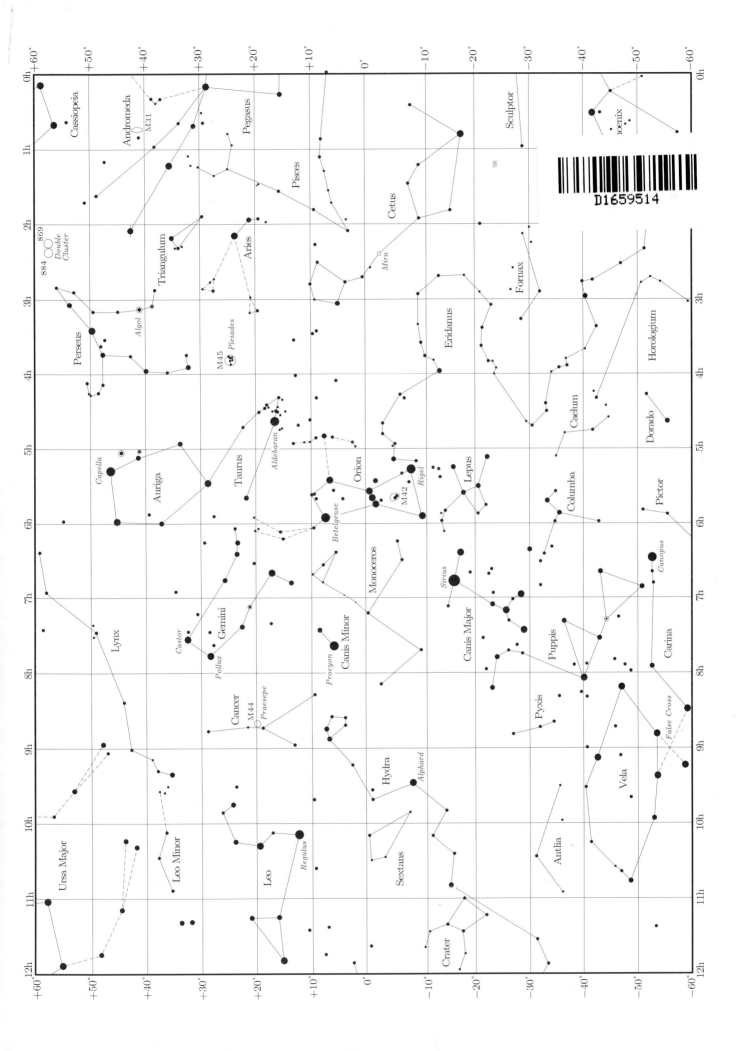

The Night Sky Observers' Guide

Volume 1
Autumn & Winter

George Robert Kepple • Glen W. Sanner

Published by

Willmann-Bell, Inc.
Publishers and Booksellers Serving
Astronomers Worldwide Since 1973

P.O. Box 35025 • Richmond, Virginia 23235 • USA • ☏(804) 320-7016 • Fax (804) 272-5920
www.willbell.com

Copyright © 1998 by George Robert Kepple and Glen W. Sanner

All rights reserved. Except for brief passages quoted in a review, no part of this book may be reproduced by any mechanical, photographic, or electronic process, nor may it be stored in any information retrieval system, transmitted, or otherwise copied for public or private use, without the written permission of the publisher. Requests for permission or further information should be addressed to Permissions Department, Willmann–Bell, Inc., P.O. Box 35025, Richmond, VA 23235.

Printed in the United States of America

Library of Congress Cataloging-in-Publication Data.

Kepple, George Robert.
   The night sky observer's guide / George Robert Kepple, Glen W. Sanner.
       p. cm.
   Includes bibliographical references and index.
   Contents: v. 1. Autumn and winter -- v. 2. Spring and summer.
   ISBN 0-943396-58-1 (v. 1). -- ISBN 0-943396-60-3 (v. 2)
   1. Astronomy--Observers' manuals. I. Sanner, Glen W. II. Title.
QB64.K46 1998   98-31044
523.8'02'16--dc21   CIP

98 99 00 01 02 03 04 05 9 8 7 6 5 4 3 2

Acknowledgements

We feel fortunate to have had the opportunity to create both a periodical magazine and this two volume publication dealing with the night sky. For both of us, observational astronomy has been an enjoyable pastime for most of our adult lives. For the past 15 years we have traveled to amateur gatherings throughout the country where we have met and made friends with hundreds of observers. We have drawn from this pool of friendship and our own observing experiences to create this publication.

A side-by-side comparison of the various issues of our magazine will reveal that there was a steady evolution in both what we covered and how it was presented. When we began this new project we choose to use our magazine experience only as a starting point and where ever possible to take advantage of the unique opportunities available only in a book. Two of which immediately come to mind: freedom from a ever constant deadline and the luxury of expandable space – at no time during this process did we or our publisher limit the page count nor set a deadline. Therefore while this work has certain likenesses to its parentage it is a unique being.

By its nature, observational astronomy is a solitary pursuit. While it is true that a large telescope can show an amazing amount of detail, this detail often does not jump out at you and say *Here I am!* To see what can be seen takes practice. If you are lucky, someone will point out what all too easily can be overlooked. So, our objective and the purpose of this publication is to show you what can be seen. It will be a success if it enables you to easily locate objects with maps and finder charts; and once you have found the objects, show in an understandable way with sketches, photographs and descriptions what can be seen with the aperture you are using.

Craig Crossen, author of *Binocular Astronomy* and several other books currently in press, has fulfilled the duties of our editor. He authored the Foreword which outlines the history of observational astronomy from its very beginning in ancient Mesopotamia to the present day. He also, in the Introduction, deals with the astrophysics of the night sky. Since we are observational astronomers, Craig's expertise in astrophysics has significantly augmented our original text. Seldom do editors have the luxury of another hand to help craft their work. This has been a long journey and we were indeed fortunate to have Craig help us finish as strongly as we began.

Appendix C of this volume is dedicated to the observers who have participated in creating the descriptions, sketches and photographs for the *Night Sky Observer's Guide*. We thank all of these individuals for their participation. A special thanks must be extended to Martin Germano who has provided nearly 75% of the photographs and to Steve Gottlieb, Steve Coe, Tom Polakis, A.J. Crayon, and Mark Stauffer who sent us thousands of observations. Where possible we have obtained photographs of these observers with their telescopes to drive home the point that most of the instruments used are just like yours.

We were also fortunate to be able to provide bound reading copies of both volumes to Brian Skiff, Richard Berry, and Harold Suiter. Their comments helped us make these better books.

Finally, we wish to express our appreciation to our wives, Barbara Kepple and Deanna Sanner to whom we dedicate this work.

George R. Kepple

Glen W. Sanner

September 1998

Foreword

Astronomy is both a very old and a very young science. For one thing, many of the star and constellation names we use today were methodically catalogued before 2000 B.C. by the Sumerians of ancient Mesopotamia, but the true nature of the stars has been understood for less than two centuries. For another thing, mathematical methods were used to predict the movements of the Moon and the planets by the Babylonians before the time of the famous pre-Socratic Greek astronomers, but a reliable distance estimate even for the Moon, the nearest of all celestial bodies, was not possible until after Columbus. Finally, the Babylonians and the Greeks gave so much time and effort to mathematical astronomy that the great Alexandrian astronomer Ptolemy was able to produce a system in the *Syntaxis* for describing lunar and planetary motion that would not be superceded for nearly a millennium and a half; yet the basic fact that our Milky Way is just one star system in a universe of independent star systems has been know for less than a century!

We who pursue astronomy are therefore fortunate to be living in the present era. However, we would be very wrong to dismiss our Babylonian, Greek, and Roman forbearers for what they did not know about the cosmos. Indeed, we should give them credit for their ability to reason — and sometimes guess – their way to truths for which they lacked hard observational evidence. For example, the Roman architect Vitruvius wrote during the time of Augustus that, "Just as the Bears (Ursa Major and Ursa Minor) turn round the pivot of the axis (the north celestial pole) without ever setting or sinking under the Earth, there are likewise stars that keep turning round the southern pivot, which on account of the inclination of the firmament lies always under the Earth, and, being hidden there, they never rise and emerge above the Earth. Even more impressive was the intuition of Vitruvius' contemporary, the astronomer Marcus Manilius, that in the Milky Way a greater host of stars has woven itself into a dense circlet.

Nor should we dismiss the cosmologies of the ancients, which (when purely mythological elements are removed) were consistent with the facts at their disposal. There was nothing inherently irrational about the Egyptians' and Babylonians' flat Earth overspread by a vaulted heaven upon which move the Sun, the Moon, the five planets, and the stars. Indeed, the Mesopotamian idea that the raw material of the Creation had been a primeval saltwater ocean, preserved in the Hebrew book of *Genesis* and in the *Theogony* of Hesiod, was based upon the observed fact of sedimentary island building at the mouths of the Tigris and Euphrates rivers along the head of the Persian Gulf.

Of course the ancient Mesopotamians had their faults (just as we do). One of them was the belief that the heavenly bodies are deities. Unfortunately, this belief lingered even into High Classical civilization. In his *Laws* Plato recommended public worship of the Sun, Moon, and stars on the grounds that it fostered public unity. Even such a sophisticated and urbane an individual as the great Roman jurist and statesman Cicero could state that the stars are "endued with such a degree of sense and understanding as places them in the rank of gods." From the pen of someone as eloquent as Cicero the notion of the deity of the stars sounds almost intellectual. But as the Roman empire aged, star worship gradually degenerated, by the very modern method of mass mania, into the artificialities of horoscope astrology and the perversions of the mystery cults of Mithras (the sun-god) and of Isis/Aphrodite (the planet Venus).

By then worship of the heavenly bodies had already been in a millennium-long decline. "History's first monotheist," as he had been called, the eccentric Egyptian Pharaoh Ikhnaton (1375–1358 B.C.), identified a universal, all-powerful deity with the Sun alone. The Hebrews, even before their Exodus from Egypt under one of Ikhnaton's successors, probably believed Jehovah to be pure spirit, and nothing in the Hebrew Old Testament reads like a residual of the astral cults of Mesopotamia, from whence Abraham had emigrated around 1700 B.C. The Christians followed the Hebrews in believing the godhead to be spirit, and with the political rise of Christianity during the 4th and 5th centuries A.D. came the decline of astrology and of the cults of Mithras and Isis. Islam, which spread by sword across the Near East and North Africa during the 7th and 8th centuries A.D., was almost an intensification of the monotheistic idea – in part in reaction to the lingering presence of Sabaean star worshippers (spiritual descendants of the Mesopotamian astral cults) in the vicinity of Mecca.

Meanwhile in both Medieval Christian Europe and the Islamic Near East the great book on true astronomy remained Ptolemy's *Syntaxis* – which Arabian astronomers did in fact call "The Greatest Book," *Al Kitab al Majisti*, now shortened to *Almagest*. The mathematics of the *Syntaxis/Almagest* assumes a geocentric cosmos, but is more of a mathematical convenience than a philosophical postulate: after all, even before 400 B.C. the Greek Pythagoreans had speculated about a heliocentric universe. Similarly, Copernicus' revolution was less a rebellion against the cosmology of the *Almagest* than simply an attempt to construct a more accurate description of the motions of the Sun, Moon and planets, which after 1,400 years had gotten out of sync with the Ptolemaic model. Indeed, even after Copernicus (1473–1543) the Danish astronomer Tycho Brahe (1546–1601) devised a system that could account for planetary motions as accurately and as easily as the new heliocentric Copernican model of the Solar System but with the Earth back at the center of the universe.

The true revolution in astronomy required the telescope of Galileo (1564–1642) and the physics of Newton (1642–1727), the latter based in part upon the laws of planetary motion of Kepler (1571–1630). What Galileo saw through his "glazed optic tube" not only knocked the Ptolemaic Earth out of the center of the universe but the Copernican Sun as well; for Galileo confirmed the 1,600-year-old speculation of Marcus Manilius that the Milky Way is a "host of stars" in a "dense circlet" – and we are manifestly *not* near the center of that circlet, for the Milky Way is much brighter toward Sagittarius than it is on the opposite side of the sky toward Auriga and Gemini. Unfortunately, because Sagittarius and its brilliant star clouds do not culminate very high above the horizon for mid-northern latitude observers, Europe is badly situated to see this. Hence the early star counts of selected regions around the northern Milky Way by William Herschel (1738–1822) resulted in a disk-like star system some five times wider than it is thick with the Sun near its center.

Despite the progress made in observational astronomy during the 18th and early 19th centuries, the problem of the nature of the Milky Way Galaxy and of the Sun's position within it remained stubbornly insoluble. Even before the end of the 18th century the German philosopher Immanuel Kant, concerning the numerous tiny fuzzy nebulae visible around the sky, had suggested astronomers perhaps should "consider these elliptical spots as systems of the same order as our own – in a word, to be Milky Ways." But decisive evidence pro or con concerning Kant's brilliant intuition remained obstinately elusive. And things only worsened with the discovery of the "spiral nebulae" by Lord Rosse with his big reflector telescopes in the mid-1800s: Were these "spiral nebulae" simply glowing gases within the Milky Way Galaxy itself, or were they indeed independent star systems so remote that their individual members could not be resolved even in Lord Rosse's huge instruments? Despite the evidence of the spectroscope, first used to analyze the light of astronomical objects later in the century, that the spectra of the spiral nebulae more resemble those of open and globular star clusters than those of such gaseous nebulae as M42 in Orion and M8 in Sagittarius, conviction grew among astronomers that these peculiarly-symmetrical objects were indeed members of the Milky Way Star System. Exactly how bad things still were in studies of the Milky Way in particular and in cosmology in general, was demonstrated in the first decade of the 20th century when J. C. Kapteyn, counting from photographic plates of selected regions around the entire Milky Way, nevertheless derived a Milky Way Star System practically identical to Herschel's, again with the Sun near its center. By then results that showed the Sun near the center of anything were taken almost as *proof* of faulty reasoning or bad data.

The problem with Kapteyn's count was that he, like Herschel before him, had ignored the effects of interstellar dust. Indeed, most astronomers – even the great E. E. Barnard, whose wide-field Milky Way astrophotos remain some of the best ever obtained – still believed that the dark patches in the star fields of the Milky Way were actual star vacancies rather that clouds of obscuring matter. In fact, not until the early 1930s and R. J. Trumpler's work on the distances of open clusters was the existence of interstellar dust taken to be definitely proven.

The first solid evidence that our Sun is out near the edge rather than near the center of the Milky Way Star System was offered by Harlow Shapley in a series of papers published in 1917-18. Shapley suggested that the Milky Way's 100-plus globular clusters constitute a single structural unit of our Galaxy, and that, because the majority of the globulars are toward the adjacent constellations Sagittarius, Scorpius, and Ophiuchus, that must be the direction of the Milky Way Galaxy's center. He derived a distance (using the globulars' RR Lyrae variables, the true brightness of which were known) from the Sun to the galactic center of nearly 50,000 light years – almost twice the actual truth because Shapley too did not believe in interstellar dust. But that still left unsolved the problem of the "spiral nebulae" and their fuzzy elliptical cousins. However, in the early 1920s, using the new 100-inch telescope on Mount Wilson, Edwin Hubble photographed individual stars in the Andromeda Nebula M31 – which henceforth could be called the Andromeda Galaxy. Some of these stars proved to be Cepheid variables, the intrinsic luminosities of which had been determined by the studies of Henrietta Leavitt: they confirmed that M31 is indeed a sizable star system hundreds of thousands of light years beyond our Milky Way Galaxy.

Thus for less than a century we have known that we live in an independent star system – a galaxy – in a universe with uncountable billions of independent star systems. We are fortunate, indeed, to be living at this point in time: the Ptolemaic and Copernican universes by comparison with reality seem claustrophobic! But all the mysteries of the cosmos are far from unraveled. We still do not have a clue to how the universe began or how it will end – or even if it is reasonable to speak about the universe beginning and ending! Black holes, mere mathematical theory less than half a century ago, have been observationally verified; but their properties strain the limits of modern physics, and we cannot claim to have a very good understanding of those objects thought to contain black holes, which include quasars, the centers of active galaxies, and the center of our own Galaxy. Many of the less exotic aspects of astronomy also remain problems: the mechanisms that trigger star formation in giant interstellar dust clouds are still more theoretical speculation than observed fact; the evolution of stars is well understood, but the evolution of galaxies is not; and we do not even have a complete knowledge of the basic spiral structure of our Galaxy in our very own neighborhood! For the first time in history humanity has some concept of the true size and the aesthetic richness of the universe into which we have come into being; but profound mysteries remain to be explored by those who have the hearts of adventurers!

As amateur astronomers we are also exceptionally fortunate in our time, an era when high quality, and very large, optics are so affordable. In the first half of the 20th century the telescope deluxe for the amateur was the 6-inch refractor (and even today the 6-inch Clark refractor remains a coveted instrument). However, such telescopes were so expensive that very few amateurs could afford them: the majority of stargazers had to content themselves with instruments in the 60mm range. Consequently, most observing guides published during that time emphasized double and multiple stars, with honorable mention for variable stars and planetary nebulae, objects which do well in long focal length refractors. Webb's 1858 *Celestial Objects for Common Telescopes* and Olcott's 1936 *Field Book of the Skies* were not superceded for so many decades simply because the average amateur instrument did not dramatically improve during the century after Webb.

By the 1950s the mass-produced or homemade 6-inch parabolic mirror brought medium-sized optics into the price range of the average amateur, and with it the emission nebulae, open clusters, and galaxies that had been seen only as amorphous blobs – if seen at all – in small refractors. The 1948 *Skalnate Pleso Atlas of the Heavens* had already displaced the classic *Norton's Star Atlas* as the frontline sky-chart for amateurs, but the observing guides badly needed rewriting. Dover reissued Webb in 1962. However, not until the late 1970s and *Burnham's Celestial Handbook* was there an observing guide worthy of the 6-inch Newtonian reflector or of the more expensive, but increasingly popular, 8-inch Schmidt-Cassegrain telescope (SCT).

By the early 80s another revolution in amateur optics was underway thanks to the inexpensive and easily-constructed mounting for large aperture Newtonian reflectors invented by John Dobson. Today's amateurs may purchase or build telescopes with mirrors ranging from 12-inch to 36-inch at a cost comparable to that of the 6-inch Newtonian or the 8-inch SCT of thirty years ago. In these big "light buckets" on their Dobsonian mounts one can see scores of emission nebulae, hundreds of star clusters, and thousands of galaxies, and with details visible in virtually all of them. Truly it is a splendid time to be an amateur astronomer!

But once again observing literature has failed to keep pace with the optics. The purpose of *The Night Sky Observer's Guide* is to close this rewidened gap by providing the owner of a medium or large aperture telescope with some idea of what to look for in such instruments – both what *objects* can be seen, and what *details* may be seen within these objects. This book endeavors to assist the observer in the act of observing – in truly seeing what there is to see in each of the objects described in these pages – because the first step in astronomy is to actually look with attention at what is in the night sky.

At various times and places during the creation of this book the manuscript has been shown to amateur astronomers. Almost without exception the first thing said was "oh, another Burnham's" – no doubt because like *Burnham's Celestial Handbook* it is physically massive. However, though similar in size, this book is not an update of the incomparable *Burnham's Celestial Handbook*. The late Robert Burnham Jr. wrote a guidebook covering virtually all that can be seen in the standard 2- to 6-inch refractor and the 6- to 8-inch reflector. He did not stop there, he also compiled a sourcebook of astronomical, astrophysical data and historical information. He did this with a literary style so engaging and so conversational that, as you looked through the eyepiece of your telescope, it seemed like the *Handbook's* author himself was at your side speaking directly to you. Burnham's astronomical and astrophysical data, being of the 1960s and 1970s (and sometimes even older), is beginning to age; and his constellation and star name history did not fully credit how much the Greeks owed to the Babylonians because that information was just not available to him. Bold indeed would be the author or authors who would claim to be the successors of Burnham!

The Night Sky Observer's Guide is uniquely a visual guide of and by amateur astronomers. It began in a doctor's office when nurse Deanna Sanner noticed

George Kepple reading a copy of *Sky & Telescope* magazine. Deanna mentioned to George that her husband Glen was interested in astronomy and had a telescope. Eventually Glen and George met and a lasting friendship developed.

In addition to astronomy both Glen and George shared an interest in computers, specifically the Macintosh, which among other things, created the modern day concept of desktop publishing. Publishing was not entirely foreign to George since he had, for years, produced and sold *AstroCards* and as a result had become friends with a wide group of observers. From this convergence of people and technology sprang *The Observer's Guide*, a bimonthly publication in which George and Glen, with the help of its readers, set out to describe, constellation by constellation, the night sky.

Burnham's Celestial Handbook had entered the scene in the 1970s to fulfill the need for an observing guide to the objects visible in a 6-inch Newtonian or an 8-inch SCT. However, in his tables Burnham gave only thumbnail descriptions of sources of the fainter objects which became available to amateur observers with the Dobsonian revolution of the 1980s. George and Glen's *The Observer's Guide* was designed to meet the demand for fuller descriptions of these objects. It was an immediate success when it appeared in 1987. Over 3000 observers subscribed and over the next 6 years the magazines covered all the constellations visible from mid-northern latitudes. Contemporary with the aperture revolution was the revolution in amateur astrophotography thanks to hypersensitizing and Kodak TP 2415 film. Thus, whereas *Burnham's Celestial Handbook* had to use mostly observatory astrophotos, *The Observers Guide* could draw from an ever-increasing abundance of high-quality amateur astrophotography.

The object descriptions in *The Night Sky Observer's Guide* derive from those in the original *Observer's Guide*, but George and Glen have reviewed and edited each so it will conform to a set style. (My own contribution to these object descriptions was merely to give them the best possible literary flow.) In those instances where inconsistencies arose the editors re-observed the object and rewrote the original *Observer's Guide* description. *The Night Sky Observer's Guide* also includes many photographs and maps that did not appear in the magazine.

Though both *The Observer's Guide* and now *The Night Sky Observer's Guide* were aimed at amateurs especially interested in observing galaxies, nebulae and clusters, neither the magazine nor these volumes have neglected double and variable stars. Data tables for doubles and variables within a constellation are provided near its beginning, and these stars are labeled on maps and finder charts. Moreover, the most famous or visually impressive doubles and variables are given written descriptions similar to those for other deep-sky objects. Splitting doubles and plotting variable star light curves are not nearly as popular with amateurs today as they were thirty or forty years ago, so doubles and variables are not emphasized in these volumes. Nevertheless, double stars in particular offer the observer many fine, and even spectacular, sights in the eyepiece.

For those who are curious, the number of celestial objects of each type covered in the two volumes of *The Night Sky Observer's Guide* are:

Double Stars	2,104
Variable Stars	433
Galaxies	2,030
Planetary Nebulae	127
Bright Nebulae	131
Dark Nebulae	69
Open Clusters	550
Globular Clusters	92
Misc. (QSOs, Asterisms)	5
Total Objects	5,541

In addition there are:

Photographs	446
Drawings (Eyepiece Impressions)	827
Star Charts	431
Tables	143

During this past spring, while writing the introduction to these volumes, I had the opportunity to observe with a pair of Russian 15 x 110 giant binoculars. I decided to test the binoculars, the charts, and descriptions in these volumes, on the most difficult type of object I could: the Coma-Virgo Supercluster galaxies scattered from Ursa Major on the north down to Centaurus on the south. With the aid of the finder charts and photographs in Volume II (in which are to be found the galaxy-rich constellations of Spring), I was able to positively identify, even at only 15x, over ninety galaxies, including several of the 12th magnitude. Moreover, the object descriptions helped me perceive details in the tiny images of nearly all those ninety galaxies. Thus there is no doubt in my mind that if you are using these volumes you will indeed know where to look and what to look for. Thank you George Kepple and Glen Sanner!

Craig Crossen
July 1998

Table of Contents

Acknowledgements --------------------------------- iii

Foreword --- v

Introduction ------------------------------------- xi
    I.1 Perspective on Our Galaxy ------------------------------ xi
    I.2 The Stars --- xii
    I.3 Stellar Magnitudes and Luminosities ------------------- xiii
    I.4 Stellar Spectra --------------------------------------- xiv
    I.5 Stellar Evolution ------------------------------------- xvii
    I.6 Variable Stars -- xxi
    I.7 Intrinsic (Pulsating) Variables --------------------- xxii
    I.8 Eruptive Variables ----------------------------------- xxiv
    I.9 Eclipsing Variables ---------------------------------- xxvi
    I.10 Double Stars --------------------------------------- xxvii
    I.11 Stellar Groups ------------------------------------- xxviii
    I.12 Globular Clusters ---------------------------------- xxix
    I.13 Open Clusters -------------------------------------- xxx
    I.14 Stellar Associations and Stellar Streams --------- xxxiv
    I.15 Nebulae -- xxxvi
    I.16 Dark Nebulae --------------------------------------- xxxvi
    I.17 Bright Nebulae ------------------------------------- xxxvii
    I.18 Galaxies -- xli
    I.19 Elliptical Galaxies -------------------------------- xliv
    I.20 Lenticular Galaxies (S0 or SB0) ------------------- xliv
    I.21 Spiral Galaxies ----------------------------------- xlv
    I.22 Irregular Galaxies (I or Irr) --------------------- xlvi
    I.23 Galaxy Groups and Clusters ----------------------- xlvi

Chapter 1 Observation of Deep-Sky Objects - 1
    1.1 Catalogs–Celestial Inventories ---------------------- 1
    1.2 Object Visibility ---------------------------------- 2
    1.3 Field Orientation ---------------------------------- 2
    1.4 Position Angles ------------------------------------ 2
    1.5 Visual Impressions --------------------------------- 3
    1.6 Dark Adaptation ------------------------------------ 3
    1.7 Visual Telescopic Descriptions --------------------- 4
    1.8 Visual Rating Guide -------------------------------- 4
    1.9 Using the Star Charts ------------------------------ 4
    1.10 Getting Started ----------------------------------- 5
    1.11 Keeping Records ----------------------------------- 5
    1.12 Magnification and Eyepiece Selection -------------- 8
    1.13 Sketching Made Easy ------------------------------- 8
    1.14 Estimating Field of View -------------------------- 9
    1.15 Enjoy the Night Sky ------------------------------- 10

Chapter 2 Andromeda, the Princess -------- 11
    2.1 Overview --- 11
    2.2 Interesting Stars ---------------------------------- 11
    2.3 Deep-Sky Objects ----------------------------------- 14

Chapter 3 Aquarius, the Water-Bearer ---- 25
    3.1 Overview --- 25
    3.2 Interesting Stars ---------------------------------- 25
    3.3 Deep-Sky Objects ----------------------------------- 28

Chapter 4 Aries, the Golden Ram ---------- 39
    4.1 Overview --- 39
    4.2 Interesting Stars ---------------------------------- 39
    4.3 Deep-Sky Objects ----------------------------------- 41

Chapter 5 Auriga, the Charioteer ---------- 59
    5.1 Overview --- 59
    5.2 Interesting Stars ---------------------------------- 59
    5.3 Deep-Sky Objects ----------------------------------- 63

Chapter 6 Camelopardalis, the Giraffe ----- 63
    6.1 Overview --- 63
    6.2 Interesting Stars ---------------------------------- 63
    6.3 Deep-Sky Objects ----------------------------------- 66

Chapter 7 Cancer, the Crab ----------------- 77
    7.1 Overview --- 77
    7.2 Interesting Stars ---------------------------------- 77
    7.3 Deep-Sky Objects ----------------------------------- 79

Chapter 8 Canis Major, the Big Dog -------- 85
    8.1 Overview --- 85
    8.2 Interesting Stars ---------------------------------- 85
    8.3 Deep-Sky Objects ----------------------------------- 88

Chapter 9 Canis Minor, the Little Dog ----- 101
    9.1 Overview --- 101
    9.2 Interesting Stars ---------------------------------- 101
    9.3 Deep-Sky Objects ----------------------------------- 101

Chapter 10 Cassiopeia, the Queen ---------- 105
    10.1 Overview -- 105
    10.2 Interesting Stars --------------------------------- 105
    10.3 Deep-Sky Objects ---------------------------------- 108

Chapter 11 Cepheus, the King -------------- 131
    11.1 Overview -- 131
    11.2 Interesting Stars --------------------------------- 131
    11.3 Deep-Sky Objects ---------------------------------- 135

Chapter 12 Cetus, the Whale — 151
12.1 Overview — 151
12.2 Interesting Stars — 151
12.3 Deep-Sky Objects — 155

Chapter 13 Columba, the Dove — 171
13.1 Overview — 171
13.2 Interesting Stars — 171
13.3 Deep-Sky Objects — 171

Chapter 14 Eridanus, the River — 175
14.1 Overview — 175
14.2 Interesting Stars — 175
14.3 Deep-Sky Objects — 178

Chapter 15 Fornax, the Furnace — 191
15.1 Overview — 191
15.2 Interesting Stars — 191
15.3 Deep-Sky Objects — 191

Chapter 16 Gemini, the Twins — 203
16.1 Overview — 203
16.2 Interesting Stars — 203
16.3 Deep-Sky Objects — 206

Chapter 17 Lacerta, the Lizard — 215
17.1 Overview — 215
17.2 Interesting Stars — 215
17.3 Deep-Sky Objects — 217

Chapter 18 Lepus, the Hare — 221
18.1 Overview — 221
18.2 Interesting Stars — 221
18.3 Deep-Sky Objects — 224

Chapter 19 Lynx, the Lynx (Bobcat) — 229
19.1 Overview — 229
19.2 Interesting Stars — 229
19.3 Deep-Sky Objects — 231

Chapter 20 Monoceros, the Unicorn — 237
20.1 Overview — 237
20.2 Interesting Stars — 237
20.3 Deep-Sky Objects — 237

Chapter 21 Orion, the Hunter — 279
21.1 Overview — 279
21.2 Interesting Stars — 279
21.3 Deep-Sky Objects — 282

Chapter 22 Pegasus, the Winged Horse — 291
22.1 Overview — 282
22.2 Interesting Stars — 291
22.3 Deep-Sky Objects — 294

Chapter 23 Perseus, the Hero — 293
23.1 Overview — 293
23.2 Interesting Stars — 293
23.3 Deep-Sky Objects — 297

Chapter 24 Pisces, the Fishes — 315
24.1 Overview — 315
24.2 Interesting Stars — 315
24.3 Deep-Sky Objects — 319

Chapter 25 Piscis Austrinus, the Southern Fish — 333
25.1 Overview — 333
25.2 Interesting Stars — 333
25.3 Deep-Sky Objects — 333

Chapter 26 Puppis, the Ship's Stern — 339
26.1 Overview — 339
26.2 Interesting Stars — 339
26.3 Deep-Sky Objects — 343

Chapter 27 Pyxis, the Mariner's Compass — 361
27.1 Overview — 361
27.2 Interesting Stars — 361
27.3 Deep-Sky Objects — 361

Chapter 28 Sculptor, the Sculptor — 365
28.1 Overview — 365
28.2 Interesting Stars — 365
28.3 Deep-Sky Objects — 367

Chapter 29 Taurus, the Bull — 375
29.1 Overview — 375
29.2 Interesting Stars — 375
29.3 Deep-Sky Objects — 378

Chapter 30 Triangulum, the Triangle — 389
30.1 Overview — 389
30.2 Interesting Stars — 389
30.3 Deep-Sky Objects — 389

Appendix A The Local Galaxy Group — 399

Appendix B Meteor Showers — 401

Appendix C The Editors and Contributors — 403

Bibliography — 411

Index — 413

Introduction
Comprehending Our Fascinating Universe

by Craig Crossen

I.1 Perspective on Our Galaxy

In the early 1920s astronomers finally proved that we live in an independent star system, the Milky Way Galaxy, in a universe of independent star systems. The natural follow-up question was: What is the structure of our Milky Way Galaxy? Fortunately the answer was easy to infer: Because much of what we see in the Galactic vicinity of the Solar System – hot blue giant and supergiant stars, cool red supergiant stars, Cepheid variables, emission nebulae, dark lanes (like the Great Rift), and open clusters – are also seen in the arms of the nearest spiral galaxies, M31 in Andromeda and M33 in Triangulum, we must live in an outer arm of a giant spiral similar to those systems.

However, the details of our Galaxy's structure are difficult to disentangle, even in the immediate neighborhood (Galactically speaking) of our Sun. The problem is that our Sun is located almost precisely on the plane of our Galaxy's spiral disk, along which interstellar dust tends to congregate, and therefore our view through the Milky Way is dimmed or, in certain directions, almost completely obscured. Nevertheless, by analogy with other spiral galaxies we know that our Galaxy must have a large central *hub* or *bulge*, in the shape of a flattened sphere, around which wind the spiral arms. (In fact the Milky Way's bulge shows up fairly well on wide-field photographs that cover Sagittarius, Scorpius, and SE Ophiuchus.) The spiral arms are actually simply the most conspicuous feature of a spiral galaxy's *disk*, which is two or three times thicker than the spiral arms themselves. (The disk component is very evident in photos of such edge-on spiral galaxies as NGC 4565 in Coma Berenices and NGC 891 in Andromeda.) Centered upon a spiral galaxy's bulge is its family of globular clusters, probably 200 strong in the case of the Milky Way, distributed in a large sphere that extends out well beyond the edges of the galaxy's disk – though the number-density of globulars decreases exponentially with increasing distance from the galaxy's center. The globulars are the most conspicuous constituent of the *halo* or *spheroidal* component of a galaxy. The actual star density in most spiral galaxy halos is quite small. However, M104, the "Sombrero Galaxy," has an unusually extensive, populous, and therefore conspicuous halo.

The spiral disk of the Milky Way Galaxy is estimated to be nearly 150,000 light years in diameter. The Galaxy's central bulge has an equatorial diameter of perhaps 10,000 light years and a polar diameter of roughly 8,000 light years. The Sun is thought to be a little less than 30,000 light years out from the Galaxy's center on the inner edge of a spiral feature called the Orion-Cygnus Spiral Arm. In the Solar region the spiral arms seem to be about 1,000 light years, and the disk itself around 2,500 light years, thick. (A spiral galaxy's disk, like its central bulge and its halo, does not have well-defined boundaries.) The most remote globular clusters are about 100,000 light years out from the Galaxy's center, with a handful, perhaps better though of as "intergalactic wanderers," that have strayed into our Galaxy's gravitational influence, as much as 300,000 light-years from the center.

The stars associated with the different structural components of our Galaxy have different characteristics. This was first discovered not by observations of stars within our Galaxy itself but by photographs of the Andromeda Galaxy obtained by Walter Baade in the mid-1940s. The wartime blackout of Los Angeles permitted Baade to take extremely long-exposure photographs of M31 with the 100-inch telescope. Hubble had been able to resolve blue supergiants and Cepheid variables in the spiral disk of M31, but the galaxy's central bulge had remained obstinately hazy. Baade's long-exposure red-sensitive plates, however, succeeded in resolving red and yellow giants in the M31 bulge, as well as similar stars in M31's satellite galaxies M32 and NGC 205, elliptical systems which had the appearance of spiral galaxy bulges minus the disk. Baade therefore proposed two stellar populations: *Population I* stars are in the spiral arms of galaxies and include blue and red supergiants, blue O- and

B-type main sequence stars, and Cepheid variables; and *Population II* stars are found in the central bulges of spiral galaxies, in globular clusters, and in elliptical galaxies like M32 and NGC 205, and include most red and yellow giant stars and RR Lyrae cluster variables. From the first there was strong presumptive evidence that these two populations are distinguished as much by age as by location: supergiants, because of their spendthrift luminosity, are necessarily short-lived; and they and other Population I stars are usually found in very close proximity to the clouds of dust and gas which provide the raw material for star formation. By contrast, elliptical galaxies and globular clusters, the natural habitat of Population II stars, are virtually dust and gas free.

Unfortunately reality is never quite this cut-and-dried, and astronomers were soon forced to sophisticate Baade's simple system. The objects which lie right along the spiral arms are now called *Extreme Population I* and include blue and red supergiants, O and B main sequence stars, and the complexes of bright emission nebulae and dark dust clouds where such stars are presently being formed. *Intermediate Population I* includes the objects found in a spiral galaxy's disk, but not necessarily along its spiral arms: Cepheid variables, less luminous yellow and red supergiants, old open clusters, solar type stars, long-period variables and planetary nebulae. The *Bulge Population* are the stars typical of galaxy bulges, the brightest of which are red and yellow low-mass giants. Finally, *Extreme Population II* are the stars typical of globular clusters and found not only within globulars but, like the globular clusters themselves, are spread throughout the Galaxy's halo: the brightest of these stars are low-mass red and yellow giants similar to those found in galactic bulges, but this population also includes W Virginis Cepheids, RV Tauri variables, and long-period RR Lyrae stars.

But even this more elaborate star population scheme has its inadequacies, for certain types of stars and objects still resist neat classification. Planetary nebulae, for example, are abundant in the Galaxy's disk: but at least four are in globular clusters; and planetaries as a whole are, like the globulars themselves, concentrated toward the Galaxy's interior. Long-period variables (LPVs) of shorter than average period can be found in globular clusters; but long-period LPVs are not, and the longer an LPV's period, the younger the star population of which it is a part. Extreme Population II stars are much poorer in metals (the astrophysicists' term for all elements heavier than helium) than Extreme Population I stars, but some Bulge Population stars are even richer in metals than Extreme Population I, and Population I stars of the inner spiral arms are more metal-rich than Population I stars of the outer spiral arms. To accomodate these complications Intermediate Population I has been further divided into the Thin, Intermediate, and Thick Disk populations, depending upon how far above and below the Galactic plane objects representative of each of those populations extend; and a transitional Disk/Halo Population has also been suggested. But at some point the temptation to further refine the star population system must be resisted, otherwise the scheme will lose its simplicity and therefore usefulness. Obviously the stars themselves form a continuum of populations, and any classification scheme, however useful, will never be entirely free of artificiality.

I.2 The Stars

Stars are luminous spheres of extremely hot gas held together by their own gravity. They radiate by energy derived from the thermonuclear reactions that occur deep within their interiors. The Sun is the nearest star and the only one astronomers can study thoroughly since all others lie at huge distances from us: the second nearest star, the Alpha Centauri triple, lies some 270,000 times farther than the Sun. Astronomers thus use the Sun as the standard of comparison by which to measure other stars: in particular the luminosities, radii, and masses of other stars are usually expressed in solar units.

Stars exist in a remarkable variety of sizes and luminosities, from brilliant supergiants hundreds of times larger and tens of thousands times brighter than the Sun to faint dwarfs just a fraction of the Sun's size, mass, and luminosity. Stars also come in a variety of colors – that is, temperatures – from hot bluish-white, through cooler white and yellow, to orange, reddish-orange, and even, in the case of the extremely cool carbon stars, poppy red. However, stellar temperatures and luminosities are not directly related: some cool reddish stars, such as Antares and Betelgeuse, are extremely bright simply because they are extremely large; and some very hot bluish stars – specifically, the white dwarfs – are extremely faint simply because they are extremely small.

Individual stars are usually involved in one (or more) of the various types of stellar groupings. Indeed, all stars were probably born in groups: stars like the Sun, which has no known star-group affiliation (other than being a member of the Milky Way Galaxy), have become loners no doubt only because they have outlived their parent group's cohesiveness. The simplest type of stellar group is the binary or multiple star, which consists of two or more components in actual orbit around their mutual center of gravity. Star clusters, in which the members orbit the whole aggregation's center of gravity, are of two types: open (in older literature sometimes also called galactic) clusters, found mainly in and around the Galaxy's spiral arms and consisting of anywhere from a few dozen to a few thousand members; and globular clusters, concentrated toward the Galaxy's central bulge but found right out to the limits of the galaxy's gravitational field and containing usually hundreds of thousands of stars. Open clusters are for the most part rather young members of our Galaxy, few being more than a billion, and most less than few hundred million, years old. Globular clusters, by contrast, are all very ancient, 9 or 10 billion years and older: they can become so old because of their gravitational stability. New open clusters are being constantly born within the spiral arms' giant clouds of gas and dust; but for all we can tell, our Galaxy ceased making globular clusters billions of years ago.

Two types of stellar groupings that are not gravitationally bound but distinguished and identifiable principally by the shared motion through the Galaxy of their individual members are *associations* and *streams* or *moving groups*. Associations are aggregations of recently formed stars and are usually involved with bright nebulae and dark dust clouds – the remnants, in which star formation often is still occurring, of the interstellar matter which gave birth to the members of the association. Most of the bright stars of the constellation Orion are members of the Orion Association, at the heart of which is the Orion Nebula (M42) and the famous Trapezium multiple star. Moving groups, or stellar streams, are in effect decayed associations: they preserve their parent association's original motion through the Galaxy but lack any really hot and luminous members, which long since have exploded as supernovae. The middle five stars of the Big Dipper are the core of the Ursa Major Stream, of which Sirius itself is a member.

The largest type of stellar grouping (though this is to stretch the meaning of the term a bit) is the galaxy, which comes in a variety of sizes and shapes. Galaxies can contain anywhere from only a few million stars, in the case of the dwarf ellipticals and dwarf irregulars, to hundreds of billions of stars, in the case of the supergiant ellipticals. Spiral, irregular, lenticular, and some elliptical galaxies also contain, in addition to their stars, large clouds of gas and dust, usually concentrated along the galaxy's disk or plane.

Each of the above stellar groupings – from double stars to supergiant galaxies – will be described in greater detail in the course of this introduction.

I.3 Stellar Magnitudes and Luminosities

Ancient Greek astronomers rated star brightness by magnitude, 1st magnitude stars being the brightest and 6th magnitude those just visible to the unaided eye. Because, as modern photometry demonstrates, the difference in brightness between the typical 1st magnitude and the typical 6th magnitude star (using the magnitudes in Ptolemy's star-catalogue in the *Almagest*) is 100 times, the brightness ratio for a difference in 1 magnitude is 2.512 (because 2.512 multiplied by itself 5 times is 100. This brightness ratio, though an awkward number, is not entirely accidental but related to the chemical physiology of the retina of the human eye.) The ancient Greek magnitude system proved easy to apply to objects visible only in telescopes, which have been assigned magnitudes greater than 6, and to objects very much brighter than the stars – such as Venus, which at its brightest has a magnitude of -5. Extended objects like emission nebulae, star clusters, and galaxies have integrated magnitudes, which is simply the magnitude they would have if all their light was concentrated into a single starlike point.

Supplementing this system of apparent magnitudes is a scale of absolute magnitudes used to measure and compare the true luminosities of celestial objects. An object's absolute magnitude is simply the apparent magnitude it would have if it was 32.6 light years = 10 parsecs from us. Similarly, the integrated absolute magnitude of an extended object is the absolute magnitude it would have if all its light was concentrated at a stellar point 32.6 light years away (in the case of galaxies, a rather mind-boggling thought!). The absolute magnitude of the Sun is a mere $+4.8$; that of Sirius, by contrast, is $+1.4$, and that of Rigel, intrinsically the brightest of the twenty-two 1st magnitude stars, is -7.1. Our Milky Way Galaxy's integrated absolute magnitude is thought to be around -20.5. The Andromeda Galaxy's absolute magnitude is -21.5.

At first glance this magnitude system might seem cumbersome. But after you use it awhile, it proves to be quite easy and natural. What you must keep in mind is that when you add magnitudes, you *multiply* luminosities. For example: The difference between the absolute magnitudes of the Sun and of Rigel is $+4.8 - (-7.1) = 11.9$ magnitudes. Because a difference of 5 magnitudes corresponds to a ratio of 100 times in brightness, and a difference of 1.9 magnitudes is a ratio of about 6 in brightness, the difference in true luminosity between the Sun and Rigel is $11.9 = 5+5+1.9$ magnitudes $= 100 \times 100 \times 6 = 60,000$ times. In other words, Rigel is about 60,000 times more luminous than the Sun. For rapid conversion of magnitude difference into brightness ratios use the following approximations:

Difference of magnitude =	(approx.) brightness ratio
0.7	2
1	2.5
2	6.25
3	16
4	40

To compute the brightness ratio of one object that is fainter than another you divide. For example: The famous white dwarf companion of Sirius has an absolute magnitude of just +11.5, which means that it is 11.5 − 4.8 = 6.7 = 5.0 + 1.0 + 0.7 magnitudes fainter than, or $1/100 \times 1/2.5 \times 1/2 = 1/500$ as bright as, the Sun.

I.4 Stellar Spectra

The color differences among stars which are so beautiful in binoculars and telescopes correspond to true differences in surface temperatures: bluish stars like Spica are in fact hotter than blue-white stars like Rigel which in turn are hotter than white stars like Sirius which are hotter than yellow-white stars like Procyon, yellow stars like our Sun, orange stars like Aldebaran, and finally reddish-orange stars like Antares. Cool stars like Antares radiate principally in the invisible infrared, and hot stars like Spica radiate principally in the invisible ultraviolet: in fact only about 2.5% of the total energy output of a star with a surface temperature of 50,000 K (kelvin), like Zeta Puppis, is in visible light.

The spectra of stars, including of course our Sun, show dark absorption lines (called Fraunhofer lines, after their discoverer), and sometimes bright emission bands, superimposed upon the background continuum of red, orange, yellow, green, blue, indigo, and violet. In reddish stars the red end of the background continuum is more intense than the blue end, and in bluish stars the blue end of the continuum is stronger. The individual absorption lines and emission bands are from the individual types of ions, atoms, and molecules in a star's atmosphere and therefore can be used to determine the star's chemical composition. But these lines and bands are also an index of the star's temperature, for in the atmosphere of hotter stars the atoms cannot combine into molecules and therefore the spectra of those stars show no absorption lines of molecules, and in the atmosphere of cooler stars few atoms lose any of their electrons and therefore the spectra of these stars show few, if any, ions.

In the late 19th century E.C. Pickering classified stellar spectra in sixteen types from A to Q based strictly upon observed features. In 1888, however, Antonia C. Maury, in a brilliant leap of intuition, rearranged Pickering's groups into the order that is still used today, which only later was understood to be a temperature sequence. Thus we find ourselves with a sequence of spectral types, from the hottest to the coolest stars, which follows the non-alphabetical order O, B, A, F, G, K, and M. And, to complicate matters, four more spectral types were later added to Maury's sequence. To the red end were added three spectral types for stars with normal M-type temperatures but abnormal chemical compositions: N stars are extremely abundant in carbon, R stars are moderately abundant in carbon, and S stars are abundant in such rare earths as zirconium and barium. Eventually types N and R were rolled up together into a new type, designated C; but even today observing lists still variously quote N or C for the spectral type of individual carbon stars. At the blue end of the spectral sequence has been added the peculiar Wolf-Rayet stars, extremely luminous objects undergoing almost explosive mass loss due to their powerful stellar winds and radiation pressure: Wolf-Rayets with strong emission bands of nitrogen are designated by WN and those with strong emission features of carbon are designated by WC.

The term "early" is often applied to O and B, and sometimes even to A and F, stars and the term "late" to K and M stars. These expressions have only the vaguest connection with stellar evolution. It is true that the youngest open clusters are rich in O and B stars; but they also often contain late M-type supergiants (to mention but two examples: NGC 663 in Cassiopeia, and the NGC 884 component of the Perseus Double Cluster). Moreover, white dwarfs, which have early O-, B-, or A-type surface temperatures, are in fact highly evolved stars – indeed, they are essentially dead stars radiating only by residual heat. The words "early" and "late" were attached to the spectral sequence before the true course of stellar evolution was understood but have become so deeply rooted in astronomical nomenclature that they are now impossible to eradicate.

Each spectral type is further split into ten subdivisions from 0 to 9, 0 being earlier – that is, hotter – than 9. (The earliest O-type stars, however, are just O3. But that is just an accident of the spectral scale: O3 stars are simply as hot as stars can get.) Furthermore, Maury began adding single-letter suffixes and prefixes to the basic spectral notation to indicate unusual characteristics of individual stellar spectra. Her shorthand was later elaborated and is still in common use. The prefixes "c," "g," "d," and "D" respectively mean supergiant, giant, dwarf, and white dwarf. (Occasionally one also sees an sd = subdwarf.) The suffixes "e" and "p" mean that the star's spectrum has emission features or other peculiarities. The suffix "f" indicates strong emission lines of ionized helium and nitrogen (usual in the spectra of O-type supergiants); and "m" implies a stellar spectrum uncommonly abundant in *metals* – elements heavier than hydrogen and helium.

In the early years of the 20th century two astronomers, Ejnar Hertzsprung and Henry Norris Russell, independently plotted diagrams of stellar luminosities versus temperature/spectral type – thereafter known as the Hertzsprung-Russell, or H-R, diagram (Figure I-1). They discovered that the vast majority of stars fall along a band which runs from hot, luminous O-type stars on the upper left of the diagram to faint, cool M-

Table I-1. Stellar Spectral Types

Type	Temperature (°K)	Color	Examples	Spectral Features (Absorption Lines)
WN, WC	Above 50,000	Bluish	Gamma (γ) Velorum (WC8)	"Wolf-Rayet" stars. Strong emission bands of nitrogen designated by WN and strong emission features of carbon designated by WC.
O	Over 25,000	Bluish	Zeta (ζ) Puppis (O5), Iota (ι) Orionis	Ionized Helium; hydrogen weak; doubly or triply ionized metals.
B	25,000 to 11,000	Bluish	Spica (B1), Rigel (B8)	No ionized helium; neutral helium strong; hydrogen stronger; fewer ionized metals.
A	11,000 to 7,600	Blue-white to white	Sirius (A1), Altair (A7)	Helium absent; hydrogen strong (strongest at A0); Calcium II present but weak; other singly ionized metals strong.
F	7,600 to 6,000	Yellowish-white	Canopus (F0), Procyon (F5)	Hydrogen weaker; calcium II stronger; at F0 singly ionized metals as strong as neutral metals, but in later F neutral metals predominate.
G	6,000 to 4,500	Yellow	Sun (G2) Capella (G8)	Hydrogen weak; calcium II at maximum; many lines of neutral metals; lines of molecules CH and CN appear and become stronger toward late G.
K	5,100 to 3,200	Orange	Arcturus (K2) Aldebraran (K5)	Hydrogen faint; neutral metals very strong; CH and CN stronger.
M	Less than 3,700	Orange-red	Antares (M1) Betelgeuse (M2) Mira (M5e-M9e)	Neutral metals and CH & CN strong; titanium oxide (TiO) appears and becomes stronger toward late M; "Me" stars have hydrogen emission lines.
C	Same as K & M	Red	19 Psc (C6) Y CVn (C5)	"Carbon stars": strong lines of such carbon molecules as C2, CN, CO; TiO absent.
S	Same as K & M	Orange-red	(None discussed in this book.)	Resembles type-M but with oxides of such exotic elements as zirconium and ytterbium replacing TiO.

type stars on the lower right. This band is called the main sequence, and the Sun falls in its lower middle. Several other areas of the H-R diagram contain noticeable concentrations of stars. In its lower left, well below the luminous O- and B-type main sequence stars, are the hot but faint white dwarfs. Toward the diagram's upper right, well above faint K- and M-type red dwarfs (as they are called) of the main sequence's lower end, are the red giants (amongst which are G and K as well as M stars). Scattered along the upper edge of the diagram are the numerically few but visually conspicuous supergiants. (Blue and white B and A, and red M-type, supergiants are decidedly more numerous than yellow F-, G-, and K-type supergiants, leaving a conspicuous gap in the supergiant sequence. The reason for the gap is simply that massive stars expand very quickly from relatively compact B/A supergiants into bloated M supergiants.)

Obviously spectral type (that is, temperature) alone does not unambiguously fix a star's location on the H-R diagram; if it did, all stars would fall somewhere on the main sequence. Hence in the 1940s W.W. Morgan and P.C. Keenan refined stellar spectral notation by appending to each star's spectral type a luminosity class; and the combination of spectral type and luminosity class together, called the star's MK spectral type, uniquely determines its place on the H-R diagram. The MK luminosity classes are:

```
    IaO or Ia+ = hyper-supergiants
            Ia = luminous supergiants
           Iab = moderately luminous supergiants
            Ib = less-luminous supergiants
            II = bright giants
           III = normal giants
            IV = subgiants
             V = main sequence
```

The Sun's complete MK spectral type is G2 V. Rigel's is B8 Ia, Procyon's F5 IV-V, Arcturus' K2p III, Betelgeuse's M2 Iab, and Zeta Puppis' O4fn Ia (the rather rare "n" indicating the broadening of its spectral lines –

another O-supergiant peculiarity). Fortunately it is not always necessary to know a star's distance to determine its luminosity class. Supergiants in particular are distinguished by the thin, sharp absorption lines in their spectra, the consequence of their thin, tenuous atmospheres.

The H-R diagram (called a color-magnitude diagram, or CMD, if stars are plotted by simple color – a more easily-measured property – than precise spectral type) is a powerful astrophysical tool, for it helps astronomers determine the distances, ages, and chemical compositions of open and globular clusters. Distance determination using the H-R or CM diagram is in principal straightforward (though in practice it is beset by numerous observational pitfalls). Assume that you know the distance to one cluster and want to know the distance to a second. If you plot the stars of both clusters on the same H-R or CM diagram using apparent magnitude, you will find that you end up with two main sequences, one below the other. Obviously the lower main sequence will be that of the more remote cluster, which in most cases will be the cluster with the unknown distance. All you need to do is measure how many magnitudes lower the more remote cluster's main sequence is than the nearer cluster's. Let us assume that it turns out to be exactly 3 magnitudes. Because 3 magnitudes corresponds to a brightness ratio of just about exactly 16, the stars in the more remote cluster appear 1/16 as bright as the stars in the nearer cluster. Now, brightness decreases with the square of the distance: therefore, since the square root of 16 is 4, the second cluster must be 4 times farther than the first cluster. If you know the distance to the nearer cluster, you simply multiply by 4 to get the distance to the more remote cluster.

When the H-R or CM diagrams of a variety of open and globular clusters are superimposed, they display a remarkable feature: the main sequences of different clusters go up to different points. However, the main sequences of similar clusters always go up to a similar point. For example, the main sequences of open clusters embedded in complexes of bright and dark nebulae (such as M16 in Serpens, NGC 6530 in the Lagoon Nebula in Sagittarius, and NGC 2244 in the Rosette Nebula in Monoceros) invariably go all the way up to spectral type O. On the other hand, the main sequences of open clusters with luminous blue giants and subgiants, and perhaps an M or K Ib supergiant or two, but well away from any nebulosity (such a M6 in Scorpius, the Alpha Persei Cluster, and – given the fact that their reflection nebulosity is not actually physically related to them – the Pleiades) go up only to the mid-B range. Open clusters without any bright blue stars but a well-populated red giant region (such as the Hyades, M44 in Cancer, and M11 in Scutum) have main sequences up just to mid- or late-A. And the main sequences of globular clusters do not extend even beyond mid-F. Because open clusters embedded within clouds of gas and dust are obviously very young objects near their place of birth, and because globular clusters, most of which are tens of thousands of light years away from any interstellar matter, are obviously very old objects, the main sequence turnoff of a cluster is a close index of its age. This goes back to the very simple fact, discussed in the following section on stellar evolution, that the hotter and more luminous O- and B-type main sequence stars squander their energy resources much more quickly than even solar G-type, let alone the extremely faint M-type, dwarfs and therefore have but a fraction of the main sequence life-expectancy of those less showy stars. Thus as a cluster ages, its main sequence gradually "peels" away toward the right, its more massive stars becoming red supergiants and since red giants are short lived they soon explode as supernovae or shrink into white dwarfs.

When the H-R or CM diagram of even one individual cluster is plotted, its main sequence is never a perfectly straight line. Indeed, often a cluster's main sequence will look downright fat. Even after observational errors and the effects of binary stars are deducted, cluster main sequences perniciously persist in possessing width. Things are even worse when two different clusters are plotted on the same H-R or CM diagram: the main sequences of the two groups – especially if one is an open and the other a globular cluster – actually run along different lines! The reason for all this is in the differences in chemical composition not only between the stars of different clusters, but even among the stars of the same cluster. In a nice, neat, orderly universe, all stars would have exactly the same chemical composition and a star's position on the main sequence would be determined solely by its initial mass and in all H-R diagrams the main sequence would be a perfectly straight line. However, it is a dirty universe and its primordial hydrogen and helium contaminated in different places by varying amounts of carbon, oxygen, nitrogen, titanium, sodium, iron, aluminum, and other elements. These elements collectively called metals, exist only in mere trace amounts, and have an influence disproportionate to their quantity. When in a star's atmosphere, these metals absorb a significant percentage of the blue light emitted by the star's surface, reradiating the energy at longer red and yellow wavelengths. Thus a metal-rich star with exactly the same mass and surface temperature as a metal-poor star will look redder. (The actual difference between metal-rich and metal-poor is minute, involving less than 2% of the two stars' masses.) If our Sun lacked metals (it is classified as a metal-rich star), it would look a little less yellow than it does. Consequently a main sequence of metal-poor stars would be shifted a little to the left (bluer) than a main sequence of metal-rich stars. Subdwarfs (excluding stars in the process of shrinking into true white dwarfs) are metal-poor stars that appear less luminous for their color/spectral

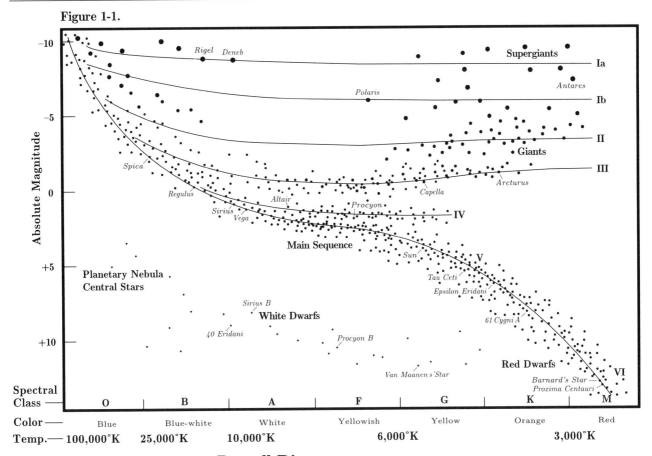

Figure 1-1.

Hertzsprung-Russell Diagram

type but are in fact too blue for their luminosity. (Hence the term "subdwarf" is actually a misnomer.) If two star clusters – for example an open cluster and a globular cluster – with stars of significantly different chemical compositions are plotted on the same absolute magnitude versus color diagram, the main sequence of the relatively metal-rich cluster – in this case the open cluster – will appear to lie above the main sequence of the metal-poor cluster – here the globular.

I.5 Stellar Evolution

Stars are born not individually but collectively in cool, dense, massive clouds of hydrogen gas and interstellar dust. The larger, more massive clouds – called giant molecular clouds, GMCs – produce not only more stars, but also more giant and supergiant stars than smaller molecular clouds – SMCs. An SMC like Bernes 157 in Corona Australis, described and illustrated in Chapter 38 in Volume Two, is generating only modest-mass, modest-luminosity stars (like the nebular variables R, S, and TY Corona Australis). But GMCs such as those toward Orion produce giant and supergiant stars (like Rigel and Betelgeuse) as well as multitudes of modest-mass, modest-luminosity stars. The stellar nursery in the Sword of Orion has hatched not only the very hot, luminous Theta-1 (θ^1) C and Theta-2 A Orionis, but a whole cloud (still obscured to us by dust) of hundreds of solar-mass stars.

Numerous dark clouds of interstellar matter are silhouetted against the glow of the Milky Way, the most conspicuous and famous of them being the Great Rift in the northern Milky Way and the Coalsack in the southern Milky Way. Of course star formation is not occurring in every dark cloud we see along the Milky Way. Some trigger is necessary to initiate the first stage of star formation in the cloud, the gravitational contraction of its denser pockets. This trigger could be a supernova shock wave or radiation pressure from a nearby cluster of hot, luminous stars or some other mechanism.

Whatever the trigger, shortly after a dense section of an interstellar cloud begins contracting its own gravitational field takes over and it begins gravitationally collapsing. As it collapses, it fragments, first into individual cloudlets of several thousand solar masses each, which probably should be thought of as protoclusters, and finally into individual globules =

protostars. Gravitational collapse releases energy (the kinetic energy gained by a falling rock, for example, is what enables it to break a windshield), which heats the interior of the contracting globule (with some of the energy diverted first to dissociate the hydrogen molecules into atoms and then to ionize the hydrogen atoms). During the latter stages of its gravitational collapse, a globule will become self-radiant; but the energy it radiates is merely the by-product of gravitational contraction. Some nucleosynthesis of hydrogen nuclei (the free protons) into deuterium (a hydrogen nucleus with one proton and one neutron) and lithium occurs during the latter phases of gravitational contraction.

Eventually, however, the globule's central temperature reaches 20 million kelvin (K), at which point the fusion of four hydrogen nuclei (protons) into one helium nucleus (two protons plus two neutrons) proceeds with a net release of energy. A star has been born. The infant star still will be swathed in remnants of its natal dust cloud – indeed it will be observable only by the infrared radiation its visible light excites in the surrounding dust cloud – but this cloud will soon be dissipated by the star's increasing stellar winds and radiation pressure. (M78 in Orion and NGC 1333 in Perseus are two examples of emission + reflection nebulae around young stars which have just dispersed a section of the obscuring dust cloud in which they have been hitherto embedded.) The new star's two energy sources – the last stages of its gravitational contraction and the first stages of its nuclear hydrogen-burning – complement and conflict, with the consequence that the newly-formed star will be erratically variable in light output and often even in surface temperature = spectral type. Such stars are called either nebular variables, because they are found only in complexes of bright and dark nebulae, or T Tauri and RW Aurigae stars, after two of their earliest-recognized specimens. After several tens of millions of years (in the case of a solar-mass star – less for more massive stars) of erratic behavior, the infant star's gravitational contraction ceases and it settles down onto the stable zero-age main sequence. Astrophysicists say that main sequence stars are in hydrostatic equilibrium: this simply means that the inward-directed force of gravity is precisely counterbalanced by the outward-directed forces of gas and radiation pressures.

The new star's mass determines exactly where upon the main sequence it will settle: the more massive a protostar, the hotter and more luminous its main sequence descendant shall be. The range of stellar masses actually is not very large. The minimum mass required to generate sufficient energy from gravitational contraction to ignite a protostar's interior hydrogen is only 8% the Sun's. The result is an M7 or M8 red dwarf with a total energy output of less than 1/2,000 the Sun's. On the other hand, the hottest and brightest stars (for example Zeta Puppis, a rapidly-evolving O4 supergiant) radiate more than 2 million times more energy each second than the Sun but contain only about 100 times the mass. The main sequence lifetime of the Sun is calculated to be about 10 billion years. Thus the main sequence lifetime of an O-type star of 100 solar masses which is radiating energy 2 million times faster than the Sun is

$$\frac{10,000,000,000 \text{ yrs} \times 100 \text{ times the solar mass}}{2,000,000 \text{ times the energy burning rate}} = 500,000 \text{ years}.$$

By contrast, a red dwarf of $1/10 = 0.1$ solar mass (an M6 V star) radiating only 1/800 as much energy each second as the Sun will have a main sequence lifetime of

$$\frac{10,000,000,000 \times 0.1}{1/800} = 800 \text{ billion years}.$$

These numbers, crude approximations though they be, demonstrate quite convincingly that the most luminous members of our Galaxy are also its most transient: they literally burn themselves out in what is but an instant of our Galaxy's lifetime (though, as we shall see, they do not go quietly!) The red dwarfs, on the other hand, could, if such a thing was possible, outlive the universe itself.

M- and K-type red dwarfs in effect have no evolutionary history: they will continue to glow feebly until the very end of time. Solar-type and high-mass stars, by contrast, not only are evolving toward inevitable extinction, they have radically different evolutionary histories that climax in dramatically different deaths.

When the hydrogen is exhausted in the core of a solar-mass star, hydrogen-burning continues in a shell around the inert helium core. The time required for core hydrogen exhaustion is, however, the bulk of the star's main sequence existence, requiring, in the case of the Sun, some 7 billion years. As the hydrogen-burning shell slowly eats upward toward the star's surface, it leaves in its wake an increasingly more massive helium core, which slowly gravitationally contracts and heats. The increase in energy from hydrogen shell burning and from helium core contraction increases the star's overall luminosity and expands its outer layers, which cool. (Cooling of the star's surface can occur despite the increase in luminosity because the radiating surface area increases proportionately with the *square* of the radius: thus a relatively small increase in radius results in a surprisingly large increase in radiating surface.) Hence the star begins evolving up and off the main sequence toward the upper right quadrant of the H-R diagram, becoming first a luminosity class IV subgiant and finally a class III red giant. This too is a very leisurely process in a solar-mass star, requiring around 2 billion years.

The density in the helium core eventually becomes so great that the core becomes *degenerate* – that is, its electrons (a star's internal gas consists of atomic nuclei in a sea of free electrons) are as packed together as they can be. However, the core continues absorbing heat from the hydrogen-burning shell above and its temperature keeps rising until it reaches 80 million K, at which point helium can be nucleosynthesized into carbon and oxygen with a net release of energy. The helium ignition, however, occurs explosively in a *helium flash*, which for a brief time generates each second as much energy as 100 billion Suns – in fact as much energy as our entire Galaxy! But none of this incredible energy output ever reaches the star's surface: it is *all* absorbed in raising the helium core out of electron degeneracy.

In fact after the helium flash, which takes place when the star is at the very tip of the red giant branch as luminous and as red as it will get, the star actually *decreases* in luminosity. Moreover, its outer layers shrink in response to the decrease in internal energy generation, and the star quickly evolves back down the red giant branch to its base, at which is the *horizontal branch*, a conspicuous feature of the H-R and CM diagrams of globular clusters. A horizontal branch star has a hydrogen-burning shell, a helium-burning core, a spectral type of A or F, an absolute magnitude of around +1 to +0.5, and a mass of only about 70% the Sun's. The decrease in mass had occurred while the star was near the tip of the red giant branch, where strong stellar winds of particles had escaped from its bloated, tenuous atmosphere.

After some 70 or 80 million years, the helium at the very center of the star's core is exhausted and helium burning continues in a shell around an inert, but *extremely* hot core of oxygen and carbon. As mass is added to it from the helium-burning shell, the oxygen/carbon core gravitationally contracts. Thus the star now has, beneath its exterior hydrogen envelope, an outer shell in which hydrogen is being nucleosynthesized into helium, an inner shell in which helium is being nucleosynthesized into oxygen and carbon, and an inert, but gravitationally contracting, oxygen/carbon core. This increases the star's overall energy output, and once again its outer layers expand and cool. But this is a situation fraught with potential instabilities, and the star experiences the first of them on its way back to the red giant region as it passes through the RR Lyrae *instability strip* and undergoes strong, but extremely regular, pulsations.

After successfully negotiating the RR Lyrae instability strip, the star evolves back up the red giant region by the *asymptotic giant branch*, a slightly different route from its original foray into this area of the H-R diagram. As the now extremely old, highly evolved star continues to expand and cool and approaches for the final time the far upper right of the red giant region, its pulsations increase in violence and it becomes a *long-period*, *semiregular*, or *irregular* M-type red variable. Moreover, in at least some asymptotic giant branch stars internal mixing seems to occur and core carbon is dredged up to the stellar surface: these are the intensely red, irregularly variable *carbon stars*. In other asymptotic branch giants such exotic elements as zirconium, ytterbium, barium, strontium, and technetium are mixed to the surface: these are stars of spectral type S, and their existence implies that at least some exotic nuclear reactions must be occurring in the oxygen/carbon cores of highly evolved red giants.

However, the core never does succeed in reaching a sufficiently high temperature for the onset of core-wide nucleosynthesis of the oxygen and carbon into heavier elements. Before that point the pulsations of the star's outer layer become so extreme that they reach escape velocity and are puffed out into space as a *planetary nebula*. The multiple shell structure of most planetaries suggests that an asymptotic branch red giant usually suffers several shell ejection events. After only a few ten thousand years the expanding planetary nebula dissipates itself into the near nothingness of interstellar space. Though the planetary nebula phase of a star's evolution is exceedingly brief by astronomical standards – on such scales it is comparable in duration to the beat of a mayfly's wing – it occurs to such a large number of stars that over 1,200 planetary nebulae are known in our region of the Galaxy alone.

Meanwhile, the loss of the star's hydrogen-rich envelope puts an end to its hydrogen-burning shell, and the helium-burning shell soon fizzles out as well. With no internal energy source to withstand gravitation the ex-red giant's inner layers begin to contract, becoming even hotter. Thus the expanding planetary exposes a stellar object which appears as an underluminous O-type subdwarf – very blue, very hot, but very small and consequently relatively faint. And this stellar object continues to get even smaller, losing luminosity but not temperature (which is sustained by energy derived from gravitational contraction and by residual helium-burning of infalling planetary material). The contraction is rather rapid, and sometimes even before the surrounding planetary dissipates this stellar core becomes sufficiently dense to drop into electron degeneracy: at that point it has officially become a *white dwarf*. The typical white dwarf packs nearly the mass of the Sun into a sphere the size of the Earth.

For any stellar remnant of less than about 1.25 solar mass no further contraction is possible, so the white dwarf has no further energy source and simply radiates its residual energy away. However, these objects have such small surface areas that billions of years must pass before a classic white dwarf with the surface temperature of an O, B, or A star will cool down to a white dwarf with the surface temperature of a yellow G-type star. Nevertheless several G-and K-type white dwarfs have been identified, the brightest of them being the magnitude 12.4 Wolf 28, "Van Maanen's Star," in Pisces. (See Chapter 24.) Given sufficient time a white dwarf will cool down to the 3K ambient temperature of interstellar space.

The evolution of a high-mass star is significantly – a better word is *drastically* – different from that of a solar mass star. Things are different from the start because, as it contracts, a high-mass protostar generates a great deal more energy than a solar-mass protostar. (Energy released by gravitational contraction = the amount of mass × the distance which that mass falls – and a high-mass protostar not only has more mass than a solar-mass protostar, it initially occupies a greater volume for that mass to contract from.) Thus a high-mass protostar heats up much more quickly than a solar-mass protostar and reaches the critical 20 million K central temperature at which hydrogen is effectively nucleosynthesized into helium long before its center is anywhere near as dense as a solar-mass protostar's.

The simple fact of a high-mass star's low central density changes everything about its subsequent evolution. To begin with, initial hydrogen-burning in such stars does not take place in a compact core but throughout an extended central region containing some 20% of the star's mass. When the hydrogen fuel gives out in this extended core, hydrogen burning shifts into a thick shell around the now inert helium core, which rapidly contracts. All this extra energy is absorbed in expanding the star's outer layers, and quickly (in the case of a 9 solar-mass star, a mere quarter million years) evolves from B-type subgiant status to a red M-type supergiant. A huge amount of energy is required to drive even only the outer layers of a really massive star out a couple hundred million miles (as we experience for ourselves even in the simple act of climbing stairs, gravity is a very energy-intensive phenomenon): hence the overall luminosity of the star remains relatively constant as it expands despite the much greater energy generation in its interior from thick-shell hydrogen burning and from helium core contraction.

Thus a red supergiant has a hydrogen-burning shell around a contracting helium core that is becoming ever denser and hotter. However, because of the initially lower central density of a massive star, a red supergiant's contracting helium core will not drop into degeneracy before becoming sufficiently hot – 80 million K – to ignite its helium. The helium is nucleosynthesized into carbon-twelve by the triple-alpha process (so called because three alpha particles – helium nuclei of two protons and two neutrons – combine to make one C^{12} nucleus – 6 protons plus 6 neutrons). In a 9 solar-mass star helium ignition occurs after the star has been a red supergiant for less than 100,000 years; however, in extremely massive stars – 15 solar masses and more – helium ignition in fact takes place while the star is still expanding into a red supergiant.

The onset of core helium-burning in a 9 solar-mass star upsets its internal and external balance. The hydrogen-burning shell thins, overall internal energy generation decreases, and the outer layers contract, thus forcing the star back to the left across the K, G, and F Ib supergiant regions of the H-R diagram. The contraction preserves the star's total luminosity, but at this point it crosses the Cepheid instability strip, suffering the clockwork pulsations characteristic of such variables. Cepheids, then, have an upper hydrogen-burning shell above a lower helium-burning shell encasing a contracting, and increasingly hot and massive, inert carbon core.

In stars of 9-10 solar masses the carbon core eventually gets sufficiently dense to drop into degeneracy and the star's subsequent evolution is so complicated that it has not yet been worked out by the theoreticians. However, in more massive stars, which started their main sequence lifetime with even less dense interiors than their 9-10 solar mass siblings, the contracting inert carbon core reaches ignition temperature before dropping into degeneracy, nucleosynthesizing oxygen. In extremely massive stars the resulting oxygen core does not drop into degeneracy either, but in its turn ignites, nucleosynthesizing neon. This cycle – core exhaustion, core contraction, core ignition – will be repeated at the center of an extremely massive star until it has an "onion-shell" structure of superimposed layers nucleosynthesizing successively heavier elements (though the layers might not all be firing simultaneously).

But with iron the cycle ends – catastrophically. The nucleosynthesis of iron into still heavier elements does not release energy, it absorbs energy. Hence core contraction accelerates, and the core rapidly increases in density and temperature. Obviously this cannot go on indefinitely. In fact at a temperature of 5 billion K the iron nuclei disintegrate into helium nuclei and free neutrons, absorbing so much more energy in the process that it is as if the core has been bodily removed from the interior of the star. With in effect nothing holding them up, the star's upper layers violently implode.

In the resulting cataclysm huge quantities of neutrinos are produced which, because they are transparent to the upper layers of the star, carry out enormous amounts of energy. (This preliminary neutrino outflow was actually observed from the 1987 supernova in the Large Magellanic Cloud prior to its visual brightening.) Meanwhile the outer layers rebound in the terrific explosion called a supernova. The energy produced by a supernova is something like 2×10^{51} ergs. Because the energy production rate of the sun is about 3.8×10^{33} ergs/second and there are 3.2×10^7 seconds/year, the amount of time necessary for the Sun to radiate as much energy as is generated in a single supernova explosion is

$$\frac{\text{Supernova energy}}{\text{Solar energy production per year}} =$$

$$\frac{2 \times 10^{51} \text{ ergs}}{(3.9 \times 10^{33} \text{ ergs/sec}) \times (3.2 \times 10^7 \text{ sec/yr})} = 15 \text{ billion years}$$

But the Sun's main sequence lifetime is only 10 billion years!

The fate of the supernova's collapsed core depends upon its mass. If it is less than about 2.25 solar masses, its gravitational collapse will end when its protons and electrons have been gravitationally crushed into neutrons and this neutron gas has been squashed into degeneracy: the core is now a neutron star. Such objects are only about a dozen miles in diameter but have incredibly rapid rotational velocities because the gravitational collapse carried the now-destroyed star's rotational energy with it down into this tiny dozen-mile sphere. (The same phenomenon – on a rather more modest scale! – is seen when a spinning skater draws in their arms.) The original star's magnetic field has also become concentrated into this minute volume and is so intense that the neutron star's radiation is channeled along the field's lines of force and beamed out its magnetic poles. The star's rapid rotation sweeps its magnetic poles around much like a lighthouse beam (unless the magnetic and rotational poles coincide), and if the observer happens to lie in a direction swept by the beam, they see the star rapidly turn on and off as a pulsar. The Crab Pulsar at the center of the Crab Nebula flashes on and off in X-rays, radio waves, and visible light about 33 times each second.

But a collapsed supernova remnant of more than about 2.25 solar masses has an even more bizarre end. Its gravity overwhelms even its degenerate neutrons, crushing them into a Schwartschild discontinuity – better known as a black hole – which is so dense that the escape velocity on its surface is greater than the speed of light. Such objects contain several solar masses of material (the state of which is not understood) in a sphere only about 4 miles across, and have the nasty habit of sucking up anything unfortunate enough to get into their gravitational well. Indeed, some black holes – such as that orbiting Cygnus X-1, the O9.7 Ib supergiant star HDE 226868 – are thought to be ingesting their companion stars, the outer layers of the latter vortexing into the black hole and being superheated to such high temperatures that the material, in something of an astrophysical death-cry, is emitting extreme amounts of X-rays.

Meanwhile the supernova shell continues to expand. Planetary nebulae have expansion velocities of around 10 to 20 kilometers per second. But the early expansion velocities of supernova shells are several thousand km/second: indeed, the Crab Nebula even now, 950 years after the original explosion, is still expanding at nearly 1,000 km/second. The Crab seems to be an unusual type of supernova remnant because it radiates by the synchrotron process in which electrons ejected by the Crab Pulsar near the speed of light accelerate or decelerate along the nebula's magnetic lines of force, emitting highly polarized light. (The electrons are carrying away some of the pulsar's rotational energy, the consequence being a measurable increase in its rotational period during the decades since its discovery.) However, neither Tycho's Star of 1572, nor Kepler's Star of 1604 have left as impressive remnants as the Crab Supernova: optically all that can be seen, even on long-exposure red-sensitive photographs, at the locations of Tycho's and Kepler's stars are a few nebulous shreds. The advanced stage of supernova shell evolution is represented by the Veil Nebula in Cygnus (the NGC 6992/5 segment of which is visible in large binoculars) and IC 443 in Gemini (a difficult telescope object), both of which are thousands of years old.

I.6 Variable Stars

Several of the many types of variable stars were mentioned in the preceding precis of stellar evolution. The behavior of the different variable star types differs dramatically. Some variables have ranges of 10 magnitudes or more in periods measured in years; others have ranges of only a few hundredths of a magnitude in periods of just hours. Some variables are as regular as clockwork; others are only vaguely periodical, and many are hopelessly erratic. In some variables the spectrum and color change with the light curve. Several types of variable display explosive behavior; but only supernovae can destroy themselves entirely.

The basic tool for studying variable stars is the light curve, a simple plot of the star's magnitude changes with time. Anyone with graph paper and a comparison chart showing the magnitudes of the (nonvariable) stars in a variable's field can plot a variable star's light curve. Making the requisite magnitude estimates for plotting light curves is so expensive of telescope time that, excluding some of the most exotic variables, this is work the professionals have left to the amateurs. Indeed, making magnitude estimates of variable stars is one of the most scientifically useful things an amateur can do. The American Association of Variable Star Observers gathers variable star estimates made by amateurs around the world, processes them, and forwards the data to the concerned professionals. The AAVSO also provides interested amateurs with the advice and comparison charts necessary for this work. Variable star observing is also an excellent way to develop an understanding of some of the physics of astronomy: not only will it familiarize you with the many variable types, each of which (excluding eclipsing variables) occupies a specific location on the H-R diagram and in stellar evolution, but by plotting your own light curves from your own magnitude estimates you can get a better feel for the mechanics of stellar variations, whether pulsations, explosions, or eclipses. The three basic types of variable star are:

Intrinsic or *pulsating variables* – stars in which the light changes are the consequence of actual changes in the star itself, commonly pulsations of its outer layers.

Eruptive variables – stars which exhibit explosive behavior, often periodic and frequently the influence of a near companion.

Eclipsing variables – stars in which the light changes are the consequence of the partial or total eclipses of one another by the components of a binary star system.

I.7 Intrinsic (Pulsating) Variables

At several points in their evolution most, if not all, stars are vulnerable to internal instabilities that manifest themselves in actual expansions and contractions, often periodic, of their outer layers. The following survey of the types of intrinsic variable stars is in approximate order of stellar evolution (not necessarily the same thing as actual age), the first types described being stars near their birth and the last types stars near their death.

Nebular variables are newly-formed stars still within the residuals of the cloud of gas and dust from which they have just condensed. Their light changes are completely erratic, both in range and in period, and are usually accompanied by changes in the star's spectrum and color. The spectrum always has emission lines from the gas and dust still falling into the star from the residual nebular cloud around it. The light changes can reach several magnitudes in amplitude. These variables are also often called RW Aurigae or T Tauri stars, after their two prototypes. Three splendid examples of nebular variables are R, T, and TY Coronae Australis embedded in the small molecular cloud Bernes 157 and surrounded by the tiny emission + reflection nebulae NGCs 6726, -27, and -29. (See Chapter 38.) R Monocerotis, the star at the tip of Hubble's Variable Nebula, NGC 2261 in Monoceros (Chapter 20), is also a nebular variable. Many faint nebular variables are embedded in the Orion Nebula M42.

Flash stars are essentially nebular variables without the nebula: they are pre-main sequence K and M red dwarfs which are still unstable because they are still gravitationally contracting; but in such low-mass stars the process of contraction onto the main sequence takes so long that the nebula within which they were formed has long since dispersed. A number of flash stars have been observed even in the 70 million-year-old Pleiades. The flashes can brighten the star from 1 to almost 4 magnitudes in just a couple of hours.

Beta Canis Majoris variables are high-mass, high-luminosity B1 to B3 subgiants just beginning their evolutionary expansion off the main sequence toward the right in the H-R diagram. Their variations, though only a few hundredths of a magnitude in range and a few hours in period, result from the shift from core to shell hydrogen-burning in the star's deep interior. In addition to Beta Canis itself, the brightest examples of this type of variable include Lambda Scorpii, Beta Cephei, and Gamma Pegasi.

Alpha Canum Venaticorum variables resemble Beta Canis Majoris variables as stars just beginning their evolutionary expansion off the stable main sequence and therefore suffering internal instabilities. However, they are distinguished by strong and variable magnetic fields and by peculiar abundances of such elements as silicon, magnesium, chromium, and europium. They are therefore also known as **magnetic spectrum variables**. Alpha Canum stars are of spectral types late-B to mid-A and have absolute magnitudes normal for main sequence stars of those spectral types, -1 to $+2.5$. Their ranges are under 0.2 magnitude and their periods under 6 hours. Their variations possibly are not due to pulsations at all but perhaps simply to stellar rotation sweeping magnetic "hot spots" on the star's surface across our line of sight. Their periods are consistent with the rapid rotation that we would expect from stars with strong magnetic fields and with peculiar chemical abundances on the surface caused by rotationally-induced mixing with the star's deep interior. In addition to Alpha Canum, other bright representatives of these variables include Epsilon Ursae Majoris (the brightest star in the Big Dipper) and Beta Coronae Borealis.

Delta Scuti variables, in effect the lower-mass version of the Beta Canis Majoris stars, are A2 to F6 subgiants, normal giants, and bright giants with absolute magnitudes from -2 to $+2$ but only two or three solar masses of material. They are just beginning their evolutionary expansion up and off the main sequence in response to the interior shift from core hydrogen burning to shell hydrogen burning around an inert but contracting helium core. Like the Beta Canis stars, their ranges are under 0.2 magnitude and their periods under 6 hours. But because they are much less massive than Beta Canis stars, the main sequence lifetimes of the Delta Scuti variables had been much longer and they therefore are now much older: Beta Canis stars are only around 20-30 million years old, but the Delta Scuti variables are up to several hundred million years old. The brightest of these variables are Rho Puppis and Beta Cassiopeiae. Delta Scuti stars are also common in such intermediate-age open clusters as the Hyades.

Classical Cepheid variables are yellow supergiants of spectral types F, G, and early K which pulsate in remarkably regular periods. Their light curves are characterized by a rapid rise to maximum followed by a slow decline back to minimum. Their spectra change with their brightness, being earliest (hottest) near maximum and latest (coolest) near minimum, and Cepheids of the longest periods often cool by a whole spectral type during the course of their light cycle. Cepheids can have periods from as little as 1 day to more than 50 days and light ranges between 0.5 and 1.5 magnitudes; but periods of around a week and ranges of about 0.7 magnitude are typical. The most important property of Cepheid variables is the period-luminosity law: the longer a Cepheid's period, the greater its luminosity. The period-luminosity law, coupled with their high intrinsic brightness as supergiants, makes Cepheids invaluable for estimating distances to other galaxies (out to about 30 million light years). Among the naked-eye Cepheids are Eta Aquilae, Zeta Geminorum, and l Carinae.

W Virginis Cepheids, though also yellow F, G, and early K supergiants with precisely repeated light curves characterized by a rapid rise to maximum followed by a slower decline to minimum, are in fact physically very different stars from the classical Cepheids: they are ancient Population II objects (hence they are sometimes called Type II Cepheids) conspicuous in globular clusters and in our Galaxy's halo whereas classical Cepheids are thin disk Population I stars found in and near the Galaxy's spiral arms. W Virginis Cepheids also follow a period-luminosity law, but on the average are nearly 2 magnitudes fainter than classical Cepheids and therefore of more limited value as distance-finders. The brightest W Virginis Cepheid is no less than the primary of Polaris, the North Star; however, its light range is too slight to be detected visually.

Dwarf Cepheids are low-mass pulsating variables of spectral types B8 to F2, periods of 1 to 5 hours, light ranges of a few tenths of a magnitude, and Cepheid-like curves. Despite the name they are not even remotely related to classical Cepheids; and the only property they share with the W Virginis Cepheids is in being highly-evolved Population II objects. They are underluminous subdwarfs, their absolute magnitudes of $+1$ to $+5$ locating them on the H-R diagram well below Population I main sequence stars of the same spectral types. No dwarf Cepheid is a naked-eye object. Two of the earliest-discovered specimens of this class of variable are CY Aquarii and SX Phoenicis.

Long-period variables, or LPVs, are pulsating red giants with periods from a few dozen up to several hundred days and light ranges from 3 to over 10 magnitudes. However, neither the period nor the light range of an individual LPV is the same from one cycle to the next: the numbers cited in the tables for these quantities are mere averages. The archetype LPV, Mira = Omicron Ceti, has an average period of 330 days and usually varies between magnitudes 3.5 and 9. The spectrum of an LPV also changes during its light cycle, being earliest near maximum and latest around minimum: the spectral range of Mira, for instance, is M5.5e–M9. The light curves of these stars display a more rapid rise to maximum than fall to minimum. Typically the longer an LPV's period, the greater its light range and the later its spectral type. The peak absolute magnitudes of LPVs are thought to be between -1 and -2 (the longer-period stars probably with the greater luminosity), but their light decrease is something of an optical illusion: as an LPV cools, most of its radiation shifts from visible red to invisible infrared wavelengths, hence its total energy output actually drops by only 1 magnitude or less. LPVs are evolved stars of one or two solar masses which have two internal energy sources: an outer shell where hydrogen is being nucleosynthesized into helium and an inner shell where helium is being nucleosynthesized into oxygen. The conflict between these two energy sources causes the internal instabilities that are externally manifested by the pulsations of the star's outer envelope. Most LPVs seem to be thick disk Population I stars; but because some of the shorter-period LPVs are found in globular clusters, a few of them must be highly-evolved Population II objects. Besides Mira, the brightest LPVs include Chi Cygni, R Hydra, and R Leonis.

Semiregular and **irregular variables** are pulsating red giants and supergiants with only marginally recognizable, or completely erratic, light cycles. These stars are of two completely different groups: (1) Most semi-regular and irregular red variables are simply LPV-type stars with extreme cycle-to-cycle fluctuations in period and magnitude range – though their periods are almost always under 200 days, their ranges less than 3 magnitudes, and their luminosity class II rather than the typical LPV III. Many of these stars seem to be evolved Population II objects. Examples are Beta Pegasi, Rho Pegasi, and the primary of the Alpha Herculis system. (2) The second group of semi-regular and irregular red variables are young, luminous, but quickly evolving red supergiants with not just two, but several internal shell energy sources – the disruptive multiplicity of which no doubt explains their greater irregularity as a group as opposed to the less luminous M II and III semiregulars and LPVs. Examples include Betelgeuse and Antares.

RR Lyrae variables are highly-evolved solar-mass

stars of exceptionally regular light curves resembling those of Cepheids, though often with an even steeper rise to maximum. The periods of RR Lyrae stars are between 0.1 and 1.4 days and their light ranges between 0.5 and 1.25 magnitudes. Their spectral types are from early-A through mid-F, and, like those of the Cepheids, change during the course of their light cycles, being earliest at maximum light and latest at minimum light. They are conspicuously located on the horizontal branch of the H-R diagrams of globular clusters and therefore are often referred to as cluster variables (though globular clusters also contain W Virginis Cepheids, short-period LPVs, and RV Tauri stars). Because they are on the appropriately named horizontal branch, all RR Lyrae stars have about the same median absolute magnitude, +0.7, and therefore are a powerful distance-finding tool for our Galaxy's globulars. Though most RR Lyrae variables are ancient, highly-evolved Population II members of our Galaxy's halo component, those with periods under 0.5 day seem to be somewhat younger than the rest, for they can be found in our Galaxy's thick disk.

RV Tauri variables are pulsating yellow or red G, K, and M supergiants with absolute magnitudes around -4.5 or -5 and a basic light curve of several dozen days similar in shape to that of the Cepheids but sometimes superimposed upon a much longer LPV-like cycle. The prototype of the class, RV Tauri itself, has a 79-day short period and a 1,300-day long period. The light ranges of these stars are between 2 and 3.5 magnitudes. Their spectra undergo complicated changes during their light cycle: at maximum these variables have the color of early-G type stars but their spectra are superimposed with bright emission bands reminiscent of those of LPVs; and at minimum their color is early-K, but their spectra have M-type titanium oxide absorption lines. RV Tauri variables, despite the shape of their light curve, are neither young Cepheid-like stars nor middle aged LPVs but in fact extremely old Population II objects found in globular clusters and throughout our galaxy's halo. The brightest RV Tauri star is R Scuti.

I.8 Eruptive Variables

Nova, from the Latin word for "new," was applied by Renaissance European astronomers to several "new" stars that appeared and then faded from sight during the 16th, 17th, and 18th centuries, including Tycho's Star of 1572, Kepler's Star of 1604, and the star now designated as P Cygni. However, in modern astrophysical usage the term nova does not refer to any of these objects, the first two of which were in fact supernovae and the third a recurrently eruptive supergiant star, but rather to a type of violent but not catastrophic explosive binary star system. The major groups of eruptive variable stars, from intrinsically faintest to intrinsically brightest, are:

Flare stars in a sense hardly qualify as "eruptive," for they are simply very faint red dwarfs which at erratic intervals emit what seems to be no more than a grandiose version of the typical solar flare: the event is so conspicuous, brightening the star by a magnitude or two in just a second, only because it takes place on such an intrinsically faint object. All flare stars are M-type dwarfs with hydrogen-emission lines in their spectra. The most notorious of them are Proxima Centauri (the nearest star after the Sun to the Earth), Krueger 60B in Cepheus (see Chapter 11), and UV Ceti (which once flared by 5.5 magnitudes).

Cataclysmic variables, also known either as dwarf novae or, after their prototypes, SS Cygni and U Geminorum stars, are very close binary systems that couple a yellow G-type subdwarf with a compact bluish subdwarf. The two stars are in such tight orbits around each other that matter is lost from the G-type star's extended atmosphere to the blue subdwarf, upon the surface of which it accumulates until it becomes explosively unstable. The outbursts recur in irregular intervals of anywhere from two weeks to three months (longer in some SS Cygni stars) and brighten the system by about 3 to 6 magnitudes – though even at maximum such variables reach only absolute magnitudes of +4 or +5 and therefore are only about as luminous as the Sun. Often these stars have two types of maxima, one more extended than the other; and a subclass of the cataclysmic variables, Z Camelopardalis stars, have prolonged periods of nearly constant brightness intermediate between maximum and minimum.

Symbiotic variables are similar to SS Cygni stars except a standard K- or M-type red giant is paired with the compact B-type subdwarf. The large star loses mass from its tenuous, bloated envelope into a ring around the subdwarf. The outbursts, 4 or 5 magnitudes in amplitude, are thought to occur both in the ring and on the subdwarf's surface among material accreted from the ring. The period between outbursts is generally between 1 and 3 years and the peak absolute magnitude reached is around -5, a luminosity of 10,000 Suns. Two examples of symbiotic systems are Z Andromedae and R Aquarii, in both of which the red giant component also exhibits normal LPV or semi-regular pulsations.

Recurrent novae are eruptive stars that have suffered novalike outbursts more that once. RS Ophiuchi exploded four times from 1898 to 1967 and T Pyxidis five times from 1890 to 1967. T Coronae Borealis, on the other hand, is a recurrent nova with the minimum of two outbursts, in 1866 and 1946: the amplitude of its explosions, however, has been 7–8 magnitudes whereas RS Ophiuchi's have been only 6–7 magnitudes. The light curves of the recurrent novae, other than their modest 6 to 9 magnitude rise, resemble those of the more violent classical novae, with a very rapid rise to maximum followed by a leisurely decline back to "normal" brightness. Spectroscopic studies suggest that recurrent novae consist of a yellow to red G- or M-type subgiant in close orbit with a bluish A- or B-type subdwarf. Thus they seem to be similar to the dwarf novae and presumably the explosions are generated in a similar manner: material from the bloated subgiant accumulates on the surface of the subdwarf until it becomes explosively unstable. Recurrent novae, however, achieve much higher absolute magnitudes than dwarf novae, 0 all the way up to -9 as opposed to the latter's $+4$ or $+5$.

Classical novae seem to be physically similar to dwarf novae, symbiotic variables, and recurrent novae, though in their case the bloated K or M giant or subgiant is paired with a near-white dwarf. The mechanism of the explosions once again seems to be the accumulation of material from the extended star onto the compact star's surface until it becomes unstable and spontaneously erupts into nucleosynthesis. However classical nova outbursts brighten their system by 10 to 15 magnitudes (a factor of 10,000 to 1 million), reach absolute magnitudes of up to -10, and – if they really are just extreme recurrent novae – must have periods measured in millennia. Their light curves are remarkable for the rapidity of the rise, maximum light being achieved in just a few days. A subclass of classical novae appropriately termed slow novae require a couple weeks to reach maximum, may have two or three distinct peaks, and require years, not just months, to fade back to prenova brightness. Despite the brilliance of nova explosions, only a few millionths solar mass of matter is actually ejected (at velocities of a couple thousand kilometers per second – less than half that of supernovae shells) and the two stars of the system return to their preoutburst state apparently unscathed. Two of the more famous classical novae were Nova Aquilae 1918, which brightened to apparent magnitude -1.4, and Nova Cygni 1975, a magnitude 1.8 object with the astonishing range of 19 magnitudes.

P Cygni stars are blue supergiants that display continuous and almost explosive mass loss in expanding circumstellar shells, with occasional more serious, near-supernova-luminosity, eruptions. P Cygni itself, presently a 5th magnitude object, twice during the 17th century brightened to magnitude 3 or 3.5. Similarly, Eta Carinae, now famous for the Hubble Space Telescope photo of its expanding bipolar nebula, though presently only a 7th magnitude object, reached magnitude -1 in 1843. At their maximum P Cygni and Eta Carinae probably achieved absolute magnitudes near -11 and -14, respectively – the latter value corresponding to a luminosity of nearly 40 million Suns! P Cygni stars are perhaps hyper-supergiants in the first stage of their evolutionary expansion into red supergiants, a process that seems to require the shedding of a considerable fraction of the original blue hyper-supergiant's mass. P Cygni stars are probably related to the Wolf-Rayet stars, which are also hyper-supergiants experiencing ongoing explosive mass-loss in expanding shells – though the Wolf-Rayets exhibit no variability in luminosity.

Supernovae are catastrophic stellar explosions of such violence that they virtually annihilate the stars involved, leaving in the wake of the disaster an expanding debris cloud around a stellar corpse in the form of a neutron star or a black hole. Supernovae can reach the luminosity of an entire galaxy. They are of two radically different types: Type II supernovae are the eruptions of massive Population I supergiants which began nucleosynthesizing iron into even heavier elements in their core – a disastrously energy-absorbing process; and Type I supernovae occur among the highly-evolved Population II stars of galaxies and apparently take place when two white dwarfs in close, mutually decaying orbits literally "fall" into each other. Type II supernovae are the most common; but Type I are the most violent, reaching absolute magnitudes of -18 to -19, two magnitudes greater than Type II. The two types are also distinguished by their light curves: Type I explosions display a very rapid rise to maximum followed by an initially rapid but later slow, steady decline into invisibility; but Type II supernovae rise to a maximum in a more moderate manner and a have a peculiar near-standstill about 80 days into their leisurely decline after which the decline becomes more precipitous. The 1987 supernova in the Large Magellanic Cloud was a Type II event (not surprising: the LMC contains only Population I stars), and Tycho's Star of 1572 was a Type I supernova.

Appendix: Though **R Coronae Borealis stars**, sometimes called **reverse novae**, are always included among the eruptive variables, they are not really explosive stars at all. Their outstanding characteristic is their sudden, unpredictable, unperiodic decrease in brightness by 5 magnitudes and more. Some R Coronae Borealis stars display moderately carbon-rich R-type spectra; others have the spectra of F, G, K, or even M supergiants with peculiar chemical abundances. All this suggests that these stars are asymptotic branch giants undergoing severe internal mixing. If the spectral features are not misleading, R Coronae Borealis stars would have maximum absolute magnitudes of around -5 and minima in the range of $+2$ or $+3$. The classic theory has been that the abrupt drop in brightness of these stars is the consequence of the formation of what amounts to a "soot cloud" of carbon molecules in their outer atmosphere. But a more recent, and more exotic, hypothesis is that R Coronae Borealis stars result from the gentle coalescence of carbon/oxygen and helium white dwarfs that had been a close binary system: the consolidated star should have a helium envelope, a planetary-nebula-like carbon-rich shell providing the "soot cloud," and a resuscitation of nuclear-burning processes – hence the supergiant-class luminosity.

I.9 Eclipsing Variables

As the name suggests, eclipsing variables are simply binary star systems so aligned with respect to our line of sight that one, or both, of the components partially or totally eclipses the other. Complications occur when the components are so close that their mutual gravitational attraction has distorted them into ellipsoids and/or has induced mass exchange between the stars. (A mass-exchange eclipsing binary with a subdwarf or white dwarf component technically speaking is *both* an eclipsing *and* an SS Cygni type variable. U Geminorum is an example of such a system.) The effects of mass exchange are revealed only in a binary system's composite spectrum; but the effects of gravitational distortion can be seen in its light curve. Eclipsing variables come in five major types:

Algol systems, named after the famous "Demon Star," Beta Persei, are the most straightforward type of eclipsing binary, involving two normal stars orbiting each other at a fairly safe distance and the two components mutually eclipsing. In such stars maximum light is simply a more-or-less flat plateau between eclipses and the decline to minimum and return to maximum nice straight lines. Minima will either be flat, if eclipses are total, or pointed, if eclipses are only partial. Usually two distinct minima occur. In the case of Algol, for instance, primary eclipse, 1.3 magnitudes deep, takes place when the hot blue B-type primary of the system is obscured by the cool K-type subgiant, and secondary eclipse, a mere 0.1 magnitude deep, occurs as the B star transits and partially occults the disk of the larger but much less luminous K star. Such systems are not always uncluttered by mass exchange, for even in Algol itself the B-type primary is gaining material from the outer layers of the K-type secondary's extended atmosphere. U Cephei and U Sagittae are in that respect two purer examples of Algol type eclipsing binaries than Algol itself.

Zeta Aurigae systems are basically the same as Algol-type eclipsing binaries except the cool star is a massive, luminous, extremely large K- or M-type class II bright giant or class I supergiant. In Zeta Aurigae stars the two components are usually very similar in luminosity, but mass loss from the red star is unavoidable because the atmospheres of such giants and supergiants are hundreds of millions of miles across and their outer layers very tenuously bound by gravity to the star. Nevertheless the light curve will be similar to those of Algol systems, though secondary minimum is virtually nonexistent because the binary's blue component occults very little of the extensive disk of the red supergiant as it passes in front of it. As the blue star orbits behind the bloated red star, it fades very gradually and provides a splendid opportunity to study the progressively deeper strata of the peculiar atmospheres of such objects. The periods of Zeta Aurigae stars, by virtue of the size of the red components, are necessarily measured in years. VV Cephei, for instance, has a period of 20.4 years. Three other representatives of Zeta Aurigae eclipsing binaries are KQ Puppis and 31 and 32 Cygni.

Beta Lyrae stars, also known as Lyrid eclipsing binaries, combine two young B- or early A-type giants in such extremely close orbit around one another that not only are they eclipsing, but their mutual gravitational pull has distorted both into egg-shaped ellipsoids. Thus as they orbit around one another they display an ever-changing amount of surface to our line of sight. Hence, the light curves of Lyrid systems are never flat nor linear like normal Algol-system light curves: maxima (between eclipses) are "humped," minima are valleys, and declines to minima and rises back to maxima are arced. Such light curves are called sinusoidal. In Beta Lyrae binaries not only is mass exchange taking place between the components, but such is the turbulence in the system that some matter is lost from the two stars into an expanding ring around them. The spectra of these variables are complications of absorption and emission lines from each of the components plus emission lines from the turbulent gas between and around them. Two of the brighter Lyrid-type eclipsing binaries are UW = 29 Canis Majoris and AO Cassiopeiae, though the primaries of both these systems are O rather than B giants.

Ellipsoidal variables are fundamentally Beta Lyrae-type systems without the eclipses (or at most grazing eclipses), though the stars involved need not be, as in true Lyrids, B or early A giants or subgiants. The light curves of ellipsoidal variables are sinusoidal but so shallow in amplitude that photometers are necessary to follow them. Pi-five Orionis, magnitude 3.7-3.8, is one of the few naked-eye ellipsoidal variables.

W Ursae Majoris variables, also called dwarf eclipsing systems, couple yellow F- or G-type dwarfs that revolve around one another in such close proximity the two components virtually touch and their orbital periods, hence the intervals between eclipses, are extremely short. The period of W Ursae Majoris itself, for example, is only 8 hours. (By contrast Lyrid and ellipsoidal variables, because of the size of their components, have periods of several days.) The stars' mutual gravitational pull distorts them into ovoids and the system's light curve is therefore sinusoidal like those of the Beta Lyrae variables. Moreover, like Lyrids, mass exchange is occurring between the stars and an expanding ring of matter encircles the whole system – though of course all this is on a considerably smaller scale than in the Beta Lyrae binaries! The component accreting mass will rapidly evolve into a hot, compact subdwarf, at which point the binary will resemble the SS Cygni cataclysmic variables: indeed, some W Ursae Majoris systems, including the prototype, exhibit minor eruptions of up to 0.3 magnitude. Dwarf eclipsing systems are remarkably common, but none are very bright. W Ursae Majoris itself ranges between magnitudes 7.9 and 8.6.

I.10 Double Stars

The numerous double and multiple stars that can be resolved in telescopes and binoculars are of two types:

1. Optical pairs or multiples are merely the chance alignment of nearer with more distant stars.

2. Physical pairs or multiples are gravitationally-bound stars in actual orbit around one another.

Double or multiple stars in which actual orbital motion has been observed (by plotting the stars' position with respect to one another over periods which usually must be decades long) are called visual binaries. Comparatively few visual binaries are known because the orbital speed of binaries wide enough to be resolved by Earth-based telescopes is necessarily extremely slow.

Many stars which stubbornly remain single even at the highest possible magnification under the best possible sky conditions in the largest telescopes are found to be binary systems with two or more members. Such binaries are of three types, depending upon how the secondary member(s) manifest themselves: (1) *Eclipsing binaries* are those in which a faint companion gives itself away by totally or (more commonly) partially obscuring the system's bright primary. (2) *Astrometric binaries* are those in which the faint companion's presence is revealed by its influence upon its bright primary's motion across the sky: the primary does not move in a straight line on the celestial sphere, but arcs first over and then under the line of its motion as it and its unseen companion orbit around their mutual center of gravity. (Of course, the amount of this deviation is minuscule – a mere fraction of a second of arc. Technically speaking, it is the binary system's center of gravity which traces the straight line on the sky above and below which the primary arcs.) (3) *Spectroscopic binaries* are apparently single stars with either a composite spectrum – that is, a spectrum that displays sets of absorption lines from two or more stars – or a spectrum with lines of only one star which periodically shift their position by the Doppler effect as that star orbits the binary system's center of gravity and first approaches and then recedes from us. The latter is called a single-line spectroscopic binary. The secondary of a binary system cannot be more than 2.5 magnitudes fainter than the primary for its spectrum to be visible.

Frequently the two components of a double star – the true status of which (optical or physical) is in doubt – are found to be moving in the same direction at the same speed (seconds of arc per year) on the celestial sphere. Such systems are called common proper motion (CPM) pairs. Because no actual orbital motion is implied by a common proper motion, such stars are not necessarily binaries. Indeed, some CPM pairs are quite widely separated from one another. An extreme, but very intriguing, case is that of Alpha and Beta Aquarii. These two stars are virtually identical in apparent magnitude and spectrum (Alpha is a magnitude 2.96 G2 Ib star and Beta magnitude 2.91 and G0 Ib) and therefore must just be about the same distance from us, around 700 light years. Both stars have a comparable motion on the sky, roughly 0.20″ per year toward the ESE yet they are 10° apart! Even if the two stars are both precisely 700 light years away, they are not less than 120 light years apart. Obviously they are not orbiting one another! However, they are unquestionably related: probably they are the remnants of a decayed stellar association. (See section I.14 for more about associations.)

The brightest member, the primary, of a double or multiple star (whether a physical or merely an optical pairing) is designated its "A" component, the secondary its "B" component, and any fainter associates "C," "D," and so forth by decreasing brightness. (The one major exception to this rule is the famous Trapezium multiple, Theta-one Orionis, at the heart of the Orion Nebula, the four members of which are designated A to D in west-to-east order.) In the double star data of this book the A component's magnitude is given first, then the B component's, then the separation between the two stars in seconds of arc, and finally the direction from the primary to the secondary – its position angle – in degrees, north being 0°, east 90°, south 180°, and west 270°. Multiple stars will have two or more sets of data lists for different pairings within the system: BC, for example, means that the magnitudes, separation, and position angle are for the multiple's B and C components, the separation and P.A. being measured from the B to the C star.

The best thing double and multiple stars have to offer the observer – and it is not a small thing – is their aesthetic appeal. A close pair of nearly equally-bright star gems is a beautiful sight. Beta Monocerotis in fact offers three comparably bright stars within a few arc seconds of one another. The Trapezium has four similarly bright members in a 20″×15″ area. But other brightness combinations are also attractive. The minute star-spark of a faint C component of a triple, even when the A and B components are fairly widely separated, is always a remarkably delicate, fragile object: photographs simply cannot capture the visual effect such a tiny light-chip makes in contrast to its brighter associates! Multiples such as Sigma Orionis, Rho Ophiuchi, and Burnham 1 in NGC 281 of Cassiopeia present several stars in a rich disorganization of separations and brightnesses. Once again the thick disk-images on photographs are dull compared to the glittering crystalline sparkle of the half dozen point-like star gems competing for the eye's attention in these multiple systems.

Double and multiple stars offer the observer even more: colors and color-contrasts. Beta Monocerotis is one of the finest sights in the entire sky not only because it presents three magnitude 4.5 -5.5 stars in a mere 8″ area, but all three stars have an astonishing electric blue-white sparkle. Gamma Leonis' reputation as one of the best of all visual doubles is based in part upon the beautiful yellow-orange hues of its components. And the famous Albireo in Cygnus pits an orangish K-type primary against a blue-white B-type secondary. These three far from exhaust the truly spectacular color and color-contrast doubles and multiples. The wide binary 61 Cygni, for example, though most famous as the first star with a measured parallax, consists of two mid-K dwarfs with intense chrome-orange colors. In the southern Milky Way is the neglected k Puppis, a splendid, moderately-close pair of magnitude 4.5 silver-blue B-type stars.

As Albireo dramatically demonstrates, the bluish tones of B- and A-type, and the oranges of K- and M-type, stars are enhanced when they are put next to each other. But even two stars of similar spectral types often show distinctly different hues: Gamma Leonis' components have K0 and G7 spectra, but the secondary is decidedly less orange than the primary. And color contrasts are usually enhanced when the secondary is considerably fainter than the primary: the faint B-type companion of a K- or M-type primary will look almost green (Antares B, for example); and an M-type dwarf next to a bright primary of any spectral type often appears purple. Star colors are notoriously subjective, conditioned not only by the star's intrinsic color, and not only by the sky conditions at the time of the observation, but especially by the observer's color-sensitivity in general and physical state at the moment of observation in particular. (Aperture makes a big difference too: in larger telescopes the brighter stars appear merely as subtly color-tinted whites. Binoculars are the best color-detectors for naked-eye stars.) However, all this relativism is no reason that whatever colors are perceived should not be enjoyed in and of themselves.

In order to eliminate the subjectivity of colors, astronomers have created an artificial number called color index. A star's, or other celestial object's, color index is simply its apparent magnitude as it appears on a blue-sensitive photographic plate minus its apparent magnitude as it appears on a yellow-sensitive photographic plate: c.i. = $m_B - m_V$ = B–V ("V" standing for "visual," the human eye being most sensitive to yellow light since we live under a yellow-tinted star). The color index of A0 V stars like Vega chanced to be near 0.0, so the color of Vega has been arbitrarily chosen as the zero point for the scale. Stars bluer than Vega have a negative B–V and stars yellower a positive B–V. (The reason for this is simply that the smaller or more negative a magnitude, the brighter the object: hence in B-type stars, for example, $m_B < m_V$ and therefore $m_B - m_V$ = B–V < 0.) The color index of the Sun is +0.65. The color indices of the "red" supergiants M1 Antares and M2 Betelgeuse are +1.83 and +1.85, respectively. The famous "Garnet Star," Mu Cephei, has a B–V of +2.38. At the other end of the scale the color index of the O4 supergiant Zeta Puppis is −0.26, just about as "blue" as a star can get.

Just as the apparent brightnesses and true luminosities of extended objects can be condensed into integrated apparent and absolute magnitudes, so too the colors of clusters, nebulae, and galaxies can be integrated into overall color indices. The color index of the Andromeda Galaxy's disk, for example, is +0.8, a surprisingly red value about equal to the B–V of late-G main sequence stars. One of the bluer galaxies is M101 in Ursa Major: its B–V of +0.46 corresponds to that of late-F main sequence stars. The relative redness of the disks of spiral galaxies is probably the result of a combination of red supergiants and of older solar-type thick and thin disk stars. Nevertheless the colors of such giant ellipticals as M87 in Virgo, systems which lack the blue giants and supergiants which are in the arms of spiral galaxies, are decidedly redder than those of M31- and M101-type spirals: their B-V's of +0.9 to +1 correspond to the color of K0 normal giants. The color indices of globular clusters range from about +0.6 to around +0.85 and depend upon the strength of a cluster's horizontal branch: older globulars with more RR Lyrae variables and other bluish horizontal branch stars have bluer color indices.

I.11 Stellar Groups

No star is born alone: star-formation always occurs in interstellar clouds of gas and dust even the smallest of which are sufficiently large to generate dozens of at least solar-mass stars. The Sun seems to be a single star only because it is 5 billion years old and therefore has had plenty of time to escape from, or to outlive, the stellar group of its birth. Massive giant and supergiant stars, on the other hand, seldom survive sufficiently long to escape their natal group. But how long a solar-mass star stays within its original group is directly related to how star-dense and massive that group is: the less massive and concentrated a star group, the shorter time it will be able to maintain its identity against the disruptive influence of our galaxy's tidal pull. We therefore should not be surprised that, of the major types of star group, the most populous and star-dense is also the type with the oldest star population and that the type which is the most scattered is also that with the youngest star population. In the next three sections star-groups are described in order of decreasing gravitational coherence, which amounts to order of decreasing average age. For the observer, this is also these groups' order of decreasing concentration and conspicuousness.

I.12 Globular Clusters

Globular clusters contain typically hundreds of thousands of stars in (sometimes slightly flattened) spheres usually over 100 light years in diameter. They are therefore the most conspicuous and visually impressive type of star group – though, because even the nearest of them are several thousand light years distant, they require at least moderate-aperture telescopes to be seen at best advantage. A globular's star density usually is very high in its core and decreases very gradually toward its periphery. Indeed, the star density of globular clusters decreases *so* gradually toward their edges that is always hard to tell where the globular ends and the surrounding star field begins: hence the ambiguously-edged halos these clusters display in the eyepiece of telescopes of all sizes. Even in smaller instruments a globular will look larger (in actual arc minutes) at higher than at lower powers because of the enhanced contrast higher powers give between the globular's tenuous outer halo and its sky background.

The degree of star-crowding toward a globular's center is expressed by its Shapley-Sawyer concentration class, class I globulars being the most star-dense and class XII the lowest. Some class XI and XII globulars – M71 in Sagitta and NGC 5897 in Libra, to name two – are hardly richer than the most populous open clusters. Several globular cluster pairs around the sky give the observer the chance to conveniently contrast the opposing ends of the concentration class scale. The class XI NGC 5053 in Coma Berenices is only 1° SE of class V M53. In Sagittarius the extremely concentrated class I globular M75 is just 10° NNE of the extremely loose class XI M55. And about 4.5° due west of the famous M3 in Canes Venatici, a moderately concentrated class VI object, is NGC 5466 in Boötes, an extremely low surface brightness class XII globular. M3 and NGC 5466 are a particularly good pair for contrasting in binoculars.

Despite a reputation for generic similarity in appearance, individual globular clusters are visually distinguished from one another by more than just concentration class. Some globulars with the dense cores of the lower concentration classes have large, straggling halos. The profiles of some globulars – the great Omega Centauri, for one – are not circular but noticeably flattened, which implies that the cluster as a whole is rotating (though the individual stars have highly elongated comet-like orbits around the cluster center). A couple globulars, particularly M62 in Scorpius, even have asymmetric profiles. The halos of many globulars contain long, curiously conspicuous star chains that look like they can hardly be chance alignments. Each globular cluster has its own individual characteristics and should be carefully scrutinized for structural peculiarities. The only thing all globulars really have in common is their star-richness.

Our Galaxy's family of 150-plus globular clusters is distributed in a great sphere a couple hundred light years in radius centered upon the Galactic center. However, the number-density of globulars increases precipitously toward the Galactic center: the number-density of globulars at the Sun's distance from the center, for example, is only 1/300 of what it is on the outer edge of the Galaxy's bulge. Thus the constellations toward the Galaxy's bulge, Sagittarius, Scorpius, and Ophiuchus, are exceptionally rich in globular clusters. Two subgroups have been identified in our Galaxy's globular cluster family: halo globulars are found from the very outer edge of the globular cluster system all the way into the bulge itself and are distinguished by the extreme poverty of their stars in metals (elements heavier than hydrogen and helium), a symptom of age; and bulge globulars are found only within the circuit of the Sun's orbit, are only moderately metal-poor, and are in a somewhat flattened distribution. A few globulars can be found more than 100,000 light years from the Galactic center: they include NGC 2419 in Lynx, the famous "Intergalactic Wanderer." However, none of these clusters is less than 200,000 light years from the Galactic center. Though they are indeed gravitationaly-bound to our galaxy, their original membership in our Galaxy's globular cluster family is questionable: they perhaps should be considered as part of a separate, truly intergalactic, system that includes the dwarf spheroidal galaxies (the Sculptor System, the Fornax System, Leo Dwarfs I and II, etc.) that have been captured by and are in orbit around the Milky Way.

The H-R diagrams of globular clusters share certain special features. First, the main sequence is always very short, extending up only through G-type stars. This is a consequence of these clusters' great age: all the stars in them earlier than type G or late-F have already evolved off the main sequence up toward the red giant region. Thus globulars have a well-populated arm of subgiants and red giants extending up and to the right off the tip of their truncated main sequence. However, many of those red giants are asymptotic branch giants on their second visit to this area of the H-R diagram. The stage intermediate between a star's two sojourns in the realm of the red giants is the horizontal branch extending toward the left (that is, toward the blue side of the H-R diagram) at about absolute magnitude +0.5 from the base of the red giant branch. The length toward the left of a globular cluster's horizontal branch is a measure of its age (though there are exceptions to this generality, particularly among globulars in the outer halo): bulge globulars, which are a mere 9–10 billion years old, have only short stubs for horizontal branches; but halo globulars boast horizontal branches that reach almost as far left as the original main sequence, crossing the RR Lyrae instability strip. Thus ancient globular clusters are both richer in RR Lyrae variables than younger globulars (bulge globulars often do not contain even one single RR Lyrae star), and bluer in integrated color. Two of the oldest globulars in our Galaxy's family are M92 in Hercules and Omega Centauri. Among the youngest galactic globulars are M69 in Sagittarius, M71 in Sagitta, and the large, bright, and unfortunately far southern 47 Tucanae.

Globular clusters have a large range of intrinsic luminosities. Omega Centauri seems to be our Galaxy's most brilliant globular, its absolute magnitude of −10.3 corresponding to a true luminosity of 1.1 million Suns. A close second is NGC 6388 in the tail of Scorpius with an absolute magnitude of −10.0. On the other hand M71 has an absolute magnitude of just −5.5, a luminosity of 13,000 suns. But even fainter is the magnitude 13.6 Palomar 1 in Cepheus, a class XII object with an absolute magnitude of a mere −2.5. The average absolute magnitude for all globulars is around −8.0.

Even the nearest globular clusters, M4 in Scorpius and NGC 6397 in Ara, are not less than 6,500 light years from us. Hence globulars are not good small telescope objects: the best northern sky globulars, M13 in Hercules, M3 in Canes Venatici, and M5 in Serpens, require 8-inch instruments to begin to look like the showpiece objects they are, with scores of individual stars resolved right into the clusters' blazing cores. In larger telescopes dozens of globulars can be significantly resolved: moderate to high powers, when seeing conditions permit, are best. The middle concentration classes are perhaps the best aesthetically: the cores of class I to III globulars resist resolution, and classes X to XII are hardly more than glorified open clusters, not the star-swarms globulars should be. The middle concentration class globular clusters display a pale, partially resolved, granular core that seems to hover against a glittering screen of background star-specks – a breathtaking three-dimensional illusion.

I.13 Open Clusters

Unlike globular clusters, which have a certain uniformity (though not a monotony) of appearance, open clusters come in an impressive variety of sizes, star numbers, concentrations, and textures. Some open clusters (such as M11 in Scutum and NGC 6259 in Scorpius) are nearly as populous and concentrated as class XI and XII globulars; others (M39 in Cygnus, NGCs 2244 and 2264 in Monoceros, and the Hyades) appear as little more than an enhancement of the background star field. Some open clusters are a swarm of similarly-bright stars (M46 and NGC 2477 in Puppis, NGC 2506 in Monoceros, M37 in Auriga, NGC 7789 in Cassiopeia); others consist only of half a dozen bright stars with a mere handful of supporting cast (M29 in Cygnus, M47 in Puppis). Some open clusters are scattered over more than a degree of sky (the Coma Star Cluster, M44 in Cancer, Stock 2 in Cassiopeia, M7 in Scorpius); others are tiny star knots, hardly more than gloried multiple stars (NGC 6823 in Vulpecula, IC 4996 in Cygnus, NGC 1444 in Perseus, NGC 1502 in Camelopardalis). Some have conspicuous star chains (M38 in Auriga, NGC 6603 in Sagittarius, NGCs 2252 and 2301 in Monoceros); others have noticeable star vacancies (M93 in Puppis, M50 in Monoceros, NGC 663 in Cassiopeia). R.J. Trumpler attempted to make sense out of the chaos of open cluster appearances with the classification scheme detailed in Table I-2.

Globular clusters can be found in practically every direction in the sky (though they are certainly not abundant in some directions!) However, of the more than a thousand open clusters that have been catalogued in our galaxy, only a mere handful can be found more than 20°–25° off the galactic equator. Certain directions through the Milky Way are extraordinarily rich in open clusters – specifically toward Cassiopeia, Monoceros, Puppis, and Crux + NE Carina. These are relatively dust free windows through which we can see thousands of light years along the spiral plane of our Galaxy, hence past accumulations of open clusters. However, open clusters are not especially concentrated within the spiral arms themselves: they are also abundant between as well as just above and below the arms. A few open clusters – but not many – can be found more than about 2,000 light years above or below the spiral plane of our Galaxy.

The variety in appearance of open clusters derives from the individual circumstances of their birth, for the size and other properties of an interstellar cloud of dust and gas determines both the types and the numbers of stars than can be generated within it. Small molecular clouds (SMCs) will form open clusters containing several score of solar-type stars but no luminous B-type members. (The Bernes 157 dark cloud in Corona Australis and the

Taurus Dark Cloud Complex beyond the Pleiades are presently forming just such clusters.) On the other hand, giant molecular clouds (GMCs) can produce both solar-type dwarfs and O- and B-type giants. But different mass GMCs, and different conditions (densities, turbulence, magnetic fields) within similar-mass GMCs, will generate rather different looking clusters – including some with luminous stars but few solar-type stars, and others with both in abundance. Certain unusual, but apparently not uncommon, conditions can result in a cluster with one or two O-type luminaries dominating a swarm of much more modest B-type and fainter companions: examples are NGC 2362 in Canis Major, NGCs 2264 and 2353 in Monoceros, and NGC 6383 in Scorpius.

All the stars of an open cluster are not born simultaneously: a massive protostar contracts onto the main sequence in a much shorter time than does a solar-mass protostar. Thus in open clusters with luminous O and early-B main sequence lucidae, such as the clusters at the centers of the Rosette, Lagoon, and Eagle (or Star Queen) nebulae, the lower-mass members are still gravitationally contracting. An even earlier stage in the life of an open cluster is represented in the Sword of Orion: the famous Trapezium includes only the most massive and therefore the first-exposed stars of an open cluster that eventually will contain hundreds of A-, F-, and G-type members; but as of yet many of the group's lower-mass pre-main sequence stars are still hidden within the thick dust cloud behind the Trapezium and the Orion Nebula, their presence known only from the infrared radiation they excite in the dust cocooning them.

After a couple million years the O-type luminaries of an infant open cluster like the Trapezium will have blown away most of the gas and dust residual from the cluster's formation, and we will see the group in the "clear." (As the nebulosity-free but very young open clusters NGC 2362 in Canis Major and NGC 6231 in Scorpius show, the process need not take even that long. Indeed, the O-type stars of NGC 2244 have needed only 300,000 years to sweep out the central hole of the Rosette Nebula.) But O-type stars are rapidly evolving objects and require only a few million years to become B Ia supergiants. The NGC 884 half of the Perseus Double Cluster is just such a group – a few million years old, free of nebulosity, its lucidae early-B Ia supergiants. After another few million years the early-OB Ia blue supergiants all have exploded as supernovae, leaving the cluster with the less massive, more slowly-evolving, late-B/early-A white supergiants and the M-type red supergiants and the cluster will resemble the NGC 869 half of the Double Cluster or NGC 4755, the Jewel Box Cluster, in Crux.

But A- and M-type Ia supergiants do not have a very long life expectancy either: they too are doomed to die as supernovae. Thus the most brilliant members of an open cluster will blow themselves into extinction one by one. However, open clusters like NGC 2362, NGC 2353, and NGC 6383, with only one or two bright members will not suffer aesthetically from this process: they will remain well-populated, well-concentrated star groups with hopes of many hundreds of millions of years further existence. But to those unfortunate open clusters with a few bright lucida and not much else – such as M29 and IC 4996 in Cygnus and NGC 6823 in neighboring Vulpecula (that three such similar clusters are in the same region of the Milky Way is no accident but the consequence of conditions in the interstellar medium toward that direction) – the loss of their lucidae means the end of their identity as star groups. Even NGC 663 in Cassiopeia, which looks rich because of its six nearly equally-bright B and M Iab supergiants, will not be all that impressive visually after losing those stars.

However even without its most massive members a populous open cluster after 60 or 70 million years will resemble the Pleiades – not such a bad fate! – or, after 100 million years, M6 in Scorpius, its brightest members mid-B main sequence, subgiant, and giant stars, with perhaps one or two K or M Ib supergiants. By then its K-type dwarfs have finally finished their gravitational contraction and are members in good standing of the cluster's main sequence. However, also by then the cluster will have begun to feel the disruptive effects of the Galaxy's tidal force and the cumulative gravitational influence of close encounters with other open clusters, globular clusters, and massive clouds of interstellar matter. The longer the cluster lives the more such encounters it will suffer. The effect of these encounters and of the Galaxy's tidal force is to "stir up" the cluster's less massive members, often to escape velocity. Thus a cluster's K- and M-type dwarfs, no sooner than they succeed in reaching the main sequence, begin to "evaporate" out of the group.

Therefore as a cluster enters intermediate age it is losing members from both ends of the luminosity hierarchy: its brightest stars keep blowing up or shrinking into white dwarfs, and its fainter stars keep escaping from the group's ever-dwindling gravitational field. These effects are so hard on open clusters that half of them never celebrate their 200 millionth birthday. If a cluster has begun with sufficient mass – particularly in solar-mass stars – it can hang on to the lower end of its main sequence for a good long time: the 660 million-year-old Hyades, for example, seems to have most of its original complement of K and M dwarfs. But the 500 million-year-old Coma Star Cluster, at present weighing in at only about 100 solar masses, has lost virtually all its members below absolute magnitude +6.0, equivalent to spectral type K0 V.

Even if an open cluster is massive enough to hold on to most of its low luminosity members as it ages, it keeps losing its higher luminosity stars to stellar evolution – by now particularly to planetary nebula ejections, the residual of which is a white dwarf. Because clusters begin with more A-type main sequence stars than B-type main sequence stars, and with more F-type main sequence stars than A-type main sequence stars, as time goes on its brightest stars are less bright but there are more of them. This evolutionary effect gives such youngish intermediate-age open clusters as M38 in Auriga, M35 in Gemini, and M41 in Canis Major (all 100–200 million years old) their peculiarly rich appearance. The appearance of richness gets even better with age (always providing that the cluster can hold on to the majority of its K and M dwarfs), as can be seen in the 250 million-year-old M11 in Scutum, the 2 billion-year-old NGC 7789 in Cassiopeia, the 3 billion-year-old NGC 2158 in Gemini, and the 9 billion-year-old NGC 6791 in Lyra. Of course only the initially most populous open clusters have any hope of surviving to a billion years anyway.

NGC 6791, one of the oldest known open clusters in our Galaxy, is only about as old as the youngest of the globular clusters. But there the resemblance ends: our Galaxy is still making open clusters as rich as NGC 6791; but the era of globular cluster production is long past because whatever special conditions in the Galaxy encouraged the formation of clusters with hundreds of thousands of members no longer exist. Some of the younger globulars have greater abundances of metals than many intermediate-age and ancient open clusters; but they are the bulge globulars which presumably had been formed near the center of the Galaxy after it had become metal-enriched by the earliest stellar populations.

Open cluster diameters range over about a factor of 10, the smallest clusters (M39 in Cygnus, NGC 7160 in Cepheus, NGC 225 in Cassiopeia, and NGCs 1513 and 1528 in Perseus, to name only a few) being just half a dozen light years across and the largest (among which are each component of the Perseus Double Cluster) 60–70 light years in extent. Many open clusters have a core of bright stars several light years across embedded in something of a halo of fainter members some 20–30 light years in diameter: examples include the Pleiades, the Hyades, M34 in Perseus, and M48 in Hydra. Populous intermediate-age open clusters with more-or-less even stellar distribution, such as M23 in Sagittarius, M37 and M38 in Auriga, M35 in Gemini, M50 in Monoceros, M46 in Puppis, and M41 in Canis Major, are typically 15–25 light years across. However, open clusters, like globular clusters, merge imperceptibly into their stellar background, so these values are minima. A splendid visual example of the ambiguity of open cluster boundaries is the Perseus Double Cluster, each component of which blends seamlessly into the surrounding Perseus OB1 association.

Open cluster luminosities extend over about the same range as do globular cluster luminosities, 8–9 magnitudes. But globular clusters all have the same types of stars, the brightest of which are modest-luminosity K and M giants, so differences between globular cluster absolute magnitudes are simply a matter of differences between numbers of stars. But open cluster luminosities depend both upon numbers and upon cluster age – upon age because the younger a cluster the more high-luminosity stars it is likely to contain. A young open cluster can be poor in numbers but very high in luminosity if it has a couple supergiants or a handful of bright giants. M29 in Cygnus, for instance, has little more than its five B0 giants; but these stars are sufficient to give the cluster an absolute magnitude of -8.2, greater than that of the average globular cluster. NGC 2129 in Gemini is even more extreme: its absolute magnitude of -7.2 depends largely upon its two B3 Ib supergiants, which have absolute magnitudes of -6.2 and -5.6. Old open clusters, on the other hand, boast only modest-luminosity red giants, and therefore their star numbers must make up for what such clusters lack in single-star candlepower. But they do not succeed: even the most populous evolved open clusters – which include NGC 2158 in Gemini, NGC 2477 in Puppis, and NGC 7789 in Cassiopeia – do not exceed absolute magnitude -6 or -6.5 (which indeed seems to be a cap on the potential luminosity of such objects).

The most luminous open clusters are the rare supergiant-rich aggregations like the two components of the Perseus Double cluster, the absolute magnitudes of NGC 869 and NGC 884 being respectively -8.0 and -8.6. However, the most luminous open cluster in our part of the galaxy apparently is the O-supergiant-rich NGC 6231, which has an absolute magnitude of -10.2, a luminosity of one million Suns and virtually equal to that of Omega Centauri. At the opposite end of the open cluster luminosity scale is the Ursa Major Moving Group with an absolute magnitude of only -1.4, a ridiculously low luminosity of 310 Suns. The Coma Star Cluster has an absolute magnitude of only -2.0 and therefore is just twice as luminous as the Ursa Major Group.

The Ursa Major Moving Group, which consists of the five central stars of the Big Dipper and is centered some 70 light years away, is generally considered to be the nearest open cluster to the Solar System. However this star-poor group, some 24 light years long and therefore most definitely not gravitationally bound, is perhaps better thought of simply as the core of the Ursa Major Stream (see Section I-14). The next nearest open cluster is the 150 light year distant Hyades, which indeed does look like a true open cluster (if a trifle loose). Beyond the Hyades is the Coma Star Cluster, 260 light years away, and the Pleiades, 410 light years from us. These three clusters are at their best in binoculars. They are a contrast in ages and structures: the Pleiades, 70 million years old, is a compact, rather populous younger intermediate-age group dominated by silver-blue mid-B giants; but Coma and the Hyades, respectively 500 and 660 million years old, are older intermediate-age clusters showing the wear-and-tear of the aeons – Coma is on the verge of disruption, and even the better-populated Hyades is getting seriously scattered. The 525 light year distant Praesepe in Cancer, the twin of the Hyades and a Taurus Stream cluster with it, is best in high-power giant binoculars. The extremely loose NGC 2451 in Puppis, about 710 light years away (if it actually exists, over which there has been some debate), is also a giant binocular object and, like many other younger intermediate-age open clusters, is dominated by a single (very attractive) orange K-type Ib supergiant. Beyond it are the large but loose M7 in Scorpius, M39 in Cygnus, and the compact but rather star-poor Delta Lyrae cluster (Stephenson 1), all between about 800 and 830 light years from us and requiring low power in telescopes: M7 and M39 are average intermediate-age clusters a couple hundred million years old, but M7 is more populous and therefore will much outlive the already borderline M39; and the Delta Lyrae cluster is slightly younger than the Pleiades, its lucida a blue-white B3 main sequence star contrasting with a ruddy-orange M4 II giant.

Most of the frontline Messier open clusters, which begin to be at their best at moderate powers (50x-100x) in moderate-aperture telescopes, are rather well-populated intermediate-age groups between roughly 2,000 and 4,000 light years away: they include M23 and M25 in Sagittarius; M52 in Cassiopeia; M34 in Perseus; M36, M37, and M38 in Auriga; M35 in Gemini; M50 in Monoceros; M93 in Puppis; and M41 in Canis Major. Somewhat more distant, about 5,000 to 6,000 light years from us, are several highly-populated Messier and NGC groups that benefit from somewhat higher aperture and power: M11 and M26 in Scutum; NGC 7789 in Cassiopeia; M46 and NGC 2477 in Puppis; NGC 2362 in Canis Major. Most run-of-the-mill NGC clusters are extremely remote, or intrinsically small and faint, and require moderately high power in 12-inch and larger telescopes to be seen at their best.

Open clusters offer the observer several aesthetically pleasing features. First of course is their star-richness: indeed, even the mere sense of condensation in the general star field (which is all some of them give) is attractive. Most open clusters at practically every magnification have interesting doubles and multiples: the Pleiades, for example, has an exquisite triple immediately west of Alcyone and a double right in the middle of the "dipper bowl;" the Praesepe even at 15x displays several doubles and multiples; the lucida of the isolated NGC 1502 is a stunning double of nearly equally-bright blue-white B0 giants; and the brightest stars of IC 4996 in Cygnus are involved in a Trapezium-like multiple. In many open clusters the brightest stars offer some splendid colors and color contrasts: the brightest Pleiades are all silver-blue (use binoculars for the best effect); in both the Hyades and Praesepe the lucidae are a mix of cream-white mid-A stars with orange K0 giants; NGC 663 in Cassiopeia combines five bluish-white late-B supergiants with a ruddy-orange M-type supergiant (a good color-contrast group for larger telescopes); and M41 in Canis Major pits blue-white late-B main sequence stars against chrome-orange early-K giants. Many open clusters are dominated by a single striking yellow or orange G, K, or M giant or supergiant (M6 in Scorpius, NGCs 2439 and 2451 in Puppis); others imitate the Pleiades with a bright-star population of only silver-blue or blue-white B-type objects (M21 in Sagittarius, M29 in Cygnus, IC 4665 in Ophiuchus, M36 in Auriga). Finally, the stars in many open clusters are distributed in chains, often arcing around conspicuous, nearly starless voids, that are so long and so regular they can hardly be accidental but must have been the consequence of some physical condition – perhaps a magnetic field? – in the giant molecular cloud where the cluster formed. Some of the more striking star chains are in M38 in Auriga, M41 in Canis Major, M50 and NGC 2301 in Monoceros, M93 in Puppis, NGC 6866 in Cygnus, and NGC 6603 in Sagittarius. Indeed, the open clusters NGC 2252 and Collinder 104 on the east edge of the Rosette Nebula are in fact just two very long, north-south star chains.

The open cluster data in these two volumes is from *The Deep Sky Field Guide to Uranometria 2000.0* and is organized for each cluster as follows: After the cluster's NGC designation – for example NGC 1960 for M36 in Auriga (Chapter 5) – and its Messier number (where applicable – M36 in this case) comes its star count (60⋆ for M36) and its Trumpler (Tr) Type. The Tr Type for M36 is II 3 m, which translates (see Table I-2): II = detached from the surrounding field but only weakly concentrated toward its center; 3 = contains stars of a large brightness range; m = moderately populous. The second line of data specifies the cluster's official diameter ϕ (12′ for M36), its integrated apparent magnitude m (6.0v for M36, where "v" = visual; sometimes only a "p" = photographic magnitude is available), and the magnitude of its brightest star Br⋆ (8.86v in M36).

But all this data is not without its problems. I have always found it difficult to visualize what a cluster will look like based upon its Trumpler type. For one thing, the apparent richness or poorness of a cluster's population is a relative thing and depends critically upon how much light-gathering power and magnification one brings to bear upon the group: the Trumpler richness criteria and the *Deep Sky Field Guide* star counts are based upon photographic plates, which frequently relate very badly to the real world of the eyepiece. For example, many open clusters that look very rich and promising in high-power

> **Table I-2. Star Cluster Trumpler Types**
>
> *The Trumpler classification is a three-part code that characterizes the cluster's degree of concentration, the range in brightness of its stars, and the degree of richness, as follows:*
>
> Concentration
>   I. Detached; strong concentration toward center.
>   II. Detached; weak concentration toward center.
> III. Detached; no concentration toward center.
> IV. Not well detached from surrounding star field.
>
> Range in brightness
>   1. Small range in brightness.
>   2. Moderate range in brightness.
>   3. Large range in brightness.
>
> Richness
>   p Poor (less than 50 stars).
>   m Moderately rich (50–100 stars).
>   r Rich (more than 100 stars).
>   n Nebulosity is associated with the cluster.

binoculars – NGC 6633 in Ophiuchus, NGC 7209 in Lacerta, NGCs 2244 and 2264 in Monoceros – are virtually blown out of existence even at low powers in telescopes. On the other hand, such groups as NGC 957 in Perseus, NGC 2301 in Monoceros, and NGC 6866 in Cygnus promise very little in small telescopes but with sufficient aperture and magnification are discovered to be impressively rich. Even in the same telescope low power sometimes will make a relatively scattered group look "rich" but high power might be necessary to resolve a compact cluster's fainter members and thereby bring out its "richness."

"Detachment" is relative as well, for it depends entirely upon a cluster's setting – and unless you know the setting you cannot beforehand visualize what "detachment" means with respect to the specific cluster. The Praesepe in Cancer looks very impressive where it is in front of the darkness of intergalactic space, and it in fact has been officially assigned to the same Trumpler concentration class as M36, II: but if the Praesepe was, like M36, in front of the heart of the Auriga Milky Way, it would not be quite so conspicuous. On the other hand, some fairly rich and concentrated clusters have the misfortune of populous Milky Way settings. NGC 2335 in Monoceros at the northern tip of the IC 2177 emission nebula is only a Tr concentration class III group because of the star density of its Milky way background; but if it was at the position of Praesepe, NGC 2335 would be one of the best-known clusters in the sky. Three open clusters in Cassiopeia – NGC 1027, Melotte 15, and the huge Stock 2 – also suffer in reputation only because of their background competition: they are all rated as Tr class III groups, which does an injustice to their true natures.

Open cluster integrated magnitudes can also be misleading because a cluster can be bright either because it has a handful of very bright stars or a multitude of faint ones, or even one bright star assisted by a multitude of faint ones (as in NGCs 2264 and 2353 in Monoceros, NGC 2362 in Canis Major, and NGC 6383 in Scorpius). Cluster apparent diameters are similarly ambiguous because they have been photographically determined and all open clusters, like globular clusters, simply thin out into the surrounding star fields. A glance at the Perseus Double cluster is sufficient to illustrate the point.

The lesson of these seemingly cynical comments about open cluster data is don't rely upon the official numbers. Look for yourself!

I.14 Stellar Associations and Stellar Streams

Stellar associations are loose aggregations of recently formed stars, often, but not necessarily always, still within or near the clouds of interstellar gas and dust from which their stars have just condensed. They frequently have one or two open cluster cores, and their involved clouds of matter usually are still bringing forth more stars. Associations are of three types:

OB associations are those whose most luminous members are O and B main sequence and (usually) giant and supergiant stars.

B associations are those with B, but no O, main sequence and giant stars. They often are simply aged OB associations that have lost their massive, quickly-evolving O-type members to supernova explosions.

T associations are aggregations of nebular T Tauri type variable stars still gravitationally contracting on their way to becoming modest-mass, modest-luminosity A-, F-, and G-type main sequence stars. T associations, because of the extreme youth of nebular variables, are always still partially embedded in dark dust clouds, and their individual members, which have just been revealed from the thickest of the dust, are still surrounded by small patches of emission and reflection nebulae.

OB associations are huge, usually hundreds of light years across. This is inevitable, for only huge giant molecular clouds themselves hundreds of light years across can generate the numerous massive O and early-B type stars that compose OB associations. T associations are always small, sometimes only a few light years across. A T association can be part of an OB association complex; but the nearest T associations to the Solar System, those several hundred light years away in small dust clouds toward Corona Australis (its dust cloud catalogued as Bernes 157), Chameleon, Lupus (dust cloud B228), and just beyond the Pleiades, are isolated groups. Most B associations, as was said, are probably just ex-OB associations – the huge association containing M103 and NGC 663 in Cassiopeia, for example, would be a B association but for one remaining O-type member – but associations that began only with B-type stars are easily conceivable.

Associations are the ideal naked-eye deep sky object because the nearest of them are many degrees in extent and are composed of from several to dozens of naked-eye stars. The Orion Association, centered about 1,600 light years away, includes all the stars in the constellation down to magnitude 3.5 (except Gamma and Pi-3), many of Orion's 4th, 5th, and 6th magnitude stars, and the Orion Nebula, M42. All the many emission, reflection, and dark nebulae in Orion visible in telescopes are embedded in the giant molecular clouds from which Orion's bright O-and B-type supergiant, giant, and main sequence stars have been formed. And star formation continues in the molecular cloud behind the Orion Nebula. The entire Orion Association complex is probably 700–800 light years N–S and perhaps 1,000 light years deep.

The Belt and Sword of Orion illustrate the process called sequential association subgroup formation. The supergiants in the Belt are estimated to be several million years old and are surrounded by a loose cluster, catalogued as Collinder 70, of B-type main sequence stars (a splendid binocular field – but it does not take magnification) which formed with them. Stellar winds and radiation pressure from the Belt supergiants have cleared the region around them of most of the residual gas and dust that remained of the original molecular cloud in which they and Collinder 70 formed. However their radiation pressure and stellar winds – plus supernova shock waves from any now-destroyed Belt supergiants – rammed into the giant molecular cloud of the Sword region, initiated contraction of the denser segments of those clouds, and the result has been the O and early-B stars of the Sword (specifically, the Iota and two Theta multiples). But radiation pressure and stellar winds from those O and early-B stars have in their turn initiated gravitational contraction further back in the Sword GMC, where a third generation of Orion Association stars is now coming into being.

Much nearer, even larger in apparent size, but older than the Orion Association is the Scorpius-Centaurus Association, which includes the majority of the 1st, 2nd, 3rd, and even 4th magnitude stars from Scorpius on the NE through Lupus and Centaurus to Crux on the SW. The Scorpius-Centaurus group technically is a B association, for it has no O-type stars: all its brightest members, except for the red supergiant Antares and a couple evolving late-B giants, are early-B main sequence, subgiant, and giant stars, many of them Beta Canis Majoris variables just beginning their evolutionary expansion. The association's lack of O-type stars might be the consequence of its age,
already at least 20 million years; but the group may well have lacked O-type stars from the start.

The apparent size of the Scorpius-Centaurus Association is $70° + 25°$. Because its center (located between Alpha Lupi and Zeta Centauri) is about 550 light years distant, its true extent is 700×250 light years. It is roughly 400 light years deep. Its highly elongated ellipsoidal shape is a consequence of galactic rotational sheer: an association might begin as a sphere, but because all its stars are orbiting around the Galactic center with the same velocity, those farther from the Galactic center, having larger orbits to cover at the same velocity, fall behind those nearer the center. Hence with time an association elongates. Given that the Scorpius-Centaurus group is already about twice as long as it is deep but can have covered only about 10% of its full orbit around the Galaxy's center (the "Galactic" year in our region being at least 220 million years), the process of rotational sheer is obviously very rapid.

Other OB or B associations in the Solar region of the Galaxy include the 1,300 light year distant Zeta Persei Group (Per OB2), involving Zeta, Xi, and Omicron Persei as well as the California Nebula, NGC 1499, and its related dust clouds; Lacerta OB1, 1,900 light years away toward southern Lacerta (making that otherwise unpretentious area a fine binocular field) and a group well-populated with Beta Canis Majoris variables; the evolved-supergiant-rich Canis Major Association, 2,500 light years from the Sun, its brightest members F8 Ia Delta, B5 Ia Eta, M0 Iab Sigma, and K3 Iab and B3 Ia Omicron's 1 and 2 Canis Majoris; and the Mu Cephei Association (Cep OB2), about 3,000 light years distant and a very populous group that includes most of the 4th, 5th, 6th, and 7th magnitude stars in the rich southern Cepheus Milky Way, plus the huge emission nebula IC 1396 and its central star cluster.

As the Scorpius-Centaurus Association illustrates, associations age rather rapidly: their massive, luminous O-type stars are soon lost to supernova events, and differential galactic rotation quickly sheers them into ever-lengthening ellipsoids. Even from birth associations are so scattered they lack the gravitational cohesion necessary to resist disruption: their identity persists only as long as a significant number of their brighter members remain in the same general area of a spiral arm and still move with something of the original space motion through the Galaxy of the association's parent giant molecular cloud.

As the association loses its B-type members to stellar death mechanisms, its gets longer and looser and its remaining A-type and later stars, now dispersed over a very large region, have only a generally similar space motion. The group has now degenerated into a *stellar stream*. The central five stars of the Big Dipper, though often called an open cluster, are in reality merely the most concentrated part of the Ursa Major Stream – also known as the Sirius Supercluster after its brightest member (in apparent magnitude). The Sun presently lies in the very midst of the Ursa Major Stream, for stream members can be seen in every direction on the celestial sphere and include such widely-distributed stars (in addition to Sirius and the five Dipper stars) as Alpha Coronae Borealis, Delta Leonis, Beta Eridani, Delta Aquarii, and Beta Serpentis. Because the Ursa Major Stream's earliest members are A0 and A1 main sequence objects, its age must be about 300 million years.

The Ursa Major Stream (and the Sun) is within a still larger and older stellar stream: The Hyades in Taurus and M44, the Praesepe, in Cancer are the dual open cluster core of an extremely extensive, loose aggregation of stars called the Hyades Stream. This group, which includes among its brighter members Capella, Alpha Canum Venaticorum, Delta Cassiopeiae, and Lambda Ursae Majoris, extends over 200 light years beyond the Hyades and 300 light years behind us (as we look toward the Hyades). The Hyades Stream has a main sequence up only to the mid-A range and therefore is over twice as old as the Ursa Major Stream. Older yet – indeed, several billion years old – is a stream of low-luminosity, metal-poor subdwarfs paralleling the space motion of the Population II orange giant Arcturus. The Arcturus Group proves that stellar streams sometimes can maintain, against the chilly encouragement of the Galaxy's gravitational field, something of their original identity for a significant fraction of the Galaxy's lifetime.

I.15 Nebulae

The word nebula derives from the Latin for "mist" or "vapor," which in its turn was derived from the Greek *nephelion*, "little cloud." Ancient Greek and Latin astronomers mentioned just four *nepheloeides*, "cloudy spots:" Praesepe, our M44, in Cancer; M7 in Scorpius; the Double Cluster in Perseus; and M41 in Canis Major. (Strange to say, the Andromeda Galaxy was not noticed in astronomical literature until 964 A.D. and the Arabian astronomer Al Sufi's *Description of the Fixed Stars*.) All these are open clusters that appear hazy only because of their distances from us; but even with the slightest optical aid all resolve into scores of stars. After Galileo first pointed a telescope at the sky, the term nebula became applied to anything that looked hazy. But as telescopes got larger and larger, more and more nebula were discovered to be, like the ancient four, merely distant star clusters. Many astronomers believed that all nebulae would turn out to be merely distant groups of faint stars.

The plot thickened in the middle of the 19th century when Lord Rosse discovered the peculiarly symmetrical "spiral nebulae." He thought he could partially resolve some of them, and this misapprehension seemed to be confirmed a few years later when the newly-introduced spectroscope showed that spiral nebulae have spectra resembling those of star clusters. But the spectroscope also proved that large numbers of nebulae are not merely unresolved star clusters but in fact clouds of glowing gases. Thus the hot question around the turn of the 20th century was whether the spiral nebulae were, despite their spectra, gas clouds within our Milky Way Star System or remote extragalactic star groups – perhaps Milky Ways in and of themselves. Therefore were coined the expressions "galactic nebulae," meaning gas clouds within our own galaxy, and "extragalactic nebulae," meaning independent star systems.

Finally in the early 1920s Edwin Hubble discovered Cepheid variables in the Andromeda Spiral Nebula M31 and in the Triangulum Spiral Nebula M33 and thereby proved for once and for all that these two objects are in fact independent extragalactic star systems. Thus the spiral nebulae were conclusively demonstrated to be spiral galaxies. The term nebula henceforth was reserved for the clouds of gas and dust within our own galaxy (and, when they were discovered, to such objects in other galaxies).

Nebulae are of two basic types, dark and bright. We "see" dark nebulae only because they chance to be silhouetted against a luminous background – either against a bright nebula or against a Milky Way star cloud or star field. Bright nebulae, on the other hand, shine either by reflected light or by their own fluorescence.

I.16 Dark Nebulae

Dark nebulae are simply clouds of interstellar dust silhouetted against bright nebulae or against the Milky Way itself. They contain light-absorbing grains of graphites, silicates, ices, and possibly metals like iron and aluminum. Despite their seeming opacity, the density of these clouds is very low: most of their mass is in molecular hydrogen, H_2, and even that has a number density well below one thousand molecules per cubic centimeter. (This is, however, very high by the standards of the interstellar medium.) The darkness of the clouds, then, is the consequence not of density, but of depth – for these objects are scores, sometimes hundreds, of light years thick. Their temperature is very low even for interstellar matter, 10–20°K. They give themselves away not only by their darkness but also by the low-energy radio-wavelength emission of their carbon monoxide molecules.

The most conspicuous dark nebula is the Great Rift, a series of dust clouds only a few hundred light years away that bisects the Milky Way from Deneb in Cygnus SW into the spaces of central Ophiuchus. Because the Great Rift cloud chain is slightly tilted with respect to the plane of our Galaxy, it arcs out of the Milky Way in Ophiuchus, passes behind the Antares region of Scorpius, and rejoins the Milky Way in Ara, where it begins another conspicuous, though shorter, rift that extends as far as Alpha and Beta Centauri. The most famous dark nebula is the Coalsack in Crux, which is around 550 light years away, some 50–60 light years across, and simply a detached Great Rift dust cloud. The so-called "Northern Coalsack" is an area of obscuration between Deneb and the constellation-figure of Cepheus: it is, however, neither as pronounced as the true Coalsack nor is it a part of the Great Rift chain. The 7°-long Pipe Nebula, a splendid binocular object, in the Theta Ophiuchi region is a detached Great Rift feature. The Great Rift and the Coalsack are not intrinsically very large dust features: they look so big simply because they are relatively nearby.

The shape of a dark cloud is conditioned by its environment: whatever predisposition it might have toward a Coalsack-like circularity is usually disrupted by radiation pressure and stellar winds from nearby O and early-B giant and supergiant stars, by expanding supernova shells (70% of interstellar space is occupied by the hollows left from old supernova shock waves), by the gravitational effects of recently-passed globular clusters or of other massive dark clouds, and probably by our Galaxy's magnetic field. And many dark nebulae have been, or are being, eaten into by star-formation regions. All these effects contribute to the interesting shapes of these objects.

Some dark nebulae actually display presently-occurring interaction with their interstellar environment. The 1°-long north-south reef of very pale emission nebula upon which is superimposed the Horsehead Nebula in Orion is the zone of excitation along which the radiation pressure and stellar winds of Orion's Belt supergiants are crashing into the dense, resistant, dark cloud east and southeast of Zeta Orionis. Similar rim nebulae are conspicuous on photos of the Lagoon Nebula, M8 in Sagittarius, and the NGC 6188 nebula in Ara. The famous Hubble Space Telescope photograph of M16 in Serpens dramatically demonstrates how denser globules within dust clouds resist stellar winds and radiation pressure, resulting in long "tails" on their lee side.

Appearances to the contrary notwithstanding, dark nebulae are not truly opaque. Even the Coalsack lets through about 16% of the background starlight. The darkness of a dark nebula can also be affected by how many foreground stars are between the cloud and us (though the cloud must be thousands of light years away for this to appreciably affect its darkness). The Parrot's Head Nebula, B87 in the Great Sagittarius Star Cloud, is considerably less well defined and distinct than the dark little B86, 4.5° north of the Parrot's Head in the Star Cloud, simply because more foreground stars are superimposed upon it (which also implies that it is more distant than B86). The relative opacity of dark nebulae is graded on a scale from 1, those which are little more than diminutions of the background Milky Way brightness, to 6, those which are nearly black. The most common form of dark nebula designation is its Barnard number, "B," given to it early in the 20th century by the great Milky Way astrophotographer E.E. Barnard (who, by a cruel irony, suspected these objects to be voids rather than obscuring masses). LDN = Lynds Dark Nebula numbers are also often used.

The general rule for observing dark nebulae is "less" is more: use the lowest possible magnification to enhance the contrast between the dark nebula – which is, after all, only relatively dark – and its sky and Milky Way background. If you use too high a power, you will tenuate the dark nebula's bright surroundings and thereby lose the very thing by which it can be seen at all. The reason the Horsehead Nebula in Orion is so notoriously difficult is that it is very small, only about 6′ across, and consequently requires so much power that usually the pale glow of the IC 434 emission nebula upon which it is silhouetted is simply magnified out of existence. However, some small class 6 nebulae take magnification well: B86, B92, and B93 in Sagittarius are three of the most interesting because the first adjoins a small, tight (but unrelated) open cluster and the other two are on the NW edge of the brilliant Small Sagittarius Star Cloud, M24.

I.17 Bright Nebulae

Bright nebulae are divided into four types based upon the process by which they shine: reflection, emission, and planetary nebulae (emission and planetary nebulae have a certain superficial kinship), plus supernova remnants. They are rated by their photographic depth and by their color response on the Palomar Observatory Sky Survey (POSS) plates. In this observing guide both ratings are given (where available) for each nebula. Photographic brightness (abbreviated "Photo Br" in the second line of each nebula's data heading) is measured from 1 to 6, nebulae rated as 6 being the most conspicuous on the plates and nebulae rated as 1 being those which are just detectable. The color response ("Color" in the second line) compares the nebula's brightness on the blue sensitive and on the red sensitive POSS plates and is indicated by 1 through 4 where

1 = nebulae that are brightest on blue plates
2 = nebulae that are equally bright on both blue and red plates
3 = nebulae that are brightest on red plates
4 = nebulae that are visible only on red plates.

The "color" is an indirect indication of the nebula type: nebulae that are prominent on the blue plates are usually reflection nebulae because most such nebulae surround bluish late-B and early-A type stars; and nebulae that are prominent on red plates are almost always emission nebulae because such nebulae radiate strongly, usually principally, in the H-alpha line at wavelength 6562Å at the red end of the visible light spectrum. However, these rules have their exceptions: some reflection nebulae are around yellow or red stars (Antares is within a reflection nebula); and some emission nebulae radiate strongly in the blue-green "forbidden" lines of doubly-ionized oxygen at 4959 and 5007Å. And supernova remnants are a law unto themselves: the Crab Nebula, M1 in Taurus, radiates strongly in both blue and red light.

Reflection nebulae are exactly what the name implies – nebulae that shine by light reflected from stars near, or embedded within, them. They are composed of the same material as dark nebulae: minute grains, about 0.00001 cm in size, of graphite, silicates, ices, and possibly metals within a very cool, relatively dense gas of molecular hydrogen (though sometimes this gas can be ionized atomic hydrogen and glow independently as an emission nebulae: M78 in Orion is a classic emission + reflection nebula). Like the dust grains in the Earth's atmosphere, interstellar matter selectively scatters blue light: hence color photos of the Pleiades reveal that the reflection nebulosity in which the four stars of the Pleiades "dipper" is embedded is in fact bluer than the stars themselves. Stars seen through dust clouds or reflection nebulae are reddened because of the scattering of their blue light. Indeed, the famous Garnet Star, Mu Cephei, is so red – much redder than even an M2 star has any right to be – because it shines through a dense peripheral dust cloud of the huge IC 1396 emission nebula.

As a class, reflection nebulae are small, low surface brightness objects best searched with moderate powers in larger telescopes. However, a few reflection nebulae have photographic brightnesses of 1 or 2: among them are NGC 7023 in Cepheus, NGCs 2170 and 2183 in Monoceros, and the northern lobe of the Trifid Nebula, M20 in Sagittarius, a reflection nebula around a yellowish F5 star. At the opposite end of the visibility scale is the large, extremely tenuous, extremely pale Witch Head Nebula, IC 2118 in Eridanus, feebly glowing by reflected light from Rigel 3° to the ESE: it is strictly a binocular and richest-field telescope object. The most famous reflection nebula undoubtedly is that around the Pleiades, part of a dust cloud through which the Pleiades just chances to be passing at present. The NGC 1435 section of the Pleiades Nebula, which extends south from Merope, is just visible with averted vision under excellent skies in 10x50 binoculars. By their very nature reflection nebulae cannot be improved by O III and H-alpha filters; but they can be somewhat enhanced by filters which block city light and natural sky glow.

Emission nebulae are interstellar clouds of ionized gas glowing by what amounts to simple fluorescence. These nebulae are always near or around hot O and early-B stars because it is the copious ultraviolet radiation from such stars that ionizes the atoms in the gas, which is the essential first step in the mechanics of fluorescence. When the freed electrons recombine with the ionized atoms and cascade back down the atoms' electron orbital shells, they emit characteristic wavelengths of radiation at every step. Hydrogen is far and away the most abundant element in all the interstellar medium (even including the dark nebulae, though it contributes nothing to their opacity): thus most emission nebula glow by the light given off by electrons cascading down the orbital shells of hydrogen atoms. Particularly numerous are electron jumps from the third down to the second lowest orbital shell of hydrogen, a transition which yields the H-alpha line of red light at wavelength 6562Å. Hence most emission nebulae are red in color – though the intensity of the color is too low to be seen in the eyepieces of even large telescopes. In some emission nebulae, particularly those which are unusually dense, such as the Orion Nebula, radiation from doubly-ionized oxygen, the O III ion, at the blue-green wavelengths of 4959 and 5007Å is especially strong. The surface brightness of the Orion Nebula is so great that its blue-green glow can be glimpsed even with moderate-aperture instruments. As all this suggests, all emission nebulae will be enhanced with an H-alpha filter and many with an O III filter as well.

The size of an emission nebula depends not only upon the extent of the gas cloud providing the hydrogen and oxygen to be ionized, but more particularly upon the cloud's density and upon the number and luminosity of the O and B stars providing the ultraviolet photons necessary for ionizing the gas. The more luminous the stars within the gas cloud, the more ultraviolet photons are available for ionizing the gas and therefore the further out the gas can be ionized. But the denser the gas, the more quickly with distance from the central star or stars will their ultraviolet photons be used up – but also, from the standpoint of the observer, the higher will be the nebula's surface brightness.

If left to their own devices, emission nebulae would tend to be simple spheres centered upon their hot illuminating star or star cluster. Among the few emission nebulae that appear circular in profile and which therefore must be more-or-less spherical in space are M20, the Trifid Nebula in Sagittarius, IC 1396 in Cepheus, NGC 2174 in Orion, and NGC 2237, the Rosette Nebula in Monoceros. In practice, however, the cool dust clouds from which their fluorescing stars have just been born usually hem in emission nebulae. The Orion Nebula, for example, is bounded on its north by a dark dust cloud and therefore fades out gradually to the south but to the north ends rather abruptly at the so-called "Fish Mouth" dark feature. The Lagoon Nebula, M8 in Sagittarius, has an almost rectangular profile because of bordering dark clouds. The huge 2° × 1° low surface brightness IC 1848 in Cassiopeia likewise has a distinctly rectangular outline clearly visible in 10x50 binoculars. And the peculiar shape which has given M17 in Sagittarius such flamboyant names as the "Swan," the "Omega," and the "Horseshoe" is the consequence of a lobe of dust jutting over the nebula's central glow from the west coupled with a dark bordering cloud along the nebula's north.

The Orion Nebula, bright and impressive though it looks to us, is not especially large for an emission nebula. It is about 30 light years in size; but the Lagoon measures 60 × 38 light years, the Rosette is over 100 light years across, and even the apparently compact Trifid is 40 light years in diameter (omitting its northern reflection lobe). The Eta Carina Nebula, NGC 3372, in the far southern Milky Way and centered by an impressive cluster of more than a dozen O-type stars, is 200 light years in extent. But much larger emission complexes are known in other galaxies. The Tarantula Nebula, NGC 2070 in the Large Magellanic Cloud, is a truly astonishing 900 light years in diameter – 30 times the size of the Orion Nebula!

Though emission nebulae have very low surface brightness, many of them are not completely featureless in the eyepiece. Indeed, several of the brighter emission nebulae offer actual internal detail to the visual observer. The intricate wispiness of the Orion Nebula can be glimpsed with apertures of only 4.5-inches. The same filamentary structure appears in M17 (as well as in the Eta Carina and Tarantula nebulae in the far southern heavens). The Hourglass near 9 Sagittarii in the western half of the Lagoon Nebula is easy in 6-inch telescopes. The central "hole" of the Rosette Nebula is accessible to large telescopes at very low powers. And with richest-field telescopes the large S-shaped nebula IC 2177, which spans the Canis Major/Monoceros frontier, can be seen to have a brighter patch in its middle, separately catalogued as NGC 2327.

Several emission nebulae are silhouetted with striking dark nebulae: dark lanes bisect the Lagoon Nebula and NGC 2024 in Orion; three dark rifts trisect the Trifid Nebula; dark lanes more miscellaneously oriented can be glimpsed with large telescopes in the Rosette and in IC 1396 of Cepheus; the Rosette also displays small dark globules; and the photogenic "Black Pillar" or "Tornado" feature superimposed upon the glow of M16 in Serpens Cauda is accessible to 12-inch telescopes at moderately high power. And all these internal or superimposed details far from exhausts the list of what can be seen in these and other emission nebulae – particularly by an experienced observer armed with an O III or H-alpha filter.

Finally, a large number of small nebulae around the sky are both emission and reflection objects. Examples

Table I-3. Planetary Nebulae Types
Based on the Vorontsov-Velyaminov Classification System.
1. Stellar
2. Smooth disc
    a. Brighter toward the center
    b. Uniform brightness
    c. Traces of ring structure
3. Irregular disc
    a. Very irregular brightness distribution
    b. Traces of ring structure
4. Ring structure
5. Irregular form
6. Anomalous form
    Complex structure may have two codes, such as 4+2 (ring & disc)

include M78 and its associated NGC nebulae in NW Orion, NGC 1999 south of the Orion Nebula, the NGC 1973 complex just north of the Orion Nebula, Hubble's Variable Nebula NGC 2261 in Monoceros, NGC 1333 in Perseus, and the NGC 6726 complex in Corona Australis. What all these (and kindred) emission + reflection nebulae have in common is that they are around extremely young stars, often actual nebular variables (such as R Monocerotis in NGC 2261, and R, T, and TY Coronae Australis in the NGC 6726 complex) still gravitationally contracting. These stars have only recently been unveiled from within the dark cloud which gave them birth and, because they are only modest luminosity B-type and later stars rather than high-powered O-type giants, are relatively slow and inefficient at sweeping the residual dust and gas out of their environs. Eventually, however, they will succeed, and in several cases (M78, NGC 1333, NGC 6276) what will be revealed when the dust and gas clears are entire open clusters of modest-luminosity stars.

Planetary nebulae are, as was described in Section I.5, the death shrouds of highly evolved asymptotic branch red giants in which the mechanism of atmospheric pulsations got out of hand and the star's outer layers reached escape velocity. The multiple shells or rings of most planetaries suggests that the ejection of the red giant's entire envelope usually requires several episodes. The expansion velocities are a rather gentle 10–20 kilometers per second.

The name "planetary" was given to these objects by Herschel because many of them display in the eyepiece small greenish images reminiscent of the disks of Jupiter, Saturn, and particularly Uranus. However, they actually come in an impressive array of sizes, some so small that they look stellar even at high powers in large telescopes, and others so large that they are comparable in size to the

Moon. The Helix Nebula, NGC 7293 in Aquarius, measures 16′ × 12′ – about half the apparent diameter of the Moon! – and is so tenuous that it can be seen only at very low powers: any kind of magnification, even in large telescopes, simply enlarges it right out of visual existence. By contrast, the Dumbbell Nebula, M27 in Vulpecula, is a very healthy 8′ × 5′ but has sufficiently high surface brightness that it takes high powers well.

The blue-green color visually evident in many planetary nebulae (especially easy to see in the Ring Nebula, M57 in Lyra; the Cat's Eye Nebula, NGC 6543 in Draco; NGC 3242 in Hydra; and NGC 3132 in Vela) is from the 4959 and 5007Å radiation of doubly-ionized oxygen. The oxygen is ionized by ultraviolet photons from the planetary's central star, the condensed core of the ex-red giant, no longer producing energy from nucleosynthesis but still very hot. The nebula's hydrogen is likewise ionized and radiates rather strongly in the red H-alpha and blue-green H-beta lines. Thus planetary nebulae respond best to O III filters but can also be enhanced with H-alpha filters and Light Pollution Reduction filters.

Planetary nebulae offer many interesting visual details, particularly in larger telescopes at higher powers. The Vorontsov-Velyaminov classification scheme (Table I-3) sorts planetaries into morphological types, but hardly does justice to the rich variety of their appearances. For example, the classic Ring Nebula structure is found both in compact, bright planetaries like M57 itself and in huge ghostly planetaries like NGC 7293, and entirely different optics are necessary to the see the annularity of the two extremes: M57 may easily be seen with 100x through small telescopes; however, the "hole" in NGC 7293 can be glimpsed only at extremely low power with large telescopes or RFTs. Ring structure also appears in multiple-shell planetaries (such as NGC 6543 in Draco and NGC 3242 in Hydra – though large telescopes are again necessary). As visually interesting as ring nebulae are the double-lobed planetaries like the Dumbbell Nebula in Vulpecula, the Little Dumbbell Nebula, M76, in Perseus, and the Bug Nebula, NGC 6302 in Scorpius. These objects apparently are true bipolar nebulae in which some physical process, probably the parent red giant's strong magnetic field, funneled the star's ejected outer envelope in opposite directions.

As interesting as the nebulae themselves are planetaries' central stars. These are highly compact bluish subdwarfs with colors equivalent to O and B main sequence stars but, because they are no longer generating any energy by nucleosynthesis, are much smaller and therefore much fainter. Planetary central stars are in fact shrinking into true white dwarfs: an object like the magnitude 13.6 sdO central star of the Helix Nebula, which must have an absolute magnitude of +10, probably is very close to true electron degeneracy. Because of their intrinsic faintness, the central stars of most planetary nebulae are very difficult even in large telescopes – particularly because the overlaying nebula-glow competes for the eye's attention. Very few planetary nebula central stars are as bright as magnitude 10, or even magnitude 11. The central star of NGC 1514 in Taurus is a magnitude 9.4 object; but its spectrum is a composite A0 + sdO. Likewise the magnitude 10.1 central star of the Eight-burst Nebula, NGC 3132 in Vela (not discussed in this guide because of its southerly declination), has as A2 V + sdO spectrum. However, the magnitude 10.2 central star of IC 418 in Lepus, the magnitude 10.5 central star of the Eskimo Nebula, NGC 2392 in Gemini, and the magnitude 10.6 central star of NGC 6826 in Cygnus are all probably the real thing, for all three have simple O6 or O7 spectra unsullied by any evidence of brighter companions.

1,340 planetary nebulae were known in our Galaxy as of 1991. Their distances have not been known with good accuracy individually; but as a group they show a clear distribution in a thick disk around the Galaxy's bulge, their number density increasing toward the Galactic interior. Hence planetary nebulae are decidedly more numerous in such Galactic interior constellations as Sagittarius, Scorpius, and Aquila (and even in the little dust-rimmed Serpens Cauda) than they are in such Galactic exterior constellations as Auriga, Gemini, and Monoceros (though planetary nebulae are not anywhere near as rare as globular clusters in the Galactic exterior part of the celestial sphere). A few planetary nebulae can be found well out of the Milky Way (M97 in Ursa Major, NGC 246 in Cetus, and NGCs 7009 and 7293 in Aquarius, to name a few); but these are relatively nearby planetaries, and thick disk population objects like planetaries can lie as much as 3,000 light years off the Galactic plane.

Many planetary nebulae have been identified in neighboring galaxies (including the Large Magellanic Cloud and the Andromeda Galaxy), where the brightest of them are found to have absolute magnitudes of −4 to −4.5. Because of the uncertainties in the distances of the Galactic planetaries, this value cannot be confirmed as a peak for the Milky Way's nebulae. The planetary nebula members of the open cluster NGC 2818 in Pyxis, and the globulars M15 in Pegasus and M22 in Sagittarius, among the very few planetaries with accurately known distances, have absolute magnitudes of −2, 0, and 1.5 (photographic). The true sizes of these three cluster planetaries are from 0.5 to 2 light years. Presumably these diameters, and absolute magnitudes are somewhere in the mid-range for planetary nebulae.

Supernova remnants (SNRs) are the expanding debris clouds of supernova explosions. The most famous of them is also the most unusual: the Crab Nebula, M1 in Taurus, is the remnant of a supernova that occurred in 1054 A.D. and was recorded by Chinese astronomers as being as bright as Venus. However, it is an easily-observed, high surface brightness object not because it is so young but because it is radiating by the synchrotron process: electrons ejected by the Crab Pulsar, the collapsed neutron-star core of the supergiant that exploded, at velocities near the speed of light emit radiation (in many wavelengths) as they decelerate or accelerate in the nebula's magnetic field. The Gum Nebula in Vela, the huge (36° = 1,000 light years) remnant of a supernova that erupted around 9000 B.C., is also radiating in part by the synchrotron process – but not with such intensity as the Crab Nebula and over a much larger area. Perhaps the Vela SNR once resembled the Crab Nebula; but now it consists only of areas of pale glow and of thin filamentary arcs vaguely centered upon the Vela Pulsar.

Certainly not all supernova remnants go through a Crab phase: neither Tycho's Star of 1572, nor Kepler's Star of 1604, nor the Lupus SN of 1006 AD, nor the Cassiopeia SN of c. 1670 has left behind anything more than pale shreds around the periphery of an expanding debris cloud. All they share with the Crab is their strong emission in radio waves and X-rays – the consequence of the tremendous energy of the original explosion, half of which would have gone into the kinetic energy of the expanding debris cloud and the other half into heating the thin gas left within the debris cloud to a temperature of one million degrees Kelvin, at which the gas becomes a strong emitter of both radio waves and X-rays.

I.18 Galaxies

Galaxies are gravitationally bound aggregations of millions, often billions, of stars, usually accompanied by massive clouds of gas and dust. For observers they are the most numerous type of deep sky object (excluding of course individual stars), several thousand of them being brighter than the 13th magnitude and therefore accessible to moderately large optics. However, galaxies are extremely distant – even the nearest large galaxy, the Andromeda Galaxy, is 2.2 million light years from us – hence they are faint and small and relatively few of them offer much visual detail even in larger telescopes. In the eyepiece most galaxies appear merely as a stellar nucleus within a tiny, low-surface-brightness, ambiguously-edged halo.

Of course intrinsically galaxies are neither small nor faint. However, they come in an impressive range of true sizes and luminosities. The most brilliant galaxies are supergiant ellipticals with integrated absolute magnitudes of -23 and -24 – luminosities approaching one trillion Suns and diameters of several hundred thousand light years. (An example is NGC 4889 in the Coma Galaxy Cluster: see Chapter 37.) At the opposite extreme are the dwarf spheroidals, probably the most numerous type of galaxy in the universe but with absolute magnitudes as little as -8.5 to -9 (hardly greater than that of the average globular cluster) and diameters of just a few thousand light years. Our Milky Way Galaxy has an absolute magnitude of -20.5 and a diameter approaching 150,000 light years (not easy quantities to measure given our position well within the Galaxy's disk), and therefore is significantly above average in size and luminosity. The Andromeda Spiral, however, is both larger and brighter than the Milky Way.

As unpromising as the small, faint images of galaxies might initially seem, the observer soon learns that each has certain distinctive figures that it shares with other "faint fuzzies." Some galaxies persist in appearing even to a trained eye as nothing more than amorphous or vaguely circular blobs of low surface brightness haze. But they are in fact the exception rather than the rule. Most galaxies have a distinct, starlike nucleus embedded in the center of their haze, the nucleus in some galaxies sharp and bright, but in others a barely discernible twinkling. Sufficient magnification often will persuade the nucleus to become a tiny bright disk-like core. Galaxies without a stellar nucleus usually have at least a broad central brightening, if not a true extended moderate surface brightness core.

The halos of galaxy images have as much variety in appearance as the nuclei and cores. Some galaxy halos are compact and of only moderately low surface brightness; others – even some around particularly intense stellar nuclei – are relatively large but exceedingly diaphanous. Some galaxies (M94 in Canes Venatici is an example) with extended, moderate surface brightness cores, have only faint, narrow halos. Not all galaxy halos are amorphous/circular: many are oval, and some are impressively elongated – a few (which are spiral galaxies viewed edge-on) to such an extreme that they are mere slivers (examples include NGC 4565 in Coma Berenices, NGC 4631 in Canes Venatici, NGC 5907 in Draco, and NGC 891 in Andromeda). Most of the galaxies with highly elongated halos also can be seen with sufficient magnification to have tiny elongated cores. Both the sliver galaxies and the more oval galaxies usually contain at least a stellar nucleus.

Many moderately bright galaxies display a tripartite structure: a stellar nucleus within a small, moderate surface brightness core surrounded by a tenuous halo. The brightest galaxies have mottled halos and sometimes even dust lanes and/or hints of spiral structure. Many of the sliver galaxies can be seen, with sufficient aperture and magnification, to be bisected by a dust lane reminiscent of our own Milky Way's Great Rift.

Even in small telescopes and 10x50 and larger binoculars many galaxies can be seen with the above-described details (excluding of course the mottling, dust lanes, and spiral arms). These details reflect the actual structure of the galaxies in which you see them and are

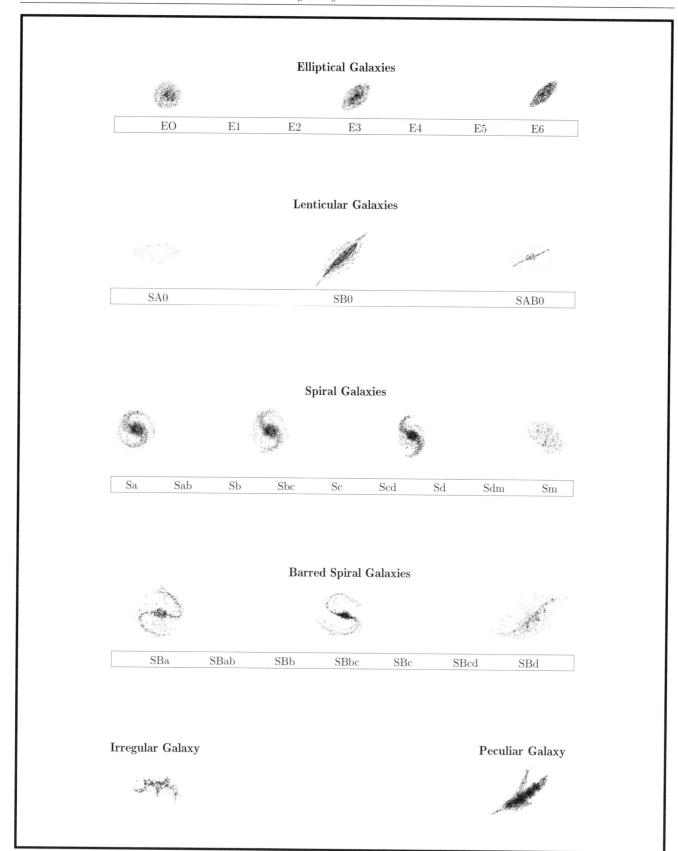

Figure 1-2. Major Morphological Galaxy Types

consistent with those galaxies' morphological types. The presently-followed system of galaxian morphological types was invented by Hubble, who began with four basic galaxy forms: elliptical (E), spiral (S), barred spiral (SB), and irregular (Irr or I). At first he regarded lenticular (S0) galaxies as a "more or less hypothetical class" at the junction of the most oval of the ellipticals (E7) with the most tightly wound spirals (Sa and SBa), but later he upgraded them into a full-fledged official morphological class. However Hubble did us all a disservice by calling the tightly wound, spiral-arm-dominated Sc and SBc galaxies "late spirals." The problem with this nomenclature is that "early"-type Sa and SBa spirals are almost entirely composed of late-type Population II stars (G, K, and M giants and dwarfs) and "late"-type Sc and SBc spirals have large numbers of conspicuous early-type Population I O and B giants and supergiants.

But just as Baade's Population I and II proved too simple for the reality about star populations, so too Hubble's system proved too simple for the reality about galaxy morphology. Consequently more complex systems of galaxy classification have been introduced. The one followed in this observing guide combines Gerard de Vaucouleurs' extension of the original Hubble system with S. van den Bergh's DDO (David Dunlap Observatory) spiral galaxy luminosity classes.

As can be seen in Table I-4 (Table 1 of *The Deep Sky Field Guide to Uranometria 2000.0*), the de Vaucouleurs scheme preserves Hubble's original division of galaxies into ellipticals, spirals and barred spirals, and irregulars – plus the afterthought lenticulars – but adds notation for bars in general (B), for the absence of bars (A), for inner and outer ring structures (r and R), and for a global "S" design (s). (The "T" column in Table I-4 gives the so-called Hubble stage for the various galaxy types.) In this system the Andromeda Galaxy, usually simply described as an Sb spiral ("b" in the original Hubble system, implying a moderately large bulge and moderately tightly wound spiral arms) is classified as SA(s)b, the "A" indicating that the galaxy has no central bar and the "(s)" that its global structure is S-shaped. Something of the weakness of the deVaucouleurs' rather Byzantine scheme is revealed in the case of the face-on spiral M101, classified as SAB(rs)cd, meaning that it has an incipient bar, its spiral arms are not quite as straggling as they could be, and that the galaxy has an overall S-shape and its inner arms close in a ring around the central bulge. Without a photo of M101 it would be very difficult to envision what an "SAB(rs)cd" galaxy might look like!

The DDO luminosity class of a spiral galaxy is indicated by a Roman numeral following its deVaucouleurs morphological type – SAB(rs)cd I in the case of M101. It applies only to spirals and strictly speaking is only a morphological description of a galaxy's spiral arms, class

Table I-4. de Vaucouleurs Revised Morphological Galaxy Classification System					
Classes	Families	Varieties	Stages	T	Types
Ellipticals		Compact		-6	cE
			Elliptical (0-6)	-5	E0
			Intermediate	-5	E0-1
		"cD"		-4	E+
Lenticulars			S0		
	Non-barred				SA0
	Barred				SB0
	Mixed				SAB0
		Inner ring			S(r)0
		S-shaped			S(s)0
		Mixed			S(rs)0
			Early	-3	S0-
			Intermediate	-2	S0°
			Late	-1	S0+
Spirals	Non-barred				SA
	Barred				SB
	Mixed				SAB
		Inner ring			S(r)
		S-shaped			S(s)
		Mixed			S(rs)
			0/a	0	S0/a
			a	1	Sa
			ab	2	Sab
			b	3	Sb
			bc	4	Sbc
			c	5	Sc
			cd	6	Scd
			d	7	Sd
			dm	8	Sdm
			m	9	Sm
Irregulars	Non-barred				IA
	Barred				IB
	Mixed				IAB
		S-shaped			I(s)
			Non-Magellanic	90	I0
			Magellanic	10	Im
		Compact		11	cI
Peculiars				99	Pec
Peculiarities		Peculiarity		pec	
(All types)			Uncertain		:
			Doubtful		?
			Spindle		sp
			Outer ring		(R)
			Pseudo outer ring		(R')

I spirals having thick, well-developed arms and class V spirals anemic, poorly-developed arms. However, these classes prove to correlate with actual galaxy luminosities. M31 is a DDO luminosity class I-II spiral and its absolute magnitude of −21.6, though very high, is indeed a little below the observational maximum for Sb spirals. The absolute magnitude of the luminosity class I M101, −21.5, is in fact at the peak of Sc spiral brilliance.

I.19 Elliptical Galaxies

The main morphological characteristic of all elliptical galaxies is a smooth decline in brightness = star density from a luminous central core out to a gradually fading periphery. In their surface brightness distribution elliptical galaxies therefore resemble globular clusters – globular clusters on a grand scale. Though they have no distinct boundaries (another property they share with globular clusters), the profiles of elliptical galaxies are very distinct, ranging from circular (E0) to highly elongated (E7). Their bright star population is the same as that of globular clusters – highly evolved, subsolar mass, Population II K- and M-type giants. A few ellipticals contain patches of dust; but any indication of a true dust lane banishes a galaxy from the ellipticals to the lenticulars.

Elliptical galaxies come in the most extreme range of sizes and luminosities, from supergiants hundreds of thousands of light years across and nearly a trillion times brighter than the Sun – the "cD" ellipticals – to dwarfs only a couple thousand light years across and barely brighter than the average globular cluster. Two varieties of dwarf elliptical galaxies exist. The dwarf ellipticals proper are merely smaller versions of the giant and supergiant ellipticals: they display the same gradual decline in star density and brightness from a dense core out to a diffuse periphery, and they come in the same range of elongations, as their giant cousins – they simply are much smaller in size. The four satellites of the Andromeda Galaxy – M32 and NGCs 147, 185, and 205 – are typical dwarf ellipticals, the smallest and faintest of the quartet being NGC 147, an elongated dE4 system some 7,200 light years across and with an absolute magnitude of -14.9. The dwarf spheroidals are an extreme class of dwarf elliptical: they have the bright star types and star distribution profiles of normal ellipticals, but their star density, from core to periphery, is a mere 1% of the normal dwarf elliptical. Consequently the typical dwarf spheroidal galaxy in the vicinity of the Milky Way is only 2–3 thousand light years across and has an integrated absolute magnitude of just -11: the Draco System in fact has an absolute magnitude of a mere -8.6. Dwarf spheroidals are probably extremely numerous throughout the universe, but we cannot see them beyond our Local Galaxy Group because they are too spread out to produce a consolidated haze: they can be identified only by the loose gathering of their brightest stars, (which have absolute magnitudes of only around -2) on photographic plates.

Dwarf spheroidals obviously are objects only for large telescopes. However, a few dwarf ellipticals are accessible to smaller amateur instruments, the best being the Andromeda Galaxy's four satellites. The vast majority of elliptical galaxies seen optically are of the giant or supergiant persuasion, though the nearest giant ellipticals are M105 and NGC 3379 in the 30 million light year distant M96 galaxy group. Nevertheless these two galaxies are easily visible in high power binoculars – as are several Messier-numbered giant ellipticals in the 70 million light year distant Coma-Virgo Cluster, Messiers 49, 59, 60, 84, 86, 87, and 89.

The appearance of elliptical galaxies in the eyepiece is always the same: a bright stellar nucleus or (if magnification is sufficient) a bright core within a halo of ambiguous extent but (if aperture is sufficient) distinct shape. There are a number of good elongation-contrast sets of elliptical galaxies around the sky. For example, even in 10x50 binoculars the difference between the almost circular E2 M32, just south of the Andromeda Spiral's hub, and the very elongated E5 NGC 205, somewhat north of the hub, is obvious. In the core of the Coma-Virgo Galaxy Cluster are two close pairs of contrasting ellipticals: the round type E1 M84 is only 17′ from the slightly but distinctly more elongated E3 M86; and M59, a highly elongated E5 elliptical, is less than 0.5° WNW of the much rounder E2 M60. Finally, in central Leo are M105 and NGC 3377, about 1.5° apart in the north-south direction, the former a circular E1 system and the latter a highly-extended E5–6 galaxy.

I.20 Lenticular Galaxies (S0 or SB0)

The main difference between ellipticals and lenticular galaxies is that the latter contains a disk. Morphologically lenticulars are such a smooth extension of the ellipticals that sometimes it is difficult to discern an S0 from an E7 system. Lenticulars are permitted to contain dust – even a complete dust lane – but any evidence of luminous gas or of star-formation regions marks a galaxy as an Sa or SBa rather than as an S0 or SB0 system spiral. In the Hubble classification system nonbarred lenticulars are classified from $S0_1$ to $S0_3$ depending on the amount of dust they contain – none in $S0_1$, a complete lane in $S0_3$ – and barred lenticulars are classed by the strength of the central bar: in $SB0_1$ lenticulars the bar is manifested only by little "knobs" at each end of the central bulge; but in $SB0_3$ systems the bar is a conspicuous feature cutting boldly across the entire bulge. The newer deVaucoulers system used in this observing guide uses the notation SAO$^-$, SAO$^\circ$, and SAO$^+$ (or SBO$^-$, SBO$^\circ$, and SBO$^+$) for lenticulars, S0$^-$ galaxies lacking conspicuous discs and therefore resembling E7 ellipticals and S0$^+$ systems with discs almost as prominent as those of Sa spirals. Like the ellipticals, S0 galaxies are composed of evolved Population II red and yellow giant and dwarf stars.

Lenticulars are rather rare outside of rich galaxy groups, in which fully half the bright members might be

S0 systems. Therefore few lenticulars are found in the vicinity of the Milky Way and its Local Galaxy Group, situated way out on the perimeter of the Coma-Virgo Supercluster. The nearest lenticular to us might be NGC 404 near Beta Andromedae, an S0⁻ galaxy 16 million light years distant and with a rather modest absolute magnitude of −18.0. NGCs 5128 and 5102 in the 22 million light year distant Centaurus Galaxy Group are usually classified as S0 systems and have very respectable absolute magnitudes of −22.5 and −20.4, respectively; but both are marred by structural peculiarities. NGC 3384, a near companion to the giant elliptical M105 in the M96 galaxy group, is an SB0⁻ system with an absolute magnitude of −20.0 and a highly elongated image that splendidly contrasts with its circular neighbor. Even better than NGC 3384 is the Spindle Galaxy, NGC 3115 in Sextans, a perfect lens-shaped S0⁻ lenticular some 30 million light years away with an absolute magnitude of nearly −21.

I.21 Spiral Galaxies

Spiral galaxies, the galaxy type par excellence, are constructed of a disk containing the spiral arms surrounding a central bulge or hub. The disk is composed of Population I stars, the youngest of which – blue and red high-luminosity supergiants, star-formation regions with emission and dark nebulae, OB and T associations – are concentrated along, indeed define, the spiral arms. A spiral galaxy's hub, and the huge spheroidal component that includes the system's family of globular clusters, consist of Population II red and yellow dwarfs and modest-luminosity giants.

As Hubble originally conceived it, the Sa-Sb-Sc (or SBa-SBb-SBc) sequence of spiral galaxy types expressed three things: (1) the degree of openness of the spiral arms, in Sa the arms being tightly wound and in Sc the arms loose and straggling; (2) the bulge-to-disk brightness ratio, greatest in Sa and least in Sc; and (3) the degree of resolution into stars of the spirals arms, least in Sa and greatest in Sc (because Sc systems have more active regions of supergiant star formation). The spirit of these criteria are preserved in the de Vaucouleurs designations, but transitional types (Sab, Sbc, etc.) have been introduced, a greater degree of spiral-arm openness is accommodated by a new type, "Sd," and "spirals" with barely detectable arms or segments of arms are given their own type, "Sm" (where m = "Magellanic" after the Large Magellanic Cloud, which has an incipient spiral feature).

In only the largest and brightest spiral galaxies, and only with moderate-to-large telescopes, are spiral features optically visible. However, even when the spiral arms are not resolved, each of the different types of spiral galaxy has its signature appearance in the eyepiece. Face-on Sc systems are always amorphous or circular, very-low-surface brightness objects with only slight central brightening: examples include M33 in Triangulum, M83 in Hydra, M101 and NGC 3184 in Ursa Major, NGC 6946 in Cepheus, M74 in Pisces, and IC 342 in Camelopardalis (though sufficient aperture will bring out spiral features and a core in all these). Sb galaxies, face-on or otherwise oriented, can be counted on to have a small bright core, stellar if the galaxy is distant or a tiny disk (given enough magnification) if the galaxy is relatively nearby. Around the core will be a moderately high surface brightness halo that is elongated if the galaxy is tilted to our line of sight. The essential visual difference between Sb spirals and elliptical galaxies is that the former do not gradually diminish in brightness away from the core; their halos tend to be uniformly bright (or faint), and often the brightness difference between core and halo is very abrupt. (Both of these visual effects are particularly conspicuous in the small telescopic image of the Whirlpool Galaxy M51 in Canes Venatici. Other classic examples of Sb spirals include the Andromeda Galaxy, M81 in Ursa Major, M66 in Leo, and NGC 7331 in Pegasus.) Sa spirals are the most difficult type to distinguish at the eyepiece from elliptical and lenticular galaxies: they tend to have very bright cores, like E and S0 systems; and their halos are very populous and extensive and virtually continuations of their very ample central bulges and consequently fade gradually outward much in the manner of E and S0 halos. M104, the Sombrero Galaxy in Virgo, is a splendid example of how lenticular an Sa spiral can appear. M65 in Leo is one of the less ambiguous-looking Sa systems. Two Sa galaxies in Leo which fit in the same field of view are NGCs 3190 and 3185. Several of the brighter Messier galaxies are specimens of the transitional Sab type of spiral, which has a large central hub but also a fairly extensive spiral disk: M81 in Ursa Major, M96 in Leo, M64 and M98 in Coma Berenices, and M63 in Canes Venatici.

One of the strangest things about spiral galaxies, including our own Milky Way, is that the orbital velocity around the center of the galaxy of objects in the disk is more-or-less the same from near the bulge all the way out to the disk's rim. But the law of gravitation states that rotational velocity should *decrease* with distance from the center of the mass. This conundrum has resulted in much speculation about a "dark population" composed of what no one really knows lying out beyond the disks of spiral galaxies and influencing their rotational velocity curves. As puzzling as this rotational velocity problem, and related to it, is how the arms of spiral galaxies are formed and – even more amazing –

can be sustained. The phenomenon of differential rotational sheer described in Section I-14 that elongates stellar associations into nonexistence should do the same to spiral arms. But it does not: indeed, photographs of spiral galaxy disks taken on red-sensitive plates, which bring out the several-hundred-million-year-old thick disk population of evolved G, K, and M giant stars, shows that this population is also distributed in spiral arcs – smoother and broader than the blue-light spiral arms of O- and B-type giants and supergiants, but following the same curves. Hypothetically, the only way a spiral pattern could be maintained is if the rotational velocity in a spiral galaxy actually *increased* with distance from the galaxy's center!

Spiral galaxies have an impressive range of luminosities (though not as great as the luminosity range of elliptical systems). The most brilliant spirals, such as M58, M61, M99, and M100 in the core of the Coma-Virgo Supercluster, have absolute magnitudes of around -22 (nearly 60 billion Suns). The peculiarly-rich Sa system M104 might have an absolute magnitude nearer -23.5, equal to that of many supergiant ellipticals. At the opposite end of the spiral luminosity hierarchy is the SBc NGC 5229 in the M101 galaxy group, which has an absolute magnitude of around -16.5, a luminosity of just 330 million suns (less than that of the Small Magellanic Cloud). Sm systems can be even feebler in light output: D127 in the NGC 4449 galaxy group in Canes Venatici has an absolute magnitude below -15.

I.22 Irregular Galaxies (I or Irr)

Irregular galaxies are almost amorphous intergalactic patches of gas and dust within which stars are presently being formed. Thus they are practically pure extreme Population I systems. But they are always small and faint: the Small Magellanic Cloud, which in fact is one of the very largest and brightest Im galaxies, has an absolute magnitude of -17.0 (a luminosity of nearly 1 billion Suns) and a diameter of about 15,000 light years (which is, however, only half the distance from the Sun to the center of our Galaxy). The Local Galaxy Group contains perhaps as many as a dozen of these dwarf irregulars, the faintest of which, the Sagittarius Dwarf Irregular Galaxy (SDIG) has an absolute magnitude of only -10.5, comparable to that of the brightest Galactic open clusters. The Small Magellanic Cloud is a naked-eye object, of course; but at least two other Local Group dwarf Irregulars can be seen in amateur instruments: NGC 6822 in Sagittarius can be glimpsed even with 10x50 binoculars; and IC 1613 in Cetus is available to large telescopes.

I.23 Galaxy Groups and Clusters

As is the case for stars, clustering is the rule rather than the exception for galaxies. The Milky Way and the Andromeda Spiral are the two dominant members of a small aggregation of perhaps three dozen (mostly dwarf) galaxies called the Local Galaxy Group. (See Appendix A at the end of Volume 2.) Similarly, M81 and M82 in Ursa Major, and NGC 2403 in Camelopardalis, are the brightest members of about a score of (again mostly dwarf) galaxies centered about 10 million light years away and including NGCs 2366, 2976, 3077, and 4236. In the southern heavens is the Sculptor Galaxy Group, at 8 million light years probably the closest group to the Local Group: it consists entirely of loose-armed Sc and Sd systems and dwarf irregulars, and its brightest members – NGCs 45, 55, 247, 253, 300, and 7793 – are distributed in a peculiar 20° diameter ring. Also peculiar in shape and constitution is the 22 million light year distant Centaurus Galaxy Group, the members of which are strung in a 30° long chain that includes the supergiant radio galaxy NGC 5128, the supernova-rich M83, the strange lenticular NGC 5102, and the amorphous NGC 5253.

All these galaxy groups are themselves on the fringe of the huge Coma-Virgo Supercluster, the center of which is 60–70 million light years away toward the M84–M86–M87 region. Because of our vantage from one edge of our supercluster, we see it as a rich band of galaxies extending from the area of the Big Dipper on the NW down through Canes Venatici, Coma Berenices, and Virgo into Centaurus on the SE. The flattening of this band implies that the entire supercluster is rotating. The core of Coma-Virgo, like other supercluster cores, is well-populated with giant ellipticals and lenticulars.

Coma-Virgo is but one of many known superclusters. The nearest, its edge perhaps 70 million light years from us, includes most, if not all, the NGC galaxies in Eridanus, Fornax, Cetus, and Pisces. Beyond this are many other extremely rich galaxy clusters, including the Perseus I Galaxy Cluster, the Coma Galaxy Cluster, the Hercules Galaxy Cluster, and the extremely distant Abell 2065 in Corona Borealis, all of which (plus several others) are described in these two volumes. Several remote galaxy clusters are dominated by one or two supergiant E+ or cD ellipticals, thought to have gotten as big as they are by cannibalizing other galaxies in their cluster.

Chapter 1

Observation of Deep-Sky Objects

1.1 Catalogs – Celestial Inventories

Mankind has kept records from the very beginning of recorded history, and the heavens certainly were not without notice. The first known star catalogues were compiled by Hipparchus who observed the naked eye sky at Rhodes between 146 and 127 B.C., and Ptolemy who observed at Canopus in Egypt from A.D. 127 to 151. Tycho Brahe catalogued 777 stars by 1590, and Hevelius compiled 1500 stellar positions before his death in 1690.

After the invention of the telescope in the early seventeenth century, mankind's concept of the heavens has slowly changed. Galileo was the first to see nebulous naked eye objects resolved into tiny stars.

Charles Messier (1730–1817) was the first to keep a record of nebulous objects but only because they interfered with his comet hunting; he discovered 13 comets during his lifetime. When he found a nebulous object that did not turn out to be a comet, he added it to his list of fixed objects so that he would not mistake it again in future searches. In addition to objects that were a nuisance too him, he almost certainly added objects simply because they were new, and for completeness sake (M42, M44, and M45, for instance). Though his telescope was rather small and inferior by today's standards, his catalog turned out to be a list of the very finest objects in the sky. Observing his catalog of objects is a good starting place for beginners.

In the nineteenth century, William and Caroline Herschel systematically catalogued thousands of deep-sky objects. John Herschel continued his father's work by observing the southern skies from South Africa. In 1864, John published a catalog of 5,097 objects, 4,630 of them discovered by the Herschels.

Due to historical interest and the popularity of the Herschel Club, we have included Herschel numbers for your convenience. Herschel sorted his discoveries into eight general categories according to their appearance (see Table 1-1). Although he used Roman numerals, it is

Table 1-1. Herschel Classifications
(*Shown as a superscript after the number:* $H78^8$)
1. Bright nebulae
2. Faint nebulae
3. Very faint nebulae
4. Planetary nebulae
5. Very large nebulae
6. Very Compressed, rich clusters of stars
7. Compressed clusters of small & large stars
8. Coarsely scattered clusters of stars

less cumbersome to identify the category by a superscript numeral after the object's catalogue number—example: $H476^3$ instead of Herschel III 476.

By 1886, so many new lists had emerged that something had to be done. The Royal Astronomical Society proposed that all such lists needed merged into one master list, a New General Catalogue, and assigned John Dreyer to this task. When it was completed, Dreyer had a total of 13,226 objects listed in order of right ascension with telescopic descriptions. Dreyer followed the format of John Herschel's General Catalogue of Nebulae, with abbreviated descriptions providing useful information about the object's telescopic appearance, its brightness, size, shape, brightness variations and field descriptions.

Our descriptions expand upon the NGC's basic shorthand descriptions by Dreyer and that of the Herschels. Some of the terminology has been updated to make the observations more readable. Adhering to the spirit of the original NGC, the descriptions are all visual telescopic observations made by the editors and a host of contributors. We have edited these observations, grouping them by the size of the instrument used. (See section 1.7, part IV.) To ensure the description's accuracy, the observations were compared to the original NGC

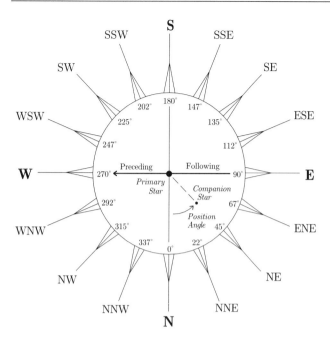

Figure 1-1. Compass Directions & Position Angles

Catalogue descriptions. Photographs and star maps were used to verify position angles, directions and distances to nearby stars, and estimate stellar magnitudes.

1.2 Object Visibility

The published magnitudes of deep-sky objects, galaxies in particular, are not a true indication of how bright they may appear. Photographic magnitudes are measured in the blue part of the spectrum, but the eye is more sensitive toward the center of the spectrum near yellow and green where visual magnitudes are measured. A galaxy with a photographic magnitude of 13.0 may have a visual magnitude of 12.4. If only the photographic magnitude is available, you may expect the galaxy to appear somewhat brighter. Besides seeing conditions and transparency, other factors affecting visibility are a galaxy's surface brightness, its morphological type, and its orientation to our line of sight. Tightly wound spirals appear brighter than the looser variety, and galaxies that are positioned edge-on to us appear more concentrated than those oriented face-on.

Although most galaxies are difficult to observe, there are some fine, bright examples visible in binoculars and small telescopes. Medium-size instruments from 8-inches to 14-inches will detect several thousand galaxies brighter than 13th magnitude and reveal some details in the brighter galaxies. In telescopes with primary mirrors of 16 inches or larger, the showpiece galaxies become beautiful objects revealing delicate structural details, dark lanes, and spiral arms. The larger instruments allow the observer to detect faint 14th and 15th magnitude objects invisible in smaller scopes.

1.3 Field Orientation

Early observers gave directions using the words "preceding" for west and "following" for east. Since they had no drive mechanisms, these terms were quite apparent as the star field drifted by. If your scope has a right ascension drive, the compass directions may easily be determined by shutting off the drive motor to see the direction of field drift. Center a star, or any object that you are observing, and watch where it exits the field–this point is west. If you are using a Newtonian reflector, north is 90° counterclockwise from west. If you are using a Schmidt-Cassegrain Telescope (SCT) or a refractor with a star diagonal, the direction of north is 90° clockwise from west.

1.4 Position Angles

In addition to field orientation, position angles are necessary for scientific accuracy and usefulness. When a celestial object is extended, the ratio of elongation between the major and minor axis is indicated, followed by the compass directions or the position angle. Position angles given for the elongation of the galaxy, or any other deep-sky object, do not exceed 179°; however, the location of a star relative to another star or a deep-sky object may be expressed by a position angle up to 359°. The object's catalog size is listed in the data line ahead of the visual description. An object's visual diameter is usually different than its catalog size due to surface brightness. In our descriptions, we have estimated the visual size to within a quarter (0.25) of an arc minute ('). Example: 6.75' × 2.25' NNW-SSE. Diameters smaller than one arc minute are given in seconds (") of arc.

When there is a star embedded in the halo, or lying nearby, the direction and distance given is from the center of the object unless otherwise stated. When there are companions in the field, positions are given for the companion from the center of brighter object. If the object being discussed is part of a group, it may be indicated as being the 1st of 3, or 2nd of 4 objects, etc. as they drift through the field with the drive turned off.

Make a photocopy of Figure 1-1 and keep it in your notebook for estimating compass directions and position angles when you make observations. Align west on the chart with the direction of field drift, and you will be able to read the position angle in degrees. The direction of drift is west, 270°, or the preceding direction. An object's elongation may be stated as: 4' × 2' NW–SE or 4' × 2' in position angle 135°. Occasionally, we may state preceding for west and following for east. In ordinary conversation it is customary to say north–south and east–west so we have followed suit when giving compass directions.

1.5 Visual Impressions

In our task to edit descriptions, we found that one observer would describe a particular object as fairly bright while another would call it fairly faint. There is a fine line between pretty bright and pretty faint. We received many conflicting descriptions so we had to set some guide lines as to what we would allow to be called bright, faint, large, or small. We considered the object's surface brightness, the observer's instrument size, sky conditions, and other factors. We have substituted "fairly" or "moderately" for the word "pretty" which we will reserve to describe aesthetically pleasing or beautiful objects.

Table 1-2 shows the shorthand code that Dreyer used to describe the telescopic appearance of objects in the NGC Catalogue. Beside these descriptions we have indicated an arbitrary magnitude range that seems to give these impressions when viewed through telescopes of two popular sizes having 12″ to 14″ mirrors. Obviously, the object's surface brightness plays a major factor in its appearance. If you use a smaller instrument, objects will appear fainter, and if you use a larger telescope, objects will appear brighter than our example. Eyepiece magnification also affects your impression of an object's brightness and size, so it is a good idea to compare objects with one particular eyepiece before inspecting them with higher powers. Quite often, especially when the object has low surface brightness as is the case with galaxies and nebulae, the perceived diameter will appear smaller than values given in catalogues. A better indication of a galaxy's visibility is surface brightness, calculated by dividing its magnitude by its area. In addition, the visual extent of the halo or disk is limited by the luminosity of the night sky itself. This is why a clear, dark observing site with minimal light pollution is essential; and, of course, you should observe deep-sky objects with the moon absent from the sky. Nearly all of our descriptions were made under the best possible sky conditions.

1.6 Dark Adaptation

Your eye is a refractive device working much like a camera, each has a lens to focus the image. The iris of the eye and the diaphragm of a camera each regulate the amount of light received. The retina of the eye and the film of the camera are each sensitive to light. The retina has two kinds of photosensitive cells, cones and rods attached by nerve fibers which come together at the center to form the optic nerve. The cones lie near the center of the retina, and the rods lie toward the edges. The cone cells process most of our vision in the daylight and are responsible for our perception of color. The rods are extremely sensitive to very dim light, but produce a rather blurred image and are not sensitive to color. The observer may take advantage of the eye's sensitive peripheral area by using a technique called averted vision. Simply look to the side of the object rather than directly at it. Shift your gaze slowly to each side, or up and down if you prefer. This is a technique that should be used any time one views a faint object. It is surprising how much extra detail becomes visible.

It is important to "dark-adapt" your eyes before viewing faint objects. After 15 minutes in the darkness, the eye becomes six to seven times more sensitive to low light levels than it was after just leaving a lighted room. However, as the eye remains in darkness, a slow chemical change requiring a minimum of 30 minutes takes place in the photoreceptive cells of the retina making the rod cells in our peripheral vision a thousand times more sensitive. Once your eyes have adapted to the darkness, you should take care not to look at white light or even a bright red light while you are still observing. Use a deep red filter over a flashlight to read star maps.

Table 1-2. Visual Impression with 12/14″ Scopes

Apparent Brightness (v)

NGC Description		Galaxies	Star Clusters
eF	extremely faint	13.5 or fainter	13th or fainter
vF	very faint	13.0–13.4	11th-12th mag.
F	faint	12.5–12.9	9th-10th mag.
cF	considerably faint	12.0–12.4	8th mag.
pF	pretty (fairly) faint	11.5–11.9	7th mag.
pB	pretty (fairly) bright	11.0–11.4	6th mag.
cB	considerably bright	10.5–10.9	5th mag.
B	bright	09.5–10.4	4th mag.
vB	very bright	08.0–09.4	3nd mag.
eB	extremely bright	07.9 & brighter	2nd mag.

Apparent Size (′) in minutes of arc

NGC Description		Galaxies	Star Clusters
eS	extremely small	0.5 or smaller	1 or smaller
vS	very small	0.5–0.9	2–3
S	small	1.0–1.4	4–5
cS	considerably small	1.5–1.9	6–7
pS	pretty (fairly) small	2.0–3.9	8–9
pL	pretty (fairly) large	4.0–5.9	10–14
cL	considerably large	6.0–7.9	15–29
L	large	8.0–9.9	30–44
vL	very large	10.0–19.9	45–59
eL	extremely large	20.0 or larger	60 or larger

1.7 Visual Telescopic Descriptions

The two bold lines preceding the telescopic description give the following data:

I. Line 1
   A. NGC Number
   B. Numbers from other catalogues:
      M = Messier Catalogue
      H = William Herschel's Catalogue.
      IC = Index Catalog
      (See Table 1-4 for other Catalogues)
   C. Number of stars in open clusters
   D. Object's type or class
   Note: Brackets indicate estimates made from observations where no data exists.

II. Line 2
   A. ø = Object's Diameter
      ′ = Minutes of arc
      ″ = Seconds of arc
   B. m = Apparent Magnitude
      v = Visual magnitude
      p = Photographic magnitude
   C. Br★ = Brightest star in cluster
      SB = Integrated surface brightness
      CS = Central star's magnitude
         (Planetary Nebula)
   D. Celestial Coordinates
      Right Ascension (Equinox 2000)
      Declination (Equinox 2000)

III. Line 3 *(Shown with Italic fonts)*
   A. Finder or Constellation Chart number. Indicates chart that plots the object. In some instances, two charts are listed.
   B. Figure Number. Indicates that there is an illustration, either an eyepiece impression (drawing) or a photograph.
   C. Name of Object
   D. Visual Rating (See Section 1.8 & Table 1-3)

IV. Line 4 Visual Telescopic Descriptions
   A. Telescope size & recommended best magnification for each particular object. Use less power to locate objects then switch to the recommended magnification. When seeing conditions permit, inspect object with higher power. Descriptions are grouped into categories as shown below with our recommendation for the overall best magnification taking the object's size into account. Below is the recommended magnification for average-size objects.
      2/4″ Scopes–50x: (majority 3″)
      4/6″ Scopes–75x: (majority 6″)
      8/10″ Scopes–100x: (majority 8″)
      12/14″ Scopes–125x: (majority 12.5–13″)
      16/18″ Scopes–150x: (majority 17.5″)
      20/22″ Scopes–175x: (majority 20″)

   B. Visual Description of Object
      Descriptions include the following information:
      1. Galaxies & Nebulae
         a. Brightness
         b. Size
         c. Position Angle
         d. Shape
         e. Brightness variations across disk or halo
         f. Field description noting other objects in the field of view, such as companions, nearby stars, or stars actually involved with the object (prominent stars that aid in locating the object are mentioned first.)
      2. Star Clusters
         a. Size
         b. Richness & compression
         c. Number of stars
         d. Brightness of the stars
         e. Field description noting other objects or stars in the field of view

1.8 Visual Rating Guide

Our rating system (shown in Table 1-3) is a combination of the object's visibility and interest. Each object is assigned one to five asterisks (stars), the more asterisks, the more highly visible or interesting the object is. Objects that show interesting features but are of similar brightness to others often are rated in a higher category. The rating system allows the observer to judge the visibility of an object. If you use a small scope or the transparency is not very good on a particular night, you may avoid objects having a rating of only one or two asterisks. Beginners should look for objects rated with five or four asterisks then graduate to fainter objects as observing skills improve.

Please note that in the rating of each object, that object is judged within its own class; galaxies are compared to other galaxies, and star clusters are compared to other star clusters, etc. A star cluster called bright will seem brighter than a bright galaxy or a bright planetary nebula.

1.9 Using the Star Charts

I. Large Constellation Map

This map plots 5th magnitude stars normally visible to the naked eye under clear, dark skies. Interesting double and variable stars listed in the tables, even though they may be fainter, are plotted to show their general location. These stars are normally visible in an 8×50 viewfinder. If the double star listed in the tables falls within the star field of a more detailed finder chart, the chart's number will be indicated in the notes column.

II. Master Finder Chart Areas

These maps show a wide view of the feature constellation and surrounding constellations. The finder chart areas are indicated along with arrows pointing to recommended guide stars. After aiming your viewfinder at the guidestar, you are ready to "starhop" using the bright stars and other star patterns shown in the finder chart. The finder chart number is given next to the arrow.

III. Finder Charts

The finder charts plot stars to about 8th magnitude and show the location of deep-sky objects in our survey. The scale of the star field in finder charts are easily recognized in a typical 8 × 50 viewfinder having a three degree field. A degree scale is plotted so that you may judge the chart's area. After you use the charts awhile, you will be able to judge the area visible in your viewfinder.

The charts have north at the top so they will need to be turned upside-down (as do all other star charts) when used with a telescope, such a Newtonian reflector, that inverts the image. We do not recommend the use of right angle diagonals because they invert north and south but not east and west making star charts difficult to read. If you prefer a diagonal, then we recommend an Amici-type image erector which will allow reading star charts directly.

Each chart shows the same naked eye guidestar indicated in the master finder chart. Using the guidestar as a starting point, starhopping to a deep-sky object is easily accomplished by following the brighter stars and other star patterns to the area where the object is plotted. Use low power to locate objects and then switch to the recommended magnification. When sky conditions permit, study the object with higher powers to reveal finer details. For the users of mechanical or digital setting circles, the guide star's right ascension and declination are given.

1.10 Getting Started

We assume that most of our readers already own a telescope or have access to one. For those who are contemplating purchasing a telescope, let us review a few things we deem important:

★Get Advice

Beginners should become familiar with different types of telescopes before purchasing their own instruments. Joining an astronomy club and attending their star parties are two of the best ways to get advice and to actually see through the equipment other amateurs are

---

Tabel 1-3 Visual Rating Guide

★★★★★ *Showpiece Objects*
These are the most impressive objects–the finest examples of their type or class, visible in small telescopes.

★★★★ *Bright or Interesting Objects*
Bright or generally more interesting objects visible in nearly all telescopes. 4″ to 6″ telescopes generally provide a good view

★★★ *Average Objects*
Moderate in brightness or size, these objects may appear fairly bright in larger telescopes but fairly faint in smaller instruments. Faint objects showing detail may be advanced to the category. 8″ to 10″ telescopes generally provide a good view.

★★ *Faint Objects*
Low surface brightness objects or clusters with faint members. Averted vision may be needed with smaller telescopes and is helpful with larger instruments. A 12″ telescope generally provides a good view.

★ *Very Faint Objects*
These objects require clear, dark skies and are best viewed near culmination. Averted vision is needed or helpful even with good seeing and transparency. 16″ or larger telescopes usually are required.

---

using. Going to astronomy conventions is the perfect forum for discovering what is available in the marketplace. These regional meetings are listed in various astronomy magazines. At any of these meetings, you may ask for advice from both the vendors and experienced amateurs.

★Telescopes

Purchase the largest aperture you can afford or easily transport. The aperture of a telescope is the diameter of its objective lens or its primary mirror. The single most important feature of a telescope is its light gathering ability. A telescope's function is to collect light and focus it to a precise focal point at the eyepiece. Assuming the optics of a telescope are of good quality, the larger the aperture of the instrument the better the image will be. The power at which an object is viewed has little to do with its clarity. Resolution is not only affected by the quality of the optics, but the number of photons the instrument collects–the bottom line is APERTURE. If quality is equal, the larger telescope will resolve more detail and detect fainter objects than a smaller instrument. We recommend Newtonian type telescopes because these are the least expensive instruments for any given aperture.

★Mountings

The mounting must be solid. Equatorial or fork mounts must have a clock drive to track the stars if you plan to do astrophotography. One of the most stable and economical mounts is the Dobsonian mount. This wonderful altazimuth mounting, developed by John Dobson, enables amateurs to afford and enjoy viewing with very large Newtonian reflectors. The mount is very stable, easily transported, and easily made by amateur telescope makers. The Dobsonian mount is not usually motor driven, but Poncet platforms or stepper motors can be added if tracking is desired.

★Eyepieces

Purchase the best eyepieces you can afford. The image seen through a good telescope will only be degraded by a poor eyepiece. Sorry to say, most commercial telescopes come with mediocre eyepieces so upgrading is often necessary. Here again, star parties and conventions are the perfect place to view and compare one eyepiece brand to another. At the major star parties, some of the vendors offer discounts or specials, and, in addition, there are usually flea markets where you can purchase used eyepieces and other equipment at bargain prices.

Generally, the more elements an ocular has in its design, the better the eyepiece performance. These additional elements give better image quality and a wider apparent field.

Kellner eyepieces, with 3-elements, give an acceptable quality at low and medium powers for the lowest cost but have poor eye relief and narrow fields (40°). These eyepieces are fine on long focal length refractors but are not suitable on short focal length reflectors. Avoid Ramsden or Huygenian designs.

Orthoscopic eyepieces are some of the most popular eyepieces because they give good overall performance, good eye relief, and a moderately wide apparent field of view (45°) in a 4-element design.

Plossl eyepieces have extremely sharp designs with pinpoint images right to the edge. They are a 4-element design and give wider fields (50°) than Orthoscopics.

Erfle eyepieces use a 6-element design for a very wide field of view but suffer from poor edge sharpness. These are best used at low to medium powers.

Wide-Angle & Ultra-Wide Field type eyepieces come in various designs with 6 to 8 elements giving breathtaking apparent fields of view of 80° or more. In some brands the eye relief is poor and the whole view cannot be seen without moving the eye around. Eye relief varies between brands so try these out before you make a purchase. These oculars are expensive, but owning at least one good quality ultra-wide is well worth the money. We have found the Nagler eyepiece and similar designs to be best suited for the short focus Newtonian reflector telescope.

★Viewfinders

The telescope should have an adequate view finder. We recommend at least an 8×50 viewfinder even if another sighting device is used. A 6×30 viewfinder is okay for locating the Moon, planets, and bright stars but is inadequate when searching for deep-sky objects. The serious deep-sky observer should have a viewfinder ranging from 12×60 to 15×80, which may also be used as a second telescope to enjoy large extended objects such as the Pleiades.

★Digital Setting Circles

If you have not already used a telescope equipped with a miniature computer known as a digital setting circle (DSC), you are in for a pleasant surprise. These devices enable amateurs to locate and observe deep-sky objects fast and easy. The DSC package comes with encoders which are installed on both of the telescope's axis. As you turn the telescope the encodes register the movement of the axis telling the computer where the telescope is pointed. To find an object, the observer enters the object's NGC or other number from the computer's database and turns the telescope while watching the computer's display until the coordinates are "zeroed." Some telescope manufacturers also offer models with motorized axis that will point the telescope to the object after it is entered. These model work best on a permanent pier so that the telescope remains properly polar-aligned.

Digital setting circles are great, but don't dare throw away your star charts – DSCs can be temperamental. Rarely is the object you want to view centered within eyepiece's field of view. If the object is a galaxy lying within a group of galaxies you will need a star chart to sort it out. Another common plight is that the encoders won't work properly after traveling some distance to a star party. I know of several cases where amateurs did not bother to equip their telescopes with viewfinders, and when the DSC did not work, they couldn't even resort to star hopping. Don't let this happen to you, include a viewfinder as standard equipment on you scope – we mean viewfinder, not just a sighting device that has no magnification).

★Star Charts

Star maps and celestial guides are as important to the astronomer as sheet music is to the musician. You need a set of star charts showing stars down to magnitude 8 or 9, and a sky guide, such as this book, showing where to find interesting objects. The amateur astronomer will find both *Sky Atlas 2000* and *Uranometria 2000.0* indispensable.

★Books and Magazines

Subscribe to astronomy magazines and journals. These publications will keep you informed on important upcoming celestial events such as: eclipses, comet and asteroid passages, meteor showers, occultations involving the Moon, planets, and bright stars. These publications feature countless articles on all aspects of astronomy, including viewing interesting celestial objects. In addition to the articles, these magazines are full of ads for all types of telescopes and equipment which, in most cases, are not available at a local store. In the USA, *Astronomy* magazine and *Sky & Telescope* may be found at major newsstands and bookstores. Astronomy books are another important resource for the amateur astronomer. There are books available for both the beginner and the advanced amateur on any astronomical subject. Again, if you can't find what you need at a local bookstore, phone a vendor advertising in one of the current journals.

★Internet Resources

For computer users with internet access, the World Wide Web ("the Web" or WWW) has become an important access to information. To reach a site all you need do is enter its Web address, called a uniform resource locator (URL). We won't publish any of the current "addresses" because they are too subject to change.

No matter what astronomical interests you have you are certain to find it on the "Web." Information is available on discoveries made by professional and amateur astronomers, astronomical almanacs of celestial events (eclipses, planetary and lunar conjunctions, occultations, asteroids, meteor showers, etc.), and periodic and new comet discoveries. Images from JPL, NASA, and other sources may be downloaded. Specific areas of any portion of the sky may be downloaded from the National Geographic Society-Palomar Observatory Sky Survey (POSS) plates through the Digitized Sky Survey. There are lists of observatories, planetariums, and astronomy clubs. Astronomical vendors post entire catalogs of items which may be ordered by e-mail. There are even on-line want ad bulletin boards where you can purchase used telescopes and other equipment or sell your own items. "Chat rooms" allow you to talk with other amateur astronomers world wide.

★Constellations

Learn the constellations; the naked eye star map found inside the front cover is ideal for this. Start with the brighter, more easily recognized constellations, these will help you find the fainter ones lying nearby.

★Advice for Neophytes

A telescope is like a musical instrument, you can not become proficient at playing it without practice. First, read your telescope's manual and assemble your instrument with care. The viewfinder must be aligned to show the same view seen through the eyepiece. Aim the telescope at a bright star centering it in a low power eyepiece; then adjust the viewfinder so that the star is also centered in its field. Next, fine tune the adjustment with a high power eyepiece. The greatest obstacle confronting the new telescope owner is learning to locate objects. Using 50x to 75x, practice aiming at bright stars until you are able to readily sweep them up. Now you are ready to try your hand at finding the brighter deep-sky objects. Start with objects that we have rated with five stars. If this is your first attempt, choose an object that lies in a field of fairly bright stars so that you can easily identify the star patterns. Start by finding the designated guide star; then identify other bright stars shown on the chart. Starhop by repositioning the viewfinder's crosshairs along a path past each bright star to the area of the deep-sky object. If the viewfinder is in proper alignment, the object will be visible in a low power eyepiece.

1.11 Keeping Records

Long before photography or electronic imaging devices were developed, the only way to record a telescopic image was with pencil and paper. This method may seem old fashioned and inaccurate, but it offers the amateur many advantages. Writing and sketching are far less expensive than cameras and other equipment. The observer can make a sketch in several minutes compared to a half hour or longer needed to make a photograph, and there is no guarantee that the photograph will be any good. Contrary to what you may think, you don't have to be an artist to make good sketches, celestial objects are much easier to draw than everyday terrestrial objects because they are mostly circular shapes and dots. Follow the simple sketching procedure discussed in section 1.13.

Those who consider themselves serious amateur astronomers should keep observation records for several good reasons. Records enable the observer to compare observations made at different times to see what changes may have occurred. Written records or drawings will also enable the observer to compare results when using different instruments or to exchange notes with other observers. Keeping records or making drawings will help a person become a more skilled observer than the casual viewer. The main reason for keeping notes, however, is for enjoyment and the satisfaction gained when an observer has compiled his or her own catalog. Beginners may start by viewing the objects in Messier's Catalog;

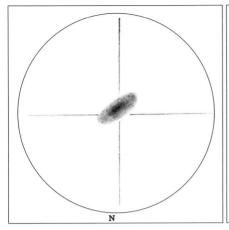 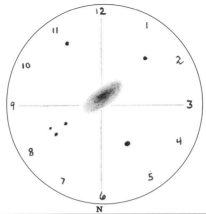 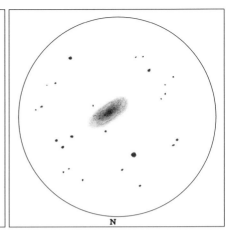

Figure 1-2. Sketching Procedure

Step 1. *Divide the field into imaginary quarters, center the object, then draw its size and shape. You may use an illuminated reticle normally used for photographic guiding.*

Step 2a. *Record the brightest stars first, estimating each star's postion from the center in hour angles as if the field of view were the face of a clock.*

Step 2b. *Use the bright stars as an additional point of reference to help place the fainter stars. Finish by marking the direction of drift (west) with an arrow, north is 90 degrees counterclockwise.*

these are the brightest and finest objects in the sky. The Astronomical League awards certificates for this accomplishment. The Herschel Club has a similar but more ambitious program.

An accurate drawing of an object and the surrounding star field may be used to compare objects on a photograph or star chart. There may be an extra star in the halo of a galaxy that is not on a photograph–you could be the discoverer of a supernova! You might also spot an asteroid or a comet in the star field that is not visible in the photograph or plotted on the star chart.

Rather than using a tablet or notebook, we recommend loose-leaf paper stored in a three-ring binder or index cards kept in a file box so that you can keep observations filed in numerical or right ascension order for later reference. The single leaflets can be used on a clipboard which has a hard surface suitable for writing and drawing. If you own a computer, you may record the observations within a software database program that will allow you to arrange or sort the data.

1.12 Magnification and Eyepiece Selection

At the beginning of each description, we indicate the best overall magnification for each aperture telescope range. Dawes limit is about $30\times$ per inch of aperture but we have found from experience that less than half this power gives the better performance. Use a low power eyepiece to locate the object; then switch to the recommended magnification. Choose eyepieces that have a comfortable eye relief, especially if you wear glasses. Higher powers require placement of the eye close to the lens in order to see the entire field of view. Often, this means removing eyeglasses to get close enough to the lens. Most eyeglass wearers will find satisfactory viewing after readjusting the focus. If your eyeglasses have plastic lenses, removing them prevents scratching. A Barlow lens helps increase eye relief. Instead of using a 12mm eyepiece, a 24mm eyepiece with a 2x Barlow retains the same eye relief that the lower power eyepiece normally provides.

After viewing the object with the recommended magnification, inspect the object at higher powers to see if any additional details become visible. Smaller deep-sky objects such as planetary nebulae require higher powers. View the object with increasing powers until a limit is reached where the stars become objectionably blotted, then back off to the last satisfactory view. By doing this, you have determined which eyepiece gives the best compromise between image size, contrast, and sharpness. You may be surprised how much more of a faint, distended object is visible at higher powers because the loss in sharpness is offset by an increase in contrast between the object and the sky background. When viewing a small planetary nebula, the higher magnification makes the tiny disk more visible, and because the nebula's disk is diffuse anyway, the loss in sharpness is not as objectionable as it is with stellar objects. Globular clusters and open star clusters containing very faint stars also benefit from the use of higher powers, but some star clusters such as the Pleiades are so huge that they may only be seen in their entirety at very low power.

1.13 Sketching Made Easy

Only paper and pencil are needed for sketching. Any blank paper suitable for drawing will do; however, if you want to do quality work, visit your local art store and purchase sketching paper and at least three pencils with hard, medium, and soft lead. A compass or circle template for drawing circles is optional; a jar lid or any round object with a diameter between 2 and 2.25 inches will do.

Stars are simple to draw; they are represented by small dots for faint stars and progressively larger dots for brighter stars up to about 1/8th inch for first magnitude stars. To draw nebulae or the halo of a galaxy, shade the paper by holding the pencil nearly flat (parallel) to the paper. To make a diffuse appearance, rub the shaded area lightly with your finger tip or a Q-tip. Since a pencil drawing is actually a negative representing the bright areas with dark shadings, dark lanes will be shown by light areas made by erasing some of the shading with a sharply pointed eraser.

Selecting an eyepiece for sketching has different considerations. Ideally, you should use the same eyepiece most of the time so that your drawings have a standard scale. A 10 or 12mm eyepiece is a good compromise; it is powerful enough to see detail in all but the smallest objects and still allows a reasonable span of field. Objects in or near the Milky Way have so many stars in the field that they overwhelm you. The trick here is to reduce the star field and the number of stars to be drawn by using higher powers; and, of course, it is not necessary to place every faint star in the field. Just be sure to include all of the brighter stars and apparent star patterns. A motorized right ascension drive is helpful when sketching due to star field drift; however, do not let the lack of a drive deter you.

Sketching – Step 1

A. Divide the field of view into imaginary quarters like the crosshairs in your viewfinder by drawing very light lines on the observing form circle. Although not necessary, you may use an eyepiece with crosshairs or an illuminated reticle normally used for photographic guiding. Also, envision the field as the face of a clock with numbers around the edge of the field.
B. Center the object; or if there are two or more, center the objects at a point between them to include the group. You should include any bright stars or patterns of stars that help identify a faint object or enhance the setting. When framing two deep-sky objects, try to center a star, even a faint one, at the center of the field for a point of reference.
C. Judge the size of the object in relation to the apparent field of view. If a galaxy's halo spans a 1/3 of the field, it should be represented on your sketch accordingly. Be careful to represent the position angle and elongation accurately. Look for variations in brightness. Is the center brighter or fainter? Are the edges uniform or irregular? Are there any bright spots, streaks, lanes, condensations, or mottling? Are there stars or nebulosity involved?

Sketching – Step 2

A. Record the brightest field stars first. Estimate each star's position as accurately as possible. Judge all star positions from the center of the field to the outside in clock hour angles. Example: The brightest star may be half the distance from the center in the 5 o'clock position. The second brightest star may be 75% of the distance from the center in the 11 o'clock position. Continue recording all the brighter stars.
B. Place the fainter stars between the bright stars using the bright stars as reference points. Two methods of checking accuracy are the hour angles and distances between the brighter stars. Accurate star field sketches may be drawn using this simple method. Don't be discouraged by an abundant star field; make your sketch in quarter fields from brightest to faintest stars.
C. After your drawing is finished, indicate the compass directions. Turn off the drive motor and indicate the direction of drift by placing an arrow in the drawing showing this direction. The direction of field drift is west or the "preceding" direction while the opposite direction is east or "following." As we discussed before, most telescopes have an inverted image which places north toward the bottom of the field 90° counterclockwise from the direction of drift, and, therefore, south is at the top. Please note that south will only be exactly at the top when you are pointed directly at the meridian; when an object is east or west of the meridian, the field will be tilted accordingly. As mentioned before, the use of a star diagonal prism complicates matters by inverting only north and south, making star maps nearly impossible to use. If you prefer using a right angle prism, use the type that gives an erect view.

1.14 Estimating Field of View

An easy method of measuring eyepiece field diameter is to time a star's transit across the field of view. Select a star near the celestial equator and position it at the very edge of the field. Time, in seconds, how long it takes for the star to drift through the center of the field to the opposite edge. The time divided by four gives the field diameter in minutes of arc.

To become familiar with the size of celestial objects, select objects of different sizes and view them with the same eyepiece. Use an eyepiece that you will be using to make most of your observations. 24mm to 15mm eye-

Table 1-4. Abbreviations

I. Object Data

Abbreviations and symbols used in the data lines and tables.

Br★	Brightest star (In open star clusters)
CS	Central star (Planetary Nebulae)
Dec.	Declination (Equinox 2000)
°	Degree(s)
φ	Diameter
>	Greater than
<	Lesser than
m	Apparent visual magnitude (v = visual) (p = photographic)
Max.	Maximum
Min.	Minimum
′	Minutes of arc
P.A.	Position Angle
R.A.	Right Ascension (Equinox 2000)
″	Seconds of Arc
Sep.	Separation
Spec.	Spectral type
★	Star(s)
SB	Integrated surface brightness
Tr Type	Trumpler Type (See Table I.2)
[ ]	Brackets indicate estimates made from observations when no data exists

II. Double Stars Catalogs

β	S. W. Burnham
Δ	J. Dunlop
Σ	Wilhelm Struve
ADS	Aitken Double Star Catalogue
A	R. G. Aitken
CorO	Cordoba Observatory
Es	T. E. H. Espin
Frk	W. S. Franks
h	John Herschel
HD	Henry Draper Catalogue
S	James South
S, h	James South & John Herschel
OΣ	Otto Struve, Pulkovo Catalogue
OΣΣ	Otto Struve, Pulkovo supplement

III. Star Clusters & Deep-Sky Catalogs

Bar	Barkhatova
Bas	Basel
Be	Berkeley
Bi	Biurakan
Bl	Blanco
Bo	Bochum
Ced	Cederblad
Cr	Collinder
Cz	Czernik
Do	Dolidze
DoDz	Dolidze/Dzimselejsvili
Haf	Haffner
H	Harvard
H	William Herschel - When number is followed by a superscript numeral: H78^s See Table I-1.
Ho	Hogg
IC	Index Catalogue
Isk	Iskudarian
J	Jonckheere
K	King
Lo	Loden
Mrk	Markarian
Mel	Melotte
M	Messier
NGC	New General Catalogue
PK	Perek & Kohoutek (Planetary Nebula)
Pi	Pismis
Ro	Roslund
Ru	Ruprecht
Sh(2)	Sharpless
Ste	Stephenson
St	Stock
Tom	Tombaugh
Tr	Trumpler
U	(UGC) Uppsala General Catalogue
Up	Upgren
vdB	van den Bergh-Waterloo
vdB-HA	van den Bergh-Hagen
Wa	Westerlund

pieces are best for most types of deep-sky objects except for small planetary nebulae which require higher powers. Locate at least four or five bright star clusters varying from 5′ to 20′ diameter, paying attention to how much of the field each object takes up. A word of caution is necessary at this point in our discussion; if you try this experiment with galaxies, the entire halo may not be visible. The halo of a galaxy fades toward the edges, and you may not be seeing the entire length measured from a photographic plate. With a little practice, judging an object's size will become easy.

1.15 Enjoy the Night Sky

The intent of this book is to provide the telescope user, both experienced amateur astronomers and beginners alike, a useful celestial guide to the night sky's many treasures conveniently grouped by constellation. We hope that each of you will enjoy becoming acquainted with new objects each time you observe. Nothing is more splendid than nature; and the grandest phenomenon of all is the universe visible through the eyepiece.

Table 1-5. The Greek Alphabet

Alpha	α	Iota	ι	Rho	ρ
Beta	β	Kappa	κ	Sigma	σ
Gamma	γ	Lambda	λ	Tau	τ
Delta	δ	Mu	μ	Upsilon	υ
Epsilon	ε	Nu	ν	Phi	φ
Zeta	ζ	Xi	ξ	Chi	χ
Eta	η	Omicron	ο	Psi	ψ
Theta	ϑ	Pi	π	Omega	ω

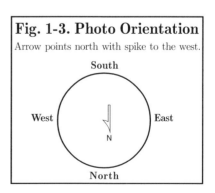

Fig. 1-3. Photo Orientation

Arrow points north with spike to the west.

Chapter 2

Andromeda, the Princess

2.1 Overview

Andromeda, which for mid-northern observers occupies the zenith during the early evenings of autumn, extends from the NE corner of the Great Square of Pegasus. With an area of 722 square degrees, it ranks as the 19th largest constellation. Due to her position well away from the galactic plane, few nebulae can be found within the constellation; however, the area is a window to deep space allowing us to see external galaxies of all types. The constellation's most famous object is M31, the Great Andromeda Galaxy, a "sister" to the Milky Way. M31 is similar in size and shape to our own galaxy. In addition to galaxies, Andromeda contains a fine assortment of double stars of varying brightness, separation, and colors as well as a couple open star clusters. Included among its treasures is the lovely bluish planetary nebula, NGC 7662.

The tale of Andromeda is one of the most famous Greek myths. The beautiful Princess was chained to a rock by her royal father, Cepheus, as a sacrifice to appease the avenging sea monster, Cetus. The hero in this story, Perseus, slew the monster and married Andromeda. Adjacent constellations represent all the characters in this myth: her parents Cepheus and Cassiopeia, her rescuer, Perseus, and the sea monster, Cetus.

Andromeda: An-DROM-eh-da
Genitive: Andromedae, An-DROM-eh-dee
Abbrevation: And
Culmination: 9pm–Nov. 23, midnight–Oct. 9
Area: 722 square degrees
Showpieces: 57–γ And, M31 (NGC 224), M32 (NGC 221), M110 (NGC 205), NGC 891, NGC 7662
Binocular Objects: 56 And, M31 (NGC 224), M32 (NGC 221), M110 (NGC 205), NGC 752, NGC 891, NGC 7686

2.2 Interesting Stars

Groombridge 34 (ADS 246) Double Star   Spec. M2
m8.2, 10.6; Sep. 40.0″; P.A. 62° $00^h17.9^m$ +44°00′
Constellation Chart 2-1   ★★★

Located 1/4° north of 26 Andromedae, this red dwarf binary system is one of the closer stars to our solar system lying only 11.7 light years away from us. The pair has a large proper motion of 2.89″ per year in P.A. 82°. A careful sketch made at high power once a year will show the movement of Groombridge 34 in relation to the surrounding star field. This reddish pair is composed of 8.2 and 10.6 magnitude stars separated by 40″.

R Andromedae Variable Star   Spec. S3-S8 (M7e)
m5.8 to 14.9, Per. 409.33 days $00^h24.0^m$ +38°35′
Finder Chart 2-4   ★★★

R Andromedae, located near the bright triangle of Theta, Rho, and Sigma Andromedae, is the brightest long-period variable in the constellation. It is a pulsating red giant similar to Mira (Omicron Ceti), and has the very large range of nine magnitudes (5.8 to 14.9) and the long period of 409.3 days. Its distance is uncertain, but probably is on the order of 1,000 light years.

Pi (π) = 29 Andromedae Multiple Star   Spec. B3
AB: m4.4, 8.6; Sep. 35.9″; P.A. 173° $00^h36.9^m$ +33°43′
Constellation Chart 2-1   ★★★★

Pi Andromedae is a multiple star with two visual companions. The 13th magnitude "C" star is only a chance alignment. Pi's absolute magnitude is −1.1, a luminosity of about 170 suns. Its distance is 390 light years.

8/10″ Scopes–100x: This unequal double has a bright white primary and a faint blue companion.

Andromeda, the Princess

Constellation Chart 2-1

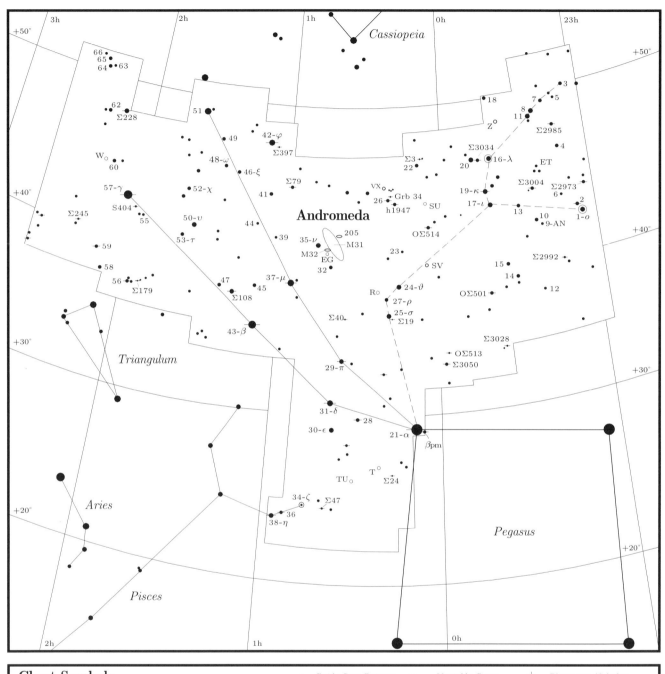

Table 2-1. Selected Variable Stars in Andromeda

Name	HD No.	Type	Max.	Min.	Period (Days)	F*	Spec. Type	R.A. (2000)	Dec	Finder Chart No. & Notes
SV And	225192	M	7.7	14.3	316.21	0.38	M5-M7	00^{h}04.3^m	+40°43′	2-1
VX And	1546	SRa	7.8	9.3	369.00		C4.5(N7)	19.9	+44 43	2-1
T And	1795	M	7.7	14.5	280.76	0.46	M4-M7.5	22.4	+27 00	2-3
R And	1967	M	5.8	14.9	409.33	0.38	S3-S8(M7)	24.0	+38 35	2-4 Red Giant
TU And	2890	M	8.7	13.1	316.77	0.48	M5	00^{h}32.4^m	+26 02	2-3
EG And	4174	Z And	7.08	7.8			M2	44.6	+40 41	2-4
W And	14028	M	6.7	14.6	395.93	0.42	M4-M1	17.6	+44 18	2-1
Z And	221650	Z And	8.0	12.4			M2+B1	23^{h}33.7^m	+48 49	2-6 Recurrent Nova?

F* = The fraction of period taken up by the star's rise from min. to max. brightness, or the period spent in eclipse.

Table 2-2. Selected Double Stars in Andromeda

Name	ADS No.	Pair	M1	M2	Sep. ″	P.A.°	Spec. Type	R.A. (2000)	Dec	Finder Chart No. & Notes
OΣ514	30		6.1	8.7	5.2	168	A2	00^{h}04.6^m	+42°06′	2-1
β pm	69	AB	6.3	8.9	158.6	250	K0	06.6	+29 01	2-3
	69	AxCD			153.2	185				
	69	CD	10.0	10.5	3.3	210				
21-a And	94		2.1	11.3	81.5	280	A0	08.4	+29 05	2-3
Σ3	119		8.1	9.1	5.0	84	A3	10.0	+46 23	2-1
h1947	215		6.2	10.2	9.2	76	A0	16.4	+43 36	2-1
Σ19	220		7.1	9.6	2.4	138	A0	16.7	+36 38	2-1
Grb 34	246	AB	8.2	10.6	40.0	62	M2	17.9	+44 00	2-1 Red Binary
Σ24	252		7.6	8.4	5.2	248	A2	18.5	+26 08	2-3
26 And	254		6.0	9.7	6.2	240	B9	18.7	+43 47	2-1
Σ40	486		6.6	9.2	11.7	313	K0	35.2	+36 50	2-4 Both stars yellow
29-π And	513	AB	4.4	8.6	35.9	173	B3	36.9	+33 43	2-1 White, blue
31-δ And	548	AB	3.3	12.4	28.7	298	K2 M2	39.3	+30 52	2-3
Σ47	562	AB	7.4	9.2	16.5	205	A3	40.3	+24 03	2-3
	562	AC		11.1	42.4	230				
36 And	755	AB	6.0	6.4	w0.9	*313	K0	55.0	+23 38	2-1 Close yellow pair
38-η And			4.4	11.3	133.4	232	G5	57.2	+23 25	2-1
Σ79	824		6.0	6.8	7.8	192	B9 B0	01^{h}00.1^m	+44 43	2-1
Σ397	926	AB	7.6	9.8	8.7	142	K2	07.9	+46 51	2-1
42-φ And	940		4.6	5.5	0.5	*127	B8	09.5	+47 15	2-1
43-β And	949	AB	2.1	14.1	27.0	202	M	09.7	+35 37	2-4 Orangish
	949	Bb		13.4	24.6	126				
Σ108	1055		6.4	9.2	6.2	62	A3	18.8	+37 23	2-4
Σ179	1500		7.4	8.4	3.5	160	F5	53.2	+37 19	2-5
56 And	1534	AB	5.7	6.0	190.4	300	K0 K2	56.2	+37 15	2-5 Binocular double
S404	1560	AB	7.6	9.6	25.0	78	G5	58.1	+41 23	2-5
57-γ And	1630	AB	2.3	5.5	9.8	63	K0	02^{h}03.9^m	+42 20	2-5 Yellow, blue-green
	1630	BC	5.5	6.3	c0.4	*103	A0			
59 And	1683		6.1	6.8	16.6	35	A0 A2	10.9	+39 02	2-5
Σ228	1709		6.6	7.1	c1.0	*282	F0	14.0	+47 29	2-1
Σ245	1763	AB	7.2	8.2	11.0	292	F2	18.6	+40 04	2-1
Σ2973	16472		6.4	9.6	7.4	41	B3	23^{h}58.4^m	+44 04	2-6
Σ2985	16557		6.6	8.6	15.4	254	G5 G5	10.0	+47 58	2-6
Σ2992	16599		7.6	9.3	14.2	286	A3	13.1	+40 00	2-1
Σ3004	16685		6.1	9.6	13.2	178	A3	20.7	+44 07	2-6
Σ3028	16894		7.1	9.6	16.0	201	A2	38.6	+35 02	2-1
OΣ501	16913		6.5	9.9	14.9	164	F0	40.0	+37 39	2-1
Σ3034	16965		7.6	9.8	5.4	103	A0	44.5	+46 23	2-6
OΣ513	17136		6.8	9.3	3.6	22	A3	58.4	+35 01	2-1
Σ3050	17149	AB	6.6	6.6	1.7	335	F8	59.5	+33 43	2-1 Deep yellow stars

Footnotes: *= Year 2000, a = Near apogee, c = Closing, w = Widening. Finder Chart No: All stars listed in the tables are plotted in the large Constellation Chart, but when a star appears in a Finder Chart, this number is listed. Notes: When colors are subtle, the suffix *-ish* is used, e.g. *bluish*.

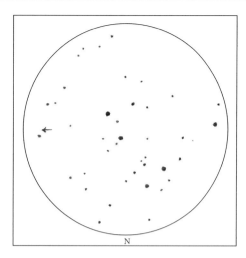

Figure 2-1. *Gamma (γ) = 57 Andromedae 12.5″, f5–225x, by G. R. Kepple*

36 Andromedae = Σ73 Double Star   Spec. K0
m6.0, 6.4; Sep. 0.9″; P.A. 313° $00^h55.0^m$ +23°38′
Constellation Chart 2-1   ★★★

This close binary is a test for large amateur telescopes, its components varying in separation from 1.4″ to only 0.6″ in a 165 year period. The primary is a subgiant several times more luminous than the Sun. The companion is probably similar in spectral type because it has the same color, a lovely, brilliant yellow. Because this system is around 390 light years away, the 0.6″ minimum separation corresponds to the distance between the Sun and Pluto.

Gamma (γ) = 57 Andromedae Multiple Star   Spec. K0
AB: m2.3, 5.5; Sep. 9.8″; P.A. 63° $02^h03.9^m$ +42°19′
Finder Chart 2-5, Figure 2-1   Almach   ★★★★★

Almach is a very beautiful multiple star with a lovely golden primary and a greenish-blue companion which is also a close double. The A-B pair can be resolved in small telescopes; but the B-C stars are only 0.5″ apart, requiring large aperture to be split. The A component is also a spectroscopic binary; hence Almach is in fact a quadruple system. The primary is about 650 times as luminous as the Sun while the combined luminosity of the other three stars equals 50 suns. This system is nearly 120 light years away.

4/6″ Scopes–125x: Gamma is a beautiful yellow and blue pair with no hint of separation between the B-C stars.

12/14″ Scopes–150x: Medium power shows a lovely yellow primary. The bluish B-C pair appears oblong.

Σ3050 Double Star   Spec. F8
m6.6, 6.6; Sep. 1.7″, P.A. 335° $23^h59.5^m$ +33°43′
Constellation Chart 2-1   ★★★★

8/10″ Scopes–150x: This lovely double star has two close yellow components in contact.

12/14″ Scopes–175x: Struve 3050 is a tight pair of equally matched deep yellow suns.

2.3 Deep-Sky Objects

NGC 80 U203 Galaxy Type SA0−:
ø2.2′ × 2.0′, m12.1v, SB 13.5 $00^h21.2^m$ +22°21′
Finder Chart 2-3, Figure 2-2   ★★

12/14″ Scopes–125x: NGC 80 forms a trio with NGC 83 and NGC 91 and is the brightest in a group of a dozen faint galaxies situated between two 8th magnitude stars. Although it is the most obvious galaxy of the group, NGC 80 appears rather faint, small, and round with a stellar nucleus.

16/18″ Scopes–150x: NGC 80 exhibits a prominent core and stellar nucleus surrounded by a moderately faint, circular 1.5′ diameter halo. Three 14th magnitude stars lie to the north, the closest 1.5′ NNE.

NGC 83 Galaxy U206 Type E
ø1.3′ × 1.2′, m12.5v, SB 12.8 $00^h21.4^m$ +22°26′
Finder Chart 2-3, Figure 2-2   ★★

12/14″ Scopes–125x: This is the second brightest galaxy in a trio with NGC 80 and NGC 91. Situated on the west side of a triangle of 11th magnitude stars, NGC 83 is visible as a very faint, small, round smudge.

16/18″ Scopes–150x: NGC 83 is situated at the southern end of a group of a dozen faint galaxies. It has a fairly faint 1′ diameter halo with a gradual brightening toward center. NGC 80 lies 5′ SW.

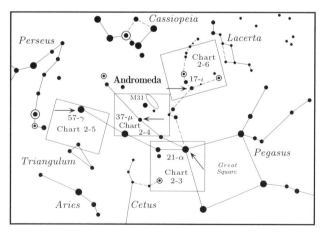

Master Finder Chart 2-2. Andromeda Chart Areas
Guide stars indicated by arrows.

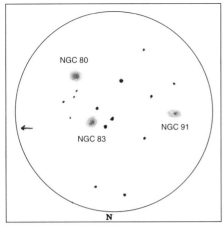
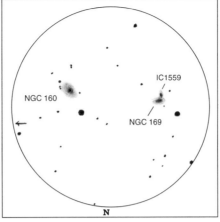
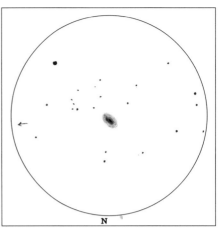

Figure 2-2. NGC 80, NGC 83, & NGC 91
17.5", f4.5–135x, by G. R. Kepple

Figure 2-3. NGC 160 & NGC 169
17.5", f4.5–135x, by G. R. Kepple

Figure 2-4. NGC 214
17.5", f4.5–200x, by G. R. Kepple

NGC 91 U208 Galaxy Type SAB(s)c pec I
⌀2.2' × 0.8', m13.7v, SB 14.2 $00^h21.8^m$ +22°25'
Finder Chart 2-3, Figure 2-2   ★

12/14" Scopes–125x: This faint galaxy is the third brightest in a trio with NGC 80 and NGC 83 and lies near the center of a larger group of a dozen faint galaxies. NGC 91 is extremely faint, small, and round, and requires averted vision.

16/18" Scopes–150x: NGC 91 is very faint, small, and elongated 1' × 0.5' NW–SE with a slight brightening at the core. There is a 13th magnitude star 1' SW. NGC 93 lies 3' east.

NGC 160 H476³ U356 Galaxy Type (R)SA0+ pec
⌀2.7' × 1.5', m12.7v, SB 14.0 $00^h36.1^m$ +23°57'
Finder Chart 2-3, Figure 2-3   ★

12/14" Scopes–125x: Lying 4' SSW of a 7th magnitude star, NGC 160 appears fairly faint, small, and diffuse with a bright center.

16/18" Scopes–150x: NGC 160 has a fairly faint halo, elongated 2' × 1' NE–SW with a sudden brightening to a stellar nucleus. A 13th magnitude star is visible 1' ENE, a 12.5 magnitude star lies 2' SW, and a 15" pair of 15th magnitude stars aligned nearly N–S lies 1.5' WSW. NGC 169 is located 11' ENE.

NGC 169 U365 Galaxy Type SA(s)ab: sp II-III
⌀3.2' × 1.0', m12.4v, SB 13.5 $00^h36.9^m$ +23°59'
Finder Chart 2-3, Figure 2-3   ★★

12/14" Scopes–125x: NGC 169 is situated 3.75' WSW of a 6th magnitude star which forms a wide pair with a 7th magnitude star further west. This galaxy appears as a faint, fairly small blob with a twin nucleus, the southern nucleus being IC 1559.

16/18" Scopes–150x: NGC 169 is moderately faint, elongated 1.5' × 1' E–W. Appearing as a faint nebulous star, companion galaxy IC 1559 makes contact with it to the south. A 14th magnitude star lies 45" NE.

NGC 183 U387a/b Galaxy Type E
⌀2.1' × 1.6', m12.7v, SB 13.9 $00^h38.3^m$ +29°30'
Finder Chart 2-3   ★

12/14" Scopes–125x: This galaxy is the brightest and northernmost of three galaxies, including NGC 181 and NGC 184, visible in the same field about 12' north of 4.4 magnitude star Epsilon (ϵ) = 30 Andromedae. NGC 183 appears faint, small, and round with a well concentrated core.

16/18" Scopes–150x: NGC 183 has a fairly faint halo elongated 1.5' × 1' NW–SE containing a bright core.

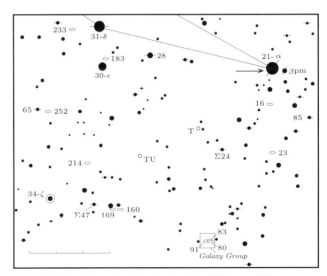

Finder Chart 2-3. 21–a And: $00^h08.2^m$ +29°05'

Figure 2-5. *The Great Galaxy in Andromeda and its companions M32 (above center) and NGC 205 (at lower left) are at the opposite end of the Local Galaxy Group from our own Milky Way Galaxy and its two Magellanic Cloud satellites. At a distance of 2.2 milion light-years, M31 is the nearest spiral galaxy to our own spiral system. Some of the brighter features are labeled in the photo above. The prefix A and C are stellar associations and star clusters, G designates globular clusters. William Harris made this 60 minute photograph with an 8″, f4 Wright Newtonian reflector telescope on Kodak 2415 Tech Pan film.*

NGC 205 Messier 110 H18⁵ Galaxy E5 pec
ø19.5′ × 12.5′, m8.1v, SB 13.9   00ʰ40.4ᵐ +41°41′
Finder Chart 2-4, Figure 2-5   ★★★★★

Throughout space the satellite is a common phenomenon. Our Milky Way Galaxy has the Magellanic Clouds as satellites. Similarly, the Andromeda Galaxy (Messier 31) has Messier 32 and NGC 205, two elliptical galaxies, as companions. NGC 205 is only about 12,000 light-years across, less than one-tenth the Andromeda Galaxy's span of 150,000 light years. The structure of NGC 205 shows an odd twist caused by the gravitational perturbations on it by M31. Many hot, young O and B type stars are mixed with the older stars typical of elliptical systems, suggesting past interactions between the two systems. Even though outclassed by the nearby Andromeda galaxy, NGC 205 is a fine object.

8/10″ Scopes–125x: NGC 205 is visible as a fairly faint, large nebulosity oval-shaped NNW–SSE with a slight brightening at center. There are two 9.5 magnitude stars off the SW edge and an 8.5 magnitude star to the NE. NGC 205 lies broadside to M31 on its NW side, opposite M32, which is much closer and smaller. These three galaxies make an impressive sight at low power.

12/14″ Scopes–125x: This elliptical galaxy is fairly bright, very large, elongated 10′ × 4′ with a broad prominent core but no stellar nucleus. Numerous faint stars (not true members of the galaxy) are visible in its halo, the most noticeable being a 14th magnitude object at the edge of the halo on the SSE edge.

16/18″ Scopes–150x: NGC 205 is a bright 12′ × 5′ oval-shaped elliptical galaxy with a bright mottled core displaced slightly west of center. A 12th magnitude star is just inside the east edge and a 12.5 magnitude star lies just north of the nucleus. This superb object is certainly a showpiece elliptical system.

NGC 214 H209² U438 Galaxy Type SAB(r)c I
ø2.0′ × 1.6′, m12.2v, SB13.4   0ʰ41.5ᵐ +25°30′
Finder Chart 2-3, Figure 2-4   ★★

12/14″ Scopes–125x: Lying 6′ NE of a 10th magnitude star, NGC 214 is a small, faint, round 1.25′ glow with a fairly low surface brightness.

16/18″ Scopes–150x: This galaxy is fairly faint and elongated 1.5′ × 1′ NE–SW with a moderately concentrated, highly extended core having no visible nucleus. Extending half its length, the core is large in comparison to the faint surrounding halo. A 14.5 magnitude star lies 1.5′ SE.

NGC 221 Messier 32 U452 Galaxy Type cE2
ø11.0′ × 7.3′, m8.1v, SB 12.7   0ʰ42.7ᵐ +40°52′
Finder Chart 2-4, Figure 2-5   ★★★★

This object is a satellite of the Great Andromeda Galaxy. It is composed of older faint red and yellow stars with virtually no gas or dust, typical of elliptical galaxies. It is 6,000 light years in diameter. M32's total mass of only two billion solar masses has had a profound impact on the parent galaxy, M31, for M31's structure near M32 is distorted. Astronomers speculate that M31 has pulled some of the outer stars away from M32, leaving it smaller and less massive.

8/10″ Scopes–125x: This elliptical is a large circular nebulosity with a gradual brightening toward the center, at which is a bright stellar nucleus.

12/14″ Scopes–125x: Located at the SE edge of M31's halo, it is quite obvious but small compared to M31. M32 is brighter but smaller than NGC 205 on the opposite side of M31. It is oval-shaped 3′ × 2′ N–S pointing to the core of M31. Its brightness increases gradually to a stellar nucleus but there is no core.

16/18″ Scopes–100x: M32 appears brighter than its assigned magnitude with a blazing nucleus surrounded by a 5′ × 3′ halo elongated NNW–SSE. The light from M31 floods the SE corner of the field like an approaching sunrise. The galaxy is very tolerant of high powers even though no additional details are revealed. It fills a third of the field at 300x.

NGC 224 Messier 31 U454 Galaxy Type SA(s)b I-II
ø185.0′ × 75.0′, m3.4v, SB 13.6   0ʰ42.7ᵐ +41°16′
Finder Chart 2-4, Figure 2-5   ★★★★★
Great Andromeda Galaxy

The Great Andromeda Galaxy, Messier 31, is our Milky Way Galaxy's closest galactic neighbor in space. On a crisp, dark autumn night it appears as a faint, hazy

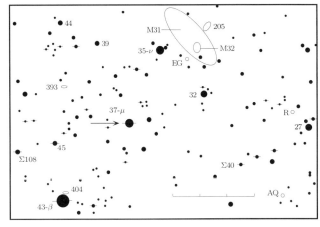

Finder Chart 2-4. 37-μ And: 00ʰ56.7ᵐ +38°29′

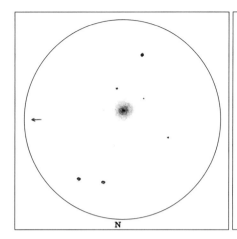
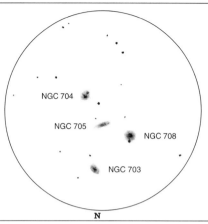
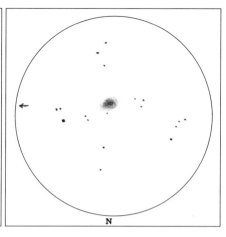

Figure 2-6. NGC 679
17.5″, f4.5–135x, by G. R. Kepple

Figure 2-7. NGC 708 Group
13″, f5.6–135x, by Steve Coe

Figure 2-8. NGC 753
12.5″, f5–75x, by Andrew D. Fraser

patch of light several times as wide as the full moon! Around 9:00p.m. during mid-November evenings, M31 lies nearly overhead for observers in the USA and Europe. Lying east of the Great Square of Pegasus, it may be found by following the bright stars from the upper left of the Great Square to Beta Andromedae and then taking a right angle turn from Beta NW through Mu to Nu (See Constellation Chart 2-1). Light from the Andromeda Galaxy began its journey to us while woolly mammoths were still roaming our planet during the Pleistocene epoch 2.2 million years ago. M31 has an apparent magnitude of 3.4 and an absolute magnitude of −21.6, corresponding to a luminosity of 40 billion Suns.

A mirror image of our own Galaxy, this huge aggregation of stars, gas, and dust spans 150,000 light years and contains a mass of approximately 200–300 billion suns. M31 is a classic example of an "Sb" spiral galaxy with moderately thick and wound-up arms. In a typical spiral galaxy the arms contain the younger stars, gas, and dust while the central condensation, or nucleus, contains older stars concentrated in a much smaller volume. The full extent of M31 cannot be seen because its galactic plane is oriented approximately thirteen degrees from our line of sight and therefore it is nearly edge-on. Its neighboring satellite galaxies, M32 and NGC 205, are visible as smudges in small telescopes.

The Andromeda Galaxy's hazy disk was noticed as early as 905 A.D. when it was mentioned by the Persian astronomer, Al Sufi. It was labelled on star charts and catalogues as the "Little Cloud" prior to its first telescopic observation in the early seventeenth century by Simon Marius. But the true extra-galactic nature of this object was not proven until the larger telescopes and technologically advanced instruments of the twentieth century came into use.

10 × 50 Binoculars: M31 has a bright, large core surrounded by an elongated glow at least three degrees long.

4/6″ Scopes–25x: At low power M31 stretches beyond a two degree field while M32 and M110 command attention at opposing sides. At 40x, a deep-sky filter reveals many light or patchy areas along the length of its arms. At 100x, M31 is a large flat oval of grayish light. It is very elongated and suddenly brighter toward the center with a round core about 8′ in diameter. Spiral arms extend from the core to the NE and SW with the SW arm being brighter. A dark lane divides the SW extension starting at the nucleus and running about halfway out. The arms in the NE region are not as prominent, but there is a slight brightening on the outer edges to the east. Many faint stellar objects are visible throughout M31; these are globular clusters and HII regions that deserve to be studied in greater detail with a large telescope.

8/10″ Scopes–50x: This bright, extremely large galaxy extends 120′ × 40′ beyond the field of view. The field also includes the two bright companion galaxies, M32 and NGC 205. Two dark lanes are obvious and the galaxy's entire body is mottled. At higher power the bright central core looks somewhat like a globular cluster. At 125x, NGC 206, a stellar association, is easily visible in the SE arm of M31. It is elongated 2:1 and fades off into the spiral arm.

12/14″ Scopes–75x: M31 is an extremely large, very bright galaxy, elongated 125′ × 50′ NE–SW with a large broadly concentrated core. The outer halo is diffuse with no definite edge and the sky transparency dictates how much is visible. At 100x, a dark lane is visible NW of center, extending past two stars

15' from center. Several more dark lanes are located in the western and SW portion of the halo. Many other objects are visible within the halo at 150x; most are stellar objects of 14th and 15th magnitude which are actually globular clusters. The brightest globular is G76 in the eastern half of the SW arm. Also in the SW arm is the most noticeable stellar association, designated A78. This feature was assigned an NGC number of its own, NGC 206 (See Figure 2-5).

NGC 233   H149³   U464   Galaxy   Type E
⌀2.0' × 1.7', m12.4v, SB 13.6   00ʰ43.4ᵐ +30°35'
Finder Chart 2-3   ★★

12/14" Scopes–125x: NGC 233 is situated nearly one degree ESE of 3.3 magnitude Delta (δ) Andromedae and 15' east of two 9th magnitude stars. It appears faint, small, and round with a slight brightening to a faint stellar nucleus or knot.

16/18" Scopes–150x: Large instruments show a bright core and a stellar nucleus surrounded by a diffuse 1.25' diameter outer halo.

NGC 252   H609²   U491   Galaxy   Type (R)SA(r)0+
⌀1.7' × 1.3', m12.3v, SB 13.1   00ʰ48.0ᵐ +27°38'
Finder Chart 2-3   ★★★

12/14" Scopes–125x: NGC 252 is the brightest and westernmost of three galaxies in a group 12' west of 6.3 magnitude 65 Piscium. It appears fairly bright but small and round, with good concentration toward the faint nucleus at its center.

16/18" Scopes–150x: This is a moderately bright galaxy with a bright core and stellar nucleus surrounded by a 1.25' × 1' halo slightly elongated ENE–WSW.

NGC 404   H224²   U718   Galaxy   Type SA(s)0−:
⌀6.1' × 6.1', m10.3v, SB 14.0   01ʰ09.4ᵐ +35°43'
Finder Chart 2-4, Figure 2-9   ★★★

8/10" Scopes–100x: Located 6.5' NW of 2nd magnitude star Beta (β) = 43 Andromedae, this galaxy makes an interesting contrast to the star's yellowish-orange glow. Even with Beta Andromedae in the field, NGC 404 is visible with direct vision as a large, round nebulosity slightly brighter at the center. A faint star lies close to the NNW.

12/14" Scopes–100x: This is a beautiful object making a fine pair with the bright star Beta Andromedae. It is fairly bright, small, round, and slightly brighter in the center. During moments of good seeing, a stellar nucleus may be glimpsed.

16/18" Scopes–125x: NGC 404 shows a bright stellar

Figure 2-9. *NGC 404 lies in the brilliant orange glow of Beta Andromedae. Martin C. Germano made this 35 minute photo on Kodak 2415 Tech Pan film with an 8", f5 Newtonian reflector at prime focus.*

nucleus in a small, round, mottled 2' diameter glow. A 14th magnitude star is embedded in northern edge.

NGC 679   H175³   U1283   Galaxy   Type S0−:
⌀1.8' × 1.8', m12.3v, SB 13.5   01ʰ49.7ᵐ +35°47'
Finder Chart 2-5, Figure 2-6   ★★

12/14" Scopes–125x: NGC 679 is located 20' SW of a 7th magnitude star near the border of Triangulum. It is faint, small and round with some brightening at the center.

16/18" Scopes–150x: At medium power it is moderately bright with a round 1.5' halo growing gradually brighter to a core containing a stellar nucleus.

NGC 687   H563³   U1283   Galaxy   Type S0
⌀1.4' × 1.4', m12.3v, SB 12.8   01ʰ50.6ᵐ +36°21'
Finder Chart 2-5   ★★

12/14" Scopes–125x: This galaxy is located 10' NW of a 9th magnitude star which is the northernmost star in a chain that includes magnitude 7.5 to 8.5 stars to the south. This galaxy is a faint, small, round object with considerable brightening toward center to a faint stellar nucleus.

16/18" Scopes–150x: In larger instruments, NGC 687 exhibits a bright core and a stellar nucleus surrounded by a round 1' diameter halo.

Figure 2-10. *NGC 752 is a large, loose cluster five degrees south of Gamma (γ) Andromedae. Use low power for best separation from the surrounding star field. Lee C. Coombs made this five minute exposure on 103a-O film with a 10″, f5 Newtonian reflector at prime focus.*

NGC 704   H563³   Galaxy   Type ?
ø–, m13.1v $01^h52.7^m$ +36°06′

NGC 703   H562³   Galaxy   Type S0–:
ø1.2′ × 0.9′, m13.3v, SB 13.2 $01^h52.8^m$ +36°09′

NGC 705   H564³   Galaxy   Type S0/a
ø1.1′ × 0.2′, m13.6v, SB 11.8 $01^h52.8^m$ +36°08′

NGC 708   H565³   U1348   Galaxy   Type E
ø3.3′ × 2.6′, m12.7v, SB 14.8 $01^h52.8^m$ +36°10′
Finder Chart 2-5, Figure 2-7   ★/★/★/★

12/14″ Scopes–125x: NGC 708 is the most obvious object lying in a tiny cluster of at least five galaxies, four of which are within a 3′ circle. It is visible as a small, round glow with even surface brightness.

16/18″ Scopes–125x: NGC 708 is situated near the center of a 10′ diameter group of seven galaxies within the Abell 262 galaxy cluster. This galaxy is the most obvious of the group, being visible as a faint, circular 1′ diameter glow with a 13.5 magnitude star touching its NNW edge. NGC 703, 1.75′ NW of NGC 708, has a faint halo elongated 45″ × 30″ NE–SW with a stellar nucleus. NGC 705, 1′ SW of NGC 708, is faint and elongated 1′ × 0.25′ ESE–WNW. NGC 704, 50″ SW of NGC 705, has a faint 30″ halo with a stellar nucleus and what appears to be a faint star at its SSE edge, but is in fact the nucleus of a very faint interacting companion.

NGC 710   U1349   Galaxy   Type Scd:
ø1.6′ × 1.6′, m13.7v, SB 14.5 $01^h53.0^m$ +36°02′
Finder Chart 2-5   ★

12/14″ Scopes–125x: Positioned at the southern edge of the Abell 262 galaxy cluster, NGC 710 is visible as a very faint, small, round glow with some brightening at center.

16/18″ Scopes–150x: This galaxy is a fairly faint 1′ diameter oval, slightly elongated E–W. There is a 15th magnitude star 30″ SW and a wide 15th magnitude pair 45″ ESE.

NGC 753   U1437   Galaxy   Type SAB(rs)bc I-II
ø3.0′ × 1.9′, m12.3v, SB 14.0 $01^h57.7^m$ +35°55′
Finder Chart 2-5, Figure 2-8   ★

8/10″ Scopes–100x: Lying near the Triangulum border, NGC 753 is located 1.5° south of the large star cluster NGC 752. It is visible 6′ ESE of a 10th magnitude star as a very faint, small, circular nebulosity with a slight brightening at center.

12/14″ Scopes–125x: The halo is faint, small, and round with a gradual brightening at center. The field has only a scattering of faint stars.

16/18″ Scopes–150x: NGC 753 has a faint 1.5′ diameter halo slightly elongated NW–SE with a slight brightening at center. A 13.5 magnitude star lies 45″ north.

NGC 752   Collinder 33   Open Cl. 60★   Tr Type III 1 m
ø50′, m5.7v, Br★ 8.96 $01^h57.8^m$ +37°41′
Finder Chart 2-5, Figure 2-10   ★★★★

8/10″ Scopes–50x: This cluster is scattered NNE of the bright pair 56 Andromedae (5.7, 6.0; 190″; 300°). A 5.9 magnitude star is 9.5′ SW. NGC 752 is a very large, loose collection of 60 to 70 stars. A 6th magnitude star marks the east edge and a close trio is visible south of center. The cluster is full of short chains, irregular clumps and a large number of double stars. NGC 752 is a fine sight at low power and is at its best in giant binoculars.

12/14″ Scopes–75x: NGC 752 is bright, very large, irregularly round, and scattered with no central compression. Seventy-five stars are visible in a 50′ area, fifteen 9th magnitude, almost twice as many 10th magnitude, and the remaining fainter. All 12th magnitude and fainter stars are not true cluster members. A jagged star chain wraps around a peanut-shaped void elongated NE–SW near the center. A concentrated knot of ten stars is visible at the void's SW edge, but the northern and eastern quadrants are the most concentrated. The lovely

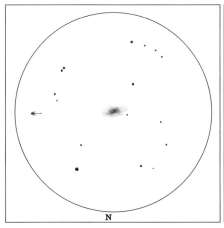

Figure 2-11. NGC 818
17.5", f4.5–250x, by G. R. Kepple

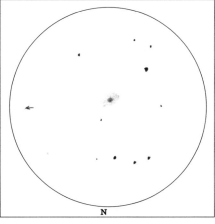

Figure 2-12. NGC 828
17.5", f4.5–250x, by G. R. Kepple

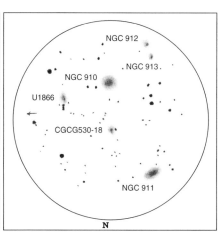

Figure 2-13. NGC 910 & NGC 911 Area
17.5", f4.5–175x, by G. R. Kepple

double star, 56 Andromedae (see Table 2-1), both components of which have a beautiful yellow-orange color, lies immediately SSW. In the surrounding area many other pairs are visible, as are several long star chains running NW–SE.

NGC 759 U1440 Galaxy Type E
ø1.8' × 1.8', m12.7v, SB 13.9   01ʰ57.8ᵐ +36°20'
Finder Chart 2-5   ★

12/14" Scopes–125x: This galaxy has a faint, round 1' diameter halo with a bright core.

16/18" Scopes–150x: At medium power NGC 759 is moderately faint with a round 1.5' diameter halo having considerable brightening at the core.

NGC 797 H566³ U1541 Galaxy Type SAB(s)a
ø1.7' × 1.3', m12.7v, SB 13.9   02ʰ03.4ᵐ +38°07'
Finder Chart 2-5   ★

12/14" Scopes–125x: This galaxy has a very faint, small, round 1' diameter halo displaying a sudden brightening at the core.

16/18" Scopes–150x: NGC 797 has a fairly faint 1.25' diameter halo slightly elongated ENE–WSW. The core displays considerable brightening to a stellar nucleus. There is a 14th magnitude star 1' WNW while NGC 801 lies 9' NNE.

NGC 818 H604² U1633 Galaxy Type SABc:
ø3.2' × 1.3', m12.5v, SB 13.9   02ʰ08.7ᵐ +38°47'
Finder Chart 2-5, Figure 2-11   ★

12/14" Scopes–125x: NGC 818 appears fairly faint, elongated 1.5' × 0.75' ESE–WNW and exhibits a much brighter core. A faint star chain to the NW points the way to the galaxy.

16/18" Scopes–150x: This galaxy is moderately bright with a 2.5' × 1.25' halo having a large bright center. A 14th magnitude star is visible 35" ENE and a 13th magnitude star lies 1.5' SSE.

NGC 828 H605² U1655 Galaxy Type Sa: pec
ø3.2' × 2.5', m12.2v, SB 14.4   02ʰ10.2ᵐ +39°12'
Finder Chart 2-5, Figure 2-12   ★

12/14" Scopes–125x: NGC 828 is located 15' NW of the bright double star, 59 Andromedae (6.1, 6.8; 16.6"; 35°). Another double composed of 11th magnitude stars is positioned between 59 Andromedae and NGC 828. The galaxy appears faint and small with an irregularly round 1.5' diameter halo having considerable brightening at the core.

16/18" Scopes–150x: NGC 828 is a moderately bright, fairly small galaxy, oval-shaped 2' × 1.5' NW–SE with a bright core. A wide double of 11th magnitude stars lies 3' east and a 12th magnitude star is visible 1.5' SE.

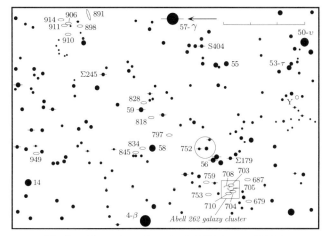

Finder Chart 2-5. 57-γ And: 02ʰ03.5ᵐ +42°20'

Figure 2-14. *NGC 891 is a fine edge-on Sb galaxy in a rich star field. High power on a large instrument will show the dark lane. Martin C. Germano made this 30 minute exposure on hypered Kodak 2415 Tech Pan film with an 8″, f5 Newtonian reflector.*

NGC 834 H567³ U1672 Galaxy Type S?
ø1.1′ × 0.5′, m13.1v, SB 12.3   02ʰ11.0ᵐ +37°40′
Finder Chart 2-5   ★

12/14″ Scopes–125x: This faint galaxy is situated just ESE of a triangle of 9th magnitude stars and about 45′ SE of 58 Andromedae (m4.8). The halo is a faint, small, slightly elongated smudge with a slight brightening at center.

16/18″ Scopes–150x: This small galaxy is fairly faint and elongated 1′ × 0.5′ NNE–SSW with a slight brightening at the core.

NGC 845 H604³ U1676 Galaxy Type SB II
ø1.6′ × 0.5′, m12.5v, SB 13.1   02ʰ12.3ᵐ +37°29′
Finder Chart 2-5   ★★

12/14″ Scopes–125x: NGC 845 is positioned just north of the middle star in a row of three 8th magnitude stars about one degree SE of 58 Andromedae (m4.8). It is quite faint, small, and elongated 1′ × 0.5′ NW–SE with a stellar nucleus.

16/18″ Scopes–150x: This galaxy is moderately faint, elongated 1.5′ × 0.5′ NNW–SSE with a very bright core and a stellar nucleus. There are two wide, faint double stars to the north. NGC 834 lies 11′ NNW.

NGC 891 H19⁵ U1831 Galaxy Type SA(s)b? sp III
ø13.0′ × 2.8′, m9.9v, SB 13.7   02ʰ22.6ᵐ +42°21′
Finder Chart 2-5, Figure 2-14   ★★★★

8/10″ Scopes–100x: NGC 891 appears fainter than its assigned magnitude due to low surface brightness. The halo is large and very elongated 9′ × 2′ NNE–SSW. The dust lane revealed in photographs is not visible. An 11th magnitude star lies inside the galaxy's west edge, a 9th magnitude star is situated just SW of center, and a 10.5 magnitude star is located at its eastern edge. A 7.5 magnitude yellow star is located at the northern edge of the field. Several faint stars are embedded in the envelope.

12/14″ Scopes–100x: This is a fine needle suspended in a rich star field! It is moderately faint, large, elongated 10′ × 2′ NNE–SSW with a moderate brightening to a slightly bulged core. At 125x there are bright patches visible along the major axis on each side of the core separated by a very faint, indistinct dark lane needing averted vision. There is a 13th magnitude star at the SSW tip. The SE field is moderately populated with faint stars near or touching the halo.

16/18″ Scopes–100x: At this power NGC 891 is an absolutely stunning galaxy despite its low surface brightness! It is very large and slender, elongated 12' × 2.5' NNE–SSW. It seems to float three-dimensionally among the stars. At 125x the center is noticeably brighter, and there is a dark equatorial lane bisecting the full length of the galaxy. The lane is prominent near the nucleus but becomes difficult in the extensions. At 175x the envelope fills the field of view.

NGC 906   U1868   Galaxy   Type SBab II
ø1.9' × 1.8', m12.9v, SB 14.1 $02^h25.2^m$ +42°06'
Finder Chart 2-5   ★

12/14" Scopes–125x: NGC 906 is situated in the northern portion of the Abell 347 galaxy cluster about 15' from a 7th magnitude star while NGC 891 is located 20' NW from the 7th magnitude star. NGC 906 is visible as a faint, diffuse, round 1' diameter patch.

16/18" Scopes–150x: This galaxy is fairly faint with a round 1.5' halo. NGC 909 lies 3.5' to the south.

NGC 910   H571³   U1875   Galaxy   Type E+
ø2.0' × 2.0', m12.2v, SB 13.5 $02^h25.4^m$ +41°50'
Finder Chart 2-5, Figure 2-13   ★★

12/14" Scopes–125x: NGC 910 is a member of the Abell 347 galaxy cluster positioned between two bright stars, 5.5' east of a 10th magnitude star and 4' SW of an 11th magnitude star. It is visible as a faint, small, round 1' diameter spot.

16/18" Scopes–150x: NGC 910 has a fairly faint, round 1.5' diameter halo with a small core. 13th magnitude stars are visible 1.25' WNW and 1.5' SW. Other members of the Abell 347 galaxy cluster are visible nearby. Lying 4' ESE of NGC 910 is NGC 912, a faint, circular 30" diameter smudge with a faint stellar nucleus. NGC 913, 1.25' north of NGC 912, is just detectable with averted vision. Other averted vision objects in the group are UGC 1866, 4' WNW of NGC 910, and CGCG 539-18, 4.5' north of NGC 910. NGC 911, 10' north of NGC 910 is much brighter.

NGC 911   Galaxy   U1878   Type E
ø1.7' × 0.9', m12.7v, SB 13.0 $02^h25.8^m$ +41°59'
Finder Chart 2-5, Figure 2-13   ★★

12/14" Scopes–125x: This galaxy is situated nearly at the center of the Abell 347 galaxy cluster about 20' from a 7th magnitude star. At medium power it is very faint, small, and slightly elongated 1' × 0.5' ESE–WNW with a slight brightening at center. There is a 12th magnitude star 1' north.

16/18" Scopes–150x: NGC 911 is fairly faint, small, and elongated 1.5' × 0.75' ESE–WNW with a moderately concentrated core. 13th magnitude stars are visible 45" WNW and 1' SSW.

NGC 914   Galaxy   U1887   Type SA(s)c
ø1.9' × 1.3', m13.0v, SB13.8 $02^h26.1^m$ +42°09'
Finder Chart 2-5   ★

12/14" Scopes–125x: NGC 914 lies north of the main concentration in the Abell 347 galaxy cluster 35' east of a 7th magnitude star. It is visible as an extremely faint, small, diffuse spot.

16/18" Scopes–150x: This galaxy is very faint, fairly small, slightly elongated 1.5' × 1.25' ESE–WNW, and has a diffuse halo.

NGC 7640   H600²   U12554   Galaxy   Type SB(s)c II
ø10.0' × 2.2', m11.3v, SB 14.5 $23^h22.1^m$ +40°51'
Finder Chart 2-6, Figure 2-15   ★★★★

8/10" Scopes–100x: NGC 7640 appears faint and elongated 7' × 2' N–S with a slight central brightening. The galaxy lies within a prominent triangle of 11.5 magnitude stars.

12/14" Scopes–125x: This galaxy is faint and very elongated 7' × 2' nearly N–S with a mottled halo. A faint star was glimpsed SE of the core.

16/18" Scopes–125x: NGC 7640 is rather faint, elongated 8' × 2' NNW–SSE, and has a moderately concentrated oval-shaped core. Besides a 13.5 magnitude star at the core's SE edge, two 13th magnitude stars are embedded in the northern extension.

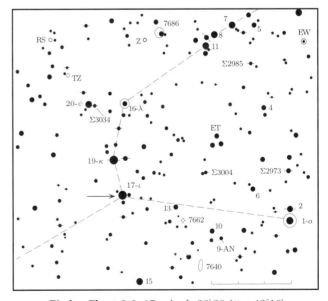

Finder Chart 2-6. 17-ι And: $23^h38.1^m$ +43°16'

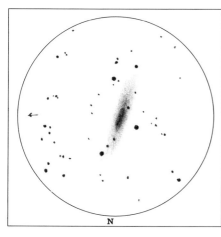 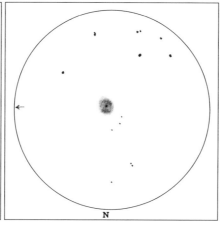 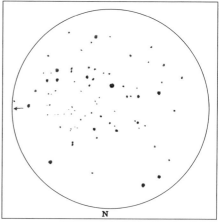

Figure 2-15. NGC 7640
17.5", f4.5–250x, by G. R. Kepple

Figure 2-16. NGC 7662
12.5", f5–250x, by G.R. Kepple

Figure 2-17. NGC 7686
12.5", f5–125x, by G.R. Kepple

The halo stretches beyond an 11.5 magnitude star to the south and touches a 12.5 magnitude star 3.75′ SSE of center. An 11.5 magnitude star lies at the edge 3′ NNW of the center.

NGC 7662 H18⁴ PK106-17.1 Planetary Neb. Type 4+3
ø>12″, m8.3v, CS 13.17 (var)   23ʰ25.9ᵐ +42°33′
Finder Chart 2-6, Figure 2-16   ★★★★
                The Blue Snowball

8/10″ Scopes–225x: A lovely bluish planetary! NGC 7662 is bright, large, and round with apparent dimming at center. The annular structure is more obvious with averted vision. The faint central area is surrounded by a suddenly brighter middle ring and a faint outer ring which does not encompass the entire periphery. The central star is not visible.

12/14″ Scopes–250x: NGC 7662 is bright, moderately large, and somewhat elongated with a very faint central star. There are two lovely bluish-green rings surrounding a slightly darker center resembling the "CBS" eye. A 13th magnitude star lies beyond the east edge.

16/18″ Scopes–250x: This is a fine bluish colored planetary nebula with a high surface brightness! The disk is elongated 30″ × 25″ NW–SE with a slightly darker center. There is some brightening along the NE and SW sides but the edges are diffuse. The central star is very difficult but a twinkle may be glimpsed in periods of good seeing. A trio of 13th and 14th magnitude stars lies just beyond the NNE edge.

NGC 7686 H69⁸ Open Cluster 20★ Tr Type IV 1 p
ø14′, m5.6v, Br★ 6.17   23ʰ30.2ᵐ +49°08′
Finder Chart 2-6, Figure 2-17   ★★★★

8/10″ Scopes–75x: This cluster is 15′ in diameter and rather poorly concentrated. Low power shows ten 10th magnitude, ten 11th magnitude, and two 8th magnitude stars extending ENE–WSW.

12/14″ Scopes–75x: NGC 7686 is a bright, large, loose, and irregularly scattered cluster. There are about fifty stars in a 25′ area, most being 10th to 11th magnitude, lying north and west of a 6th magnitude reddish-orange star.

16/18″ Scopes–75x: This loose group has twenty-five 10th to 13th magnitude stars scattered around an 8th magnitude star. An interesting chain of stars snakes off WSW of the bright star. Another loose E–W star chain skirts the northern perimeter. NW of the cluster's brightest star, a granular haze of very faint stars occupies the space between the two chains.

Chapter 3

Aquarius, the Water-Pourer

3.1 Overview

Aquarius, the Water-Pourer, is the 11th sign of the Zodiac, representing a man pouring water from an urn. The star pattern originated in early Babylonian times when its faint stars were more visible to the naked eye than in our light polluted skies. The most prominent star pattern in the constellation is the Y-shaped asterism centered on Zeta Aquarii, which the ancients called "The Water Jar." Aquarius is only one of several constellations in this region of the heavens with watery associations. Indeed, the Babylonians knew this area as a celestial sea, and from them, the Greeks inherited not only Aquarius but also Pisces, the Fishes, Capricornius, the Goat-Fish, Piscis Austrinus, the Southern Fish, and Eridanus, the River. The "Age of Aquarius" heralded by astrologers will not take place for another 600 years when the vernal equinox will be located in Aquarius.

Galaxies are the most numerous type of deep-sky objects in Aquarius, but most are faint. Since the constellation lies far from the plane of the Milky Way, it is naturally lacking in star clusters and diffuse nebulae. Aquarius contains three globular clusters, two of which are Messier objects, M2 and M72. A third Messier object, M73, is only an asterism. There are many fine double and multiple stars within Aquarius, and the constellation contains two notable planetaries, Saturn and Helical Nebulae. Although the Helical Nebula is quite faint, it is enormous and impressive when viewed through a large aperture telescope with an O-III filter.

3.2 Interesting Stars

12 Aquarii = Σ2745 Double Star   Spec. F5 & A3
m5.9, 7.3; Sep. 2.8″; P.A. 192° $21^h04.1^m$ $-05°49'$
Finder Chart 3-4   ★★★★

8/10″ Scopes–175x: 12 Aquarii is a close yellowish and pale blue pair of stars.

Aquarius: AK-WARE-ee-us
Genitive: Aquarii, AK-WARE-ee-eye
Abbrevation: Aqr
Opposition: 9pm–July 15, midnight–May 30
Culmination: 9pm–Oct. 9, midnight–Aug. 25
Area: 980 square degrees
Showpieces: 12 Aqr, 41 Aqr, 55–$\zeta^{1,2}$ Aqr, 94 Aqr, 107 Aqr, M2 (NGC 7089), M72 (NGC 6981), NGC 7009 (Saturn Nebula)
Binocular Objects: M2 (NGC 7089), M72 (NGC 6981), M73 (NGC 6994), NGC 7009 (Saturn Nebula), NGC 7293 (Helix Nebula)

Σ2838 Double Star   Spec. F8
m6.3, 9.1; Sep. 17.6″; P.A. 184° $21^h54.6^m$ $-03°18'$
Finder Chart 3-5   ★★★★

4/6″ Scopes–100x: Struve 2838 is fine a yellow and pale blue pair in a nice field.

29 Aquarii = S 802 Double Star   Spec. A2
m7.2, 7.4; Sep. 3.7″; P.A. 244° $22^h02.4^m$ $-16°58'$
Finder Chart 3-6   ★★★★

8/10″ Scopes–175x: 29 Aquarii is a close pair of white stars requiring moderately high power for a clean split.

41 Aquarii = H N 56 Double Star   Spec. G8
m7.1, 7.1; Sep. 5.0″; P.A. 114° $22^h14.3^m$ $-21°04'$
Finder Chart 3-6   ★★★★

8/10″ Scopes–100x: 41 Aquarii is a close, equally bright pair of gold and blue stars.

53 Aquarii = S, h345 Multiple Star   Spec. G0 & G0
AB: m6.4, 6.6; Sep. 3.1″; P.A. 343° $22^h26.6^m$ $-16°45'$
Finder Chart 3-6   ★★★★

8/10″ Scopes–175x: 53 Aquarii has a close pair of lovely

Aquarius, the Water-Bearer

Constellation Chart 3–1

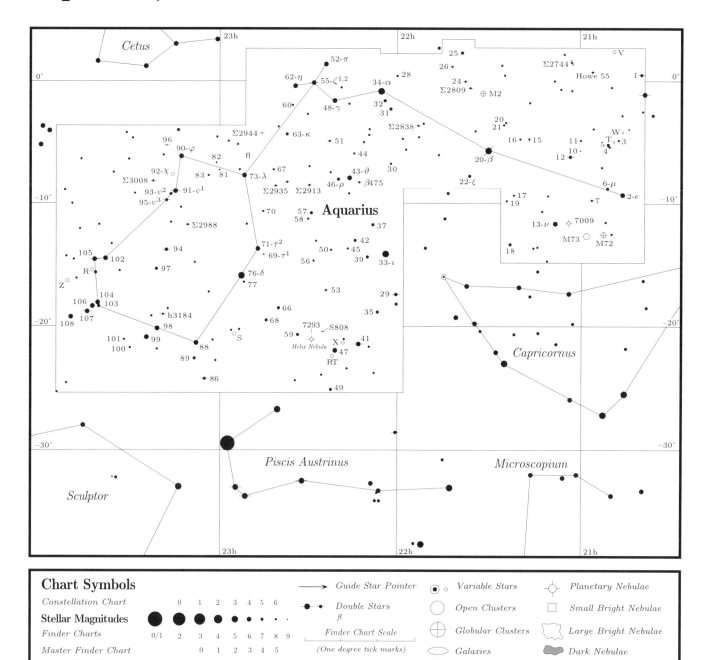

Table 3-1. Selected Variable Stars in Aquarius

Name	HD No.	Type	Max.	Min.	Period (Days)	F*	Spec. Type	R.A. (2000)	Dec.	Finder Chart No. & Notes
W Aqr	225192	M	8.4	14.9	381.10	0.42	M6–M8	$20^h46.4^m$	$-04°05'$	3-3
V Aqr	197942	SRa	7.6	9.4	244.00		M6	46.8	+02 26	3-1
T Aqr	198373	M	7.2	14.2	202.10	0.48	M2–M5	49.9	−05 09	3-3
X Aqr	211610	M	7.5	14.8	311.65	0.42	M4–M6	$22^h18.7^m$	−20 54	3-6
S Aqr	216907	M	7.6	15.0	279.27	0.39	M4–M6	57.1	−20 21	3-1
RT Aqr	212243	M	8.8	13.1	246.3		M5–M6	23.2	−22 03	3-6
R Aqr	222800	M	5.8	12.4	386.96	0.42	M5–M8	$23^h43.8^m$	−15 17	3-9
Z Aqr	223737	SRa	9.5	12.0	135.50	0.33	M1–M7	52.2	−15 51	3-1

F* = The fraction of period taken up by the star's rise from min. to max. brightness, or the period spent in eclipse.

Table 3-2. Selected Double Stars in Aquarius

Name	ADS No.	Pair	M1	M2	Sep."	P.A.°	Spec	R.A. (2000)	Dec	Finder Chart No. & Notes
4 Aqr	14360	AB	6.4	7.2	c0.4	*57	F2	$20^h51.4^m$	$-05°38'$	3-4
	14360	AC		12.9	68.7	316				
	14360	AD		9.4	131.3	329				
Howe 55	14457	AB	6.1	9.8	26.2	72	K2	57.2	+00 28	3-3
	14457	AC		12.3	35.4	116				
Σ2744	14573	AB	6.7	7.2	1.3	137	F5	$21^h03.0^m$	+01 32	3-1
	14573	AC		12.5	89.1	102				
12 Aqr	14592		5.9	7.3	2.8	192	F5 A3	04.1	−05 49	3-4 Yellowish, pale blue
Σ2809	15142		6.3	8.7	31.1	163	A2	37.6	−00 23	3-5 White, light blue
24 Aqr	15176	AB	7.2	7.6	0.5	264	F8	39.5	−00 03	3-5
	15176	ABxC		10.8	37.6	160				
Σ2838	15432		6.3	9.1	17.6	184	F8	54.6	−03 18	3-5 Yellow, light blue
29 Aqr	15562	AB	7.2	7.4	3.7	244	A2	$22^h02.4^m$	−16 58	3-6 Close pair of white stars
β475	15725		7.1	9.9	1.3	203	F2	12.6	−08 01	3-1
41 Aqr	15753	AB	5.6	7.1	5.0	114	G8	14.3	−21 04	3-6 Yellow, blue companions
	15753	AC		9.0	212.1	43	F5			
β171	15760	CD		11.7	12.0	256				
51 Aqr	15902	AB	6.5	6.5	0.5	324	A0	24.1	−04 50	3-1
S808	15926		7.1	8.3	6.8	152	F8	25.8	−20 14	3-6
53 Aqr	15934	AB	6.4	6.6	3.1	334	G0 G0	26.6	−16 45	3-6
(S, h345)	15934	BC		12.9	46.7	339				
	15934	CD		13.9	1.8	101				
55–ζ^{1,2} Aqr	15971		4.3	4.5	w2.1	*192	F2	28.8	−00 01	3-1 Yellowish stars
Σ2913	15994		7.7	8.7	8.2	329	F0	30.5	−08 07	3-1
Σ2935	16208	AB	6.9	7.9	2.3	308	A2	43.1	−08 19	3-1
69–τ^1	16268		5.8	9.0	23.7	121	B9	47.7	−14 03	3-7 Bluish, yellowish-orange
Σ2944	16270	AB	7.5	8.0	2.3	282	G0 G0	47.8	−04 14	3-1
	16270	AC		8.5	49.7	106				
Σ2988	16579		7.8	7.8	3.6	279	K0	$23^h12.0^m$	−11 56	3-8
91–ψ^{1}Aqr	16633	AxBC		4.5	49.6	312	K0	15.9	−09 05	3-8 Yellowish-orange, blue
	16633	BC	10.8	11.5	0.3	105	K			
94 Aqr	16672		5.3	7.3	12.7	350	G5	19.1	−13 28	3-9 Yellow, bluish-green
96 Aqr	16676		5.6	10.5	10.6	20	F2	19.4	−05 07	3-8
h3184	16688		7.2	9.2	5.4	283	G5	21.0	−18 32	3-1
Σ3008	16725		7.2	8.2	4.0	176	K0	23.8	−08 28	3-8 Optical pair
h316	16878		5.7	9.6	33.1	93	G5	37.7	−13 04	3-9
105–ω^{2}Aqr	16944		4.5	10.5	5.7	86	A0	42.7	−14 33	3-9
107 Aqr	16979		5.7	6.7	6.6	136	A5	46.0	−18 41	3-1 White, bluish

Footnotes: *= Year 2000, a = Near apogee, c = Closing, w = Widening. Finder Chart No: All stars listed in the tables are plotted in the large Constellation Chart, but when a star appears in a Finder Chart, this number is listed. Notes: When colors are subtle, the suffix -ish is used, e.g. bluish

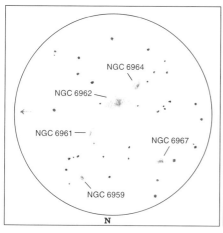

Figure 3-1. NGC 6962 Group
17.5″, f4.5–175x, by G. R. Kepple

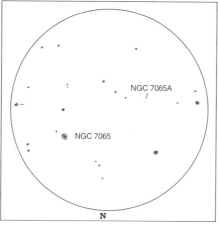

Figure 3-2. NGC 7065 & NGC 7065A
17.5″, f4.5–275x, by G. R. Kepple

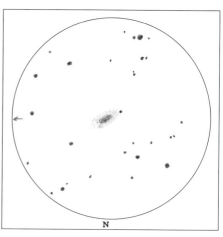

Figure 3-3. NGC 7171
13″, f5.6–165x, by Steve Coe

yellows suns; however these stars are only an optical alignment with no physical connection. Two fainter members at 12.9 and 13.9 magnitude separated only by 1.8″ in P.A. 101° lie 46.7″ from the A-B pair.

Zeta (ζ) = 55 Aquarii (Σ2909) Double Star   Spec. F2
m4.3, 4.5; Sep. 2.1″; P.A. 192° $22^h28.8^m$ $-00°01'$
Constellation Chart 3-1   ★★★★

2/3″ Scopes–175x: Zeta Aquarii is a very close, but elegant, pair of white stars.

8/10″ Scopes–175x: Moderately high power will just split these bright, nearly equal yellowish stars.

**Tau-1 (τ¹) = 69 Aquarii (Σ2943) Double Star
Spec. B9**
m5.8, 9.0; Sep. 23.7″; P.A. 121° $22^h47.7^m$ $-14°03'$
Finder Chart 3-7   ★★★★

8/10″ Scopes–100x: The colors are subtle, various observers reporting them to be: blue-white and greenish; yellowish and orange; white and yellow; white and pale blue; and white and pale red. What do you see? Tau-1 and Tau-2 also make a nice pair in binoculars, Tau-2 having a fine orange hue.

94 Aquarii = Σ2998 Double Star   Spec. G5
m5.3, 7.3; Sep. 12.7″; P.A. 350° $23^h19.1^m$ $-13°28'$
Finder Chart 3-9   ★★★★

2/3″ Scopes–100x: Smyth called the colors of 94 Aquarii "pale rose and emerald." Indeed, the stars are nicely tinted yellowish-red and light green.

Psi-1 (ψ¹) = 91 Aquarii Double Star   Spec. K0 & K
m4.5, 10.8; Sep. 49.6″, P.A. 312° $23^h15.9^m$ $-09°05'$
Finder Chart 3-8   ★★★★

2/3″ Scopes–75x: Low power reveals a wide, unequal pair of orange stars with a greenish field star that reminds one of Uranus. A very pretty color contrast.

107 Aquarii = H II 24 Double Star   Spec. A5
m5.7, 6.7; Sep. 6.6″; P.A. 136° $23^h46.0^m$ $-18°41'$
Constellation Chart 3-1   ★★★★

2/3″ Scopes–100x: 107 Aquarii is a close double with subtle colors of yellowish-white and bluish-white.

3.3 Deep-Sky Objects

NGC 6959 Galaxy Type (R:)S0°: pec
ø0.7′ × 0.5′, m13.7v, SB12.4 $20^h47.1^m$ $+00°27'$
Finder Chart 3-3, Figure 3-1   ★

16/18″ Scopes–150x: NGC 6959 is the fourth brightest in the NGC 6962 galaxy group, lying 7′ NNW of NGC 6962. It has a faint halo elongated 30″ × 15″

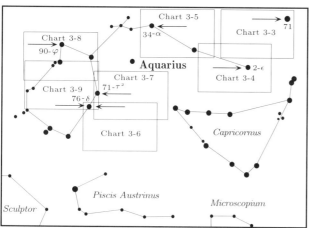

Master Finder Chart 3-2. Aquarius Chart Areas
Guide stars indicated by arrows.

NE–SW with a slight brightening through its center. 2′ SSW of the galaxy is a short NE-SW line of two 13th magnitude stars flanking one of the 14th magnitude. A 12.5 magnitude star lies 1.5′ WSW.

NGC 6961 Galaxy Type E2/S0− pec?
ø0.7′ × 0.4′, m13.7v, SB 12.2   20^{h}47.2^m +00°23′
Finder Chart 3-3, Figure 3-1   ★

16/18″ Scopes–150x: NGC 6961, lying 3.25′ NW of NGC 6462, has a very faint 30″ halo with a faint stellar nucleus. A 14.5 magnitude star lies at its NE edge. It is the fifth brightest object in the NGC 6962 galaxy group.

NGC 6962 H426² Galaxy Type (R′)SAB(s)a:
ø2.7′ × 2.1′, m12.1v, SB 13.9   20^{h}47.3^m +00°19′
Finder Chart 3-3, Figure 3-1   ★★

16/18″ Scopes–150x: NGC 6962 is located 2.5′ NE of an 11th magnitude star. It has a moderately faint 2.5′ × 2′ halo elongated ENE–WSW with a much brighter center and a stellar nucleus. It is the largest and brightest system in the NGC 6962 galaxy group.

NGC 6964 H427² Galaxy Type SA:0° pec
ø1.6′ × 1.1′, m13.0v, SB 13.5   20^{h}47.4^m +00°18′
Finder Chart 3-3, Figure 3-1   ★★

16/18″ Scopes–150x: NGC 6964 touches the SE edge of NGC 6962. It has a bright core and a nonstellar nucleus surrounded by a faint halo elongated 1.25′ × 1′ N–S. A 13th magnitude star lies 40″ ESE of center.

NGC 6965 IC 5058 Galaxy Type Sa:
ø0.9′ × 0.6′, m14.0v, SB 13.2   20^{h}47.4^m +00°30′
Finder Chart 3-3   ★

16/18″ Scopes–150x: NGC 6965 is located in a triangle of three stars and is the faintest of six galaxies 10′ north of NGC 6962. NGC 6965 is very faint but with direct vision can be seen as a tiny 30″ diameter spot, slightly elongated N-S with a gradual brightening to a prominent core and faint stellar nucleus.

NGC 6967 Galaxy Type SB?(rs:)0+:
ø0.9′ × 0.6′, m13.1v, SB 12.3   20^{h}47.6^m +00°25′
Finder Chart 3-3, Figure 3-1   ★

16/18″ Scopes–150x: NGC 6967, 45″ west of a 10th magnitude star which interferes with viewing, has a faint halo elongated 45″ × 30″ ESE–WNW and a conspicuous core. It is the third brightest in the NGC 6962 galaxy group and is situated 6.5′ NE of the flagship object.

NGC 6976 = NGC 6975? Galaxy Type SAB?(r:)bc II:
ø1.3′ × 1.1′, m14.0v, SB 14.2   20^{h}52.4^m −05°46′
Finder Chart 3-3   ★

16/18″ Scopes–150x: NGC 6976 is the southernmost of a NNE–SSW string of three faint galaxies. It is visible as an extremely faint, round, diffuse 30″ diameter spot.

NGC 6977 Galaxy Type SB(r)a pec
ø1.3′ × 1.0′, m13.2v, SB 13.4   20^{h}52.5^m −05°44′
Finder Chart 3-3   ★

16/18″ Scopes–150x: NGC 6977 is the center of three galaxies, NGC 6976 lying 2′ SSW and NGC 6978 lying 2′ NNE. It has a very faint, round, diffuse 45″ diameter halo with uniform illumination.

NGC 6978 Galaxy Type Sb II
ø1.5′ × 0.7′, m13.3v, SB 13.2   20^{h}52.6^m −05°43′
Finder Chart 3-3   ★

16/18″ Scopes–150x: NGC 6978 is fairly faint and elongated 1.5′ × 0.75′ NW–SE with a conspicuous core. It is the brightest and northernmost of three galaxies with NGC 6976 and NGC 6977.

NGC 6981 Messier 72 Globular Cluster Class IX
ø5.9′, m9.3v   20^{h}53.5^m −12°32′
Finder Chart 3-4, Figure 3-4   ★★★★

Messier 72 was discovered in August, 1780 by M. Mechain. The cluster is rather remote at about 56,000 light years distant. Though the cluster is rather loosely concentrated, its 15th magnitude stars are quite difficult to resolve. It lies 3° WSW of the Saturn Nebula.

8/10″ Scopes–125x: This fine globular is fairly bright and moderately large with a broad, unevenly bright core. The 2′ diameter halo has a grainy texture with

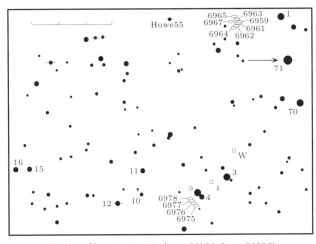

Finder Chart 3-3. 71 Aqr: 20^{h}38.2^m −01°06′

Figure 3-4. *M72 (NGC 6981) is a loose Class IX globular cluster but its faint 15th magnitude stars require high power and larger telescopes for resolution. Lee C. Coombs made this ten minute exposure on 103a-O film with a 10″, f5 Newtonian.*

Figure 3-5. *Messier 73 (NGC 6994) is a neat little Y-shaped asterism composed of four stars visible in small telescopes at low power. Lee C. Coombs made this five minute exposure on 2415 film with a 10″, f5 Newtonian.*

some small dark areas visible. A few outlying stars may be resolved.

16/18″ Scopes–175x: Messier 72 is moderately well concentrated with a bright compressed core and a ragged, nonsymmetrical halo. Faint strings of stars protrude from a 3′ diameter disk that on a good night may be resolved into swarms of faint 15th and 16th magnitude objects. However, under less than perfect conditions, only a dozen stars may be resolved around the cluster's edges. There are several field stars just to the south, a 9th magnitude star 5′ ESE, and an abundance of scattered faint stars.

NGC 6994 Messier 73 Asterism 4★ Tr Type IV 1p
ø2.8′, m8.9p, Br★ 10.0p $20^h59.0^m$ $-12°38'$
Finder Chart 3-4, Figure 3-5   ★★★

Messier 73 is merely an asterism of four faint stars having magnitudes of 10.5, 10.5, 11.0 and 12.0. Messier described it in October, 1780, as "three or four small stars which looks like a nebula at first sight."

8/10″ Scopes–100x: Messier 73 is a Y-shaped asterism of faint stars. The brightest star, on the south side, is 10th magnitude and two fainter stars are to the west.

NGC 7009 H1[4] Planetary Nebula Type 4+6
ø>25′, m8.3p, CS 12.78 $21^h04.2^m$ $-11°22'$
Finder Chart 3-4, Figure 3-6   Saturn Nebula ★★★★

The Saturn Nebula is a small, bright planetary nebula located one degree west of 4.5 magnitude star Nu (ν) = 13 Aquarii. It was first seen by Sir William Herschel in 1782. In 1850, Lord Rosse gave it the name "Saturn" Nebula when he saw extensions projecting from its disk. These projections may be seen with a medium-sized scope on a good night. NGC 7009 lies about 2,900 light years away.

8/10″ Scopes–100x: NGC 7009 is a fine, bright, bluish-green, oval-shaped planetary nebula, elongated 25″ × 20″ E-W with a uniform illumination. 175x shows very faint, diffuse extensions about 1′ long to the east and west. A 9th magnitude star is 10′ NNW.

12/14″ Scopes–150x: The Saturn Nebula has a bright bluish-green disk elongated 30″ × 25″ E-W with faint antennae extending from the ends of the major axis. At 200x, faint "bulbs" are visible at the end of each antenna. Some central brightening is apparent, but the central star is not visible.

16/18″ Scopes–200x: This bright planetary nebula is light bluish-green and elongated 40″ × 30″ E-W with a faint outer shell and two very faint projections to the east and west. At 250x, the central star is obvious and the antennae stand out more clearly with spots visible along them. The west antenna is brighter. A 14th magnitude star lies 1.75′ NNW.

Figure 3-6. *NGC 7009, The Saturn Nebula, is a bright planetary but its faint extensions require good viewing conditions. Martin C. Germano made this 15 min. exposure on 2415 film with a 14.5", f5 Newtonian.*

Figure 3-7. *Messier 2 (NGC 7089) is a fine bright globular cluster, the showpiece of the constellation Aquarius. Evered Kreimer made this 10 minute exposure on Tri-X film with a cold camera and a 12.5", f7 Newtonian reflector.*

NGC 7065 Galaxy Type SB:(r)b: II-III
ø1.1′ × 0.8′, m13.3v, SB 13.0 $21^h26.7^m$ $-07°00′$

NGC 7065A Galaxy Type SAB(r)c pec
ø1.4′ × 1.4′, m13.1v, SB 13.7 $21^h27.0^m$ $-07°01′$
Finder Chart 3-4, Figure 3-2   ★★/★

16/18″ Scopes–150x: NGC 7065, 4.5′ WSW of a 9th magnitude star, has a conspicuous stellar nucleus surrounded by a very faint halo slightly elongated 45″ × 30″ NE–SW. A 13th magnitude star lies 30″ WSW of center. Situated 4′ SE of NGC 7065 and 2.5′ SSW of the 9th magnitude star, NGC 7065A is a very faint, diffuse 30″ diameter glow with a slight central brightening. A 7.5 magnitude star 6′ east interferes with observation. 3′ east of NGC 7065A, a star of 11th magnitude forms the NE apex of a 1.25′ × 0.75′ diameter isosceles triangle with two 12th magnitude stars.

NGC 7077 U11755 Galaxy Type S0°? pec
ø0.6′ × 0.6′, m13.1v, SB 11.9 $21^h29.9^m$ $+02°25′$
Finder Chart 3-5   ★

16/18″ Scopes–150x: This galaxy is a faint, diffuse 30″ spot appearing fainter than NGC 7081 21′ to the ENE. A star brighter than the galaxy lies 2′ NE.

NGC 7081 Galaxy Type Sb? pec
ø1.0′ × 1.0′, m12.7v, SB 12.5   21h31.4m +02°30′
Finder Chart 3-5   ★★

16/18″ Scopes–150x: NGC 7081 has a faint 45″ diameter halo with a gradual brightening to a small, faint core. A 15″ pair of 13th and 14th magnitude stars centered 45″ SE of the galaxy's center point toward it. Lying 4.5′ SE, UGC 11760 is a very faint haze requiring averted vision. NGC 7077 is 21′ west.

NGC 7089 Messier 2 Globular Cluster Class II
ø12.9′, m6.4v $21^h33.5^m$ $-00°49′$
Finder Chart 3-5, Figure 3-7   ★★★★★

M2 was discovered by Maroldi in 1746 and added to Messier's catalogue in 1760. It is about 36,800 light years distant, and a fine object for amateur telescopes.

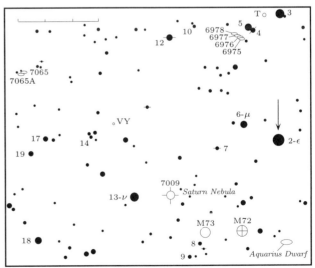

Finder Chart 3-4. 2-ε Aqr: $20^h47.7^m$ $-09°30′$

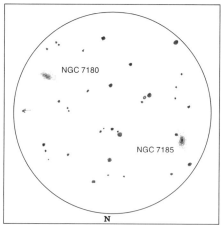

Figure 3-8. NGC 7180 & NGC 7185
17.5", f4.5–250x, by G. R. Kepple

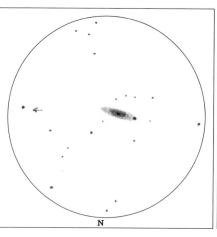

Figure 3-9. NGC 7184
13", f4.5–165x, by Tom Polakis

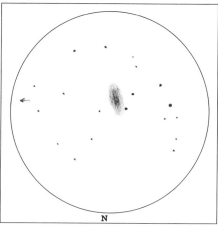

Figure 3-10. NGC 7218
17.5", f4.5–135x, by Steve Coe

8/10" Scopes–100x: Messier 2 has a bright, well compressed 5′ diameter halo with a suddenly brighter center. Though it is unresolved, its perimeter is mottled and shows two stars on the SE side. A 10th magnitude star lies 4.5′ NE of center.

16/18" Scopes–150x: A most impressive object, this is the showpiece of the constellation! Messier 2 is a bright, large globular cluster well compressed to an intense core. The 8′ diameter halo is symmetrical with a slight N–S elongation. There is a profusion of stars in the outer corona, and perhaps, a hundred stars can be resolved across the disk against a background haze. Star chains meandering out from the core extend the halo's span to a diameter of 12′. Several dark lanes are visible, the most prominent located in the NE portion.

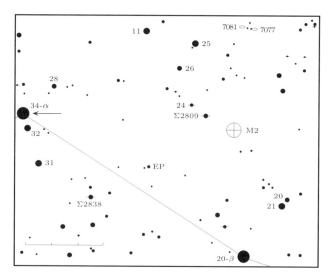

Finder Chart 3-5. 34-α Aqr: $22^h05.7^m$ $-00°19'$

NGC 7171 H692³ Galaxy Type SAB:(rs)bc II
⌀2.1′ × 1.3′, m12.2v, SB 13.1 $22^h01.0^m$ $-13°16'$
Finder Chart 3-7, Figure 3-3   ★★

8/10" Scopes–100x: NGC 7171 is located SW of a 4′ row of 10th magnitude stars. It is very faint and elongated 1.5′ × 0.75′ NW–SE with a slight brightening at center. A 12th magnitude star is at the SE tip.

12/14" Scopes–150x: This galaxy has a fairly faint halo elongated 2′ × 1.25′ NW–SE with a broad, slightly more concentrated central area extended 1′ × 0.75′. The field is well populated with faint to moderately bright stars.

NGC 7180 H693³ Galaxy Type S0-:
⌀1.5′ × 0.7′, m12.6v, SB 12.5 $22^h02.3^m$ $-20°33'$
Finder Chart 3-6, Figure 3-8   ★★★

16/18" Scopes–150x: NGC 7180 is fairly small with a moderate surface brightness. The oval core is bright and well condensed surrounded by a faint halo elongated 1.5′ × 0.75′ NE–SW. It lies 10′ WSW of NGC 7185, and 16′ NNW of NGC 7184.

NGC 7183 Galaxy Type SA: sp
⌀4.1′ × 1.1′, m11.9v, SB 13.4 $22^h02.4^m$ $-18°56'$
Finder Chart 3-6   ★★★

16/18" Scopes–150x: NGC 7183 is fairly faint and elongated 3.5′ × 1′ ENE–WSW with a bright core. It is located among four 11th to 13th magnitude stars.

NGC 7184 H1² Galaxy Type SB(r)b I-II
⌀6.5′ × 1.4′, m11.2v, SB 13.4 $22^h02.7^m$ $-20°49'$
Finder Chart 3-6, Figure 3-9   ★★★

8/10" Scopes–100x: NGC 7184 is a nice bright edge-on galaxy elongated 4′ × 1′ NE–SW with a small, bright

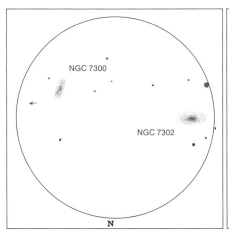

Figure 3-11. NGC 7300 & NGC 7302
13″, f5.6–165x, by Steve Coe

Figure 3-12. NGC 7371
17.5″, f4.5–135x, by Steve Coe

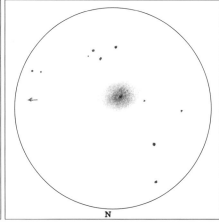

Figure 3-13. NGC 7377
20″, f4.5–175x, by Richard W. Jakiel

core. An 11.5 magnitude star, appearing as bright as the nucleus, lies at the NE tip, another star of similar brightness is placed 3′ WSW, and a 10.5 magnitude star is located 4′ west. It is the southern-most galaxy in group with NGC 7180, NGC 7185, and NGC 7188.

16/18″ Scopes–150x: A bright, circular core is surrounded by a well concentrated 3′ × 0.75′ inner region and a fainter halo spanning a diameter of 5.5′ × 1.5′ elongated NE–SW. A 12.5 magnitude star is embedded in the halo's edge 1′ ENE of center.

NGC 7185 Galaxy Type S0-? pec
ø2.1′ × 1.4′, m12.1v, SB 13.1   22ʰ03.1ᵐ −20°28′
Finder Chart 3-6, Figure 3-8   ★★

16/18″ Scopes–150x: NGC 7185 is just slightly fainter than NGC 7180 lying 10′ WSW. It displays a fairly conspicuous core containing a stellar nucleus surrounded by a faint envelope elongated 2′ × 1.25′ NNE-SSW. A 13th magnitude star is embedded in the halo 20″ SW of center. To the west of the galaxy is an irregularly spaced 6.5′ string of unequally bright stars aligned NNW–SSE.

NGC 7188 Galaxy Type SB(r)b: II-III
ø1.6′ × 0.7′, m13.2v, SB 13.2   22ʰ03.5ᵐ −20°20′
Finder Chart 3-6   ★

16/18″ Scopes–150x: NGC 7188 has a very faint, diffuse halo elongated 1′ × 0.5′ NE-SW with a slight brightening at center. It is located 10′ NNE of NGC 7185 in the NGC 7184 group.

NGC 7218 H897² Galaxy Type SB(rs)cd II-III
ø2.8′ × 1.0′, m12.0v, SB 13.0   22ʰ10.2ᵐ −16°40′
Finder Charts 3-6 & 3-7, Figure 3-10   ★★

8/10″ Scopes–100x: NGC 7218 is fairly obvious, elongated 2′ × 1′ NNE–SSW with a slightly brighter core. A 12th magnitude star touches the NE end while another star of similar brightness lies 1′ east.

16/18″ Scopes–125x: The halo is a fairly bright, thin oval elongated 2.5′ × 1′ NE–SW. A bright extended core sits in a mottled, irregularly concentrated central region.

NGC 7251 Galaxy Type (R′?)SA(rs)0/a:
ø1.8′ × 1.5′, m11.8v, SB 12.7   22ʰ20.4ᵐ −15°46′
Finder Chart 3-7   ★★

12/14″ Scopes–150x: NGC 7251 is faint with a circular 1.5′ halo gradually brightening to center.

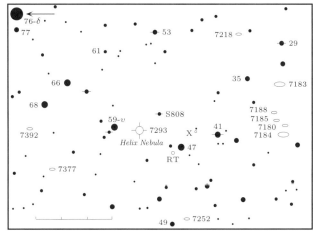

Finder Chart 3-6. 76-δ Aqr: 22ʰ54.6ᵐ −15°49′

Figure 3-14. *NGC 7293 "The Helix Nebula" is the largest and perhaps nearest planetary nebula. Because of its extremely low surface brightness, it is surprisingly easy to spot in 10 x 50 binoculars but nearly impossible to observe in telescopes without the help of an O-III or nebular filter. Martin C. Germano made this 65 minute photograph on hypered 2415 Kodak Tech Pan film with a 14.5", f5 Newtonian reflector.*

NGC 7252 H458³ Galaxy Type SAB0°? pec
ø2.0' × 1.6', m11.4v, SB 12.5   22ʰ20.7ᵐ −24°41'
Finder Chart 3-6   ★★

8/10" Scopes–100x: NGC 7252 is located 40' WNW of the 6th magnitude star 49 Aquarii in a field of faint stars. It has a faint 1' diameter halo with a stellar nucleus.

16/18" Scopes–150x: This galaxy shows a fairly conspicuous core with a stellar nucleus surrounded by a circular 1.5' halo. 13th magnitude stars are visible 2' NNE and 2' south.

NGC 7293 PK36-57.1 Planetary Nebula Type 4+3
ø>769", m7.3v, CS 13.62v   22ʰ29.6ᵐ −20°48'
Finder Chart 3-6, Figure 3-14   Helix Nebula   ★

The Helix Nebula is perhaps the nearest planetary to the Solar System, its distance probably less than 300 light-years. It is in any case the largest planetary in apparent size, its diameter being one-half that of the full moon. NGC 7293 has an extremely low surface brightness, and its ring structure, so conspicuous in photos, is not easy to detect visually. In binoculars and rich field scopes, the planetary is a large ghostly disk, vaguely circular. In larger instruments it is very challenging, so use a very low power eyepiece. The O-III and other nebula filters help quite substantially.

4/6" Scopes–50x: The Helix Nebula is visible as a faint, very large smoke ring. The center is plainly darker, and the central star is visible with averted vision.

8/10" Scopes–50x: The Helix Nebula is very large and faint with a uniformly illuminated annular-shape having a much fainter center. A UHC filter and averted vision reveals two slightly brighter spots in the ring. Three 10th magnitude stars are in the center; and at least five faint stars are involved in the disk.

16/18" Scopes–75x: The Helix Nebula is extremely faint when viewed without a filter, but an Oxygen-III filter shows a huge, glowing 15' diameter wreath with a dark center. The ring is rather faint around its outer edge. At least ten stars are embedded in the nebulosity, including the central star and a close pair on the southern side of the ring. 100x reveals two brighter areas in the wreath's outer area at the NNE and SSW sides.

NGC 7298 Galaxy Type SAB:(rs?)c I-II:
⌀1.5′ × 1.2′, m13.7v, SB 14.2   22ʰ30.8ᵐ −14°11′
Finder Chart 3-7   ★

16/18″ Scopes–150x: Located in a star poor field 22′ NNE of 6.4 magnitude star 56 Aquarii, NGC 7298 is an extremely faint 1′ diameter nebulous patch. NGC 7300 lies 11′ NNE.

NGC 7300 Galaxy Type SB(r:)bc I-II
⌀2.1′ × 1.0′, m12.8v, SB 13.5   22ʰ31.0ᵐ −14°00′
Finder Chart 3-7, Figure 3-11   ★

16/18″ Scopes–150x: This galaxy, located north of a large trapezoid of 12th magnitude stars, is a very faint, small, diffuse glow elongated 1.5′ × 0.75 NNW–SSE without central brightening. A 14th magnitude star lies 1′ SW. NGC 7298 lies 11′ SSW and NGC 7302 is 24′ ESE.

NGC 7302 H31⁴ Galaxy Type (R′?)SA?0°:
⌀2.1′ × 1.3′, m12.2v, SB13.0   22ʰ32.4ᵐ −14°07′
Finder Chart 3-7, Figure 3-11   ★★

8/10″ Scopes–100x: NGC 7302 is 4′ north of a 9th magnitude star and 40′ NE of 6th magnitude 56 Aquarii. It is visible as a dim, circular 1′ diameter glow with a stellar nucleus.

16/18″ Scopes–150x: The halo is faint and slightly elongated 1.5′ × 1.25′ E–W with a small core and a conspicuous stellar nucleus. The galaxy forms a triangle with NGC 7298 20′ WSW and NGC 7300 24′ WNW.

NGC 7309 H476² Galaxy Type SB(rs)c I-II
⌀1.6′ × 1.6′, m12.5v, SB13.3   22ʰ34.3ᵐ −10°21′
Finder Chart 3-7   ★★

8/10″ Scopes–100x: NGC 7309 is situated 15′ NE of an 8th magnitude star and 1° ENE of Sigma (σ) = 57 Aquarii (m4.8v). It is a very dim, circular 1′ diameter object with even surface brightness.

16/18″ Scopes–150x: Larger instruments show a fairly bright core surrounded by a poorly concentrated 1.25′ diameter halo. A 13th magnitude star lies 1.5′ east.

NGC 7371 H477² Galaxy Type (R′?)SAB:(rs)b: III
⌀1.9′ × 1.9′, m11.5v, SB12.8   22ʰ46.1ᵐ −11°00′
Finder Chart 3-7, Figure 3-12   ★★

8/10″ Scopes–100x: NGC 7371, located 10′ north of a 7th magnitude star which hinders viewing, is a very faint, round 1′ diameter object with uniform illumination.

16/18″ Scopes–150x: NGC 7371 has a fairly faint, round 1.5′ diameter halo with a stellar nucleus. 2′ SE is a wide double of 11.5 and 12.5 magnitude stars separated by 20″.

NGC 7377 H598² Galaxy Type (R′)SAB(rs)0+?
⌀4.4′ × 3.6′, m10.4v, SB 13.2   22ʰ47.8ᵐ −22°19′
Finder Chart 3-6, Figure 3-13   ★★

8/10″ Scopes–100x: NGC 7377 has a faint, round, diffuse 1′ diameter halo. Two 9th magnitude stars aligned N–S lie 6′ NNW and 10′ north.

16/18″ Scopes–150x: The 2′ diameter halo appears moderately faint and diffuse with a faint 30″ core. A 2′ × 1′ triangle composed of 12.5 magnitude stars pointing SE is centered 2.5′ SSW of the galaxy's core.

NGC 7378 Galaxy Type (R′?)SB(r)ab
⌀1.5′ × 1.0′, m12.7v, SB 13.0   22ʰ47.9ᵐ −11°49′
Finder Chart 3-7   ★★

12/14″ Scopes–125x: Located 4′ NW of an 8.5 magnitude star, NGC 7378 is a faint object elongated 1′ × 0.5′ N-S with a large, bright core.

NGC 7392 H702² Galaxy Type SB(s)b? I-II
⌀2.1′ × 1.1′, m11.8v, SB 12.6   22ʰ51.8ᵐ −20°36′
Finder Chart 3-6, Figure 3-15   ★★★

8/10″ Scopes–100x: This moderately faint galaxy is elongated 1′ × 0.5′ with a slight brightening at center. 12th magnitude stars lie 1.5′ north and 2.5′ ESE.

16/18″ Scopes–150x: NGC 7392 is faint and elongated 1.5′ × 1′ ESE–WNW with a bright core.

NGC 7416 Galaxy Type SB?(rs)b: II-III
⌀3.0′ × 0.7′, m12.3v, SB 13.0   22ʰ55.7ᵐ −05°30′
Finder Chart 3-8   ★★

8/10″ Scopes–100x: NGC 7416 lies south of a 5th magnitude star and just north of 9th and 8th magnitude stars aligned N-S. It is faint and elongated 2.0′ × 0.5′

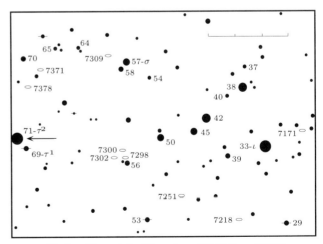

Finder Chart 3-7. 71-τ² Aqr: 22ʰ49.5ᵐ −13°36′

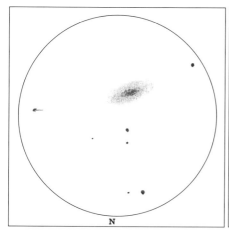 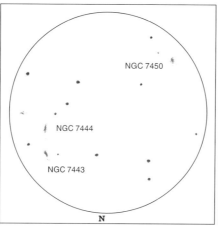 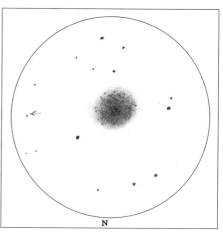

Figure 3-15. NGC 7392
20″, f4.5–175x, by Richard W. Jakiel

Figure 3-16. NGCs 7443, 7744 & 7750
17.5″, f4.5–250x, by G. R. Kepple

Figure 3-17. NGC 7492
13″, f4.5–165x, by Tom Polakis

ESE–WNW with some brightening at center.

16/18″ Scopes–150x: This galaxy is fairly faint, elongated 2.5′ × 0.7′ ESE–WNW with a gradual brightening to a stellar nucleus. A 10.5 magnitude star lies 2.5′ ENE.

NGC 7443 H450² Galaxy Type SB:(s:)0°:
ø1.5′ × 0.6′, m12.6, SB 12.3   23^{h}00.1^m −12°48′
Finder Chart 3-9, Figure 3-16   ★★

16/18″ Scopes–150x: This galaxy forms a faint but distinctive duo with NGC 7444 1.5′ SSE. NGC 7450 is located 12′ SE. NGC 7443 is faint, elongated 1.25′ × 0.5′ NE–SW with a sudden brightening at center. A 13th magnitude star lies 1.25′ WSW.

NGC 7444 H451² Galaxy Type SB?(r?)0°
ø1.4′ × 0.6′, m12.8v, SB 12.5   23^{h}00.1^m −12°50′
Finder Chart 3-9, Figure 3-16   ★★

16/18″ Scopes–150x: Nearly its twin, this galaxy forms

a close pair with NGC 7443 1.5′ north. NGC 7444 is faint, elongated 1′ × 0.5′ N-S with a sudden brightening at center. 13th and 12th magnitude stars lies 1′ and 2′ SE respectively. NGC 7450 lies 11′ ESE.

NGC 7450 Galaxy Type (R)SB(r)a
ø1.4′ × 1.4′, m12.4v, SB 13.0   23^{h}00.9^m −12°56′
Finder Chart 3-9, Figure 3-16   ★★

16/18″ Scopes–150x: This is the easternmost of three galaxies, NGC 7443 and NGC 7444 forming a close pair 11.5′ NW. NGC 7450 is faint with a circular 1′ diameter halo gradually brightening to center. A 13th magnitude star lies 2′ SW.

NGC 7492 H558² Globular Cluster Class XII
ø6.2′, m11.4v   23^{h}08.4^m −15°37′
Finder Chart 3-9, Figure 3-17   ★★

8/10″ Scopes–150x: NGC 7492 is an extremely faint, round 2.5′ blob requiring averted vision.

12/14″ Scopes–200x: NGC 7492 shows a pale 3.5′ disk with a uniform glow. There is no hint of resolution.

16/18″ Scopes–225x: This faint globular has a round, diffuse 4′ diameter halo. Four 12.5 magnitude stars lie around the periphery to the NW, SW, east, and NE, along with half a dozen 13th magnitude stars.

NGC 7576 H454² Galaxy Type (R′)SA?0°:
ø1.5′ × 1.1′, m12.9v, SB 13.3   23^{h}17.4^m −04°44′
Finder Chart 3-8, Figure 3-19   ★

16/18″ Scopes–150x: NGC 7576 is located 11′ SW of NGC 7585. It is fainter than its companion but displays a prominent core surrounded by a faint envelope elongated 1′ × 0.75′ N-S.

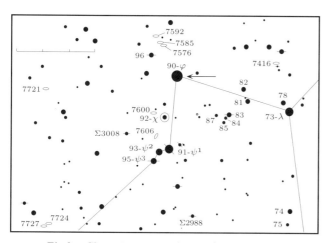

Finder Chart 3-8. 90-φ Aqr: 23^{h}14.3^m −06°03′

NGC 7585 H236² Galaxy Type SA?0°pec
ø3.0' × 2.6', m11.4v, SB 13.5
23ʰ18.0ᵐ −04°39'
Finder Chart 3-8, Figure 3-19   ★★★

8/10" Scopes–100x: This galaxy lies 10' south of a 9.5 magnitude star and 30' NW of 96 Aquarii, a nice white and blue double star (see Table 3-2). It is fairly faint, small, and round with a faint stellar nucleus. NGC 7576 lies 11' SW while NGC 7592 is 15' NW; both galaxies are considerably fainter than NGC 7585.

16/18" Scopes–150x: NGC 7585 is a moderately bright galaxy in larger instruments, its halo elongated 2.5' × 2' ESE–WNW with a small core containing a bright stellar nucleus.

NGC 7592 H186³ Galaxy Type S0+ pec
ø1.1' × 0.9', m14.2v, SB 14.0
23ʰ18.4ᵐ −04°25'
Finder Chart 3-8   ★

12/14" Scopes–150x: NGC 7592 is located 15' NNE of NGC 7585 and 14' east of an 8.5 magnitude star. It is an extremely faint, round 1' diameter smudge.

NGC 7600 H431² Galaxy Type E5/S0−
ø2.0' × 1.4', m11.9v, SB 13.4   23ʰ18.9ᵐ −07°35'
Finder Chart 3-8   ★★

8/10" Scopes–100x: NGC 7600 lies 40' ENE of 5th magnitude star Chi (χ) = 92 Aquarii and SE of two 8th magnitude stars. It has a faint 1' × 0.5' diameter halo elongated ESE–WNW with a tiny core.

16/18" Scopes–150x: This galaxy is fairly faint and elongated 1.5' × 1' ESE–WNW with a small, conspicuous core containing a faint stellar nucleus.

NGC 7606 H104¹ Galaxy Type SB:(rs)b II
ø4.4' × 2.0', m10.8v, SB 13.0   23ʰ19.1ᵐ −08°29'
Finder Chart 3-8, Figure 3-18   ★★★★

8/10" Scopes–100x: This galaxy is faint but obvious. It has a 3' × 1' diameter halo elongated NW–SE with a slight brightening at center. 11th magnitude stars lie 2.5' south and 3' NNW.

12/14" Scopes–125x: The halo is fairly bright, elongated 3.5' × 1.5' with a brighter center and a stellar nucleus.

16/18" Scopes–150x: NGC 7606 appears fairly bright and large. A broad core containing a stellar nucleus is surrounded by a fairly well defined and moderately concentrated halo elongated 4' × 1.5' NNW–SSE. 200x shows some mottling, and averted vision gives the impression of a dark lane west of the core. A threshold star is embedded in the halo 40" north of center.

Figure 3-18. *NGC 7606 is a bright, SB galaxy with a large, extended core and a stellar nucleus. Harvey Freed made this 55 minute exposure on 2415 film with a 10", f6.3 Newtonian at prime focus.*

NGC 7721 H432² Galaxy Type SAB(rs:)c I-II
ø3.3' × 1.3', m11.6v, SB 13.0   23ʰ38.8ᵐ −06°31'
Finder Chart 3-8   ★★★

8/10" Scopes–100x: This galaxy appears faint, elongated 2' × 0.5' N–S with some brightening at center.

12/14" Scopes–125x: NGC 7721 is fairly faint and elongated 2.5' × 0.75' NNE–SSW with a bright center.

16/18" Scopes–150x: The halo is moderately bright,

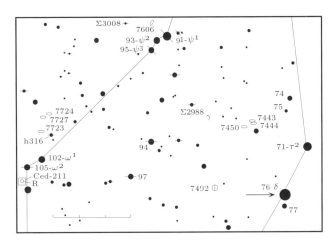

Finder Chart 3-9. 76-δ Aqr: 22ʰ54.6ᵐ −15°49'

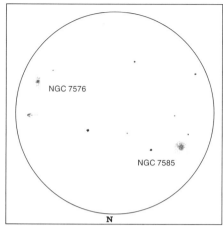 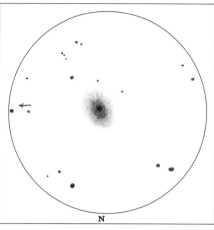 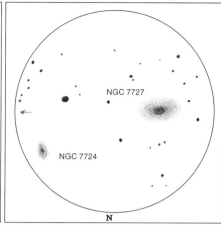

Figure 3-19. NGC 7576 & NGC 7585
17.5", f4.5–225x, by G. R. Kepple

Figure 3-20. NGC 7723
17.5", f4.5–175x, by G. R. Kepple

Figure 3-21. NGC 7724 & NGC 7727
17.5", f4.5–175x, by G. R. Kepple

elongated 3' × 1' NNW–SSE with an extended core. Both the core and the central area are mottled, and some stellar spots lie along the major axis. The surrounding star field is bleak.

NGC 7723   H110¹   Galaxy   Type SB(r)bc II
ø2.8' × 1.9', m12.2v, SB 12.9   23ʰ38.9ᵐ −12°58'
Finder Chart 3-9, Figure 3-20   ★★★

8/10" Scopes–100x: This galaxy lies 15' ENE of h5355, a wide triple star with components of 7.6, 8.9, and 10.1 magnitude. The halo is fairly bright, and elongated 2.5' × 1.5' NE–SW with a brighter center. Two 11th magnitude stars are located 5' NNW and 5.5' NE, each having a faint, widely separated companion.

16/18" Scopes–150x: NGC 7723 has a moderately bright, oval-shaped halo elongated 3' × 2' NE–SW with a prominent oval-shaped core and a faint stellar nucleus. At 200x, the core is mottled and extended 1' × 0.5' ENE–WSW. A 13th magnitude star lies 2.25' SW.

NGC 7724   Galaxy   Type (R')SB(r)b? pec
ø1.4' × 0.9', m13.5v, SB 13.6   23ʰ39.1ᵐ −12°14'
Finder Chart 3-9, Figure 3-21   ★

8/10" Scopes–100x: Situated 12' NW of NGC 7727 and 5.5' NNW of a 10th magnitude star, this object is an extremely faint, small, round smudge.

16/18" Scopes–150x: NGC 7724 has a faint, diffuse halo elongated 1' × 0.75' NNE–SSW with a poorly concentrated core.

NGC 7727   H111¹   Galaxy   Type SAB:(S?)0/a pec.
ø5.6' × 4.0', m10.6v, SB 13.8   23ʰ39.9ᵐ −12°18'
Finder Chart 3-9, Figure 3-21   ★★★

8/10" Scopes–100x: NGC 7727, located 9' west of a 10th magnitude star, has a fairly faint, circular 2.5' diameter halo brightening at center to a stellar nucleus.

12/14" Scopes–125x: NGC 7727 appears fairly bright with a circular 3' diameter halo and a bright core. The galaxy lies within, and SE of the center of, an irregular 8' diameter circlet of 12.5 and 13th magnitude stars.

16/18" Scopes–150x: Larger instruments show a bright, circular 45" diameter core with a nonstellar nucleus. The halo is faint and oval-shaped 5' × 3' ENE–WSW, fading smoothly to the edges. The faint galaxy NGC 7724 lies 12' NW.

Chapter 4

Aries, the Golden Ram

4.1 Overview

Aries is the first sign of the Zodiac. In ancient Greek times, when the Zodiac was organized from even more ancient Babylonian constellations, the Sun lay in Aries on the vernal equinox, the first day of spring. Thus Greek and Roman astronomers called the vernal equinox the "first point of Aries." The term persists though precession has moved this point from Aries into Pisces.

Aries rescued Phryxus and Helle, children of the king of Thessaly, from their cruel stepmother. Phryxus later sacrificed the ram, and when he hung its fleece in a sacred grove it turned to gold. Jason and his Argonauts sought and found the fleece.

The ram has played an important role in the religion and mythology of many different cultures. In ancient Egypt, Ammon Ra, the God of Fertility and Creative Life, was depicted with the body of a man and the head of a ram.

Aries is a small constellation with only 441 square degrees of sky. Its western portion has the brightest stars and the most distinctive asterism, the head of the Ram, Alpha (α) = 13, Beta (β) = 6, and Gamma (γ) = 5, the individual names of which are Hamal, Sheratan, and Mesarthim. The constellation offers only double stars and external galaxies, none of which are showpieces.

> **Aries:** AY-ri-eez
> **Genitive:** Arietis, AY-ri-e-tis
> **Abbrevation:** Ari
> **Opposition:** 9pm–Aug. 25, midnight–Jul 10
> **Culmination:** 9pm–Dec. 14, midnight–Oct. 30
> **Area:** 441 square degrees
> **Showpieces:** 1 Ari, 5-γ Ari, 9-λ Ari, 30 Ari, 33 Ari, 48-ϵ Ari, Σ326
> **Best Deep-Sky Objects:** NGC 772, NGC 877, NGC 972
> **Binocular Objects:** 5-γ Ari, 9-λ Ari, 30 Ari, 41 Ari, Σ366

4.2 Interesting Stars

1 Arietis = Σ174   Double Star Spec. F5
m6.2, 7.2; Sep. 2.8″; P.A. 166° $01^h50.1^m$ +22°17′
Finder Chart 4-3 ★★★★

4/6″ Scopes–150x: 1 Arietis is a fine pair of yellow and pale blue stars.

Gamma (γ) = 5 Arietis (Σ180) Double Star Spec. A0
AB: m4.8, 4.8; Sep. 7.8″; P.A. 0° $01^h53.5^m$ +19°18′
Finder Chart 4-4   Mesarthim   ★★★★★

4/6″ Scopes–125x: Gamma Arietis is a beautiful pair of equal, bluish-white stars dominating a nice star field.

Lambda (λ) = 9 Arietis (H V 12) Double Star A5
m4.9, 7.7; Sep. 37.4″; P.A. 46° $01^h57.9^m$ +23°36′
Finder Chart 4-3 ★★★★

4/6″ Scopes–75x: Lambda Arietis is a wide, easy pair of pretty yellowish-white and pale blue stars.

30 Arietis = Σ I 5   Double Star Spec. F5 F5
m6.6, 7.4; Sep. 38.6″; P.A. 274° $02^h37.0^m$ +24°39′
Finder Chart 4-6 ★★★★

4/6″ Scopes–75x: 30 Arietis is a pleasing pair of yellow stars.

33 Arietis = Σ289   Double Star Spec. A2
m5.5, 8.4; Sep. 28.6″; P.A. 0° $02^h40.7^m$ +27°04′
Finder Chart 4-6 ★★★★

2/3″ Scopes–100x: This double is an easily split pair of yellowish-white and blue stars.

Σ326 Double Star Spec. K0 & K5
AB: m7.6, 9.8; Sep. 5.9″; P.A. 220° $02^h55.6^m$ +26°52′
Finder Chart 4-6 ★★★★

4/6″ Scopes–100x: Struve 326 is a beautiful pair of orange and dull red stars.

39

Aries, The Golden Ram

Constellation Chart 4-1

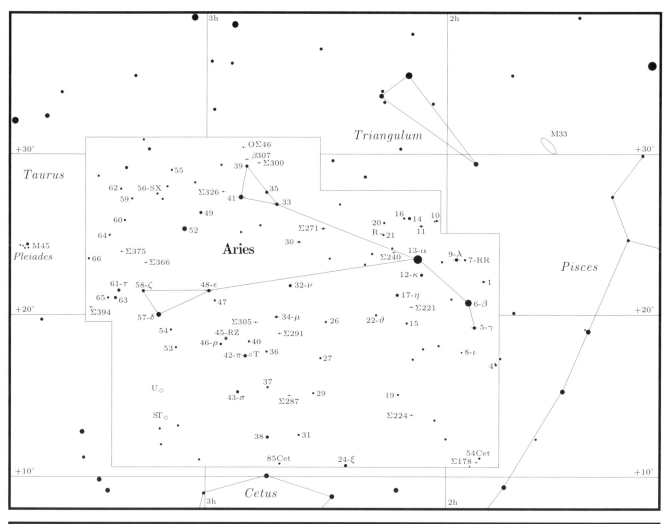

Table 4–1. Selected Variable Stars in Aries

Name	HD No.	Type	Max.	Min.	Period (Days)	F*	Spec. Type	R.A. (2000)	Dec.	Chart No. & Notes
7–RR Ari	11763	EA?	6.42	6.84:	47.9	0.08	K0	01ʰ55.9ᵐ	+23°35′	4-3 30′ W of λ Ari
V Ari		SR	8.0	8.6	75.0:		R4	02ʰ14.9ᵐ	+12 15	4-4 Deep Orange
R Ari	13913	M	7.4	13.7	186.78	0.45	M3-M6	16.1	+25 03	4-3
T Ari	17446	SRa	7.5	11.3	316.6	0.49	M6-M8	48.3	+17 31	4-5 15′ WNW of π Ari
45–RZ Ari	18191	SRb	5.62	6.01	30.0		M6	55.8	+18 20	4-5 25′ NNW of ρ Ari
X Ari	19510	RRab	8.97	9.95	0.65	0.13	A8-F4	03ʰ08.5ᵐ	+10 27	4-5
ST Ari		SRb	9.0	10.6	99.0		M4	10.1	+13 26	4-5
U Ari	19737	M	7.2	15.2	371.13	0.40	M4-M9	11.0	+14 48	4-5

F* = The fraction of period taken up by the star's rise from min. to max. brightness, or the period spent in eclipse.

Table 4–2. Selected Double Stars in Aries

Name	ADS No.	Pair	M1	M2	Sep."	P.A.°	Spec	R.A. (2000)	Dec.	Finder Chart No. & Notes
1 Ari	1457		6.2	7.4	2.8	166	F5	01^{h}50.1^m	+22°17'	4-3 Yellow & bluish
Σ178	1487		8.5	8.5	2.1	201	F0	52.2	+10 49	4-4 Both stars white
5–γ Ari	1507	AB	4.8	4.8	7.8	0	A0	53.5	+19 18	4-4 Matched white stars
	1507	AC		9.6	221.3	84				
9–λ Ari	1563	AB	4.9	7.7	37.4	46	A5	57.9	+23 36	4-3 Yellowish-white & blue
10 Ari	1631	AB	5.9	7.3	w1.3	*339	F5	02^{h}03.7^m	+25 56	4-3 White & light blue
Σ221	1678	AB	8.3	9.5	8.4	146	A3	09.7	+20 21	4-3
Σ224	1689		8.4	8.9	5.5	243	G5	10.9	+13 41	4-4
Σ240	1749		8.4	8.9	4.8	51	F0	17.4	+23 53	4-3
Σ271	1904	AB	5.9	10.4	12.4	182	F5	30.5	+25 14	4-1 Yellow & lilac
30 Ari	1982	AB	6.6	7.4	38.6	274	F5 F5	37.0	+24 39	4-6 Yellow pair
Σ287	2008		7.4	9.7	6.8	74	K0	39.0	+14 52	4-5
33 Ari	2033		5.5	8.4	28.6	0	A2	40.7	+27 04	4-6 Topaz & Blue
Σ291	2042		7.7	8.0	3.4	118	B9	41.1	+18 48	4-5
	2042	AC		9.3	65.7	241				
Σ300	2091		7.8	8.0	3.1	312	F0	44.6	+29 28	4-6
Σ305	2122		7.4	8.2	w3.7	*308	G0	47.5	+19 22	4-5
	2122	AC		12.6	88.6	34				
β307	2126		7.2	11.6	15.4	316	B5	47.6	+29 41	4-6
42–π Ari	2151	AB	5.2	8.7	3.2	120	B5	49.3	+17 28	4-5 White & gray
	2151	BC		10.8	25.2	110				
OΣ46	2158	AB	6.8	10.0	4.9	74	F0	50.0	+30 32	4-6
h656	2158	AC		12.7	20.1	168				
41 Ari	2159	AB	3.6	10.7	24.6	277	B8	50.0	+27 16	4-6
H V 116	2159	AC		10.5	31.3	213				
H VI 5	2159	AD		9.0	124.9	232				
Σ326	2218		7.6	9.8	5.9	220	K0 K5	55.6	+26 52	4-6 Orange & red
48–ε Ari	2257	AB	5.2	5.5	1.5	208	A2	59.2	+21 20	4-1 White pair
Σ366	2414	AB	6.9	9.6	47.2	40	K0	03^{h}14.3^m	+22 57	4-1
β530	2414	BC		10.3	1.7	192				
Σ375	2473	AB	7.5	9.6	2.3	315	A5	20.4	+23 41	4-6
	2473	AC		12.6	64.9	290				
Σ394	2546		7.1	8.1	6.8	162	A3	28.0	+20 28	4-1 Yellowish, light blue

Footnotes: *= Year 2000, a = Near apogee, c = Closing, w = Widening. Finder Chart No: All stars listed in the tables are plotted in the large Constellation Chart, but when a star appears in a Finder Chart, this number is listed. Notes: When colors are subtle, the suffix -ish is used, e.g. bluish.

Epsilon (ε) = 48 Arietis (Σ333) Double Star Spec. A2
m5.2, 5.5; Sep. 1.5"; P.A. 208°   02^{h}59.2^m +21°20'
Constellation Chart 4-1   ★★★★

8/10" Scopes–150x: Almost equal in magnitude, this bright pair of white stars may be split at 150x when seeing conditions are good.

4.3 Deep-Sky Objects

NGC 673 H589² Galaxy Type SAB(s)c
⌀2.1' × 1.6', m12.6v, SB 13.8   01^{h}48.4^m +11°32'
Finder Chart 4-4   ★

16/18" Scopes–150x: This moderately faint galaxy is 50' NW of a 6th magnitude star and 50' NNE of a 7th magnitude star. It is extended 1.5' × 1.25' N–S with a slight brightening at the center. An 11th magnitude star lies 3.25' ENE.

NGC 678 H228² Galaxy Type SB(s)b:
⌀7.6' × 1.0', m12.2v, SB 13.7   01^{h}49.4^m +22°00'
Finder Chart 4-3, Figure 4-1   ★★

8/10" Scopes–100x: NGC 678 is one of six NGC objects in the NGC 697 galaxy group clustered around 1 Arietis, a fine double star (see Table 4-2). With averted vision, this galaxy appears elongated 2' × 1' nearly E–W with some brightening at center.

16/18" Scopes–150x: NGC 678, located 18' SSW of 5.9 magnitude star 1 Arietis, is the larger but fainter of a pair with NGC 680 lying 5.5' ESE. It is moderately faint and elongated 4' × 1' ENE–WSW with a bright core. A 13th magnitude star lies 3' west.

NGC 680 H229² Galaxy Type E+ pec:
⌀2.7' × 2.4', m11.9v, SB 13.8   01^{h}49.8^m +21°58'
Finder Chart 4-3, Figure 4-1   ★★★

8/10" Scopes–100x: NGC 680 is one of the six galaxies in a group around 1 Arietis, NGC 697 being the

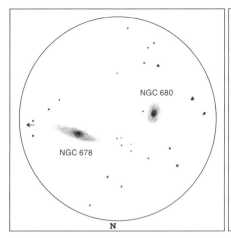

Figure 4-1. NGC 678 & NGC 680
17.5″, f4.5–200x, by G. R. Kepple

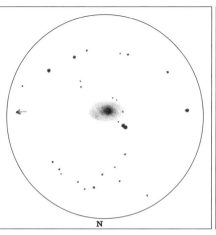

Figure 4-2. NGC 691
17.5″, f4.5–200x, by G. R. Kepple

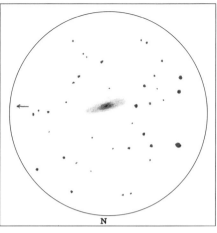

Figure 4-3. NGC 697
17.5″, f4.5–200x, by G. R. Kepple

brightest. NGC 680 has a faint round 1.5′ halo growing considerably brighter at center.

16/18″ Scopes–150x: This galaxy makes a nice pair with NGC 678 5.5′ WNW. At medium power, NGC 680 has a 30″ diameter core surrounded by a moderately faint, circular 2′ diameter halo slightly elongated N–S.

NGC 691 H617² Galaxy Type SA(rs)bc
⌀3.5′ × 2.5′, m11.4v, SB 13.6   01ʰ50.7ᵐ +21°46′
Finder Chart 4-3, Figure 4-2   ★★★

8/10″ Scopes–100x: NGC 691 is the southernmost object in the NGC 697 galaxy group scattered around the double star 1 Arietis. It is 15′ SSE of NGC 680. 100x shows a faint, diffuse 2′ diameter halo slightly elongated E–W with a slight central brightening.

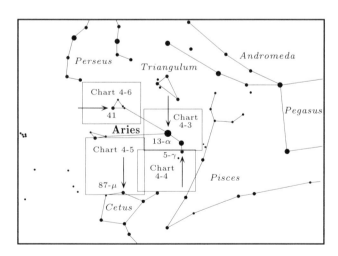

Master Finder Chart 4-2. Aries Chart Areas
Guide stars indicated by arrows.

16/18″ Scopes–150x: NGC 691 lies 1.75′ SW of a 7″ pair of nearly matched 10.5 magnitude stars aligned ENE–WSW. It has a conspicuous 30″ diameter core surrounded by a fairly faint, diffuse 2.5′ diameter halo slightly elongated E–W.

NGC 694 Galaxy Type S0? pec
⌀0.5′ × 0.3′, m13.7v, SB 11.5v   01ʰ51.0ᵐ +22°00′
Finder Chart 4-3   ★

16/18″ Scopes–150x: NGC 694 lies 25′ SE of double star 1 Arietis in the NGC 697 galaxy group. It has a faint, diffuse halo appearing larger than the catalog size. With averted vision it seems to be elongated 1′ × 0.5′ NW–SE with a uniform surface brightness. A wide pair of 11th magnitude stars aligned NNW–SSE lies 5.5′ NNW.

NGC 695 H618² Galaxy Type S0? pec
⌀0.5′ × 0.4′, m12.8v, SB 10.9   01ʰ51.2ᵐ +22°35′
Finder Chart 4-3   ★★

12/14″ Scopes–125x: NGC 695, 23′ NE of double star 1 Arietis, is the northernmost galaxy in the NGC 697 galaxy group. NGC 695 is 13′ north of NGC 697 in the same low power field of view. NGC 695 has a poorly concentrated 30″ diameter halo with a brighter center. It lies at the east edge of a 3′ row of three 13th magnitude stars. A 13.5 magnitude star lies 1′ SSW.

NGC 697 H179³ Galaxy Type SAB(r)c:
⌀4.7′ × 1.6′, m12.0v, SB14.1   01ʰ51.3ᵐ +22°21′
Finder Chart 4-3, Figure 4-3   ★★★

8/10″ Scopes–100x: This galaxy is 16′ east of double star 1 Arietis in a group of faint galaxies. NGC 697 is faint and elongated 1.5′ × 0.75′ E–W with a slightly brighter center. It is in a rich star field NW of a dozen 11th to 13th magnitude stars.

16/18″ Scopes–125x: Larger instruments reveal a fairly bright galaxy with a somewhat mottled halo elongated 3′ × 1.25′ ESE–WNW. A small, faint extended core containing a faint stellar nucleus is just visible. A 12.5 magnitude star lies 2′ east.

NGC 770 Galaxy Type E3:
ø1.0′ × 0.7′, m12.9v, SB 12.4   01ʰ59.2ᵐ +18°57′
Finder Chart 4-4   ★★

8/10″ Scopes–125x: This faint object is a companion of NGC 772 3.5′ SSW. NGC 770 is visible as a very faint 30″ diameter spot.

16/18″ Scopes–150x: This galaxy is a faint oval-shaped object elongated 45″ × 30″ N–S with a somewhat brighter center.

NGC 772 H112¹ Galaxy Type SA(s)b
ø7.3′ × 4.6′, m10.3v, SB 14.0   01ʰ59.3ᵐ +19°01′
Finder Chart 4-4, Figure 4-4   ★★★

8/10″ Scopes–100x: NGC 772, 1.5° east of the 4.8 magnitude double star Gamma (γ) = 5 Arietis, has a small, bright core surrounded by a very faint, diffuse halo elongated 3′ × 1.5′ NW–SE.

16/18″ Scopes–125x: A conspicuous 30″ diameter core with a stellar nucleus is surrounded by a mottled, fairly well concentrated, 1.5′ diameter core. The outer halo is much fainter, extending to 4′ × 2′ NW–SE. The immediate area around the galaxy is well sprinkled with faint stars, including 13th magnitude stars 2.25′ NE and 3′ west. A threshold star is embedded in the halo 1′ NE of center. NGC 770 lies 3.5′ SSW.

NGC 803 H208³ Galaxy Type SA(s)c
ø3.2′ × 1.5′, m12.6v, SB 14.1   02ʰ03.8ᵐ +16°02′
Finder Chart 4-4, Figure 4-5   ★★

16/18″ Scopes–125x: NGC 803, situated in a lightly adorned star field 1′ ENE of a 10.5 magnitude star, is fairly faint and elongated 2.5′ × 0.75′ N–S with a small, faint core. A 14th magnitude star lies 45″ NE of the galaxy's center and a 13th magnitude star is 1.5′ NE.

NGC 821 H152¹ Galaxy Type E6?
ø3.3′ × 2.3′, m10.7v, SB 12.7   02ʰ08.4ᵐ +11°00′
Finder Chart 4-4, Figure 4-6   ★★★

8/10″ Scopes–100x: NGC 821 is 12′ south of an 8th magnitude star, and nestled between a 9th magnitude star 1′ to the galaxy's NW and an 11th

Figure 4-4. *NGC 772, the brightest galaxy in Aries, displays a bright core surrounded by a faint halo. Lee C. Coombs made this 10 minute exposure on 103a-0 film with a 10″, f5 Newtonian.*

magnitude star 1.75′ to its south. It is fairly faint, and elongated 1.5′ × 0.5′ NE–SW with considerable brightening at its center.

16/18″ Scopes–150x: Larger instruments show a small but conspicuous stellar nucleus surrounded by a faint, diffuse oval-shaped halo elongated 2′ × 1′ NNE–SSW. A 13.5 magnitude star touches the SSW tip.

NGC 871 H201³ Galaxy Type SB(s)c:
ø1.0′ × 0.3′, m13.6v, SB 12.2   02ʰ17.2ᵐ +14°33′
Finder Chart 4-4, Figure 4-7   ★★

8/10″ Scopes–100x: NGC 871 is 5′ north of the westernmost star of an 8th magnitude pair aligned E–W about 11′ apart. It makes a interesting pair with NGC 877, lying nearly the same distance north of the eastern 8th magnitude star. NGC 871 has a fairly faint halo elongated 45″ × 20″ N–S.

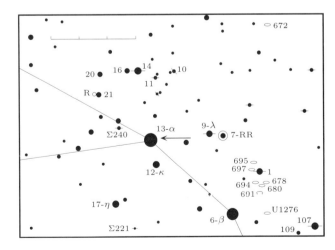

Finder Chart 4-3. 13–α Ari: 02ʰ07.2ᵐ +23°28′

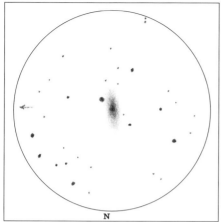 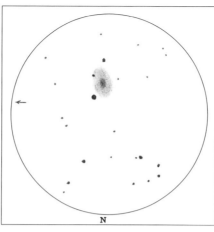 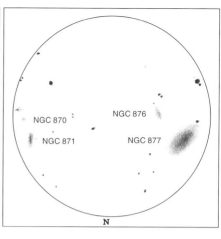

Figure 4-5. NGC 803
17.5", f4.5–250x, by G. R. Kepple

Figure 4-6. NGC 821
17.5", f4.5–250x, by G. R. Kepple

Figure 4-7. NGC 877 Group
17.5", f4.5–175x, by Steve Coe

16/18" Scopes–150x: NGC 871 is fairly faint and elongated 1′ × 0.25′ N–S with a long, thin core running almost the length of the major axis. A 13th magnitude star lies along the east flank 20″ from center. A pair of widely separated 12th magnitude stars is 2.5′ SSW. NGC 870, 1.5′ SSW of NGC 871, is just detectable with averted vision as a faint blur.

NGC 877 H246² Galaxy Type SAB(rs)bc I-II
⌀2.1′ × 1.7′, m11.9v, SB13.1   02ʰ18.0ᵐ +14°33′
Finder Chart 4-4, Figure 4-7   ★★★

8/10" Scopes–100x: 1.25° SE of magnitude 5.7 19 Arietis is a pair of 8th magnitude stars aligned E–W, each with a faint galaxy pair 5′ to its north. NGC 877, NNW of the easternmost 8th magnitude star, is accompanied by the tiny faint galaxy NGC 876 2′ SW. 12′ west of NGC 877, 5′ north of the westernmost 8th magnitude star, is NGC 871 and its faint

companion galaxy NGC 870. NGC 877 is faint and elongated 1.5′ × 0.75′ NW–SE with uniform surface brightness.

16/18" Scopes–150x: NGC 877 has a fairly bright halo elongated 2′ × 1.5′ NW–SE with a uniform surface brightness except for a brighter streak running along the center of the major axis. The NW extension seems longer. A 13.5 magnitude star is off the galaxy's SE tip 1′ from its center. A pair of 14th magnitude stars separated by 20″ lies 3′ SSW. 1.5′ north of this star-pair is NGC 877's companion, NGC 876, a very faint streak elongated 1′ × 0.25′ NNE–SSW.

NGC 918 Galaxy Type SAB(rs)c:
⌀3.3′ × 1.9′, m12.2v, SB 14.0   02ʰ25.9ᵐ +18°30′
Finder Chart 4-4, Figure 4-8   ★

16/18" Scopes–150x: NGC 918, 3′ NNW of a 10.5 magnitude star, has a diffuse, oval-shaped, very low surface brightness halo elongated 2′ × 1.25′ NNW–SSE with a slight brightening at its center. A very faint knot or star is embedded just NW of the galaxy's center.

NGC 927 Galaxy Type SB(r)c I
⌀1.2′ × 1.2′, m13.4v, SB 13.7   02ʰ26.6ᵐ +12°09′
Finder Chart 4-4   ★

16/18" Scopes–150x: NGC 927, 18′ NE of an 8th magnitude star, has a very faint 45″ diameter halo with moderate brightening at its center. A 13.5 magnitude star touches the north edge while a star of similar brightness is visible 2′ WSW. A small but conspicuous Y-shaped asterism of 12th and 13th magnitude stars with its stem pointing toward the galaxy lies 7′ south. A 9th magnitude star is 9′ SE.

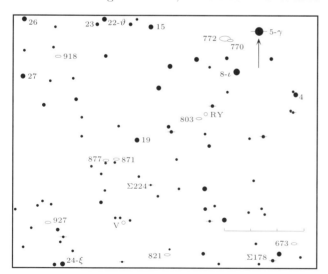

Finder Chart 4-4. 5–γ Ari: 01h53.5m +19°18′

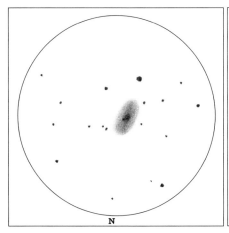
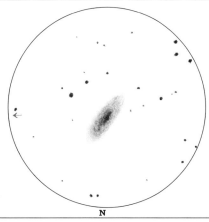
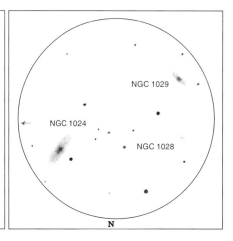

Figure 4-8. NGC 918
17.5", f4.5–250x, by G. R. Kepple

Figure 4-9. NGC 972
20", f4.5–175x, by Richard W. Jakiel

Figure 4-10. NGC 1024, 1028, & 1029
20", f4.5–175x, by Richard W. Jakiel

NGC 972 H211² Galaxy Type Sab
ø3.4' × 1.6', m11.4v, SB 13.1 $02^h34.2^m$ +29°19'
Finder Chart 4-6, Figure 4-9   ★★★★

8/10" Scopes–150x: This galaxy has a fairly faint halo elongated 1.5' × 0.75' NNW–SSE with some brightening at the center. To the SW three stars form a row, each star toward the NW being brighter than the next.

16/18" Scopes–150x: NGC 972 is a fine oval-shaped object! It has a bright halo elongated 2.5' × 1' NNW–SSE with a thin mottled core and a faint stellar nucleus. At 200x, faint wispy plumes are visible extending from a bar-like center. The star field to the south is well sprinkled with a mixture of both bright and faint stars, but the field to the north is barren.

NGC 1012 H152³ Galaxy Type S0/a?
ø2.7' × 1.4', m12.0v, SB 13.3 $02^h39.3^m$ +30°09'
Finder Chart 4-6   ★★

16/18" Scopes–150x: NGC 1012, 22' NNW a wide double of 7th and 9th magnitude stars, appears faint and elongated 1.5' × 0.5' NNE-SSW with a bright center. A 13th magnitude star touches the halo's edge 20" SE of the galaxy's center and a 14th magnitude star is embedded in the SSW tip 35" from the center.

NGC 1024 H592² Galaxy Type (R')SA(r)ab III
ø4.4' × 1.6', m12.1v, SB 14.0 $02^h39.2^m$ +10°51'
Finder Chart 4-5, Figure 4-10   ★

12/14" Scopes–125x: NGC 1024 is found 1.5° WNW of 4.3 magnitude star Mu (μ) – 87 Ceti and 12' north of a 7th magnitude star. It has a faint halo elongated 2' × 1' NNW–SSE with a some brightening at the center. An 11th magnitude star is 45" NE.

16/18" Scopes–150x: NGC 1024 has a 30" diameter core surrounded by a very tenuous halo extended 3' × 1' NNW–SSE. A 13th magnitude star does not quite touch the SSE tip. Lying 7' SE of NGC 1024, NGC 1029 is visible as a faint streak elongated 1' × 0.25' ENE–WSW with a 13.5 magnitude star lying beyond the tip 1' ENE of center. 3' north of NGC 1029 and 6' west of NGC 1024, is NGC 1028, visible with averted vision as a hazy smudge.

NGC 1036 H475³ Galaxy Type Pec?
ø1.5' × 1.0', m13.2v, SB13.5 $02^h40.5^m$ +19°18'
Finder Chart 4-5   ★

16/18" Scopes–150x: NGC 1036 is located between two 7th magnitude stars which form a triangle with a 6th

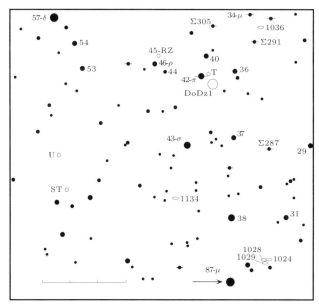

Finder Chart 4-5. 87–μ Cet: 02h44.9m +10°07'

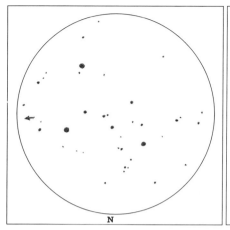 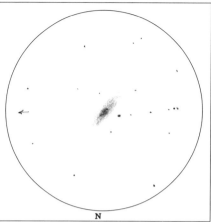 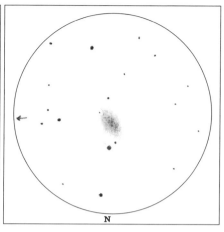

Figure 4-11. DoDz 1
12.5", f5–200x, by G. R. Kepple

Figure 4-12. NGC 1134
17.5", f4.5–250x, by G. R. Kepple

Figure 4-13. NGC 1156
17.5", f4.5–250x, by G. R. Kepple

magnitude star to the NE. This galaxy is visible as a faint, circular 1' diameter haze slightly elongated N–S with a slight central brightening.

Dolidze-Dzimselejsvili 1   Open Cl. 12★   Tr Type III 2 p
⌀12', m –, [Br★ 9:]   02ʰ47.4ᵐ   +17°12'
Finder Chart 4-5, Figure 4-11   ★★

12/14" Scopes–100x: This loose open cluster is found 0.5° SW of 5.2 magnitude Pi (π) = 42 Arietis. At low power it appears fairly bright and surprisingly well concentrated for an object not listed in the NGC catalog. The cluster is triangular, its the two brightest stars, 9th magnitude objects, lying along the western edge. The majority of its 30 stars are spread E–W along the northern side in a stream measuring 12' × 5'. A 10th magnitude star is in the NE portion, and three 11th magnitude stars are near the center.

NGC 1134   H254²   Galaxy   Type S
⌀2.3' × 0.8', m12.1v, SB 12.6   02ʰ53.6ᵐ   +13°00'
Finder Chart 4-5, Figure 4-12   ★★

16/18" Scopes–150x: NGC 1134 may be found 15' WNW of the 8th magnitude star at the southern angle of a large 30' triangle of 8th magnitude stars. It has a faint, diffuse halo elongated 2' × 0.75' NNW–SSE with a poorly concentrated 30" diameter core. A 13th magnitude star lying 50" NE of center is the beginning of a string of faint stars trailing eastward.

NGC 1156   H619²   Galaxy   Type IB(s)m IV-V
⌀3.4' × 2.8', m11.7v, SB 14.0   02ʰ59.7ᵐ   +25°14'
Finder Chart 4-6, Figure 4-13   ★★★

8/10" Scopes–100x: Centered 1' south of an 11.5 magnitude star, this galaxy has a faint, diffuse disk elongated 1.5' × 0.75' NE–SW.

12/14" Scopes–125x: NGC 1156 appears fairly faint and elongated 2' × 1' NNE–SSW. The envelope is somewhat mottled but generally uniform in brightness.

16/18" Scopes–150x: The halo is fairly faint and elongated 2.5' × 1.5' NNE–SSW with a more concentrated 1' × 0.5' central area. The central portion is uneven in brightness with several spots and a slightly darker area in the southern portion. A threshold star is embedded in the halo 45" NNE of center, and a 13th magnitude star lies 1' south of center.

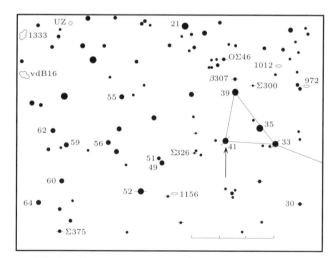

Finder Chart 4-6.   41 Ari: 02ʰ49.2ᵐ +27°16'

Chapter 5

Auriga, the Charioteer

5.1 Overview

Auriga is known as the Charioteer, but ancient Greek and Roman descriptions of the constellation do not mention his chariot. The ancient Babylonians, on the other hand, saw these stars as a chariot but, as far as we know, it had no charioteer. The name of Auriga's brightest star, Capella, literally means "she-goat," her kids being the stars Zeta and Eta Aurigae. One classical myth identifies Auriga with Erichthonius, the fourth of the early kings of Athens, whose lameness inspired him to invent the chariot.

Auriga is a bright pentagon-shaped constellation lying in a rich, interesting section of the winter Milky Way. The constellation has a wide variety of open star clusters and nebulae. Three very fine open clusters are part of Messier's famous catalog: M36, M37, and M38. Many fainter star clusters abound, hidden like gemstones among the Milky Way's quarry of stars.

Auriga: aw-REE-ga
Genitive: Aurigae, aw-RYE-je
Abbrevation: Aur
Opposition: 9pm–Sep. 20, midnight–Aug 5
Culmination: 9pm–Feb. 4, midnight–Dec. 21
Area: 657 square degrees
Showpieces: 14 Aur, 41 Aur, M36 (NGC 1960), M37 (NGC 2099), M38 (NGC 1912)
Binocular Objects: Cr 62, M36 (NGC 1960), M37 (NGC 2099), M38 (NGC 1912) NGC 1664, NGC 1778, NGC 1857, NGC1893, NGC 1907, NGC 2281, Stock 10

5.2 Interesting Stars

Omega (ω) = 4 Aurigae (Σ616) Double Star Spec. A0
m5.0,8.0; Sep. 5.4″; P.A. 359° $04^h59.3^m$ +37°53′
Finder Chart 5-5   ★★★★

2/3″ Scopes–100x: In smaller instruments the stars of Omega Aurigae appear white and bluish.

5 Aurigae = OΣ92 Double Star   Spec. F5
m6.0, 9.7; Sep. 3.7″; P.A. 265° $05^h00.3^m$ +39°24′
Finder Chart 5-4   ★★★★

12/14″ Scopes–200x: Moderately high power is needed for a clean split of this double. Its components are both yellow and very unequal in magnitude.

Σ644 Double Star   Spec. B2
AB: m6.7,7.0; Sep. 1.6″; P.A. 221° $05^h10.3^m$ +37°18′
Finder Chart 5-5   ★★★★

12/14″ Scopes–200x: Σ644 appears single at 100x and notched at 200x, a split occurring in moments of good seeing. The stars are yellowish and light blue.

14 Aurigae = Σ653 Triple Star   Spec. A2
AB: 5.1,7.4; Sep. 14.6″; P.A. 352° $05^h15.4^m$ +32°31′
Finder Chart 5-5   ★★★★

2/3″ Scopes–100x: The two brightest stars appear pale yellow and blue.

12/14″ Scopes–100x: The primary is yellowish while the brighter companion is a pale blue. These three stars form a crooked row in a sparse star field.

AE Aurigae Variable Star   Spec. O9
m5.4 to 6.1, Period Irr. $05^h16.3^m$ +34°19′
Finder Chart 5-5   ★★★

Located some 1,500 light-years away, AE Aurigae is an unusual O-type star with irregular light variations. Its average absolute magnitude is −5, a luminosity of nearly 10,000 Suns. The star illuminates IC 405, the "Flaming Star Nebula," a turbulent cloud of gas and dust some 9 light-years in extent. The connection between the star and the nebula is a chance encounter: AE Aurigae has only recently entered the nebula, and photographs suggest that the star has swept clear the nebulosity in its path.

AE Aurigae is one of three "Runaway Stars" receding from the Orion Association of O- and B-type stars, the other two Orion escapees being 53 Arietis and Mu Columbae. Different theories have been suggested as the

Auriga, The Charioteer

Constellation Chart 5-1

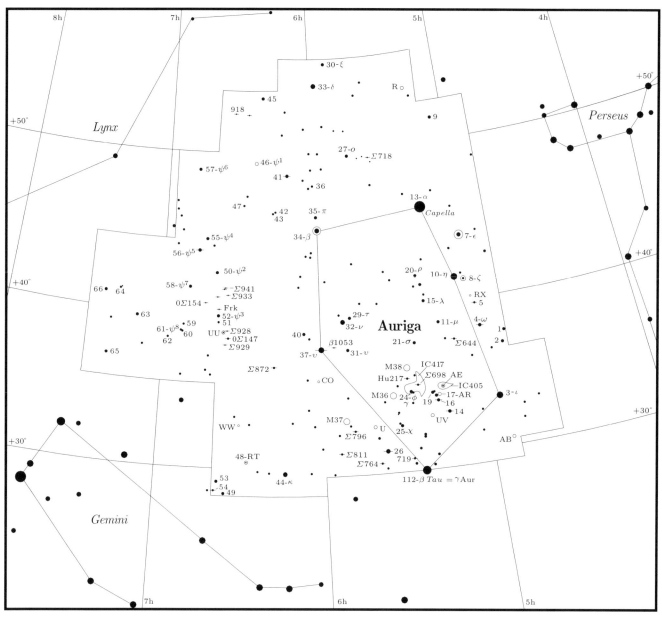

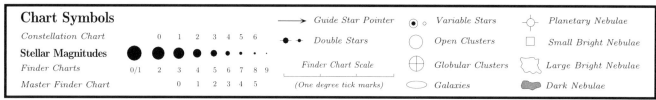

Table 5–1. Selected Variable Stars in Auriga

Name	HD No.	Type	Max.	Min.	Period (Days)	F*	Spec. Type	R.A. (2000)	Dec.	Finder Chart No. & Notes
AB Aur	31293	Ina	6.9	8.4			B9	04^{h}55.8^m	+30°33′	5-3 In vdB31
RX Aur	31913	Cδ	7.2	8.0	11.62	0.49	F6-G2	05^{h}01.4^m	+39 58	5-4
7–ε Aur	31964	EA/GS	2.9	3.8	9892.00	0.08	A8-F2	02.0	+43 49	5-4
AE Aur	34078	Ina	5.4	6.1			O9	16.3	+34 19	5-5 In IC405
R Aur	34019	M	6.7	13.9	457.51	0.51	M6-M9	17.3	+53 35	5-1
17–AR Aur	34364	EA/DM	6.1	6.8	4.13	0.07	A-B9	18.3	+33 46	5-5
UV Aur	19737	M	7.4	10.6	394.42		C6-C8	21.8	+32 31	5-5 Burnt-orange, blue
U Aur	37724	M	7.5	15.5	408.09	0.39	M7-M9	42.1	+32 02	5-6
CO Aur	40457	SRd	7.4	8.0	39.00		F5	06^{h}00.5^m	+35 19	5-6
46–ψ^1 Aur	44537	Lc	4.7	5.0			K5-M0	24.9	+49 17	5-7 Lovely orange star
48–RT Aur	45412	Cδ	5.0	5.8	3.72	0.25	F4-G1	28.6	+30 30	5-1 Classic Cepheid Var.
WW Aur	46052	EA/DM	5.7	6.5	2.52	0.10	A3+A3	32.5	+32 27	5-1

F* = The fraction of period taken up by the star's rise from min. to max. brightness, or the period spent in eclipse.

Table 5–2. Selected Double Stars in Auriga

Name	ADS No.	Pair	M1	M2	Sep."	P.A.°	Spec	R.A. (2000)	Dec	Finder Chart No. & Notes
4–ω Aur	3572		5.0	8.0	5.4	359	A0	04^{h}59.3^m	+37°53′	5-5 Pale yellow & orange
5 Aur	3589		6.0	9.7	3.7	265	F5	05^{h}00.3^m	+39 24	5-4 Unequal yellow stars
Σ644	3734	AB	6.7	7.0	1.6	221	B2	10.3	+37 18	5-5 Yellowish & light blue
	3734	AC		9.3	72.6	15				
14 Aur	3824	AB	5.1	11.1	11.1	352	A2	15.4	+32 31	5-5 Yellowish & blue
(Σ653)	3824	AC		7.4	14.6	226				
	3824	AD		10.4	184.0	321				
Σ698	4000		6.6	8.7	31.2	346	K0 K	25.2	+34 51	5-5 Yellowish & blue
Hu 217	4072		6.8	8.3	0.6	251	B5	29.7	+35 23	5-5 LY Aur
Σ719	4086	AB	7.3	9.8	1.0	330	G5	30.1	+29 33	5-3
	4086	AC		9.3	15.0	351				
Σ718	4119	AB	7.5	7.5	7.7	74	F5	32.4	+49 24	5-1 A-B: Elegant white pair
	4119	AC		9.2	119.4	185				
26 Aur	4229	AB	6.0	6.3	w 0.2	*03	A2	38.6	+30 30	5-6 Yellowish & blue
Σ753	4229	ABxC		8.0	12.4	267				
β90	4229	ABxD		11.5	33.1	114				
Σ764	4262		6.9	7.5	26.0	14	A A	41.3	+29 29	5-6
Σ796	4421	AB	7.0	8.1	3.8	62	A3	49.9	+31 47	5-6
	4421	AC		10.5	207.5	324				
β1053	4472		6.9	8.9	1.4	350	F5	53.5	+37 20	5-1
Σ811	4483		7.7	9.2	5.0	231	B5	54.3	+30 30	5-6
37–υ Aur	4566		2.6	7.1	3.6	313	A0	59.7	+37 13	5-1 White & bluish
	4566	AC		10.6	50.0	297				
41 Aur	4773		6.3	7.0	7.7	356	A0	06^{h}11.6^m	+48 43	5-7 White & Lilac
Σ872	4849	AB	6.9	7.9	11.3	217	F0	15.6	+36 09	5-1 Yellow & Lilac
Σ918	5178	AB	7.2	8.2	4.7	333	A3	34.0	+52 28	5-1
OΣ147	5188	AB	6.6	10.0	43.2	73	K0	34.3	+38 05	5-1 Yellowish primary with two bluish companions
	5188	AxCD			46.3	117				
	5188	CD	10.6	11.0	0.5	109				
Σ928	5191	AB	7.6	8.2	3.5	133	F5	34.7	+38 32	5-1 Charming white stars
	5191	BC		11.0	128.6	123				
Σ929	5208		7.2	8.3	6.0	25	G5	35.4	+37 43	5-1 Nice! Yellowish & blue
Σ933	5233		8.4	8.9	25.6	75	A2 A2	36.8	+41 08	5-7 Easy duo of white stars
Frk			7.7	9.4	48.9	253	K0	38.6	+40 20	5-7 Yellowish primary with two bluish companions
				9.2	61.5	90				
Σ941	5269	AB	7.2	8.2	2.0	81	B9	38.7	+41 35	5-7 White primary with two bluish companions
	5269	AC		10.2	82.8	134				
54 Aur	5289		6.0	7.8	0.9	36	B8	39.6	+28 16	5-1
OΣ154	5379		6.9	9.4	23.8	110	M	44.3	+40 37	5-1
56–ψ^5 Aur	5425		5.3	8.3	36.2	31	G0	46.7	+43 35	5-6 Fine yellow & blue pair

Footnotes: *= Year 2000, a = Near apogee, c = Closing, w = Widening. Finder Chart No: All stars listed in the tables are plotted in the large Constellation Chart, but when a star appears in a Finder Chart, this number is listed. Notes: When colors are subtle, the suffix -ish is used, e.g. *bluish*.

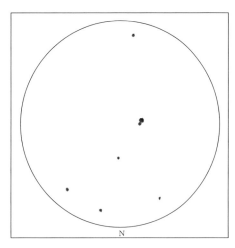

Figure 5-1. 41 Aurigae
3.2″, f11–60x, by Mark T. Stauffer

cause of these stars' "escape." Perhaps a binary companion went supernova, and the explosion freed the stars. With abundant numbers of massive and rapidly evolving stars in the region, supernovae in the Orion complex are highly likely.

UV Aurigae Variable & Double Star   Spec. C6-C8
m7.4 to 10.6 in 394.4 days $05^h21.8^m$ +32°31′
Finder Chart 5-5   ★★★★

12/14″ Scopes–175x: UV Aurigae is a variable carbon star paired with an 11.5 magnitude B-type giant star 3.4″ away in P.A. 4°. This gorgeous double of burnt-orange and blue stars is well worth finding.

Σ698 Double Star   Spec. K0 & K
m6.6, 8.7; Sep. 31.2″; P.A. 345° $05^h25.2^m$ +34°51′
Finder Chart 5-5   ★★★★

2/3″ Scopes–75x: This luminary is a nice, easy target for small scopes. Low power shows a wide double of yellowish-orange components in a beautiful star field.

26 Aurigae = Σ1240 Double Star   Spec. A2
ABxC: m6.0, 8.0; Sep. 12.4″; P.A. 267° $05^h38.6^m$
+30°30′
Finder Chart 5-6   ★★★★

2/3″ Scopes–100x: The A–B pair is much too close (0.2″) for separation in amateur instruments, but ABxC is an easy, wide double of unequal yellowish and blue stars.

Theta (ϑ) = 37 Aurigae (OΣ545) Double Star Spec. A0
AB: m2.6, 7.1; Sep. 3.6″; P.A. 313° $05^h59.7^m$ +37°13′
Constellation Chart 5-1   ★★★★

Theta Aurigae is a "silicon star" with strong lines of that element in its spectrum. It is a binary with a primary 80 times as luminous as the Sun and a companion nearly equal to our Sun in luminosity and spectral type. The pair has an orbital period of seven or eight centuries. A second companion 52″ away is not a true physical member of the system: its separation is increasing due to Theta's proper motion. Theta Aurigae lies over 80 light years from us.

2/3″ Scopes–150x: This is a very difficult binary. The secondary may just be glimpsed within the primary's diffraction ring. Both stars appear bluish-white.

12/14″ Scopes–175x: Theta appears elongated at low power and requires a magnification of about 175x to be split. The stars are white and light blue.

41 Aurigae = Σ845 Double Star   Spec. A0
m6.3, 7.0; Sep. 7.7″; P.A. 356° $06^h11.6^m$ +48°43′
Finder Chart 5-7, Figure 5-1   ★★★★

2/3″ Scopes–100x: 41 Aurigae is a lovely white and blue-white pair. 3° ENE is 46 Aurigae, a beautiful orange colored variable star.

12/14″ Scopes–125x: This double has a white primary with a lilac attendant.

RT Aurigae Variable Star   Spec. F4–G1
m5.0 to 5.8 in 3.72 days $06^h28.6^m$ +30°30′
Constellation Chart 5-1   ★★★★

RT is a classical Cepheid with a light variation of 5.0 to 5.8 magnitude in 3.72 days. The rise to maximum takes 1.5 days and the decline back to minimum about 2.5 days. The variations seem to be due to actual stellar pulsation. RT is a supergiant with a luminosity of about 2,300 suns, an absolute visual magnitude of −3.1, at maximum. During the cycle its spectral class changes from F4 to G1.

There is a relationship between the periods and luminosities of Cepheid variables: the brighter the Cepheid, the longer it takes for the star to complete its cycle. The true luminosities of distant Cepheids may be calculated from their periods, thus allowing the distance to be inferred. RT Aurigae is estimated about 1,600 light years away.

Σ872 Double Star   Spec. F0
AB: m6.9, 7.9; Sep. 11.3″; R.A. 217° $06^h15.6^m$ +36°09′
Constellation Chart 5-1   ★★★★

12/14″ Scopes–100x: This splendid pair has a fine contrast of yellow and lilac stars.

OΣ147   Quadruple Star   Spec. K0
AB: m6.6, 10.0; Sep. 43.2″; P.A. 73°
AC: –   10.6; Sep. 46.3″; P.A. 117°
$06^h34.3^m$ +38°05′
Constellation Chart 5-1   ★★★★

2/3″ Scopes–50x: Easy at low power, this beautiful trio is an isosceles triangle of a yellowish primary with bluish companions. The C component is also a close 0.5″ pair.

Σ928   Double Star   Spec. F5
m7.6, 8.2; Sep. 3.5″; P.A. 133°
$06^h34.7^m$ +38°32′
Constellation Chart 5-1   ★ ★ ★ ★

2/3″ Scopes–125x: This is a charming pair of nearly equally bright white stars.

Σ929   Double Star   Spec. G5
m7.2, 8.3; Sep. 6.0″; P.A. 25°
$06^h35.4^m$ +37°43′
Constellation Chart 5-1   ★★★★

2/3″ Scopes–100x: Struve 929 is a very nice pair of yellowish-white and bluish stars!

Psi-5 (ψ^5) = 56 (S, h75) Aurigae Dbl. Star Spec. G0
m5.3, 8.3; Sep. 36.2″; P.A. 31° $06^h46.7^m$ +43°35′
Finder Chart 5-7   ★★★★

2/3″ Scopes–50x: Low power shows a fine contrast of yellow and blue stars in a nice field.

5.3 Deep-Sky Objects

NGC 1664 H59[8] Open Cluster Tr Type III 1 p
⌀18.0′, m7.6, Br★ 10.5 $04^h51.1^m$ +43°42′
Finder Chart 5-4, Figures 5-2 & 5-4   ★★★

8/10″ Scopes–100x: This cluster, which has thirty members visible at these apertures, is fairly bright and large but not rich and compressed. A 7th magnitude star is on the south side from which a nice chain of rather faint stars leads NW toward the cluster's center.

12/14″ Scopes–125x: This cluster appears fairly bright, large, loose, and irregular with 50 stars, including eight of the 10th and eighteen of the 11th magnitude. The area is rich and the cluster's outlying stars blend into the surrounding Milky Way. At low power the cluster may be spotted readily as an enrichment in the background star field. An irregular star chain forming a hook runs through the cluster. A 7th magnitude star at the hook's SE end

Figure 5-2. *NGC 1664 is large and loose but may be readily spotted at low power as an enrichment in the surrounding star field. Lee C. Coombs made this 10 minute exposure on 103a-O film with a 10″, f5 Newtonian reflector.*

is probably not a true cluster member. There are many rows and arcs of three or four stars each.

Barnard 26-28   Dark Nebula
⌀20′, Opacity 6 $04^h55.2^m$ +30°35′
Finder Chart 5-3, Figure 5-3   ★★★★

16/18″ Scopes–75x: This dark nebula is visible as an irregular Z-shaped starless patch NW of variable star AB Aurigae. It is silhouetted by both a rich star field and the very faint reflection nebula vdB31.

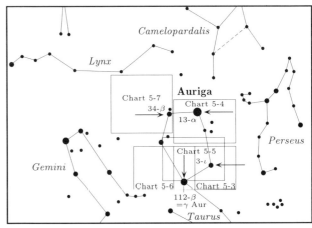

Master Finder Chart 5-2. Auriga Chart Areas
Guide stars indicated by arrows

Figure 5-3. *Dark Nebulae B26-28 stand out against the very faint reflection nebula vdB31 to the NW of variable star AB Aurigae, which interferes with observation. Martin C. Germano made this 70 minute photo on 2415 film with an 8", f5 Newtonian.*

van den Bergh 31 Reflection Nebula
ø8' × 5', Photo Br 3-5, Color 1-4   04^{h}55.7^m +30°33'
Finder Chart 5-3 ★

16/18" Scopes–125x: This difficult reflection nebula is visible as a diffuse, hazy area around the 6.8 magnitude variable star AB Aurigae. The dark patch of Barnard 26-28 stands out NW of the star.

Czernik 19 Open Cluster 50★ Tr. Type III 2 m
ø18', m –   04^{h}57.0^m +28°47'
Finder Chart 5-3 ★★

16/18" Scopes–125x: This is a large, faint, loose, irregular group of 15 stars without central concentration. It has four 12th magnitude stars, two of which are on the NE edge; four 13th magnitude stars; and eleven 13–14 magnitude stars. The cluster is flanked by brighter stars on both the preceding and following sides.

Dolidze 15 Open Cluster (Asterism?) Tr. Type IV 1 p
ø18', m –   05^{h}04.6^m +34°50'
Finder Chart 5-3 ★

12/14" Scopes–75x: This cluster is a faint, large, poor scattering of stars.

Barnard 29 Dark Nebula
ø10', Opacity 6   05^{h}06.2^m +31°44'
Finder Chart 5-3 ★★★★

16/18" Scopes–75x: Barnard 29 is a prominent dark streak about two degrees south following Z Aurigae.

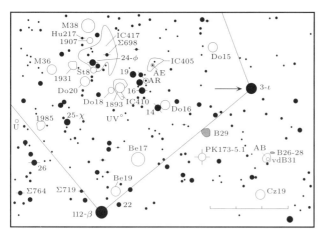

Finder Chart 5-3. 3-ι Aur: 04h56.9m +33°10'

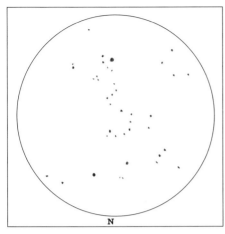
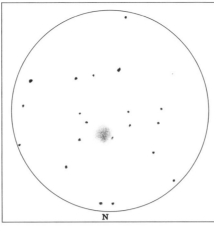
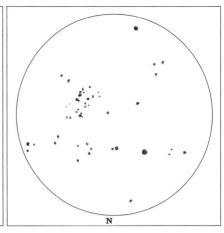

Figure 5-4. NGC 1664
13", f4.5–100x, by Tom Polakis

Figure 5-5. PK173-5.1
13", f5.6–100x, by Steve Coe

Figure 5-6. NGC 1778
8", f7–165x, by G. R. Kepple

PK173-5.1 Planetary Nebula   Type 3
ø132", m –, CS 18.2p   05^{h}08.1^m +30°48'
Finder Chart 5-3, Figure 5-5   ★

16/18" Scopes–100x: With an O-III filter, this large, round planetary appears very faint and diffuse.

NGC 1778   H61[8]   Open Cluster 25★   Tr Type III 2 p
ø6', m7.7v, Br★ 10.06v   05^{h}08.1^m +37°03'
Finder Chart 5-5, Figures 5-6 & 5-7   ★★★

8/10" Scopes–100x: This cluster lies half a degree SW of the westernmost star in an E–W row with 6th magnitude stars at both ends. NGC 1778 is a NW–SE scattering of 15 stars.

12/14" Scopes–150x: Barely noticeable against the Milky Way, this cluster is poor and uncompressed with eighteen 10th to 14th magnitude stars in scattered clumps.

16/18" Scopes–150x: Medium power shows a fairly bright cluster of thirty 10th magnitude and fainter stars extended 6' × 3' N–S. A starless streak running NW–SE divides the cluster into two groups. The northern group has more stars and most of the brighter members. A triple consisting of two 10th magnitude and one 13th magnitude star is located at the northern end. There are many star chains in the area, the most prominent one to the north. A large asterism of faint stars forming a figure-8 may be seen in the star field ENE of the cluster.

NGC 1798   Open Cluster 50★   Tr Type I 1 m
ø5', m –, Br★ 13.0p   05^{h}11.6^m +47°40'
Finder Chart 5-4   ★★

12/14" Scopes–125x: This cluster is a faint, hazy spot 5' across with a slight E–W elongation. At 125x fifteen 13.5 magnitude and fainter stars are visible. A 12th magnitude star is at the east end.

16/18" Scopes–100x: NGC 1798 is a faint, moderately compressed cluster of sixteen 13th magnitude and fainter stars in an 5' area, with a lone 12th magnitude star at the eastern edge. A 5th magnitude star lies 45' north while an 8.5 magnitude star is an equal distance to the SW. A small but nice Y-shaped asterism of 11th and 12th magnitude stars lies just beyond the 8.5 magnitude star.

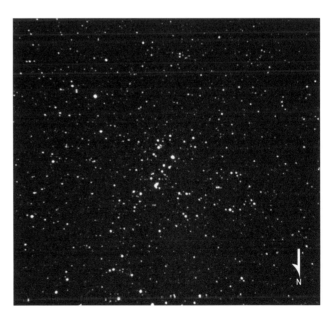

Figure 5-7. *NGC 1778 is a fairly bright cluster of several dozen 10th magnitude and fainter stars. This 10 minute exposure was made on 103a-O film with a 10", f5 Newtonian reflector by Lee C. Coombs.*

Figure 5-8. *Use an O-III or UHC filter to detect IC405, the Flaming Star Nebula, an extremely faint glow fanning north of AE Aurigae. This 45 minute photo was made on hypered 2415 film with an 8″, f5 Newtonian reflector by Martin C. Germano.*

Dolidze 16 Open Cluster 10★ Tr Type III 2 p n
ø12′, m−   05ʰ14.6ᵐ +32°43′
Finder Chart 5-5   ★★

12/14″ Scopes–100x: This cluster is faint, loose and poor, with less than a dozen stars in a 12′ area.

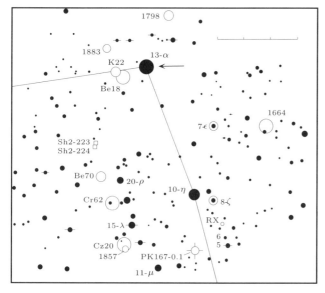

Finder Chart 5-4. 13-α Aur: 05ʰ16.7ᵐ +46°00′

IC 405 Reflection & Emission Nebula
ø30′ × 20′, Photo Br 2-5, Color 2-4   05ʰ16.2ᵐ +34°16′
Finder Chart 5-5, Figure 5-8   Flaming Star Nebula★

12/14″ Scopes–75x: A UHC filter at 75x reveals a huge, extremely faint glow fanning north of AE Aurigae, the nebula brightest around the star.

16/18″ Scopes–75x: An O-III filter and averted vision reveals an extremely faint, marginally perceptible, nebula around, and extending north of, AE Aurigae.

NGC 1857 H33⁷ Open Cluster 40★ Tr Type II 2 m
ø5′, m7.0v, Br★ 7.38v   05ʰ20.2ᵐ
+39°21′
Finder Chart 5-4, Figure 5-9   ★★★★

8/10″ Scopes–100x: This is a lovely cluster of three dozen stars concentrated in a 5′ area. There are two 7.5–8th magnitude stars aligned NW-SE of each other, two 9th magnitude stars in a NNE–SSW position angle, and thirty stars ranging from 10th to 14th magnitude. The field is extremely rich, most of its brighter stars lying north of the cluster. Dark starless voids are visible east, south and west of the cluster.

12/14" Scopes–125x: A nice, rich cluster! At least sixty stars are resolved in a 9′ area. It is mildly concentrated with two 7th and 8th magnitude, two 9th magnitude, and a host of 10th to 14th magnitude stars. Several star chains radiate from the cluster.

Berkeley 17   Open Cl. 100★   Tr Type III 1 r
⌀13′, m –, Br★ 16.0p   05^{h}20.6^m +30°36′
Finder Chart 5-3   ★★

16/18" Scopes–125x: Berkeley 17 is a large, faint unresolved haze at 125x. 175x reveals a dozen or so minute cluster members. Preceding Be 17 is a triangle of stars slightly brighter than those of the cluster.

Berkeley 18   Open Cl. 300★   Tr Type II 1 r
⌀20′, m –, Br★ 16.0p   05^{h}22.2^m +45°24′
Finder Chart 5-4   ★★

12/14" Scopes–125x: Berkeley 18 is very faint, quite large and rich but without resolution. The cluster is connected to open cluster King 22; both are misty patches with Berkeley 18 a little more concentrated than its neighbor.

16/18" Scopes–125x: This cluster forms a larger assemblage connected with open cluster King 22 to the east. At 175x Berkeley 18 shows a hundred stars with a slight condensation at each end. Both clusters contain extremely faint 15th to 16th magnitude stars.

Collinder 62   Open Cluster   Tr Type IV 3 p
⌀28′, m4.2p   05^{h}22.5^m +41°00′
Finder Chart 5-4   ★★

16/18" Scopes–75x: Viewed at low power, this is a large, rather poor, uncompressed cluster of about twenty 9th to 12th magnitude stars distributed over a 30′ area.

IC410   Emission Nebula
⌀40′ × 30′, Photo Br 2-5, Color 3-4   05^{h}22.6^m +33°31′
Finder Chart 5-5, Figure 5-10   ★★

12/14" Scopes–75x: Embedded in open cluster NGC 1893, this nebula is faint, large, irregular and very diffuse at 75x without a filter. With an Oxygen-III filter IC 410, though still rather faint, is more impressive, appearing as a C-shaped glow, the gap in the "C" opening to the south. The nebula is more obvious at the NW edge of the cluster. There are many faint notches and extensions surrounded by fainter nebulosity spanning half a degree.

Figure 5-9. *NGC 1857 is a nice, moderately rich cluster that stands out well at low to medium powers. Lee C. Coombs made this 10 minute photo on 103a-O film at prime focus with a 10", f5 Newtonian reflector.*

NGC 1893   Open Cluster 60★   Tr Type II 2 m n
⌀12′, m7.5v, Br★ 9.31   05^{h}22.7^m +33°24′
Finder Chart 5-5, Figure 5-10   ★★★

8/10" Scopes–100x: Located within a triangle of three 8th magnitude stars in a rich Milky Way field, NGC 1893 has fifteen 9th to 11th magnitude stars visible in a 10′ area.

12/14" Scopes–100x: Medium-sized instruments show forty faint stars elongated N–S lying within a triangle of 9th magnitude stars. The cluster is set in a very rich star field and is also embedded in the nebulosity of IC 410. Without a filter the nebulosity is very difficult.

16/18" Scopes–100x: Fairly bright at low power, this large 15′ x 8′ cluster has fifty 9th to 13th magnitude stars elongated N–S. The greatest concentration, located in the northern portion, is somewhat diamond-shaped with two extensions to the north and east adding another three dozen stars.

King 22   Open Cluster   Tr Type III 3 r
⌀14′, m –, Br★ 15.0p   05^{h}22.9^m +45°28′
Finder Chart 5-4   ★★

12/14" Scopes–100x: King 22 is a misty patch lying beside open cluster Berkeley 18 to the west; both are

Figure 5-10. *Open cluster NGC 1893 is embedded in the faint emission nebula IC410. An O-III or nebula filter will help detect its faint glow which is more obvious NW of the cluster. This 75 minute photo was made on hypered 2415 Kodak Tech Pan film with an 8″, f5 Newtonian reflector by Martin C. Germano.*

very faint, large, and rich but unresolved with an occasional twinkle from threshold stars. Berkeley 18 is a little more concentrated than King 22.

16/18″ Scopes–100x: Located nearly a degree SE of Capella, this large cluster of 15th to 16th magnitude stars is intertwined with open cluster Berkeley 18. At 175x, there are well over 100 stars visible with a slight compression toward the center.

Berkeley 19 Open Cluster 40★ Tr Type IV 2 p
ø6′, m11.4v, Br★ 14.7v   05^{h}24.1^m +29°36′
Finder Chart 5-3   ★★

16/18″ Scopes–150x: Medium power shows a faint, loose, circular cluster of 14th magnitude stars in a 6′ diameter area. On the SW side, the brightest stars form a triangular or arrowhead asterism. Do not mistake a brighter unnumbered group of stars lying to the SE for Berkeley 19: its stars are much too bright.

Dolidze 18 Open Cluster 15★ Tr Type IV 2 p n
ø12′, m –   05^{h}24.1^m +33°18′
Finder Chart 5-5, Figure 5-11   ★★

16/18″ Scopes–100x: This is a fairly large, moderately faint, loose, oval-shaped cluster located just SE of open cluster NGC 1893 and emission nebula IC 410. 100x shows 15 stars but higher power reveals several dozen fainter members. From the west, a row of four stars act as a pointer to the cluster.

Berkeley 70 Open Cluster 40★ Tr Type IV 3 m
ø12′, m –, Br★ 15.0p   05^{h}25.7^m +41°54′
Finder Chart 5-4   ★

16/18″ Scopes–100x: This cluster is an extremely faint, fairly large, hazy patch of unresolved stars that requires averted vision even on the best of nights. The stars are only 15th magnitude and fainter, requiring large instruments and transparent skies to be discerned.

NGC 1883 H34^7 Open cluster 30★ Tr Type II 1 p
ø3′, m12.0p, Br★ 14.0p   05^{h}25.9^m +46°33′
Finder Chart 5-4, Figure 5-12   ★★

12/14″ Scopes–150x: Easy to miss, NGC 1883 appears as a faint patch of stardust lying 3′ NNE of a 10.5 magnitude star. At 175x, it is faint, small, irregularly round and fairly compressed with half a dozen stars visible.

16/18″ Scopes–150x: At 150x, this cluster is faint, fairly small, and compressed with a dozen 14th magnitude

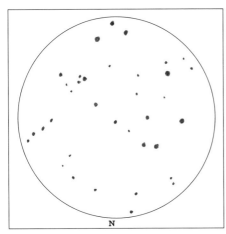
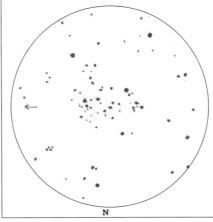
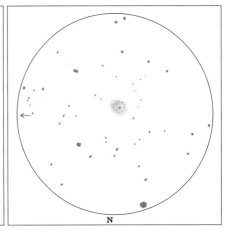

Figure 5-11. Dolidze 18
12.5", f5–70x, by G. R. Kepple

Figure 5-12. NGC 1883
12.5", f5–200x, by G. R. Kepple

Figure 5-13. IC 2149
13", f4.5–215x, by Tom Polakis

and fainter stars. Averted vision at 200x reveals 18 stars in an unresolved haze.

Sh2-224 Emission Nebula (SNR)
ø20′ × 3′, Photo Br 5-5, Color 4-4   05ʰ27.3ᵐ +42°59′
Finder Chart 5-4   ★

12/14" Scopes–100x: This nebula is found 15′ SW of a 7th magnitude star and 5′ SSE of an 8th magnitude star which should be kept out of the field when viewing the nebula. With a UHC filter at 100x, it is extremely faint and fairly large with two very elongated sections running E-W.

Stock 8 Open Cluster 40★ Tr Type I 2 p n
ø5′, m –, Br★ 9.0p   05ʰ28.1ᵐ +34°26′
Finder Chart 5-5   ★

12/14" Scopes–100x: Stock 8, lying 10′ ESE of 5.1 magnitude star Phi (φ) Aurigae, is a rich group of 40 faint stars involved with emission nebula IC 417. It is misplotted a little west of its true position. The cluster shows two strings of stars aligned NNW-SSE and a knot of stars just west of its center.

NGC 1907 Open Cluster 30★ Tr Type II 1 m n
ø6′, m8.2v, Br★ 11.26v   05ʰ28.0ᵐ +35°19′
Finder Chart 5-5, Figure 5-15   ★★★★

8/10" Scopes–100x: This is a very nice compact cluster making a fine contrast, with the much looser group Messier 38 to the north. NGC 1907 is a fairly rich and moderately compressed cluster of thirty 9th to 12th magnitude stars.

12/14" Scopes–125x: NGC 1907 is irregularly round with 30 stars of a wide range of magnitudes. Two 9.5 magnitude stars to the south do not seem to be part of the group because a small starless gap separates them from the main body.

16/18" Scopes–125x: This is a fairly bright, irregular cluster of forty 10th magnitude and fainter stars in a 6′ area. It is well concentrated toward the center, but there are outlying stars and protruding star chains in all directions. The most conspicuous chain, which includes an E-W pair of magnitude 9.5 stars, curves from the SW edge of the cluster toward the south and east. This nice cluster is overshadowed by M38 to the north.

IC 417 Emission Nebula
ø13′ × 10′, Photo Br 2-5, Color 3-4 05ʰ28.1ᵐ +34°26′
Finder Chart 5-6, Figure 5-14   ★

8/10" Scopes–100x: Requiring averted vision, IC 417 is a very faint haze involved with three 9th to 10th magnitude stars. The brightest portion is located 8′ SE of 5.5 magnitude Phi (φ) Aurigae. Both the UHC and O-III filters help considerably, but the sky must

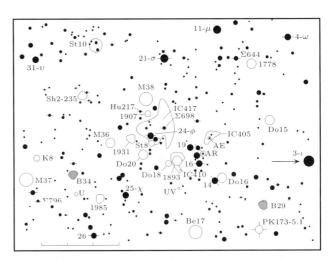

Finder Chart 5-5. 3-ι Aur: 04ʰ56.9ᵐ +33°10′

Figure 5-14. *Emission nebula IC 417, Lying just ESE 24-φ Aurigae, is embedded in open cluster Stock 8, a collection of forty faint stars. Martin C. Germano made this 60 minute exposure on hypered Kodak 2415 Tech Pan film with an 8″, f5 Newtonian.*

be very transparent.

16/18″ Scopes–125x: Viewed through an O-III filter, IC 417 is a faint nebula appearing roughly wedge-shaped with several indentations and faint wispy filaments embedded in a cluster of stars located ESE of Phi (φ)Aurigae. Because open cluster Stock 8 is plotted directly south of Phi Aurigae and there are no other clusters listed in this area, we can assume that the designation "IC 417" applies both to this nebulosity and to its involved open cluster.

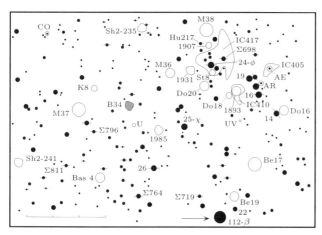

Finder Chart 5-6. 112-β Tau = γ Aur: $05^h26.3^m$ +28°36′

Dolidze 20 Open Cluster 10★ Tr Type IV 1 p
⌀12′, m – $05^h28.1^m$ +33°26′
Finder Chart 5-6   ★★

12/14″ Scopes–100x: Dolidze 20 is fairly large but loose and poor. The cluster is triangular with two 7th magnitude stars on its western side and a 9th magnitude star at the eastern apex. Many very faint stars are near each of the triangle's corners, and a tight gathering of stars lies beyond its SE apex.

NGC 1912 Messier 38 Open Cl. 100★ Tr Type III 2 m
⌀21′, m6.4v, Br★ 9.53v $05^h28.7^m$
+35°50′
Finder Chart 5-6, Figure 5-15   ★★★★★

M38 was discovered in 1749 by Le Gentil. It is a large, irregular, scattered group, its earliest member a class B5 star having an absolute magnitude of about -1.5. The cluster also contains a number of A-type main sequence stars and several G-type giants. The brightest cluster star is a yellow G0 giant with a visual magnitude of 7.9 and an actual luminosity of about 900 suns. If we could view our Sun from the distance of M38 it would be only 15th magnitude object. M38 has a diameter of 25 light years, about the same as M37.

4/6″ Scopes–75x: A truly glorious object! M38 is very

large and rich, containing twenty 9th, and fifty 10th to 12th, magnitude stars in a 20′ area. Its overall outline is irregular and somewhat elongated NW–SE with a 9th magnitude star near the center. It has many double stars and several starless lanes. Although M38's brightest members are fainter than those of nearby M36, it is a much more populous cluster.

8/10″ Scopes–100x: M38 is a bright, irregularly round, rich group with over a hundred stars, many being 9th magnitude or brighter. The cluster contains an asterism shaped like the Greek letter π and some dark lanes. M38 stands out well from the evenly distributed background star field. Visually it is the second most impressive of Auriga's Messier "big three."

12/14″ Scopes–100x: M38 is a fine bright cluster with a hundred 8th magnitude and fainter stars visible in a 20′ area. Outlying stars increase the count to 150. There are many gaps and star chains, the most prominent chain protruding to the north. The southern portion is somewhat richer, and there is a noticeable starless void in the eastern part around a triplet of 11–13th magnitude stars. Numerous doubles and other star patterns are visible.

Figure 5-15. *M38 (NGC 1912) is a fine, bright cluster at low power. Lying south of this cluster is NGC 1907, a rich but smaller cluster often overlooked. This 10 minute photo was made on 103a-O film with a 10″, f5 Newtonain by Lee C. Coombs.*

NGC 1931 H261[1] Emission & Reflection Nebula
⌀4′ × 4′, Photo Br 1-5, Color 3-4   05^{h}31.4^m
+34°15′

Finder Chart 5-6, Figure 5-16   ★★

8/10″ Scopes–100x: This nebula, located in an interesting area, is visible as a small, round glow surrounding an equilateral triangle of stars. There is a faint double star to the east and a brighter double to the NE.

12/14″ Scopes–150x: This is a small, fairly conspicuous nebula elongated 1.5′ × 1′ NE–SW with an 11th magnitude central star having four companions. The nebula is a little brighter at its center. The surrounding star field is rich, particularly just SW of the nebula.

NGC 1960 Messier 36 Open Cluster 60★ Tr Type II 3m
⌀12′, m6.0v, Br★ 8.86v   05^{h}36.1^m +34°08′

Finder Chart 5-5, Figure 5-17   ★★★★★

M36, discovered in 1749 by Le Gentil, is some 4,100 light-years away, ten times the distance of the Pleiades. If it was at the same distance as the Pleiades, it would outshine even that famous star-group; for the true luminosity of M36 is about 8,000 suns, nearly three times that of the Pleiades. M36 is a rather young cluster, for its brightest members are hot, blue B2 and B3 main sequence stars and it lacks red giants. The cluster's true size is around 14 light-years.

4/6″ Scopes–75x: M36 is a bright, large, irregular cluster of three dozen 9th to 13th magnitude stars in a 12′ area. In a 6 × 40 finder it is a bright, large, round glow.

8/10″ Scopes–100x: M36 is bright and large with 40 stars. Its brightest members are arranged in chains that give the cluster a crab-like appearance. This star-group has ten 9th to 10th magnitude, twenty 11th to 12th magnitude, and ten 13th to 14th magnitude stars.

Figure 5-16. *Emission and reflection nebula NGC 1931 is a small, round glow surrounding an equilateral triangle of stars. Martin C. Germano made this 75 minute photo on hypered 2415 film with an 8", f5 Newtonian reflector.*

Figure 5-17. *Messier 36 (NGC 1960) is a moderately rich, irregularly scattered but attractive open star cluster of 9th magnitude and fainter stars. Lee C. Coombs made this 10 minute photo on 103a-O film with a 10", f5 Newtonian reflector.*

12/14" Scopes–100x: This irregularly scattered cluster is bright and large with sixty stars in a 12′ area. There are ten 9th magnitude stars and a dozen or so 10th magnitude stars, with fainter members sprinkled throughout. M36 is smaller and not as rich as open cluster M38 but more concentrated toward its center. A chain of brighter stars crosses the cluster from NE to SW while another less prominent star chain crosses it in the opposite direction forming a distorted-X.

NGC 1985 H865³ Reflection Nebula
ø60″, m –   05ʰ37.5ᵐ +32°00′
Finder Chart 5-6   ★★

12/14" Scopes–150x: NGC 1985, a small, circular reflection nebula surrounding a 13.5 magnitude star in a fairly rich star field, is faint but visible without filters.

Stock 10 Open Cluster 15★ Tr Type IV 3 p
ø25′, m11.3v   05ʰ39.0ᵐ +37°56′
Finder Chart 5-5   ★

12/14" Scopes–75x: Stock 10 is large, loose, irregularly scattered, and unconcentrated with 25 stars having a wide range of moderately bright stars. The brighter stars form an extended patch from the east to the SW side with two 7th magnitude stars at the east edge. There are three 8.5 magnitude and three 9th magnitude stars, the other members are between 10th to 14th magnitude.

Barnard 34 Dark Nebula
ø20′, Opacity 4   05ʰ43.5ᵐ +32°40′
Finder Charts 5-6   ★★★★

12/14" Scopes–50x: This dark nebula is an easy object in a 38mm Erfle eyepiece. The dark area is roundish and about 1/2° in size with several dark lanes winding west out of the field of view. Higher magnification does not help.

Basel 4 Open Cluster 15★ Tr Type II 1 p
ø6′, m9.1v, Br★ 12.18v   05ʰ48.5ᵐ +30°13′
Finder Chart 5-6   ★★

12/14" Scopes–125x: Lying 15′ south of a 7.5 magnitude star, Basel 4 is a faint 6′ diameter patch with a dozen very faint stars mildly concentrated toward the center.

King 8 Open Cluster 30★ Tr Type I 3 m
ø7′, m11.2v, Br★ 13.48v   05ʰ49.4ᵐ
+33°38′
Finder Chart 5-6   ★★

12/14" Scopes–100x: This cluster may be found 30′ SW of a 6th magnitude star and nearly a degree NNW of open cluster M37. King 8 contains a triangle of three 11th to 12th magnitude stars and a scattering of ten very faint stars standing out against a background haze of unresolved cluster members concentrated in a 2′ area elongated ENE–WSW.

16/18" Scopes–125x: This cluster is visible as a faint 3′

diameter mist lying partly within, but extending SW from, a triangle of 10.5 to 12th magnitude stars. Ten 13.5 magnitude and fainter stars may be resolved against an unresolved background haze.

**NGC 2099   Messier 37   Open Cl.150★
Tr Type II 1 r, ø20′, m5.6v, Br★ 9.21v
05ʰ52.4ᵐ   +32°33′**

Finder Chart 5-6, Figure 5-18   ★★★★★

M37 is the finest open cluster in Auriga with at least 150 stars from 9th to 12.5 magnitude. It is older and more evolved than M36, the majority of its bright members being main sequence A-type stars of about −1 absolute magnitude whereas the brightest members of M36 are B2 and B3 stars with absolute magnitudes near −3. The cluster also contains at least a dozen red giant stars while M36 has none. Its stars have a total luminosity of about 2,500 suns spread over 25 light years. M37 was discovered by Messier in 1764.

4/6″ Scopes–75x: This is an excellent cluster, similar in appearance to open cluster M11 in Scutum. M37 is bright, large, and very rich with at least 75 stars visible. The stellar density is considerably higher in the inner regions. A lone bright star stands out near the center. Several brighter stars are also noticeable in the southern part of the cluster. M37 is noticeably richer and more compressed than either open cluster M36 or M38. All three clusters are visible to the naked eye under ideal skies.

8/10″ Scopes–75x: M37 is a bright 20′ diameter cluster with a hundred 9th to 13th magnitude stars. A 9th magnitude star is in the center, and the cluster has a prominent chain of ten 9th to 10th magnitude stars aligned N–S. The field is very rich and includes an orange star south of the cluster.

12/14″ Scopes–100x: Larger instruments show a bright, very large and well concentrated cluster with a 150 stars in a 24′ area. Within an area 50′ across well over 200 cluster stars can be counted. The majority of the brighter stars are arranged in a triangular formation against the fainter stars. One apex of the triangle points west; the other two SW and NW. Several dark voids notch the edge of the triangle, and a narrow, jagged dark lane snakes though the stars. A reddish-orange star near center stands out well against the other stars.

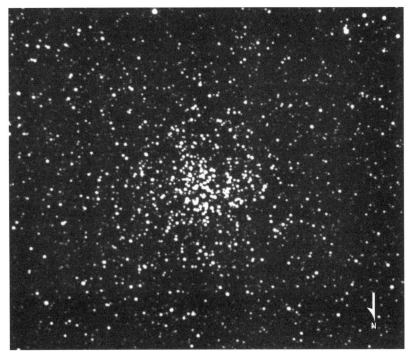

Figure 5-18. *Rich with 150 stars, M37 (NGC 2099) is the finest open cluster in Auriga being similar in appearance to M11 in Scutum. South is up in this 10 min. photo by Lee C. Coombs with a 10″ Newtonian.*

**IC 2149   PK166+10.1   Planetary Nebula   Type 3b+2
ø8″, m10.7v, CS 11.59v   05ʰ56.3ᵐ +46°07′**

Finder Chart 5-7, Figure 5-13   ★★★

12/14″ Scopes–275x: IC 2149 is bright and easily visible, appearing round and nearly stellar. High power reveals a well defined bluish disk and a central star that sparkles in and out of visibility with changing seeing conditions.

16/18″ Scopes–300x: This is a bright but small planetary with a nice bluish tint. The disk is elongated 15″ × 10″ ENE–WSW and brighter near the center, where the central star may be held with direct vision.

**NGC 2126   H68⁸   Open Cluster 40★   Tr Type II 1 p
ø6′, m10.2p, Br★ 13.0p   06ʰ03.0ᵐ +49°54′**

Finder Chart 5-7, Figure 5-19   ★★

8/10″ Scopes–100x: NGC 2126 presents a very pleasing sight–like diamond dust on black velvet! The cluster is round and quite faint. Fifteen 11th to 14th magnitude stars can be seen. A lone bright star stands out at the NE edge.

12/14″ Scopes–125x: This cluster is a rather faint, fairly small, loose, scattering of several dozen 11th magnitude and fainter stars in a 6′ area. A 6th magnitude star is at the cluster's NE edge.

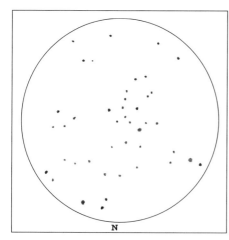

Figure 5-19. NGC 2126
12.5″, f5–70x, by G. R. Kepple

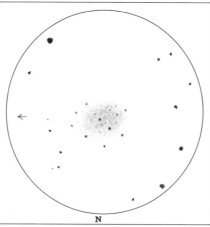

Figure 5-20. NGC 2192
12.5″, f5–70x, by G. R. Kepple

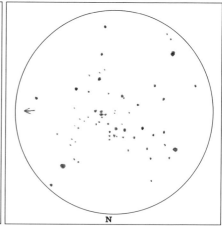

Figure 5-21. NGC 2281
12.5″, f5–70x, by G. R. Kepple

NGC 2192   H57⁷   Open Cluster 45★   Tr Type III 1 p
⌀5′, m10.9p, Br★ 14.0 $06^h15.2^m$ +39°51′
Finder Chart 5-7, Figure 5-20   ★★

12/14″ Scopes–125x: NGC 2192 is a faint, small, irregularly round 5′ patch of a dozen 14th magnitude stars scattered over a background of stardust.

16/18″ Scopes–100x: Larger instruments reveal a faint, irregularly round misty patch just south of an 8th magnitude reddish star. 150x reveals two dozen 14-15th magnitude stars against a twinkling background of fainter cluster members.

NGC 2242   PK170+15.1   Planetary Nebula
⌀22″, m15.0v, CS 15.2 $06^h34.0^m$ +44°46′
Finder Chart 5-7   ★

16/18″ Scopes–250x: This planetary is faint, small, and round with diffuse edges and a somewhat brighter center. The central star is not visible. There is a poor response to both the H-Beta and O-III filters.

NGC 2281   Open Cluster 30★   Tr Type I 3 p
⌀14′, m5.4v, Br★ 7.30v $06^h49.3^m$ +41°04′
Finder Chart 5-7, Figure 5-21   ★★★

8/10″ Scopes–75x: Located 3.4° SSW of Psi-7 (ψ^7) = 58 Aurigae, this cluster is bright and loose with two dozen 8th to 10th magnitude stars in a 15′-long "U."

12/14″ Scopes–75x: NGC 2281 is bright, large, and loose with three dozen stars. A dozen 10th to 11th magnitude stars are elongated E-W, the fainter stars spreading southward. A diamond-shaped asterism is at center, each of its four stars having a fainter companion. A faint star of about 13th magnitude is visible within the diamond. The northernmost double in the diamond displays a fine contrast of reddish and blue stars. This nice cluster is often neglected by observers.

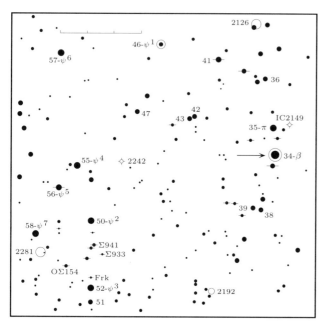

Finder Chart 5-7. 34-β Aur: $05^h59.5^m$ +44°57′

Chapter 6

Camelopardalis, the Giraffe

6.1 Overview

The northern circumpolar constellation of Camelopardalis seems to have been invented by the Dutch astronomical globe-maker Petrus Kaerius in 1613. However, it did not receive general acceptance until after it was published in 1624 in a book by the German mathematician Jakob Bartsch, a son-in-law of Johannes Kepler. Though rather substantial in area, its southeastern corner within the Milky Way, Camelopardalis has only four stars as bright as magnitude 4.5 and is not marked by any conspicuous star-patterns. However, the constellation harbors some fine open clusters, several bright, large galaxies, a nice planetary nebula, and a neat string of stars known as Kemble's Cascade.

Camelopardalis: ka-MEL-oh-pard-al-iss
Genitive: Camelopardalis, ka-MEL-oh-pard-al-iss
Abbrevation: Cam
Culmination: 9pm–Feb. 6, midnight–Dec. 23
Area: 757 square degrees
Best Deep-Sky Objects: NGC 1501, NGC 1502, NGC 2403, Σ1694
Binocular Objects: Cr464, IC 342, NGC 1502, NGC 2403, Kemble's Cascade (see NGC 1502), Stock 23

6.2 Interesting Stars

Σ390 Double Star   Spec. A2
m5.1, 9.5; Sep. 14.8″; P.A. 159° $03^h30.0^m$ +55°27′
Finder Chart 6-3   ★★★★

4/6″ Scopes–100x: Struve 390 is an unequal double with a white primary and a purple secondary.

Σ485 (SZ) Double Star   Spec. B0
m7.0, 7.1; Sep. 17.9″; P.A. 304° $04^h07.9^m$ +62°20′
Finder Chart 6-4   ★★★★

8/10″ Scopes–100x: This easy double star exhibits two equally matched components of blue-white.

1 Camelopardalis = Σ550 Double Star   Spec. B1
AB: m5.7, 6.8; Sep. 10.3″; P.A. 308° $04^h32.0^m$ +53°55′
Finder Chart 6-3   ★★★★

4/6″ Scopes–100x: This easy double star has a white primary and a light blue attendant.

11 & 12 Camelopardalis Double Star   Spec. B3 K0
AB: m5.4, 6.5; Sep. 108.5″; P.A. 8° $05^h06.1^m$ +58°58′
Constellation Chart 6-1   ★★★

2/3″ Scopes–50x: 11 and 12 Camelopardalis form a wide pair of white and deep-yellow stars for binoculars and small telescopes at low power.

29 Camelopardalis = H IV 125 Double Star   Spec. A2
m6.5, 9.5; Sep. 25.1″; P.A. 131° $05^h50.6^m$ +56°55′
Constellation Chart 6-1   ★★★★

4/6″ Scopes–100x: 29 Camelopardalis is an easy but unequal pair of yellow and light blue stars.

Beta (β) = 10 Camelopardalis Triple Star Spec. G0 A5
AB: m4.0, 8.6; Sep. 80.8″; P.A. 208°
BC: 11.2; Sep. 14.8″; P.A. 168° $05^h03.4^m$ +60°27′
Constellation Chart 6-1   ★★★★

4/6″ Scopes–75x: The AB pair is a wide, easy pair for small scopes and binoculars. At low power, this yellow and blue can be seen embedded in a streak of dark nebulosity tinting the pale Milky Way background glow.

Σ1122 Double Star   Spec. F2
m7.8, 7.8; Sep. 15.4″; P.A. 5° $07^h45.9^m$ +65°09′
Finder Chart 6-6   ★★★★

2/3″ Scopes–50x: Struve 1122 is an attractive pair of equally matched white stars.

Camelopardalis, The Giraffe
Constellation Chart 6-1

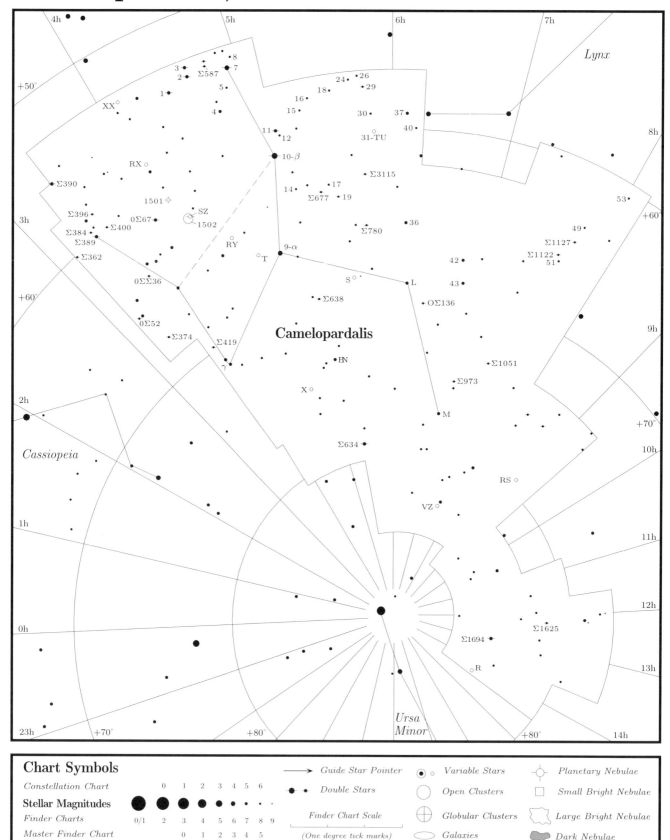

Table 6-1. Selected Variable Stars in Camelopardalis

Name	HD No.	Type	Max.	Min.	Period (Days)	F*	Spec. Type	R.A. (2000)	Dec.	Finder Chart No. & Notes
RX Cam	11763	Cδ	7.30	8.07	7.91	0.28	F6-G2	$04^h05.0^m$	$+58°40'$	6-3
SZ Cam		EA/DM	7.0	7.29	2.69	0.17	B0+B0	07.9	+62 20	6-4 SZ = Σ485
T Cam	13913	M	7.3	14.4	373.20	0.47	S4-S8	40.1	+66 09	6-4
X Cam	17446	M	7.4	14.2	143.56	0.49	K8-M8	45.7	+75 06	6-5
S Cam	18191	SRa	7.7	11.6	327.26	0.51	C7	$05^h41.0^m$	+68 48	6-5
RS Cam	19510	SRb	7.9	9.7	88.6	0.45	M4	$08^h50.8^m$	+78 58	6-7 & 6-8
R Cam		M	6.97	14.4	270.22	0.45	S2-S8	$14^h17.8^m$	+83 50	6-1

F* = The fraction of period taken up by the star's rise from min. to max. brightness, or the period spent in eclipse.

Table 6-2. Selected Double Stars in Camelopardalis

Name	ADS No.	Pair	M1	M2	Sep."	P.A.°	Spec	R.A. (2000)	Dec	Finder Chart No. & Notes
Σ362	2426	AB	8.5	8.8	7.1	142	A0	$03^h16.3^m$	$+60°02'$	6-3 AB pair both light yellow
	2426	AC		10.5	26.1	43				(In open cl. Stock 323)
	2426	AD		11.1	30.9	286				
	2426	AE		9.9	35.3	242				
OΣ52	2436	AB	6.8	7.3	0.5	84	A2	17.5	+65 40	6-4
Σ374	2494		7.8	9.3	10.9	295	F8	24.2	+67 27	6-4 White & yellow pair
Σ384	2540	AB	7.9	9.1	2.0	270	F8	28.5	+59 54	6-3
Σ390	2565	AB	5.1	9.5	14.8	159	A2	30.0	+55 27	6-3 White & purple
Σ389	2563		6.5	7.5	2.8	67	A0	30.2	+59 22	6-3
Σ396	2592	AB	6.3	8.2	20.4	*243	A2	33.5	+58 46	6-3
Σ400	2612	AB	6.8	7.6	w1.6	*264	F5	35.0	+60 02	6-3
OΣΣ36	2650	AB	6.8	8.6	46.1	69	F5	40.0	+63 52	6-4
Σ419	2678	AB	7.9	7.9	3.0	74	A3	42.8	+69 51	6-1
A984	2678	BC		10.6	0.5	154				
OΣ67	2867		5.3	8.5	1.9	44	K0	57.1	+61 07	6-4
Σ485 (SZ)	2984	AB	7.0	7.1	17.9	304	B0	$04^h07.9^m$	+62 20	6-4 AB pair blue-white
1 Cam	3274	AB	5.7	6.8	10.3	308	B1	32.0	+53 55	6-3 White & light blue
2 Cam	3358	ABxC	5.8	7.3	1.3	258	F0	40.0	+53 28	6-1
Σ587	3442	AB	7.4	8.9	21.0	185	A3	48.0	+53 07	6-1
Σ586			9.4	9.8	28.4	129				In field with Σ587
10-β Cam	3615	AB	4.0	8.6	80.8	208	G0 A5	$05^h03.4^m$	+60 27	6-1 Yellow & blue
	3615	BC		11.2	14.8	168				
11 & 12 Cam		AB	5.4	6.5	108.5	8	B3 K0	06.1	+58 58	6-1 Binocular pair
Σ638	3759		7.6	8.6	5.2	222	K0	14.3	+69 49	6-5
Σ634	3864	AB	5.1	9.1	10.4	91	F8	22.6	+79 14	6-1
Σ677	3956		7.9	8.2	w1.2	136	G0	24.7	+63 23	6-1
Σ3115	4376		6.5	7.6	1.0	5	A2	49.1	+62 49	6-1
29 Cam	4412		6.5	9.5	25.1	131	A2	50.6	+56 55	6-1 Yellow & light blue
Σ780	4405	AB	6.8	8.1	3.8	104	F8	51.0	+65 45	6-1
	4405	AC		10.0	12.3	150				
	4405	AD		13.4	19.3	56				
OΣ136	5039		6.0	9.8	5.7	80	A2	$06^h28.2^m$	+70 32	6-5
Σ973	5669	AB	7.1	8.1	12.6	31	G0	$07^h04.1^m$	+75 14	6-7
Σ1051	6028	AB	7.1	9.2	1.1	284	F0	26.6	+73 05	6-7
	6028	AC		7.8	31.5	82	F0			
Σ1122	6319		7.8	7.8	15.4	5	F2	45.9	+65 09	6-6 Equal white stars
Σ1127	6336	AB	7.0	8.8	5.3	340	A2	47.0	+64 03	6-6
	6336	AC		9.9	11.3	175				
Σ1625	8494		7.3	7.8	14.4	219	F0 F0	$12^h16.2^m$	+80 08	6-1 Both stars bluish
Σ1694	8682		5.3	5.8	21.6	326	A2 A0	49.2	+83 25	6-8 Bluish-white & greenish

Footnotes: *= Year 2000, a = Near apogee, c = Closing, w = Widening. Finder Chart No: All stars listed in the tables are plotted in the large Constellation Chart, but when a star appears in a Finder Chart, this number is listed. Notes: When colors are subtle, the suffix *-ish* is used, e.g. *bluish*.

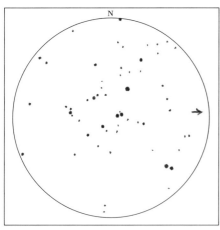

Figure 6-1. Stock 23
12.5", f4.5–70x, by G. R. Kepple

Figure 6-2. King 6
13", f5.6–100x, by Steve Coe

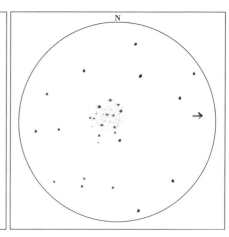

Figure 6-3. Berkeley 10
13", f5.6–100x, by Steve Coe

Σ1625 Double Star Spec. F0 F0
m7.3, 7.8; Sep. 14.4"; P.A. 219° $12^h16.2^m$ +80°08'
Constellation Chart 6-1 ★★★★

2/3" Scopes–100x: Struve 1625 is a nearly equal pair of yellowish stars.

Σ1694 Double Star Spec. A2 A0
m5.3, 5.8; Sep. 21.6"; P.A. 326° $12^h49.2^m$ +83°25'
Finder Chart 6-8 ★★★★

2/3" Scopes–75x: This double is an easy pair of whitish stars.

8/10" Scopes–100x: Subtle hues of yellow were reported for the brighter star and bluish for the fainter star, but our own observations show white stars.

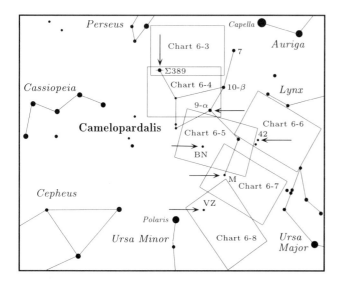

Master Finder Chart 6-2. Camelopardalis Chart Areas
Guide stars indicated by arrows.

6.3 Deep-Sky Objects

Stock 23 Open Cluster 25★ Tr Type III 3 p n
ø14', m5.6v $03^h16.3^m$ +60°02'
Finder Chart 6-3, Figure 6-1 ★★★

4/6" Scopes–50x: Stock 23 is bright and large but rather poor and not at all compressed. In its center six stars are arranged in a keystone-shaped asterism reminiscent of the Hercules Keystone. Part of the asterism keystone is Σ362, a close equal double, and a wide double of 9th magnitude stars. Located 10' west is another wide double star.

12/14" Scopes–75x: This cluster is bright, large, loose, and irregular with forty stars in a 10' diameter area. It contains a quadrilateral formed by four of the brightest stars, the westernmost being a double with 7.5 and 8th magnitude components 7" apart. Ten stars are concentrated near the easternmost star in the quadrilateral, eight in a NE–SW chain. Some 20 stars are around the western and southern stars of the quadrilateral, five of them forming another chain.

King 6 Open Cluster 35★ Tr Type IV 2 p
ø6', m –, Br★ 10.0p $03^h28.1^m$ +56°27'
Finder Chart 6-3, Figure 6-2 ★★★

12/14" Scopes–100x: King 6 is a moderately rich and somewhat compressed cluster covering a 6' area with two dozen 10th to 13th magnitude stars standing out against a granular background.

Figure 6-4. *Reflection nebulae van den Bergh 14 & 15 are a challenge to detect and require very dark, transparent skies. Martin C. Germano made this 60 minute exposure on hypered Kodak 2415 Tech Pan film with an 8", f5 Newtonian reflector.*

van den Bergh 14 Reflection Nebula
ø20' × 8', Photo Br 3-5, Color 1-4   03^{h}29.2^m +59°57'
Finder Chart 6-3, Figure 6-4 ★

16/18" Scopes–75x: Extremely faint and difficult, this nebula requires averted vision and excellent sky transparency. Because it is a reflection nebula, filters will not help you see it better. It is located SW of a 4th magnitude star extending toward a 5th magnitude star. These bright stars interfere and should be placed outside the field of view. vdB14 is a faint milky streak with many faint stars embedded in it.

van den Bergh 15 Reflection Nebula
ø25' × 10', Photo Br 3-5, Color 1-4   03^{h}30.1^m +58°54'
Finder Chart 6-3, Figure 6-4 ★

16/18" Scopes–75x: Located 3/4° south of vdB14, this nebula surrounds a bright 5th magnitude star with its most obvious area north of the star. Even on the best of nights, vdB15 is extremely faint, requires averted vision, and can be seen only as an amorphous haze. Use Martin Germano's photo to help identify the area.

Berkeley 10 Open Cluster 50★ Tr Type II 3 p
ø12', m –, Br★ 14.0p   03^{h}39.4^m +66°32'
Finder Chart 6-4, Figure 6-3 ★★

12/14" Scopes–100x: Berkeley 10 is a faint, fairly large, sparse, and uncompressed cluster of only ten stars against a faint background glow. Few more members are added at 150x.

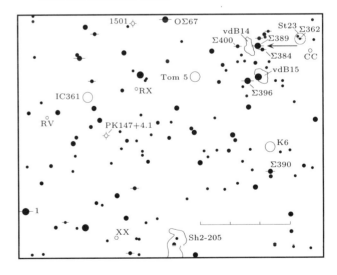

Finder Chart 6-3. Σ389: 03^{h}30.2^m +59°22'

Figure 6-5. *Face-on galaxy IC 342 has a bright core surrounded by a large, faint halo with many foreground stars superimposed. Martin C. Germano made this 90 minute exposure on hypered Kodak 2415 Tech Pan film with an 8″, f5 Newtonian reflector.*

IC 342 U2847 Galaxy Type SAB(rs)cd I-II
ø22.0′ × 22.0′, m8.4v, SB 15.0   03ʰ46.8ᵐ +68°06′
Finder Chart 6-4, Figure 6-5   ★★★

12/14″ Scopes–75x: IC 342 displays a prominent nucleus surrounded by a diffuse outer halo about 12′ across with many faint stars embedded in it.

16/18″ Scopes–100x: This galaxy is faint with an exceptionally large 14′ diameter halo around a large core and a stellar center. There are many stars superimposed on its faint disk. SW of center is a string of six stars running NW–SE and four 11th magnitude stars are visible NE of center. There is a loose clump of stars south of the halo.

Tombaugh 5 Open Cluster 60★ Tr Type III 2 m
ø17′, m8.4v, Br★ 11.62   03ʰ47.8ᵐ +59°03′
Finder Chart 6-3, Figure 6-10   ★★★

12/14″ Scopes–100x: Standing out well in the star field, Tombaugh 5 is a fairly faint, large, loose, irregular cluster of forty 12th and 13th magnitude stars embedded in an unresolved haze. Its stars are spread evenly over a NW-SE rectangular area. A fairly bright concentration of stars is visible in the southern portion. An 8th magnitude star is located to the SW of the cluster.

NGC 1501 H53⁴ PK144+6.1 Planetary Neb. Type 3
ø52′, m11.5v, CS 14.45   04ʰ07.0ᵐ +60°55′
Finder Chart 6-4, Figure 6-6   ★★★★

8/10″ Scopes–100x: This planetary nebula is bright, large, and round with a light blue tint. The edges are well defined and some dark markings in the disk make it resemble the Eskimo Nebula in Gemini. The central star fades in and out at lower power but is steady with direct vision at 250x.

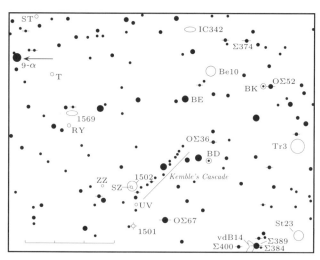

Finder Chart 6-4. 9–α: 04ʰ54.1ᵐ +66°21′

Figure 6-6. *NGC 1501 is a fine bluish planetary nebula displaying a mottled disk. Martin C. Germano made this 20 minute exposure on hypered 2415 film with a 14.5″ Newtonian.*

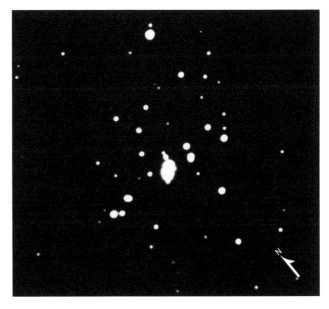

Figure 6-7. *NGC 1502 is an interesting cluster containing many double stars. Don Walton made this 30 minute photo on hypered 2415 film with an 11″, f10 SCT.*

12/14″ Scopes–250x: NGC 1501 is a lovely bluish planetary nebula! It is bright with a fairly large mottled halo nearly 1′ across. The periphery is diffuse but seems slightly elongated NE–SW. The center is slightly darker, its annularity being more noticeable at higher power. The prominent central star shines steadily.

NGC 1502 H47⁷ Open Cluster 45★ Tr Type II 3 p
ø7′, m5.7v, Br★ 6.93   04ʰ07.7ᵐ +62°20′
Finder Chart 6-4, Figure 6-7   ★★★★

While searching for NGC 1502, you cannot help but stumble upon a beautiful 2.5° long rivulet of 8th magnitude stars immediately to the NW of NGC 1502. This asterism is known as Kemble's Cascade, a moniker given it by Walter Scott Houston. Lucien J. Kemble is the Canadian amateur astronomer for whom it is named. Kemble's Cascade is best observed in 7x or 10x binoculars.

3/4″ Scopes–50x: This nice cluster stands out well in an interesting star field. It is bright, small, fairly rich and moderately compressed in a vague triangular shape. The triangle's southern base is the line along which lies the cluster's brightest members. At the midpoint of this base line is the cluster lucida, the conspicuous double Σ485.

12/14″ Scopes–100x: NGC 1502 lies near the SE end of Kemble's Cascade, a line of 8th to 10th magnitude stars with a 5th magnitude star just NW of its midpoint. NGC 1502 is bright, moderately large, fairly rich and considerably compressed with some three dozen 7th magnitude and fainter stars in a 7′ area. Σ485 = ADS 2984, a wide pair of 7th magnitude stars (the B component of which is the Beta Lyrae type variable SZ Cam), lies near its center. The cluster contains half a dozen 8th and a dozen 9th magnitude stars. A nice blue and gold double star lies to the NE.

PK147+4.1 Minkowski 2-2 Planetary Nebula Type 3
ø12′, m13.8v, CS –   04ʰ13.3ᵐ +56°57′
Finder Chart 6-3   ★★★

12/14″ Scopes–200x: This planetary nebula has a faint, small, round disk with a slight brightening at its center.

16/18″ Scopes–225x: With a UHC filter, the disk appears fairly bright, small, and round without a central star. Without the filter, the disk is readily apparent but appears fainter.

IC 361 Open Cluster 60★ Tr Type II 1 r
ø6′, m11.7:v, Br★ 14.55v   04ʰ19.0ᵐ +58°18′
Finder Chart 6-3, Figure 6-8   ★★★

16/18″ Scopes–150x: On a good night this fine but faint cluster looks like a mist of stardust! Ten 14.5 magnitude and fainter stars are irregularly scattered over a 6′ area against an unresolved haze. The two brightest stars lie on the NNW side, and a string of four stars extends southward from the east side.

Figure 6-8. *Open cluster IC 361 looks like a mist of stardust in large scopes. Lee C. Coombs made this 10 minute exposure on 103a-0 film with a 10″, f5 Newtonian reflector.*

Figure 6-9. *Edge-on galaxy NGC 1560 is a large ghostly streak of light in small scopes. Martin C. Germano made this 80 minute exposure on 2415 film with an 8″, f5 Newtonian.*

NGC 1530 U3013 Galaxy Type SB(rs)b
⌀4.8′ × 2.6′, m11.4v, SB 14.0   04ʰ23.4ᵐ +75°18′
Finder Chart 6-5   ★★★

12/14″ Scopes–150x: NGC 1530 presents a faint and very diffuse halo elongated 1.5′ × 0.8′ NNE–SSW with a slight brightening at center.

16/18″ Scopes–150x: This faint galaxy has a tiny core containing a stellar nucleus surrounded by a uniform halo elongated 2′ × 1′ NNE–SSW. There is a 12th magnitude star 3′ north, and, slightly nearer, are two 13th magnitude stars aligned with the galaxy's major axis.

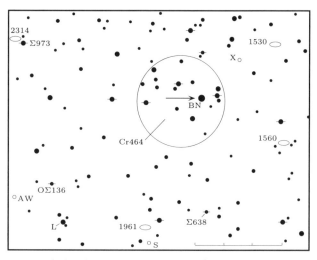

Finder Chart 6-5. BN Cam: 05ʰ12.2ᵐ +73°57′

NGC 1569 H768² Galaxy Type IBm IV-V
⌀3.0′ × 1.9′, m11.0v, SB 12.8   04ʰ30.8ᵐ +64°51′
Finder Chart 6-4   ★★★

12/14″ Scopes–150x: NGC 1569 is a nice little galaxy lying 6′ NNW of a 9th magnitude star and 20′ SW of a 7th magnitude star. A 10.5 magnitude star lies 1′ north. The galaxy's halo is fairly bright, elongated 2′ × 0.8′ ESE–WNW, and the bright, extended core has a hint of mottling. A 10th magnitude star lies 1′ north.

16/18″ Scopes–150x: This galaxy is moderately bright, fairly small, and elongated 2.5′ × 1′ ESE–WNW. The lens-shaped halo has tapered ends. The core is oval-shaped with a faint stellar nucleus. A 12th magnitude star touches the halo on the ESE edge.

NGC 1560 IC 2062 Galaxy Type SA(s)d
⌀9.2′ × 1.7′, m11.4v, SB 14.3   04ʰ32.8ᵐ +71°53′
Finder Chart 6-5, Figure 6-9   ★★★

12/14″ Scopes–100x: This galaxy is a very faint ghostly streak, highly elongated N–S with uniform surface brightness. 13th magnitude stars are embedded in the halo near each edge of the major axis, and a 12th magnitude star nearly touches the halo to the SE. The galaxy is flanked east and west by 10th magnitude stars, each having an 8th magnitude star lying further beyond.

16/18″ Scopes–125x: NGC 1560 appears rather faint, large, and elongated 8′ × 1.5′ NNE–SSW with a large prominent core.

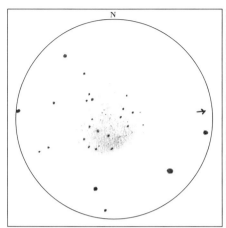

Figure 6-10. Tombaugh 5
17.5", f4.5–100x, by G. R. Kepple

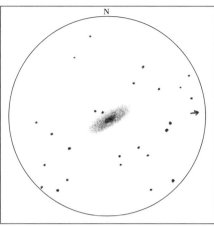

Figure 6-11. NGC 2146
13", f5.6–100x, by Steve Coe

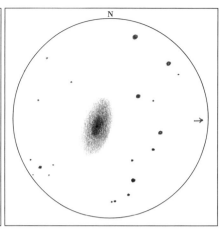

Figure 6-12. NGC 2336
8", f7–116x, by G. R. Kepple

Collinder 464 Open Cluster 50★ Tr Type IV 3 p
⌀120', m4.2p   **05ʰ22.0ᵐ +73°:**
Finder Chart 6-5   ★★★

15 × 65 Binoculars: Collinder 464 is either an extremely large, loose, irregular cluster or merely a rich Milky Way field. It is best seen at low powers or with binoculars; resemblance to a cluster is lost with any magnification at all. The brightest star is 5th magnitude, and around it are scattered four 6th magnitude, eight 7th magnitude, and thirty-five 8th to 14th magnitude stars.

NGC 1961 H747³ Galaxy Type SAB(rs)c II
⌀4.3' × 3.0', m11.0v, SB 13.9   **05ʰ42.1ᵐ +69°23'**
Finder Chart 6-5   ★★★

3/4" Scopes–50x: This galaxy is very faint and small, and requires averted vision.

12/14" Scopes–125x: NGC 1961 is a very faint galaxy, elongated 3' × 1.5' E–W with a diffuse, uniform surface brightness and a faint core. A faint star is embedded in the halo just SE of center.

16/18" Scopes–150x: This galaxy appears moderately bright in larger instruments, its a fairly large halo, elongated 3.5' × 1.5' E–W, slightly brightening to a small core. There is a 13th magnitude star 30" SE of the core and hints of a dark patch between the core and the star.

NGC 2146 U3429 Galaxy Type SB(s)ab pec II
⌀5.4' × 4.5', m10.6v, SB 13.9   **06ʰ18.7ᵐ +78°21'**
Finder Chart 6-7, Figure 6-11   ★★★

12/14" Scopes–125x: NGC 2146 has a bright envelope elongated 5' × 2' ESE–WNW with a hint of spiral structure. Located to the east are half a dozen bright stars, one of which is a close pair.

16/18" Scopes–150x: This moderately bright galaxy is elongated 5.5' × 2' ESE–WNW with a large faint core and a faint stellar nucleus. The halo is variegated with indistinct light and dark streaks.

NGC 2314 Galaxy Type E3
⌀1.8' × 1.6', m12.2v, SB 13.2   **07ʰ10.5ᵐ +75°20'**
Finder Chart 6-7   ★★

12/14" Scopes–125x: This galaxy may be found 25' east of Σ973, a double with 7.1 and 8.2 magnitude stars 13" apart. NGC 2314 is a moderately faint, tiny round spot with a stellar nucleus.

16/18" Scopes–150x: NGC 2314 is visible as a faint, small, round spot only 45" across with a prominent stellar nucleus. IC 2174 lies 6' WNW.

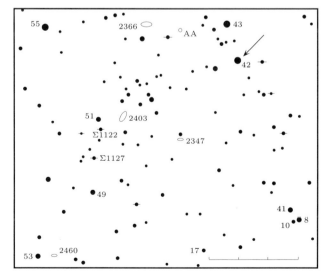

Finder Chart 6-6. 42 Cam: 06ʰ50.9ᵐ +67°34'

Figure 6-13. *NGC 2336 displays a tiny core surrounded by a faint halo while its smaller companion, IC 467 appears as a much fainter, elongated streak. Martin C. Germano made this 75 minute exposure on hypered 2415 film with an 8″, f5 Newtonian. North is up.*

NGC 2268 U3653 Galaxy Type SAB(r)bc II
ø3.4′ × 2.2′, m11.5v, SB 13.6 $07^h14.0^m$ +84°23′
Finder Chart 6-8   ★★★

12/14″ Scopes–125x: NGC 2268 is a fairly faint, small, oval-shaped galaxy elongated 1.5′ × 1′ ENE–WSW with a slight brightening at center.

16/18″ Scopes–150x: This galaxy is moderately faint and elongated 2′ × 1.3′ ENE–WSW. Its faint halo increases slightly in brightness to a large core with a faint stellar nucleus. A 13th magnitude star is at the SW edge.

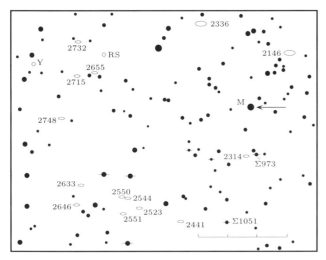

Finder Chart 6-7. M Cam: $07^h00.0^m$ +76°59′

NGC 2347 H746³ Galaxy Type (R′)SA(r)b: I-II
ø1.8′ × 1.4′, m12.4v, SB 13.3 $07^h16.1^m$ +64°43′
Finder Chart 6-6   ★★

12/14″ Scopes–125x: Located 4′ south of a 7.5 magnitude star, this galaxy is a very faint, round, diffuse smudge 1′ across with a faint, tiny core. On the opposite side of NGC 2347 from the magnitude 7.5 star, 13′ north of the galaxy, is the faint IC 2179.

16/18″ Scopes–150x: NGC 2347 is faint and small with a circular 1.5′ diameter halo containing a faint core and a nonstellar nucleus. There is a 13th magnitude star 2′ SW.

NGC 2336 U2809 Galaxy Type SAB(r)bc I
ø6.4′ × 3.3′, m10.4v, SB 13.6 $07^h27.1^m$ +80°11′
Finder Chart 6-8, Figures 6-12 & 6-13   ★★★

12/14″ Scopes–125x: This galaxy has a fairly faint halo elongated 3′ × 1′ N–S with a slight brightening in the middle. It forms a triangle with two stars to the west. IC 467 lies 20′ SSE.

16/18″ Scopes–150x: In larger instruments, NGC 2336 displays a small, faint core and a stellar nucleus surrounded by a diffuse halo elongated 4′ × 2′ N–S. There is a 15th magnitude star east of the core.

NGC 2366 H748³ Galaxy Type IB(s)m V
ø8.2′ × 3.3′, m10.8v, SB 14.3 $07^h28.9^m$ +69°13′
Finder Chart 6-6   ★★★

12/14″ Scopes–125x: This faint galaxy is located 50′ north of a 5.5 magnitude star. The halo is a very faint, diffuse haze elongated 4′ × 1.5′ NNE–SSW with an even surface brightness.

16/18″ Scopes–150x: NGC 2366 appears elongated 4′ × 2′ NNE–SSW with several areas of diffuse concentration through center, including a bright stellar knot. At the SW end, its companion, NGC 2363, is visible as a diffuse spot less than 1′ across.

IC 467 Galaxy Type SAB(s)c:
ø3.2′ × 1.3′, m12.6v, SB 14.0 $07^h30.0^m$ +79°52′
Finder Chart 6-8, Figure 6-13   ★★

16/18″ Scopes–150x: Located 20′ SSE of NGC 2336, this galaxy is much fainter and a little smaller than its companion. IC 467 appears rather faint and elongated 3′ × 1′ E-W with a slight brightening at the core. There is a 15th magnitude star near the west end.

Figure 6-14. *NGC 2403 displays a bright core surrounded by a tenuous halo with many stars superimposed. The halo exhibits some spiral structure in larger instruments when seeing conditions are good. Martin C. Germano made this 65 minute exposure on hypered Kodak 2415 Tech Pan film with a 14", f5 Newtonian reflector at prime focus.*

NGC 2403 H44⁵ Galaxy Type SAB(s)cd III
ø25.5' × 13.0', m8.5v, SB 14.6   07ʰ36.9ᵐ +65°36'
Finder Chart 6-6, Figure 6-14   ★★★★

NGC 2403 is an outlying member of the M81-M82 Galaxy Group. Allen Sandage used the 200" Hale telescope to detect Cepheids in it. This was the first galaxy beyond the Local Group to have Cepheid variables identified in it. Sandage derived a distance of 8,000 light years at the time! It is now believed to be about 8 million light years away.

8/10" Scopes–100x: This galaxy is a moderately bright 10' × 4' ovoid with a slight central brightening. It lies a little north of the 6' long E–W line between two 10th magnitude stars. Very faint stars are involved in its nebulous halo near the western 10th magnitude star. NGC 2403 may be spotted in most finders, and is easy in 10 x 50 binoculars.

12/14" Scopes–125x: Lovely! NGC 2403 is bright, very large, and elongated 12' × 5' ESE–WNW. The galaxy has a bright core with spiral arms attached at opposite ends. Both arms can be traced for more than a half-turn around the galaxy, their outer arcs shimmering and sparkling with a mottled texture. The northern arm is connected to NGC 2404, visible as a knot. Many faint stars are superimposed upon the halo.

16/18" Scopes–125x: Impressive! This fine galaxy is bright, very large, elongated 15' × 5' ESE–WNW, and has an 11th magnitude star on each side of a slightly brighter core. The halo is tenuous with many faint stars superimposed, the brightest being just south of the core. A dark area runs between the center and the 11th magnitude star on the ESE side. Another dark streak extends north of the core.

NGC 2441 U4036 Galaxy Type SAB(r)b: II
ø2.1' × 2.0', m12.2v, SB 13.6   07ʰ52.2ᵐ +73°02'
Finder Chart 6-7   ★★

12/14" Scopes–125x: This galaxy is visible as a very faint, small, round spot with a slight brightening at center.

16/18" Scopes–150x: NGC 2441 is a faint, diffuse, circular 1.5' halo slightly elongated NW–SE with a slight brightening in the core area.

NGC 2460 U4097 Galaxy Type SA(s)a III
ø3.7′ × 2.6′, m11.8v, SB 14.1   07^{h}56.9^m +60°21′
Finder Chart 6-6   ★★

12/14″ Scopes–125x: This galaxy has a prominent core surrounded by a moderately faint, diffuse 1′ diameter halo slightly elongated NE–SW. A pair of 12th magnitude stars is 4.5′ SE.

16/18″ Scopes–150x: NGC 2460 shows a large prominent core and a stellar nucleus surrounded by a relatively faint, small halo, slightly elongated 1.2′ × 1′ NE–SW. IC 2209 lies 5.5′ WSW.

NGC 2523 U4271 Galaxy Type SB(r)bc I
ø2.7′ × 2.0′, m11.9v, SB 13.6   08^{h}15.0^m +73°35′
Finder Chart 6-7, Figure 6-15   ★★

12/14″ Scopes–125x: This galaxy shows a fairly faint circular halo 1.5′ across with a slight brightening at center. There is an 11th magnitude star 1.5′ SW. NGC 2523 is just a small amorphous glow in smaller telescopes.

16/18″ Scopes–150x: Larger instruments reveal a prominent core embedded in a distinct central bar extending NW–SE surrounded by a moderately faint 2′ halo slightly elongated ENE–WSW. Companion galaxy NGC 2523B, 9′ west, appears very faint, small, and round.

NGC 2544 U4312 Galaxy Type SB(s)a:
ø0.9′ × 0.7′, m12.8v, SB 12.2   08^{h}21.7^m +73°59′
Finder Chart 6-7   ★

12/14″ Scopes–125x: This galaxy is only a faint, diffuse circular glow in a field that includes several fairly bright stars.

16/18″ Scopes–150x: NGC 2544 has a very faint, round 30″ diameter halo with a faint stellar nucleus.

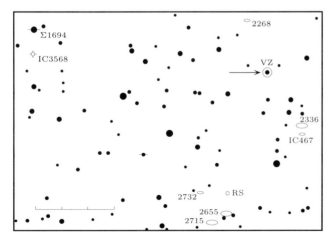

Finder Chart 6-8. VZ Cam: 7^{h}31.1^m +82°25′

NGC 2550 U4359 Galaxy Type Sb:
ø1.0′ × 0.4′, m12.8v, SB 11.7   08^{h}24.6^m +74°01′
Finder Chart 6-7   ★

12/14″ Scopes–125x: This galaxy is an extremely faint, small, elongated smudge SW of a 9th magnitude star.

16/18″ Scopes–150x: The halo is faint and elongated 1′ × 0.25′ ESE–WNW with a slight brightening toward center.

NGC 2551 U4362 Galaxy Type SA(s)0/a
ø1.6′ × 1.1′, m12.1v, SB 12.6   08^{h}24.8^m +73°25′
Finder Chart 6-7   ★★★

12/14″ Scopes–125x: This galaxy has a faint halo elongated 1′ × 0.5′ E–W with a slight brightening toward center. A 12th magnitude star is 2′ NE.

16/18″ Scopes–150x: NGC 2551 is fairly faint and elongated 1.5′ × 0.8′ ENE–WSW. The core is considerably brighter and contains a very faint stellar nucleus.

NGC 2633 U4574 Galaxy Type SB(s)b I-II
ø2.6′ × 1.6′, m12.2v, SB 13.6   08^{h}48.1^m +74°06′
Finder Chart 6-7   ★★

12/14″ Scopes–125x: NGC 2633 has a small conspicuous nucleus surrounded by faint indefinite extensions. NGC 2634, 8′ south, is nearly a carbon copy of NGC 2633.

16/18″ Scopes–150x: This galaxy is a moderately faint, diffuse glow elongated 2′ × 1′ N–S with a stellar nucleus at center.

NGC 2646 U4604 Galaxy Type SB(r)0°:
ø1.7′ × 1.7′, m12.1v, SB 13.1   08^{h}50.4^m +73°28′
Finder Chart 6-7   ★★★

12/14″ Scopes–125x: NGC 2646 has a fairly faint, round 1′ diameter halo with a stellar nucleus. A faint star is near the south end and a wide, unequal double star 2′ SSE.

16/18″ Scopes–150x: This galaxy is fairly bright and elongated 1.3′ × 1′ E–W with a slight brightening to a faint stellar nucleus. IC 520 is 14′ east while IC 2389 lies 11′ to the WNW.

NGC 2655 H288^1 Galaxy Type SAB(s)0/a
ø6.0′ × 5.3′, m10.1v, SB 13.7   08^{h}55.6^m +78°13′
Finder Chart 6-8   ★★★

8/10″ Scopes–100x: NGC 2655 forms an equilateral triangle with a 7.5 magnitude star to the SE and a 9th magnitude star to the NE. Its halo appears bright and elongated 3′ × 2′ E–W with a much

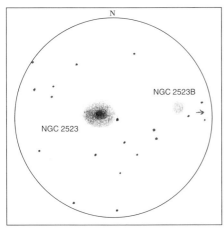

Figure 6-15. NGC 2523B & NGC 2523
13", f5.6–100x, by Steve Coe

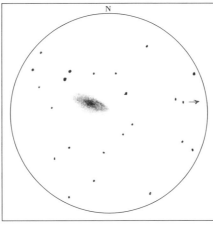

Figure 6-16. NGC 2748
13", f5.6–100x, by Steve Coe

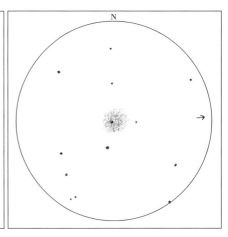

Figure 6-17. IC3568
13", f5.6–300x, by Steve Coe

brighter, extended core.

12/14" Scopes–125x: NGC 2655 has a tenuous 4' × 3' halo, somewhat brighter on the south side. The prominent core is also elongated E–W, and within it is a large nucleus.

NGC 2715   U4759   Galaxy   Type SAB(rs)c II
⌀4.6' × 1.6', m11.2v, SB 13.2 $09^h08.1^m$ +78°05'
Finder Chart 6-8   ★★★

12/14" Scopes–125x: NGC 2715 is a nice lens-shaped galaxy! The halo is elongated 3' × 1.5' NNE–SSW with uniform surface brightness.

16/18" Scopes–150x: Medium power reveals a moderately bright halo elongated 3.5' × 2' NNE–SSW with some mottled texture. The large, faint core contains a very faint stellar nucleus.

NGC 2732   U4818   Galaxy   Type S0
⌀1.8' × 0.8', m11.9v, SB 12.2 $09^h13.4^m$ +79°11'
Finder Chart 6-8   ★★★

8/10" Scopes–100x: This galaxy displays a fairly bright halo elongated 1.2' × 0.3' ENE–WSW with a bright center. A 12th magnitude star lies at the ENE tip.

12/14" Scopes–125x: NGC 2732, a fairly bright, small, edge-on galaxy, has a lenticular halo elongated 1.5' × 0.5' ENE–WSW containing a small oval core and a stellar nucleus. A 12th magnitude star is in the halo ENE of the galaxy's center. Galaxy U4832 is 4' to the east.

NGC 2748   U4825   Galaxy   Type SAbc III
⌀2.8' × 1.1', m11.7v, SB 12.7 $09^h13.7^m$ +76°29'
Finder Chart 6-7, Figure 6-16   ★★★

12/14" Scopes–125x: NGC 2748 is fairly bright and elongated 2' × 0.75' NE–SW with a large bright core.

16/18" Scopes–150x: This galaxy has a fairly bright halo elongated 3' × 1' NE–SW with a bulging core and a very faint stellar nucleus.

IC 3568   PK123+34.1   Planetary Nebula   Type 2+2a
⌀>6.0", m10.6v, CS 11.4 $12^h32.9^m$ +82°33'
Finder Chart 6-8, Figure 6-17   ★★★

8/10" Scopes–175x: IC 3568 has a bright, tiny, round disk with a 13th magnitude star in contact to the west. The disk is definite at 175x but almost stellar at 100x. The planetary forms one apex of an isosceles triangle with two stars.

12/14" Scopes–250x: A beautiful bluish disk! This planetary is fairly bright but rather small and round, its central star offset to the east.

16/18" Scopes–275x: This planetary nebula is bright but fairly small even at high power. The 10" halo is round with a bright center fading to a diffuse periphery. The central star is easily visible. The 13th magnitude star on the planetary's west edge tends to merge with the disk during bad seeing conditions.

Chapter 7

Cancer, the Crab

7.1 Overview

Cancer, the Crab, is the faintest of the twelve Zodiacal constellations, containing not one single star as bright as magnitude 3.5. A dark, moonless night is necessary to identify its dim stars scattered between the Sickle of Leo the Lion and the parallel star-lines of Gemini the Twins. The Beehive or Praesepe ("Manger"), Star Cluster is more visually striking than any of the star-patterns in the constellation. The cluster is in the middle of an irregular square formed by Gamma (γ), Delta (δ), Eta (ϵ), and Theta (φ) Cancri. The square marks the body of the Crab, the creature's legs extending out to Alpha (α), Beta (β), Phi-two (φ^2) and Iota-one (ι^1). The hazy little oval of the Praesepe has been used since at least ancient Greek times as a visual barometer or weather forecaster: if it was invisible on an otherwise dark, moonless night, the ancient mariners of the Mediterranean knew that rain and storms were on the way. Cancer, 506 square degrees in area, ranks 31st among the 88 constellations.

In mythology, the goddess Juno, who hated Hercules, sent Cancer, the Crab, to pinch and distract him as he fought with Hydra. However, Hercules simply crushed Cancer underfoot. Juno then elevated the Crab to the heavens as a reward.

Two thousand years ago, the summer solstice (June 21) occurred while the Sun was in Cancer, but today the Sun traverses the constellation from July 7 to August 14. Nevertheless, the parallel of latitude which marks the northernmost points on the Earth's surface where the Sun passes straight overhead at noon on June 21 is still called the Tropic of Cancer.

Cancer contains two large, impressive open star clusters, many worthwhile double and multiple stars, and because it is located away from the Milky Way, numerous galaxies.

Cancer: KAN-ser
Genitive: Cancri, KAN-kre
Abbrevation: Cnc
Culmination: 9pm–Mar. 16, midnight–Jan. 30
Area: 506 square degrees
Showpieces: Zeta (16–ζ) Cnc, Phi-2 (23–ϕ^2) Cnc, Iota-1 (48–ι^1) Cancri, Σ1245, M44 (NGC 2632), M67 (2682)
Binocular Objects: M44 (NGC 2632), M67 (2682), β584, Iota-1 (48-ι^1) Cancri.

7.2 Interesting Stars

Zeta (ζ) = 16 Cancri (Σ1196)   Triple Star   Spec. F7 G2
AB: 5.6, 6.0; Sep. 0.8″; P.A. 72° $08^h12.2^m$ +17°39′
ABxC: 6.2; Sep. 5.7″; P.A. 88°
Constellation Chart 7-1   ★★★★★

12/14″ Scopes–225x: Zeta Cancri is a lovely triple of yellow stars. The AB pair appears oblong but is not separated.

Phi-2 (φ^2) = 23 Cancri (Σ1223)   Dbl Star   Spec. A2 A2
m6.3, 6.3; Sep. 5.1″; P.A. 218° $08^h26.8^m$ +26°56′
Finder Chart 7-3   ★★★★

2/3″ Scopes–75x: Beautiful! Phi Cancri is an equally matched pair of white stars.

12/14″ Scopes–125x: This fine pair of white stars looks like approaching headlights.

Σ1245   Multiple Star   Spec. F6 G5
AB: 6.0, 7.2; Sep. 10.3″; P.A. 25° $08^h35.8^m$ +06°37′
Finder Chart 7-5   ★★★★

2/3″ Scopes–50x: Σ1245 is a fine triple! The closer AB pair is composed of yellowish-white and yellow stars. The C component is white.

Cancer, the Crab

Constellation Chart 7-1

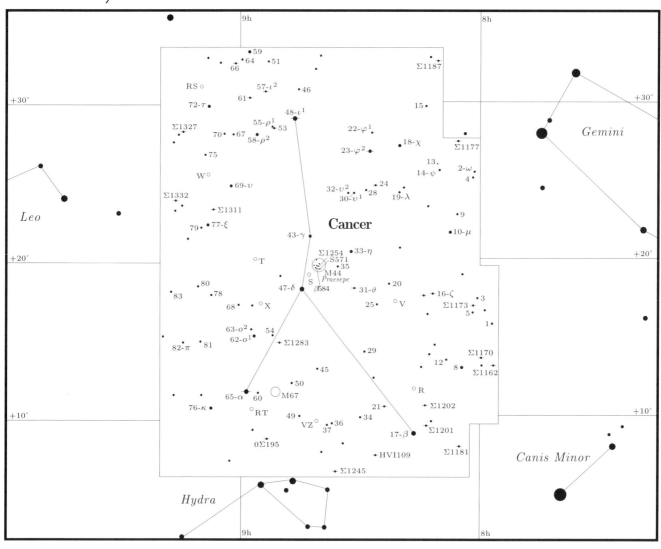

Table 7-1. Selected Variable Stars in Cancer

Name	HD No.	Type	Max.	Min.	Period (Days)	F*	Spec. Type	R.A. (2000)	Dec.	Finder Chart No. & Notes
R Cnc	69243	M	6.07	11.8	361.60	0.47	M6-M9	$08^h16.6^m$	+11°44'	7-1
V Cnc	70276	M	7.5	13.9	272.13	0.46	S0-S7	21.7	+17 17	7-1
VZ Cnc	73857	δ Sct	7.18	7.91	0.17	0.26	A7-F2	40.9	+09 49	7-5
X Cnc	76221	SRb	5.6	7.5	195.		C5,4	55.4	+17 14	7-4
T Cnc		SRb	7.6	10.5	482.0	0.35	C3,8-C5,5	56.7	+19 51	7-4
RT Cnc	76734	SRb	7.12	8.6	60.		M5	58.3	+10 51	7-5 1° south of α Cnc
W Cnc	78585	M	7.4	14.4	393.22	0.40	M6.5-M9	$09^h09.9^m$	+25 15	7-1
RS Cnc	78712	SRc?	6.2	7.7	120.		M6	10.6	+30 58	7-1

F* = The fraction of period taken up by the star's rise from min. to max. brightness, or the period spent in eclipse.

Table 7-2. Selected Double Stars in Cancer

Name	ADS No.	Pair	M1	M2	Sep."	P.A.°	Spec	R.A. (2000)	Dec	Finder Chart No. & Notes
Σ1162	6482		7.9	9.8	9.1	328	G5	07^{h}57.4^m	+13°12′	7-1 Yellowish & blue
Σ1170	6499		8.5	8.5	2.2	101	F5	59.7	+13 41	7-1 Equal pair of white stars
Σ1173	6519		7.9	9.6	10.0	51	G0	08^{h}01.4^m	+16 57	7-1 Yellowish & bluish
Σ1181	6570		8.0	9.5	5.2	140		05.4	+08 12	7-1 Nice! Whitish & bluish
Σ1177	6569		6.6	7.5	3.5	351	B9	05.6	+27 32	7-3 Boths stars are white
Σ1187	6623		7.1	8.0	2.8	27	F2	09.5	+32 13	7-1 Fine pair of white stars
16–ζ Cnc	6650	AB	5.6	6.0	6.0	*72	F7	12.2	+17 39	7-1 A,B,C: All yellow stars
	6650	ABxC		6.2	5.7	88	G2			
	6650	ABxD		9.7	287.9	108				
Σ1201	6659		7.8	9.5	6.6	183	A5	12.9	+09 35	7-1 Whitish & bluish
Σ1202	6663		7.4	9.5	2.4	311	F8	13.6	+10 51	7-1
H VI 109	6805		5.1	9.2	31.5	342	K0	25.9	+07 34	7-1
24 Cnc	6811	AB	7.0	7.8	5.8	47	A3 G	26.7	+24 32	7-3 Yellowish & bluish
23–φ² Cnc	6815		6.3	6.3	5.1	218	A2 A2	26.8	+26 56	7-3 White-like auto headlights
31–υ Cnc			5.4	9.8	63.1	61	M	31.6	+18 06	7-4
Σ1245	6886	AB	6.0	7.2	10.3	25	F6 G5	35.8	+06 37	7-5 A-B: Yellowish & bluish
	6886	AC		10.7	93.2	120				C: White
	6886	AD		12.2	117.0	282				
	6886	AE		8.8	112.4	210				
β584 (In M44)	6915	AB	6.9	11.9	1.4	292	A0	39.9	+19 33	7-4 Orange primary with
S571 (In M44)	6915	AC		7.2	45.2	156	A0 A0			two bluish companions
	6915	AD		6.7	92.9	241	K0			
Σ1254 (In M44)	6921	AB	6.4	8.9	20.5	54	G5	40.4	+19 40	7-4 Yellowish primary with
	6921	AC		8.6	63.2	342	A0			all bluish companions
	6921	AD		8.9	82.6	43				
48–ι¹ Cnc	6988		4.2	6.6	30.5	307	G5 A5	46.7	+28 46	7-3 Yellow & blue
Σ1283	7031		7.6	8.6	16.4	123	F0	50.0	+14 50	7-1
53 Cnc			6.2	9.7	43.1	333	M0	52.5	+28 16	7-1
OΣ195	7073		8.0	8.5	9.6	139	F8	54.0	+08 25	7-5 Subtle yellow & blue
57–ι² Cnc	7071	AB	6.0	6.5	1.4	316	K0	54.2	+30 35	7-1 A-B: deep yellow
	7071	ABxC		9.1	55.6	199				C: Blue
65–α Cnc	7115		4.3	11.8	11.3	325	A3	58.5	+11 51	7-5
66 Cnc	7137	AB	5.9	8.0	4.6	137	A2	09^{h}01.4^m	+32 15	7-1 A-B: White & light blue
	7137	AC		10.8	187.4	319				
Σ1311	7187	AB	6.9	7.3	7.5	200	F5	07.4	+22 59	7-4 Both stars white
	7187	AC		12.6	27.8	118				
Σ1327	7260	AB	8.2	9.4	8.0	61	F8	15.5	+27 55	7-1 Yellowish primary with
	7260	AC		9.3	26.3	22				bluish companions
Σ1332	7281		7.8	8.1	5.9	26	F5	17.3	+23 39	7-1 Matched white stars

Footnotes: *= Year 2000, a = Near apogee, c = Closing, w = Widening. Finder Chart No: All stars listed in the tables are plotted in the large Constellation Chart, but when a star appears in a Finder Chart, this number is listed. Notes: When colors are subtle, the suffix -*ish* is used, e.g. *bluish*.

Iota-1 (ι¹) = 48 Cancri (Σ1268) Dbl Star   Spec.G5 A5
m4.2, 6.6; Sep. 30.5″; P.A. 307°   08^{h}46.7^m +28°46′
Finder Chart 7-3   ★★★★★

2/3″ Scopes–50x: Iota-1 is a striking gold and blue pair.

12/14″ Scopes–100x: This lovely deep-yellow and blue pair reminds one of Beta Cygni (Albireo).

Iota-2 (ι²) = 57 (Σ1291) Cancri Triple Star   Spec. K0
AB: 6.0, 6.5; Sep. 1.4″; P.A. 316°   08^{h}54.2^m +30°35′
Constellation Chart 7-1   ★★★★

12/14″ Scopes–300x: Iota-2 is a triple with an unresolved, but visibly oblong, yellow primary and a blue companion.

7.3 Deep-Sky Objects

NGC 2535 U4264 Galaxy Type SA(r)c pec I
⌀3.3′ × 2.3′, m12.8v, SB 14.8   08^{h}11.2^m +25°12′

NGC 2536 Galaxy Type SB(rs)c pec
⌀0.7′ × 0.5′, m14.2v, SB 12.9   08^{h}11.3^m +25°11′
Finder Chart 7-3, Figure 7-1   ★/★

12/14″ Scopes–125x: Located 22′ SSE of 5.7 magnitude Psi (ψ) = 14 Cancri, NGC 2535 is extremely faint, small, round, and diffuse with uniform surface brightness. Three stars of 11.5 to 12.5 magnitude nearly touch the NW side. Companion NGC 2536, 2′ SSE, is very faint, tiny, and diffuse.

16/18″ Scopes–150x: NGC 2535 is a very faint, small spot less than 30″ in diameter, slightly elongated

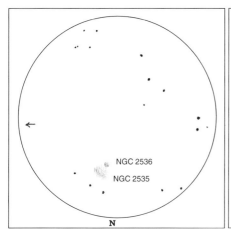

Figure 7-1. NGC 2535 & NGC 2536
17″, f4.5–255x, by Murray Cragin

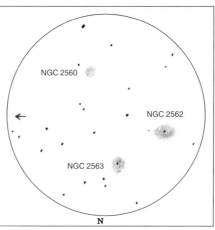

Figure 7-2. NGC 2562, 2563, & 2560
13″, f5.6–165x, by Steve Coe

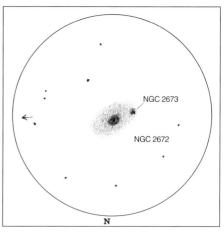

Figure 7-3. NGC 2672-73
13″, f5.6–165x, by Steve Coe

NE–SW, and slightly brighter in the center. Lying 2′ SSE, interacting companion NGC 2536 is just visible as tiny spot, much smaller than its companion but with a faint stellar nucleus.

NGC 2545 H627² Galaxy Type (R)SB(r)ab I-II
⌀2.3′ × 1.1′, m12.4v, SB 13.3   08ʰ14.2ᵐ +21°21′
Finder Chart 7-3   ★★

12/14″ Scopes–100x: This galaxy is faint, fairly small, and elongated 1.2′ × 0.4′ N–S with some brightening in the middle. An 8.5 magnitude star lies 24′ SSE.

16/18″ Scopes–150x: NGC 2545 has a moderately faint halo, elongated 1.5′ × 0.5′ N–S with a slight central brightening and a faint stellar nucleus. A star slightly fainter than the nucleus lies at the galaxy's NW edge.

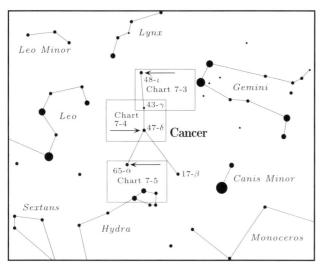

Master Finder Chart 7-2. Cancer Chart Areas
Guide stars indicated by arrows.

NGC 2554 H303² Galaxy Type S0/a
⌀3.2′ × 2.4′, m12.0v, SB 14.0   08ʰ17.9ᵐ +23°28′
Finder Chart 7-3   ★

16/18″ Scopes–150x: NGC 2554 is located 3/4° SW of 5.3 magnitude Lambda (λ) = 19 Cancri, midway between it and a 6th magnitude star to the SW. This galaxy appears faint, small, elongated 2′ × 1.5′ N–S with considerable brightening at the center.

NGC 2562 H607³ Galaxy Type S0/a:
⌀1.2′ × 0.9′, m12.9v, SB 12.8   08ʰ20.4ᵐ +21°08′
Finder Chart 7-3, Figure 7-2   ★★

12/14″ Scopes–125x: NGC 2562 has a faint halo less than 30″ in diameter with a slight elongation N–S and some brightening at its center. NGC 2562 lies 5′ NW of NGC 2563 in the Cancer I Galaxy Group.

16/18″ Scopes–150x: NGC 2562 has a fairly faint oval-shaped halo elongated 0.75′ × 0.5′ N–S with a slight brightening at center and a stellar nucleus. Lying 7′ SW, NGC 2560 has a faint, small, diffuse circular halo.

NGC 2563 H634² Galaxy Type S0°:
⌀2.0′ × 1.7′, m12.2v, SB 13.4   08ʰ20.6ᵐ +21°04′
Finder Chart 7-3, Figure 7-2   ★★

12/14″ Scopes–125x: NGC 2563 shows a faint halo elongated 1′ × 0.75′ E–W with some brightening at center. Lying 5′ NW, NGC 2562 is similar in brightness to, but much smaller than, NGC 2563. Both are members of the Cancer I Galaxy Group.

16/18″ Scopes–150x: This galaxy is elongated 1.5′ × 1′ E–W with a slight brightening at center and a faint stellar nucleus. There is a 13th magnitude star 2′ west.

Cancer

NGC 2577 H259² Galaxy Type S0–:
ø1.8′ × 1.1′, m12.4v, SB 13.0 08ʰ22.7ᵐ +22°33′
Finder Chart 7-3   ★

16/18″ Scopes–150x: NGC 2577 is located 1/2° NE of a wide pair of 7th and 8th magnitude stars. It has a very faint, oval-shaped halo elongated 1.25′ × 0.75′ ESE–WNW.

NGC 2598 U4443 Galaxy Type SBa?
ø1.1′ × 0.5′, m13.7v, SB 12.8 08ʰ30.0ᵐ +21°29′
Finder Chart 7-3   ★

16/18″ Scopes–150x: NGC 2598, 1/2° NE of a 7th magnitude star, is visible as a very faint smudge elongated 1′ × 0.5′ N–S.

NGC 2599 H234³ Galaxy Type SAa
ø2.4′ × 2.4′, m12.2v, SB 14.0 08ʰ32.2ᵐ +22°34′
Finder Chart 7-3   ★

16/18″ Scopes–150x: NGC 2599 lies 8′ north of two widely separated 9th magnitude stars in a rather sparse star field. The halo appears very faint, round, and diffuse with a 1.5′ diameter halo.

NGC 2608 H318² Galaxy Type SB(s)b: II
ø2.3′ × 1.7′, m12.3v, SB 13.6 08ʰ35.3ᵐ +28°28′
Finder Chart 7-3   ★

12/14″ Scopes–125x: NGC 2608 is located 1/2° from a wide pair of 7th and 8th magnitude stars and just 5′ north of an 11.5 magnitude double star. Its faint halo is elongated 1.25′ × 1′ ENE–WSW with a slight brightening at the center.

16/18″ Scopes–150x: This galaxy appears faint, elongated 1.5′ × 1′ ENE–WSW with considerable brightening at center and a stellar nucleus. NGC 2619 is located 33′ ENE beyond an 8th magnitude star located between it and NGC 2608.

NGC 2623 U4509 Galaxy Type Pec
ø2.5′ × 0.7′, m13.4v, SB 13.8   08ʰ38.4ᵐ +25°45′
Finder Chart 7-3   ★

12/14″ Scopes–125x: This galaxy has a very faint, circular 1.5′ halo and is slightly brighter in the center.

NGC 2632 Messier 44 Open Cluster 50★ Tr Type II 2m
ø95′, m3.1v, Br★ 6.3   08ʰ40.1ᵐ +19°59′
Finder Chart 7-4, Figure 7-4   ★★★★★
The Beehive or Praesepe

M44 has been known since at least early Greek times. It was called Praesepe, the "Manger," by the Greek astronomical poet Aratus in 270 B.C. Gamma (γ)

Figure 7-4. *Messier 44 (NGC 2632), the Beehive, is a loose, irregular cluster of a hundred stars requiring low power to take in its span of 1.5 degree. J. C. Mirtle took this 20 minute exposure on 2415 film with a 6″, f4.5 Newtonian reflector.*

and Delta (δ) Cancri were called the Aselli, "Asses," feeding at the manger. M44 cannot be resolved with the unaided eye, so its true nature remained a mystery until Galileo looked at the cluster through his newly-invented telescope in 1610.

Messier 44 is, in apparent terms, one of the largest and brightest of the open clusters. To a large degree this

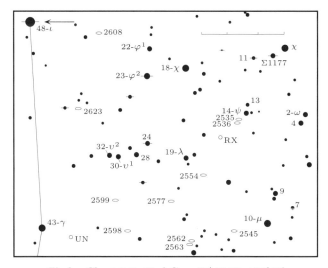

Finder Chart 7-3. 48-ι¹ Cnc: 08ʰ46.7ᵐ +28°46′

Figure 7-5. *One of the oldest known clusters, Messier 67 (NGC 2682) is an impressive sight with a hundred stars visible in small telescopes. Martin C. Germano made this 30 minute exposure on 2415 film with an 8", f5 Newtonian reflector.*

is because it is also one of the nearest, being only about 525 light-years distant. It is, however, a rather aged open cluster, probably around 700 million years old, ten times the age of the Pleiades cluster. It seems to have the same motion, velocity, and age as the Hyades cluster in Taurus, perhaps indicating some common origin.

4/6" Scopes–40x: Messier 44 is very bright and extremely large, spanning 1.5°. At least sixty 6th to 12th magnitude stars are visible, including a prominent star chain running NNW–SSE. Messier 44 may be seen with the naked eye, and even a typical viewfinder resolves about twenty stars. In binoculars it truly does suggest a swarm of bees.

8/10" Scopes–50x: This cluster is very bright, exceptionally large and only slightly compressed. 75 stars can be seen. Several star chains wind through the cluster, and a nice triple star is situated near its center.

12/14" Scopes–60x: Messier 44 is a splendid sight! It is an exceptionally bright, large, loose, and irregular cluster of a hundred 8.5 to 12th magnitude stars. The stars are arranged in doubles, triples, and star-groups. Many of the brighter stars are distinctly yellow or bluish-white.

PK208+33.1   Abell 30   Planetary Nebula   Type 2c
ø109", m15.5v, CS 14.3v   08ʰ46.8ᵐ +17°53'
Finder Chart 7-4   ★

16/18" Scopes–100x: Viewed through an O-III filter, this planetary is an extremely faint, large disk just detectable with averted vision. However, its 14th magnitude central star is rather easy.

NGC 2672   U4619   Galaxy   Type E1-2
ø3.0' × 2.8', m11.7v, SB 13.9   08ʰ49.37ᵐ +19°04'
Finder Chart 7-4, Figure 7-3   ★★★

12/14" Scopes–125x: NGC 2672 shows a moderately

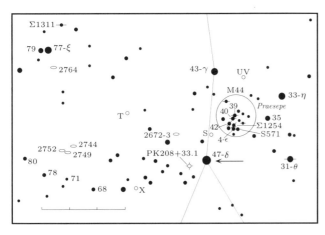

Finder Chart 7-4. 47-δ Cnc: 08ʰ44.7ᵐ +18°09'

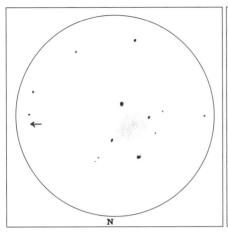
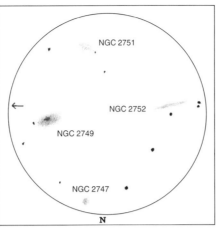
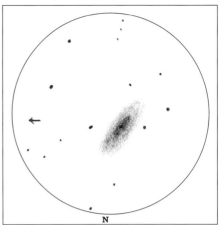

Figure 7-6. PK219-31.1
17.5″, f4.5–80x, by Dr. Jack Marling

Figure 7-7. NGC 2747, 2749, 2751, & 2752
20″, f4.5–165x, by Richard W. Jakiel

Figure 7-8. NGC 2775
13″, f5.6–165x, by Steve Coe

bright 1.5′ diameter halo slightly elongated E–W with a brighter middle. NGC 2673 makes contact at the eastern edge.

16/18″ Scopes–150x: This galaxy has a fairly bright, 2′ diameter halo slightly elongated ENE–WSW with a much brighter center and a faint, tiny nucleus. A 13th magnitude star is 1′ NNE. NGC 2673, a faint companion, lies less than 45″ ESE.

NGC 2673   U4620   Galaxy   Type E0 pec
⌀0.8′x0.8′, m13.3v, SB12.7   08ʰ49.41ᵐ +19°04′
Finder Chart 7-4, Figure 7-3   ★

12/14″ Scopes–125x: Lying on the ESE edge of NGC 2672, this object is a very faint 30″ diameter knot with a prominent core and a tiny nucleus. Two stars are just to its SE and NNE.

NGC 2682   Messier 67   Open Cl. 200★   Tr Type II 2 m
⌀29′, m6.9v, Br★ 9.69   08ʰ50.4ᵐ +11°49′
Finder Chart 7-5, Figure 7-5   ★★★★★

Messier 67, roughly 5 billion years old, is one of the most ancient star clusters known. Most open clusters are distributed along or near the plane of our Galaxy's spiral disk; but M67 is so old that it has worked its way 1,500 light-years off the plane of the Milky Way out to the dust and gas poor fringes of the spiral disk where there is little obscuring gas and dust. The cluster is 2,600 light years away and about 12 light years in diameter with over 500 members.

4/6″ Scopes–50x: Impressive! M67 is a bright cluster, with two 8th magnitude and at least fifty 9th to 12th magnitude stars spread over a 30′ diameter area. The 20 brightest stars form a chain in the eastern portion which runs southward then swings westward below most of the other members.

8/10″ Scopes–60x: This is a fine, bright cluster! It is very large and quite rich with nearly a hundred 10th to 14th magnitude stars. Its members are distributed in several rich clumps, one of which is on the cluster's southern edge near the brighter stars. 150x shows some beautiful star chains, and several obvious dark lanes meander like rivers through the cluster. A 7.5 magnitude star is near its NE edge.

12/14″ Scopes–75x: A very nice cluster! M67 is very bright, quite large and very rich with well over a hundred 9th to 14th magnitude stars visible, twenty being 10th magnitude. The main concentration of stars runs NE–SW in a bar-shaped formation that fans out at the NE end. An 8th magnitude star lies just beyond these stars.

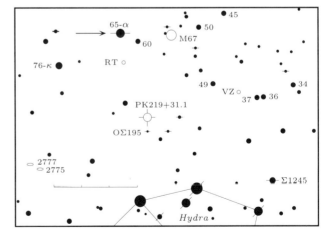

Finder Chart 7-5. 65-α Cnc: 08ʰ58.3ᵐ +11°51′

PK219-31.1 Sh2-290 Planetary Nebula Type 3a
ø>980″, m12.0v, CS 15.5v $08^h54.2^m$ +08°55′
Finder Chart 7-5, Figure 7-6   ★

8/10″ Scopes–75x: In a UHC filter, this planetary nebula is an extremely faint, large glow. It is slightly brighter around the edge near four 10th magnitude stars. There are eight 10th to 12th magnitude stars in the field of view. The planetary lies 30′ north of OΣ195, a double composed of 7.5 and 8th magnitude stars separated by 10″.

12/14″ Scopes–75x: Through a UHC filter, PK219-31.1 appears very faint, exceptionally large, and irregularly round with no brightening at center. At least three stars are involved in its dull glow.

16/18″ Scopes–75x: Viewed in an O-III filter, this giant planetary nebula is faint but obvious. The halo is circular, its edges diffuse, and contained by a parallelogram of fairly bright stars elongated 12′ × 8′ N–S.

NGC 2744 H60³ Galaxy Type SB(s)ab: pec
ø1.7′ × 1.3′, m13.5v, SB 14.2 $09^h04.7^m$ +18°28′
Finder Chart 7-4   ★

16/18″ Scopes–150x: This galaxy is the northernmost object in a group of faint galaxies located 14′ NW of NGC 2749. NGC 2744 appears faint, elongated 1.5′ × 1′ E–W with a dim stellar nucleus. A faint star is 2′ NE. Along the southern edge, a row of four stars curves westward, with the third star a double.

NGC 2749 U4763 Galaxy Type E3
ø1.8′ × 1.7′, m11.8v, SB 12.8 $09^h05.4^m$ +18°19′
Finder Chart 7-4, Figure 7-7   ★★★

12/14″ Scopes–125x: NGC 2749 has a faint, 1.5′ diameter halo slightly elongated E–W with a fairly bright core and a faint stellar nucleus. It is the brightest galaxy in a group which includes NGC 2744 14′ NW, NGC 2751 4′ SE, and NGC 2752 5′ east.

NGC 2752 U4772 Galaxy Type SBb: sp II-III
ø1.6′ × 0.4′, m13.7v, SB13.1 $09^h05.7^m$ +18°20′
Finder Chart 7-4, Figure 7-7   ★

16/18″ Scopes–150x: NGC 2752 is very faint, small, and elongated 1.5′ × 0.4′ ENE–WSW with a very slight brightening at its center. A fairly bright, close double star lies 1.5′ ENE. NGC 2752 is located in the eastern portion of a group of half a dozen faint galaxies. A row of three fairly bright stars extends NW from its northern side toward NGC 2747, visible as a faint, small, round smudge. NGC 2749 lies 5.5′ west from the SW tip of NGC 2747.

NGC 2764 H236³ Galaxy Type S0:
ø1.3′ × 0.8′, m12.9v, SB 12.8 $09^h08.3^m$ +21°27′
Finder Chart 7-4   ★★

12/14″ Scopes–150x: This galaxy is 1/2° SSW of the wide pair of bright stars Xi (ξ) = 77 Cancri (m5.14) and 79 Cancri (m6.01). NGC 2764 has a fairly faint, circular 1′ diameter halo, slightly elongated NNE–SSW with a prominent core. A very faint star lies at the northern edge. A 10th magnitude star is 2′ north and another star SE.

NGC 2775 H2¹ Galaxy Type SA(r)ab
ø4.6′ × 3.7′, m10.1v, SB 13.1 $09^h10.3^m$ +07°02′
Finder Chart 7-5, Figure 7-8   ★★★★

8/10″ Scopes–100x: This galaxy is located 1/2° south of an 8th magnitude star at the east end of an E–W row with a 7th magnitude and another 8th magnitude star. Its halo is fairly bright, 3′ × 1.5′ in diameter, and has a large, bright nucleus.

12/14″ Scopes–125x: NGC 2775 has a bright mottled halo, elongated 3.5′ × 1.7′ NNW–SSE with a gradual increase in surface brightness toward the center to an extended core containing a stellar nucleus. NGC 2777 lies 11′ NNE.

NGC 2777 U4823 Galaxy Type Sab?
ø0.6′ × 0.5′, m13.3v, SB 11.8 $09^h10.7^m$ +07°12′
Finder Chart 7-5   ★

16/18″ Scopes–150x: This tiny object may be found 30′ SE of the 7th magnitude star at the center of a 45′ long row with two 8th magnitude stars. At 225x, NGC 2777 has a faint, circular 30″ diameter halo slightly elongated NNW–SSE with a slight brightening in the core area. A stellar nucleus is suspected. NGC 2775 lies 11′ SSW.

Chapter 8

Canis Major, the Big Dog

8.1 Overview

Orion had two hunting dogs, Canis Major and Canis Minor, his faithful companions who help him track and fetch game. Southeast of Orion, Canis Major, the big dog, stands beside the hunter's feet, ready to spring into action. The Big Dog has his eye, Sirius, fixed on Lepus, the Hare, crouched at Orion's feet. With bright and easily recognized stars, Canis Major contains Sirius, the brightest star in the heavens with a magnitude of −1.4. Excluding the Sun and Moon, only the planets Venus, Jupiter, Saturn, and Mars outshine it. Sirius is a part of the Winter Triangle along with the stars Procyon, marking the little dog, and Betelgeuse indicating Orion's right shoulder.

In ancient Egypt, the rising of Sirius in the early morning sky, was an indication of the coming annual flooding of the Nile River. In Egypt, Sirius was known as Anubis, depicted as a god with a dog's head.

Canis Major covers a small area, only 380 square degrees of sky; but it certainty is not lacking in interesting objects for the telescope user. Due to its location in the winter Milky Way, Canis Major is rich in open star clusters with Messier 41 claiming top honors. There are several complex nebulosities, a planetary nebula, and even a few galaxies shining through the Milky Way's obscuring blanket of dust and gas. For the double star enthusiast, this rich area is a special treat.

8.2 Interesting Stars

Nu-1 (ν^1) = 6 Canis Majoris Double Star Spec. G5 G0
m5.8, 8.5; 17.5″; 262° $06^h36.4^m$ $-18°40'$
Finder Chart 8-3   ★★★★

2/3″ Scopes–50x: Low power will show an easy yellow pair with 6–ν^1 and 7–ν^2 (Nu) Canis Majoris in the same field of view.

8/10″ Scopes–75x: These stars display colors of subtle yellowish and dark bluish hues.

Canis Major: KAY-nis May-jer
Genitive: KAY-nis May-JOR-is
Abbrevation: CMa
Culmination: 9pm–Feb. 16, midnight–Jan. 2
Area: 380 square degrees
Showpieces: 6-ν^1 CMa, 18-μ CMa, 21-ϵ CMa, h3945, M41 (NGC 2287), NGC 2360, NGC 2362
Binocular Objects: Δ47, M41 (NGC 2287), NGC 2345, NGC 2360, NGC 2362, NGC 2367, NGC 2374, NGC 2384

Alpha (α) = 9 Canis Majoris   Spec. A1 V
m−1.5, 8.5; Sep. 4.5″; P.A. 5° $06^h45.1^m$ $-16°43'$
Finder Chart 8-3   Sirius   ★★★★

Sirius, "The Sparkling One," is also known as the "Dog Star" or "Nile Star." It is the brightest star in our sky; at −1.46 apparent magnitude it is nine times brighter than a first magnitude star. Sirius has a diameter 1.8 times that of the Sun and a mass 2.35 times the Sun's. Its surface temperature is 10,000°K with a luminosity of about 23 suns, but its brilliance in our sky is due primarily to its close proximity. At a distance of 8.7 light years, it is the fifth nearest star. (We're counting binaries as one star system.)

Sirius shares the true space motion of the widely-scattered Ursa major Moving Group, which includes Alpha Ophiuchi, Beta Aurigae, Alpha Coronae Borealis, and Delta Leonis as well as the five central stars of the Big Dipper.

Sirius has a white dwarf companion about 24 A.U. away with an orbital period of approximately 50 years. The companion has a mass equal to that of the Sun, but its diameter is 40 to 50 times smaller. A typical white dwarf star has an incredibly high density of about 2.25 tons per cubic inch compared to the Sun's density of about 0.5 ounce per cubic inch.

Canis Major, the Big Dog

Constellation Chart 8-1

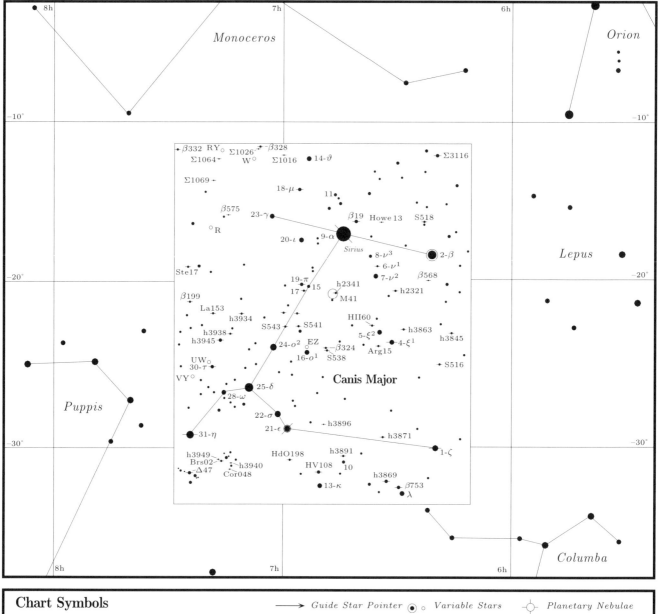

Table 8-1. Selected Variable Stars in Canis Major

Name	HD No.	Type	Max.	Min.	Period (Days)	F*	Spec. Type	R.A. (2000)	Dec.	Finder Chart No. & Notes
2-β CMa	44743	BC	1.3	2.0	0.250		B1	06ʰ22.7ᵐ	−17°57′	8-3
W CMa	54361	Lb	6.3	7.9			C6	07ʰ08.1ᵐ	−11 55	8-6
RY CMa	56450	Cd	7.7	8.4	4.678	0.24	F6–G0	16.6	−11 29	8-6
UW CMa	57060	EB/KE	4.8	5.3	4.393		O7	18.7	−24 34	8-7 22′ N of NGC 2362
R CMa	57167	EA/SD	5.7	6.3	1.135	0.15	F1	19.5	−16 24	8-6
VY CMa	58061	Unq	6.5	9.6			M5	23.0	−25 46	8-8 Reddish-orange

F* = The fraction of period taken up by the star's rise from min. to max. brightness, or the period spent in eclipse.

Table 8-2. Selected Double Stars in Canis Major

Name	ADS No.	Pair	M1	M2	Sep."	P.A.°	Spec	R.A. (2000)	Dec	Finder Chart No. & Notes
h3845			6.1	11.1	47.6	6	G0	06ʰ17.1ᵐ	−22°43′	8-4 Bright star is ruddy S516
		AB	7.2	8.3	61.5	7	A0	19.3	−24 58	8-4 Yellowish-white & blue
		AC		7.1		243	B9			
1-ζ CMa			3.0	7.6	175.5	338	B8 K0	20.3	−30 04	8-1
Σ3116	4978	AB	5.6	9.8	4.2	23	B2	21.4	−11 46	8-1
β568	5023		7.2	7.5	0.8	155	B8	23.8	−19 47	8-4 White & bluish
S518	5034		6.7	9.1	16.6	88	A3	24.3	−16 13	8-3
β753			5.9	7.9	1.3	45	B3	28.7	−32 22	8-5
h3863	5128		6.9	8.7	2.6	119	A2	29.4	−22 35	8-4 Yellow & rust colored
h2321	5164		8.2	9.2	5.6	307	F5	31.4	−20 37	8-4 Fine pair of white stars
h3869			5.7	7.7	24.9	258	B3 A0	32.6	−32 02	8-5 White & bluish
h3871	5214		7.1	8.2	7.6	354	A2	34.1	−29 38	8-5 Yellowish-white & bluish
Howe 13	5242		7.8	8.8	11.2	300	B9	35.8	−16 06	8-3
Arg 15			7.9	9.0	39.1	234	B9 F8	36.1	−24 07	8-4 White & light blue
6-ν¹ CMa	5253		5.8	8.5	17.5	262	G5 G0	36.4	−18 40	8-3 Subtle yellow & dark blue
H II 60	5260		6.4	10.0	9.1	336	B8	36.7	−22 37	8-4
β19	5358		7.0	9.3	3.6	166	B9	42.0	−16 00	8-3 White & blue pair
9-α CMa	5423		−1.5	8.5	4.5c	y 5	A0	45.1	−16 43	8-3 Alpha brilliant blue-white
h3891			5.7	8.0	4.9	223	B3	45.5	−30 57	8-5 White & bluish
h2341	(In M41)		8.3	9.1	45.0	86	A0 A0	45.6	−20 41	8-4
S538			7.2	8.5	27.2	3	A2	49.6	−24 09	8-7 White & bluish
β324	5498	AB	6.3	7.6	1.8	206	A0	49.7	−24 05	8-7 AB both white
S537	5498	AC		8.6	30.5	281	A2			C component is blue
	5498	AD		12.8	29.3	2				
H V 108		AB	5.6	8.2	42.9	66	B8 A3	50.4	−31 42	8-8 White & bluish
h3896	5527	ABxC	9.4	8.9	11.3	163	F0	51.1	−28 44	8-8 Fine white duo
17 CMa	5585	AB	5.8	9.3	44.4	147	A2 K5	55.0	−20 24	8-3 White primary
	5585	AC		9.0	50.5	184	—			Companions both bluish
	5585	AD		9.5	129.9	186	—			
19-π CMa	5602		4.7	9.7	11.6	18	F2	55.6	−20 08	8-3 Yellowish & bluish stars
18-μ CMa	5605	AB	5.3	8.6	3.0	340	G5	56.1	−14 03	8-6 Yellow & blue
S541			8.0	9.0	23.2	44	K0 F	56.6	−22 39	8-7 Yellow & blue
21-ε CMa	5654		1.5	7.4	7.5	161	B1	58.6	−28 58	8-8
HdO 198		AB	6.4	9.0	35.0	315	B6	58.7	−31 00	8-8 White & bluish
		AC		10.0	70.0	320	—			
S543			8.5	8.6	91.0	272	K0 A0	07ʰ00.0ᵐ	−22 43	8-7 Orangish & white
Σ1016	5761		7.4	9.4	5.3	150	B5	04.6	−11 31	8-6
β328	5795	AB	5.7	6.9	0.6	116	B3	06.7	−11 18	8-6
Σ1026	5795	ABxC		9.1	17.8	350	—			
h3934	5863	AB	7.0	9.4	13.6	236	B5 B9	11.3	−21 48	8-7 White & light bluish
h3940			7.9	9.9	6.8	99	B9	13.2	−30 58	8-8
h3938	5912		6.5	8.3	19.7	250	B3 A5	13.8	−22 54	8-7 White & bluish
CorO 48			8.0	9.8	7.8	10	B9	14.6	−31 30	8-8
β575	5925	AB	8.0	8.0	0.7	75	F8	14.8	−15 29	8-6 White & bluish
Σ1057	5925	AC		9.7	15.6	2	—			
h3945	5951		4.8	6.8	26.6	55	K5 F0	16.6	−23 19	8-7 Beautiful yellow & blue
BrsO 2			6.3	8.1	37.9	182	A5 A5	17.0	−30 54	8-8 White & blue-white
Σ1064	5956	AC	7.6	9.1	15.6	240	F5	17.1	−12 01	8-6 White & blue
Σ1069			8.3	8.3	25.3	194	A A	18.0	−13 42	8-6
h3949			7.7	8.0	3.1	77	B3	18.6	−30 48	8-8 Lovely white pair
30-τ CMa	5977	AB	4.4	10.5	8.2	90	O9	18.7	−24 57	8-7 Yellowish & bluish
	5977	AC		11.2	14.5	79	—			
	5977			8.8	85.0	74	—			
La153	5986		7.5	7.6	3.9	346	A2	19.3	−22 03	8-7
31-η CMa			2.4	6.9	178.7	285	B7 A0	24.1	−29 18	8-8 White & blue stars
Δ47		AC	5.5	7.6	99.2	342	K2 A0	24.7	−31 49	8-8 Binocular double
β199	6065	AB	7.1	8.1	1.8	22	B2	25.1	−21 10	8-7 White & bluish
Ho 522	6065	AC		12.8	6.7	117				
Stone 17	6078		7.9	9.9	4.9	76	A0	26.0	−18 21	8-6 Close white stars
β332	6104	AB	6.1	8.1	0.7	175	F5	27.8	11 33	8-6 Yellow & bluish
Σ1097	6104	AC		8.5	20.0	313	—			
β332	6104	AD		9.5	23.2	157	—			

Footnotes: *= Year 2000, a = Near apogee, c = Closing, w = Widening. Finder Chart No: All stars listed in the tables are plotted in the large Constellation Chart, but when a star appears in a Finder Chart, this number is listed. Notes: When colors are subtle, the suffix -*ish* is used, e.g. *bluish*.

White dwarf stars are objects of great interest. These stars have exhausted their hydrogen supply, and their stellar evolution is at an end. Although they no longer produce nuclear energy, white dwarfs are visible because contraction has left them at a high temperature. If massive enough, their ultimate fate will be a "black hole." All known post-nova stars are hot dwarfs which may be an intermediate stage between normal stars and true white dwarfs. It is believed that the contraction toward the white dwarf state causes periods of instability which result in nova outbursts. The exchange of matter between very close binary stars also hastens a star's fuel burning process.

8/10″ Scopes–300x: Remaining unresolved and not even showing elongation, half of this star's image appears white while the other half is yellowish. After many observations the companion was seen in a rare moment of near perfect seeing.

16/18″ Scopes–300x: Sirius may be resolved for a few seconds but air currents keep it blurred into an elongated star. Its companion was closest in the year 1995 at 3.1″ but it will widen again to 4.6″ by the start of the 21th century.

β324 Multiple Star   **Spec. A0 & A2**
AB: m6.3, 7.6; Sep. 1.8″; P.A. 206° $06^h49.7^m$ $-24°05'$
Finder Chart 8-4   ★★★★

2/3″ Scopes–150x: The AB pair is assigned the number β324 while the AC pair is designated S537. The A and B stars appear white while the C component (m8.6; 30.5″; 281°) appears blue. There is also another companion designated D (m12.8; 29.3″; 2°). Lying 5′ south, S538 is a wide double composed of white and blue stars aligned almost perpendicular to S537.

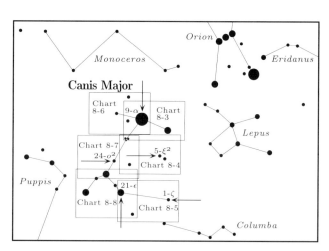

Master Finder Chart 8-2. Canis Major Chart Areas
Guide stars indicated by arrows

17 Canis Majoris = H V 65 Multiple Star Spec. A2 & K5
AB: m5.8, 9.3; Sep. 44.4″; P.A.147° $06^h55.0^m$ $-20°24'$
Finder Chart 8-6   ★★★★

2/3″ Scopes–50x: Very nice! Low power shows a lovely wide triple with a white primary and two orangish companions forming a right triangle. Another companion at 9.5 magnitude lies 130″ distant in P.A. 186°. Pi (19–π) Canis Majoris is located in the same field to the NE.

Pi (π) = 19 Canis Majoris (H V 123) Dbl Star   Spec. F2
m4.7, 9.7; Sep. 11.6″; P.A. 18° $06^h55.6^m$ $-20°08'$
Finder Chart 8-6   ★★★★

2/3″ Scopes–75x: This double has a yellowish-white primary with a bluish companion. 17 Canis Majoris lies in the same field to the SW.

Mu (μ) = 18 Canis Majoris (Σ997) Multiple Star Spec. G5
AB: 5.3, 8.6; Sep. 3.0″; P.A. 340° $06^h56.1^m$ $-14°03'$
Finder Chart 8-6   ★★★★

2/3″ Scopes–175x: This unequal pair of stars in contact presents a striking contrast of orange and blue.

8/10″ Scopes–175x: These beautiful stars of deep yellow and blue may just be separated at 175x. Two 10th magnitude companions lie much further away.

h3945 Double Star   **Spec. K5 & F0**
m4.8, 6.8; Sep. 26.6″; P.A. 55° $07^h16.6^m$ $-23°19'$
Finder Chart 8-7   ★★★★

2/3″ Scopes–50x: This intense orange and blue double rivals Albireo (β Cygni) and Almach (γ Andromedae).

Tau (τ) = 30 Canis Majoris (h3948) Multiple Star Spec.O9
AB: 4.4, 10.5; Sep. 8.2″; P.A. 90° $07^h18.7^m$ $-24°57'$
Finder Chart 8-7   ★★★★

2/3″ Scopes–50x: This fine multiple star is located within open cluster NGC 2362 in a wonderful setting profuse with stars! The primary is yellowish and the closest companion is bluish.

8.3 Deep-Sky Objects

NGC 2204 H13[7] Open Cluster 80★ Tr Type III 3 m
ø12.0′, m8.6v, Br★ 12.20 $06^h15.7^m$ $-18°39'$
Finder Chart 8-3, Figure 8-1   ★★

8/10″ Scopes–75x: Located just SSE of a 6th magnitude star, NGC 2204 is visible as a group of several dozen faint stars with a bright orange star at the northern end. The brighter stars form a wavy X-shaped pattern.

12/14″ Scopes–100x: This is a loose, irregular, 6′ diameter cluster of three dozen 12th magnitude and

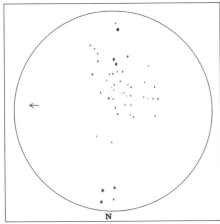

Figure 8-1. NGC 2204
12.5", f5–100x, by G. R. Kepple

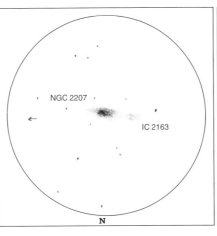

Figure 8-2. NGC 2207 & IC 2163
12.5", f5–100x, by Tom Polakis

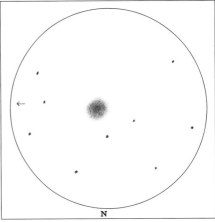

Figure 8-3. IC 2165
13", f5.6–300x, by Steve Coe

fainter stars against a background haze of unresolved stars. There are several strings of stars and a parallelogram which includes an orange star in its northern section.

NGC 2207 Galaxy Type SAB(r)c pec II
ø4.8′ × 2.3′, m10.8v, SB 13.3   06ʰ16.4ᵐ −21°22′
Finder Chart 8-4, Figure 8-2   ★★★

8/10" Scopes–100x: NGC 2207 and IC 2163 form an interacting pair lying ESE–WNW of each other. NGC 2207 appears fairly faint, elongated 1.5′ × 1′ E–W with a slight brightening at center. IC 2163 is a faint spot beyond its eastern edge.

12/14" Scopes–125x: Located on the preceding side, NGC 2207 is the brighter and larger of two interacting galaxies with a 1.5′ × 1′ halo elongated ENE–WSW and containing a bright core. IC 2163, near NGC 2207's eastern edge, is a fairly noticeable but small nodule, oval-shaped ESE–WNW.

IC 2165 PK221-12.1 Planetary Nebula Type 3b
ø4.0″, m12.9p, 10.6v, CS 14.99   06ʰ21.7ᵐ −12°59′
Finder Chart 8-3, Figure 8-3   ★★

8/10" Scopes–250x: Stellar at low power, this planetary nebula is the third star from the western end of a string of stars located 38′ west of double star Σ903 (6.1,10.8; 23.2″; 295°). A faint, small, round, bluish disk is just discerned.

12/14" Scopes–300x: This bluish planetary nebula is bright but very small and slightly elongated NE–SW with diffuse edges. The central star is not visible.

16/18" Scopes–325x: IC 2165 is a bright, tiny bluish disk slightly extended NE–SW with an irregular brightening at center, including several spots.

NGC 2217 Galaxy Type (R)SB(r)0/a
ø5.0′ × 4.5′, m10.2v, SB 13.4   06ʰ21.7ᵐ −27°14′
Finder Chart 8-5   ★★★

8/10" Scopes–100x: Located in a rich star field, NGC 2217 has a fairly faint, round 2′ diameter halo with a slight brightening at center.

12/14" Scopes–125x: A bright core is surrounded by a glow elongated 3′ × 2.5′ E–W. A stellar nucleus is visible in moments of good seeing conditions.

16/18" Scopes–150x: NGC 2217 has a fairly bright 3.5′ halo slightly extended ESE–WNW with a broad, much brighter core containing a faint stellar nucleus. The surrounding field is rich in stars.

NGC 2223 Galaxy Type SAB(r)c II
ø3.2′ × 2.3′, m11.8v, SB 13.8   06ʰ24.6ᵐ −22°50′
Finder Chart 8-4   ★★

8/10" Scopes–100x: At low power, NGC 2223 may be found 15′ east of an 8th magnitude star. Its halo is a faint, fairly small smear of light lying SW of the

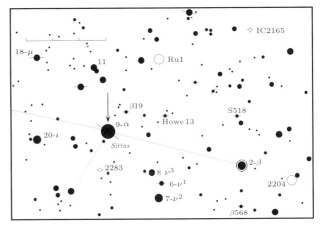

Finder Chart 8-3. 9-α CMa: 06ʰ45.1ᵐ −16°43′

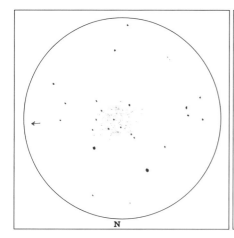
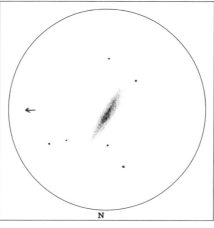
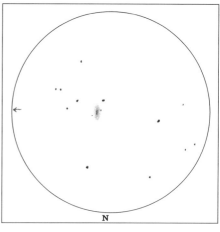

Figure 8-4. NGC 2243
17.5", f4.5–150x, by Steve Coe

Figure 8-5. NGC 2280
20", f4.5–175x, by Richard W. Jakiel

Figure 8-6. NGC 2325
13", f4.5–100x, by Tom Polakis

center star of a row of three 11th magnitude stars.

12/14" Scopes–125x: This galaxy appears fairly faint and round with some brightening in the center. A 13th magnitude star is at the northern edge.

16/18" Scopes–150x: Larger instruments reveal a prominent core, several knots, and dusty patches surrounded by a halo elongated 2.5' × 1' NNE–SSW. Two stars of the 13th–14th magnitude are involved within the halo's southern edge, and a 13.5 magnitude star is 4' north.

NGC 2243 Open Cluster 100★ Tr Type I 2 r
ø13.0', m9.4v, Br★ 11.79   06ʰ29.8ᵐ −31°17'
Finder Chart 8-5, Figure 8-4   ★★

8/10" Scopes–100x: NGC 2243 is a hazy glow located 9' SW of a 7.5 magnitude star. Five faint stars stand out against a 5' patch of stardust slightly elongated E–W.

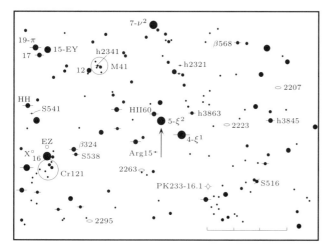

Finder Chart 8-4. 5-ξ² CMa: 06ʰ35.0ᵐ −22°58'

12/13" Scopes–125x: NGC 2243 is a moderately small sphere of 25 very faint stars embedded within an unresolved background haze. The cluster resembles a class XII globular. 250x shows good spacing between the brighter stars with many fainter members becoming visible, but resolution is subject to seeing conditions. Three brighter stars are in the middle, and a "headlight" double star is to the east.

16/18" Scopes–150x: NGC 2243 is a fairly conspicuous but rather small cluster of several dozen faint stars resolving against a background haze. The cluster appears rich at low power, but higher powers show an irregular concentration. It is surrounded by 12th magnitude stars, including a row on the preceding side and a pair on the SE side.

Ruprecht 1 Open Cluster 15★ Tr Type III 1 p
m11.0', m –, Br★ 11.0v   06ʰ36.4ᵐ −14°11'
Finder Chart 8-3   ★★

12/14" Scopes–100x: Ruprecht 1 is a faint, fairly large, sparse group of twenty stars elongated E–W. The brightest members form a kite-shaped asterism, the top of the kite facing west. A faint group of stars is detached to the NE. The cluster includes eight 11th to 12th magnitude, its other members being fainter.

NGC 2263 Galaxy Type (R)SB(r)ab
ø2.8' × 1.3', m11.9v, SB 13.2   06ʰ38.4ᵐ −24°49'
Finder Chart 8-4   ★★★

12/14" Scopes–125x: This galaxy has a fairly bright halo elongated 2.5' × 1' NNW–SSE, the slightly brighter core extending along the major axis. 12.5 magnitude stars are 2' north and south of the halo.

Ruprecht 2 Open Cluster 10★
Tr Type IV 3 p, ⌀6.0', m –, Br★ 12.0p
06ʰ41.0ᵐ –29°33'
Finder Chart 8-5   ★

12/14" Scopes–125x: This inconspicuous cluster is a faint condensation of stardust in a rich field. Less than a dozen stars are visible, the brightest being 12th and 13th magnitude. Ruprecht 3, another faint mist of stars, is located 15' NE.

Ruprecht 3 Open Cluster 15★
Tr Type II 1 p, ⌀2.8', m –, Br★ 11.0p
06ʰ42.1ᵐ –29°27'
Finder Chart 8-5   ★

12/14" Scopes–125x: This cluster is a little more compact and obvious than Ruprecht 2 located 15' SW. Ruprecht 3 has a dozen 11th to 13th magnitude stars in a 3' area and a 9th magnitude star at its SE edge.

NGC 2280 Galaxy Type SA(s)c: I-II
⌀6.3' × 2.8', m10.5v, SB 13.5
06ʰ44.8ᵐ –27°38'
Finder Chart 8-5, Figure 8-5   ★★★

12/14" Scopes–125x: NGC 2280 is rather faint, elongated 2' × 1.2' NNW–SSE, and has a weak extended core. The galaxy is contained by four 12.5 to 13 magnitude stars shaped like the Big Dipper's bowl.

16/18" Scopes–150x: The halo is fairly faint and elongated nearly 3' × 1.5' NNW–SSE. The core is inconspicuous, highly extended, and irregular in brightness with some mottling.

NGC 2283 H271³ Galaxy Type SB(s)c I-II
⌀2.8' × 2.0', m12.2v, SB 13.9   06ʰ45.9ᵐ –18°14'
Finder Chart 8-3   ★★★

12/14" Scopes–125x: Located 1.5° south of Sirius, this galaxy lies in a rich star field NE of an 18' equilateral triangle of 7–7.5 magnitude stars. NGC 2283 appears very faint, round, and diffuse with a 2' diameter halo. 12th magnitude stars are at the NE and NW edges of the halo, and a 13th magnitude star touches its southern edge.

16/18" Scopes–150x: This galaxy is faint but readily visible, its diffuse, circular 2.5' diameter halo slightly elongated N–S. Except for a slight brightening at center, the surface brightness of the galaxy is uniform.

Figure 8-7. *M41 (NGC 2287) is a fine, bright open cluster for small telescopes. Lee C. Coombs made this photograph with a 10", f5 Newtonian reflector at prime focus for five minutes on 103a-O Spectroscopic film. The image is enlarged 5.5x.*

NGC 2287 Messier 41 Open Cluster 80★ Tr Type II 3 m
⌀38', m4.5v, Br★ 6.91   06ʰ47.0ᵐ –20°44'
Finder Chart 8-4, Figure 8-7   ★★★★★

Messier 41 is a large, bright, open cluster located 4° south of Sirius. On dark clear nights it is visible to the naked eye as a hazy patch virtually the size of the full moon. Binoculars partially resolve it, and it is a fine object at low power in small telescopes. The cluster's brightest star, located near center, is a reddish K-type giant star about 700 times as luminous as the Sun. Several more K-type giants are in the cluster, but the majority of the more prominent members are blue B-type giant stars. Messier 41 lies approximately 2350 light years away. Its true size is around 24 light-years and its true luminosity 8,000 suns.

4/6" Scopes–50x: Visible to the naked-eye SE of Sirius, Messier 41 is large and bright with 50 stars visible at low power, many of them in scattered groupings. The center is marked by a 7th magnitude orange star with a 7.5 magnitude yellow companion. A row of 9th magnitude stars curves northward from the central pair. A nice triple of 7.5, 9th, and 10th magnitude stars is in the southern portion. This part of the cluster contains fainter 9th to 11th magnitude stars whereas the northern portion has relatively bright stars.

8/10″ Scopes–60x: A truly impressive cluster! It is very large, bright, and coarse with a hundred 7th to 12th magnitude stars, many of them in curved chains. Some of the star chains form together to compose a backward question mark. Other bright cluster members are in a parallelogram. The cluster has two yellow-orange stars. It is faint in a 6 × 30 finder, but an 8 × 50 finder resolves many of its stars.

12/14″ Scopes–75x: M41 is a bright, very large, loose, irregularly round cluster of a hundred 8th to 12th magnitude stars. The small clump of stars at the center is surrounded by many groupings and star chains. M41's preceding side is sprinkled with faint stars while the south side has brighter outlying stars. The 6th magnitude star 12 Canis Majoris lies near the SE edge.

NGC 2295 Galaxy Type Sab: sp
ø2.4′ × 0.7′, m12.7v, SB 13.1 $06^h47.3^m$ $-26°44'$
Finder Chart 8-4   ★★★

12/14″ Scopes–125x: NGC 2295 lies at the center of a row of four stars aligned N–S, a 12th magnitude star at each end. Its halo touches the row's two central stars, its 12.5 magnitude star to the north and a 13th magnitude star to the south. The halo is a very faint streak, elongated 1.5′ × 0.4′ NE–SW with a faint extended core. The interacting galaxies NGC 2292-93, 4′ to the east of NGC 2295, appears as one object.

NGC 2292 Galaxy Type SAB0° pec
ø4.2′ × 2.3′, m10.8v, SB 13.1 $06^h47.6^m$ $-26°45'$
NGC 2293 Galaxy Type SAB(s)0+ pec
ø4.3′ × 3.0′, m10.7v, SB 13.3 $06^h47.7^m$ $-26°45'$
Finder Chart 8-8   ★★★

12/14″ Scopes–125x: NGC 2292-93, a pair of interacting galaxies, appear as one faint common smooth halo, elongated 1.5′ × 1′ NW–SE. The SE galaxy, NGC 2293, has a tiny nucleus with its halo extending NW to NGC 2292, visible as a faint N–S streak at the halo's edge. A small asterism of 13th and 14th magnitude stars shaped like a three-bladed propeller lies 2.5′ SW. NGC 2295 is just 4′ west.

Collinder 121 Open Cluster 20★ Tr Type III 3 p
ø50′, m2.6v, Br★ 3.79 $06^h54.2^m$ $-24°38'$
Finder Chart 8-7   ★★★

8/10″ Scopes–35x: Collinder 121 is an extremely large group of 20 stars not well separated from the star field. Its five brightest stars form a crooked N–S chain with 3.79 magnitude star Omicron-1 (o^1) = 16 Canis Majoris at the northern end. The Wolf-Rayet variable EZ CMa, just north of Omicron-one, might be a true cluster member.

Ruprecht 7 Open Cluster 30★ Tr Type II 2 m
ø4′, m –, Br★ 14.0 $06^h57.7^m$ $-13°13'$
Finder Chart 8-6   ★

16/18″ Scopes–125x: Ruprecht 7 is a faint, fairly small, mildly compressed cluster of 14th magnitude and fainter stars. A dozen faint members resolve against a background haze.

Tombaugh 1 Open Cluster 45★ Tr Type III 1m
ø5′, m9.3v, Br★ 14.0 $07^h00.4^m$ $-20°28'$
Finder Chart 8-7   ★

16/18″ Scopes–125x: Tombaugh 1 is a fairly faint but rich cluster with four dozen very faint stars in a 5′ area elongated E–W without central compression. Sprinkled among its faint stars is a 10.5 magnitude star on the west side, an 11th magnitude star near center, and an 11th magnitude star on the east side. There are many pairs and groupings with a noticeable gap at the center surrounding the 11th magnitude star. The cluster lies NNW of Σ572, a double consisting of 6.6 and 10.6 magnitude stars 5.2″ apart in P.A. 143°.

Ruprecht 8 Open Cluster 10★ Tr Type IV 1 p
ø4′, m –, Br★ 12.0v $07^h01.7^m$ $-13°35'$
Finder Chart 8-6   ★

16/18″ Scopes–125x: This is a faint, unconcentrated cluster of a dozen 12th magnitude and fainter stars in a 4′ area. It is surrounded by a moderate gap in a rich star field yet is rather easy to miss.

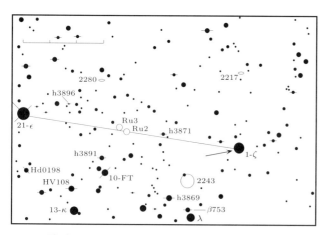

Finder Chart 8-5. 1-β CMa: $06^h20.3^m$ $-30°04'$

Figure 8-8. *IC 2177 is a large winding strip of nebulosity extending northward into Monoceros. Cederblad 90 is the most obvious concentration (upper left) while NGC 2327 is immersed in a concentration of stars (right of center) just above a dark notch. Martin C. Germano made this six minute exposure on hypered Kodak 2415 Tech Pan film with an 8", f5 Newtonian.*

NGC 2325 Galaxy Type SAB(s)0-:
ø3.7′ × 2.2′, m11.2v, SB 13.3   07ʰ02.7ᵐ −28°42′
Finder Chart 8-8, Figure 8-6   ★★★

12/14" Scopes–125x: Three magnitude 7 to 8.5 stars midway between Epsilon (ε) = 21 and Sigma (σ) = 22 Canis Majoris form a line pointing SE at the galaxy. NGC 2325 is a fairly bright 1′ diameter glow without a nucleus. An 11th magnitude star is 1.5′ SE. 13th magnitude stars are near the NW and SE ends of the galaxy's halo.

16/18" Scopes–125x: NGC 2325 appears fairly bright and elongated 1.5′ × 1′ N–S with a slightly brighter core but no stellar nucleus. There are 13th magnitude stars very near the halo NW and SE of center and 14th magnitude stars lie 1′ to the north and south.

PK242-11.1 Minkowski 3-1 Planetary Neb. Type ?+6
ø12.0″, m11.3:v, CS 15.77   07ʰ02.8ᵐ −31°35′
Finder Chart 8-8   ★★★

12/14" Scopes–225x: This planetary is moderately bright, small, and round. It is visible without a filter, but an Oxygen-III filter adds contrast. A 9th magnitude star is 2′ north.

Tombaugh 2 Open Cluster 50★ Tr Type I 1 m
ø3′, m –, Br★ 16.0   07ʰ03.4ᵐ −20°51′
Finder Chart 8-7   ★

16/18" Scopes–150x: Tombaugh 2 is a small, faint patch of haze lying 4′ south of an 8th magnitude star. Higher powers resolve only a few twinkles in the cluster haze.

NGC 2327 H25⁴ Emission & Reflection Nebula
ø1.5′ × 1.5′, Photo Br 1–5, Color 2–4 07ʰ04.3ᵐ −11°18′

Cederblad 90 Gum 3 Emission & Reflection Nebula
ø10′ × 10′, Photo Br 1–5, Color 2–4 07ʰ05.2ᵐ −12°20′

IC 2177 Gum 2 Emission Nebula
ø120′ × 40′, Photo Br 3–5, Color 3–4 07ʰ05.3ᵐ −10°38′
Finder Chart 8-6, Figure 8-8   ★★★

12/14" Scopes–100x: IC 2177 is a large winding strip of nebulosity that extends northward into Monoceros near open cluster NGC 2335 and southward to an 8th magnitude star centered on Cederblad 90. IC 2177 can be glimpsed on dark, transparent nights in 10 × 50 binoculars. In telescopes look for it using the lowest possible power and a UHC or O-III filter. The central part of IC 2177, NGC 2327 is located 35′ west of Σ1026 (5.7, 9.1; 17.8″; 350°) and 14′ NNW

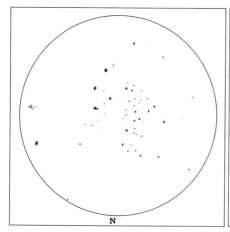

Figure 8-9. NGC 2345
13", f5.6–100x, by Steve Coe

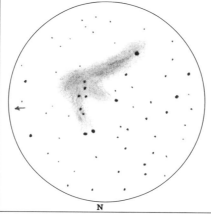

Figure 8-10. Sharpless 2-301
13", f5.6–100x, by Steve Coe

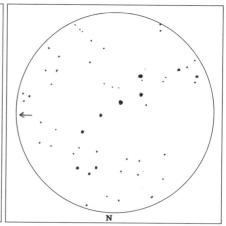

Figure 8-11. NGC 2354
8", f6–100x, by A. J. Crayon

of Σ1026 (7.4, 9.4; 5.3"; 150°): it is a faint, diffuse, but fairly concentrated, reflection nebula surrounding a 10th magnitude star. These nebulae are just slightly more visible in 16/18" instruments.

Haffner 4   Open Cluster 30★   Tr Type III 1 p
ø2.4', m –, Br★ 14.0v   **07ʰ06.2ᵐ −14°59'**
Finder Chart 8-6   ★

16/18" Scopes–150x: Haffner 4 is 5' SE of an 8.5 magnitude star, between two 9th magnitude stars to its west and NE. The cluster's brighter stars form a triangle south of a line between the two 9th magnitude stars. There are fifteen 14th to 15th magnitude stars concentrated around and east of the easternmost 9th magnitude star. Altogether three dozen very faint stars resolve against the cluster's granular haze. A faint pair stands out near its west edge.

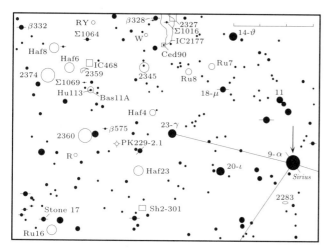

Finder Chart 8-6. 9-α CMa: 06ʰ45.1ᵐ −16°43'

Ruprecht 11   Open Cluster 20★   Tr Type III 2 p
ø2.9', m –, Br★ 11.0p   **07ʰ07.4ᵐ −20°48'**
Finder Chart 8-7   ★★

16/18" Scopes–150x: Ruprecht 11 is a faint, small, irregular cluster with two concentrations generally N–S of each other. The northern clump is larger and has more than twice the stars of the southern group. Sixteen stars of 11th and 12th magnitude resolve, including a wide pair at the northern edge.

NGC 2345   Open Cluster 70★   Tr Type I 3 m
ø12', m7.7v, Br★ 9.87   **07ʰ08.3ᵐ −13°10'**
Finder Chart 8-6, Figure 8-9   ★★★

8/10" Scopes–75x: Spreading southward from an 8.5 magnitude star, NGC 2345 is an appealing 12' diameter cluster of thirty 10th to 14th magnitude stars. The brighter members form a V-shaped asterism marked by a double at its NNE point.

12/14" Scopes–100x: This is a fairly bright, large, loose irregular cluster of about fifty stars, the majority 11th magnitude and fainter. An 8.5 magnitude star is on the northern edge and a 9th magnitude star on the southern edge. The cluster's brighter stars, SW of the main concentration, form an arrowhead pointing toward an empty region. The star at the tip of the arrowhead is a lovely 10th and 11th magnitude bluish-green and yellowish-orange double. The three stars along the arrowhead's eastern side all have companions, the central star being a triple. Overall the cluster is elongated 12' × 8' NNE–SSW.

Haffner 23   Open Cluster 40★   Tr Type III 2 m
ø11', m –, Br★ 13.0p   **07ʰ09.4ᵐ −16°57'**
Finder Chart 8-6   ★

16/18" Scopes–100x: Haffner 23, 40' south of a wide pair

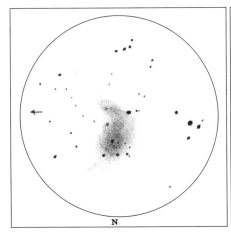 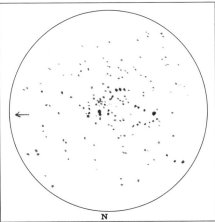 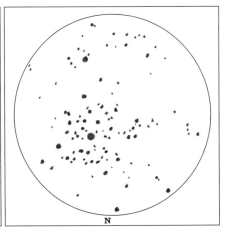

Figure 8-12. NGC 2359
13", f4.5–75x, by Tom Polakis

Figure 8-13. NGC 2360
12.5", f5–100x, by G. R. Kepple

Figure 8-14. NGC 2362
12.5", f5–100x, by G. R. Kepple

of 6th and 7th magnitude stars, is a large, loose, faint spray of 13th magnitude and fainter stars, a single 9th magnitude star near its center. The brighter stars lie south and west of this central star, three of those to the south forming a triangle. Three dozen 13th to 14th magnitude stars are resolved against a hazy background which occasionally appears granular.

Sharpless 2-301 Gum 5 Emission Nebula
⌀8' × 7', Photo Br 1–5, Color 3–4 $07^h09.8^m$ $-18°29'$
Finder Chart 8-6, Figure 8-10   ★★★

12/14" Scopes–75x: With an O-III filter, this nebula is a fairly obvious pale glow. It is located NW of an 8th magnitude star. Sharpless 2-301 has an 8' long distorted triangular shape. The east side is more distinct because several stars are embedded in the nebulosity. A star to the north has a faint, small halo.

16/18" Scopes–100x: In a UHC filter, this 8' long nebula appears fairly bright and fan-shaped, with three prongs to the north, NE, and SE. Among the many involved stars is a bright one at the SE edge and several on the north edge. The N–S prong contains at least five faint stars. A few detached sections of nebulosity extend beyond a 30' field of view.

NGC 2354 H16⁷ Open Cluster 100★ Tr Type III 2 m
⌀20', m6.5v, Br★ 9.13v $07^h14.3^m$ $-25°44'$
Finder Chart 8-8, Figure 8-11   ★★★

8/10" Scopes–100x: Located in a star-rich region of the Milky Way, NGC 2354 is a large, loose, and highly irregular cluster with a starless void near its center. Most of its members are faint and bunched into three groups over a 20' area. The brighter cluster stars are arranged in a pattern reminiscent of the constellation Scorpius: the "head" (at the SE end of the cluster) marked by three 9th magnitude stars, the "body" is three 10th magnitude stars to the WNW, and the "tail" six 9th to 11th magnitude stars looping counterclockwise to the north.

12/14" Scopes–100x: This is a bright, large, irregular cluster without central compression composed of about fifty stars of a wide range of magnitudes in at least four condensations. A bright curved star chain links several of the condensations, the brightest of which, to the west, has sixteen stars.

16/18" Scopes–100x: NGC 2354 is a loose scattering of 60 stars located between Tau (τ) = 30 Canis Majoris and Gamma (γ) = 25 Canis Majoris. Because of its loose structure, it can be very easy to pass over. A broad, loose, curved stream of some 30 stars, marked by a 9th magnitude double at its center, forms the cluster's western half. The eastern segment includes a keystone composed of two 9.5 and two 10th magnitude stars with faint members sown throughout. The star at the keystone's NW corner is a double with very close 12th magnitude components. Several other multiple groups are spread throughout the cluster.

Collinder 132 Open Cluster 25★ Tr Type III 3 p
⌀95.0', m3.6v $07^h14.4^m$ $-31°10'$
Finder Chart 8-8   ★★★

8/10" Scopes–35x: Collinder 132 is a very large condensation of the rich southern Canis Milky Way star field. It consists of several dozen stars of various brightness spread over a 1.5° area. The four brightest Cr 132 stars, one of the 5th and three of the 6th magnitude, are arranged in a box, the corners of which point N–S and E–W. As is the case with most Collinder clusters, it is best seen at very low power or with a pair of binoculars.

Figure 8-15. *NGC 2360 is a rich, attractive cluster with at least a hundred stars visible in medium-size telescopes. Lee C. Coombs made this photo with a 10″, f5 Newtonian at prime focus for ten minutes on 103a-O Spectroscopic film.*

Basel 11A Open Cluster 30★   Tr Type II 3 m
ø9′, m8.2v, Br★ 10.90v $07^h17.1^m$ $-13°58′$
Finder Chart 8-6   ★★

12/14″ Scopes–150x: Basel 11A, 12′ south of a 7th magnitude star, is a scattering of several dozen 11th to 14th magnitude stars. The cluster's center is marked by a wide pair of 8.5 and 9.5 magnitude stars. Three 8.5 magnitude stars form a 15′ triangle SW of the cluster while a another 8.5 magnitude star lies 5′ NE.

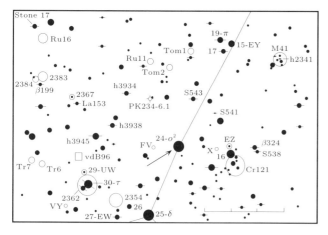

Finder Chart 8-7. $24-o^2$ CMa: $07^h03.0^m$ $-23°50′$

IC 468 Emission Nebula
ø20′ × 20′, Photo Br 3, Color 3
$07^h17.5^m$ $-13°05′$
Finder Chart 8-6   ★

12/14″ Scopes–100x: Located 15′ east of a 7.5 magnitude star, this very faint, indistinct, elongated patch of nebulosity is attached to the NW end of the more easily observed emission nebula NGC 2359.

NGC 2360 H12⁷ Open Cluster 80★
Tr Type II 2 m, ø12′, m7.2v, Br★ 10.36v
$07^h17.8^m$ $-15°37′$
Finder Chart 8-6, Fig. 8-13 & 8-15   ★★★

8/10″ Scopes–75x: This beautiful group lies 20′ east of a bright 5th magnitude field star. It is large, very rich and irregularly-shaped. Most of its 60 stars are 11th and 12th magnitude with a brighter unequal pair at its eastern boundary. Most of its stars are gathered into conspicuous clumps or strung along delicate star chains.

12/14″ Scopes–100x: This attractive cluster is bright, large, rich, and moderately compressed with at least a hundred 10.5 to 13th magnitude stars visible in a 12′ area. The western half is more star-rich and concentrated than the eastern half, in which are several conspicuous star-vacancies. To the south the cluster merges gradually into the Milky Way background, but on the north a long meandering star chain extends NE and then curves north.

NGC 2359 Gum 4 Emission Nebula
ø9′ × 6′, Photo Br 2–5, Color 3–4 $07^h18.6^m$ $-13°12′$
Finder Chart 8-6, Figures 8-12 & 8-16   ★★

8/10″ Scopes–100x: This bright nebula is called the "Duck Nebula" by some observers, who see it in the profile of a duck with a prominent head and bill. The nebula consists of two very faint N–S areas just in contact. The southern segment, forming the duck's bill, is elongated E–W. The 9th magnitude star at its SE edge marks the duck's eye. The northern area of nebulosity is larger, more diffuse, and has three 11th magnitude stars along its northern edge.

12/14″ Scopes–100x: Without a filter NGC 2359 is just a dim haze; but an O-III filter truly brings it to life. The nebula is large and obvious, and consists of two main N–S regions. The larger northern region reveals filamentary structure on the best of nights. The southern region is thinner and has a 9th magnitude star on its southern edge.

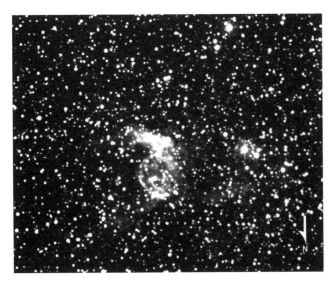

Figure 8-16. *NGC 2359's duck-shape is more obvious with nebula filters in medium and large telescopes. Martin C. Germano made this exposure with an 8", f5 Newtonian reflector at prime focus for 70 minutes on hypered 2415 film.*

Figure 8-17. *NGC 2362 is a bright, impressive cluster surrounding Tau Canis Majoris. Martin C. Germano made this photograph with an 8", f5 Newtonian reflector at prime focus for 30 minutes on hypered 2415 film.*

16/18" Scopes–150x: Viewed through an O-III filter, this nebula is fairly bright, large, and irregularly V-shaped with several obvious concentrations, the brightest lies at the southern end spreading west from a 9th magnitude star. Several filaments extend northward into a triangular asterism, three 11th magnitude stars marking its northern edge and a fourth 11th magnitude star at its southern vertex. The very faint nebula, IC 468, is attached at the NW end.

NGC 2362 H17[7] Open Cluster 60★   Tr Type I 3 p n
ø8', m4.1, Br★ 4.39v $07^h18.8^m$ $-24°57'$
Finder Chart 8-7, Figures 8-14 & 8-17   ★★★★★

NGC 2362 is a bright and beautiful group surrounding 4th magnitude Tau (τ) = 30 Canis Majoris. It is rich and compact, and its brightest stars are very young O- and B-type giants. NGC 2362 is estimated to be only about one million years old and therefore is one of the youngest known star clusters. The cluster lies some 5,000 light years away. If we could view our own Sun from this distance, its visual magnitude would be only 15.5. The actual diameter of NGC 2362 is about 8 light years. The brightest member of the group, Tau CMa, is believed to be a true cluster member, a spectroscopic binary with very massive and luminous components.

2/3" Scopes–75x: Small telescopes show a neat clump of several dozen stars surrounding Tau Canis Majoris. At lower power, the stars blend into a granular haze around Tau.

8/10" Scopes–100x: Impressive! NGC 2362 is an exceptionally bright, large, concentrated cluster centered on the bright Tau Canis Majoris. It has some thirty-five 8th to 9th magnitude and several dozen much fainter stars. Just east of Tau are two very faint companions. A 7th magnitude star is on the east edge of the cluster. A stream of stars loops around the cluster's northern side toward the east.

12/14" Scopes–125x: This beautiful and stunning cluster is bright, highly compressed, and centered upon the brilliant bluish-white 4th magnitude Tau Canis Majoris, which has two close bluish companions just to its east. The cluster contains fifty members from 8th to 11th magnitude. A relatively starless strip separates the smaller northern from the richer southern segment. The cluster's exterior boundary is roughly triangular, one apex pointing south. A 7th magnitude star lies at the cluster's east edge.

Haffner 6 Open Cluster 60★   Tr Type IV 3 p m
ø4', m9.2v, Br★ 11.11v $07^h20.1^m$ $-13°08'$
Finder Chart 8-6   ★

16/18" Scopes–150x: This cluster is just a mist of stardust with eighteen 11th to 14th magnitude stars standing out against a hazy background.

NGC 2367 Open Cluster 30★   Tr Type IV 3 p
ø3.5', m7.9:v, Br★ 9.39v $07^h20.1^m$ $-21°56'$
Finder Chart 8-7   ★★

4/6" Scopes–75x: NGC 2367 lies 10' north of two 7th magnitude stars forming a triangle with them. The cluster is a very faint glow involved with four stars forming a Y-shaped asterism.

12/14" Scopes–125x: This object is a pleasant surprise! It is a fairly bright, moderately large concentration

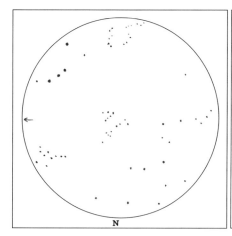
Figure 8-18. NGC 2374
8″, f6–125x, by A. J. Crayon

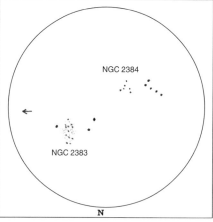
Figure 8-19. NGC 2383 & NGC 2384
4.5″, f9–35x, by Allister Ling

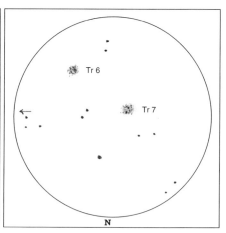
Figure 8-20. Trumpler 6 & 7
4.5″, f9–35x, by Allister Ling

of eighteen 11th to 14th magnitude stars. The cluster is elongated N–S in an 9′ × 4′ area with a conspicuous double of 9th magnitude stars 4.5″ apart at its center. The star field has a noticeable void immediately around the cluster, but 7′ north lies a nice little clump of four stars.

Ruprecht 16 Open Cluster 15★ Tr Type IV 2 p
⌀11′, m –, Br★ 13.0p   07ʰ23.2ᵐ −19°27′
Finder Chart 8-6   ★

16/18″ Scopes–125x: Ruprecht 16 is another faint group that requires a careful looking to be seen. It lies 30″ south of an irregular crescent of five 5th to 7th magnitude stars. It is a poor cluster with several dozen 13th to 15th magnitude stars visible extending NE-SW in a loose, irregular scattering. Two nearly detached groups, each having one of the cluster's brighter stars, stands out in its west and SSW portions. A loose concentration of stars runs between these two groups.

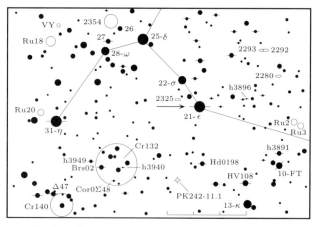
Finder Chart 8-8. 21–ε CMa: 06ʰ58.6ᵐ −28°58′

Haffner 8 Open Cluster 30★ Tr Type IV 3 m
⌀4.2′, m9.1v, Br★ 11.09v   07ʰ23.4ᵐ −12°20′
Finder Chart 8-6   ★★

16/18″ Scopes–150x: Haffner 8 is a fairly faint, small, loose, irregular cluster of about two dozen stars, most are 12th magnitude and fainter. However, one 10.5 magnitude star is on the cluster's northern side, and a curved row of 11th magnitude stars is at its southern end.

Collinder 140 Open Cluster 30★ Tr Type III 3 p
⌀42.0′, m3.5v   07ʰ23.9ᵐ −32°12′
Finder Chart 8-8   ★★★

8/10″ Scopes–50x: This very large open cluster, visible to the naked eye as a fuzzy Milky Way condensation, marks the "tuft" at the end of the Big Dog's tail. The stem of a large Y-shaped asterism composed of four 5th and 6th magnitude stars protrudes out of the cluster to the NE. The center star of the Y is Dunlap 47, a wide blue and yellow double with 5.5 and 7.6 magnitude stars separated by 99″ in P.A. 342°. The area is excellent in binoculars.

NGC 2374 H35⁸ Open Cluster 25★ Tr Type II 3 p
⌀19′, m8.0v, Br★ 10.74v   07ʰ24.0ᵐ −13°16′
Finder Chart 8-6, Figure 8-18   ★★★

8/10″ Scopes–75x: This is a rather large, irregular, and loose but easily distinguished cluster of 35 stars in a rich Milky Way field. The brightest stars lie in its north and east sections, where a dozen 10th and 11th magnitude stars may be seen in a 6′ or 7′ area. To the SW is the main concentration, 4′ across with approximately 15 stars visible against a background haze.

12/14″ Scopes–100x: NGC 2374, just NE of an 8.5

magnitude star, is a small, loose, but visually interesting gathering of three dozen stars. On its SW it has several dozen 12th magnitude stars in a 4′ area and a string of three wide 10th magnitude doubles. The NE portion is much sparser but has brighter stars, six of them in a small arc.

NGC 2383 Open Cluster 40★ Tr Type I 3 m
⌀6′, m8.4:v, Br★ 9.76v   07ʰ24.8ᵐ −20°56′
Finder Chart 8-7, Figure 8-19   ★★★

4/6″ Scopes–35x: NGC 2383 is a glow around three stars north preceding NGC 2384. At 100x, the glow seems elongated N–S.

8/10″ Scopes–100x: NGC 2383 and NGC 2384 are two fairly bright, moderately-sized clusters, separated 8′ NW–SE. The NW cluster, NGC 2383, is a small, rich cluster of three 10th and twenty 11th to 13th magnitude stars.

12/14″ Scopes–100x: NGC 2383 has several dozen 11th to 13th magnitude stars with three 10th magnitude stars, two on the NE and one on the west. It is larger but fainter than open cluster NGC 2384.

Ruprecht 18 Open Cluster 40★ Tr Type III 2 p
⌀4′, m9.4v, Br★ 10.98   07ʰ24.8ᵐ −26°13′
Finder Chart 8-8   ★★★

8/10″ Scopes–100x: This cluster is a faint, compact, nebulous knot against the star field. 150x shows a dozen 11th magnitude and fainter stars in an unresolved background haze.

12/14″ Scopes–125x: Ruprecht 18 is faint but rather nice with several dozen 11.5 magnitude and fainter stars in a compressed gathering 5′ across. A 10.5 magnitude star stands out on the SW edge.

NGC 2384 Open Cluster 15★ Tr Type IV 3 p
⌀2.5′, m7.4:v, Br★ 8.58   07ʰ25.1ᵐ −21°02′
Finder Chart 8-7, Figure 8-19   ★★★

4/6″ Scopes–75x: NGC 2384 is in the same low power field with open cluster NGC 2383 lying 8′ NW. It is the more conspicuous of the two faint hazy objects, appearing elongated E–W with several resolved stars.

8/10″ Scopes–100x: This cluster is a small, faint, compact group of a dozen 8.5 magnitude and fainter stars. NGC 2384 looks like a detached segment of NGC 2383, but both clusters are difficult and may be swept by unnoticed.

12/14″ Scopes–100x: West of NGC 2384 is a pair of 9th and 9.5 magnitude stars 4.5″ apart. The fainter cluster members are scattered eastward from this double to an 8.5 magnitude star with several faint companions.

Trumpler 6 Open Cluster Tr Type III 2 p
⌀6′, m10.0p   07ʰ26.1ᵐ −24°18′
Finder Chart 8-7, Figure 8-20   ★★★

4/6″ Scopes–75x: At low power, Trumpler 6 is a faint, round, soft glow. 150x reveals a pair of stars oriented N–S accompanied by fainter stars.

12/14″ Scopes–125x: This cluster is much fainter and smaller than Trumpler 7 across the border in Puppis. Trumpler 6 has twenty 11th magnitude and fainter stars in a loose gathering without central compression. Trumpler 7 is brighter and twice as large with several dozen 9th to 13th magnitude stars featuring a prominent string running NW–SE.

Ruprecht 20 Open Cluster 30★ Tr Type III 2 m
⌀10′, m9.5v, Br★ 11.59   07ʰ26.7ᵐ −28°50′
Finder Chart 8-8   ★★★

12/14″ Scopes–100x: Ruprecht 20 is a faint, loose group of a dozen 11.5 magnitude and fainter stars standing out against a haze of threshold stars. 9th magnitude stars are near its western and northern edges.

Chapter 9

Canis Minor, the Little Dog

9.1 Overview

Canis Minor, the Little Dog, is the smaller of the two celestial hounds of Orion the Hunter. The constellation is small and obscure, but it boasts the brilliant beacon Procyon, eighth brightest star in the sky. The name Procyon literally means "Before the Dog," and was given to this star by the ancient Greeks because it rises shortly before the Dog Star, Sirius. The constellation of the Little Dog was formed by the Romans. Procyon with Sirius and Betelgeuse in Orion together comprise the "Winter Triangle." The only other bright star in Canis Minor is 2.9 magnitude Beta Canis Minoris, a blue star named Gomeisa, the old Arabic name for Procyon itself. Canis Minor, though on the edge of the winter Milky Way, has only a few faint galaxies to offer the deep-sky observer. Even double stars are not in great supply, but some of the pairs are quite interesting and deserve attention.

Canis Minor: KAY-nis MY-ner
Genitive: Canis Minoris, KAY-nis My-NOR-is
Abbrevation: CMi
Culmination: 9pm–Feb. 28, midnight–Jan. 14
Area: 183 square degrees
Best Deep-Sky Objects: Σ1149, NGC 2470
Binocular Objects: Dolidze 26, R CMi

Σ1149 Double Star   Spec. G0
m7.9, 9.6; Sep. 21.7″; P.A. 41°   07^{h}49.5^m +03°13′
Finder Chart 9-2   ★★★

2/3″ Scopes–75x: Struve 1149, located 3° SE of Procyon at the western end of an E–W string of bright stars, is a real gem with lovely yellowish and bluish stars, a faint version of Eta (η) Cassiopeiae. At higher power the companion takes on a purplish hue.

9.2 Interesting Stars

Σ1095 Double Star   Spec. A2
m8.4, 8.9; Sep. 10.1″ P.A. 78°   07^{h}27.4^m +08°45′
Finder Chart 9-2   ★★★

2/3″ Scopes–100x: Although rather faint, Σ1095 is easily found 12′ SW of Gamma (γ) = 4 Canis Minoris and 25′ north of Beta (β) = 3 Canis Minoris. This neat little pair has a white primary and a bluish companion.

Σ1103 Double Star   Spec. B9
m7.7, 9.2; Sep. 4.4″, P.A. 243°   07^{h}30.6^m +05°15′
Finder Chart 9-2   ★★★

2/3″ Scopes–100x: To find this double, sweep 2° west of Procyon. Σ1103 will be obvious as a close pair of yellowish-white and bluish-white stars.

9.3 Deep-Sky Objects

NGC 2350 Galaxy Type S0/a
ø1.4′ × 0.7′, m12.3v, SB 12.1   07^{h}13.2^m +12°15′
Finder Chart 9-2, Figure 9-1   ★

16/18″ Scopes–150x: Located 10′ south of a 10″ pair of 10.5 magnitude stars, NGC 2350 is a very faint object requiring averted vision. A faint core is surrounded by a diffuse halo of uniform surface brightness elongated 1′ × 0.5′. A tiny Y-shaped asterism of 13th magnitude stars lies 3′ south, and a wide 12th magnitude pair is located 6′ SE.

NGC 2394 Open Cluster [14★ Tr Type III 2 p]
[ø8′], m – Asterism?   07^{h}30.1^m +11°54′
Finder Chart 9-2, Figure 9-2   ★★

12/14″ Scopes–100x: This object is probably an aster-

Canis Minor, the Little Dog

Constellation Chart 9-1

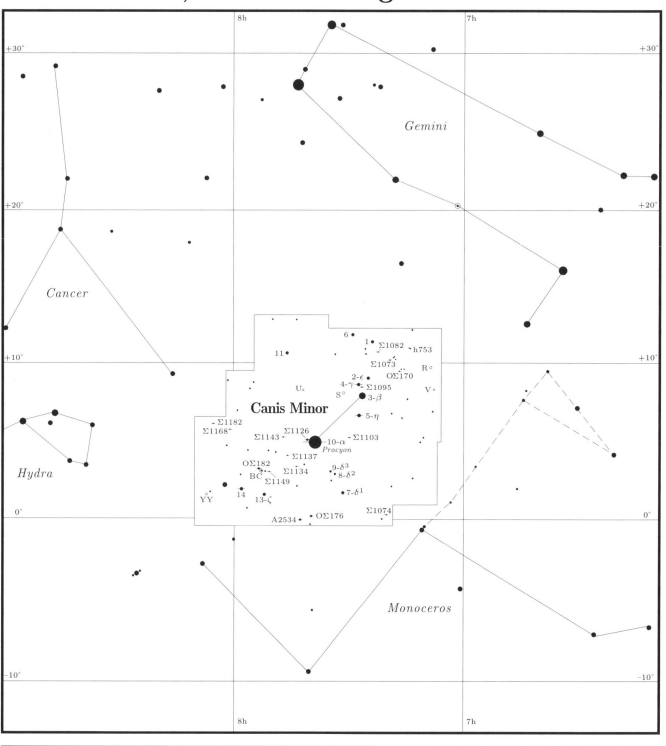

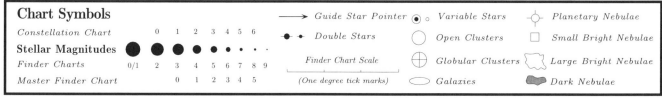

Table 9-1. Selected Variable Stars in Canis Minor

Name	HD No.	Type	Max.	Min.	Period (Days)	F*	Spec. Type	R.A. (2000)	Dec.	Finder Chart No. & Notes
V CMi	53847	M	7.4	15.1	366	0.39	M4-M10	$07^h07.0^m$	+08°53′	9-1
R CMi	54300	M	7.2	11.6	337	0.48	C7	08.7	+10 01	9-1
S CMi	59950	M	6.6	13.2	332	0.49	M6-M8	32.7	+08 19	9-2
U CMi	61789	M	8.0	14.0	413	0.52	M4	41.3	+08 23	9-2
YY CMi	67110	EB	8.33	9.13	1.09		F6	$08^h06.6^m$	+01 56	9-1

F* = The fraction of period taken up by the star's rise from min. to max. brightness, or the period spent in eclipse.

Table 9-2. Selected Double Stars in Canis Minor

Name	ADS No.	Pair	M1	M2	Sep.″	P.A.°	Spec	R.A. (2000)	Dec	Finder Chart No. & Notes
h753			8.2	9.0	22.6	4	A2	$07^h15.0^m$	+11°01′	9-2
OΣ170	5958		7.6	7.9	1.5	95	G0	17.6	+09 18	9-2
Σ1074	5996	AB	7.4	7.8	0.6	161	B9	20.5	+00 24	9-1
β577	5996	AC		12.5	12.8	101				
	5996	AD		12.0	15.3	11				
	5996	AE		9.9	53.7	278				
Σ1073	6002		7.8	9.8	8.8	67	A0	21.0	+10 12	9-2
Σ1082	6035	AB	8.5	9.2	19.9	326	A3	23.8	+10 42	9-2
Σ1095	6088		8.4	8.9	10.1	78	A2	27.4	+08 45	9-2 White & bluish pair
5-η CMi	6101		5.3	11.1	4.0	25	A5	$07^h28.0^m$	+06 57	9-2
Σ1103	6140		7.7	9.2	4.4	243	B9	30.6	+05 15	9-2 White & bluish
OΣ176	6240	AB	7.1	9.1	1.5	214	B9	38.5	+00 30	9-1
10-α CMi	6251	AB	0.4	12.9	5.2	26	F5	39.3	+05 14	9-2
Σ1126	6263	AB	6.6	6.9	1.1	158	A0	40.1	+05 14	9-2 Close yellowish pair
	6263	AC		10.8	44.4	249				
A2534	6313	ABxC	7.0	8.2	0.6	217	G5	43.1	+00 11	9-1
Σ1134	6317	AB	7.5	10.7	10.2	147	F8	43.5	+03 29	9-2
	6317	AC		10.1	85.5	347				
Σ1137	6360		8.3	9.3	2.9	132	F5	46.6	+04 08	9-2
Σ1143			7.0	11.0	9.3	152	K0	48.1	+05 25	9-2
Σ1149	6391		7.9	9.6	21.7	41	G0	49.5	+03 13	9-2 Yellow & bluish or purple
14 CMi		AB	5.4	8.4	88.6	77	K0	58.3	+02 13	9-1
		AC		9.3	120.0	150				
OΣ182	6425		7.5	8.0	1.0	21	A2	52.7	+03 23	9-2
Σ1168	6492		6.8	10.6	6.2	220	B9	58.8	+05 37	9-2
Σ1182	6571		8.0	10.0	4.5	73	B9	$08^h05.4^m$	+05 50	9-2

Footnotes: *= Year 2000, a = Near apogee, c = Closing, w = Widening. Finder Chart No: All stars listed in the tables are plotted in the large Constellation Chart, but when a star appears in a Finder Chart, this number is listed. Notes: When colors are subtle, the suffix *-ish* is used, e.g. *bluish*..

ism rather than a true open cluster. The bracketed data above is our own because no information is available on NGC 2394 outside the original NGC catalogue. NGC 2394 is situated 10′ NE of Eta (η) = 5 Canis Minoris (m5.3) and 4′ SW of a 9th magnitude star. It is a large S-shaped asterism of eight 10th-11th magnitude stars around which are sprinkled half a dozen fainter cluster "members." Four faint stars form an E–W row including the central star in the S-shaped asterism. Overall, the cluster is elongated 8′ × 5′ N–S. Five outlying stars to the north and south could be included, but they seem separated from the main group.

Dolidze 26 Open Cluster Tr Type IV 1 p
⌀23′, m – $07^h30.1^m$ +11°54′
Finder Chart 9-2, Figure 9-3   ★★

12/14″ Scopes-100x: Dolidze 26 is a fairly bright, large, loose, irregular cluster of sixty stars in a 23′ × 12′ area. 6 Canis Minoris (m4.54) at the NNW edge is probably a foreground object. A 9th magnitude star is near the center. The cluster also contains ten 10th to 11th magnitude stars, a dozen 12th magnitude members, and an indeterminate number of fainter stars. The main concentration of stars form a NE–SW stream passing between 6 CMi and the 9th magnitude star at the cluster's center. Another much fainter star stream flowing E–W passes the 9th magnitude star on the south side of

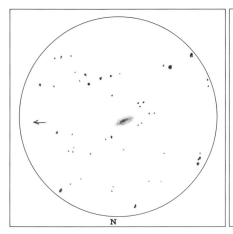
Figure 9-1. NGC 2350
17.5″, f4.5–150x, by G. R. Kepple

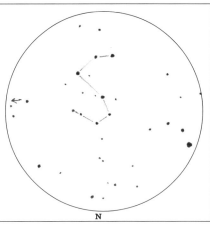
Figure 9-2. NGC 2394
12.5″, f5–150x, by G. R. Kepple

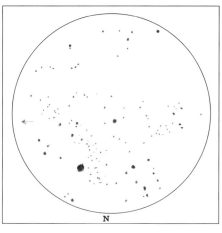
Figure 9-3. Dolidze 26
12.5″, f5–100x, by G. R. Kepple

the more prominent stream. Both star streams merge on the WSW, giving the cluster a triangular shape.

NGC 2470 Galaxy Type Sab
ø2.0′ × 0.6′, m12.7v, SB 12.7   07ʰ54.3ᵐ +04°27′
Finder Chart 9-2   ★★★

16/18″ Scopes–150x: Located 0.5° west of a 6th magnitude star, NGC 2470 appears moderately bright and elongated 1.5′ × 0.5′ NW–SE with a bright core.

NGC 2485 Galaxy Type Sa
ø1.5′ × 1.5′, m12.2v, SB 12.9   07ʰ56.7ᵐ +07°29′
Finder Chart 9-2   ★★

16/18″ Scopes–150x: NGC 2485 is situated 25′ NW of a 6th magnitude star and 7′ north of a 9.5 magnitude star. Its small 30″ halo is diffuse and round with a stellar nucleus. At low power the galaxy and a 12.5 magnitude star 30″ south masquerade as a double star.

NGC 2496 Galaxy Type E
ø1.1′ × 0.9′, m12.9v, SB 12.8   07ʰ58.6ᵐ +08°02′
Finder Chart 9-2   ★★

16/18″ Scopes–150x: NGC 2496 is located 25′ south of a 6.5 magnitude star which is part of a 2° long arc of bright stars in the field north of the galaxy. NGC 2496 appears faint, elongated 1′ × 0.75′ N–S with a slight brightening in the core area. There is a 14.5 magnitude star 30″ east. NGC 2491 lies 4′ SW.

NGC 2508 H7³ Galaxy Type E?
ø2.0′ × 1.6′, m12.7v, SB 13.8   08ʰ02.0ᵐ +08°34′
Finder Chart 9-2   ★★

16/18″ Scopes–150x: This galaxy is situated 19′ south of a 6th magnitude star and 12′ NE of a 6.5 magnitude star, the two easternmost in a 2° long arc of bright stars. NGC 2508 is a faint, oval-shaped galaxy elongated 2.0′ × 1.5′ NW–SE with a small bright core. Two 14th magnitude stars lie 30″ and 1′ west.

NGC 2538 Galaxy Type (R′)SBa
ø1.6′ × 1.2′, m12.6v, SB 13.1   08ʰ11.4ᵐ +03°37′
Finder Chart 9-2   ★★

16/18″ Scopes–150x: NGC 2538 is fairly faint, about 1.25′ in diameter with a slight elongation NNE–SSW and a well condensed core. The galaxy lies between a 13 magnitude star to the SE and a 14.5 magnitude star to the NW. There is a faint double of 14th and 15th magnitude stars 1′ east.

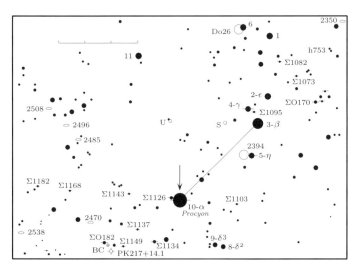
Finder Chart 9-2. 10–α CMi: 07ʰ39.3ᵐ +05°14′

Chapter 10

Cassiopeia, the Queen

10.1 Overview

The familiar "W" or "M" pattern formed by the brightest stars of Cassiopeia is superimposed upon the brilliant star fields of the autumn Milky Way, a splendid region to scan with binoculars or an RFT. Cassiopeia is a circumpolar constellation best positioned for early evening viewing during the late autumn and early winter; but those who prefer warm weather observing, and who are willing to stay up late, can watch Cassiopeia culminate after midnight in late summer.

Cassiopeia's rather modest 600 square degree area is exceptionally rich in open clusters even for a Milky Way constellation. Many of its clusters are tiny groups or knots of stars embedded in rich star fields that make them difficult to discern; but other Cassiopeia clusters are among the best in the entire sky. The constellation is also liberally garnished with an assortment of planetary nebulae, a few galaxies, and a host of colorful double and multiple stars. Double star observing in Cassiopeia is especially rewarding because a plentiful supply of colorful doubles of both bright and faint stars is crowded within the constellation's confines.

In Greek mythology, Cepheus and Cassiopeia, the king and queen of Ethiopia, were the parents of Princess Andromeda. In the sky Cassiopeia is depicted sitting in her throne. However, half the time, when the constellation is below the pole, she is upside down. This was, according to the ancient poets, the gods' way of punishing her for her vain boast that she was more beautiful than the Sea Nymphs.

Cassiopeia: Kass-ee-oh-PEE-ah
Genitive: Cassiopeiae, Kass-ee-oh-PEE-yee
Abbrevation: Cas
Culmination: 9pm–Nov.23, midnight–Oct. 9
Area: 598 square degrees
Showpieces: 24-η Cas, ι Cas, RZ Cas, M52 (NGC 7654), M103 (NGC 581), NGC 278, NGC 457, NGC 654, NGC 663, NGC 7789
Binocular Objects: IC 1848, M52 (NGC 7654), M103 (NGC 581), Mel 15, NGC 129, NGC 225, NGC 457, NGC 654, NGC 659, NGC 663, NGC 1027, NGC 7789, NGC 7790, Stock 2, Stock 5, Stock 12, Tr 3

Alpha (α) = 18 Cassiopeiae   Multiple Star   Spec. K0 II-III
AD: m2.2, 8.9; Sep. 64.4″, P.A. 280° $00^h40.5^m$ +56°32′
Finder Chart 10-6   *Schedar*   ★★★★

2/3″ Scopes–50x: Alpha Cassiopeiae has three optical companions which are easy targets for small telescopes. The 9.8 magnitude attendant is bluish and contrasts nicely with the bright orange of the primary. The other components are much fainter at 12.7 magnitude, 38″ away in P.A. 105°, and 13.7 magnitude, 19.8″ away in P.A. 275°.

Eta (η) = 24 Cassiopeiae (Σ60) Double Star Spec. G0 V
m3.4, 7.5; Sep. 12.9″, P.A. 317° $00^h49.1^m$ +57°49′
Finder Chart 10-6   ★★★★★

Eta Cassiopeia is a well known binary first discovered by Sir William Herschel in August 1779. The period is about 500 years with the separation varying between 5″ and 16″. The stars provide a lovely contrast, observers reporting gold, yellow, or topaz for the primary and orange, red, purple, or garnet for the secondary.

10.2 Interesting Stars

Sigma (σ) = 8 Cassiopeiae (Σ3049) Double Star Spec. B2
m5.0, 7.1; Sep. 3.0″, P.A. 326° $23^h59.0^m$ +55°45′
Finder Chart 10-3   ★★★★

8/10″ Scopes–125x: Sigma Cassiopeiae is a fine bluish-white and yellow pair located in a glorious star field!

Cassiopeia, the Queen Constellation Chart 10-1

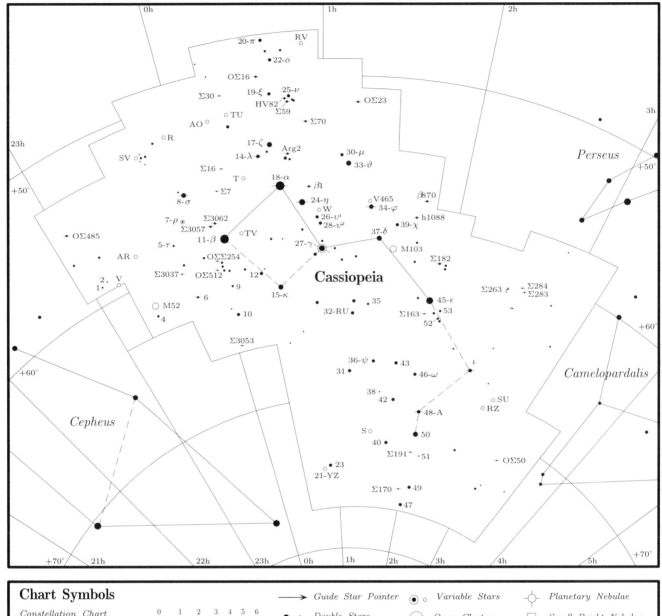

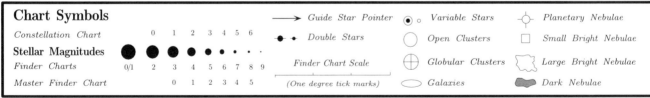

Table 10-1. Selected Variable Stars in Cassiopeia

Name	HD No.	Type	Max.	Min.	Period (Days)	F*	Spec. Type	R.A. (2000)	Dec.	Finder Chart No. & Notes
V Cas	218997	M	6.9	13.4	228	0.48	M5-M8	$23^h11.7^m$	+59°42′	10-3
7-ρ Cas	224014	SRd	4.1	6.2	320		F8-K0	54.4	+57 30	10-3
R Cas	224490	M	4.7	13.5	430	0.40	M6-M10	58.4	+51 24	10-4
TV Cas	1486	EA/SD	7.22	8.22	1.81	0.18	B9-F7	$00^h19.3^m$	+59 08	10-3
T Cas	1845	M	6.9	13.0	444	0.56	M6-M9	23.2	+55 48	10-1
TU Cas	2207	Cep	6.88	8.18	2.13	0.31	F3-F5	26.3	+51 17	10-4
21 YZ Cas	4161	EA/DM	5.71	6.12	4.46	0.15	A2-F5	45.7	+74 59	10-1
27-γ Cas	5394	gC(x)	1.6	3.0			B0	56.7	+60 43	10-6
RZ Cas	17138	EA/SD	6.18	7.72	1.19	0.17	A3	$02^h48.9^m$	+69 38	10-1
SU Cas	17463	Cds	5.70	6.18	1.94	0.40	F5-F7	52.0	+68 53	10-1

F* = The fraction of period taken up by the star's rise from min. to max. brightness, or the period spent in eclipse.

Table 10-2. Selected Double Stars in Cassiopeia

Name	ADS No.	Pair	M1	M2	Sep.″	P.A.°	Spec.	R.A. (2000)	Dec	Finder Chart No. & Notes
OΣ485	16474	AB	6.6	9.8	20.1	50	B9	$23^h02.7^m$	+55°14′	10-1
	16474	AC		10.0	56.6	260				
Σ3037	16982	AB	7.1	8.6	2.7	213	K0	46.1	+60 28	10-3 AB: Yellowish & orange
	16982	AC		9.0	29.2	186				Both C&D are bluish
	16982	AD		9.2	52.6	229				
6 Cas	17022	AB	5.5	8.0	1.6	193	A2	48.8	+62 13	10-3 Yellowish & orange
	17022	AC		10.5	62.4	309				
OΣ512	17119	AB	6.7	11.0	3.0	393	M	57.3	+61 01	10-3 All reddish stars
Arg 99	17119	CD		10.0	4.8	318				
8-σ Cas	17140	AB	5.0	7.1	3.0	326	B2	59.0	+55 45	10-3 Bluish & yellowish
OΣΣ254		AB	7.6	8.7	58.1	89	N A	$00^h01.2^m$	+60 21	10-3 Fine red & blue pair!
Σ3053	1	AB	5.9	7.3	15.2	70	G0 A2	02.6	+66 06	10-1 Yellowish & white stars
	1	AC		10.8	98.5	290				
Σ3057	36		6.6	8.7	3.7	299	B3	04.9	+58 32	10-3 Unequal white & bluish
Σ3062	61		6.4	7.2	1.2	235	G5	06.3	+58 26	10-3 Yellowish & bluish
Σ7	143		7.9	9.4	1.3	212	B8	11.6	+55 58	10-3 White & bluish
Σ16	218		7.8	9.1	5.7	38	A3	16.7	+54 40	10-4 Nice yellowish & bluish
Σ30	361	AB	6.9	8.8	15.2	308	B9	27.2	+49 59	10-4 A: blue
14-λ Cas	434		5.3	5.6	0.5	176	B8	31.8	+54 31	10-1
OΣ16	546		5.4	9.9	13.3	23	K2	39.2	+49 21	10-4
18-α Cas	561	AD	2.2	8.9	64.4	280	K0	40.5	+56 32	10-6 Lovely orange & purple
Arg 2	630	AB	8.4	9.4	11.0	115	F8	45.7	+54 59	10-4
H V 82		AB	8.0	8.3	52.3	76	K2 K0	47.5	+51 06	10-4 Both stars yellowish
Σ59	659		7.2	8.1	2.2	147	A0	48.0	+51 27	10-4 White pair
24-η Cas	671	AB	3.4	7.5	12.9w	*317	F8	49.1	+57 49	10-6 Pretty yellow & reddish
β1	719	AB	7.8	9.8	1.4	82	B2	52.8	+56 38	10-6 Triple white stars
	719	AC		8.8	3.8	133				
	719	AD		9.3	8.9	194				
Σ70	735	AB	6.3	9.3	8.0	245	A0	53.8	+52 41	10-4 Bluish-white & purple
	735	AC		9.8	78.5	148				
	735	Cc		10.2	1.7	88				
OΣ23	956	AB	8.2	8.7	14.6	192	F8	$01^h10.1^m$	+51 45	10-1 White. ds τ868 in field
h1088	1334		6.3	8.8	19.6	168	B9	42.3	+58 38	10-6
β870	1359		6.4	7.8	1.0	19	A2	44.3	+57 32	10-6
Σ163	1459	AB	6.8	8.8	34.8	35	K5	51.3	+64 51	10-5 Fine orange & blue pair
Σ170	1504		7.7	8.4	3.3	246	A5	55.4	+76 13	10-1
Σ182	1531	AB	8.2	8.2	3.6	123	A0	56.4	+61 17	10-7 Beautiful white pair
48 Cas	1598	AB	4.7	6.4	0.9	*263	A3	$02^h02.0^m$	+70 54	10-1
Σ191	1606		6.3	8.6	5.5	192	A3	03.2	+73 51	10-1
ι Cas	1860	AB	4.6	6.9	2.5w	*230	A5	29.1	+67 24	10-1 White, blue companions
	1860	AC		8.4	7.2	114				
Σ263	1877	AB	8.3	11.5	14.6	103	B	29.6	+60 40	10-7 All stars white
OΣ50	1877	A'B'	8.5	9.5	16.6	226				
Σ283	2014	AB	8.2	9.0	1.8	207	G5	40.5	+61 29	10-7 Close yellow pair
Σ284	2018	AB	7.9	9.9	5.5	195	F0	40.8	+61 17	10-7 White & bluish stars
OΣ50	2377	AB	8.4	8.4	1.3	179	F8	$03^h12.5^m$	+71 33	10-1

Footnotes: *= Year 2000, a = Near apogee, c = Closing, w = Widening. Finder Chart No: All stars listed in the tables are plotted in the large Constellation Chart, but when a star appears in a Finder Chart, this number is listed. Notes: When colors are subtle, the suffix -ish is used, e.g. bluish.

Figure 10-1. *The bright, large open cluster Messier 52 (top, center) is accompanied by two more difficult and lesser known objects, open cluster Czernik 43 lying to its SE (left of center) and emission nebula 7635 to the SW (lower, right). William Harris used an 8", f4 Wright Newtonian reflector of his own design for 30 minutes on hypered 2415 Kodak Tech Pan film.*

Iota (ι) Cassiopeiae Triple Star   Spec. A5
AB: m4.6, 6.9; Sep. 2.5"; P.A. 230°
AC: – 8.4; Sep. 7.2"; P.A. 114° $02^h29.1^m$ +67°24'
Constellation Chart 10–1   ★★★★★

Iota is one of the loveliest triple stars in the sky! A brilliant white primary is attended by yellow and blue companions. The AB pair orbits each other in a period of some 840 years. The C star is a physical component of the system but shows no definite orbital motion.

RZ Cassiopeiae Variable Star Type EA/SD Spec. A3
m6.18 to 7.72, Per. 1.19 days $02^h48.9^m$ +69°38'
Constellation Chart 10-1   ★★★★

RZ is an interesting eclipsing binary star. The period is 1.19 days, but the light drop from 6.18 to 7.72 magnitude takes only about two hours; then brightening begins immediately, and in another two hours the star is back to its normal magnitude.

10.3 Deep-Sky Objects

NGC 7635 Emission Nebula
ø15' × 8', Photo Br. 1-5, Color 3-4 $23^h20.7^m$ +61°12'
Finder Chart 10-3 , Figures 10-1 & 10-3   ★
Bubble Nebula

8/10" Scopes–75x: This nebula is an extremely faint, large, luminous shell fleetingly visible with averted vision around an 8th magnitude star. The glare from this star and from the 7th magnitude star to the west

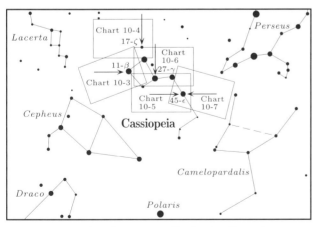

Master Finder Chart 10-2. Cassiopeia Chart Areas
Guide stars indicated by arrows.

of the nebula interfere with observation.

12/14″ Scopes–100x: This emission nebula is a very faint 2′ × 1′ elliptical glow surrounding an 8th magnitude star in a rich star field. The southern edge is more pronounced than the northern.

16/18″ Scopes–125x: NGC 7635 is quite faint, very diffuse, small, irregular, and elongated N–S. It is centered around an 8th magnitude star and is more prominent to the NW. Only about 1′ × 3′ of the nebula's 15′ × 8′ catalog size is visible. Nebulosity can also be seen around a 7th magnitude star to the WSW. Filters do not seem to help.

NGC 7654 Messier 52 Open Cluster 100★ Tr Type I 2 r
ø12′, m6.9v, Br★ 8.22v   23^{h}24.2^m +61°35′
Finder Chart 10-3, Figures 10-1 & 10-2 ★★★★★

This rich, compressed cluster was discovered by Messier in 1774. M52 is about 3,900 light-years distant, implying a true diameter of nearly 15 light-years. Its central star-density is a very respectable 50 per cubic parsec. Its brightest member, a magnitude 8.2 star on the SW edge of the group, is a yellow G-type giant. However, M52 also contains blue-white B3 main sequence stars, which imply a cluster age of around 50 million years, a little less than that of the Pleiades.

4/6″ Scopes–75x: Messier 52 is a fine, bright, well-compressed cluster rich in faint stars with a bright yellow star at its SW edge.

8/10″ Scopes–75x: A very nice cluster! It is bright, rich, and quite condensed, standing out well from the surrounding star field. It contains 80 stars, including a couple yellow stars that contrast well with the large number of relatively bright blue stars in the cluster.

12/14″ Scopes–100x: A lovely cluster! Messier 52 is bright, large, and irregular with a moderate central compression. The cluster has 150 stars from 9th to 12th magnitude with the brightest stars stretching east to west. Several 2′ large star-condensations are NE of the cluster center near its SW edge. Many very faint stars spread NE and SE of the main group. A yellow 8th magnitude star stands out on the SW side, and a group of four stars is detached at the north end. Open cluster Czernik 43 is obvious just to the south.

Czernik 43 Open Cluster 15★ Tr Type III 1 r
ø13′, m –   23^{h}25.8^m +61°19′
Finder Chart 10-3, Figure 10-1 ★★★

12/14″ Scopes–100x: This is a moderately bright, large cluster, much looser than Messier 52, 5′ north. Czernik 43 is elongated NNE–SSW and somewhat

Figure 10-2. *Messier 52 (NGC 7654) is a large, bright cluster splendid for small telescopes. Lee C. Coombs made this 10 minute exposure on 103a-O film with a 10″, f5 Newtonian.*

triangular, an 8th magnitude star marking its NNE apex. This star and three 9th magnitude stars mark the cluster's NW flank. Several more star chains to the east run in the same direction, all running toward the 8th magnitude star. The cluster contains twenty 10th to 11th and thirty-five 12th to 14th magnitude stars in a 14′ area.

Stock 11 Open Cluster Tr Type IV 2 p
ø10′, m –, Br★ 8.0p   23^{h}32.9^m +55°29′
Finder Chart 10-3 ★

12/14″ Scopes–100x: This object is a tiny asterism of eight 11th to 12th magnitude stars, five of which form a pentagon situated between two bright 7th magnitude stars. A 10th magnitude star is one pentagon-width east of the figure. St 11 is probably not a true open cluster.

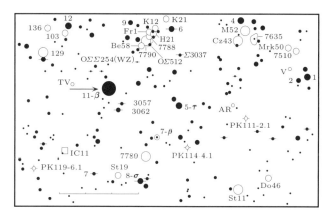

Finder Chart 10-3. 11-β Cas: 00^{h}09.2^m +59°09′

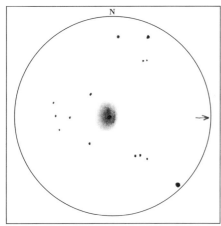 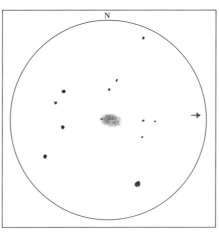 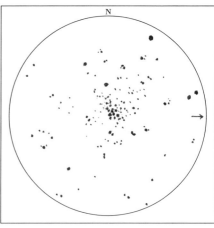

Figure 10-3. NGC 7635
13", f4.5–115x, by Tom Polakis

Figure 10-4. PK112-10.1
17.5", f4.5–105x, by Dr. Jack Marling

Figure 10-5. NGC 7788
12.5", f5–75x, by G. R. Kepple

Stock 12 Open Cluster Tr Type IV 2 p
⌀20′, m –, Br★ 8.0p   23ʰ37.2ᵐ +52°26′
Finder Chart 10-4   ★★★

12/14″ Scopes–100x: Stock 12 is a very large, loose cluster of three dozen 8th magnitude and fainter stars. The fainter stars are distributed in several chains, one of which runs E–W and protrudes west beyond the cluster's edge by two field diameters. Another star chain loops around the SE side of the cluster past a small concentration of stars detached from the main group. Another detached concentration of stars scatters southward from these chains. On the east side of the cluster is a chain of five wide double stars, the brightest pair at the southern end.

PK114-4.1 Abell 82 Planetary Nebula Type 3b
⌀95.0″, m12.7v, CS 14.92   23ʰ45.8ᵐ +57°04′
Finder Chart 10-3   ★★

12/14″ Scopes–135x: Only 10″ of this planetary's 95″ halo is visible. Viewed through a UHC filter, it is faint, fairly small, round, and uniform in brightness.

16/18″ Scopes–125x: Using an O-III filter, the 90″ diameter halo appears fairly faint, round, and uniform in brightness. The 12.7 visual magnitude was computed from photometric measurements.

PK112–10.1 Abell 84 Planetary Nebula Type 3b
⌀126.0″, m13.0v, CS 18.0   23ʰ47.7ᵐ +51°24′
Finder Chart 10-4, Figure 10-4   ★

16/18″ Scopes–125x: This planetary is large, slightly elongated E–W, and has a star on the eastern edge. The 13.0 visual magnitude was computed from photometric measurements by Dr. Jack Marling.

King 12 Open Cluster 15★ Tr Type I 2 p
⌀2′, m9.0v, Br★ 10.37v   23ʰ53.0ᵐ +61°58′
Finder Chart 10-3, Figures 10-7 & 10-16 ★★★

8/10″ Scopes–100x: This faint cluster's dozen 10th to 12 magnitude members lie just north of a line between two 7th magnitude stars. Open cluster Harvard 21 is located 10′ SE.

12/14″ Scopes–125x: King 12 is a faint, small, compact concentration of 10th magnitude and fainter stars elongated 2′ × 1′ E–W around and mostly north of the cluster's lucida which is a fairly close double (ADS 17081: 10.4, 10.7; 6.6″; 341°). The cluster's second brightest star lies 1′ NE of the lucida and twenty more stars are visible against a hazy background. A star chain trails 10′ SE from the cluster toward a 7th magnitude field star. Another 7th magnitude field star is 5′ SSW of the cluster.

Harvard 21 Open Cluster 20★ Tr Type IV 2 p
⌀6′, m9.0p   23ʰ54.1ᵐ +61°46′
Finder Chart 10-3, Figure 10-7   ★★

8/10″ Scopes–125x: Harvard 21 is a small, poorly concentrated group of five 10th magnitude stars. It is located midway between two 7th magnitude field stars. Open cluster King 12 lies 10′ NNW.

12/14″ Scopes–150x: Harvard 21 is a faint, small cluster of five 10th magnitude and two 12.5 to 13th magnitude stars. The five brighter members form an irregular pentagon about 2.5′ across.

NGC 7788 Open Cluster 20★ Tr Type I 2 p
⌀9′, m9.4p   23ʰ56.7ᵐ +61°24′
Finder Chart 10-3, Figures 10-5 & 10-7 ★★★

8/10″ Scopes–100x: NGC 7788 and NGC 7790 10′ to its

SE are a pair of moderately rich condensations of 9th magnitude and fainter stars that stand out well from a field abundant with 6th and 7th magnitude stars. NGC 7788 is a 5' diameter concentration of a dozen 9th to 12th magnitude stars.

12/14" Scopes–125x: NGC 7788 is a moderately faint, compact cluster of several dozen members, most 12th to 13th magnitude. The densest concentration, marked by a 9th magnitude star, is midway between 9th magnitude stars to its NNW and SSE. Many faint stars fan SE from the central star.

NGC 7789 H30⁶ Open Cluster 300★
Tr Type II 1 r, ø15', m6.7v, Br★ 10.70v
23ʰ57.0ᵐ +56°44'
Finder Chart 10-3, Figure 10-6   ★★★★

NGC 7789, one of the major omissions from Charles Messier's catalogue, was discovered in the late eighteenth century by Caroline Herschel. Its several hundred 11th magnitude and fainter stars are uniformly spread over half a degree of sky. It is estimated to be 5,900 light-years distant: its true diameter is therefore about 43 light-years. The cluster's brightest members are G and K orange giants with absolute magnitudes around –2. Such stars are highly evolved: indeed, NGC 7789 is estimated to be 2 billion years old. This group is very old for an open cluster, but far younger than any globular cluster in our galaxy.

4/6" Scopes–50x: NGC 7789 is visible to the unaided eye under very dark skies as a tiny hazy spot. It is one of the best open clusters for moderate-aperture telescopes, appearing as an extremely rich and very compressed concentration of uniformly faint stars. It is easily found midway between Rho (ρ) = 7 and Sigma (σ) = 8 Cassiopeiae.

8/10" Scopes–75x: NGC 7789 shows a hundred 11th to 12th magnitude stars embedded in an unresolved background glow. The uniformly dense 5' cluster core contains 30 stars embedded in the background glow and surrounded by 2' wide star-poor zones. The 5' wide band around the cluster's core is even denser, especially to the east and southwest.

12/14" Scopes–100x: This splendid cluster is large, rich, fairly dense, and well resolved. At least 150 stars are visible in a 16' area. The cluster's brighter members are 11th to 12th magnitude objects distributed in concentric rings. NGC 7789 has no distinct border:

Figure 10-6. *NGC 7789 is a magnificent cluster, rich in uniformly faint stars. Martin C. Germano made this photo using a 8", f5 Newtonian reflector at prime focus for 30 minutes on hypered Kodak 2415 Tech Pan film.*

its outlying stars seem to blend imperceptibly into the surrounding star field, only the eastern edge being at all abrupt. A small detached patch of stars is 3' south of the cluster and a tiny smudge 2' west. NGC 7789 compares favorably with many of the Messier star clusters.

Frolov 1 Open Cluster Tr Type –
ø –, m9.2v, Br★ 10.60v   23ʰ57.4ᵐ +61°38'
Finder Chart 10-3, Figure 10-7   ★★

12/14" Scopes–100x: Frolov 1 is a faint, irregularly scattered cluster of 20 stars in a 6' area. The cluster's main feature is an E–W arc of nine stars, the arc concave to the south. Other cluster stars scatter loosely to the south. A 9th magnitude star is on the west edge of the group, but most of its other members are 11th to 13th magnitude objects.

NGC 7790 H56⁷ Open Cluster 40★ Tr Type III 2p
ø17',m8.5v, Br★ 10.87v   23ʰ58.4ᵐ +61°13'
Finder Chart 10-3, Figure 10-7   ★★★

4/6" Scopes–50x: NGC 7790, 10' SE of NGC 7788, is a faint, E–W elongated mist in a field rich in 6th and 7th magnitude stars.

8/10" Scopes–75x: NGC 7790 is faint but quite rich and

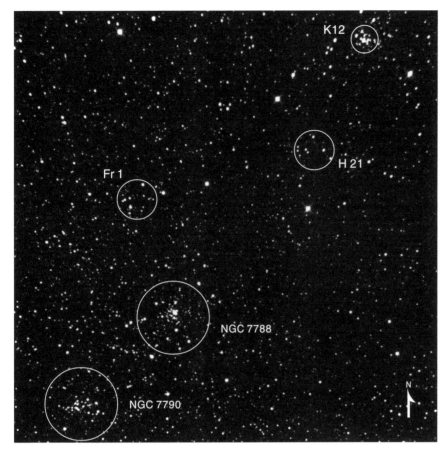

Figure 10-7. *Five open star clusters are identified in this rich area of Cassiopeia. The most prominent clusters are King 12 (at upper right), NGC 7788, and NGC 7790 (at lower left). Lee C. Coombs made this photograph using a 10″, f5 Newtonian reflector at prime focus with a ten minute exposure on Kodak 103a-O Spectroscopic film.*

concentrated, a patch of stardust extended E–W. Individual stars become more obvious with prolonged observation, the cluster's ten 11th to 12th magnitude stars standing out well against a hazy background.

12/14″ Scopes–125x: This is a bright, fairly rich, compact cluster of 30 stars in an area elongated 4′ × 2′ E–W with three clumps of stars visible. The brighter cluster members are arranged in a square in the western part of the cluster, but the main core of faint stars lies east of center, a chain of faint cluster stars extending further eastward.

Berkeley 58 Open Cluster 30★ Tr Type IV 2 p,
ø7′, m9.7v, Br★ 11.89v   00ʰ00.2ᵐ +60°58′
Finder Chart 10-3   ★★

12/14″ Scopes–100x: This cluster is located 10′ SE of NGC 7790 within a large triangle of 6–6.5 magnitude stars. At low power it is visible as a faint, moderately large patch of unresolved haze in a well populated field of bright stars. 150x shows three dozen unconcentrated 12th magnitude and fainter members scattered in a 5′ area.

Stock 19 Open Cl. 6★ Tr Type II2 p
ø3′, m –, Br★ 8.0p   00ʰ04.4ᵐ +56°02′
Finder Chart 10-3   ★★

12/14″ Scopes–100x: This sparse cluster has only a few faint stars nestled within a triangle of 9th magnitude stars pointing SW.

NGC 103 Open Cluster 30★
Tr Type I 1 2 p ø5′, m9.8v
Br★ 12.25v   00ʰ25.3ᵐ +61°21′
Finder Chart 10-6, Figure 10-8   ★★

4/6″ Scopes–75x: NGC 103 is a faint, small, foggy patch of light elongated NE–SW.

8/10″ Scopes–100x: This cluster has twenty 12th to 14th magnitude stars in a 4′ × 2′ area.

12/14″ Scopes–125x: NGC 103 is a faint but obvious cluster in a well populated field of both bright and faint stars. The brighter cluster stars form a horseshoe, the open end facing south. The western arm of the horseshoe is the richer. Two 12.5 magnitude stars are at the shoe's base, a spray of faint stars extending from them to the NW. Some two dozen stars may be counted in a 5′ area.

PK119–6.1 Hu 1–1 Planetary Nebula Type 2
ø5.0″, m12.3v, CS ?   00ʰ28.3ᵐ +55°58′
Finder Chart 10-3   ★

16/18″ Scopes–225x: Though stellar at low power, this planetary can be identified by blinking with an O-III filter. At 225x, it is a round, uniform disk less than 10″ in diameter. A 12.3 visual magnitude was computed from photometric measurements.

NGC 129 H78⁸ Open Cluster 35★ Tr Type IV 2 p
ø21′, m6.5v, Br★ 8.57v   00ʰ29.9ᵐ +60°14′
Finder Chart 10-6, Figure 10-9   ★★★

4/6″ Scopes–50x: NGC 129, just north of a 5th magnitude field star, is a fairly large, irregularly scattered, uncompressed cluster of several dozen members. A starless gap is at its center.

Figure 10-8. *NGC 103 is a small irregular cluster in a rich star field. Lee C. Coombs made this 10 minute photo on 103a-O film with a 10", f5 Newtonian at prime focus.*

Figure 10-9. *NGC 129 is a wedge-shaped cluster divided by a dark gap. Lee C. Coombs made this 10 minute exposure on 103a-O film with a 10", f5 Newtonian at prime focus.*

8/10" Scopes–75x: This is a large, irregular group of about fifty 9th to 13th magnitude stars spread over a 20' area. On the eastern edge is a N–S line of stars. The cluster does not stand out well from the surrounding field.

12/14" Scopes–100x: NGC 129 is a bright, loose, cluster of 35 stars, the more prominent in a wedge pointing south toward a 5th magnitude field star. Just south of the cluster center is a triangle of 9th magnitude stars. 125x nearly doubles the number of cluster members that can be seen. To the east of the cluster is a N–S line of 9th magnitude stars.

NGC 133   Cr3   Open Cluster 5★   Tr Type IV 1 p
ø7', m9.4p   00^{h}31.2^m +63°22'
Finder Chart 10-5, Figure 10-10   ★★★

4/6" Scopes–50x: NGC 133 is 28' NNW of Kappa (κ) = 15 Cassiopeiae in an interesting field, open cluster King 14 being only 12' ESE and NGC 146 lying 12' ESE. It appears as a N–S arc of four magnitude 9 to 9.5 stars, the arc concave to the west.

8/10" Scopes–125x: This cluster contains four 9th magnitude stars aligned N–S and several 12th to 13th magnitude stars.

12/14" Scopes–100x: The brighter stars form a Y–shaped asterism, the stem pointing northward. The Y consists of one 12th and four 9-9.5 magnitude stars. Half a dozen 13th to 14th magnitude stars are sprinkled around the Y.

NGC 147   U326   Galaxy   Type E5 pec
ø15.0' × 9.4', m9.5v, SB 14.7v   00^{h}33.2^m +48°30'
Finder Chart 10-4, Figure 10-11   ★★

8/10" Scopes–75x: This galaxy is an extremely faint, amorphous glow in a rich star field.

12/14" Scopes–100x: Only the central portion of this galaxy seems to be visible. It is a very faint, featureless, elliptical glow elongated 3' × 2' NNE–SSW with a slight brightening at the center.

16/18" Scopes–125x: NGC 147 is much fainter and more difficult than NGC 185 a degree to the east. It is similar in size and position angle to NGC 185 but more extended, roughly 15' × 8' NNE–SSW. It has a fairly prominent core surrounded by diffuse halo that fades gradually to the periphery. Very faint stars are visible at the tip of the southern extension, and on the western edge directly preceding the nucleus.

NGC 136   H35^6 Open Cluster 20★   Tr Type II 2 p
ø1.2', m –, Br★ 13.0p   00^{h}31.5^m +61°32'
Finder Chart 10-5   ★

8/10" Scopes–100x: This cluster is a small patch of stardust 6' NE of an interesting 30" pair of pale orange stars of 9th and 10th magnitudes.

12/14" Scopes–125x: This tiny, round group looks like a faint globular cluster. Ten 13th to 14th magnitude stars can be seen against a background glow.

16/18" Scopes–150x: NGC 136 is a very faint, 2' diameter patch of extremely faint stars, the brightest only 13th magnitude. With careful inspection, a

Figure 10-10. *NGC 146 (top, center), NGC 133 (lower right of center), and King 14 (lower left of center) form a fine trio of open star clusters in a rich star field north of Kappa (κ)Cassiopeiae. Lee C. Coombs made this 10 minute exposure on 103a-O Kodak Spectroscopic film with a 10″, f5 Newtonian at prime focus.*

dozen threshold stars may be resolved in the nebulous background haze. Some of the outlying stars spread NE.

King 14 Open Cluster 20★ Tr Type III 2 p
ø7′, m8.5v, Br★ 11.29 $00^h31.9^m$ +63°10′
Finder Chart 10-5, Figure 10-10   ★★★

4/6″ Scopes–75x: This cluster, 10′ NNW of Kappa (κ) = 15 Cassiopeiae, is faint but fairly rich. Eight or nine stars can be resolved against a background glow. The open clusters NGC 133 and NGC 145 are in the same low power field.

8/10″ Scopes–100x: Fifteen 11th to 13th magnitude stars are visible in a 7′ area. Most of the brighter stars are in the eastern portion. The cluster's four 11th magnitude members are in a conspicuous N–S row, but the western half is considerably richer in faint stars.

12/14″ Scopes–125x: King 14 is a fairly obvious, loose and irregular cluster of several dozen stars in a 7′ area. The brightest star, an 11th magnitude object, is near the group's center at the apex of a V-shaped asterism. The cluster has, in addition to its 11th magnitude lucida, a dozen 12th, and another dozen 13th-14th, magnitude stars.

NGC 146 Open Cluster 20★ Tr Type IV 3
ø6′, m9.1v, Br★ 11.56v $00^h33.1^m$ +63°18′
Finder Chart 10-5, Figure 10-10   ★★★

4/6″ Scopes–75x: NGC 146 is a faint and uncompressed group of several bright stars dominating a gathering of fifteen fainter cluster members. The brighter stars are in ragged E–W chains. The blue-white star Kappa (κ) = 15 Cassiopeiae and the open clusters King 14 (10′ WNW) and NGC 133 (8′ SE) are in the same low power field.

8/10″ Scopes–75x: Twenty 11th to 13th magnitude stars are visible in a 7′ area, the brighter stars in an ENE–WSW row.

12/14″ Scopes–125x: This cluster is fairly bright, loose, irregular, and elongated 10′ × 7′ E–W. Two 10th magnitude stars stand out in the southern part of the cluster, the eastern one of which is a double of comparably bright stars 7″ apart. Six 12th magnitude, and another dozen fainter, stars are concentrated around this double. The cluster's total population down to the 14th magnitude is several dozen.

NGC 185 U396 Galaxy Type E3 pec
ø14.5′ × 12.5′, m9.2v, SB 14.7 $00^h39.0^m$ +48°20′
Finder Chart 10-4, Figure 10-11   ★★★

4/6″ Scopes–75x: NGC 185 is a very faint, featureless glow between two 8th magnitude stars.

Figure 10-11. *NGC 185 and NGC 147 (left to right) may be seen at low power as two faint oval-shaped objects lying a little more than a degree apart. Martin C. Germano made this 70 minute photo on hypered 2415 film with a 8″, f5 Newtonian at prime focus.*

8/10″ Scopes–100x: This faint galaxy is a NE–SW oval, the major axis of which points toward 8th magnitude stars 15′ beyond each end of the oval. The oval is about 3′ across, and slightly brighter in the center. A much fainter galaxy, NGC 147, is located a degree west beyond 6.5 and 7.5 magnitude stars oriented ENE–WSW.

12/14″ Scopes–100x: NGC 185 has a fairly obvious halo elongated 5′ × 4′ NE–SW with a well concentrated center.

16/18″ Scopes–125x: In large instruments, NGC 185 is a diffuse but rather bright oval-shaped glow elongated 8′ × 5′ NE–SW, its moderately bright core mottled. Very faint stars are visible at the preceding and the southern edges.

NGC 189 H707² Open Cluster 15★ Tr Type III 2 p
ø3.7′, m8.8v, Br★ 10.95v   00ʰ39.6ᵐ +61°04′
Finder Chart 10-6, Figure 10-12   ★★

8/10″ Scopes–100x: NGC 189 is a small, irregularly shaped gathering of twenty faint stars about twice as concentrated as the surrounding field.

12/14″ Scopes–100x: NGC 189 is a fairly obvious, loose concentration of several dozen 11th to 13th magnitude stars in a 4′ area. Around its periphery are many wide star pairs.

16/18″ Scopes–125x: NGC 189 is a loose, irregular cluster of three dozen 11th to 14th magnitude stars in a 4′ area. On its eastern side is a double with 11.5 and 12.5 magnitude stars 11″ apart. From this star a chain extends west through and beyond the cluster.

Stock 24 Open Cluster 20★ Tr Type IV 2 p
ø4′, m8.8v, Br★ 11.05v   00ʰ39.7ᵐ +61°57′
Finder Chart 10-6   ★★

12/14″ Scopes–100x: Stock 24, 25′ west of NGC 225, is a compact group of nine 11th to 13th magnitude stars in a 3′ area.

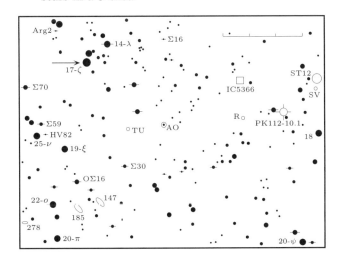

Finder Chart 10-4. 17-ζ Cas: 00ʰ36.9ᵐ +53°54″

Figure 10-12. *NGC 189 is a small, loose concentration of stars. Lee C. Coombs made this ten minute exposure on 103a-O film with a 10″, f5 Newtonian at prime focus.*

Figure 10-13. *NGC 225 is a bright, loose, irregular cluster clearly visible in small scopes. Lee C. Coombs made this ten minute photo on 103a-O film with a 10″, f5 Newtonian.*

16/18″ Scopes–125x: Stock 24 is a small cluster of eighteen 11th to 14th magnitude stars in a 4′ area, its northern portion marked by a conspicuous little star-clump. An 8th magnitude star 13′ south of the cluster forms a row with two 9th magnitude stars to the cluster's NW.

NGC 225 H78⁸ Open Cluster 15★ Tr Type III 1 p n
ø12′, m7.0v, Br★ 9.26v   00ʰ43.4ᵐ +61°47′
Finder Chart 10-5, Figure 10-13   ★★★

4/6″ Scopes–75x: NGC 225 is a fairly bright, large, and moderately rich but uncompressed cluster of 20 stars in two groups, one an arc to the NE, the other a small V-shaped asterism to the west.

8/10″ Scopes–100x: Although not rich, NGC 225 is an obvious triangular-shaped cluster with its NW side concave. The brighter stars are nearly equal in magnitude and are evenly distributed. Fourteen 10th magnitude stars and six fainter stars are visible in a 12′ area. The stars to the east, outside the triangle, appear to be part of the cluster.

12/14″ Scopes–125x: NGC 225 is a large, bright, very loose, irregular cluster of fifteen 9th to 10th magnitude, and an equal number of 11th to 14th magnitude, stars. The cluster is elongated NE–SW and triangular-shaped, its long axis pointing NE. A chain of five 9th and three 12th to 13th magnitude stars on the eastern edge looks as if it is part of the cluster, but *Uranometria 2000.0* plots it outside.

King 16 Open Cluster 35★ Tr Type II 3 p
ø3′, m10.3v, Br★ 12.49v   00ʰ43.7ᵐ +64°11′
Finder Chart 10-5, Figure 10-18   ★

12/14″ Scopes–125x: King 16, 8′ SSE of a 6.5 magnitude star, is a small Y-shaped asterism of 12th magnitude stars embedded in an unresolved haze. The northernmost star in the Y is a double. Open cluster Berkeley 4 is 15′ NNE.

Berkeley 4 Open Cluster 25★ Tr Type I 2 p
ø5.5′, m10.6v, Br★ 12.57v   00ʰ45.5ᵐ +64°24′
Finder Chart 10-5   ★

12/14″ Scopes–125x: Berkeley 4, 15′ NE of a 6.5 magnitude star, is a fairly small, faint, irregular cluster elongated E–W. Its brightest star, on the west edge of the cluster, is flanked by 13th magnitude companions. On the east edge of the group is a hazy patch of partial resolution. The northeastern stars of the cluster are arranged in a wide "V." Open cluster King 16 is 15′ to the SSW.

Dolidze 13 Open Cluster 30★ Tr Type III 1 p
ø12′, m –   00ʰ50.0ᵐ +64°08′
Finder Chart 10-5   ★★

12/14″ Scopes–100x: Dolidze 13 is a fairly obvious, loose, circular cluster spread over a 12′ area. The group contains fifteen 12th magnitude, and at least an equal number of fainter, stars. It is a good contrast to a loose asterism of brighter stars which includes a 5th magnitude star to the NE.

Figure 10-14. *The designation NGC 281 applies both to the emission nebula and its embedded open cluster. The cluster is more obvious than the nebula but careful inspection of the area will reveal many faint patches of both light and dark streaks. William Harris made this exposure with an 8", f4 Wright Newtonian reflector for 30 minutes on hypered 2415 Kodak Tech Pan film at prime focus.*

NGC 278 H159[1] Galaxy Type SAB(rs)b II–III
ø2.6′ × 2.6′, m10.8v, SB 12.8   00^{h}52.1^m +47°33′
Finder Chart 10-4   ★★★★

4/6″ Scopes–75x: NGC 278, 3′ south of a 9th magnitude star, is a very faint, small, diffuse smudge.

8/10″ Scopes–100x: This galaxy is an easily spotted round glow of uniform surface brightness.

12/14″ Scopes–125x: NGC 278 has a round 1.5′ diameter halo with a much brighter center.

16/18″ Scopes–150x: A well concentrated core containing a stellar nucleus is surrounded by a mottled 2′ diameter halo.

NGC 281 IC 1590 Open Cluster Tr Type III 1 m
ø4′, m7.4p, Br★ 9.0p   00^{h}52.8^m +56°37′
NGC 281 Emission Nebula
ø35′ × 30′, Photo Br 1-5, Color 3-4   00^{h}52.8^m +56°36′
Finder Chart 10-6, Figure 10-14   ★★★/★

8/10″ Scopes–75x: Nine 7th to 10th, and twenty 11th to 13th magnitude stars are seen against a slight haze.

12/14″ Scopes–100x: With a UHC filter the NGC 281 nebula may be seen with direct vision as a 1°-long maple leaf. The embedded NGC 281 open cluster is very poor and loose.

16/18″ Scopes–100x: With an O-III filter NGC 281 appears fairly bright, large, and irregular. Several dozen stars are concentrated at its center. The nebula is crescent-shaped with dark lanes dividing it on the north side. The brightest nebulosity, on the NW side of the object, is about 15′ across with an obvious triple star at its luminous center. The lucida of the triple had a very close companion. Other patches of nebulosity are visible to the SE.

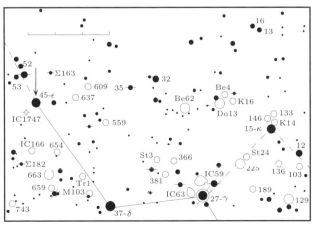

Finder Chart 10-5. 45-ε Cas: 01^{h}54.2^m +63°40′

Figure 10-15. *The faint emission and reflection nebulae IC 59 (top center) & IC 63 (left of center) lie north of the bright star, Gamma (γ) = 27 Cassiopeiae which interferes with observation. Martin C. Germano made this 70 minute exposure on hypered Kodak 2415 Tech Pan film with an 8", f5 Newtonian.*

IC 59 Emission & Reflection Nebula
⌀10' × 5', Photo Br 2-5, Color 1-4 $00^h56.7^m$ +61°04'
Finder Chart 10-6, Figure 10-15   ★

12/14" Scopes–100x: This nebula is extremely faint and fairly large and can be seen only with a UHC filter and averted vision. Three faint stars are also involved. Gamma (γ) = 27 Cassiopeiae should be kept out of the field of view.

16/18" Scopes–100x: IC 59, centered just 20' north of overly-bright Gamma (γ) = 27 Cassiopeiae, is very faint, diffuse, and elongated N–S. It appears larger but fainter than IC 63. IC 59 and IC 63 are together designated Sharpless 185.

IC 63 Emission & Reflection Nebula
⌀10' × 3', Photo Br 1-5, Color 2-4 $00^h59.5^m$ +60°49'
Finder Chart 10-6, Figure 10-15   ★

12/14" Scopes–100x: IC 63 is very faint, fairly large and elongated 8' × 4'. The rim is brighter on the south side. It should be viewed with a UHC filter and Gamma (γ) = 27 Cassiopeiae outside the field.

16/18" Scopes–100x: This nebula is very faint but more obvious than IC 59. It is fairly large, distinctly fan-shaped, and elongated E–W, the brightest apex at the west end. The southern edge of the fan is more distinct than the northern. Filters do not noticeably improve the view.

Berkeley 62 Open Cluster 50★ Tr Type III 2 p
⌀5.5', m9.3v, Br★ 10.91v $01^h01.0^m$ +63°57'
Finder Chart 10-5   ★★★

12/14" Scopes–100x: Berkeley 62 is a large, faint, irregular cluster with no central compression. It contains 11th magnitude and fainter stars against a background glow. Only eight stars can be resolved, six of them in two E–W rows of three stars each in the southern part of the cluster. Be 62 does not stand out well from the star field.

NGC 358 Open Cluster 25★
⌀3' , m – $01^h05.2^m$ +62°05'
Finder Chart 10-6   ★★

8/10" Scopes–100x: NGC 358 is relatively faint and compact, its stars evenly distributed. A bright yellow star is superimposed.

12/14" Scopes–125x: NGC 358 is a fairly faint, small, square-shaped cluster of ten very faint stars. 10th magnitude stars mark the north and south edges of the square.

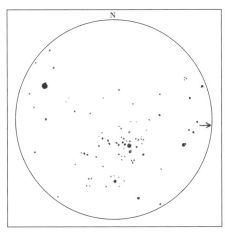 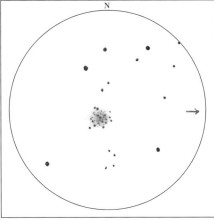 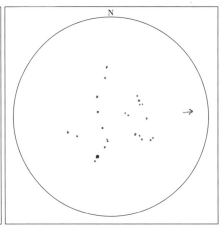

Figure 10-16. King 12
12.5″, 5–100x, by G. R. Kepple

Figure 10-17. King 16
12.5″, 5–75x, by G. R. Kepple

Figure 10-18. NGC 366
12.5″, 5.6–250x, by Alister Ling

NGC 366 Open Cluster 30★ Tr Type II 3 p
ø3′ , m –, Br★ 10.0p   01ʰ06.4ᵐ +62°14′
Finder Chart 10-6, Figure 10-18   ★★

12/14″ Scopes–125x: NGC 366 is a fairly nice compact cluster of eighteen stars, many of them doubles and triples. The west side is abundant with faint stars.

16/18″ Scopes–150x: NGC 366 is a faint, small cluster of twenty stars in a 3′ area. Three of its 10th magnitude stars are in its SE part and one is on its northern edge. A dozen very faint stars can be seen with averted vision in the cluster's western section, and a haze of unresolved stars is noticeable in the eastern portion. NGC 366 lies in a relatively sparse area of the Milky Way.

NGC 381 H64⁸ Open Cluster 50★ Tr Type III 2 p
ø6′, m9.3p, Br★ 10.0p   01ʰ08.3ᵐ +61°35′
Finder Chart 10-6, Figure 10-19   ★★★

4/6″ Scopes–75x: NGC 381, 8′ west of an 8th magnitude star, is a fairly compressed, rich gathering of faint stars.

8/10″ Scopes–100x: NGC 381 is a 5′ diameter ring of several dozen 12th to 13th magnitude stars from which an attached chain extends north. The cluster is loose but slightly more concentrated to the south. The cluster lucida, a 10th magnitude star with a 13th magnitude companion, is in the northern chain.

12/14″ Scopes–100x: NGC 381 stands out well against the star field as an irregularly round rich gathering of forty 11th to 13th magnitude stars covering a 6′ area. The cluster's 10th magnitude lucida is part of a chain of five bright, equally spaced, stars extending north from the main body. 200x reveals the 10th magnitude star to be a triple. At this magnification the cluster appears clumpy, scattered with starless patches.

Stock 3 Open Cluster 8★ Tr Type IV 1 p
ø2′ , m –, Br★ 11.0p   01ʰ12.3ᵐ +62°20′
Finder Chart 10-6   ★★

12/14″ Scopes–125x: Stock 3 is a rather faint, very small, E–W elongated cluster of seven stars in a tight 2′ group. Because of the sparse star field around it, the cluster stands out rather well. To its north and west are 10th magnitude field stars. To its east is a fairly bright gathering of field stars.

NGC 433 Stock 22 Open Cluster 15★ Tr Type III 2 p
ø2.5′ , m –, Br★ 9.0p   01ʰ15.3ᵐ +60°08′
Finder Chart 10-6   ★★

8/10″ Scopes–100x: This is a small, compact cluster of fifteen 12th to 15th magnitude stars scattered around a 9th magnitude star.

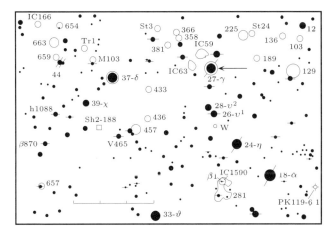

Finder Chart 10-6. 27-γ Cas: 00ʰ56.7ᵐ +60°43′

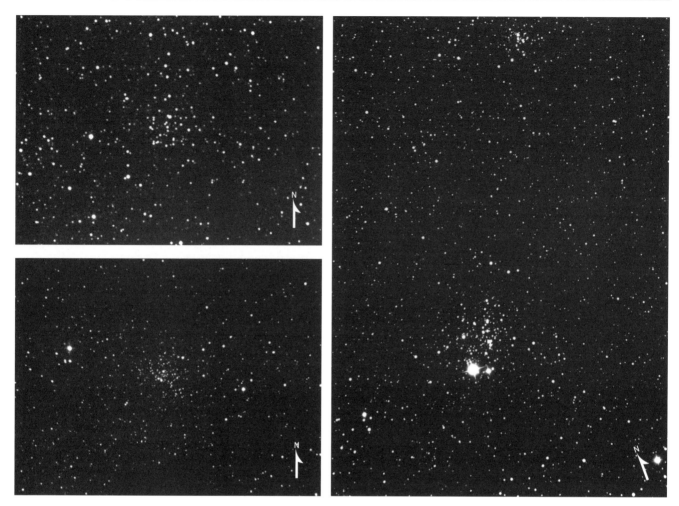

Figure 10-19 (Top). NGC 381 appears faint, circular, and fairly rich in small telescopes.
Figure 10-21 (Bottom). NGC 559 is a faint but rich patch of stardust. NGC 381 and NGC 559 were photographed by Lee C. Coombs for 10 minutes on 103a-O film with a 10″, f5 Newtonian.

Figure 10-20. NGC 436 (near top) is a small, irregular knot while NGC 457 (near bottom) is a bright impressive splash of stars extending northwest from the colorful double Phi (φ) Cassiopeiae. This 30 minute exposure was made on hypered 2415 film by J. C. Mirtle with an 8″, f6 Newtonian reflector.

12/14″ Scopes–100x: NGC 433 is a faint, loose, irregular cluster of 25 stars in several star chains. A 9th magnitude star is near its center. The cluster also has eight 11th and at least fifteen 12th-13th magnitude members.

NGC 436 H45[7] Open Cluster 30★ Tr Type I 3
ø5′, m8.8v, Br★ 11.12v   01^{h}15.6^m +58°49′
Finder Chart 10-6, Figure 10-20   ★★★

4/6″ Scopes–75x: NGC 436 is 50′ NW of magnitude 4.98 Phi (φ) = 34 Cassiopeiae and 40′ NW of open cluster NGC 457. It is a faint, small, fairly rich concentration of stars with three bright members at its center.

8/10″ Scopes–100x: NGC 436 is quite small but stands out well from the star field. It contains nine 11th to 12th magnitude stars, but many fainter members can be resolved or partially resolved. In the northern part of the cluster is a tiny triangle of two 11.5 and one 12th magnitude star. In the field 8′ east of the cluster is an 8th magnitude star. NW is a 4′ long row of three 10th magnitude field stars.

12/14″ Scopes–125x: NGC 436 is a fairly conspicuous cluster of thirty 11th magnitude and fainter stars in a 5′ area. A knot of stars is north of the center, and a string of stars extends south from the cluster. At high power several multiple stars can be resolved, but the object as a whole maintains the sense of "clusterness." In the field to the south are two widely separated 9th magnitude stars. To the east of the cluster are several more 9th magnitude stars.

NGC 457 H42⁷ Open Cluster 80★ Tr Type I 3 r
m13′, m6.4v, Br★ 8.59v   01ʰ19.1ᵐ +58°20′
Finder Chart 10-6, Figure 10-20   ★★★★

NGC 457 is a young star cluster about 9,300 light years away. The bright Phi (φ) = 34 Cassiopeiae is on the SE end of the group, but its membership is uncertain. If it is a true member, Phi Cas would have a luminosity of 275,000 suns.

8/10″ Scopes–75x: This is a beautiful cluster! The lovely yellow and blue double star Phi Cassiopeiae (5.0, 7.0; 134″; 231°) lies at its SE edge. NW from the Phi pair extends a moderately concentrated, irregular splash of 40 stars of mixed magnitudes. An arc of 11th and 12th magnitude stars curves through the center from either side of Phi, a wide pair of 12th magnitude stars marking the arc's center.

12/14″ Scopes–125x: NGC 457 is certainly one of the finer clusters in Cassiopeia. It is bright, large, quite rich, and somewhat triangular-shaped with three streams of stars protruding NNW, SW, and eastward from a moderately compressed center. It has at least fifty 8.5 to 12th magnitude, and thirty 13th to 14th magnitude, stars sprinkled within its boundaries. A gathering of bright stars is south, and a dark patch north, of the cluster center. A red star lies at the northern edge, and the beautiful yellow and blue double Phi Cassiopeiae is on the SE edge.

NGC 559 H48⁷ Open Cluster 60★ Tr Type II 2 m
⌀4.4′, m9.5v, Br★ 10.58v   01ʰ29.5ᵐ +63°18′
Finder Chart 10-5, Figure 10-21   ★★★

4/6″ Scopes–75x: This cluster is a faint, small, misty patch of stardust, three stars discernible south of its center. It is flanked by 7th and 8th magnitude field stars.

8/10″ Scopes–100x: NGC 559, easily located between two bright stars, is an evenly distributed spray of faint stars 4′ across. It has several dozen 11th and 12th magnitude stars and many more just at the threshold of visibility.

12/14″ Scopes–125x: NGC 559 is a moderately faint but nice cluster. It is a very rich and fairly compact group of forty stars in a 5′ diameter area. The cluster's three brightest members, one 10.5 and two 11th magnitude, form a triangle near the southern edge. Five 12th magnitude stars and a host of fainter stars resolve against the background haze. A chain of four 9th magnitude and two fainter stars joins the cluster from the NW. In the field 10′ to the east of the cluster is a keystone asterism that includes a 7th magnitude star. An 8th magnitude is in the field 8′ to the west.

Figure 10-22. *Messier 103 (NGC 581) is a celestial Christmas tree decorated by stars. Lee C. Coombs made this five minute photo on 103a-O Spectroscopic film with a 10″ f5 Newtonian reflector.*

NGC 581 Messier 103 Open Cluster 25★ Tr Type III 2 p
⌀6′, m7.4v, Br★ 10.55v   01ʰ33.2.ᵐ +60°42′
Finder Chart 10-6, Figure 10-22   ★★★★★

Messier 103 was discovered by Mechain in 1781 and later added to Messier's catalog. Its distance is some 9,200 light years and it has a true diameter of about 15 light years. The area is rich in star clusters, NGC 663, NGC 654, and NGC 659 located nearby. Double star Σ131, a pretty pair of 6th and 9th magnitude stars (Sep. 28.2″ in P.A. 145°) lies to the northwest. To the SE is a lovely 10th magnitude M6 red giant.

4/6″ Scopes–75x: This is an attractive cluster of moderately bright stars in a fairly rich triangular arrangement, its three brightest stars marking the angles.

8/10″ Scopes–100x: Messier 103 is an exquisite arrowhead-shaped cluster with a 7.5 magnitude star at the northern tip. The cluster is bright, large, fairly rich, moderately compressed, and has forty 8th to 12th magnitude stars. The field is liberally sprinkled with 8th and 9th magnitude stars, the brighter stars lying to the north.

12/14″ Scopes–100x: A splendid cluster! M103 is a bright, irregular, moderately rich cluster of fifty bright and faint stars in a 6′ area elongated NNW–SSE. The main body forms a wedge or Christmas tree, the three brightest stars at each apex. The brightest star (about 7.5 magnitude) is at the NNW apex — the top of the tree. The next brightest star, an 8th magnitude object is at the SE apex. The SW apex is marked by a 10.5 magnitude star. SE of the cluster's center is a pretty 9th magnitude red star.

Figure 10-23. *Trumpler 1 is a charming little clump of stars photographed by Martin C. Germano for 30 minutes on hypered 2415 film with an 8″, f5 Newtonian reflector.*

Figure 10-24. *NGC 637 is a crescent-shaped splash of fairly bright stars recorded by Lee C. Coombs with a five minute exposure on 103a-O film using a 10″, f5 Newtonian reflector.*

50 stars are within the triangle, and many faint stars are scattered beyond the wedge to the west and NE.

Trumpler 1 Cr15 Open Cluster 20★ Tr Type I 3 p
⌀4.5′, m8.1v, Br★ 9.55v   01ʰ35.7ᵐ +61°17′
Finder Chart 10-6, Figure 10-23   ★★★

4/6″ Scopes–75x: This cluster is a small, compressed knot of half a dozen stars standing out against an unresolved mass.

8/10″ Scopes–100x: This charming cluster stands out well from the field as a small clump of two dozen 10th to 14th magnitude stars, the brighter arranged in two roughly NE–SW lines. A bright star in the east portion is a double, its companion lying to its north.

12/14″ Scopes–100x: Trumpler 1 is a fairly bright, compact, irregular clump of several dozen 10th to 14th magnitude stars, the brighter stars in two short rows. The western row is the most prominent: two 10th and two 11th magnitude stars alternating along a NNE–SSW line. The NE–SW eastern row consists of two 10th and one 11th magnitude star. Many 12th magnitude and fainter stars are scattered around the rows, but most are located to the NW and SW. The star field is abundant, but the cluster's two star-rows stand out well.

NGC 609 Open Cluster Tr Type II 3 r
⌀3′, m11.0v, Br★ 14.43v   01ʰ37.2ᵐ +64°33′
Finder Chart 10-5   ★

12/14″ Scopes–125x: This cluster is a small round glow of 14th magnitude and dimmer stars. It is within a triangle of one 7th and two 8th magnitude stars, SW of the triangle's west corner. With averted vision six very faint stars may be discerned against an unresolved haze.

16/18″ Scopes–150x: NGC 609 is an extremely faint, granular patch spread over a 3′ area. Half a dozen stars stand out with averted vision, including a small arc of stars a little east of center.

NGC 637 H49⁷ Open Cluster 20★ Tr Type I 3 p
⌀3.5′, m8.2v, Br★ 9.97   01ʰ42.9ᵐ +64°00′
Finder Chart 10-5, Figure 10-24   ★★★

4/6″ Scopes–75x: This cluster is a faint, highly condensed crescent-shaped group south of a 9.5 magnitude star. A dozen members can be clearly resolved while a fair number of stars remain at the limit of visibility.

8/10″ Scopes–100x: NGC 637 is a tight, neat sprinkling of eight fairly bright and fifteen fainter stars in a NE–SW group. It stands out well from the surrounding field. A bright pair dominates the cluster center.

12/14″ Scopes–125x: NGC 637 is a bright, loose, irregular cluster of several dozen 10th to 14th magnitude stars elongated 4′ × 2′ NE–SW, a 9.5 magnitude star at the northern edge. The main concentration, south of the center, is surrounded by many faint stars. The curved row of five stars in the center includes an 8.5″ wide double of 10th and 11th magnitude stars. 3′ east of the cluster center are two 10th magnitude stars 25″ apart, aligned ESE–WNW.

Figure 10-25. *NGC 654 is a bright, rich knot, easily seen in small instruments. Lee C. Coombs made a five minute exposure on 103a-O film with a 10", f5 Newtonian reflector.*

Figure 10-26. *The faint stars of NGC 659 contrast nicely with bright stars to its south. Lee C. Coombs made this five minute exposure on 103a-O film with a 10", f5 Newtonian reflector.*

NGC 654 H46⁷ Open Cluster 60★ Tr Type II 3 m
ø5′, m6.5v, Br★ 7.36v $01^h44.1^m$ +61°53′
Finder Chart 10-6, Figure 10-25   ★★★★

4/6" Scopes–75x: This cluster is a small, rich knot of faint stars with a 7.5 magnitude star at its SSE edge. It is well defined with many faint stars nearly equal in magnitude.

8/10" Scopes–100x: NGC 654 is a 5′ diameter cluster with fifty 11th to 13th magnitude stars. One 7.5 and two 9th magnitude stars are along the southern edge. The main body is rectangular and contains many interesting close pairs and tight clumps. An 8 × 50 finder shows a round, obvious glow.

12/14" Scopes–100x: NGC 654 is a moderately rich, irregular cluster of fifty 11th to 14th magnitude stars in a 5′ area. The fainter stars are in a rich concentration, just SSE of which is a yellow magnitude 7.5 star and SSW are several 9th magnitude stars in a NNE–SSW line. The cluster's outline resembles the "teapot" of Sagittarius, the brighter stars along its south forming the handle. NGC 654 has ten 10th and forty 11th to 14th magnitude stars. South of the cluster a concentration of faint stars forms a flower-shaped asterism.

NGC 659 H65⁸ Open Cluster 40★ Tr Type III 1 p
ø5′, m7.9v, Br★ 10.43v $01^h44.2^m$ +60°42′
Finder Chart 10-5, Figure 10-26   ★★★

4/6" Scopes–75x: This nice but faint cluster lies 10′ NE of the conspicuous triangle formed by magnitude 5.9 and 44 Cassiopeia and magnitude 6.5 and 8.5 stars. NGC 659 is a granular patch in a rich star field with a dozen stars resolving against a background of stardust.

8/10" Scopes–100x: This cluster is just distinguishable as an enrichment of the Milky Way. Thirty 12th magnitude and fainter stars are visible in a 5′ area.

12/14" Scopes–125x: NGC 659 is rather faint, fairly large, moderately rich and concentrated with forty-five 12th magnitude or fainter stars. In the NE part of the cluster is a conspicuous pentagon of brighter stars, including the magnitude 10.5 group lucida. At the west of the pentagon is an in-line triple. The western and southern areas of the cluster are composed of many faint stars. The group's faint stars contrast nicely with the bright triangle involving 44 Cassiopeiae 10′ south.

NGC 663 H31⁶ Open Cluster 80★ Tr Type III 2
ø16′, m7.1v, Br★ 8.42v $01^h46.0^m$ +61°15′
Finder Chart 10-5, Figure 10-27   ★★★★

4/6" Scopes–75x: This fairly bright cluster is 23′ SE of a 6.5 magnitude star. It is large and uncompressed but well defined with 30 stars visible. A dark lane cuts into the cluster from the NNW.

8/10" Scopes–100x: This fine cluster should have made Messier's list. Four 8th magnitude and ten 9th to 10th magnitude stars are sprinkled among sixty fainter stars over a 15′ area. The cluster contains the double stars, Σ151, Σ152, and Σ153. In a viewfinder

Figure 10-27. *NGC 663 is a rich cluster divided by a dark streak. Lee C. Coombs made this five minute exposure on 103a-O film with a 10", f5 Newtonian reflector.*

NGC 663 appears as a rather large, hazy spot.

12/14" Scopes–100x: NGC 663 is bright, considerably rich and irregular with eighty stars in a 16' area. The cluster is divided into two segments by a starless void running N–S through its center. The large eastern segment grows wider toward the north, where it is rich in bright cluster members, including the magnitude 8.5 lucida. The western concentration is smaller and has only a third as many stars. The cluster contains four 8.5 magnitude, a dozen 9th to 10th magnitude, and several dozen 11th to 12th magnitude stars. In it are three prominent doubles of 9.5 to 11 magnitude stars with separations between 7" to 9". The star field is rich, several star streams flowing SW and SSW toward open cluster NGC 659 and 44 Cassiopeiae.

IC 166 Open Cluster 120★ Tr Type III 1 r
⌀4.5', m11.7v, Br★ 14.88v $01^h52.5^m$ +61°50'
Finder Chart 10-5   ★★

16/18" Scopes–150x: IC 166, 4' east of a wide 9th and 10th magnitude pair in a rich star field, is a faint mist centered on a Y-shaped asterism.

IC 1747 PK130+1.1 Planetary Nebula Type 3b
⌀13", m12.1v, CS 15.8v $01^h57.6^m$ +63°20'
Finder Chart 10-5   ★★★

12/14" Scopes–200x: This planetary nebula is located 30" SE of the 3.4 magnitude star Epsilon (ϵ) = 45 Cassiopeiae. It shows a fairly bright, smooth, round 12" disk. To the NNE are three 11th magnitude stars in a NNE–SSW line.

16/18" Scopes–250x: IC 1747 has a fairly bright, uniform 13" disk that seems slightly oval-shaped N–S. It is surrounded by a triangle of faint 13th to 14th magnitude stars. The central star is not visible.

NGC 743 Open Cluster 12★ Tr Type II 1 p
⌀5', m –, Br★ 10.0p $01^h58.7^m$ +60°11'
Finder Chart 10-7   ★★

8/10" Scopes–100x: This is a nice little cluster of 15 stars shaped like an arrowhead, an orange star at the west end. Because the brightest stars are of the 10th to 12th magnitudes, it stands out well in this bright-star-poor field. A small, curved chain lies to the SE.

12/14" Scopes–125x: NGC 743 is a faint but conspicuous cluster of twenty 10th to 12th magnitude stars concentrated in a 5' area. It is triangular, its north and east corners marked by 10th magnitude stars. In the SE portion is the wide double, h1098, 10.3 and 12.7 magnitude stars 17" apart. The star field around the cluster is interesting: a 9th magnitude star is immediately to the west, two 7th magnitude stars are 10' and 15' SE, and many 8th and 9th magnitude stars are scattered throughout.

Stock 5 Open Cluster 25★ Tr Type IV 2 p
⌀14', m –, Br★ 7.0p $02^h04.5^m$ +64°26'
Finder Chart 10-7   ★★★

8/10" Scopes–100x: This cluster is a naked eye spot in the Milky Way SW of three bright stars, one of which is 52 Cassiopeiae. Stock 5 is a bright, fairly large, uncompressed cluster of 17 stars, a circlet visible in the southern portion.

12/14" Scopes–100x: Nothing is visible at the catalogue position of Stock 5, but just to the south is a large bright circlet of stars containing many wide doubles. Two dozen stars are in and around the circlet, and another dozen stars lie to the north.

Stock 2 Open Cluster 50★ Tr Type III 1 m
⌀60', m4.4v, Br★ 8.18v $02^h15.0^m$ +59°16'
Finder Chart 10-7, Figure 10-30   ★★★

2/3" Scopes–15x: Stock 2, 2° NNW of the Perseus Double Cluster, is an ideal object for binoculars and richest-field telescopes. It is very large and uncompressed, though well-populated with 9th and 10th magnitude stars. 15x gives enough surrounding sky to set the cluster apart from the Milky Way.

8/10" Scopes–35x: This is a fairly bright, extremely large, and rich but uncompressed cluster containing many beautiful star chains and dark starless lanes.

12/14" Scopes–50x: Stock 2 is an extremely large,

Figure 10-28. *Emission nebulae NGC 896 is a small bright nodule right of center and IC 1795 is the larger, fainter nebulous patch attached to its upper left. At left, Melotte 15 is a bright cluster embedded in the faint wispy patches of emission nebula of IC 1805. Martin C. Germano made this 60 minute exposure on hypered 2415 Kodak Tech pan film with an 8″, f5 Newtonian at prime focus.*

bright, conspicuous cluster of 8th and 9th magnitude stars containing many groups and voids. Half a degree to its NE is a 63″ pair of equally matched 7th magnitude stars. At low power the catalog count of 50 stars is correct, but 100x adds enough faint stars to double the total.

NGC 886   Stock 6   Open Cluster   20★   Tr Type IV 2 p
ø20′, m –, Br★ 11.0p   02^{h}23.7^m +63°52′
Finder Chart 10-7   ★★

12/14″ Scopes–100x: At the catalogue position of NGC 886 are only a few faint stars, but to the west is a fairly large, half moon-shaped group of several dozen 11th and 12th magnitude stars. This group is NE of a line between two 9th and two 7th magnitude stars.

NGC 896   Emission Nebula
ø20′ × 20′, Photo Br 1-5, Color 3-4   02^{h}24.8^m +61°54′
IC 1795   Emission Nebula
ø40′ × 15′, Photo Br 1-5, Color 3-4   02^{h}26.5^m +62°04′
Finder Chart 10-7, Figures 10-28 & 10-30   ★★/★

12/14″ Scopes–75x: With a UHC filter, these nebulae form a footprint-shaped glow with a dark lane separating the heel from the sole.

16/18″ Scopes–75x: In an O-III filter on excellent nights these two nebulae are faint but obvious milky patches. NGC 896, the southwest patch of the pair, is a 5′ long NE–SW oval. A 30′ long E–W streak of obscuring dust cloud extends east from its east edge. The NE nebula of the pair, IC 1795, is larger but more diffuse. Both of these nebulae are indistinct and require averted vision. However, much more nebulosity is visible in this area than is plotted on *Uranometria 2000.0* or *Sky Atlas 2000*. Both of these faint milky areas lack bright involved stars.

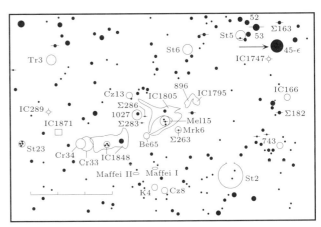

Finder Chart 10-7. 45-ε Cas: 01^{h}54.2^m +63°40′

Figure 10-29. *Melotte 15 is a bright, coarse star cluster embedded in the faint nebulous streaks of emission nebulae IC 1805. Martin C. Germano made this 60 minute exposure on hypered Kodak 2415 Tech Pan film with an 8", f5 Newtonian reflector.*

Markarian 6 Open Cluster 6★ Tr Type IV 2 p
ø4.5', m7.1v, Br★ 8.44v   02ʰ29.6ᵐ +60°39'
Finder Chart 10-7   ★★

12/14" Scopes–125x: Markarian 6, on the SW edge of the large IC 1805 nebulous complex, is a N–S row of seven 8th and 9th magnitude stars. The two stars at the southern end of the row are both doubles. A star to the north of the row might be a cluster member. In the field east of the row is a magnitude 8.5 star. A wide double lies to the north.

IC 1805 Emission Nebula
ø60' × 60', Photo Br 3-5, Color 3-4   02ʰ33.4ᵐ +61°26'

Melotte 15 Open Cluster 40★ Tr Type III 3 p
ø22', m6.5v, Br★ 7.87v   02ʰ32.7ᵐ +61°27'
Finder Chart 10-7, Figs. 10-28, 10-29, & 10-30 ★/★★★★

2/3" Scopes–25x: Melotte 15 is a loose, coarse cluster of several dozen stars, seven of which are 8th to 9th magnitude.

8/10" Scopes–50x: Melotte 15 is bright but loose, with eight 8th to 9th and twenty 10th to 12th magnitude stars in a 20' area. The nebulosity is not visible.

12/14" Scopes–60x: Through a UHC filter, IC 1805 is very faint, its most obvious part a dim 30' long streamer south of Melotte 15. The cluster is fairly compressed, its 30 stars forming a ring.

16/18" Scopes–75x: IC 1805 is an extensive, circular, hazy glow that teases averted vision with no definite edges. Located at the nebula's center, Melotte 15 contains fifty 8th magnitude and fainter stars spread over a 22' diameter area, the cluster's 7.8 magnitude lucida on its western edge. A milky area devoid of bright stars encircles Melotte 15.

King 4 Open Cluster 20★ Tr Type III 2 p
ø3', m10.5v, Br★ 12.85v   02ʰ35.7ᵐ +59°00'
Finder Chart 10-7   ★★

12/14" Scopes–125x: This cluster is a faint, tiny, irregular patch of twenty 13th magnitude and fainter stars. Low power shows it as a knot in the star field.

16/18" Scopes–150x: A worthwhile find – at least for larger instruments! King 4 is a faint but easily spotted, granular patch of light. 200x really brings it to life, clearly resolving an interesting group of several dozen stars punctuated by a few brighter luminaries mixed with star poor areas. An S-shaped asterism underlined by a string of stars may be sorted out.

Berkeley 65 Open Cluster 20★ Tr Type I 2 p
ø5', m10.2v, Br★ 10.79v   02ʰ39.0ᵐ +60°25'
Finder Chart 10-7, Figure 10-30   ★★★

12/14" Scopes–125x: Berkeley 65 is a fairly bright, loose, irregular 5' diameter cluster somewhat elon-

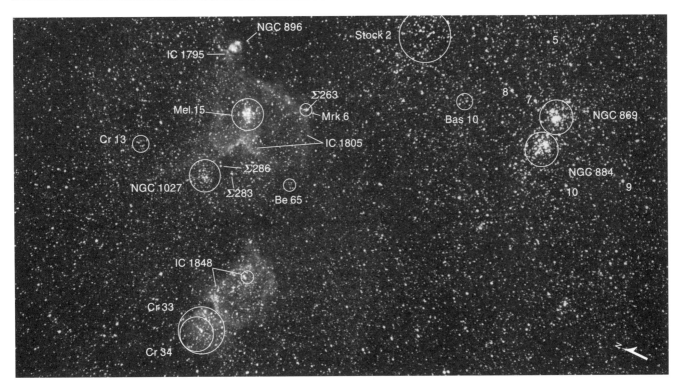

Figure 10-30. *This wide field photo may be used identify many of Casseopeia's clusters and nebulae. IC 1848 (bottom left) and the IC1795-IC1805 nebulous complex (top left). The double cluster in Perseus is prominent at right. Lying between the double cluster and the IC1795-IC1805 complex, is the large extended cluster Stock 2 (right of center at top). Martin C. Germano used a 135mm telephoto lens at f2.5 for 60 minutes on hypered Kodak 2415 Tech Pan film.*

gated N–S. The two brightest stars are on the south and SE edges, and a triangle of 11th magnitude stars is visible on the west side. The cluster contains ten 10th to 12th, and ten 13th to 14th, magnitude stars.

NGC 1027   H66⁸   Open Cluster 40★   Tr Type III 2 p n
ø20', m6.7v, Br★ 9.33v   02ʰ42.7ᵐ +61°33'
Finder Chart 10-7, Figure 10-31   ★★★

4/6" Scopes–50x: This cluster is moderately bright, large, and rich but uncondensed. A blue star is at its center, and a chain of faint stars exits its north edge and curves eastward.

8/10" Scopes–75x: NGC 1027 is a bright, large, rich cluster of 50 stars in a 20' area centered upon a 7th magnitude star. It has five 10th and twenty 11th to 13th magnitude stars, many in four prominent, mutually intersecting, star chains.

12/14" Scopes–100x: NGC 1027 is a bright, rich, irregular cluster with a 7th magnitude star just south of its center. Almost 80 stars are visible, twice the catalogue number. A relatively star-vacant lane crosses the cluster NE–SW behind the 7th magnitude lucida. A concentration of mostly 12th magnitude stars is just north of the center. The cluster is part of a profuse river of Milky Way stars meandering NW–SE.

Czernik 13   Open Cl. 10★   (Asterism?)   Tr Type II 2 p
ø5.0', m10.4v, Br★ 12.84v   02ʰ44.7ᵐ +62°21'
Finder Chart 10-7   ★

16/18" Scopes–150x: Czernik 13, 15' east of a 7th

Figure 10-31. *NGC 1027 is a highly irregular enrichment of the star field. Martin C. Germano made this 8 minute exposure on hypered 2415 film with an 8", f5 Newtonian.*

Figure 10-32. *IC 1848 is embedded with an open cluster of bright stars bearing the same designation. Faint but interesting swirls of nebula are just visible around some of the brighter stars through large instruments on a good night. Open cluster Cr 33 is located at the upper left. Martin C. Germano made this 60 minute exposure on hypered 2415 film with an 8", f5 Newtonian reflector at prime focus.*

magnitude star, is a very faint, rather small cluster of ten 13th and 14th magnitude stars in a NE–SW group. It is encircled by brighter stars, the brightest a 9th magnitude object to the south. At low power this cluster appears almost nebulous.

IC 1848 Open Cluster 10★ Tr Type IV 3 p n
⌀12′, m6.5v, Br★ 7.10v   02^{h}51.2^m +60°26′

IC 1848 Emission Nebula
⌀120′ × 55′, Photo Br 2-5, Color 3-4 02^{h}51.2^m +60°26′
Finder Chart 10-7, Figures 10-30 & 10-32   ★★★/★

12/14″ Scopes–75x: Both the cluster and nebulosity bear the designation IC 1848. The nebula is a very large, diffuse oval patch of very low surface brightness, widely scattered throughout with 9th magnitude stars. The cluster is an obvious group of several dozen stars surrounding a widely separated 7th and 8th magnitude pair.

16/18″ Scopes–100x: The cluster is moderately conspicuous, a rich but not compressed swarm of 11th magnitude and fainter stars. The assigned count of 10 stars in a 12′ area is much too low, for at least two dozen stars may be counted. And outside the main concentration are another 50 stars in a 30′ area. Both the 7th and 8th magnitude stars in the cluster core are surrounded by groups of fainter stars contained within pentagons of 11th magnitude stars. The group around the 7th magnitude cluster lucida has at least a dozen stars. Nebulosity can be glimpsed around most of the brighter IC 1848 stars.

Collinder 33 Open Cluster 25★ Tr Type II 3 m n
⌀39′, m5.9p   02^{h}59.3^m +60°24′

Collinder 34 Open Cluster Tr Type I 3 p
⌀25′, m6.8p   03^{h}00.9^m +60°25′
Finder Chart 10-7, Figure 10-32   ★★/★★

4/6″ Scopes–35x: Both of these open clusters are large, coarse scatterings of stars not well detached from the rich field around them.

8/10″ Scopes–50x: These clusters are somewhat richer and brighter than the surrounding Milky Way but are still rather difficult to distinguish.

12/14″ Scopes–50x: Collinder 33 is a bright, very large cluster of 30 stars elongated N–S, its brightest member on its west edge. A larger circular group farther west should be designated a cluster. Collinder 34 is devoid of bright stars, and has a vaguely doughnut shape. The area around both clusters is slightly milky with nebulosity.

Figure 10-33. *Emission nebula IC 1871 is a very faint patch lying south of two 10th magnitude stars. Martin C. Germano made this 50 minute photograph on hypered 2415 film with an 8", f5 Newtonian reflector.*

Figure 10-34. *Open cluster Trumpler 3 (Harvard 1) is a typical example of the many loose, irregular groups scattered throughout the Milky Way. Lee C. Coombs made this 10 minute exposure on 103a-0 film with a 10", f5 Newtonian at prime focus.*

IC 1871 Emission Nebula
ø4′ × 4′, Photo Br 4-5, Color 3-4   03ʰ06.4ᵐ +60°41′
Finder Chart 10-7, Figure 10-33   ★★★

16/18" Scopes–125x: This nebula is a small hazy patch barely detectable with averted vision. Filters are of little help.

20/24" Scopes–150x: IC 1871 is a very faint, small patch of nebulosity illuminated by a 10th magnitude star. Very slight variations in contrast across the face may be discerned.

IC 289 Planetary Nebula Type 4+2
ø>34.0′, m13.3v, CS 16.8   03ʰ10.3ᵐ +61°19′
Finder Chart 10-7, Figure 10-35   ★★

12/14" Scopes–200x: This planetary has a pale, round 35″ diameter disk of uniform surface brightness.

16/18" Scopes–225x: IC 289 is a faint planetary with a 42″ × 28″ halo elongated NW–SE and a slightly darker center about 15″ across. On a less transparent night, the annular structure was not visible. The central star was not seen. A 10th magnitude star lies just to the north.

Trumpler 3 Harvard 1 Open Cl. 30★ Tr Type III 3 p
ø23′, m7.0p   03ʰ11.8ᵐ +63°15′
Finder Chart 10-7, Figure 10-34   ★★★

4/6" Scopes–50x: This cluster is fairly bright and large, its 20 stars poorly compressed. Three N–S star chains stand out, the one at the cluster's center being the most prominent.

8/10" Scopes–75x: This is one of the better Trumpler or Harvard clusters, easily visible in an 11×80 viewfinder or large binoculars. It is bright and fairly large, but its thirty stars are not well concentrated.

Figure 10-35. *Planetary nebula IC 289 exhibits annularity in larger instruments. Martin C. Germano's 40 minute photo was taken on hypered 2415 film with an 8", f5 Newtonian.*

12/14″ Scopes–75x: Trumpler 3 is a fairly bright, large, loose, irregular cluster of approximately 35 stars, a dozen 9th to 10th magnitude. Three of the 9th magnitude stars are in a loose N–S row near the cluster's center. Two more 9th magnitude stars are on the cluster's NE edge, and a third is toward its eastern side. The cluster's stars are more concentrated to the east of the N–S star-row.

Chapter 11

Cepheus, the King

11.1 Overview

The stars of this constellation represent King Cepheus of Ethiopia, husband of Cassiopeia and father of Andromeda. Cepheus is a relatively faint constellation sandwiched between the brighter and more easily recognized Cassiopeia and Cygnus, but its house-shaped outline can be readily traced on dark nights. Alpha, Beta, Zeta and Iota form the corners of the house with Gamma at the top of the "roof." Cepheus, nearly 600 square degrees in size, is the 27th largest constellation. It is a north circumpolar group visible all year to northern observers but best seen at culmination in late autumn.

The constellation's brightest star, Alpha Cephei, named Alderamin, "Right Arm," lies near the path traced by Earth's axis in space over its 25,800-year precessional cycle and therefore periodically becomes the pole star. It will be the nearest bright star to the North Celestial Pole around 7500 A. D. Before that, however, two thousand years from now, Gamma Cephei will be the pole star.

Cepheus lies at the edge of the Milky Way. To the naked eye it seems somewhat barren, but even in binoculars it proves to be surprisingly rich in clusters, nebulae, and double stars.

Cepheus: SEE-fuss or SEE-fee-us
Genitive: Cephei, See-fee-eye
Abbrevation: Cep
Culmination: 9pm–Nov. 13, midnight–Sept. 29
Area: 588 square degrees
Showpieces: Mu(μ) Cep, Delta (27–δ) Cep, Xi (ξ) Cep, NGC 40, NGC 6939, NGC 7510
Binocular Objects: Mu(μ) Cep, Delta (27–δ) Cep, S800, Σ2893, O$\Sigma\Sigma$ 1, IC 1396, NGC 188, NGC 6939, NGC 6946, NGC 7160, NGC 7235, NGC 7261, NGC 7281, NGC 7510

11.2 Interesting Stars

OΣ440 Double Star　　　　　　　　　**Spec. M**
m6.4, 10.7; Sep. 11.4″; P.A. 181°　　$21^h27.4^m$　+59°45′
Finder Chart 11-5　　　　　　　　　　　★★★★

2/3″ Scopes–100x: OΣ440 is a beautiful pair of orange and bluish stars.

Beta (β) = 8 Cephei Σ2806 Double Star　**Spec. B2 III**
m3.2, 7.9; Sep. 13.3″; P.A. 249°　　$21^h28.7^m$　+70°34′
Finder Chart 11-4　　　*Alfirk*　　　★★★★

8/10″ Scopes–100x: This double has a brilliant white primary and a blue secondary 13″ apart Beta is a Cepheid variable with a very small light variation.

12/14″ Scopes–100x: Beta Cephei is a gorgeous double of white and emerald stars.

Mu (μ) Cephei Variable Star　　　**Spec. M2 Ia**
m3.4 to 5.1 in 730 days　　　　$21^h43.5^m$　+53°47′
Finder Chart 11-5　Herschel's Garnet Star　★★★★

Mu Cephei is Herschel's famous "Garnet Star," one of the most deeply-colored stars in the sky. It seems redder in smaller scopes and when near minimum magnitude. In medium-sized instruments Mu Cephei looks deep orange-red: in large scopes it is yellowish-orange.

The Garnet Star's period is irregular, though some authorities see a 730 day cycle in its fluctuations. Mu Cephei is a pulsating red supergiant similar to Betelgeuse in Orion. Indeed, its spectral features suggest that it might be even more brilliant than Betelgeuse with an absolute magnitude of –8.3, a luminosity of about 174,000 suns. The star's distance is estimated to be around 3,000 light-years.

Cepheus, the King

Constellation Chart 11-1

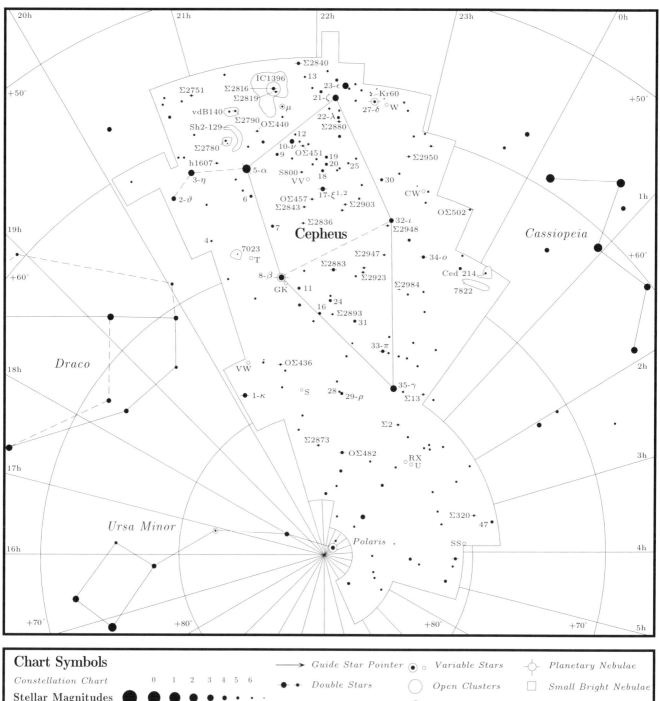

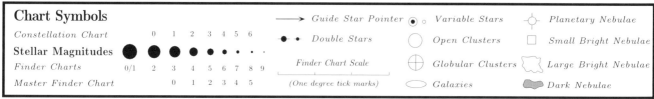

Table 11-1. Selected Variable Stars in Cepheus

Name	HD No.	Type	Max.	Min.	Period (Days)	F*	Spec. Type	R.A. (2000)	Dec	Finder Chart No. & Notes
T Cep	202012	M	5.2	11.3	388	0.54	M5-M8	21^{h}09.5^m	+68°29′	11-4
GK Cep	205372	EB/KE	6.8	7.3	0.93		A2+A2	31.0	+70 49	11-4
S Cep	206362	M	7.4	12.9	486	0.55	C7 (N8)	35.2	+78 37	11-8 Nice red color
μ Cep	206936	SRc	3.4	5.1	730		M2	43.5	+58 47	11-5 Herschel's Garnet Star
27-δ Cep	213306	Cd	3.4	4.3	5.36	0.25	F5-G1	22^{h}29.2^m	+58 25	11-6 Cepheid var. prototype
W Cep	214369	SRc	7.0	9.2	-		K0-M2	36.5	+58 26	11-6
CW Cep	218066	EA/DM	7.6	8.0	2.72	0.13	B0-B0	23^{h}04.0^m	+63 24	11-1
RX Cep	4499	SRd?	7.2	8.2	55:		G5	00^{h}50.1^m	+81 58	11-9
U Cep	5679	EA/SD	6.7	9.2	2.49	0.15	B7+G8	01^{h}02.3^m	+81 53	11-9 Eclipsing binary
SS Cep	22689	SRb	8.0	9.1	90		M5	03^{h}49.5^m	+80 19	11-9

F* = The fraction of period taken up by the star's rise from min. to max. brightness, or the period spent in eclipse.

Table 11-2. Selected Double Stars in Cepheus

Name	ADS No.	Pair	M1	M2	Sep."	P.A.°	Spec	R.A. (2000)	Dec	Finder Chart No. & Notes
1-κ Cep	13524	AB	4.4	8.4	7.4	122	B9	20^{h}08.9^m	+77°43′	11-1 White & blue
h1607	14544		7.5	10.7	11.5	82	K0	21^{h}00.3^m	+61 30	11-3
Σ2751	14575		6.1	7.1	1.5	353	B9	02.1	+56 40	11-1 Close white pair
Σ2780	14749	AB	6.0	7.0	1.0	219	B2	11.8	+59 59	11-3
OΣ436	14782		7.0	10.5	11.7	230	B9	12.1	+76 19	11-1
5-α Cep	14858	AB	2.4	10.2	206.8	22	A5	18.6	+62 35	11-3
		BxCD			19.9	172				
		CD	10.9	11.1	2.6	104				
Σ2790	14864	AB	5.7	10.0	4.5	45	K0 A0	19.3	+58 37	11-3 Orange and bluish
		AD		10.3	74.5	351				
OΣ440	14998		6.4	10.7	11.4	181	M	27.4	+59 45	11-5 Yellow & pale blue
8-β Cep	15032		3.2	7.9	13.3	249	B1	28.7	+70 34	11-4 White & emerald
β1143	15184	AB	5.6	13.3	1.6	324	O6	39.0	+57 29	11-5 AC: White & blue
Σ2816	15184	AC		7.7	11.7	121				
	15184	AD		7.8	19.9	339				
Σ2819	15214		7.5	8.5	12.4	57	F5	40.4	+57 35	11-5
Σ2836	15366		6.5	9.5	11.8	153		49.1	+66 48	11-4
OΣ451	15390	AB	7.5	8.5	4.3	221	A2	51.0	+61 37	11-5
	15390	CD	9.3	10.2	3.2	131				11-5 2″ north of brighter pair
Σ2843	15407	AB	7.1	7.3	1.5	145	A2	51.6	+65 45	11-4
15407		AC	9.9	56.0	276					
Σ2840	15405	AB	5.5	7.3	18.3	196	B3 A	52.0	+55 48	11-5 White & pale blue
S800 (EM)	15434	AB	7.0	8.7	62.6	146	B3 B8	53.8	+62 37	11-5 White pair in NGC 7160
OΣ457	15467		5.9	8.1	1.4	247	B2	55.5	+65 19	11-4
Σ2873	15571	AB	7.0	7.3	13.7	69	F5	58.2	+82 52	11-1 Matched yellow pair
17-ξ Cep	15600	AB	4.4	6.5	7.7	277	A3 G	22^{h}03.8^m	+64 38	11-5 Yellowish-white & yellow
Σ2883	15719		5.6	7.6	14.6	254	F2	10.6	+70 08	11-4
Σ2880	15729		7.8	9.7	4.2	352	K0	11.8	+59 43	11-6
Σ2893	15764		6.2	8.3	28.9	348	G5	12.9	+73 18	11-4
Σ2903	15881		6.7	6.7	4.3	96	F5 A2	21.8	+66 42	11-4
Krueger 60	15972	AB	9.8	11.3	3.3w	*137	M2 M5	28.1	+57 42	11-6 Red pair only 13 l.y. away
27-δ Cep	15987	AC	3.4	7.5	41.0	191	G0 A0	29.2	+58 25	11-6 Yellow & blue
Σ2923	16062	AB	6.4	8.7	9.4	46	A0	33.3	+70 22	11-7
OΣ482	16294		4.7	9.7	3.5	33	K0	47.5	+83 09	11-1
Σ2947	16291	AB	7.2	7.2	4.3	58	F5 F5	49.0	+68 34	11-7
Σ2948	16298		7.2	8.9	2.7	5	B9	49.6	+66 33	11-7 White duo
Σ2950	16317	AB	6.1	7.4	1.7	295	G0	51.4	+61 42	11-6 Yellow and ashy
	16317	AC		10.7	39.3	354				
Σ2984	16525	AB	7.7	10.2	4.5	295	K0	23^{h}07.4^m	+70 40	11-7
33-π Cep	16538	AB	4.6	6.6	1.2	*357	G5	07.9	+75 23	11-8
34-o Cep	16666	AB	4.9	7.1	2.8c	*223	G5	18.6	+68 07	11-7 Yellow & blue
OΣ502	16911		6.9	10.6	3.6	223	A2	39.9	+63 44	11-1
Σ2	102		6.6	6.9	0.8w	*12	A3	00^{h}09.3^m	+79 43	11-8
Σ13	207		7.0	7.3	0.9	*51	B9	16.2	+76 57	11-8
Σ320	2294		5.6	8.8	4.6	229	M	03^{h}06.1^m	+79 25	11-1 Orange & yellow stars

Footnotes: *= Year 2000, a = Near apogee, c = Closing, w = Widening. Finder Chart No: All stars listed in the tables are plotted in the large Constellation Chart, but when a star appears in a Finder Chart, this number is listed. Notes: When colors are subtle, the suffix -ish is used, e.g. bluish.

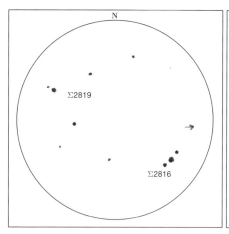
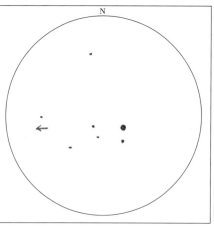
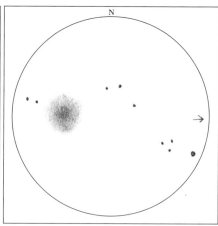

Figure 11-1. Σ2816 & Σ2819
10″, f4.5–100x, by Dave Sellinger

Figure 11-2. Delta (27-δ) Cephei
3.5″, f11–40x, by Mark Stauffer

Figure 11-3. UGC 11557
16″, f5–175x, by Bob Erdmann

Xi (ξ) = 17 Cephei Σ2863 Double Star   Type A3 & G
m4.4, 6.5; Sep. 7.7″; P.A. 277°   22ʰ03.8ᵐ +64°38′
Finder Chart 11-5   Al Kurhah   ★★★★

Xi is an attractive double star that is a true binary, its components separated by about 185 A.U. The Xi Cephei system is 120 light-years away and an outlying member of the Taurus Moving Group associated with the Hyades cluster. Webb called this pair white and tawny or ruddy.

2/3″ Scopes–100x: This lovely pair of white stars is easily resolved in small telescopes.

8/10″ Scopes–100x: Xi Cephei has subtle shades of yellowish-white and reddish.

Krueger 60 Double Star   Spec. M2-M5
m9.8, 11.3; Sep. 3.3″; P.A. 137°   22ʰ28.1ᵐ +57°27′
Finder Chart 11-6   ★★★

This faint but interesting double lies 43′ south of Delta (δ) = 27 Cephei. It is one of the nearest binaries, only 13.1 light years away. Krueger 60 is composed of 9.8 and 11.3 magnitude stars separated by 3.3″ revolving around each other in a period of only 44 years. Both stars are low-luminosity red dwarfs and are separated by only nine A.U., about the distance between Saturn and the Sun. These tiny stars have masses of only 0.25 and 0.15 suns and luminosities of less than 0.01 sun. The B component has one of the smallest stellar masses known.

A good 6″ scope with a high power ocular will usually resolve this vivid red pair. Its rapid orbital motion can be detected in just a few years.

Delta (δ) = 27 Cephei Variable Star   Spec. F5–G1
m3.48 to 4.37 in 5.4 days   22ʰ29.2ᵐ +58°25′
Finder Chart 11-6, Figure 11-2   ★★★★

Delta Cephei, the prototype of the highly regular Cepheid variables, was discovered in 1784 by the young English astronomer John Goodriche, a deaf-mute who died at age 21. Because in Cepheid variables the longer the star's light period the greater its luminosity, the distances of these stars — and most importantly the distances of the clusters and galaxies in which Cepheids are observed — can be estimated with reasonable accuracy.

Delta Cephei has a magnitude 6.3 companion 41″ distant. Though Delta is around 1,000 light-years away, implying a true separation between the stars of at least 13,000 A.U., the magnitude 6.3 star is believed to be physically involved with the variable. The secondary is a B7 star with a luminosity of some 250 suns. The primary's luminosity at maximum is around 4,000 suns. 43′ south of Delta Cephei is the faint but interesting double star, Krueger 60.

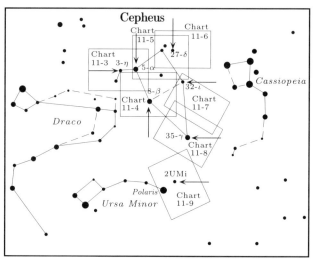

Master Finder Chart 11-2. Cassiopeia Chart Areas
Guide stars indicated by arrows.

U Cephei Variable Star   SpecB7+G8
m6.7 to 9.2, Per. 2.49 days
$01^h02.3^m$ +81°53'
Finder Chart 11-9   ★★★

U Cephei is an interesting short-period eclipsing binary. The bright B-type primary is occulted by the larger but fainter G giant every 2.5 days, the four-hour fall to minimum followed by a two hour total eclipse. Because the surface-to-surface separation of the two components is only 6.5 million miles, the compact B-type star is tearing matter from the outer layers of the G giant. This change in relative mass has caused the system's orbital period to slow by four minutes during the last century. A third star seems to be in the system but is not involved in the eclipses.

11.3 Deep-Sky Objects

UGC 11557 Galaxy Type SAB(s)dm
ø2.2' × 2.0', m13.2v, SB 14.7
$20^h24.0^m$ +60°12'
Finder Chart 11-3, Figure 11-3   ★

16/18" Scopes–150x: This galaxy is at the south end of a NE–SW line of 8th magnitude stars in a bright field, 7th magnitude stars being about 10' west and 12' NW of it. U11557 has a very faint, round 3' diameter halo with a weakly concentrated core. A tight triangle of 13th magnitude stars is 2.5' NE.

NGC 6939 Open Cluster 80★ Tr Type I 1 m
ø7.0', m7.8v, Br★ 11.91v $20^h31.4^m$ +60°38'
Finder Chart 11-3, Figure 11-4   ★★★★

4/6" Scopes–75x: NGC 6939 is a faint, round patch with a dozen tiny stars superimposed against a granular haze. The cluster is centered 10' north of a 7th magnitude star and surrounded by 10th magnitude stars to the NE, north, and west. NGC 6939 is only 39' NW of the galaxy NGC 6946: the two objects fit in the same low power field and provide an interesting contrast in appearance, the galaxy's fuzziness heightening the cluster's star-sparkling half-resolution.

8/10" Scopes–100x: This cluster is fairly bright, large, rich and compressed. Its shape is irregular rather than circular. Seventy stars can be resolved in it, many in chains that meander out into the surrounding Milky Way star field.

12/14" Scopes–125x: Quite nice! NGC 6939 is a fairly bright, rich and moderately compressed cluster of

Figure 11-4. *Open cluster NGC 6939 and galaxy NGC 6946 are only 2/3° apart and at low powers fit in the same field of view. Lee C. Coombs made this 15 minute exposure on 103a-O Spectroscopic film with a 10", f5 Newtonian.*

100 stars, most 12th magnitude and fainter. The brighter stars are gathered in the southern portion, fainter members fanning northward. The cluster spans about 10', but outliers are numerous to the south and NE. Several prominent star chains meander eastward. Three 10th magnitude stars surround the cluster to the north, west, and NE, and a 7th magnitude star is one cluster diameter away to the south.

NGC 6946 H76[4] Galaxy Type SAB(rs)cd I-II
ø13.0' × 13.0', m8.8v, SB 14.2 $20^h34.8^m$ +60°09'
Finder Chart 11-3, Figure 11-4   ★★★

4/6" Scopes–75x: NGC 6946, 39' SE of open cluster NGC 6939, is a dull, featureless haze with a smooth texture that contrasts nicely with the grainy cluster. A triangle composed of two 7th and one 8th magnitude star lies to the west.

8/10" Scopes–100x: This galaxy is faint but obvious, its diffuse oval halo brightening considerably toward the center. NGC 6946 is a nebulous object in a well populated star field and forms a nice pair with the star-rich open cluster NGC 6939 located 39' NW.

12/14" Scopes–125x: NGC 6946 is fairly bright, elongated 9' × 7' E–W, and has a well concentrated core.

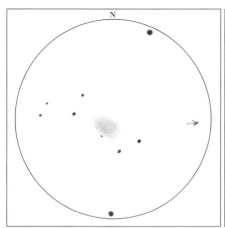

Figure 11-5. NGC 6949
16", f5–175x, by Bob Erdmann

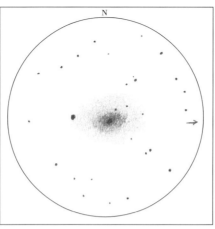

Figure 11-6. NGC 6951
18.5", f5-200x, by Glen W. Sanner

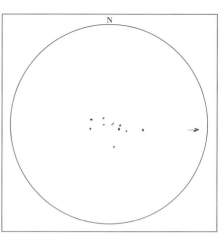

Figure 11-7. NGC 7055
12.5", f5–250x, by Alister Ling

16/18" Scopes–150x: This galaxy is impressive in larger instruments! Within its 10' × 8' halo is a multiple-arm spiral structure around a very bright core with a stellar nucleus. At 225x four spiral arms can be glimpsed: one arcs east from the central region, the second curves SE, the third (broad and blunt) extends south, and the fourth reaches from the central region west and NW. Several diffuse knots may be seen east and NE of the core. Many foreground Milky Way stars are superimposed across the halo.

NGC 6949 Galaxy Type S?
ø1.6' × 1.5', m13.5v, SB 14.3   20^{h}35.1^m +64°48'
Finder Chart 11-3, Figure 11-5   ★

16/18" Scopes–150x: NGC 6949 is very faint, elongated about 2' × 1.5' NE–SW, and slightly brighter in the center. The galaxy lies in the center of a trapezoid of four 12th magnitude yellow stars.

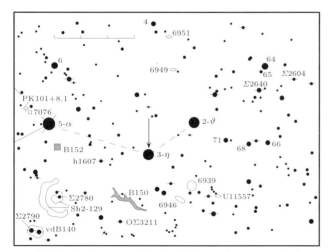

Finder Chart 11-3. Eta (3-η) Cep: 20^{h}45.2^m +61°50'

NGC 6951 Galaxy Type SAB(rs)bc I-II
ø3.8' × 3.3', m10.7v, SB 13.2   20^{h}37.2^m +66°06'
Finder Chart 11-3, Figure 11-6   ★★★

12/14" Scopes–125x: This galaxy has a moderately bright, diffuse oval halo elongated 1.5' × 1' E–W. The slightly brighter core has a stellar nucleus displaced south of center. A 12th magnitude star is at the east end.

16/18" Scopes–150x: NGC 6951 is a nice bright galaxy, elongated 3' × 2' ESE–WNW with an oval-shaped core and a bright stellar nucleus. A yellow 12th magnitude star is embedded on the east edge.

Barnard 150 Dark Nebula
ø60' × 3', Opacity 5   20^{h}50.6^m +60°18'
Finder Chart 11-3, Figure 11-8   ★★★

16/18" Scopes–50x: This dark nebula is a highly extended starless streak meandering 1.5° ENE-WSW through the Milky Way star field. Its west end passes immediately east of two 9th magnitude stars, and its east end is just south of, and parallels, a conspicuous line of three 9th-10th magnitude stars.

NGC 7023 H74^4 Open Cluster
ø5', m7.1p, Br★ 11.91v   21^{h}00.5^m +68°10'

NGC 7023 H74^4 Reflection Nebula
ø10' × 8', Photo Br 1-5, Color 1-4   21^{h}00.5^m +68°10'
Finder Chart 11-4, Figure 11-9   ★★★

8/10" Scopes–100x: NGC 7023 is a nice, fairly obvious milky nebula around a 7th magnitude star. Because this is a reflection nebula, nebular filters will not improve the view. The central cluster is a deformed-J of four fairly bright and fifteen faint stars.

Figure 11-8. *B150 is a thin dark streak meandering 1.5 degrees through the star field. Martin C. Germano made this 75 minute exposure on hypered 2415 film.*

Figure 11-9. *NGC 7023 is a fairly bright reflection nebula surrounding a 7th magnitude star. Martin C. Germano made this 30 minute exposure on hypered 2415 film. Both photos were made with an 8", f5 Newtonian.*

12/14" Scopes–125x: This is a fairly bright, irregularly-shaped reflection nebula surrounding a 7th magnitude star that interferes with observations — though very dark lanes within the nebula-glow may be discerned with careful viewing. The surrounding field has a dim glow from outlying sections of the nebula. NGC 7023 lies in a region of the Milky Way dark from the dust within which the nebula is embedded.

16/18" Scopes–150x: NGC 7023 is a moderately bright, irregular patch of nebulosity about 10' across with a bright 7th magnitude orange star embedded in its east end. A small, faint scattering of stars, one of which is yellow, lies west of the 7th magnitude star.

Barnard 152 Dark Nebula
ø15' × 3', Opacity 5   21ʰ14.5ᵐ +61°45'
Finder Chart 11-3   ★★

12/14" Scopes–100x: This highly extended starless streak is silhouetted against the diffuse haze around a 9th magnitude star. Other 9th magnitude stars are to the north.

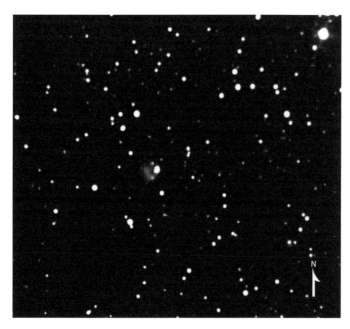

Figure 11-10. *Planetary nebula NGC 7076 is a faint object lying east of a faint star. Martin C. Germano made this 50 minute exposure on 2415 film with an 8″, f5 Newtonian.*

NGC 7055 Open Cluster
ø –   21^{h}19.4^m +57°35′

Finder Chart 11-5, Figure 11-7   ★★

12/14″ Scopes–150x: At 75x, NGC 7055 is a small, obvious clump in a nice star field. At 150x the group resolves into a handful of stars without the background glow that implies the presence of fainter members. The cluster is highly extended NE–SW, its stars arranged in pairs.

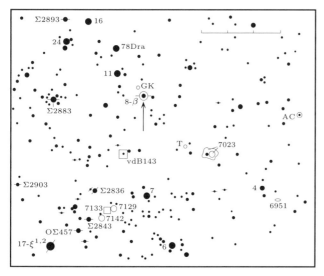

Finder Chart 11-4. Beta (8-β) Cep: 21^{h}28.6^m +70°34′

NGC 7076 PK101+8.1 Abell 75 Plan Neb Type 3 b
ø57″, m13.2v, CS 17.4v   21^{h}26.4^m +62°53′

Finder Chart 11-5, Figure 11-10   ★

16/18″ Scopes–250x: This planetary has a 30″ disk with uniform surface brightness touching a very faint star on its west side. Three 13th to 14th magnitude stars are just to the NNE.

IC 1396 Open Cluster 50★ Tr Type II m n
ø50′, m3.5v, Br★ 3.82v   21^{h}39.1^m +57°30′

IC 1396 Emission Nebula
ø170′ × 140′, Photo Br 3-5, Color 3-4 21^{h}39.1^m +57°30′

Finder Chart 11-5, Figure 11-11   ★★★★/★

2/3″ Scopes–25x: IC 1396 is visible under good sky conditions to the unaided eye, and is not difficult in 10 x 50 binoculars. This nebula with involved cluster is a large circular patch of haze spreading 3° south of the orange-red star Mu (μ) Cephei. The brightest stars of the cluster are arranged in a large "X" centered upon the triple Σ2816. 50x expands the cluster out of recognition, but reveals the two 8th magnitude companions of the bluish-white magnitude 5.5 primary of Σ2816. The pretty double of 8th magnitude stars, Σ2819, is just NE of Σ2816. The nebula is quite difficult even with a Daystar 300B filter, becoming apparent only after minutes of dark adaptation and careful study. What can be seen is a NE–SW section of nebulosity with three involved bright stars bracketed by E–W and N–S dark lanes. The remaining nebulosity was too faint to be seen unambiguously, but discerning dark lanes in it proves that it was just below the threshold of this aperture.

8/10″ Scopes–60x: This is a very pretty cluster lying south of the bright orange-red Mu (μ) Cephei, "Herschel's Garnet Star." At 40x the coarse but rich cluster is a "V," Σ2816 at its apex. At 60x this large group still looks compressed. The surrounding emission nebula responds well to a UHC filter.

12/14″ Scopes–60x: A fine sight! Lying south of the blood-red Herschel's Garnet Star, Mu (μ) Cephei, this cluster is extremely large, very bright, loose, irregular, and awash with a faint nebulous background mixed with dark lanes. At its center is the bright triple Σ2816, which has a bright white primary and two blue companions. The double Σ2819 is visible just NE. In a 25mm eyepiece the cluster spills out of the field of view. An O-III filter enhances the nebula's tenuous glow, but nebulosity and dark lanes are also faintly visible without a filter through a 15mm eyepiece.

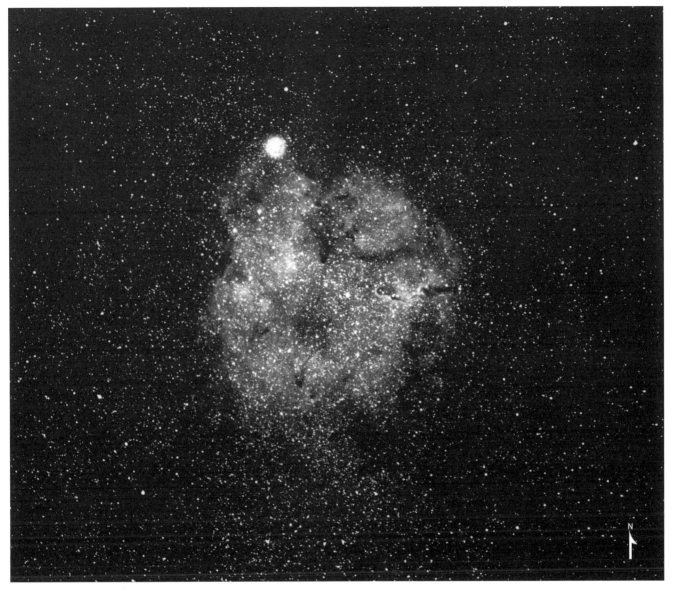

Figure 11-11. *Emission Nebula IC 1396 is very large, spanning about three degrees of sky. It is a challenge to detect visually, requiring very dark transparent skies. John Ebersole, M.D. made this photo with a 180mm telephoto at f2.8 for 70 minutes on 2415 film.*

NGC 7129 H75⁴ Open Cluster 10★ Tr Type IV 2 p n
ø7.0′, m11.5p   21ʰ42.8ᵐ +66°06′

NGC 7129 Reflection Nebula
ø7′×7′, Photo Br 1-5, Color 1-4   21ʰ42.8ᵐ +66°06′
Finder Chart 11-4, Figure 11-12   ★★/★★

12/14″ Scopes–150x: NGC 7129 is a small group containing six bright stars resembling a tiny Delphinus, nebulosity just visible around the four northernmost stars. Filers very little on the nebula.

16/18″ Scopes–150x: NGC 7129 is a obvious group composed of three wide bright pairs and a few much fainter stars. The northernmost pair (10, 10.5; 35″; 300°) and the central pair (9.5, 10; 60″; 310°) form a parallelogram. The southern pari (10, 10.5; 35″; 0°) is angled perpendicular to the other two. The nebulosity is fairly obvious, elongated 4′ × 2′ NNW-SSE, and brightest around the eastern star of the northernmost pair, fading to the north.

NGC 7133 Reflection Nebula
3′×3′, Photo Br 1-5, Color 1-4   21ʰ43.6ᵐ +66°10′
Finder Chart 11-4, Figure 11-4   ★★

16/18″ Scopes–150x: NGC 7133, lying 5′ NE of NGC 7129, is a small faint haze with a faint star touching its southern edge.

Figure 11-12. *Open cluster and reflection nebula NGC 7129-33 (lower right) lie less than half a degree from the moderately rich open cluster NGC 7142 (upper left) in an interesting star field. Martin C. Germano made this 60 minute exposure on hypered 2415 Kodak Tech Pan film with an 8″, f5 Newtonian at prime focus.*

NGC 7142 H66⁷ Open Cluster 100★ Tr Type II 2 r
4.3′, m9.3v, Br★ 12.12v   21ʰ45.9ᵐ +65°48′
Finder Chart 11-4, Figure 11-12   ★★★

8/10″ Scopes–100x: NGC 7142 is a rich cluster of three dozen 12th to 14th magnitude stars in a 10′ area against the background glow of the unresolved members. An 8th magnitude field star is 7′ north, and three outlying 10th magnitude stars are at the cluster's SE edge. Nebula NGC 7129 is 25′ NW.

12/14″ Scopes–125x: NGC 7142 is a rich, sparkling cluster of sixty stars in a roughly oval 12′-long concentration with a starless notch at its southern edge. Thirty 13th, and thirty 14th magnitude and fainter stars twinkle against an unresolved glow. Three 12.5 magnitude stars are near the cluster's center. A prominent N–S star chain almost bisects the group.

NGC 7139 H696³ Planetary Nebula Type 3b
ø78.0″, m13.3v, CS 18.1v   21ʰ45.9ᵐ +63°49′
Finder Chart 11-5, Figure 11-13   ★★

8/10″ Scopes–125x: This difficult planetary nebula is extremely faint, moderately large, and round. It lies north of a semicircle of 10th to 11th magnitude stars. Two stars of the 8th and 7th magnitudes are located 12′ and 20′ west, respectively.

12/14″ Scopes–150x: NGC 7139 is a very faint, moderately large, circular, grey smudge with a somewhat brighter center and diffuse edges. The halo, 45″ in diameter, is about the size of Jupiter's disk. A 13th magnitude star is 1′ SE of the planetary's center. To the east is a triangle of stars, from the southern

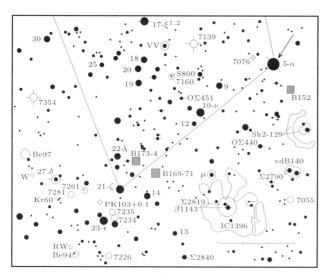

Finder Chart 11-5. Alpha (5-α) Cep: 21ʰ18.5ᵐ +62°35′

Figure 11-13. *NGC 7139 is a large but faint planetary nebula in a rich star field. Martin C. Germano made this 40 minute exposure on 2415 film with a 14.5″, f5 Newtonian.*

Figure 11-14. *NGC 7160 is an easily visible clump of bright stars. Lee C. Coombs took this five minute exposure on 103a-O film with a 10″, f5 Newtonian at prime focus.*

vertex of which a string of stars curves southwest.

16/18″ Scopes–150x: NGC 7139 appears moderately faint, quite large, and round with uniform brightness. An O-III makes the periphery more distinct. A 13.5 magnitude star touches the NE edge, and a pair of 14th magnitude stars lies 1′ NW.

NGC 7160 Open Cluster 12★ Tr Type II 3 p
ø7′, m6.1v, Br★ 7.04v   21ʰ53.7ᵐ +62°36′

Finder Chart 11-5, Figure 11-14   ★★★

8/10″ Scopes–100x: This cluster is an attractive knot of bright stars elongated 5′ E–W. 7th and 8th magnitude stars are east, and three magnitude 9.5 stars just SW, of the center. Fifteen fainter cluster members also can be seen.

12/14″ Scopes–100x: NGC 7160 is a bright, loose, irregular cluster of several dozen stars in an 8′ × 4′ ENE–WSW area. Just ENE of the center are 7th magnitude yellowish-white and 8th magnitude bluish-green stars. To the WSW of this bright pair is a trio of 9th-10th magnitude stars. More of the cluster's faint members are north than south of these five stars. A 7th magnitude field star is 5′ NW of the group.

Barnard 169 LDN 1151 Dark Nebula
ø60′, Opacity 3   21ʰ58.9ᵐ +58°47′

Barnard 170 LDN 1151 Dark Nebula
ø26′, Opacity 4   21ʰ58.9ᵐ +58°59′

Barnard 171 LDN 1151 Dark Nebula
ø19′, Opacity 5   22ʰ03.5ᵐ +58°52′

Finder Chart 11-5, Figs. 11-15 & 11-16 ★★★/★★★/★★★★

12/14″ Scopes–50x: Barnard 171 is a patch of obscuration contrasting well with the surrounding Milky Way star field. It is irregular in shape, with extensions to the NNE, NW, and SW. The NW extension is designated B169-170. The junction of B169-170 with B171 is flanked by 9th magnitude stars.

Barnard 174 LDN 1164 Dark Nebula
ø19′, Opacity 6   22ʰ07.3ᵐ +59°05′

Barnard 173 LDN 1164 Dark Nebula
ø4′, Opacity 6   22ʰ07.4ᵐ +59°10′

Finder Chart 11-5, Figure 11-16   ★★★/★★★★

12/14″ Scopes–50x: B173-4 is an S-shaped dark nebula stretched N–S. The southern half (B173) contrasts well with the surrounding Milky Way glow, but the distinctness of the northern half (B174) is compromised by the generous sprinkling of stars over it. The double Σ2872 (6.6,10.1; 117.6″;153°) is just NE. The field to the east and south is exceptionally rich in bright stars.

Figure 11-15. *B171 is the large dark area to the left and B169-170 is the finger extending to the right. Martin C. Germano made this 75 minute exposure on hypered 2415 film with an 8″, f5 Newtonian reflector at prime focus. North is up.*

Figure 11-16. *B173 is the dark streak at left. The eastern edge of B171 is visible at the lower right corner. North is up. Photographic information is the same for both photos on this page.*

NGC 7226 Open Cluster 25★ Tr Type I 1 p
ø1.4′, m9.6v, Br★ 10.80v $22^h10.5^m$ +55°25′
Finder Chart 11-6   ★★

16/18″ Scopes–150x: This easily overlooked cluster is a tiny 1.5′ mist of 14th magnitude stars in a very well populated Milky Way field. Ten pinpoints of light are sprinkled around and SW of an 11″ wide pair of magnitude 11.7 and 12.7 stars aligned NNE–SSW. A magnitude 10.8 star lies on the cluster's NW edge.

NGC 7234 Open Cluster
ø –, m – $22^h12.1^m$ +56°58′
Finder Chart 11-6, Figure 11-19   ★★

12/14″ Scopes–125x: This cluster, just north of a 5th magnitude star, is a condensation of eight faint stars. It is much less obvious than open cluster 7235, 15′ north.

NGC 7235 Open Cluster 30★ Tr Type III 2 p
ø4′, m7.7v, Br★ 8.80v $22^h12.6^m$ +57°17′
Finder Chart 11-6, Figure 11-17   ★★★

8/10″ Scopes–100x: This cluster is 23′ NW of a NE–SW line of two 6th magnitude stars flanking one of the 5th. NGC 7235 is a fairly obvious group of twenty 9th to 12th magnitude stars in a 4′ × 2′ ENE–WSW area.

12/14″ Scopes–125x: NGC 7235 is a fairly bright, loose cluster like an arrowhead pointing west. It is moderately concentrated with respect to the surrounding star field. Several dozen members can be discerned in a 5′ × 3′ NE–SW area. A 9th magnitude star is just SE. Nice, but not a showpiece.

PK103+0.1 Min 2-51 Planetary Nebula Type 2+3
ø>41″, m13.5v, CS – $22^h16.1^m$ +57°29′
Finder Chart 11-6, Figure 11-20   ★★

12/14″ Scopes–175x: This planetary is 14′ north of a 6th magnitude star, the northernmost in a NE–SW row with a 5th magnitude and another 6th magnitude star. In a Daystar 300 filter it appears fairly large, faint, and elongated N–S.

16/18″ Scopes–200x: This planetary's 30″ disk and 15th magnitude central star are just detectable without a filter. The UHC filter makes the nebula bright and obvious, but the central star disappears.

NGC 7261 Open Cluster 30★ Tr Type III 1 p
ø5′, m8.4v, Br★ 9.61v $22^h20.4^m$ +58°05′
Finder Chart 11-6, Figure 11-18   ★★★

8/10″ Scopes–100x: This cluster is a degree east of Zeta (ζ) = 21 Cephei (m3.35) and 18′ south of a 7th magnitude star. Several dozen faint members are loosely scattered in a 7′ × 4′ N–S area. The brighter stars line the eastern edge, the brightest being

Figure 11-17. *NGC 7235 is a fairly bright, compact, triangular-shaped open cluster of several dozen stars. Martin C. Germano made this 35 minute exposure on hypered 2415 film with a 14.5", f5 Newtonian reflector at prime focus. North is up.*

Figure 11-18. *Open cluster NGC 7261 is an irregular cluster of moderately faint stars elongated N-S. Lee C. Coombs made this ten minute exposure on 103a-O Spectroscopic film with a 10", f5 Newtonian at prime focus. North is up.*

easternmost. Several star chains are visible, one extending beyond the main body to the north.

12/14" Scopes–125x: NGC 7261 is an irregular scattering of thirty faint stars extended 8′ × 5′ N–S. The cluster's brighter stars form a "g"-shaped asterism. The southern region has a nearly starless rectangular void. The majority of the stars are 11.5 magnitude and fainter. Outlying stars straggle to the north.

Berkeley 94 Open Cluster 10★ Tr Type I 1 p n
ø2.3′, m8.7:v, Br★ 9.65v   22^{h}22.7^m +55°51′
Finder Chart 11-6   ★★★

12/14" Scopes–125x: Berkeley 94 lies 6′ SW of the orange variable star RW Cephei. It is an obvious group forming a tight 2′ long isosceles triangle with a dozen 11th to 14th magnitude stars. An isolated 9.5 magnitude star is 2′ north.

NGC 7281 Open Cl. or Asterism? 20★ Tr Type IV 2 p
ø12′, m –   22^{h}24.7^m +57°50′
Finder Chart 11-6, Figure 11-21   ★★

8/10" Scopes–100x: This cluster, 3/4° SW of Delta (δ) = 27 Cephei, is a loose, triangular group of a several dozen 11th magnitude and fainter stars.

12/14" Scopes–125x: NGC 7281 is a loose gathering of three dozen stars, most 11th magnitude and fainter, in a 10′ area. On the north side of the cluster is an E–W arc of three magnitude 9.5 members.

Berkeley 97 Open Cluster 12★ Tr Type III 1 p
ø5′, m –, Br★ 11.0p   22^{h}39.5^m +59°01′
Finder Chart 11-6   ★★

12/14" Scopes–125x: Berkeley 97 is a rather poor cluster of two 10th magnitude stars 1′ apart with two fainter members nearby. 225x adds five more stars of magnitude 13.5 to 14.

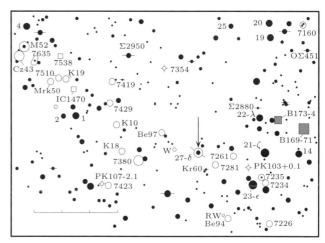

Finder Chart 11-6. Delta (27-δ) Cep: 22^{h}29.1^m +58°25′

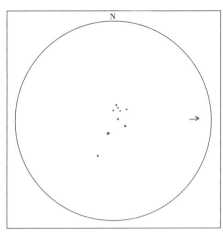 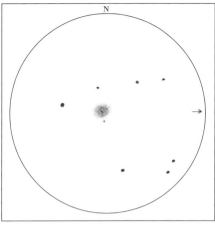 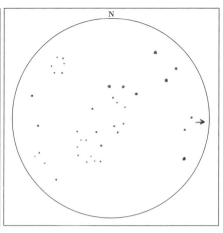

Figure 11-19. NGC 7234
12.5", f5.6–250x, by Alister Ling

Figure 11-20. PK103+0.1
16", f5–185x, by Bob Erdmann

Figure 11-21. NGC 7281
13", f5.6–165x, by Steve Coe

NGC 7354 PK107+2.1 Planetary Nebula Type 4+3b
ø>20', m12.2v, CS 16.1v   22ʰ40.4ᵐ +61°17'
Finder Chart 11-6, Figure 11-22   ★★★

8/10" Scopes–200x: This planetary nebula lies between a 10th magnitude star 4' to its SSE and two 10th magnitude stars, 30" apart E–W, 3' to its NNW. It has a faint, small, diffuse, uniform disk without a central star. An O-III filter enhances the outer edges.

12/14" Scopes–225x: At low power this planetary is a faint, small, soft uniform disk. 225x shows a grey 30" disk with a fairly sharp periphery. Two 13th magnitude stars lie just outside the disk to its west and SW.

16/18" Scopes–250x: NGC 7354 displays a fairly bright, ghostly grey 30" disk slightly fainter on its NE and SW edges. The central star is suspected but not confirmed. An O-III filter enhances the nebula's contrast with the sky background, but its fainter sections are no longer apparent.

IC 1454 Abell 67 Planetary Nebula Type 4
ø33", m13.8v, CS 18.0v   22ʰ42.6ᵐ +80°27'
Finder Chart 11-8   ★★★

12/14" Scopes–225x: This planetary lies just west of the westernmost in half a degree-long E–W arc of three 7th magnitude stars. It is visible as a faint, round 30" disk with well defined edges. A very faint star lies inside the western edge.

NGC 7380 H77⁸ Open Cluster 40★ Tr Type III 3 p n
ø12', m7.2v, Br★ 8.58v   22ʰ47.0ᵐ +58°06'

NGC 7380 H77⁸ Emission Nebula
ø25'×30', Photo Br 1, Color 3   22ʰ47.6ᵐ +58°04'
Finder Chart 11-6, Figure 11-23   ★★★/★★

12/14" Scopes–100x: This cluster is centered 15' ESE of one 5.5 magnitude star and 20' south of another. 100x shows thirty 8.5 magnitude and fainter stars in a triangular arrangement. Nebulosity is visible without a filter, but a UHC filter considerably enhances the nebula's contrast with the sky background.

16/18" Scopes–125x: NGC 7380 is a bright, irregular, little concentrated cluster of 45 fainter stars sprinkled around a prominent "V" formed by the group's magnitude 8.5–9 members. The brightest star is located at the SW edge of the V, and double star OΣ480 (8.0, 9.0; 30.9"; 117°) is on its west edge. The

Figure 11-22. *Planetary nebula NGC 7354 diplays a faint but obvious disk without a central star. Martin C. Germano used a 14.5", f5 Newtonian to make this 12 minute exposure on hypered 2415 film.*

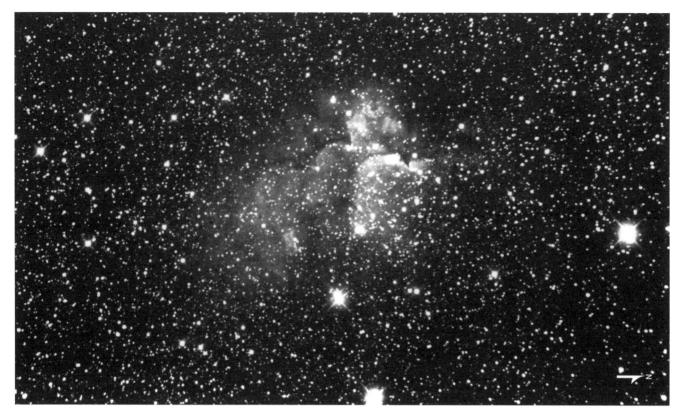

Figure 11-23. *The faint emission and open cluster NGC 7380 lies in rich star field which includes several 6th magnitude stars. Martin C. Germano made a 45 minute exposure on hypered 2415 film with an 8", f5 Newtonian at prime focus.*

cluster is surrounded by a Milky Way field rich in bright stars. Very faint wisps of nebulosity are embedded in the cluster, the most obvious lying east of the V.

King 18 Open Cluster
ø4.0′, m – $22^h52.1^m$ +58°17′
Finder Chart 11-6   ★★

16/18″ Scopes–125x: King 18 is a fairly faint, small, cluster of sixteen 11th magnitude and fainter stars.

NGC 7419 H43⁷ Open Cluster 40★ Tr Type II 3 r
ø2′, m13.0p, Br★ 10.0p $22^h54.3^m$ +60°50′
Finder Chart 11-6, Figure 11-24   ★★★

12/14″ Scopes–100x: NGC 7419, 10′ ESE of a 6th magnitude star, is a faint, rich, highly concentrated clump of two dozen 10th to 14th magnitude stars in a 2′ × 1.5′ N–S group. A 9th magnitude star lies at the NW edge.

16/18″ Scopes–150x: NGC 7419 is faint but rich, with 30 stars in a NW–SE 6′ × 4′ area. Most of its stars are 13th to 15th magnitude objects, but five 10-11th magnitude stars form a chain through the cluster center with a 9th magnitude star at its NW end.

King 10 Open Cluster 40★ Tr Type II 3 m
ø3′, m –, Br★ 11.0p $22^h54.9^m$ +59°10′
Finder Chart 11-6   ★★

8/10″ Scopes–125x: King 10, located just west of a 9th magnitude star, is a hazy 3′ long N–S mist in which are embedded a dozen faint stars. 175x resolves fifteen 13th to 14th magnitude cluster members against the mist.

12/14″ Scopes–150x: King 10 is a very faint, granular patch in a lovely star-rich setting. Twenty 11th and 12th magnitude members are resolved in a misty patch 3′ across. West of the cluster is a N–S row of three bright field stars. To the east a NW–SE chain of 10th magnitude stars leads to a wide triangle of bright field stars. King 10 is somewhat triangular, its longest dimension N–S. For an "obscure" cluster, it is a nice object.

NGC 7423 Open Cluster
ø6′, m15.0p $22^h55.3^m$ +57°08′
Finder Chart 11-6   ★

12/14″ Scopes–125x: Although listed as "non-existent," NGC 7423 lies at the precise location of the open cluster Berkeley 57. The NGC description reads,

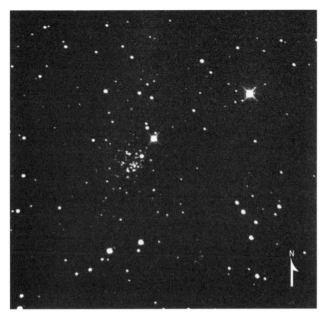

Figure 11-24. *NGC 7419 is a faint but rich triangular-shaped cluster. Lee C. Coombs recorded NGC 7419 for 10 minutes on 103a-O Spectrosopic film with a 10", f5 Newtonian.*

Figure 11-25. *NGC 7510 is a fine, bright wedge-shaped cluster. Lee C. Coombs made this 10 minute exposure on 103a-O Spectroscopic film with a 10" f5 Newtonian.*

"very faint, pretty large, irregular figure, easily resolvable." The observed group consists of two fairly bright stars surrounded by a scattering of twenty very faint ones.

NGC 7429 Open Cluster 15★ Tr Type III 2 p
ø14', m –, Br★ 11.0p $22^h55.9^m$ +59°59'
Finder Chart 11-6   ★★

12/14" Scopes–100x: This cluster is nestled between the two westernmost of a 2° long group of 5th and 6th magnitude stars. It has fifteen stars in a 12' area, half 10th to 11th magnitude and the others 12th to 13th magnitude. The cluster is oblong with crab claw-like extensions, and includes an easy triple star.

16/18" Scopes–125x: NGC 7429 is a loose, irregular 14' wide patch of fifteen 12th to 13th magnitude stars, a green-tinted 9th magnitude star on its NW edge.

PK107-2.1 Min 1-80 Planetary Nebula Type 2
ø8", m14.4v, CS – $22^h56.3^m$ +57°09'
Finder Chart 11-6   ★

12/14" Scopes–225x: This planetary nebulae is a close pair with a 12th magnitude star to its ENE. Viewed through a Daystar 300 filter, it is an extremely faint, diffuse disk only 8" across.

16/18" Scopes–250x: In an O-III filter PK107-2.1 blinks well, appearing almost as bright as the star on the nebula's following side. Its disk is very faint and tiny but definitely nonstellar. The open cluster NGC 7423 lies 7' SW.

IC 1470 Emission Nebula
ø1.2' × 0.8', Photo Br 2-5, Color – $23^h05.2^m$ +60°15'
Finder Chart 11-6   ★★★

12/14" Scopes–125x: IC 1470, 2' SE of a 10th magnitude star, is a uniformly bright oblong patch. A star at the SE end gives the nebula a cometary appearance.

16/18" Scopes–150x: A very neat object distinctly comet shaped. The small highly concentrated central knot is distinctly stellar, and extending southward from it is the short and stubby "tail." A star on the SE edge has a faint companion; but no stars are visible in the nebula.

King 19 Open Cluster 25★ Tr Type II 2 m
ø6', m9.2v, Br★ 10.37v $23^h08.3^m$ +60°31'
Finder Chart 11-6   ★★★

12/14" Scopes–100x: King 19, situated within a crescent of five 10th magnitude stars, is a sparse irregular cluster of fifteen 11th to 13th magnitude stars. NGC 7510, a lovely cluster with an arrowhead shape, lies 22' east.

16/18" Scopes–125x: King 19 is a faint, compact open cluster elongated 6' × 4' N–S with 20 resolved members surrounded by hints of unresolved stars. At the north edge of the cluster is a triangle of 10th magnitude stars; but the main concentration of the cluster is to the south.

NGC 7510 H44[7] Open Cluster 60⋆ Tr Type II 2 m n
ø4′, m7.9v, Br⋆ 9.68v   23ʰ11.5ᵐ +60°34′
Finder Chart 11-6, Figure 11-25   ★★★★

8/10″ Scopes–100x: This is a bright, rich, beautiful cluster. It is a highly concentrated 4′ diameter wedge of a dozen stars embedded within the haze of faint, unresolved cluster members. The group's lucida is on its eastern tip. 150x resolves several dozen stars down to about 13th magnitude

12/14″ Scopes–125x: A real surprise–nice at all powers! NGC 7510 is a bright, arrowhead-shaped cluster, the tip facing SW. Thirty 10th to 12th magnitude stars can be counted within the 4′ × 3′ arrowhead, the brightest along its southern edge. The multitude of 12th to 14th magnitude stars fanning northward bring the star count to well over 60 in an 8′ × 6′ area. NGC 7510 is highly concentrated, standing out well in the rich Milky Way star field. The cluster is framed within an equilateral triangle, 16′ long on a side, of 9 to 9.5 magnitude stars.

NGC 7538 H706[2] Emission Nebula Type vF 1 R
ø9′ × 6′, Photo Br 1-5, Color 3-4   23ʰ13.5ᵐ +61°31′
Finder Chart 11-6, Figure 11-26   ★★

12/14″ Scopes–100x: This nebulae is faint but can be seen without averted vision as a haze surrounding two 11th magnitude stars aligned NNE–SSW. The stars suggest ghostly eyes peering from the nebula, and make it a miniature version of Messier 78 in Orion. To the northeast a much brighter star with a wide faint companion is embedded in more nebulous haze.

16/18″ Scopes–125x: NGC 7538 is an easily seen circular nebulosity, 2′ in diameter, with diffuse edges and a slight central brightening. An 11th magnitude pair, 30″ apart, lies in the southern portion. A very faint 14th magnitude star is on the nebula's preceding edge. An O-III filter slightly enhances the nebula's glow.

Markarian 50 Open Cluster 5⋆ Tr Type 2 p n
ø5′, m8.5v, Br⋆ 9.83v   23ʰ15.3ᵐ +60°28′
Finder Chart 11-6   ★★★

4/6″ Scopes–125x: This open cluster of six stars is involved with a faint triangular nebula highly extended NE–SW, 10th magnitude stars marking each tip.

12/14″ Scopes–150x: 150x reveals a dozen stars, many at the threshold of resolution. Five 10th to 12th

Figure 11-26. *Emission nebula NGC 7538 is a haze surrounding a pair of stars, a miniature version of Messier 78 in Orion. Martin C. Germano captured the nebula with a 40 minute exposure on hypered 2415 film with an 8″, f5 Newtonian.*

magnitude stars are aligned nearly N–S. Another bright star lies to the SW one cluster width away.

PK116+8.1 Min 2-55 Planetary Nebula Type 3
ø40″, m14.2v, CS –   23ʰ31.9ᵐ +70°23′
Finder Chart 11-7   ★★

12/14″ Scopes–225x: This planetary lies 7′ west of an 8th magnitude star and 23′ east of a 5.5 magnitude star. In an O-III filter it is a fairly faint, round 35″ disk of uniform surface brightness.

NGC 7748 Nebula Type ?
ø –, Photo Br –   23ʰ45.0ᵐ +69°45′
Finder Chart 11-7   ★★

12/14″ Scopes–100x: NGC 7748 is a faint, circular 3′ haze surrounding a bright 7th magnitude star. An O-III filter improves the view.

NGC 7762 H55[7] Open Cluster 40⋆ Tr Type II 2 p
ø11′, m10.0p, Br⋆ 11.0p   23ʰ49.8ᵐ +68°02′
Finder Chart 11-7, Figure 11-27   ★★

12/14″ Scopes–100x: NGC 7762, centered 15′ NE of a 5th magnitude star, is a faint, loose, irregular group of thirty 11th and 12th magnitude stars in a 12′ area.

16/18″ Scopes–125x: NGC 7762 is loose and irregular with twenty 11th to 12th and thirty 13th to 15th magnitude stars.

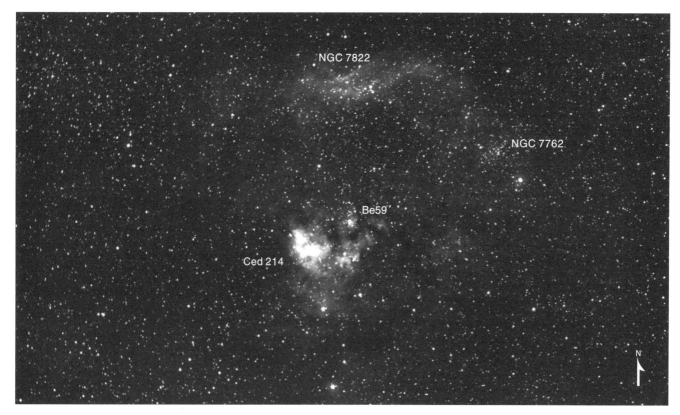

Figure 11-27. *This wide field photograph includes emission nebula Cederblad 214 (below center) with open cluster Berkeley 59 embedded in its northern portion. Emission nebula NGC 7822 lies to the north (top center) and NGC 7762 is visible to the west (right). Chris Schur made this 30 minute exposure through a 92 Wratten filter on hypered 2415 film with an 8", f1.5 Schmidt camera.*

The heaviest concentration of stars is along a NW–SE stream. A chain of bright stars is east of the cluster.

Berkeley 59 Open Cluster 40★ Tr Type III 2 p n
⌀10′, m –, Br★ 11.0p   00ʰ02.6ᵐ +67°23′
Finder Chart 11-7, Figure 11-27   ★★

12/14″ Scopes–125x: Berkeley 59 is a faint but obvious group of 40 stars embedded in the northern portion of emission nebula Cederblad 214. The majority of stars are sprinkled east and west along a 10′ area around a N–S pair of 11th magnitude stars. A detached circlet of 13th and 14th magnitude stars lies 3′ north of the northernmost of the 11th magnitude stars.

NGC 7822 Emission Nebula (SNR?)
⌀65′ × 20′, Photo Br 3-5, Color 3-4   00ʰ03.6ᵐ +68°37′
Finder Chart 11-7, Figure 11-27   ★

12/14″ Scopes–100x: NGC 7822 is a very faint, extremely large milky patch in a well populated star field. It is highly elongated E–W, the most obvious portion NE of a pair of 8th magnitude stars.

Cederblad 214 Emission Nebula
⌀50′ × 40′, Photo Br 2-5, Color 3-4   00ʰ04.7ᵐ +67°10′
Finder Chart 11-7, Figure 11-27   ★

12/14″ Scopes–75x: Cederblad 214, centered around a keystone of four 6th to 8th magnitude stars, is extremely faint and very large. On good nights irregular swirls and streaks are just visible in it. An O-III filter makes some of the streaks more obvious.

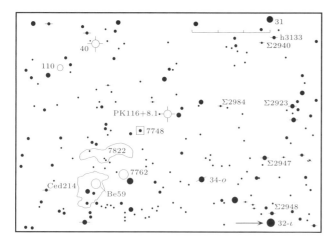

Finder Chart 11-7. Iota (32-ι) Cep: 22ʰ49.7ᵐ +66°12′

Figure 11-28. *Planetary nebula NGC 40 is a fine bright oval with a conspicuous central star. Lee C. Coombs made this five minute exposure on 103a-E Spectroscopic film with an 8″, f5 Newtonian reflector at prime focus.*

Figure 11-29. *NGC 188 is one of the oldest known open clusters. Its faint stars require large instruments for resolution: in modest-size telescopes it is a faint granular patch. Lee C. Coombs made this five minute exposure on 103a-E film with an 8″, f5 Newtonian.*

NGC 40 PK120+9.1 Planetary Nebula Type 3b+3
ø37″, m12.4v, CS 11.61v $00^h13.0^m$ +72°32′
Finder Chart 11-8, Figure 11-28   ★★★★

12/14″ Scopes–225x: A spectacular object! This planetary lies between a 9th magnitude star 2′ to its NE and two 10th magnitude stars 2-3′ SW. It is a bright, slightly oval-shaped disk with a conspicuous central star. Brighter areas along the eastern and western edges mimic the appearance of the polar caps of Mars. The western "cap" seems to run off the disk. The "polar cap" effect is only visible on the best of nights.

16/18″ Scopes–250x: NGC 40 is a bright bluish-green disk elongated 45″ × 35″ NNE–SSW with a prominent central star. It is somewhat annular. The disk is moderately brighter along its eastern edge and significantly brighter along its western edge. A 13th magnitude star lies just outside the SW rim.

NGC 188 Open Cluster 120★ Tr Type II 2 r
ø13′, m8.1v, Br★ 12.09v $00^h44.4^m$ +85°20′
Finder Chart 11-9, Figure 11-29   ★★

NGC 188 is one of the oldest known open clusters. Its estimated age, 9 billion years, is about that of the youngest globular clusters. NGC 188's brightest stars, 12th to 13th magnitude objects, are yellow luminosity class III giants with spectra of G8 to K4. The cluster completely lacks white main sequence stars.

8/10″ Scopes–100x: This cluster is a rather faint, rich granular patch, a dozen 12th to 13th magnitude stars embedded in the background haze. The field, especially to the SW, is well sprinkled with 8th and 9th magnitude stars.

12/14″ Scopes–100x: This cluster is moderately faint with three dozen pinpoint stars resolved in a rich, concentrated background glow spanning a 13-14′ area.

16/18″ Scopes–150x: NGC 188 is a nice but faint, round cluster of fifth to sixty 12th to 15th magnitude stars twinkling in and out of resolution against a granular background. Several dark gaps lie west of the cluster's center. Several wide star-pairs stand out.

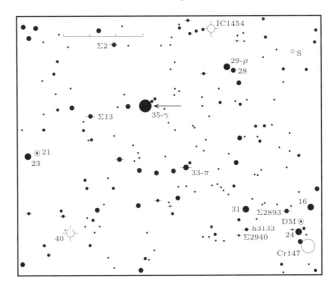

Finder Chart 11-8. Gamma (35) = γ Cep: $23^h39.0^m$ +77°38′

NGC 1184   H704² Galaxy   Type S0a
⌀2.6′ × 0.7′, m12.4v, SB 12.9   03ʰ16.6ᵐ +80°48′
Finder Chart 11-9   ★★

16/18″ Scopes–150x: NGC 1184 is a faint galaxy highly elongated 2.5′ × 0.5′ NNW–SSE with a well concentrated central area.

NGC 1544   Galaxy   Type S?
⌀1.6′ × 1.0′, m13.2v, SB 13.5   05ʰ03.0ᵐ +86°13′
Finder Chart 11-9   ★★

16/18″ Scopes–150x: NGC 1544 is 30′ NW of a 6th magnitude star about 3° from Polaris. It is quite faint, elongated 2′ × 1′ NW–SE, and slightly brighter in the center. Near one edge is a pair of 14th and 15th magnitude stars. The galaxy is just south of a row of three 13th magnitude stars aligned with its major axis.

NGC 2276   Galaxy   Type SAB(r)bc II
⌀2.6′ × 2.3′, m11.4v, SB 13.6   07ʰ27.0ᵐ +85°45′
Finder Chart 11-9   ★★★

12/14″ Scopes–125x: This galaxy lies 2′ ENE of an 8.5 magnitude star. It has a faint 2′ round halo of uniform surface brightness.

16/18″ Scopes–150x: NGC 2276 is a fairly obvious round, diffuse 2′ diameter halo without central brightening. It lies 2′ ENE of an 8.5 magnitude star which is part of an E–W row of four stars. Galaxy NGC 2300, 6′ SE of NGC 2276, is at the east end of the star-row.

NGC 2300   UGC 3798   Galaxy   Type SA0°
⌀3.2′ × 2.8′, m11.0v, SB 13.2   07ʰ32.0ᵐ +85°43′
Finder Chart 11-9   ★★★

12/14″ Scopes–125x: NGC 2300 is the brighter and larger of a pair with NGC 2276. It is fairly bright and round, and its 1.5′ halo has diffuse edges and a bright center.

16/18″ Scopes–150x: NGC 2300 is a bright but rather small galaxy elongated 2′ × 1.5′ ESE–WNW with a broad core containing a stellar nucleus. It lies at the eastern end of an E–W chain of four stars. NGC 2276 is 6′ NW.

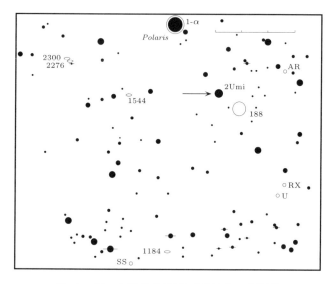

Finder Chart 11-9. 2 UMi: 01ʰ08.7ᵐ +86°15′

Chapter 12

Cetus, the Whale

12.1 Overview

Cetus is the Sea Monster of the Andromeda myth, turned to stone when Perseus exposed to its sight the severed head of the snake-haired Medusa. Modern guides often call it the Whale; but the constellation's figure on the Fornese Globe (1st century B.C.) is of a creature with a horned canine head, forepaws, and a long curved fishtail. Cetus, with 1,231 square degrees of sky, is the fourth largest constellation. Its claim to fame is the star Omicron (*o*) Ceti, named Mira "the Wonderful," prototype of the red giant long-period variables.

Cetus lies well away from the Milky Way. Except for its rather conspicuous star-pattern, it occupies a rather blank part of the sky. It contains no open clusters or diffuse nebulae, and, despite its large size, boasts just one Messier object, the Seyfert galaxy M77. However, it does have a fine planetary nebula, a number of good double stars, and a host of galaxies.

Cetus: SEE-tus
Genitive: Ceti, SEE-ti
Abbrevation: Cet
Culmination: 9 pm–Nov. 30, midnight–Oct 15
Area: 1,231 square degrees
Showpieces: 66 Cet, 84 Cet, 86-γ Cet, M77 (NGC 1068), NGC 157, NGC 246, NGC 247, NGC 578, NGC 1087
Binocular Objects: T Cet, 37 Cet, 54-χ Cet, 55-ζ Cet, 68-*o* Cet, M77 (NGC 1068), NGC 246, NGC 247

12.2 Interesting Stars

26 Ceti = Σ84   Double Star   Spec. F0
m6.2, 8.6; Sep. 16.0″; P.A. 253° $01^h03.8^m$ $+01°22'$
Finder Chart 12-5   ★★★★

8/10″ Scopes–100x: 26 Ceti is a pretty double of yellow and lilac stars.

Tau (τ) = 52 Ceti   Spec. G8 V
m3.50 $01^h44.0^m$ $-15°56'$
Constellation Chart 12-1   ★★★

Tau Ceti, 11.5 light-years away, is the seventh nearest naked eye star. (The six nearest, and their distances in light-years, are: Alpha Centauri, 4.3; Sirius, 8.6; Epsilon Eridani, 10.7; Epsilon Indi, 11.3; 61 Cygni, 11.4; and Procyon, 11.4.) Tau Ceti is a G-type yellow dwarf similar to the Sun but, with an absolute magnitude of +5.7, only 45% as bright.

66 Ceti = Σ231   Double Star   Spec. G0
m5.7, 7.5; Sep. 16.5″; P.A. 234° $02^h12.8^m$ $-02°24'$
Finder Chart 12-10   ★★★★

2/3″ Scopes–75x: 66 Ceti is a fine pair of yellow and blue stars.

8/10″ Scopes–100x: 66 Ceti is a nice color-contrast pair of topaz and violet stars.

Omicron (*o*) Ceti   Variable Star   Spec. M5e–M9e
m2.0 to 10.1 in 331.9 days $02^h19.3^m$ $-02°59'$
Finder Chart 12-10, Figure 12-1   Mira   ★★★★

David Fabricius, a Dutch amateur astronomer, discovered Mira on August 13, 1596 when he observed a star of the 3rd magnitude in the neck of Cetus unlisted in any of the current, or ancient, star catalogues. When it faded Fabricius assumed that it had been a nova like Tycho's Star of 1572. However he reobserved this supposed "new star" on February 15, 1609. Meanwhile the German astrocartographer Johann Bayer had plotted the object on the map of Cetus in his 1603 *Uranometria*, showing it to be of 4th magnitude and labelling it Omicron Ceti. The name Mira, "Wonderful," was given the star later in the 17th century after the periodic nature of its variations had been realized.

Mira has an average period of roughly 330 days. Its

Cetus, the Whale

Constellation Chart 12-1

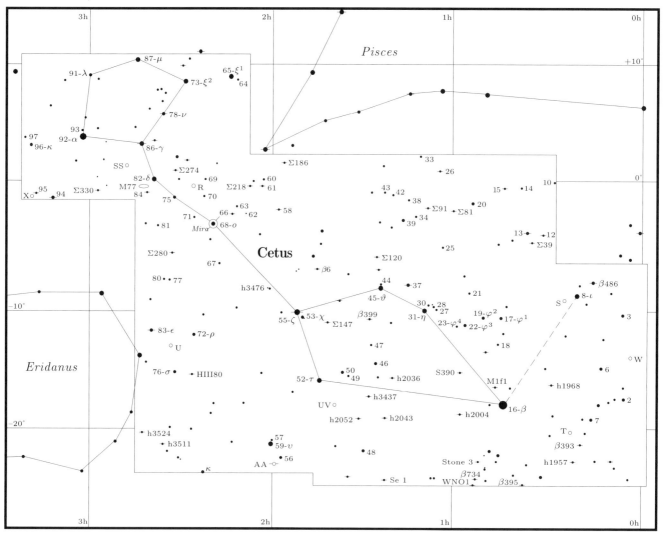

Chart Symbols

Constellation Chart — Stellar Magnitudes: 0 1 2 3 4 5 6	→ Guide Star Pointer ⊙ ○ Variable Stars	⌀ Planetary Nebulae
Finder Charts: 0/1 2 3 4 5 6 7 8 9	•–• Double Stars ○ Open Clusters	▢ Small Bright Nebulae
Master Finder Chart: 0 1 2 3 4 5	Finder Chart Scale ⊕ Globular Clusters	Large Bright Nebulae
	(One degree tick marks) ⬭ Galaxies	Dark Nebulae

Table 12-1. Selected Variable Stars in Cetus

Name	HD No.	Type	Max.	Min.	Period (Days)	F*	Spec. Type	R.A. (2000)	Dec.	Finder Chart No. & Notes
W Cet	224960	M	7.1	14.8	351		S6–S9	00^{h}02.1^m	−14°41'	12-3
T Cet	1760	SRc	5.0	6.9	158		M5–M6	21.8	−20 03	12-4
S Cet	1987	M	7.6	14.7	320	0.47	M3–M6.5	24.1	−09 20	12-3
UV Cet		UV	6.8	12.9			M5.5	01^{h}38.8^m	−17 58	12-7
AA Cet	12180	EW	6.2	6.7	0.53		F2	59.0	−22 55	12-7
o Cet	14386	M	2.0	10.1	331	0.38	M5–M9	02^{h}19.3^m	−02 59	12-10 Mira the wonderful
R Cet	15105	M	7.2	14.0	166	0.43	M4–M9	26.0	−00 11	12-10
U Cet	15971	M	6.8	13.4	234	0.44	M2–M6	33.7	−13 09	12-8

F* = The fraction of period taken up by the star's rise from min. to max. brightness, or the period spent in eclipse.

Table 12-2. Selected Double Stars in Cetus

Name	ADS No.	Pair	M1	M2	Sep."	P.A.°	Spec	R.A. (2000)	Dec	Finder Chart No. & Notes
β486	180		5.1	11.1	3.0	4	M	00^{h}14.5^m	−07°47′	12-3
β393	251		7.1	7.9	0.7	22	A0	18.3	−21 08	12-4
h1957	302	AxBC	7.6	9.9	6.1	24	G0	21.9	−23 00	12-4
h1968	366		7.3	9.8	17.3	230	F8	27.7	−16 25	12-4
12 Cet	410		5.7	10.5	10.3	194	K5	30.0	−03 57	12-5
(Dem 2)	475	AB	7.6	8.1	0.6	252	G0	34.5	−04 33	12-5
Σ39	475	ABxC		9.3	19.7	45				
β395	520		6.3	6.4	0.8y	109	K0	37.3	−24 46	12-4 Difficult orange pair
M1f 1	636		6.5	9.5	2.5	195	F0	45.7	−16 25	12-4
Stone 3	716	AB	7.6	8.3	1.8	256	G0	52.2	−22 37	12-4
	716	AC		11.9	32.6	196				
β734	726		5.5	9.7	10.8	346	K0	52.7	−24 00	12-4
WNO 1	733		6.5	8.4	5.4	9	F2	53.2	−24 47	12-4
h2004	799		7.2	10.2	3.4	240	A0	57.6	−19 00	12-1
S390	806		7.9	7.9	6.4	216	F5	58.2	−15 41	12-6
Σ81	825		7.3	10.1	17.4	68	F5	01^{h}00.1^m	−02 01	12-5
26 Cet	875		6.2	8.6	16.0	253	F0	03.8	+01 22	12-5 Pretty yellow & Lilac
Σ91	923		7.4	8.2	4.2	318	F5	07.2	−01 44	12-5
37 Cet	1003		5.2	8.7	49.7	331	F0 G	14.4	−07 55	12-6
h2036	1087		7.5	7.7	1.8	353	G0	20.0	−15 49	12-1 Close pair of yellow suns
h2043	1106		6.5	8.8	5.0	74	F5	22.5	−19 05	12-1
Se 1	1113		6.9	8.9	2.7	85	A5	23.6	−24 21	12-1
Σ120	1131		6.8	10.6	7.3	278	A0	25.0	−05 57	12-6
β399	1162		6.1	9.8	1.6	302	K0	27.8	−10 54	12-6
h3437	1171		7.3	8.9	12.0	247	F0	28.1	−17 16	12-1
h2052			7.1	7.4	78.4	117	A3 K0	31.6	−19 01	12-7
Σ147	1339		6.1	7.4	2.1	90	F5	41.7	−11 19	12-6
β6	1376	AB	6.6	9.4	2.6	167	G0	44.7	−06 46	12-1
53-χ Cet			4.9	6.9	183.8	250	F0	49.6	−10 41	12-8 Binocular double
55-ζ Cet			3.7	9.9	187.0	41	K0 K0	51.5	−10 20	12-8
Σ186	1538		6.8	6.8	1.3z	57	G0	55.9	+01 51	12-1 Close, equal yellow stars
H II 58	1581		7.4	7.7	8.4	304	F2	59.0	−22 55	12-7 Variable star AA Cet
h3476			5.5	9.6	61.7	193	M	02^{h}00.4^m	−08 31	12-8
61 Cet	1634	AB	5.9	10.4	43.0	194	G5	03.8	−00 20	12-10
	1634	AC		11.8	83.0	326				
Σ218	1673		8.3	9.3	4.9	248	F0	08.7	−00 26	12-10
66 Cet	1703		5.7	7.5	16.5	234	G0	12.8	−02 24	12-10 Topaz & violet
68-o Cet	1778	AB	2.0v	12.0	73.1	85	M	19.3	−02 59	12-10 Mira. Red var. star
H VI 1	1778	AC		9.3	118.7	78				12-8
H III 80	1849	AB	5.9	8.9	12.2	293	A2	26.0	−15 20	12-8
	1849	AC		10.8	105.8	30				
Σ274	1924		7.3	7.8	13.5	219	A2	31.5	+01 06	12-10
Σ280	1953		7.9	8.1	3.6	346	K2	34.1	−05 38	12-8
78-ν Cet	1971		4.9	9.5	8.1	81	G5	35.9	+05 36	12-9 Deep yellow & ashy
h3511	1978		7.4	9.2	14.8	98	G0	36.0	−21 24	12-1 Yellow & pale blue
84 Cet	2046		5.8	9.0	4.0	310	F5	41.2	−00 42	12-10 Lovely yellow & reddish
h3524	2079		7.5	9.3	19.4	152	G0	43.0	−20 17	12-1
86-γ Cet	2080		3.5	7.3	2.8	294	A2	43.3	+03 14	12-9 Brillant white & yellow
Σ330	2237		7.3	9.3	8.8	192	G5	57.2	−00 34	12-1 Deep yellow & bluish
95 Cet	2459		5.6	7.5	1.1*	250	K1 G8	03^{h}18.4^m	−00 56	12-1

Footnotes: *= Year 2000, a = Near apogee, c = Closing, w = Widening. Finder Chart No: All stars listed in the tables are plotted in the large Constellation Chart, but when a star appears in a Finder Chart, this number is listed. Notes: When colors are subtle, the suffix *-ish* is used, e.g. *bluish*.

maximum is usually around magnitude 3.5 and its minimum nearly 9.5. Thus the star is 6 magnitudes = 250 times brighter at maximum than at minimum. The rise from minimum to maximum takes only 110 days; the fade back to minimum is much more leisurely, occupying fully two-thirds of the light-cycle. Mira's spectrum varies as well, being M5e near maximum and a very cool M9 at minimum. Mira-type long-period variables (LPVs) are cool red giants that have strong absorption lines of oxides — particularly titanium oxide — in their spectra. They also typically have emission lines, mainly of hydrogen, in their spectra as they brighten. (In Mira the emission lines appear when the star approaches 7th magnitude.) These emission features disappear soon after maximum light is reached.

Mira lies at a distance of about 200 light years. Because it is relatively close, its diameter can be measured interferometrically. The results suggest that at

Figure 12-1. *Mira is seen near minimum, about 10.1 magnitude (top) and near maximum, about 2.0 magnitude (bottom). Lee C. Coombs made these two five minute exposures using a 150mm telephoto lens at f4 with a minus violet filter.*

maximum the star has 400–500 times the diameter of the Sun. If it was at the center of the Solar System, Mira's outer surface would be beyond the orbit of Mars! The large size of red giants like Mira is a consequence of the onset of helium burning (the fusion of helium nuclei into carbon nuclei) in their cores: the extra energy (hydrogen burning continues in a shell around the helium-burning core) expands the star's outer envelope. The envelope of such stars is so bloated that material can escape the star's relatively weak surface gravity: hence red giants

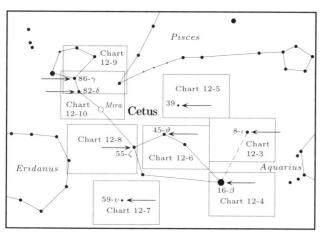

Master Finder Chart 12-2. Cetus Chart Areas
Guide stars indicated by arrows.

in general, and Mira-type LPVs in particular, undergo small, but continuous, mass loss. One theory of planetary nebula formation is that the pulsations of an LPV become so strong that they can overcome the star's weak surface gravity and expel the star's outer layers as a planetary shell. The rather modest expansion velocities of planetaries, 10 to 20 kilometers per second, are consistent with this theory.

Thousands of Mira-type long period variables have been discovered. They are all cool red M (or late K) giants. Though at maximum brightness they have a luminosity of around 200 suns, they are thought to contain only about one solar mass of material. LPVs can have periods of as little as a few dozen, or as many as several hundred, days, and light ranges of only 2–3, or as much as ten, magnitudes. In general, the shorter an LPV's period, the smaller its light range. Unlike Cepheids, however, there is no correlation between an LPV's period and its luminosity. Because short-period LPVs have been identified in globular clusters, they must be older than their longer-period cousins.

Mira is also a binary, its companion, 73.1″ distant toward the ENE, is the faint variable VZ Ceti, magnitude range 9.5 to 12. This star is a blue subdwarf that orbits the red giant in a period of perhaps 1,800 years.

84 Ceti = Σ295 Double Star   Spec. F5
m5.8, 9.0; Sep. 4.0″; P.A. 310° $02^h41.2^m$ $-00°42'$
Finder Chart 12-10   ★★★★

8/10″ Scopes–150x: 84 Ceti is a lovely pair of yellow and reddish stars.

Gamma (γ) = 86 Ceti Triple Star (Σ299)   Spec. A2V, dF3
m3.5, 7.3; Sep. 2.8″; P.A. 294° $02^h43.3^m$ $+03°14'$
Finder Chart 12-9   ★★★★

Gamma Ceti is a fine but close double, a true binary with an orbital period of several thousand years. The system is about 70 light-years away, so the true separation between the components is at least 60 A.U. A 10th magnitude red dwarf 14′ away in PA 315° is an actual physical member of the system, its projected separation from the close pair being 18,000 A.U.

8/10″ Scopes–125x: Gamma is a beautiful triple star with a close brilliant white and yellow pair and a reddish dwarf star 14′ away.

Alpha (α) = 92 Ceti   Spec. M2 III, B7 III
m2.53v $03^h02.2^m$ $+04°05'$
Constellation Chart 12-1   *Menkar*   ★★★★

2/3″ Scopes–25x: Beautiful! This orange giant star forms a very wide binocular double with the 5.6 magnitude blue star 93 Ceti 15′ north. The two stars are not, however, physically related.

Figure 12-2. *UA444 is a large, faint, diffuse galaxy requiring larger instruments and dark skies for a good view. Martin C. Germano made this 80 minute exposure on hypered 2415 film with an 8", f5 Newtonian at prime focus.*

Figure 12-3. *NGC 45 is a faint, diffuse, oval-shaped galaxy lying next to a 7th mag. star that interferes with observation. Martin C. Germano made this 75 minute exposure on hypered 2415 film with an 8", f5 Newtonian.*

12.3 Deep-Sky Objects

UA444 Galaxy Type IB(s)m V
ø11.5′ × 4.0′, m10.6v, SB 14.6 $00^h02.0^m$ −15°28′
Finder Chart 12-3, Figure 12-2   ★

16/18″ Scopes–125x: UA444, 25′ SSW of an 8th magnitude star, is very faint, diffuse, and elongated about 8′ × 3′ N–S. Its surface brightness is uniform, but several stars are embedded near its center. A 13th magnitude star is at the galaxy's southern tip, and a wide 13th and 14th magnitude pair 1′ west.

NGC 45 Galaxy Type SAB(s)dm IV
ø6.3′ × 4.6′, m10.8v, SB 14.3 $00^h14.1^m$ −23°11′
Finder Chart 12-4, Figure 12-3   ★★

8/10″ Scopes–100x: NGC 45 is 5′ ENE of a 7th magnitude star and 1′ NNW of a 10th magnitude star. It is extremely faint, small, and oval-shaped NNW–SSE, and can be observed only with averted vision.

12/14″ Scopes–125x: This galaxy is very faint, elongated 4′ x 3′ NNW–SSE, and slightly brighter in its center. A 10th magnitude star touches the SSW tip.

16/18″ Scopes–150x: NGC 45 is faint, diffuse, and elongated 6′ x 4′ NNW–SSE with a moderately concentrated oval core. A 10th magnitude star is embedded inside the SSE tip, and at least half a dozen faint stars are superimposed upon the galaxy's halo.

NGC 145 Galaxy Type (R′)SB(s)c pec
ø1.8′ × 1.3′, m12.7v, SB 13.5 $00^h31.7^m$ −05°09′
Finder Chart 12-5   ★★

8/10″ Scopes–100x: NGC 145 is 5′ east of the northernmost star of a NNE–SSW 9th magnitude pair. It is faint, small, and circular with uniform surface brightness.

12/14″ Scopes–125x: The halo is faint, elongated 1.5′ × 1′ ESE–WNW, and gradually brighter toward its center.

16/18″ Scopes–150x: NGC 145 is fairly faint, elongated about 1.75′ × 1.25′ ESE–WNW, and slightly brighter at its center. A string of stars, the brightest 8th magnitude, is 5′ east of the galaxy.

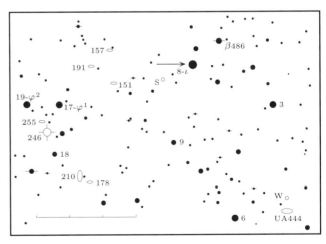

Finder Chart 12-3. 8-ι Ceti: $00^h19.5^m$ −08°49′

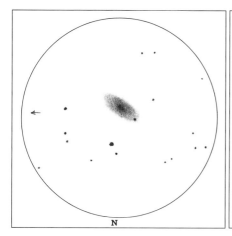
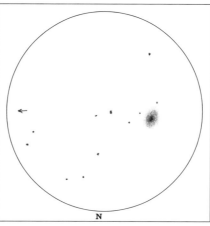
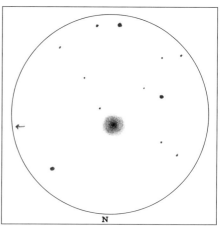

Figure 12-4. NGC 157
13", f5.6–100x, by Steve Coe

Figure 12-5. NGC 210
8", f6–100x, by A. J. Crayon

Figure 12-6. NGC 255
12.5", f5–125x, by G. R. Kepple

NGC 151 Galaxy Type SB(rs)bc I-II
⌀3.3′ × 1.4′, m11.6v, SB 13.1   00ʰ34.0ᵐ −09°42′
Finder Chart 12-3   ★★★

8/10″ Scopes–100x: NGC 151, just SW of an 11th magnitude star, is faint, small, and circular, its core prominent. The field is a poor sprinkling of 11th and 12th magnitude stars.

12/14″ Scopes–125x: This galaxy displays a prominent core surrounded by a fairly faint halo elongated 1.5′ × 0.8′ ENE-WSW.

16/18″ Scopes–150x: NGC 151 is a moderately bright 2′ × 1′ ENE-WSW oval with a large, bright core. A 13th magnitude star is 2.5′ SSE.

NGC 157 H3² Sc Galaxy Type SAB(s)c: Pec I-II
⌀4.0′ × 2.4′, m10.4v, SB 12.7   00ʰ34.8ᵐ −08°24′
Finder Chart 12-3, Figure 12-4   ★★★★

8/10″ Scopes–100x: This nice bright galaxy lies between two bright stars 10′ apart, an 8th magnitude star to the north and a 9th magnitude star to the south. The halo is a 3′ × 2′ NE-SW oval with a brighter center. A 12.5 magnitude star is at the NE edge.

12/14″ Scopes–125x: In medium-sized telescopes the galaxy's halo increases to 3.5′ × 2′ NE-SW and engulfs the magnitude 12.5 star near the NE edge. The core is well concentrated, and in moments of good seeing displays some mottling.

NGC 175 SBb Galaxy Type SB(r)a
⌀2.0′ × 1.7′, m12.1v, SB 13.3   00ʰ37.4ᵐ −19°56′
Finder Chart 12-4   ★★★

12/14″ Scopes–125x: This galaxy is 25′ NW of BB Ceti (m6.7) and 20′ south of an 8th magnitude star. It is fairly faint, its round, diffuse 1′ halo slightly brighter at center.

16/18″ Scopes–150x: NGC 175 has a moderately faint, diffuse 1.5′ halo slightly elongated ESE-WNW and slightly brighter toward the center. The galaxy forms a triangle with 10.5 magnitude stars lying 5′ east and 5′ SSE.

NGC 178 Galaxy Type IB:(s:)m: pec III?
⌀2.1′ × 0.9′, m12.6v, SB 13.2   00ʰ39.1ᵐ −14°10′
Finder Chart 12-3   ★★★

8/10″ Scopes–100x: NGC 178, 5′ NE of an 8th magnitude star, appears rather faint, elongated 1′ × 0.5′ N-S, and slightly brighter toward its center.

12/14″ Scopes–125x: NGC 178 is fairly faint, elongated 1.5′ × 0.75′ N-S, and slightly brighter at its center. NGC 210 is 27′ NE.

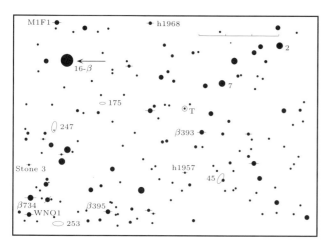

Finder Chart 12-4. 16-β Ceti: 00ʰ43.6ᵐ −17°59′

Figure 12-7. *NGC 210 is a galaxy with an extended core surrounded by a faint halo. Martin C. Germano made this 70 minute exposure on hypered 2415 film with an 14.5", f5 Newtonian reflector at prime focus.*

Figure 12-8. *NGC 246 is a bright, large planetary nebula involved with four stars. Martin C. Germano made this 80 minute exposure on hypered 2415 film with an 14", f5 Newtonian reflector at prime focus.*

NGC 210 H452² Galaxy Type (R')SAB(rs)a
⌀4.0' × 2.9', m10.9v, SB 13.4   00ʰ40.6ᵐ −13°52'

Finder Chart 12-3, Figures 12-5 & 12-7   ★★★

8/10" Scopes–100x: This galaxy lies 5' east of a 7.5 magnitude star and 1.5' ENE of an 11th magnitude star. It is a moderately faint, diffuse object elongated 2'× 1' NNW–SSE.

16/18" Scopes–150x: NGC 210 has a moderately concentrated, extended core with a tiny nonstellar nucleus surrounded by a much fainter outer halo elongated 3.5' × 2' NNW–SSE. Four stars lie along the arc of the halo's edge: a 12th magnitude star to the WSW, a 13th magnitude star to the NNW, a 13.5 magnitude star to the NW, and a threshold star to the SSE.

NGC 227 H444² Galaxy Type S0+: pec
⌀1.7' × 1.4', m12.2v, SB 12.9   00ʰ42.6ᵐ −01°32'

Finder Chart 12-5   ★★★

8/10" Scopes–100x: This galaxy has a faint, round 1' halo and a stellar nucleus.

12/14" Scopes–125x: NGC 227 appears moderately bright, is slightly elongated NNW–SSE, and has a stellar nucleus or a very suddenly brighter core.

NGC 237 Galaxy Type SB?(rs?)b pec I-II
⌀1.8' × 1.1', m13.0v, SB 13.6   00ʰ43.5ᵐ −00°07'

Finder Chart 12-5   ★

12/14" Scopes–125x: This galaxy lies 7' NE of the easternmost in a NW–SE string of four 9th magnitude stars. It is a very faint, tiny, diffuse smudge.

16/18" Scopes–150x: NGC 237 is fairly faint and elongated 1.25' × 0.75' NNW–SSE. Its weakly concentrated core has a faint stellar nucleus.

NGC 245 H445² Galaxy Type SAB(r)bc pec II:
⌀1.2' × 1.2', m12.2v, SB 12.4   00ʰ46.1ᵐ −01°44'

Finder Chart 12-5   ★★★

8/10" Scopes–125x: NGC 245 is faint, small, and irregularly round but not difficult to see. It lies in a sparse star field.

16/18" Scopes–150x: NGC 245 is moderately bright and circular, or perhaps slightly elongated 1.2' × 1.0' NW–SE. Its has a faint stellar nucleus. Two faint stars, magnitudes 13.5 and 14, are just to its SW.

NGC 246 H25⁵ Planetary Neb. Type 3b
⌀225", m10.9v, CS 11.95v   00ʰ47.0ᵐ −11°53'

Finder Chart 12-3, Figure 12-8   ★★★★

8/10" Scopes–125x: This nice planetary nebula is fairly bright, large and round. The northern and western

Figure 12-9. *NGC 247 is a bright, large, highly elongated galaxy. In larger instruments both core and halo appear mottled and many faint foreground field stars are resolved on its image. Martin C. Germano made this 45 minute exposure on hypered 2415 Kodak Tech Pan film with an 8", f5 Newtonian at prime focus.*

edges are more sharply defined than the southern and eastern, and the eastern quadrant of the nebula is fainter than the rest. Three 12th–13th magnitude stars embedded in the disk, and 12th and 13th magnitude stars NNW and SW of the disk, together form a distorted three-blade propeller. The planetary's 12th magnitude central star is the hub of the propeller.

12/14" Scopes–150x: This beautiful planetary nebula has a bright 3' disk but gives the impression of transparency because of the sprinkling of stars across it. The disk is mottled, and the notch at its SE edge makes the planetary look like a doughnut with a bite taken from it. Within the disk, about 45" SW and 65" ESE of the 12th magnitude central star respectively, are 12th and 13th magnitude field stars. Another 12th magnitude field star lies near the NNW edge, and 1' SW of the disk is a second 13th magnitude star.

NGC 247 H20⁵ Galaxy SAB(s)dm III-IV
⌀19.0' × 5.5', m9.2v, SB 14.1
00ʰ47.1ᵐ −20°46'
Finder Chart 12-4, Figure 12-9   ★★★★

8/10" Scopes–100x: This galaxy is faint, extremely elongated 12' × 3' N–S, and has a faint elongated core. A 9th magnitude star is at the south edge.

12/14" Scopes–100x: This galaxy has a faint halo elongated 15' × 3' N–S with a weakly concentrated core. A 9th magnitude star at the southern tip. At times of good seeing the halo appears mottled.

16/18" Scopes–100x: Nice! NGC 247 is a fairly bright galaxy elongated 19' × 5' in position angle 170°. The halo is teardrop-shaped, its northern portion broader and more diffuse and its southern end brighter and pointed. The 9th magnitude star at the southern edge clearly lies within the halo. The central area is mottled and brightens gradually to an irregularly extended core. At least seven faint stars are embedded in the halo. The surrounding star field is somewhat sparse.

NGC 255 H472² Galaxy Type SB(rs)bc II
⌀3.1' × 2.4', m11.9v, SB 13.9   00ʰ47.8ᵐ −11°28'
Finder Chart 12-3, Figure 12-6   ★★

8/10" Scopes–100x: NGC 255 lies 20' east of the southern apex of an equilateral triangle of magnitude 7.5 stars. It is a very faint, irregularly round milky nebulosity in a sparse star field.

12/14" Scopes–125x: NGC 255 is fairly faint, elongated 2' × 1.5' NNE–SSW, and has a large granular core. It rather resembles the smaller, fainter globular clusters. Planetary nebula NGC 246 lies 24' SW.

NGC 268 H463³ Galaxy Type SB(s)c I-II
⌀1.3' × 0.9', m13.1v, SB 13.1   00ʰ50.2ᵐ −05°12'
Finder Chart 12-5   ★★★

12/14" Scopes–125x: This galaxy lies just SW of the NW angle of a fairly large diamond or kite-shaped asterism with a 7th magnitude star at its NE angle, and 8th magnitude star at its SW angle, and 9th

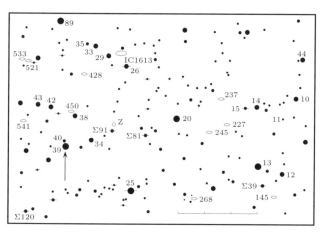

Finder Chart 12-5. 39 Ceti: 01ʰ16.6ᵐ −02°30'

magnitude stars at its SE and NW angles. NGC 268 is fairly faint, diffuse, and slightly elongated E–W, its 1' halo slightly brighter at the center.

NGC 274 H429³ Galaxy Type SAB(r:)0° pec
ø1.3' × 1.2', m11.8v, SB 12.1   00ʰ51.0ᵐ −07°03'

NGC 275 Galaxy Type S pec or P(c)
ø1.6' × 1.3', m12.5v, SB 13.1   00ʰ51.1ᵐ −07°04'
Finder Chart 12-6   ★★★/★

8/10" Scopes–100x: At this aperture and power the two faint halos of the NGC 274 and NGC 275 galaxy-pair appear to be in contact. NGC 274, to the NW, is the brighter but smaller of the two and has a fairly obvious core. NGC 275 to the SE is larger but fainter.

12/14" Scopes–125x: These two small galaxies are a fine but faint pair only 45" NW–SE of each other. Both are circular and less than 1' across. NGC 274 on the NW is the brighter of the two and has a sharp, tiny core. NGC 275 on the SE is somewhat larger than its companion but is more diffuse and has only a faint stellar nucleus.

Figure 12-10. *IC 1613, a dwarf member of the Local Galaxy Group, is visible even in larger telescopes only as a large irregular haze. Martin C. Germano made this 75 minute exposure on 2415 film with an 8", f5 Newtonian at prime focus. South is up.*

NGC 309 Galaxy Type SB(r)c I
ø2.6' × 2.0', m11.9v, SB 13.6   00ʰ56.4ᵐ −09°56'
Finder Chart 12-6   ★★

12/14" Scopes–125x: NGC 309 is 7' SSW of the southernmost of three 9th magnitude stars and 2' SSW of an 11th magnitude star. Its faint halo is slightly elongated 2' N–S. The central area is irregular in brightness, with slightly brighter spots.

NGC 337 H433³ Galaxy Type SAB:(s)cd II-III
ø3.0' × 1.5', m11.6v, SB 13.1   00ʰ59.8ᵐ −07°35'
Finder Chart 12-6   ★★★

8/10" Scopes–100x: This galaxy is fairly faint, elongated 2' × 0.75' WNW–ESE, and slightly brighter at its center. A 10th magnitude star is 9' SSW.

12/14" Scopes–125x: NGC 337 appears fairly bright, elongated 2.5' × 1.25' ESE–WNW, and has tapered ends. The core is large, extended, and weakly concentrated.

NGC 357 H434² Galaxy Type SB(rs)0°
ø2.7' × 1.8', m12.0v, SB13.6   01ʰ03.4ᵐ −06°20'
Finder Chart 12-6   ★★

8/10" Scopes–100x: This galaxy is 10' north of one 9th magnitude star and 12' NW of a second. Its halo is faint, elongated 1' × 0.75' ESE–WNW, and suddenly brighter at the middle. A 12th magnitude star lies 30" ENE.

12/14" Scopes–125x: NGC 357 is fairly bright, its 2.5' × 1.5' ESE–WNW halo enclosing a bright core containing a stellar nucleus. NGC 357 is the brightest in a group of about ten galaxies in a 12' area, including NGC 355 lying 3' WNW. This region repays careful galaxy searching.

IC 1613 Galaxy Type IB(s)m V
ø20.0' × 18.5', m9.2v, SB 15.2   01ʰ04.8ᵐ +02°07'
Finder Chart 12-5, Figure 12-10   ★

IC 1613 is a member of our Milky Way's Local Galaxy Group. It is a dwarf irregular system similar to, but much smaller and less luminous than, the Magellanic Clouds. There are more faint dwarf galaxies in space than any other type but their presence is certainly not pronounced. The Cepheid variables observed IC 1613 implies that it is about 2.8 million light-years away. It therefore has a total luminosity of about 130 million suns and a diameter of 16,000 light-years.

16/18" Scopes–100x: This difficult object is 10' south of a 7th magnitude star at the NW angle of a triangle 1.75° long with 6th magnitude stars at its south and east corners. IC 1613 is a very faint, irregular haze spread over a 20' area. With averted vision two patches of galaxy can be seen, separated by 7' in a

NE–SW direction. The NE patch is easier; the SW patch is larger but more diffuse

New 1 MCG-01-03-85 Galaxy Type ?
⌀4.4′ × 3.3′, m11.6v, SB 14.4   01ʰ05.1ᵐ −06°13′
Finder Chart 12-6   ★★

12/14″ Scopes–125x: This galaxy is faint and nearly round with a diffuse outer area and a faint stellar nucleus. It is located in a sparse star field.

16/18″ Scopes–125x: New 1 is a rather faint, diffuse galaxy with a 3.5′ × 3′ halo elongated E–W and a faint nonstellar nucleus. A pair of 12th magnitude stars lies 7′ north, and a 13th magnitude star is 3′ east.

NGC 428 H6222 Galaxy Type SAB(s)dm III-IV
⌀4.6′ × 3.4′, m11.5v, SB 14.3   01ʰ12.9ᵐ +00°59′
Finder Chart 12-5   ★★★

8/10″ Scopes–100x: This galaxy forms a triangle with two 9th magnitude field stars lying 5′ NNE and 5′ west. Its diffuse 2′ diameter halo is very faint. Averted vision helps.

12/14″ Scopes–125x: NGC 428 is considerably faint, fairly small, and elongated 3′ × 2.5′ ESE–WNW. Though it is slightly brighter in the center, it lacks a nucleus. A pair of 12th magnitude stars is 2′ south and a slightly brighter star 2′ NW.

NGC 450 H440³ Galaxy Type SAB(s:)cd III
⌀3.2′ × 2.6′, m11.5v, SB 13.7   01ʰ15.5ᵐ −00°52′
Finder Chart 12-5   ★★★

8/10″ Scopes–100x: NGC 450, 10′ NE of 38 Ceti (m5.7), appears very faint, diffuse, and circular, its halo 2.5′ in diameter.

12/14″ Scopes–125x: NGC 450 is faint, elongated 2.5′ × 2′ ENE–WSW with uniform surface brightness and a faint stellar nucleus. 2.75′ WSW of the galaxy is a 17″ wide pair of 13th magnitude stars, and a 12.5 magnitude star is 3.25′ south.

NGC 521 H461² Galaxy Type SB(r)b I-II
⌀3.5′ × 3.3′, m11.7v, SB 14.2   01ʰ24.6ᵐ +01°44′
Finder Chart 12-5   ★★

8/10″ Scopes–100x: NGC 521 has a very faint 1′ diameter halo with a slightly brighter center. Galaxy NGC 533 lies 15′ east. The star field to the west is fairly rich and includes a 6th magnitude star 28′ away from the galaxy.

12/14″ Scopes–125x: NGC 521 is faint and diffuse, its 2′ diameter halo slightly elongated NNE-SSW. The central area is concentrated but lacks a core or nucleus. A faint, tiny galaxy lies 3′ WNW. A faint N–S string of 13th and 14th magnitude stars passes the galaxy's western flank.

NGC 533 H462² Galaxy Type E+3
⌀4.1′ × 3.1′, m11.4v, SB 14.5   01ʰ25.5ᵐ +01°46′
Finder Chart 12-5   ★

8/10″ Scopes–100x: This galaxy is faint and rather small, just slightly larger than NGC 521, 15′ west. Its 1′ diameter halo is slightly elongated NE–SW and gradually brighter toward the center.

12/14″ Scopes–125x: NGC 533 is fairly faint, its 2′ × 1.5′ NE–SW halo enveloping a well concentrated core with a faint stellar nucleus. An 8th magnitude star lies 16′ south.

NGC 541 Galaxy Type E0 or S0−
⌀2.4′ × 2.4′, m12.1v, SB 13.8   01ʰ25.7ᵐ −01°23′
Finder Chart 12-5   ★

8/10″ Scopes–100x: This galaxy is the brightest of a group of otherwise very faint galaxies, most of its companions visible only on the best of nights with averted vision. NGC 541 is faint, small, and round with a brighter center.

16/18″ Scopes–150x: NGC 541 is fairly faint, circular or slightly elongated NNE–SSW with a brighter center. It is slightly brighter than NGC 545 and NGC 547 which lie 4.5′ ENE.

NGC 578 Galaxy Type SB(s)c I-II
⌀3.9′ × 2.2′, m11.0v, SB 13.2   01ʰ30.5ᵐ −22°40′
Finder Chart 12-7   ★★★★

8/10″ Scopes–100x: NGC 578, 8′ SSE of an 8th magnitude star, is fairly faint, elongated 2.5′ × 1.5′ ESE–WNW, and has a slightly brighter center.

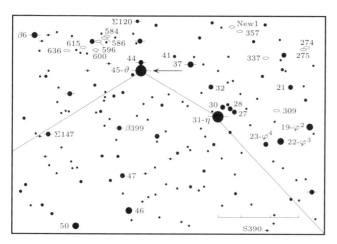

Finder Chart 12-6. 45-ϑ Ceti: 01ʰ24.0ᵐ −08°11′

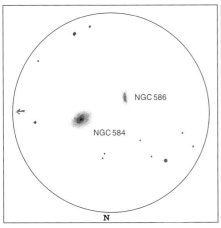

Figure 12-11. NGC 584 & NGC 586
13″, f5.6–135x, by Steve Coe

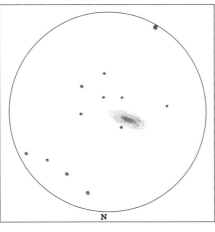

Figure 12-12. NGC 681
16″, f5–160x, by Bob Erdmann

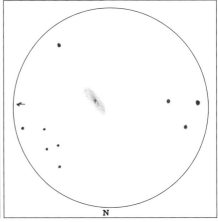

Figure 12-13. NGC 701
16″, f5–100x, by Bob Erdmann

12/14″ Scopes–125x: NGC 578 has a fairly bright halo elongated 4′ × 2′ ESE–WNW fading rapidly toward the tips. The core is broad and just slightly brighter. The triple star Innes 445, NNW of the galaxy, consists of a magnitude 7.8 primary with a secondary 67″ NNE that is itself a close double, magnitude 10.1 and 10.3 stars separated by 2.2″ in P.A. 356°.

NGC 584 H100¹ Galaxy Type SAB?0−
ø3.2′ × 1.7′, m10.5v, SB 12.2   **01ʰ31.3ᵐ −06°52′**
Finder Chart 12-6, Figure 12-11   ★★★

8/10″ Scopes–100x: NGC 584 is the brighter of a galaxy-pair with NGC 586 4.5′ SE. Two 7.5 magnitude stars 20′ apart flank the pair to the NNE and SSW. NGC 584 is fairly bright with a 1.5′ × 0.75′ halo elongated ENE–WSW enclosing a bright core.

12/14″ Scopes–125x: NGC 584 is a very bright galaxy with a 2.5′ × 1.25′ halo elongated ENE–WSW within which is a relatively bright oval core and stellar nucleus. NGC 584's companion, NGC 586, 4.5′ SE, is much fainter and smaller.

NGC 586 Galaxy Type S0+:
ø1.7′ × 0.9′, m13.2v, SB 13.5   **01ʰ31.6ᵐ −06°54′**
Finder Chart 12-6, Figure 12-11   ★

8/10″ Scopes–100x: NGC 586, 4.5′ SE of NGC 584, is fairly faint, diffuse, and elongated 1′ × 0.5′ N–S. It is slightly brighter in its core.

12/14″ Scopes–125x: NGC 586 is moderately faint, elongated 1.5′ × 0.75′ N–S, and suddenly brighter at its center.

NGC 596 H4² Galaxy Type SAB(s?)0− pec
ø2.7′ × 2.0′, m10.9v, SB 12.6   **01ʰ32.9ᵐ −07°02′**
Finder Chart 12-6   ★★★

8/10″ Scopes–100x: NGC 596, 12′ west of a 5.7 magnitude star, has a bright 1′ diameter circular halo. The core is prominent and contains a stellar nucleus.

12/14″ Scopes–125x: NGC 596 has a fairly bright halo elongated 1.5′ × 1′ NNE–SSW with a bright, large core and a stellar nucleus.

NGC 600 Galaxy Type (R′)SB(rs)cd II-III
ø2.9′ × 2.7′, m12.4v, SB 14.5   **01ʰ33.1ᵐ −07°19′**
Finder Chart 12-6   ★

12/14″ Scopes–125x: NGC 600, 20′ SW of a 5.7 magnitude star, is an extremely faint, round, diffuse object less than 2′ across with no central brightening. A 13th magnitude star lies 3′ south.

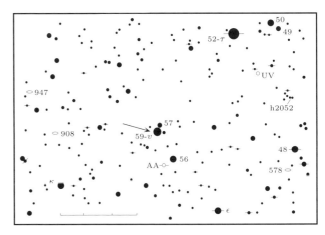

Finder Chart 12-7. 59-υ Ceti: 02ʰ00.0ᵐ −21°05′

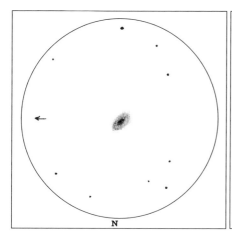
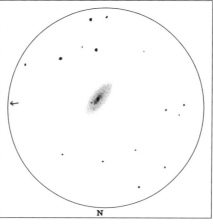
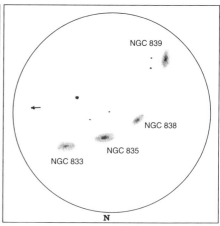

Figure 12-14. NGC 720
8", f6–100x, by A. J. Crayon

Figure 12-15. NGC 779
13", f5.6–135x, by Steve Coe

Figure 12-16. NGC 833, 835, 838, & 839
20", f4.5–175x, by Richard Jakiel

NGC 615 H282² Galaxy Type (R) SB:(r)ab
ø2.5' × 1.3', m11.6v, SB 12.7   01ʰ35.1ᵐ −07°20'
Finder Chart 12-6   ★★★

8/10" Scopes–100x: This galaxy is 25' SE of a 5.7 magnitude star and 6' ENE of an 8.5 magnitude star. A 40' long E–W arc of three 7.5 magnitude field stars is to its south. The galaxy is fairly bright, elongated 1.5' × 0.5' NNE–SSW and moderately concentrated to a stellar nucleus.

12/14" Scopes–125x: NGC 615 is bright, elongated 2.5' × 1.25' NNE–SSW, and has a tiny, faint core with a stellar nucleus offset to the north.

NGC 636 H283² Galaxy Type SA?(r:)0°: pec?
ø2.7' × 1.9', m11.5v, SB 13.1   01ʰ39.1ᵐ −07°31'
Finder Chart 12-6   ★★★

8/10" Scopes–100x: This galaxy is SW of three 9th magnitude stars and 3' east of an 11th magnitude star. It has a circular 1.25' diameter halo with a suddenly brighter core, and is fairly obvious.

12/14" Scopes–125x: NGC 636 has a tiny core with a stellar nucleus embedded within a fairly bright 1.75' diameter halo slightly elongated NW–SE.

NGC 681 H481¹ Galaxy Type SA:a sp
ø3.0' × 1.8', m12.0v, SB 13.7   01ʰ49.2ᵐ −10°26'
Finder Chart 12-8, Figure 12-12   ★★★

8/10" Scopes–100x: NGC 681 is easily located 15' NNW of Chi (χ) = 53 Ceti and 7' WSW of a magnitude 7.5 star. It is fairly faint, small, elongated 1.5' × 1.25' roughly E–W, and gradually brightens toward its center. A 12th magnitude star is less than 1' NW. The surrounding field is rather rich in bright stars for an off-Milky Way region.

12/14" Scopes–125x: NGC 681 appears moderately faint and is elongated 2' × 1.5' ENE–WSW. Its halo has diffuse edges and gradually brightens to a broad core without a nucleus. A 12.5 magnitude star is embedded just inside the halo's NW edge. The star field is rich and interesting: four 13th magnitude stars form a small diamond-shaped asterism to the SW and many other faint stars are sprinkled about.

NGC 701 H62¹ Galaxy Type SB(s)cd III
ø2.5' × 1.5', m12.2v, SB 13.5   01ʰ51.1ᵐ −09°42'
Finder Chart 12-8, Figure 12-13   ★★★

8/10" Scopes–100x: This faint galaxy is 36' north of the 3.7 magnitude orange star Zeta (ζ) = 55 Ceti and 15' west of a 7.5 magnitude star. The halo is elongated 2' × 1' NE-SW with an extended core and a stellar nucleus.

12/14" Scopes–125x: NGC 701 is fairly faint, elongated 2.25' × 1' NE–SW, and becomes slightly brighter toward an extended core with a faint stellar nucleus. At 175x an indistinct dark streak can be seen cutting across the SE side of the core. 5' SSE appears the fairly faint, small, and circular galaxy IC 1738.

NGC 720 H105¹ Galaxy Type E4 or SA(rs)0−
ø4.3' × 2.0', m10.2v, SB 12.4   01ʰ53.0ᵐ −13°44'
Finder Chart 12-8, Figure 12-14   ★★★

8/10" Scopes–100x: This galaxy displays a bright oval core with a stellar nucleus embedded within a 1.5' × 1' NW–SE halo that rapidly fades to its outer edges.

12/14" Scopes–125x: NGC 720 is a bright galaxy elongated 1.75' × 1.25' NW–SE with a prominent oval core containing a stellar nucleus.

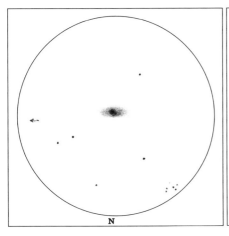

Figure 12-17. NGC 908
8", f6–100x, by A. J. Crayon

Figure 12-18. NGC 936 & NGC 941
13", f5.6–100x, by Steve Coe

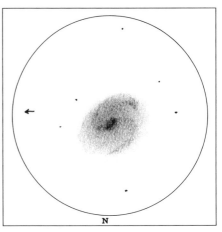

Figure 12-19. NGC 1042
20", f4.5–175x, by Richard Jakiel

NGC 779 H10¹ Galaxy Type SA?(r:)ab
⌀3.4′ × 1.2′, m11.2v, SB 12.5   01ʰ59.7ᵐ −05°58′
Finder Chart 12-8, Figure 12-15   ★★★★

8/10″ Scopes–100x: This fine galaxy lies 22′ north of an 8th magnitude star. It is bright, considerably elongated 2.5′ × 1′ NNW–SSE, and its much brighter center shows some mottling along the major axis.

12/14″ Scopes–125x: Nice! NGC 779 is a bright edge-on galaxy elongated 3′ × 1′ NNW–SSE with no central bulge. The bright core has bar-like extensions and a stellar nucleus. Just NW of the core is a knot or faint star. An 11th magnitude star is 4′ south of the galaxy.

NGC 788 H435² Galaxy Type (R′)SA:(rs)0/a
⌀2.7′ × 1.5′, m12.1v, SB 13.5   02ʰ01.1ᵐ −06°49′
Finder Chart 12-8   ★★★

8/10″ Scopes–100x: NGC 788, located between two 8th magnitude stars 35′ apart, stands out well in a sparse star field. The halo is faint and elongated 1.5′ × 0.75′ E–W. Within it is a bright, extended core.

12/14″ Scopes–125x: NGC 788 is moderately bright, elongated 2′ × 1′ ENE–WSW, and has a prominent, extended core with a nonstellar nucleus. The galaxy IC 184 lies 18′ west.

NGC 833 Galaxy Type SB?(R:)0°pec
⌀1.4′ × 0.7′, m12.7v, SB 12.5   02ʰ09.3ᵐ −10°08′
Finder Chart 12-8, Figure 12-16   ★★

8/10″ Scopes–100x: This galaxy is part of the Arp 318 galaxy group centered within a triangle of bright stars, a 6th magnitude star at the east vertex, a 7th magnitude star at the east vertex, and an 8th magnitude star at the southern vertex. NGC 833 is faint, small, and round. It is the westernmost of the four "faint fuzzy" Arp 318 galaxies arranged in a line curving east and south.

16/18″ Scopes–150x: NGC 833 is moderately bright, elongated 1.25′ × 0.7′ E–W, and has a brighter center. It is the westernmost of the curved row of galaxies with NGC 835, NGC 838, and NGC 839.

NGC 835 Galaxy Type (R)Sa: pec
⌀1.9′ × 1.6′, m12.1v, SB 13.2   02ʰ09.4ᵐ −10°08′
Finder Chart 12-8, Figure 12-16   ★★★

8/10″ Scopes–100x: This is the brightest object in the Arp 318 galaxy group and is second from the west in the NW-SE arc of NGCs 833, 835, 838, and 839. Its halo is faint, small, round, and somewhat brighter at its center.

16/18″ Scopes–150x: Moderately bright in larger instruments, NGC 835 appears round with a 1.5′ diameter halo and a bright core.

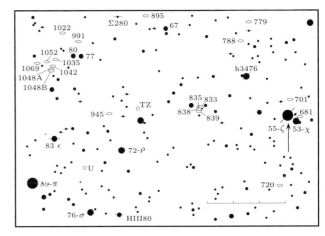

Finder Chart 12-8. 55-ζ Ceti: 01ʰ51.5ᵐ −10°20′

NGC 838 Galaxy Type SA?(rs)0°: pec
⌀1.7′ × 1.0′, m12.9v, SB 13.4   02ʰ09.6ᵐ −10°09′
Finder Chart 12-8, Figure 12-16   ★★

8/10″ Scopes–100x: NGC 838 is the third from the west in a NW-SE arc within the Arp 318 galaxy group: NGCs 833, 835, 838, and 839. It is very faint, its halo a 1′ circle.

16/18″ Scopes–150x: NGC 838 is a fairly faint, 1.5′ × 1′ E–W oval with a suddenly brighter center. A stellar nucleus is suspected.

NGC 839 Galaxy Type I0? pec
⌀1.4′ × 0.7′, m13.1v, SB 13.0   02ʰ09.7ᵐ −10°11′
Finder Chart 12-8, Figure 12-16   ★★

8/10″ Scopes–100x: NGC 839 is at the SE end of the string of four Arp 318 galaxies that includes NGCs 833, 835, and 838.

16/18″ Scopes–150x: NGC 839 is similar in appearance to NGCs 833 and 835 to its NW. It is moderately bright, elongated 1′ × 0.5′ E–W, and has a brighter center.

NGC 864 H457³ Galaxy Type SAB(rs)c II-III
⌀4.4′ × 3.2′, m10.8v, SB 13.6   02ʰ15.5ᵐ +06°00′
Finder Chart 12-9   ★★★

8/10″ Scopes–100x: Nestled against a 10.5 star at its ESE edge, this galaxy is fairly faint, diffuse, and elongated 2′ × 1.5′ NE–SW with a slight brightening at center.

12/14″ Scopes–125x: NGC 864 is fairly faint, elongated 3.5′ × 2′ NE-SW, and slightly brightens to an extended, irregular core. The diffuse halo touches a 10.5 magnitude star to the ESE.

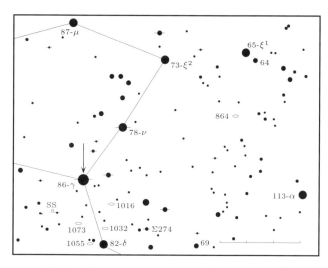

Finder Chart 12-9. 86-γ Cet: 02ʰ43.3ᵐ +03°14

NGC 895 H438² Galaxy Type SA(s)c I-II
⌀3.4′ × 2.4′, m11.7v, SB 13.9   02ʰ21.6ᵐ −05°31′
Finder Chart 12-8   ★★

8/10″ Scopes–100x: This galaxy is very faint and diffuse with a 2′ × 1.5′ E–W halo. The core is slightly brighter and contains a stellar nucleus.

16/18″ Scopes–150x: NGC 895 is rather faint and elongated 3′ × 2′ ENE–WSW. The weakly concentrated core contains a nonstellar nucleus and a faint knot just north of center. A 13th magnitude star is 2′ east of the galaxy.

NGC 908 H153¹ Galaxy Type SAB(s)c: IC
⌀5.5′ × 2.3′, m10.4v, SB 13.1   02ʰ23.1ᵐ −21°14′
Finder Chart 12-7, Figure 12-17   ★★★

8/10″ Scopes–100x: This fine, bright galaxy lies in a rather rich star field. The 4.5′ × 1.5′ E–W halo has a circular core with a stellar nucleus.

12/14″ Scopes–125x: NGC 908's bright, considerably large halo is elongated 5.5′ × 2′ E–W and contains a prominent core with a stellar nucleus. The surrounding star field is well populated with faint stars including a 12th magnitude pair (35″, 90°) 3.5′ south of the galaxy and a 13th magnitude star touching its east edge.

NGC 936 H23⁴ Galaxy Type (R′:)SB(rs)0+
⌀5.7′ × 4.6′, m10.2v, SB 13.5   02ʰ27.6ᵐ −01°09′
Finder Chart 12-10, Figure 12-18   ★★★

8/10″ Scopes–100x: This galaxy is 22′ south of a 7th magnitude star, with several 8.5 magnitude stars between. NGC 936's fairly faint, 3′ × 2′ E–W halo has a large, bright core.

12/14″ Scopes–125x: NGC 936 is fairly bright, moderately large, elongated 5′ × 4′ NW–SE, and has a prominent, mottled or unevenly illuminated, core without a stellar nucleus. NGC 941 lies 13′ east.

NGC 941 H261³ Galaxy Type
⌀3.2′ × 2.4′, m12.4v, SB 14.5   02ʰ28.5ᵐ −01°09′
Finder Chart 12-10, Figure 12-18   ★★★

8/10″ Scopes–100x: NGC 941, 13′ east of NGC 936, is only an extremely faint, circular smudge.

12/14″ Scopes–125x: NGC 941 is very faint and diffuse, its 2′ diameter halo slightly elongated N–S and slightly brighter in its center.

NGC 947 Galaxy Type SA(r)c I-II
⌀2.1′ × 1.2′, m12.7v, SB 13.6   02ʰ28.5ᵐ −19°04′
Finder Chart 12-7   ★★★

8/10″ Scopes–100x: NGC 947 is between, but somewhat

closer to the southern of, two N–S 7.5 magnitude stars. A 9th magnitude star lies between the galaxy and the northern 7.5 magnitude star. The galaxy is very faint, small, elongated 1.5′ × 0.75′ NE–SW and gradually brighter in the center.

12/14″ Scopes–125x: NGC 941 appears much the same in larger as in smaller telescopes. The halo looks a little larger, about 2′ × 1′ NE–SW, and the central brightness is more concentrated.

NGC 945 Galaxy Type SB(rs)bc I-II
ø2.2′ × 2.2′, m12.1v, SB 13.6   02^{h}28.6^m −10°32′
Finder Chart 12-8   ★★★

8/10″ Scopes–100x: NGC 945, 5′ north of a 9th magnitude star, appears very faint, small, round, and diffuse.

12/14″ Scopes–100x: NGC 945 is fairly faint, its round 1.5′ diameter halo showing only a slight brightening at the center. A 13th magnitude star lies 1.5′ SE.

NGC 955 H278² Galaxy Type Sa: sp
ø3.1′ × 0.9′, m12.0v, SB 12.9   02^{h}30.6^m −01°07′
Finder Chart 12-10   ★★★

8/10″ Scopes–100x: NGC 955, 22′ WSW of 75 Ceti (m5.4), is a fairly obvious 1.5′ × 0.5′ NNE–SSW spindle with a brighter center. An 11th magnitude star lies 3′ SE.

12/14″ Scopes–125x: NGC 955 is fairly bright, elongated 2′ × 0.75′ NNE–SSW with an extremely extended, slightly brighter, core. Averted vision brings out a very faint stellar nucleus.

NGC 958 H237² Galaxy Type SB:(rs)bc II
ø2.3′ × 0.8′, m12.2v, SB 12.7   02^{h}30.7^m −02°57′
Finder Chart 12-10   ★★★

8/10″ Scopes–100x: This galaxy is a fairly obvious, thin 2′ × 0.25′ N–S streak with a tiny core.

12/14″ Scopes–125x: NGC 958 is fairly bright, and highly elongated 2.25′ × 0.5′ nearly N–S. Its slightly brighter, irregularly oval core is more obvious on the SE side.

NGC 991 H434³ Galaxy Type SAB(rs)cd II-III
ø2.9′ × 2.5′, m11.7v, SB 13.7   02^{h}35.5^m −07°09′
Finder Chart 12-8   ★★★

8/10″ Scopes–100x: NGC 991, 40′ north of 77 and 80 Ceti, is a very faint unconcentrated circular haze. A 12th magnitude star is just 1.5′ NNW.

12/14″ Scopes–125x: NGC 991 has a very faint, small 1.5′ halo with a diffuse, even glow.

NGC 1015 U2124 Galaxy Type SB(r)0/a
ø3.0′ × 3.0′, m12.1v, SB 14.3   02^{h}38.1^m −01°19′
Finder Chart 12-10   ★★★

16/18″ Scopes–150x: NGC 1015, 5′ NNW of an 8th magnitude star, shows a prominent core embedded within a faint, 2′ diameter, slightly N–S elongated halo. A 9th magnitude field star lies 9′ NNE.

NGC 1016 Galaxy Type E+0
ø2.5′ × 2.5′, m11.6v, SB 13.4   02^{h}38.3^m +02°08′
Finder Chart 12-9   ★★★

8/10″ Scopes–100x: NGC 1016, 6′ SE of a 9th magnitude star, is the largest and most conspicuous member of a group of faint galaxies. It is fairly faint, but its 1.75′ diameter halo suddenly brightens towards its center.

12/14″ Scopes–125x: NGC 1016 is fairly faint, though with a much brighter center in its round 2′ halo. Three dim companion galaxies are in the field.

NGC 1022 H102¹ Galaxy Type (R′)SB(r)0/a pec
ø2.7′ × 2.7′, m11.3v, SB 13.3   02^{h}38.5^m −06°40′
Finder Chart 12-8   ★★★

8/10″ Scopes–100x: This galaxy is 12′ east of a 9th magnitude star and 6′ west of a triangle of 10.5 magnitude stars. It is an obvious circular, 1′ diameter halo surrounding a prominent core and a stellar nucleus.

12/14″ Scopes–125x: NGC 1022 is fairly bright with a 1.5′ diameter, circular halo and a broad, well concentrated core containing a stellar nucleus. A 13th magnitude star lies 2′ NE.

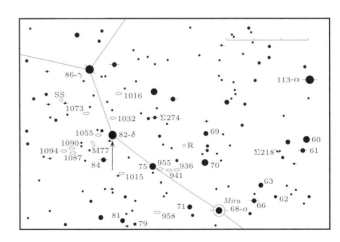

Finder Chart 12-10. 82-δ Cet: 02^{h}39.5^m +00°20′

Figure 12-20. *This wide angle photo shows seven galaxies with NGC 1042 and NGC 1052 forming an interesting pair. Martin C. Germano made this 65 minute exposure on hypered 2415 Kodak Tech Pan film with an 8", f5 Newtonian at prime focus.*

NGC 1032 Galaxy Type S0+ sp
⌀3.5' × 1.2', m11.6v, SB 13.0   02ʰ39.4ᵐ +01°06'
Finder Chart 12-9   ★★★

8/10" Scopes–100x: NGC 1032, 42' north of the 4.1 magnitude star Delta (δ) = 82 Ceti, is faint, elongated 2.5' × 0.75' ENE–WSW, and has a slightly brighter center. Three faint stars are on its north side.

16/18" Scopes–150x: NGC 1032 is fairly faint, elongated 3.5' × 1' ENE–WSW, and has a barely discernible core. It is at the SW corner of a square it forms with three 12th magnitude stars.

NGC 1035 H284² Galaxy Type Scd sp II-III
⌀2.2' × 0.7', m12.2v, SB 12.5   02ʰ39.5ᵐ −08°08'
Finder Chart 12-8, Figure 12-20   ★★★

8/10" Scopes–100x: This galaxy is a fairly obvious spindle 7' NNE of a 9th magnitude star. It is elongated 1.75' × 0.4' NNW–SSE with a faint star touching the SSE tip.

12/14" Scopes–125x: NGC 1035 is a moderately bright slash, elongated 2.5' × 0.75' NNE–SSW with an extended, irregular core. A faint star is just within the SSE tip. At 175x, the halo appears irregular in brightness. NGC 1052 lies 25' ESE.

NGC 1042 Galaxy Type SAB(rs)c I-II
⌀4.2' × 3.3', m11.0v, SB 13.7   02ʰ40.4ᵐ −08°26'
Finder Chart 12-8, Figures 12-19 & 12-20   ★★★

8/10" Scopes–100x: NGC 1042, 14' SW of NGC 1052 in a scattered group of faint galaxies, appears as a very faint, round, diffuse glow about 2' across without central brightening.

12/14" Scopes–125x: This galaxy has a faint, diffuse, halo elongated 2' × 1.75' E–W with a slightly brighter center. 12th magnitude stars are 1' ESE and 2.5' north.

16/18" Scopes–150x: In larger instruments, this galaxy displays a fairly faint, diffuse, round 3.5' halo with a faint stellar nucleus. The halo is somewhat mottled and has a slightly brighter knot at its east end.

NGC 1048A Galaxy Type SBb pec sp
⌀0.7' × 0.4', m14.5v, SB 13.0   02ʰ40.6ᵐ −08°33'
NGC 1048B Galaxy Type SBc:
⌀1.0' × 0.3', m14.5v, SB 13.0   02ʰ40.6ᵐ −08°32'
Finder Chart 12-8, Figure 12-20   ★/★

8/10" Scopes–100x: These two galaxies, 7' SSE of NGC 1042, are just visible with averted vision as two extremely faint, tiny, round spots.

Figure 12-21. *Messier 77 (NGC 1068), at left, is a bright circular Sab galaxy with a brilliant core. NGC 1055, lower right, is a fairly bright Sb galaxy exhibiting a dark lane. The very faint 14th magnitude galaxy, NGC 1027, lies at top center. Martin C. Germano made this 75 minute exposure on hypered 2415 Kodak Tech Pan film with an 8", f5 Newtonian at prime focus.*

16/18" Scopes–150x: Even in instruments of this range, these two galaxies are very faint featureless ovals. NGC 1048A, the southernmost object, is slightly elongated N–S while NGC 1048B is slightly extended E–W.

NGC 1052 H63¹ Galaxy Type SA?(r?)0−
⌀2.5' × 2.0', m10.5v, SB 12.1   02ʰ41.1ᵐ −08°15'
Finder Chart 12-8, Figure 12-20   ★★★

8/10" Scopes–100x: NGC 1052 is a bright, circular object, its 1' diameter halo surrounding a brilliant core with a stellar nucleus. NGC 1042 lies 14' SW and NGC 1035 is 22' WNW.

12/14" Scopes–125x: NGC 1052 has a prominent core with a stellar nucleus surrounded by a fainter 1.5' halo slightly elongated NW–SE. A tiny triangle of 12.5 to 13.5 magnitude stars lies 1' SW. NGC 1052 is the brightest in a group of visually contrasting galaxies that includes NGCs 1035, 1042, and 1069. NGC 1047, 10' NW, is a small pip with a stellar nucleus.

NGC 1055 H6² Galaxy Type Sb: sp II-III
⌀7.3' × 3.3', m10.6v, SB 13.9   02ʰ41.8ᵐ +00°26'
Finder Chart 12-10, Figures 12-21 & 12-22   ★★★

8/10" Scopes–100x: NGC 1055 is at the southern vertex of an equilateral triangle with two magnitude 7 to 7.5 stars 6'–7' to its NE and NW. It is a fairly faint E–W streak with a 4' × 1' halo enclosing a faint, thin, core.

12/14" Scopes–125x: NGC 1055 is a 4.5' × 1.5' ESE–WNW spindle with a slightly brighter core. NW of its center, just outside its halo is a tiny triangle of one 12th and two 13th magnitude stars, the 12th magnitude star almost touching the edge of the halo near the galaxy's core.

16/18" Scopes–150x: NGC 1055 is fairly bright, very elongated 7' × 3' ESE-WNW, and slightly brighter along its major axis. A dark lane running along the northern flank is just visible with averted vision.

NGC 1068 Messier 77 Galaxy Type (R)SAB(rs)ab
⌀8.2' × 7.3', m8.9v, SB 13.2   02ʰ42.7ᵐ −00°01'
Finder Chart 12-10, Figures 12-21 & 12-23   ★★★★

Messier 77 is the prototype of the peculiar Seyfert galaxies, systems with bright, starlike nuclei, emission-line spectra, and moderately-strong radio output. Ap-

Figure 12-22. *NGC 1055 is an obvious streak in medium-size telescopes. Larger insruments reveal its dark lane. Lee C. Coombs made this 10 minute exposure on 103a-O film with a 10", f5 Newtonian. South is up.*

Figure 12-23. *Messier 77 displays a brillant core surrounded by a mottled oval halo. Alexander Brownlee made this 60 minute exposure on Kodak 103a-O Spectroscopic film with a 16", f5 Newtonian. South is to the upper left.*

parently Seyferts are suffering serious explosions in their nuclei. M77 lies some 65 million light-years away and is four times as luminous as our Milky Way Galaxy.

4/6" Scopes–75x: M77, only 1.5′ WNW of a 10th magnitude star, is a quite bright E–W oval with a relatively large nucleus and several variegated areas in its outer regions.

8/10" Scopes–100x: M77 has a bright 3′ diameter halo enclosing a large luminous core. The core covers nearly half of the disk and is about as bright as the 9th magnitude star just to the galaxy's ESE.

12/14" Scopes–125x: M77 has a bright 6′ × 5′ halo elongated NNE–SSW containing a very bright, exceptionally large, core. The halo is mottled with dark lanes and luminous fragments of spiral arms. At 175x, a stellar nucleus becomes visible in the core.

NGC 1069 Galaxy Type SAB(s)c I-II
⌀1.5′ × 1.0′, m13.7v, SB 14.0   02ʰ42.8ᵐ −08°18′
Finder Chart 12-8, Figure 12-20   ★

16/18" Scopes–150x: NGC 1069 lies between a 9th magnitude star 6′ to its east and an 8th magnitude star 16′ to its SW. With averted vision it appears as an extremely faint spot, slightly oval-shaped NNW–SSE.

NGC 1073 H455³ Galaxy Type SB(rs)c II-III
⌀5.0′ × 5.0′, m11.0v, SB 14.3   02ʰ43.7ᵐ +01°23′
Finder Chart 12-9, Figure 12-24   ★★

12/14" Scopes–125x: NGC 1073, 7′ NE of a large equilateral triangle of 9th and 10th magnitude stars, appears faint, round, and 3.5′ across. It is slightly brighter in its center and on its northern edge.

16/18" Scopes–150x: NGC 1073 has a faint, circular 5′ halo gradually brightening to a NE–SW bar-like core. Several very faint stars or knots are visible on the galaxy's south and SW edges.

NGC 1087 H466² Galaxy Type SB:(s)c: pec II-III
⌀3.7′ × 2.8′, m10.9v, SB 13.3   02ʰ46.4ᵐ −00°30′
Finder Chart 12-10, Figure 12-25   ★★★★

8/10" Scopes–100x: This galaxy is at the western vertex of a triangle with 11th magnitude stars 6′ to its NE and 7′ to its ESE. It has a smooth, bright 2′ diameter halo.

12/14" Scopes–125x: The 2.5′ diameter halo is bright and diffuse with an indistinct, irregular outline and a slightly brighter center. NGC 1090 lies 14′ north.

16/18" Scopes–150x: NGC 1087 is bright, elongated 3′ × 2′ NNE–SSW, and has a slightly brighter center. The halo is irregular both in outline and in surface brightness. A dark lane is suspected ESE–WNW across the halo. The SSW side is somewhat broader and brighter.

Figure 12-24. *NGC 1073 has a faint, large, circular halo with a central bar visible in larger instruments. Alexander Brownlee made this 60 minute exposure on 103a-O Kodak Spectroscopic film with a 16", f5 Newtonian at prime focus.*

Figure 12-25. *An interesting triangle of galaxies is formed by NGC 1087 (top), NGC 1090 (bottom center) and NGC 1094 (bottom right). Lee C. Coombs made this 10 minute exposure on 103a-O film with a 10", f5 Newtonian.*

NGC 1090 H465² Galaxy Type SB(r)bc II
⌀3.9′ × 2.0′, m11.8v, SB 13.9 $02^h46.6^m$ −00°15′

Finder Chart 12-10, Figure 12-25   ★★★

8/10" Scopes–100x: NGC 1090, 3′ south of a magnitude 10 star, is a very faint, small, oval of uniform surface brightness. It lies 14′ north of galaxy NGC 1087 and 13′ WNW of galaxy NGC 1094.

12/14" Scopes–125x: NGC 1090 is fairly faint, elongated 2.5′ × 1.5′ ESE–WNW, and has a prominent core. It is somewhat dimmer than NGC 1087, 14′ south.

16/18" Scopes–150x: Larger instruments show a fairly bright core surrounded by a much fainter halo elongated 3.5′ × 2′ ESE–WNW. A 13th magnitude star is at the south edge.

NGC 1094 Galaxy Type Sbc I-II
⌀1.2′ × 0.9′, m12.5v, SB 12.5 $02^h47.5^m$ +00°17′

Finder Chart 12-10, Figure 12-25   ★★★

12/14" Scopes–100x: This galaxy lies 3′ south of a wide N–S pair of magnitude 9 and 10 stars. It is a faint, small, round object with a well concentrated center.

16/18" Scopes–125x: NGC 1094 is fairly faint, with a circular 1′ diameter halo and a prominent core. This galaxy is located 17′ NE of NGC 1087 and 13′ ESE of NGC 1090.

Chapter 13

Columba, the Dove

13.1 Overview

Columba was invented during the 1500s to represent the Dove that Noah sent to search for dry land after the Deluge. Johannes Bayer and other early celestial mapmakers depicted Columba perched upon Puppis, the Stern of Argo the Ship, which sometimes during the 16th and 17th centuries was identified with Noah's Ark.

Columba is a small constellation of only 270 square degrees south of the ancient Greek group Lepus the Hare. Its southerly situation makes it and its contents difficult viewing for northern hemisphere observers: you have to wait until it transits the meridian for the best possible views of its few objects. Columba contains a handful of good double stars, a few modestly bright galaxies, and the fine globular cluster NGC 1851.

Columba: Kol-LUM-ba
Genitive: Columbae, Kol-LUM-be
Abbrevation: Col
Culmination: 9pm–Feb. 1, midnight–Dec. 18
Area: 270 square degrees
Showpieces: NGC 1792, NGC 1808, NGC 1851
Binocular Objects: h3849, h3857, NGC 1851

13.2 Interesting Stars

h3857 Triple Star   Spec. G5 & G5
AB: m5.7,10.8; Sep. 12.9″; P.A. 256°
AC: 6.9; Sep. 64.8″; P.A. 72°   06^{h}24.0^m −36°42′
Finder Chart 13-4   ★★★★

4/6″ Scopes–75x: Medium power shows a fine wide pair of yellowish-orange stars. The brighter star has a faint greyish-white companion.

β755 Double Star   Spec. B9
AB: m6.0, 6.8; Sep. 1.3″; P.A. 258°   06^{h}35.4^m −36°47′
Finder Chart 13-4   ★★★

4/6″ Scopes–175x: Although not split, the brilliant white primary appears elongated with its ashy companion. The surrounding star-field is rather rich.

13.3 Deep-Sky Objects

NGC 1792 Galaxy Type SAB(s)b: II-III
⌀5.5′ × 2.5′, m9.9v, SB 12.6   05^{h}05.2^m −37°59′
Finder Chart 13-3, Figure 13-1   ★★★★

12/14″ Scopes–125x: NGC 1792 is bright and elongated 3′ × 1.25′ NW–SE. The halo appears nearly uniform, with no evident nucleus or spiral arcs. But close inspection at 175x reveals a few faint knots or stars embedded near the ends of the major axis. Galaxy NGC 1808 lies 30′ SW.

NGC 1800 Galaxy Type IB(s)m IV
⌀2.1′ × 1.0′, m12.6v, SB 13.3   05^{h}06.4^m −31°57′
Finder Chart 13-3, Figure 13-2   ★★

12/14″ Scopes–125x: NGC 1800 appears rather faint, its halo elongated 1′ × 0.5′ ESE–WNW and shaped like a checkmark or "V," the opening facing ESE. Some granularity can be seen at the point of the "V" and at the SE end of its southern leg.

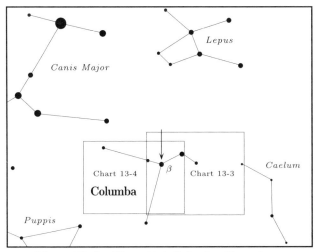

Master Finder Chart 13-1. Columba Chart Areas
Guide stars indicated by arrows.

Columba, the Dove
Constellation Chart 13-2

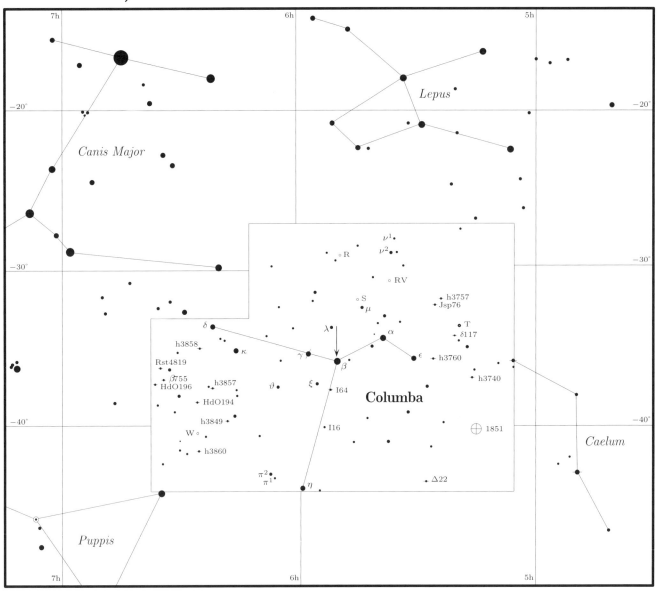

Table 13-1. Selected Variable Stars in Columba

Name	HD No.	Type	Max.	Min.	Period (Days)	F*	Spec. Type	R.A. (2000)	Dec.	Finder Chart No. & Notes
T Col	34897	M	6.6	12.7	225	0.50	M3-M6	05^{h}19.3^m	−33°42′	13-3
RV Col		SRd	9.3	10.3	105	0.40	G5	35.7	−30 50	13-3
S Col		M	8.9	14.2	325	0.46	M6-M8	46.9	−31 42	13-3
R Col	39324	M	7.8	15.0	327	0.39	M3-M7	50.5	−29 12	13-2
W Col		M	9.3	11.5	327		M6	06^{h}27.8^m	−40 06	13-2

F* = The fraction of period taken up by the star's rise from min. to max. brightness, or the period spent in eclipse.

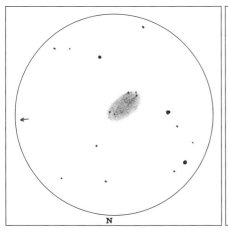
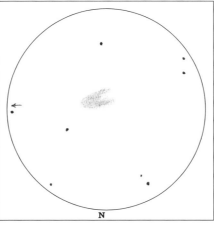
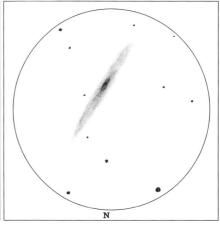

Figure 13-1. NGC 1792
13", f4.5–125x, by George deLange

Figure 13-2. NGC 1800
13", f4.5–125x, by George deLange

Figure 13-3. NGC 1808
13", f4.5–125x, by George deLange

NGC 1808 Galaxy Type (R)SB(s)a pec:
ø5.2′ × 2.3′, m9.9v, SB 12.4 $05^h07.7^m$ $-37°31'$
Finder Chart 13-3, Figure 13-3   ★★★★

12/14″ Scopes–125x: NGC 1808, half a degree NE of NGC 1792, is a beautiful barred-spiral galaxy with a bright, tiny core and a stellar nucleus surrounded by a bright halo elongated 5′ × 1.5′ NW–SE. The halo is distinctly brighter through the center of its major axis. A 12th magnitude star lies to the SW of the galaxy, and a yellow 10th magnitude star is at the northern edge of the field of view.

NGC 1851 Dun 508 Globular Cluster Class II
ø11″, m7.2v, Br★ 13.2 $05^h14.1^m$ $-40°03'$
Finder Chart 13-3, Figure 13-4   ★★

12/14″ Scopes–125x: Beautiful! NGC 1851 appears bright and large with a granular 5′ diameter halo. The bright, dense, unresolvable core is surrounded by progressively fainter and looser concentric rings of stars. The periphery is irregular and surrounded by a field nicely adorned with 10th and 11th magnitude stars.

Table 13-2. Selected Double Stars in Columba

Name	ADS No.	Pair	M1	M2	Sep.″	P.A.°	Spec	R.A. (2000)	Dec	Finder Chart No. & Notes
h3740			6.8	8.5	23.9	287	G5	$05^h15.1^m$	$-36°39'$	13-3
δ117			6.1	10.9	2.2	7	B5	21.3	$-34\ 21$	13-3
h3757			7.6	11.6	15.1	315	F0	23.2	$-31\ 45$	13-3
h3760		AB	7.7	8.3	7.4	222	F5	25.9	$-35\ 21$	13-3 1° west of Epsilon (ε) Col
		AC		10.6	26.0	282				
Jsp 76			7.0	13.0	13.6	50	K0	26.0	$-32\ 13$	13-3
Δ22			7.5	8.0	7.5	169	A5	31.0	$-42\ 18$	13-2
α Col			2.6	12.3	13.5	359	B8	39.0	$-34\ 04$	13-3
I 64		AB	5.6	11.7	16.5	250	K0	52.6	$-37\ 38$	13-4
I 16			6.8	10.3	1.3	126	K0	52.8	$-38\ 32$	13-4
γ Col			4.4	12.7	33.8	110	B3	57.5	$-35\ 17$	13-4
h3849			6.7	8.3	39.6	53	K0 G5	$06^h19.8^m$	$-39\ 29$	13-4
h3857		AB	5.7	10.8	12.9	256	G5	24.0	$-36\ 42$	13-4 Brighter stars deep yellow,
		AC		6.9	64.8	72	G5			faint greyish companion
h3858		AB	6.3	7.4	132.3	48	K0 A3	25.5	$-35\ 04$	13-4
		BC		8.5	3.8	311				
h3860			6.9	8.7	8.6	227	A3	25.8	$-40\ 59$	13-2
HdO 194			6.5	13.5	32.8	56	F0	27.1	$-37\ 54$	13-4
β755		AB	6.0	6.8	1.3	258	B9	35.4	$-36\ 47$	13-4 Brillant white & ashy
		AC		11.5	21.4	301				
IIdO 196		AB	6.0	12.0	18.0	270	G5	37.0	$-38\ 09$	13-4
		AC		10.0	25.0	180				
		AD		13.0	35.0	100				
Rst 4819			6.1	6.9	0.6	356	B9	37.2	$-36\ 59$	13-4

Footnotes: *= Year 2000, a = Near apogee, c = Closing, w = Widening. Finder Chart No: All stars listed in the tables are plotted in the large Constellation Chart, but when a star appears in a Finder Chart, this number is listed. Notes: When colors are subtle, the suffix *-ish* is used, e.g. *bluish*.

Chapter 13

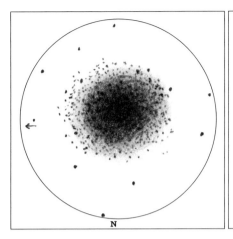

Figure 13-4. NGC 1851
13", f4.5–195x, by George deLange

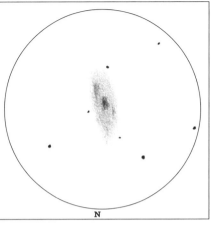

Figure 13-5. NGC 2090
13", f4.5–195x, by George deLange

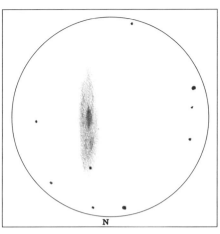

Figure 13-6. NGC 2188
13", f4.5–125x, by George deLange

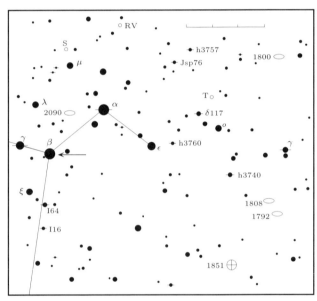

Finder Chart 13-3. Beta (β) Col: 05ʰ50.9ᵐ −35°46′

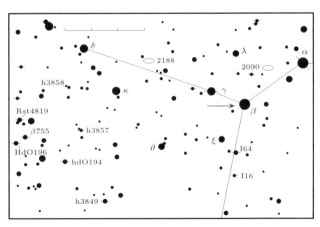

Finder Chart 13-4. Beta (β) Col: 05ʰ50.9ᵐ −35°46′

NGC 2090 Galaxy Type SA(s)c: II-III
ø4.5′ × 2.3′, m11.0v, SB 13.4   05ʰ47.0ᵐ −34°14′
Finder Chart 13-3, Figure 13-5   ★★

12/14″ Scopes–125x: NGC 2090 lies 1.75° east of magnitude 2.6 star Alpha (α) Columbae. Its moderately bright core is surrounded by a faint halo elongated 2.5′ × 1′ NNW–SSE. At 200x, tiny spiral arcs are visible NE and SW of the center. 14th magnitude stars lie 45″ WSW of the galaxy's center and just east of its southern tip, and a 15th magnitude star lies just east of the northern tip.

NGC 2188 Galaxy Type SB(s)m? Sp III
ø5.5′ × 0.8′, m11.6v, SB 13.1   06ʰ10.1ᵐ −34°06′
Finder Chart 13-4, Figure 13-6   ★★

12/14″ Scopes–125x: NGC 2188, 8′ NE of an 8th magnitude star, is a bright streak highly elongated 4′ × 0.5′ N–S. A bar is visible south of the center, and definite hints of spiral structure can be discerned. A 13.5 magnitude star is embedded in the NNW tip.

Chapter 14

Eridanus, the River

14.1 Overview

Eridanus, the River, is the sixth largest of the constellations. Its faint stars meander from Orion southwestward to nearly −60° declination. In Greek mythology the Eridanus was the river into which the foolhardy young Phaeton was thrown as he inexpertly drove the chariot of his father, the Sun god, perilously close to the Earth. Ancient writers identified the ancient river with several different terrestrial streams, including the Po in Italy, the Nile in Egypt, and the Euphrates in Babylonia. Probably the celestial river Eridanus was formed by the Babylonians and originally represented the Euphrates, for the name of the ancient Babylonian city at the mouth of the Euphrates was Eridu. The Greek (and no doubt the Babylonian) celestial River flowed through the stars of our constellation Columba to the brilliant 1st magnitude Canopus; but the Medieval Arabian astronomers stopped the river short at Theta Eridani, whose name, Acamar, is from the Arabic Achernar, meaning "River's End," the name and location of the blue 1st magnitude star Alpha (α) Eridani. The constellation is sprinkled with many galaxies, the best two visually being the spirals NGC 1232 and NGC 1300.

Eridanus: Eh-RID-an-us
Genitive: Eridani, Eh-RID-a-ni
Abbreviation: Eri
Culmination: 9 pm–Dec. 25, midnight–Nov. 10
Area: 1,138 square degrees
Best Deep-Sky Objects: NGC 1300, NGC 1332, NGC 1421, NGC 1532, NGC 1535
Binocular Objects: β1042, h3628, 40–O^2 Cet, 62 Cet, NGC 1232, NGC 1291, NGC 1332, NGC 1535

14.2 Interesting Stars

Theta ($\vartheta^{1,2}$) Eridani = Pz 2   Double Star   Spec. A2 & A2
m3.4, 4.5; Sep. 8.2″; PA 88° $02^h58.3^m$ −40°18′
Constellation Chart 14-1   ★★★★

8/10″ Scopes–100x: This is a nice pair, both components appearing white.

Rho (ϱ^2) = 9 Eridani (Σ11)   Double Star   Spec. G5
m5.3, 9.5; Sep. 1.8″; P.A. 75° $03^h02.7^m$ −07°41′
Finder Chart 14-3   ★★★★

8/10″ Scopes–150x: Rho Eridani is a very close double of lovely deep yellow suns in a nice star field.

32 Eridani = Σ470   Double Star   Spec. G5 A2
AB: m4.8, 6.1; Sep. 6.8″; P.A. 347° $03^h54.3^m$ −02°57′
Finder Chart 14-6   ★★★★

4/6″ Scopes–75x: 32 Eridani appears yellow and white.

8/10″ Scopes–100x: This double has a very beautiful deep yellow primary and a greenish-white secondary.

39 Eridani = Σ516   Double Star   Spec. K0
AB: m5.0, 8.0; Sep. 6.4″; P.A. 146° $04^h14.4^m$ −10°15′
Finder Chart 14-7   ★★★★

4/6″ Scopes–75x: 39 Eridani is a beautiful orangish-yellow and white pair lying in a well sprinkled star field.

8/10″ Scopes–100x: This pair exhibits deep-yellow and light blue stars.

Omicron (o^2) = 40 Eridani (Σ518)   Triple Star Spec. G5 A
AC: m4.4, 11.2; Sep. 7.6″; P.A. 347° $04^h15.2^m$ −07°39′
Finder Chart 14-6   ★★★★

4/6″ Scopes–100x: A yellowish primary is attended by two blue companions of magnitudes 11.2 and 9.4 lying 7.6″ and 83.4″ away, respectively.

8/10″ Scopes–125x: Omicron Eridani appears yellowish-white with two bluish companions.

Eridanus, the River

Constellation Chart 14-1

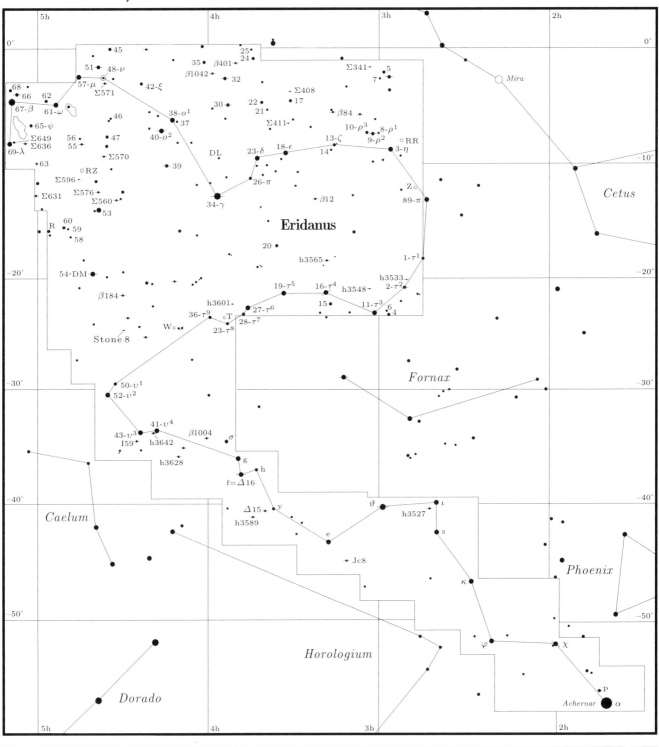

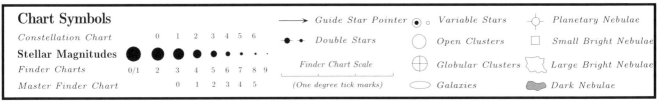

Table 14-1. Selected Variable Stars in Eridanus

Name	HD No.	Type	Max.	Min.	Period (Days)	F*	Spec. Type	R.A. (2000)	Dec.	Finder Chart No. & Notes
Z Eri	17491	SRb	7.0	8.6	80		M4	$02^h47.9^m$	$-12°28'$	14-3
RR Eri	17895	SRb	7.4	8.6	97		M5	52.2	$-08\ 16$	14-3
T Eri	24754	M	7.4	13.2	252	0.45	M3-M5	$03^h55.2$	$-24\ 02$	14-8
W Eri	26601	M	7.5	14.5	376	0.40	M7	$04^h11.5^m$	$-25\ 08$	14-8
RZ Eri	30050	EA/RS	7.79	8.71	39.28	0.05	A5-F5	43.8	$-10\ 41$	14-10

F* = The fraction of period taken up by the star's rise from min. to max. brightness, or the period spent in eclipse.

Table 14-2. Selected Double Stars in Eridanus

Name	ADS No.	Pair	M1	M2	Sep."	P.A.°	Spec	R.A. (2000)	Dec	Finder Chart No. & Notes
p Eri			5.8	5.8	w11.5	*191	G5 G5	$01^h39.8^m$	$-56°12'$	14-1
χ Eri			3.7	10.7	5.0	202	G5	56.0	$-51\ 37$	14-1 Golden primary
h3527			7.1	7.3	2.0	41	A0 A0	$02^h43.3^m$	$-40\ 32$	14-1
h3533			7.8	8.8	39.1	271	K5 G5	49.8	$-20\ 15$	14-4
ϑ Eri			3.4	4.5	8.2	88	A2 A2	58.3	$-40\ 18$	14-1 Brilliant white pair
9–o^2 Eri	2312		5.3	9.5	1.8	75	G5	$03^h02.7^m$	$-07\ 41$	14-3 Lovely yellow pair
Σ341	2316		7.7	9.7	8.6	226	F5	03.0	$-02\ 05$	14-1
h3548	2326		7.7	11.5	12.2	123	G0	03.8	$-21\ 22$	14-4
h3556		ABxC		8.9	3.5	200				
β84	2440		6.9	7.1	1.0	14	B9	16.0	$-05\ 55$	14-1
h3565	2465		5.7	9.1	7.2	118	F0	18.7	$-18\ 34$	14-4
16–$τ^1$ Eri	2472	AB	3.7	9.2	5.7	288	M2	19.5	$-21\ 45$	14-4
	2472	AC		10.5	39.8	112				
β12	2523		6.9	9.8	2.3	274	A0	24.4	$-14\ 00$	14-3
Σ408	2581		8.3	8.5	1.4	331	A3	30.7	$-04\ 17$	14-6
Σ411	2596	AB	7.8	8.8	19.1	88	F8	32.3	$-07\ 05$	14-6
	2596	AC		11.2	38.2	28				
Δ15			7.3	8.4	7.8	328	A2	39.8	$-40\ 21$	14-5
h3589			6.6	9.2	5.2	349	K0	44.1	$-40\ 40$	14-5
Δ16			4.8	5.3	7.9	212	B8 A0	48.6	$-37\ 37$	14-5 Pale yellow pair
β401	2803	AB	6.5	10.5	4.5	254	F2	50.3	$-01\ 31$	14-6
	2803	AC		11.2	40.6	289				
h3601	2825		7.7	9.7	10.6	300	G5	51.7	$-22\ 56$	14-8
30 Eri	2832		5.5	10.6	8.2	135	B8	52.7	$-05\ 22$	14-6
32 Eri	2850	AB	4.8	6.1	6.8	347	G5 A2	54.3	$-02\ 57$	14-6 Yellow & white
β1042	2909	AD	7.1	10.8	39.2	250	G5	58.6	$-02\ 39$	14-6
	2909	AxBC			55.6	93				
	2909	BC	8.7	9.5	1.2	34				
β1004		AB	7.2	7.8	1.5	106	G0	$04^h02.1^m$	$-34\ 29$	14-9
h3628			7.1	8.0	50.3	50	F5 F5	12.5	$-36\ 09$	14-9
39 Eri	3079	AB	5.0	8.0	6.4	146	K0	14.4	$-10\ 15$	14-7 Orangish & light blue
40–o^2 Eri	3093	AB	4.4	9.4	83.4	104	G5 A	15.2	$-07\ 39$	14-6 Yellowish-orange & blue
	3093	BC		11.2	7.6	347				Blue
h3642			6.4	8.4	6.0	159	A2	19.0	$-33\ 54$	14-9
I59		AB	6.6	10.0	42.4	198	F5	24.9	$-34\ 45$	14-9
		BC		11.0	3.7	281				
β184	3247		7.3	7.7	1.5	252	F5	27.9	$-21\ 30$	14-1
Stone 8	3257	AC	7.8	9.3	7.0	351	G5	28.9	$-25\ 12$	14-1
Σ560	3284	AC	6.2	9.1	29.8	44	A2	41.4	$-13\ 39$	14-1
Σ570	3318		6.7	7.7	12.8	259	A0 A0	35.2	$-09\ 44$	14-10 Bluish-white
Σ571	3328		6.3	11.0	17.5	258	B9	36.0	$-03\ 37$	14-10 White & blue
Σ576	3355		6.8	8.1	12.4	172	A0	38.0	$-13\ 02$	14-1
55 Eri	3409		6.7	6.8	9.2	317	F5	43.6	$-08\ 48$	14-10 Yellow & pale yellow
Σ596	3428	ABxC	7.7	10.2	10.4	292	F0	45.8	$-11\ 57$	14-1 Yellow & pale yellow
Σ631	3606		7.5	9.0	5.5	106	A0	$05^h00.7^m$	$-13\ 30$	14-1
Σ636	3640		7.1	8.2	3.7	102	A0	03.0	$-08\ 40$	14-10
66 Eri	3698		5.2	8.4	52.2	9	B9	06.8	$-04\ 39$	14-10
Σ649	3722		5.8	9.8	21.6	81	B8	08.3	$-08\ 40$	14-10

Footnotes: *= Year 2000, a = Near apogee, c = Closing, w = Widening. Finder Chart No: All stars listed in the tables are plotted in the large Constellation Chart, but when a star appears in a Finder Chart, this number is listed. Notes: When colors are subtle, the suffix -*ish* is used, e.g. *bluish*.

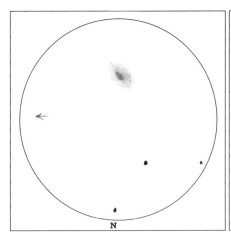

Figure 14-1. NGC 1084
8", f4.5–100x, by Dennis E. Hoverter

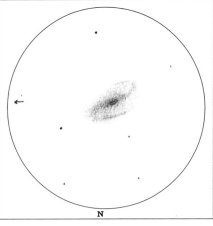

Figure 14-2. NGC 1187
17.5", f4.5–175x, by Richard W. Jakiel

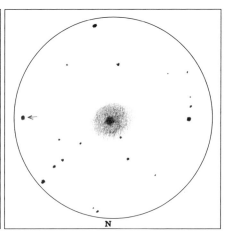

Figure 14-3. NGC 1232
13", f5.6–100x, by Steve Coe

55 Eridani = Σ3409 Double Star   Spec. F5
m6.7, 6.8; Sep. 9.2″ ; P.A. 317°   04ʰ43.6ᵐ   −08°48′
Finder Chart 14-10   ★★★★

4/6″ Scopes–75x: 55 Eridani is a fine pair of vivid yellow and pale yellow stars.

14-3. Deep-Sky Objects

NGC 1084 H64¹ Galaxy Type SB:(rs)bc pec II?
⌀3.2′ × 1.9′, m10.7v, SB 12.5   02ʰ46.0ᵐ −07°35′
Finder Chart 14-3, Figure 14-1   ★★★

8/10″ Scopes–100x: This galaxy is a fairly bright 1.75′ × 1′ NE–SW oval.

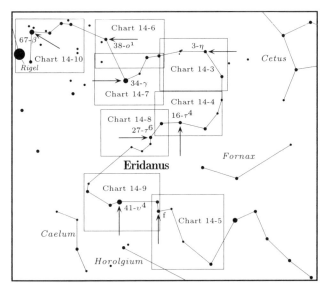

Master Finder Chart 14-2. Eridanus Chart Areas
Guide stars indicated by arrows.

12/14″ Scopes–125x: NGC 1084 is a bright even glow elongated 2′ × 1′ NE-SW with a broad oval core.

16/18″ Scopes–150x: NGC 1084 is a very bright galaxy with a 2.5′ × 1′ NE-SW oval halo. The large oval core is nearly half the size of the halo.

NGC 1140 H470² Galaxy Type IB:m pec
⌀1.9′ × 1.2′, m12.5v, SB 13.2   02ʰ54.6ᵐ −10°02′
Finder Chart 14-3   ★★★

12/14″ Scopes–125x: NGC 1140, 22′ NNE of a 7th magnitude star, is a fairly bright but tiny circular object only 30″ across with a stellar nucleus.

16/18″ Scopes–150x: NGC 1140 shows a bright halo elongated 1′ × 0.75′ N-S. The large core has a stellar nucleus.

NGC 1172 H502² Galaxy Type SA0−
⌀2.2′ × 1.6′, m11.9v, SB 13.1   03ʰ01.6ᵐ −14°50′
Finder Chart 14-3   ★★

12/14″ Scopes–125x: This galaxy is a small, very faint glow 2′ WSW of a 9.5 magnitude star.

16/18″ Scopes–150x: The halo, though faint, is not difficult. It is about 1′ across with a diffuse, uniform brightness.

NGC 1179 Galaxy Type SB(rs)cd II-III:
⌀3.6′ × 2.7′, m11.9v, SB 14.2   03ʰ02.6ᵐ −18°54′
Finder Chart 14-4   ★★

12/14″ Scopes–125x: This galaxy is an extremely faint, round 1′ diameter smudge with a 13th magnitude star nearly touching its ESE edge.

16/18″ Scopes–150x: NGC 1179 is very faint and diffuse, its circular 3′ diameter halo touching a 13th magnitude star at the ESE edge.

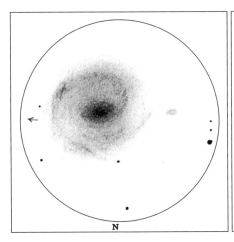
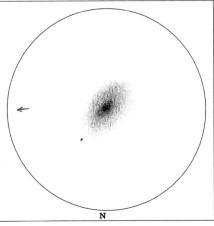
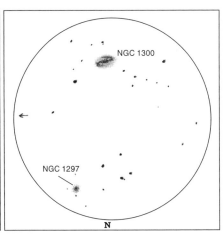

Figure 14-4. NGC 1232–32A
20″, f4.5–175x, by Richard W. Jakiel

Figure 14-5. NGC 1291
20″, f4.5–175x, by Richard W. Jakiel

Figure 14-6. NGC 1297 & NGC 1300
13″, f5.6–100x, by Steve Coe

NGC 1187   H245³   Galaxy   Type (R′)SB(s)c: I-II
ø5.2′ × 3.0′, m10.7v, SB 13.5   03ʰ02.6ᵐ −22°52′
Finder Chart 14-4, Figure 14-2   ★★★★

8/10″ Scopes–100x: This galaxy is a large, faint, unconcentrated disk 4′ SE of an 8th magnitude star. The halo is elongated 3.5′ × 3′ NW–SE.

12/14″ Scopes–125x: In medium-size telescopes the halo appears nearly the same size as in smaller instruments, but some vague hints of spiral structure can be glimpsed with averted vision.

16/18″ Scopes–150x: NGC 1187 appears fairly faint, is elongated about 4′ × 3′ NW–SE, and has an unevenly illuminated halo with a star or stellar nucleus at its center and a faint star on its NE edge.

20/22″ Scopes–150x: Nice! NGC 1187, a giant barred-spiral galaxy, displays a broad but rather weak bar with an oval-shaped core and a stellar nucleus surrounded by two arms that form a wispy S.

NGC 1199   H503²   Galaxy   Type E3: pec
ø2.2′ × 1.7′, m11.3v, SB 12.6   03ʰ03.6ᵐ −15°37′
Finder Chart 14-3   ★★★

12/14″ Scopes–125x: This fairly faint galaxy lies 3′ SW of a 10th magnitude star. It has a rather small 1′ diameter halo with diffuse edges and a slightly brighter center. Galaxy NGC 1209 lies 35′ east.

16/18″ Scopes–150x: NGC 1199 has a fairly bright 1.5′ diameter halo slightly elongated NE–SW with a broad core and a faint stellar nucleus. Four faint companion galaxies, all within 5′, lie to the west, south, and SW.

NGC 1209   H504²   Galaxy   Type S0−: sp
ø2.1′ × 1.0′, m11.4v, SB 12.1   03ʰ06.0ᵐ −15°37′
Finder Chart 14-3   ★★★

12/14″ Scopes–125x: NGC 1209 is a moderately faint 1′ × 0.5′ E–W oval with a circular core and stellar nucleus. It lies 35′ east of NGC 1199.

16/18″ Scopes–150x: This is a fairly obvious galaxy elongated 1.5′ × 1′ E–W with an unusual twin nucleus, the western nucleus clearly the brighter.

NGC 1232   H258²   Galaxy   Type SAB(rs)c I or I-II
ø6.8′ × 5.6′, m10.0v, SB 13.8   03ʰ09.8ᵐ −20°35′
Finder Chart 14-4, Figures 14-3, 14-4, & 14-7   ★★★★

4/6″ Scopes–75x: NGC 1232, 8′ west of a 9th magnitude star, appears very faint and diffuse with a round 3′ diameter halo.

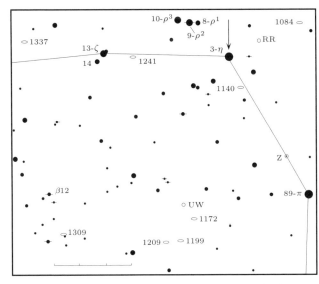

Finder Chart 14-3. 3-η Eri: 02ʰ56.4ᵐ −08°54′

Figure 14-7. *NGC 1232 is a fine, bright face-on spiral galaxy. Knots and condensations are visible in larger instruments. Harvey Freed, D.D.S. took a 55 minute exposure on 2415 film with a 10", f6.3 Newtonian reflector.*

8/10" Scopes–100x: NGC 1232 shows a faint 3.5' halo with several faint knots and a gradually brighter center.

12/14" Scopes–125x: This galaxy is fairly bright with a 5' diameter halo and an oval core containing a faint stellar nucleus. With averted vision the tiny bright knots of several H-II regions are clearly visible, and spiral arms are suspected. A 13th magnitude star touches the NE edge.

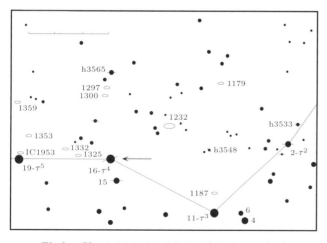

Finder Chart 14-4. 16-τ^4 Eri: $03^h19.5^m$ $-21°45'$

16/18" Scopes–150x: NGC 1232 is a bright, large 7' diameter galaxy displaying an oval core and a faint stellar nucleus. The halo contains many knots and condensations in a very faint counterclockwise spiral pattern.

NGC 1241 H286² Galaxy
Type SAB(rs)bc II
ø2.8' × 1.8', m12.0v, SB 13.6
$03^h11.3^m$ $-08°55'$
Finder Chart 14-3 ★★

12/14" Scopes–125x: NGC 1241, 3' south of a 9th magnitude star, has a circular 1.5' diameter halo of uniform surface brightness.

16/18" Scopes–150x: NGC 1241 is a fairly faint object with a 2' × 1.5' halo elongated NW–SE. A double of two 13th magnitude stars aligned 12" N–S lies 3' ESE. Galaxy NGC 1242, 1.5' to the NE, is a very faint, tiny spot.

NGC 1291 Galaxy Type (R)SB(s)0/a
ø11.0' × 9.5', m8.5v, SB 13.4
$03^h17.3^m$ $-41°08'$
Finder Chart 14-5, Figure 14-5 ★★★

12/14" Scopes–125x: NGC 1291, 10' NE of an 8th magnitude star, has a circular 7' diameter halo and a bright core containing a nearly stellar nucleus. A faint star is just to the NW.

20/22" Scopes–150x: NGC 1291 has a very bright halo elongated 9' × 7' NNW–SSE with a well concentrated core and a dense nucleus.

NGC 1297 Galaxy Type SA:(s?)0° pec
ø2.2' × 1.9', m12.0v, SB 13.4 $03^h19.2^m$ $-19°06'$
Finder Chart 14-4, Figures 14-6 & 14-8 ★★

12/14" Scopes–125x: NGC 1297, 14' SW of a 7th magnitude star, is a faint 1' diameter round spot with a well concentrated core and a stellar nucleus. A 13th magnitude star 1' NNE has the same brightness as the nucleus.

16/18" Scopes–150x: NGC 1297 shows a faint, circular 1.5' diameter halo that intensifies suddenly to a large, bright core with a stellar nucleus.

NGC 1300 Galaxy Type SB(s)bc I
ø5.5' × 2.9', m10.4v, SB 13.3 $03^h19.7^m$ $-19°25'$
Finder Chart 14-4, Figures 14-6 & 14-8 ★★★★

4/6" Scopes–75x: Small telescopes show a very faint, fairly large diffuse disk elongated E–W.

Eridanus

Figure 14-8. *NGC 1300, NGC 1297, and NGC 1301 all lie within a one degree field. NGC 1300 is an impressive barred spiral galaxy showing distinct arms in larger instruments. Martin C. Germano made this 65 minute exposure on hypered 2415 Kodak Tech Pan film with an 8″, f5 Newtonian reflector at prime focus. North is to the right.*

8/10″ Scopes–100x: This galaxy is very faint, elongated 4′ × 1.5′ E–W, and gradually brighter toward the center. A 10th magnitude star lies 6′ WNW while a chain of five 12th to 13th magnitude stars extends ENE 18′ from the galaxy.

12/14″ Scopes–125x: NGC 1300 is fairly bright, elongated 6′ × 4′ E–W, and has a faint bar that extends the entire 6′ length of the major axis. At the center of the bar is an oval core.

16/18″ Scopes–150x: This galaxy is rather faint even in larger instruments. The halo displays a reverse S-shaped spiral, the central bar of the S extending 6.5′ E–W. The core of the central bar is well concentrated but diffuse.

20/22″ Scopes–175x: NGC 1300 has a large 6.5′ long bar in P.A. 100° with faint but distinct spiral arms attached to each end, the arm attached to the west end curving north of the bar to the east and the arm attached to the east end curving south of the bar to the west. A faint knot or star lies just NE of the bar's prominent 1′ diameter core.

NGC 1309 H106¹ Galaxy Type SB?(s:)c II
⌀1.7′ × 1.7′, m11.5v, SB 12.5   03ʰ22.1ᵐ −15°24′
Finder Chart 14-3, Figure 14-9   ★★★

8/10″ Scopes–100x: This fairly bright galaxy lies 4′ NE of a 7.5 magnitude star. The halo is a round, uniform glow about 1.5′ across.

12/14″ Scopes–125x: NGC 1309 has a bright circular

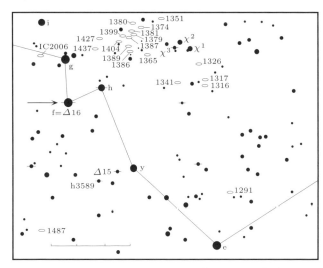

Finder Chart 14-5. f Eri: 03ʰ47.8ᵐ −36°06′

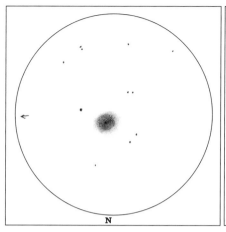

Figure 14-9. NGC 1309
12.5″, f5–125x, by Andrew D. Fraser

Figure 14-10. NGC 1337
13″, f5.6–175x, by Steve Coe

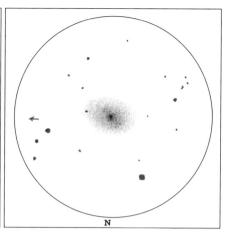

Figure 14-11. NGC 1357
17.5″, f4.5–200x, by G. R. Kepple

halo about 2′ in diameter that, except for its very faint stellar nucleus, is evenly illuminated.

NGC 1325 H77⁴ Galaxy Type SB(rs)bc: II
ø4.6′ × 1.5′, m11.5v, SB 13.4   03ʰ24.4ᵐ −21°33′
Finder Chart 14-4   ★★

12/14″ Scopes–125x: NGC 1325 is a very faint, diffuse glow elongated 2′ × 1′ NE-SW. Its NE tip touches a magnitude 11.5 star. Galaxy NGC 1319 lies 7′ east.

16/18″ Scopes–150x: NGC 1325 is a uniformly faint oval elongated 2.5′ × 1.25′ NE-SW. A 13th magnitude star lies 1.5′ ESE. The galaxy NGC 1319, 7′ west, has a faint halo elongated 45″ × 20″ NNE–SSW.

NGC 1332 H60¹ Galaxy Type S0−:
ø5.0′ × 1.8′, m10.5v, SB 12.7   03ʰ26.3ᵐ −21°20′
Finder Chart 14-4   ★★★★

12/14″ Scopes–125x: This galaxy has a bright core containing a stellar nucleus surrounded by a considerably fainter halo elongated 2.5′ × 1′ E–W. Not difficult.

16/18″ Scopes–150x: NGC 1332 displays a very bright core and a stellar nucleus in a fairly large 3′ × 1.25′ halo elongated ESE–WNW. A faint star is just SW of the core. The companion galaxy NGC 1331 is visible as a faint stellar object lying 3′ ESE.

NGC 1337 Galaxy Type SB(s)cd II–III
ø5.0′ × 1.4′, m11.9v, SB 13.9   03ʰ28.1ᵐ −08°23′
Finder Chart 14-3, Figure 14-10   ★★★

12/14″ Scopes–125x: This galaxy is fairly faint, highly elongated 3.5′ × 0.5′ NW-SE, and slightly brighter along its major axis. A 13th magnitude star lies 2.5′ ESE. A 9th magnitude star 5′ to the WSW is at the NW end of a 1/4° long row with two more 9th magnitude stars.

16/18″ Scopes–150x: Although moderately faint, NGC 1337 is a nice glowing streak elongated 4.5′ × 0.75′ NNW–SSE, its broad mottled core spanning nearly half the length of the major axis. 175x reveals an irregular halo with bright knots and dark blotches.

NGC 1353 H246³ Galaxy Type SAB(rs)b II
ø3.5′ × 1.5′, m11.4v, SB 13.1   03ʰ32.1ᵐ −20°49′
Finder Chart 14-4   ★★★

8/10″ Scopes–100x: This galaxy is fairly faint, elongated 1.5′ × 0.75′ NW–SE, and has a prominent core. An 11.5 magnitude star lies 3′ SE.

16/18″ Scopes–150x: NGC 1353 has a fairly bright halo, elongated 2.5′ × 1′ NW-SE, within which is a large circular core with a nonstellar nucleus. The halo is much more tapered at the SE than at the NW end.

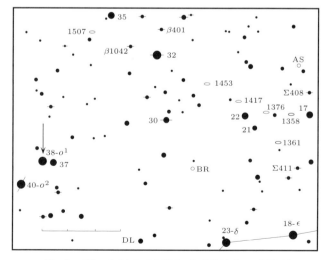

Finder Chart 14-6. 38 Eri–o¹: 04ʰ11.8ᵐ −06°50′

Eridanus 183

NGC 1357 H290² Galaxy Type (R′)SA:(r)a
ø3.6′ × 2.5′, m11.5v, SB 13.7   03ʰ33.2ᵐ −13°40′
Finder Chart 14-7, Figure 14-11   ★★★

8/10″ Scopes–100x: NGC 1357 is at the southern vertex of a triangle with an 8.5 magnitude star 3.75′ NNE and a 9.5 magnitude star 4′ WNW. It is a fairly faint oval elongated 1.5′ × 0.75′ ENE–WSW with a stellar nucleus.

16/18″ Scopes–150x: NGC 1357 is fairly bright, elongated 3′ × 2′ ENE–WSW, and has a small core with a faint stellar nucleus. A threshold star is embedded in the galaxy's halo 1.25′ west of the nucleus.

IC 1953 Galaxy Type SB(rs)cd pec II
ø2.5′ × 2.0′, m11.7v, SB 13.3   03ʰ33.7ᵐ −21°29′
Finder Chart 14-4   ★★

12/14″ Scopes–125x: IC 1953, only 6′ north of the 4.3 magnitude star Tau-2 (τ^2) Eridani, is a diffuse, round 1′ diameter object.

16/18″ Scopes–150x: Larger telescopes show IC 1953 as a 2′ circular glow surrounding a slightly brighter core elongated 30″ × 15″ NNW–SSE.

NGC 1358 H446³ Galaxy Type (R′)SB(rs:)a pec
ø2.7′ × 2.5′, m12.1v, SB 14.0   03ʰ33.7ᵐ −05°05′
Finder Chart 14-6   ★★

8/10″ Scopes–100x: This galaxy lies between two 11.5 magnitude stars, the star to the east a nearly equal 14″ double. NGC 1358 is a very faint, small, round smudge with uniform surface brightness.

12/14″ Scopes–125x: The halo is rather faint, elongated 1.5′ × 1.25′ N–S, and has a large faint core.

16/18″ Scopes–150x: In large telescopes NGC 1358 has a fairly obvious halo elongated 2.25′ × 1.75′ N–S with a bloated core. The halo looks somewhat more extended east of the core. The galaxy NGC 1355, 7′ NNW, appears highly elongated 1′ × 0.25′ E–W with a tiny bright nucleus.

NGC 1359 Galaxy Type SB(s)dm pec III:
ø2.5′ × 1.3′, m12.1v, SB 13.2   03ʰ33.8ᵐ −19°29′
Finder Chart 14-8   ★★

8/10″ Scopes–100x: NGC 1359, 5′ SSW of a 10.5 magnitude star, appears rather faint with a round 1′ diameter halo.

12/14″ Scopes–125x: NGC 1359 is a fairly faint 1.25′ diameter object with a faint stellar nucleus.

NGC 1376 H288² Galaxy Type SAB(r)c I-II
ø2.2′ × 1.7′, m12.1v, SB 13.3   03ʰ37.1ᵐ −05°03′
Finder Chart 14-6   ★

8/10″ Scopes–100x: NGC 1376, 15′ ENE of a 7th magnitude star, is a very faint, round 1′ diameter featureless blob.

16/18″ Scopes–150x: NGC 1376 is little better with increased aperture, the halo still just a faint, circular 1.25′ diameter glow only slightly brighter in its center. 12th magnitude stars are 3.75′ SSE and 4.25′ NE.

NGC 1395 Galaxy Type E2 pec
ø5.4′ × 4.6′, m9.7v, SB 13.0   03ʰ38.5ᵐ −23°02′
Finder Chart 14-8, Figure 14-12   ★★★

8/10″ Scopes–100x: This fairly bright galaxy lies 3′ north of a 9th magnitude star. The halo is elongated 1.25′ × 0.75′ E–W and has a bright circular core with a stellar nucleus.

12/14″ Scopes–125x: NGC 1395 is a bright fat oval, elongated 2.25′ × 1.75′ ESE–WNW, with a well concentrated core and a brilliant stellar nucleus. One 14th magnitude star is embedded in the halo west of the core, and another lies just 1′ north. Galaxy NGC 1401 lies 21′ NE.

NGC 1400 H593² Galaxy Type E0+ (or SA0-)
ø2.8′ × 2.5′, m11.0v, SB 13.0   03ʰ39.5ᵐ −18°41′
Finder Chart 14-8, Figure 14-13   ★★★

8/10″ Scopes–100x: NGC 1400, 16′ north of a 7.5 magnitude star, displays a conspicuous stellar nucleus surrounded by a round, faint 1′ diameter halo.

12/14″ Scopes–125x: NGC 1400 has a fairly bright, circular 1.5′ diameter halo around a slightly brighter

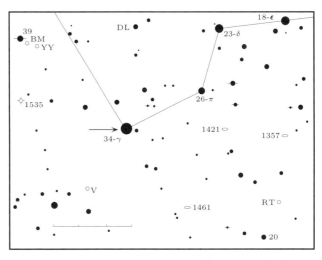

Finder Chart 14-7. 34–γ Eri: 03ʰ58.0ᵐ −13°31′

core with a prominent stellar nucleus. It is fainter and smaller than NGC 1407 lying 12′ NE.

NGC 1407 H107¹ Galaxy Type E+0
ø6.0′ × 5.8′, m9.7v, SB 13.4   03ʰ40.2ᵐ −18°35′
Finder Chart 14-8, Figure 14-13   ★★★

8/10″ Scopes–100x: This galaxy is the brightest of a group which includes NGC 1400 lying 12′ SW, the other, much fainter, members being to the NW. NGC 1407 is a brighter, slightly larger version of NGC 1400. It has a 1.25′ halo with a prominent stellar nucleus.

12/14″ Scopes–125x: NGC 1407 displays a round 2′ diameter halo growing gradually brighter toward a 30″ core containing a tiny bright nucleus.

NGC 1415 H267² Galaxy Type (R)SAB(rs)0/a: pec
ø3.2′ × 1.5′, m11.5v, SB 13.1   03ʰ41.0ᵐ −22°34′
Finder Chart 14-8   ★★

8/10″ Scopes–100x: NGC 1415, 8.5′ ESE of a 9th magnitude star, has a very faint halo elongated 1′ × 0.25′ NW–SE with tapered ends and an oval core.

12/14″ Scopes–125x: The halo appears fairly faint, is elongated 2′ × 1′ NW–SE, and has a well concentrated core with a stellar nucleus. 175x brings out some knots in the halo and some darkening in the core. An 11th magnitude star lies 2.5′ NNW, and a 13th magnitude star is 3′ SSE.

NGC 1417 Galaxy Type SB(rs)bc I-II
ø2.7′ × 1.5′, m12.1v, SB 13.5   03ʰ42.0ᵐ −04°42′
Finder Chart 14-6   ★★★

8/10″ Scopes–100x: NGC 1417, 1.5′ NW of an 11th magnitude star, is a faint, small object elongated 1.25′ × 0.75′ N–S.

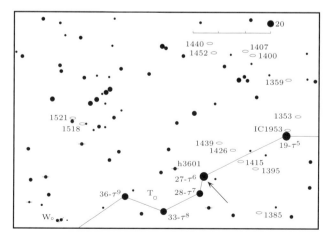

Finder Chart 14-8. 27-τ⁶ Eri: 03ʰ46.8ᵐ −23°15′

12/14″ Scopes–125x: NGC 1417 has a fairly faint halo elongated 1.25′ × 1′ N–S around a tiny, moderately concentrated core. Its companion galaxy, NGC 1418, 5′ to the ESE, is a very faint, diffuse, round 30″ diameter glow.

NGC 1421 H291² Galaxy Type Sbc: pec
ø3.1′ × 1.0′, m11.4v, SB 12.5   03ʰ42.5ᵐ −13°29′
Finder Chart 14-7, Figure 14-14   ★★★★

8/10″ Scopes–100x: This is a nice edge-on galaxy 3′ ESE of a 10.5 magnitude star. Its fairly bright halo is elongated 3′ × 0.5′ N–S and has a thin 1′ long core. Some mottling is visible in the halo. A 12th magnitude star west of the northern tip of the galaxy is at the south end of a string of stars running toward the NNW.

12/14″ Scopes–125x: NGC 1421 is a pleasing bright 3.5′ × 1′ N–S streak with a mottled, irregular surface and a highly extended core. Half a dozen knots and some indistinct dark streaks can be glimpsed along the major axis.

NGC 1426 H248³ Galaxy Type (R′)SA(rs)0−:
ø3.0′ × 2.0′, m11.2v, SB 13.0   03ʰ42.8ᵐ −22°07′
Finder Chart 14-8   ★★★

8/10″ Scopes–100x: NGC 1426, 7′ south of a 7.5 magnitude star, is fairly bright, its 1′ diameter halo slightly elongated NW–SE. At the center there appears to be an unusual double stellar nucleus.

12/14″ Scopes–125x: Larger telescopes reveal that the northern spot at the center of NGC 1426 is merely a bright knot but the southern spot is the galaxy's actual nucleus. The size of the NW-SE halo is increased slightly to 1.25′ × 1′.

NGC 1439 H249³ Galaxy Type E0:
ø2.8′ × 2.6′, m11.2v, SB 13.2   03ʰ44.8ᵐ −21°55′
Finder Chart 14-8   ★★

8/10″ Scopes–100x: This galaxy exhibits a bright stellar nucleus surrounded by a much fainter, circular 1′ diameter halo. A 9th magnitude star lies 5′ SE.

12/14″ Scopes–125x: NGC 1439 has a faint 1.25′ × 1′ halo elongated NNE–SSW with a prominent stellar nucleus. A faint star is embedded in the halo just north of the nucleus.

NGC 1440 H458² Galaxy Type SB(r:)0°
ø2.3′ × 1.8′, m11.5v, SB 12.9   03ʰ45.0ᵐ −18°16′
Finder Chart 14-8   ★★★

8/10″ Scopes–100x: NGC 1440, 5′ east of a 9th magnitude star, is a moderately faint circular object 1′

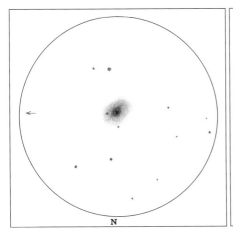

Figure 14-12. NGC 1395
13″, f4.5–165x, by Tom Polakis

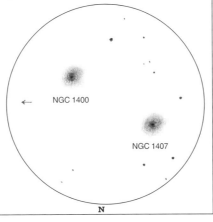

Figure 14-13. NGC 1400 & NGC 1407
13″, f4.5–165x, by Tom Polakis

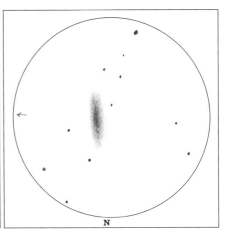

Figure 14-14. NGC 1421
13″, f5.6–165x, by Steve Coe

across with a stellar nucleus.

12/14″ Scopes–125x: The halo is rather faint, elongated 1.25′ × 1′ NNE-SSW, and has a bright nonstellar nucleus. NGC 1440 is slightly smaller than, but otherwise similar in appearance to, NGC 1452, lying 20′ south.

NGC 1452 H459² Galaxy Type SB(r)0/a
ø2.5′ × 1.7′, m12.1v, SB 13.5   03ʰ45.4ᵐ −18°38′
Finder Chart 14-8   ★★★

8/10″ Scopes–100x: This object is nearly a carbon copy of NGC 1440, 20′ north: it is moderately faint, with a 1′ diameter halo and a stellar nucleus.

12/14″ Scopes–125x: NGC 1452 has a fairly faint halo elongated 1.5′ × 1′ NE–SW with a nonstellar nucleus.

NGC 1453 Galaxy Type SAB:0−
ø2.2′ × 2.2′, m11.5v, SB 13.1   03ʰ46.4ᵐ −03°58′
Finder Chart 14-6   ★★★

8/10″ Scopes–100x: This galaxy is bright but small, its stellar nucleus embedded in a prominent core surrounded by a 1′ diameter halo. NGC 1453 is the brightest object in a galaxy group which includes NGC 1441, NGC 1449, and NGC 1451.

12/14″ Scopes–125x: NGC 1453 is bright with a circular 1.25′ diameter halo containing a bright tiny core and a stellar nucleus.

NGC 1461 H460² Galaxy Type SB:(r:)0°:
ø2.3′ × 0.8′, m11.8v, SB 12.3   03ʰ48.5ᵐ −16°24′
Finder Chart 14-7   ★★★

8/10″ Scopes–100x: NGC 1461, located 3′ SE of a 10th magnitude star, is a fairly bright object elongated 1.25′ × 0.5′ NNW–SSE with a thin, conspicuous core.

12/14″ Scopes–125x: NGC 1461 appears lens-shaped 1.5′ × 0.5′ NNW–SSE with a bright, extended core containing a stellar nucleus.

IC 2006 Galaxy Type (R)SA0−
ø2.1′ × 1.9′, m11.4v, SB 12.8   03ʰ54.1ᵐ −35°59′
Finder Chart 14-5   ★★★

12/14″ Scopes–125x: This galaxy appears diffuse and fairly faint, its round 1.25′ diameter halo containing a prominent core. A double of 9th and 10th magnitude stars 20″ apart lies 5′ SW.

16/18″ Scopes–150x: IC 2006 is a moderately faint object with a 1.5′ diameter halo, possibly elongated NE–SW, and a stellar nucleus. The halo and core are mottled and irregular in brightness.

NGC 1487 Galaxy Type Im: pec
ø2.2′ × 2.0′, m11.4v, SB 12.9   03ʰ55.8ᵐ −42°22′
Finder Chart 14-5   ★★

12/14″ Scopes–125x: NGC 1487 forms a triangle with two 11th magnitude stars 1.5′ NNE and 1.5′ WNW. This galaxy is a fairly faint 1′ diameter glow slightly brighter toward the center. Galaxies ESO 249-31A & B, embedded in the halo, are inseparable.

NGC 1507 U2947 Galaxy Type SB:(S?)dm? sp III
ø3.3′ × 1.0′, m12.3v, SB 13.5   04ʰ04.5ᵐ −02°11′
Finder Chart 14-6, Figure 14-15   ★★★

8/10″ Scopes–100x: NGC 1507 is at the northern apex of a 15′ equilateral triangle with 7th and 8th magnitude stars. It is a fairly obvious thin streak elongated 3′ × 0.5′ N–S with a slight central brightening.

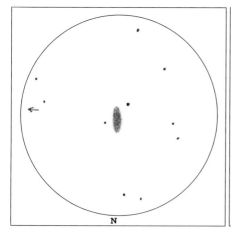

Figure 14-15. NGC 1507
12.5″, f5-175x, by Thomas Jager

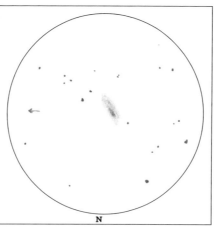

Figure 14-16. NGC 1518
12.5″, f5-125x, by Andrew D. Fraser

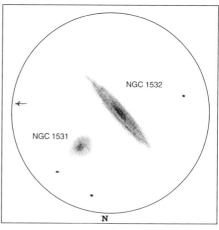

Figure 14-17. NGC 1531 & NGC 1532
20″, f4.5-175x, by Richard Jakiel

12/14″ Scopes–125x: NGC 1507 is moderately bright and highly elongated 3.5′ × 0.5′ N–S. The halo has a mottled texture, and half a dozen tiny bright spots are scattered along the major axis, but the galaxy has no distinct nucleus. A 12th magnitude star lies 2′ WNW.

NGC 1518   Galaxy   Type SB(s)m III
ø3.5′ × 1.3′, m11.7v, SB 13.2   04^{h}06.8^m −21°11′
Finder Chart 14-8, Figure 14-16   ★★★

8/10″ Scopes–100x: NGC 1518, 2′ NE of a 9th magnitude star, is a faint streak elongated 2.25′ × 0.75′ NE–SW. Galaxy NGC 1521 is 22′ ENE.

12/14″ Scopes–125x: NGC 1518 is a fairly bright even glow elongated 3.25′ × 1′ NE–SW surrounding a faint stellar nucleus. At 175x, the halo shows some brighter streaks and hints of a dust lane.

NGC 1521   Galaxy   Type SAB0°
ø3.0′ × 2.0′, m11.3v, SB 13.1   04^{h}08.3^m −21°03′
Finder Chart 14-8   ★★

8/10″ Scopes–100x: NGC 1521, 4′ north of an 8th magnitude star, is a very faint, poorly defined round smudge.

12/14″ Scopes–125x: This galaxy has a fairly faint, 1.5′ diameter halo brightening slightly to a weak core. At 175x, the halo appears somewhat elongated NE–SW. A 12th magnitude star lies 1.5′ WSW.

NGC 1531   Galaxy   Type S0° pec
ø1.3′ × 0.8′, m12.1v, SB −   04^{h}12.0^m −32°51′

NGC 1532   Galaxy   Type Sb pec sp
ø11.2′ × 3.2′, m9.9v, SB 13.6   04^{h}12.1^m −32°52′
Finder Chart 14-9, Figure 14-17   ★★★/★★★★

8/10″ Scopes–100x: This galaxy-pair is separated 1.75′ NW–SE. The brighter NGC 1532 to the SE is elongated 3.5′ × 0.5′ NE–SW, its halo suddenly brightening to a very distinct core. The much smaller and fainter NGC 1531 to the NW is elongated only 1′ × 0.75′ NW–SE; its diffuse halo has little central brightening.

12/14″ Scopes–125x: NGC 1532 is an impressive object, highly elongated 6′ × 1′ NE–SW with pointed ends and a bulging core. The halo is mottled especially toward center, and a thin dark lane extends along the major axis, passing just SE of the galaxy's core. NGC 1531 is much smaller and less elongated than its companion, its halo measuring 1.25′ × 1′. At 175x a faint stellar nucleus can be seen.

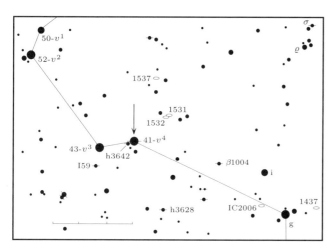

Finder Chart 14-9. 41-v^4 Eri: 04^{h}17.8^m −33°48′

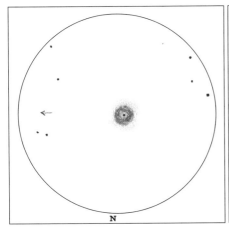
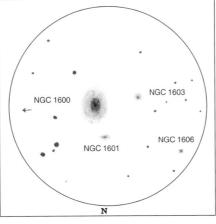
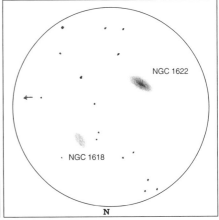

Figure 14-18. NGC 1535
17.5″, f4.5–175x, by G. R. Kepple

Figure 14-19. NGC 1600 Group
18.5″, f5–200x, by Glen W. Sanner

Figure 14-20. NGC 1618 & NGC 1622
12.5″, f5–125x, by Andrew D. Fraser

NGC 1537 Galaxy Type SAB0° pec:
⌀4.0′ × 2.6′, m10.6v, SB 13.0   04ʰ13.7ᵐ −31°39′
Finder Chart 14-9   ★★★

12/14″ Scopes–125x: This bright galaxy is easily found by aiming the viewfinder just north of the center of a 20′ × 40′ parallelogram (diagonals oriented NNW–SSE and ESE–WNW) formed by the two 7th and two 8th magnitude stars. The halo is elongated 1.25′ × 0.75′ E–W with a prominent stellar nucleus.

16/18″ Scopes–150x: NGC 1537 displays a bright non-stellar nucleus within an extended core surrounded by a fainter halo elongated 1.5′ × 1′ ESE–WNW.

NGC 1535 H26⁴ PK206-40.1 Planetary Neb. Type 4+2c
⌀>18″, m9.6p, CS 11.59v   04ʰ14.2ᵐ −12°44′
Finder Chart 14-7, Figures 14-18 & 14-21   ★★★★★

2/3″ Scopes–100x: Small scopes will reveal a faint, small, round, diffuse glow 30′ east of an equilateral triangle of 9th magnitude stars.

4/6″ Scopes–125x: This planetary nebula has a bright, small, round disk with a nice blue color.

8/10″ Scopes–100x: A nice diversion from all the galaxies in Eridanus–a real showpiece! NGC 1535 has a bright blue disk with a well concentrated central area surrounded by a misty outer ring. The central star is clearly visible.

12/14″ Scopes–125x: This fine bright, bluish planetary displays two shells surrounding a bright central star. The 20″ inner shell is bright with a dark ring immediately surrounding the central star. The outer shell is much fainter and uneven in brightness, extending to 35″.

NGC 1600 Galaxy Type SA:(s:)0−
⌀2.3′ × 1.5′, m10.9, SB 12.1   04ʰ31.7ᵐ −05°05′
Finder Chart 14-10, Figure 14-19   ★★★

8/10″ Scopes–100x: NGC 1600 is south of a 20′ × 8′ isosceles triangle of magnitude 7.5 stars. It is a fairly bright 1′ × 0.75′ N–S oval slightly brighter in the core. Its nearest companion, NGC 1601, 1.5′ to the NNE, is a very faint, small diffuse object.

12/14″ Scopes–125x: This galaxy has a bright, well concentrated halo elongated 2′ × 1.5′ N–S with a diffuse circular core. NGC 1600 is by far the brightest and largest in a group of galaxies that includes NGC 1601, NGC 1603, and NGC 1606.

Figure 14-21. *NGC 1535 is a bluish planetary nebula with two shells surrounding its bright central star. The photo above approximates the view at low power in small telescopes. Lee C. Coombs made this five minute exposure on 103a-O film with a 10″, f5 Newtonian reflector at prime focus.*

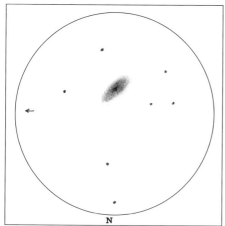

Figure 14-22. NGC 1625
8", f4.5-150x, by Dennis E. Hoverter

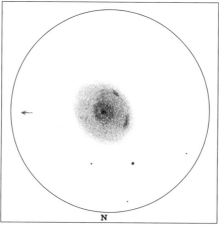

Figure 14-23. NGC 1637
20", f4.5-175x, by Richard W. Jakiel

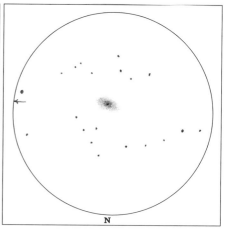

Figure 14-24. NGC 1638
12.5", f5-125x, by Andrew D. Fraser

16/18" Scopes–150x: This galaxy is a fine, bright, oval elongated 2.5' × 1.5' NNE–SSW with a well-defined core and a stellar nucleus. NGC 1601, lying 1.5' NNE, is a tiny spindle elongated 1' × 0.5' E–W. NGC 1603, 2.5' east, has a faint 30" diameter halo with a stellar nucleus. The most remote of NGC 1600's companions, NGC 1606 lying 6.5' to its ENE is a very faint, nearly stellar object.

NGC 1618 Galaxy Type SB(rs:)bc: pec II:
ø2.3' × 0.8', m12.7v, SB 13.2   04ʰ36.1ᵐ −03°09'
Finder Chart 14-10, Figure 14-20   ★★

8/10" Scopes–100x: NGC 1618, 13' north of 3.9 magnitude star Nu (ν) Eridani, is a faint, but not difficult, diffuse glow with a 1.75' × 0.75' halo elongated NNE–SSW. Galaxy NGC 1622 lies 7' ESE.

12/14" Scopes–125x: NGC 1618 has a faint, uniform halo elongated 2' × 1' NNE–SSW with a faint stellar nucleus. A threshold star lies on the north edge. An interesting parallelogram of 13th and 14th magnitude stars lies just to the east.

NGC 1622 Galaxy Type (R':)SA:(rs)ab:
ø3.5' × 0.8', m12.5v, SB 13.5   04ʰ36.6ᵐ −03°11'
Finder Chart 14-10, Figure 14-20   ★★

8/10" Scopes–100x: NGC 1622, 11' NNE of Nu (ν) Eridani, is in the same field as the slightly fainter and smaller NGC 1618 lying 7' to its WNW. NGC 1622 has a 2.5' × 0.75' halo elongated NE–SW with a large, diffuse core extending to near half the halo's width.

12/14" Scopes–125x: NGC 1622 is fairly faint, elongated 3' × 1' NNE–SSW, and has a large unevenly concentrated core with a nonstellar nucleus and a generous sprinkling of bright spots or knots.

NGC 1625 Galaxy Type (R':)SA?(r:)bc: II-III:
ø2.0' × 0.6', m12.3v, SB 12.3   04ʰ37.1ᵐ −03°18'
Finder Chart 14-10, Figure 14-22   ★★

8/10" Scopes–100x: NGC 1625, ENE of Nu (ν) = 48 Eridani, is the brightest of a small galaxy group that includes NGCs 1618 and 1622 to the NW. All three objects look about the same. NGC 1625 is fairly faint, elongated 1.75' × 0.5' NW–SE, and has a mottled, slightly brighter center.

12/14" Scopes–125x: NGC 1625 shows a 2.25' × 0.75' halo elongated NW–SE with an irregularly concentrated, extended core containing several bright spots and knots. A very faint star is at the NW tip.

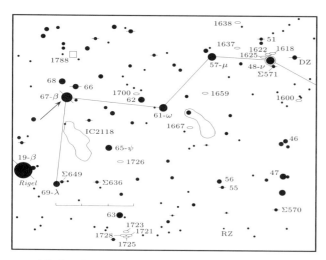

Finder Chart 14-10. 67-β Eri: 05ʰ07.8ᵐ −05°05'

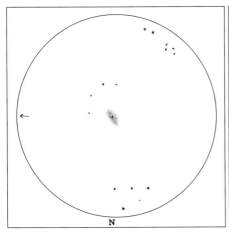

Figure 14-25. NGC 1659
12.5", f5–125x, by Andrew D. Fraser

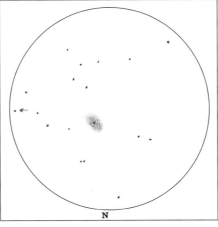

Figure 14-26. NGC 1667
12.5", f5–125x, by Andrew D. Fraser

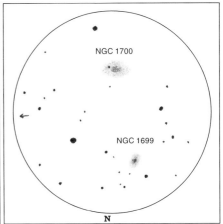

Figure 14-27. NGC 1699 & NGC 1700
17.5", f4.5–200x, by G. R. Kepple

NGC 1637 Galaxy Type SAB(rs)bc pec II
ø3.2' × 2.6', m10.8v, SB 13.0   04ʰ41.5ᵐ −02°51'
Finder Chart 14-10, Figure 14-23   ★★★

8/10" Scopes–100x: This galaxy is fairly bright, elongated 2.5' × 2' NE-SW, and has a broad, well concentrated halo surrounded by a thin, faint outer rim.

12/14" Scopes–125x: NGC 1637 is relatively bright. Its 3' × 2' NNE–SSW halo contains a circular core with a stellar nucleus. A 13th magnitude star lies 2' NE of center.

20/22" Scopes–150x: The halo is unevenly bright with hints of spiral structure in the NE and SE portions. The core is moderately large but unevenly bright and has a stellar nucleus.

NGC 1638 H525² Galaxy Type SAB(rs?)0°
ø2.3' × 1.7', m12.0v, SB 13.4   04ʰ41.6ᵐ −01°49'
Finder Chart 14-10, Figure 14-24   ★★

8/10" Scopes–100x: This galaxy is rather faint, elongated about 1' × 0.5' NE–SW, and has a very faint stellar nucleus.

12/14" Scopes–125x: NGC 1638 shows a prominent stellar nucleus embedded in a faint 1.5' × 0.75' halo elongated ENE–WSW.

NGC 1659 H589³ Galaxy Type Sbc: pec II?
ø1.6' × 0.9', m12.5v, SB 12.7   04ʰ46.5ᵐ −04°47'
Finder Chart 14-10, Figure 14-25   ★★★

8/10" Scopes–100x: NGC 1659 appears fairly faint, elongated 1' × 0.5' NE–SW, and slightly brighter in the center. The galaxy is in a rather well populated star field.

12/14" Scopes–125x: NGC 1659 has a moderately faint, 1.5' × 1' NE–SW halo with a faint stellar nucleus.

NGC 1667 Galaxy Type SB:(s?)bc I-II
ø1.8' × 1.5', m12.1v, SB 13.0   04ʰ48.6ᵐ −06°19'
Finder Chart 14-10, Figure 14-26   ★★★

8/10" Scopes–100x: NGC 1667 is a faint, round, diffuse glow about 1' in diameter.

12/14" Scopes–125x: This galaxy appears fairly bright, its 1.5' × 1' NE–SW halo slightly brightening to a weak core with a faint stellar nucleus. Galaxy NGC 1666 is 15' south.

NGC 1700 H32⁴ Galaxy Type E3/SA:0-pec
ø2.9' × 1.7', m11.2v, SB 12.8   04ʰ56.9ᵐ −04°52'
NGC 1699 Galaxy Type SAB?(s:)cd II-III
ø0.9' × 0.6', m12.9v, SB 12.1   04ʰ57.0ᵐ −04°45'
Finder Chart 14-10, Figure 14-27   ★★★/★★

8/10" Scopes–100x: This galaxy is quite obvious even though it lies only 6' SE of a 7th magnitude star. The halo is elongated 1' × 0.75' E–W and has uniform surface brightness, well defined edges, and a nearly stellar nucleus.

16/18" Scopes–150x: NGC 1700 is a bright but fairly small object elongated 2' × 1.5' E–W with a tiny bright core containing a stellar nucleus. A faint star is 30" WSW of the nucleus, and an 11th magnitude star lies 2.5' SW. The companion galaxy, NGC 1699, is 6.5' north of NGC 1700 and 4' ENE of the 7th magnitude field star. It is a faint object elongated 45" × 30" NNW–SSE with a faint stellar nucleus.

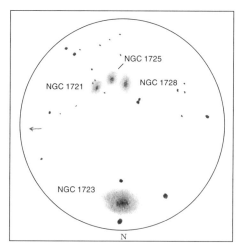

Figure 14-28. NGC 1723 Group
17.5", f4.5-200x, by G. R. Kepple

NGC 1721 Galaxy Type (R′)SAB(s)0° pec
⌀2.5′ × 1.4′, m12.3v, SB 13.5   04ʰ59.2ᵐ −11°08′

NGC 1725 Galaxy Type (R′)SA(r)0-
⌀1.9′ × 1.2′, m12.3v, SB 13.0   04ʰ59.3ᵐ −11°08′

NGC 1728 Galaxy Type Sa? pec sp
⌀1.8′ × 0.9′, m12.9v, SB 13.3   04ʰ59.2ᵐ −11°08′
Finder Chart 14-10, Figure 14-28   ★★/★★/★★

16/18" Scopes-100x: These three faint galaxies lie along an E-W arc, open to the north, 8′ due south of the bright NGC 1723. The westernmost galaxy, NGC 1721, situated between two 12th magnitude stars, is a very faint, diffuse, 1.5′ × 1′ NW-SE oval smudge. The central galaxy, NGC 1725, probably the easiest of the three, appears faint, elongated 1.5′ × 1′ NNE-SSW, but with a well concentrated core along its major axis. The easternmost galaxy, NGC 1728, is very faint, circular or slightly elongated 1.25′ × 1′ N-S, and has a very faint, tiny, core.

NGC 1723 Galaxy Type SB(r)a pec
⌀3.2′ × 2.2′, m11.7v, SB 13.7   04ʰ59.4ᵐ −10°59′
Finder Chart 14-10, Figure 14-28   ★★

16/18" Scopes-100x: This galaxy lies between a 9th magnitude star to its north and a 10th magnitude star to its south. A third star to the east completes a triangle with the other two. NGC 1723's halo is elongated 2.5′ × 1.75′ NE-SW and has diffuse edges and a much brighter center. The E-W trio of galaxies NGC 1721, 1725, and 1728 are 8′ south.

NGC 1726 Galaxy Type E4/S0−
⌀2.0′ × 1.5′, m11.7v, SB 12.7   04ʰ59.7ᵐ −07°45′
Finder Chart 14-10   ★★

12/14" Scopes-100x: This is a rather faint, diffuse object less than 1′ across slightly elongated N-S with a stellar nucleus at its center. A 12th magnitude star is just south of center, and two 9th magnitude stars are 5′ away, one to the SE and the other to the WSW. Galaxy NGC 1720 lies 8′ SW.

IC 2118 SNR/Reflection Nebula
⌀180′ × 60′, Photo Br 3-5   05ʰ06.9ᵐ −07°13′
Finder Chart 14-10   Witch Head Nebula   ★★★

10x50 Binoculars: This nebula is a very dim ribbon covering a 50′ × 10′ area. It shines by Rigel's reflected light.

12/14" Scopes-50x: IC 2118 is a very faint streamer with a uniform glow spanning a 30′ × 10′ NW-SE area. The brightest part is centered 70′ ENE of 4.8 magnitude star Psi (ψ) = 65 Eridani.

Chapter 15

Fornax, the Furnace

15.1 Overview

In 1752, Nicolas Louis de Lacaille introduced Fornax under the name Fornax Chemica, the Chemical Furnace. He dedicated his new constellation to the famous French chemist Antoine Laurant Lavoisier (A.D. 1743–1794), the father of modern chemistry. Fornax is little more than a small inconspicuous rectangle of sky: it is marked by no prominent star-pattern, and even its Alpha star is only a magnitude 3.9 object. However, it is rich in faint galaxies, most of them members of the 100 million light year distant Fornax I Galaxy Cluster. The constellation also contains a member of the Local Galaxy Group, the extremely low surface brightness Fornax Dwarf Galaxy.

Fornax: FOR-naks
Genitive: Fornacis, FOR-na-cis
Abbrevation: For
Culmination: 9pm–Dec. 17, midnight–Nov. 2
Area: 398 square degrees
Showpieces: NGC 1097, NGC 1316, NGC 1360 NGC 1365, NGC 1380, NGC 1406, NGC 1425
Binocular Objects: NGC 1097, NGC 1316, NGC 1360, NGC 1365, NGC 1380, NGC 1398

15.2 Interesting Stars

Omega (ω) Fornacis = h3506   Double Star   Spec. B9
m5.0, 7.7; Sep. 10.8″; P.A. 244° $02^h33.8^m$ $-28°14'$
Finder Chart 15-4   ★★★★

8/10″ Scopes–75x: Low power shows a white primary with a blue companion. A faint unequal double lies 2′ west.

Alpha (α) Fornacis = h3555   Double Star   Spec. F8
m4.0, 7.0; Sep. 5.1″; P.A. 299° $03^h12.1^m$ $-28°59'$
Finder Chart 15-5   ★★★★

8/10″ Scopes–125x: Both stars of this 314 year binary appear yellowish. The star's separation will widen to 5.8″ by the year 2020.

15.3 Deep-Sky Objects

NGC 922 H239³ Galaxy Type SB(s)m: pec III
ø2.2′ × 1.4′, m12.0v, SB 13.1 $02^h25.1^m$ $-24°47'$
Finder Chart 15-4   ★★

8/10″ Scopes–100x: NGC 922, 2′ SSE of a 12th magnitude star, has a fairly faint, round 1′ diameter halo slightly brighter at its center.

16/18″ Scopes–125x: Larger instruments show a prominent core with a stellar nucleus embedded in a 1.5′ halo slightly elongated N–S.

NGC 986 Galaxy Type (R′)SB(rs)b II
ø3.8′ × 1.9′, m10.9v, SB 12.9 $02^h33.6^m$ $-39°02'$
Finder Chart 15-3, Figure 15-1   ★★★

8/10″ Scopes–100x: NGC 986, 8′ north of a 9th magnitude star, is faint but obvious. Its 1.5′ × 0.75′ NE–SW halo fades evenly to the outer edges and has a prominent core.

16/18″ Scopes–125x: NGC 986 has a fairly bright halo elongated 2.5′ × 1.25′ NE–SW with a tiny core and a faint stellar nucleus.

NGC 1049 Globular Cluster
ø0.4′, m12.9v $02^h39.7^m$ $-34°17'$

ESO 356-G4 Fornax Dwarf Galaxy
ø12.0′ × 10.2′, m8.12v, SB 13.2 $02^h39.9^m$ $-34°32'$
Finder Chart 15-3, Figures 15-2 & 15-4   ★★/★

12/14″ Scopes–200x: NGC 1049 is the brightest of five globular clusters visible within the halo of the Fornax Dwarf Galaxy, a member of the Local Galaxy Group. Although because of its extremely low surface brightness the galaxy itself cannot be seen, its globular clusters, being highly concen-

Fornax, the Furnace

Constellation Chart 15-1

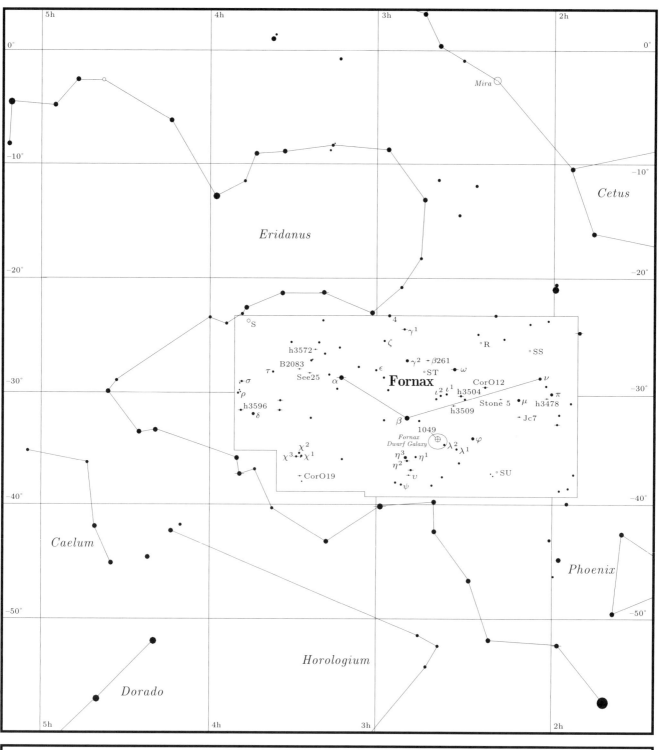

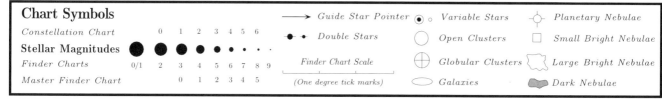

Table 15.1 Selected Variable Stars in Fornax

Name	HD No.	Type	Max.	Min.	Period (Days)	F*	Spec. Type	R.A. (2000)	Dec.	Finder Chart No. & Notes
SS For		RRab	9.45	10.60	0.49	0.14	A3-G0	$02^h07.9^m$	$-26°52'$	15-4
SU For	14729	EA	9.5	11.0	2.43	0.12	A2	21.6	$-37\ 13$	15-3
R For		M	7.5	13.0	387	0.52	Ce	29.3	$-26\ 06$	15-4
ST For	17166	SRa	8.7	10.2	277		Mb(e)	44.4	$-29\ 12$	15-4
S For	23686	cst?	5.6	8.5			F8	$03^h46.2^m$	$-24\ 24$	15-5

F* = The fraction of period taken up by the star's rise from min. to max. brightness, or the period spent in eclipse.

Table 15.2 Selected Double Stars in Fornax

Name	ADS No.	Pair	M1	M2	Sep."	P.A.°	Spec	R.A. (2000)	Dec	Finder Chart No. & Notes
h3478			8.0	8.8	42.0	146	G5 G5	$02^h02.7^m$	$-30°20'$	15-4
Jc 7		AB	8.0	10.2	6.7	280	G0	12.1	$-32\ 17$	15-4
		AC		8.7	149.9	180	F0			
Stone 5			8.1	8.8	2.7	201	F5	17.5	$-30\ 43$	15-4
CorO 12	1816		7.9	10.3	9.9	79	G0	22.0	$-29\ 21$	15-4
h3504			8.2	9.0	6.7	269	F8	30.3	$-30\ 21$	15-4
ω For	1954		5.0	7.7	10.8	244	B9	33.8	$-28\ 14$	15-4 White & blue
h3509			7.5	11.5	23.5	59	F0	34.2	$-31\ 31$	15-4
β261	2092	AB	7.9	9.4	3.0	100	G5	43.8	$-27\ 54$	15-4
		AC		9.8	70.2	131				
υ For			7.0	8.2	5.4	145	F2	48.6	$-37\ 24$	15-3
η² For			5.9	10.1	5.0	14	K0	50.2	$-35\ 51$	15-3
α For	2402		4.0	7.0	w5.1	*299	F8	$03^h12.1^m$	$-28\ 59$	15-5 Both stars yellowish
h3572			8.6	8.8	20.7	95	F5 F5	24.0	$-26\ 13$	15-5
See 25	2543		7.7	11.7	10.2	18	K0	26.7	$-28\ 34$	15-5
CorO 19			8.0	9.7	12.9	310	F5	27.1	$-37\ 53$	15-1
χ³ For			6.5	10.5	6.3	248	A0	28.2	$-35\ 51$	15-1
B2083			7.8	14.0	8.0	nf	F5	30.7	$-27\ 55$	15-5
h3596			8.2	8.5	9.2	137	A3	48.5	$-31\ 47$	15-6

Footnotes: *= Year 2000, a = Near apogee, c = Closing, w = Widening. Finder Chart No: All stars listed in the tables are plotted in the large Constellation Chart, but when a star appears in a Finder Chart, this number is listed. Notes: When colors are subtle, the suffix *-ish* is used, e.g. *bluish*.

---

trated objects, are generally within the grasp of 12" and larger instruments. NGC 1049 appears as a fairly faint, tiny, round, well concentrated object with a stellar nucleus.

16/18″ Scopes–200x: NGC 1049 is moderately bright and small, its faint halo containing a tiny bright core. NGC 1049 is the brightest of five globular clusters in the Fornax Dwarf Galaxy. Globular E356-SC08, 30′ NE of NGC 1049, is stellar in appearance. E356-SC05, 15′ SSE of NGC 1049, is smaller and fainter.

NGC 1079 Galaxy Type (R)SAB(rs)a
ø3.2′ × 2.1′, m11.3v, SB 13.2 $02^h43.7^m$ $-29°00'$

Finder Chart 15-4   ★★★

12/14″ Scopes–125x: NGC 1079 is a fairly bright but small galaxy located 12′ south of a 9th magnitude and 3.5′ NE of a 10.5 magnitude star. Its bright circular core has a stellar nucleus and is surrounded by a much fainter 2.5′ × 0.5′ halo elongated E–W.

16/18″ Scopes–150x: The halo is elongated 2′ × 1′ E–W, its west extension slightly brighter and a little shorter but broader than its east extension. The core and nucleus are quite bright.

NGC 1097A Galaxy Type E5 Pec:
ø0.9′ × 0.5′, m13.6v, SB 12.6 $02^h46.2^m$ $-30°14'$

NGC 1097 H48⁵ Galaxy Type SB(rs)bc I-II
ø10.5′ × 6.3′, m9.2v, SB13.6 $02^h46.3^m$ $-30°16'$

Finder Chart 15-4, Figure 15-3   ★/★★★★

12/14″ Scopes–125x: In medium-size telescopes no spiral structure, only the galaxy's bright 4.5′ × 1.5′ NW–SE bar, can be seen. At the center of the bar is a small oval core. NGC 1097A lies 4′ north.

16/18″ Scopes–150x: At the center of the long bright 5′ × 2′ bar is a large bright 50″ × 30″ oval core with a stellar nucleus. From the northern tip of the bar a

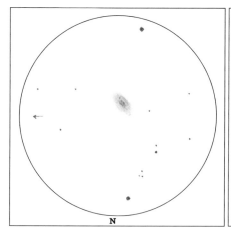
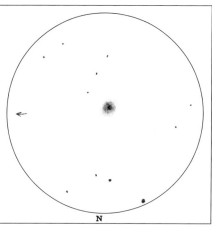
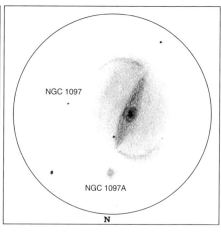

Figure 15-1. NGC 986
13", f4.5–166x, by Tom Polakis

Figure 15-2. NGC 1049
13", 4.5–166x, by Tom Polakis

Figure 15-3. NGC 1097 & NGC 1097A
20" f4.5–175x, by Richard W. Jakiel

very faint spiral arm sweeps east and then curls slightly back toward the core. A very faint, short arm fans west from the southern tip of the bar. SW of the core another small detached narrow glow is just visible. The companion galaxy, NGC 1097A, is a 30" diameter glow with a stellar nucleus about as bright as the bar of the big spiral.

NGC 1201 H109[1] Galaxy Type SAB–
ø3.5' × 1.9', m10.8v, SB 12.7   03h04.1m −26°04'
Finder Chart 15-5   ★★★

8/10" Scopes–100x: NGC 1201 is on the western edge of the 8' × 3.5' triangle formed by one magnitude 10.5 and two magnitude 11.5 stars, the magnitude 10.5 star at the triangle's NE vertex. The galaxy is moderately bright, elongated 2' × 1' N–S, and slightly

brighter in its core.

16/18" Scopes–150x: This galaxy displays a tiny bright core with a stellar nucleus embedded in a diffuse halo elongated 2.5' × 1.25' in P.A. 8°.

NGC 1255 A60 Galaxy Type SAB(s)bc II
ø4.0' × 2.5', m11.0v, SB 13.3   03h13.5m −25°44'
Finder Chart 15-5, Figure 15-6   ★★★

12/14" Scopes–125x: NGC 1255 is a faint, diffuse object with a 3' × 2' halo elongated ESE–WNW. It is situated among several 9th magnitude stars just 2' NE of a 12th magnitude star.

16/18" Scopes–150x: The halo is fairly faint, elongated 3.5' × 2.5' ESE–WNW, and has a granular texture of uneven brightness. Its edges are diffuse and irregular, and it slightly brightens toward the center.

NGC 1288 Galaxy Type SAB(rs)c IC
ø2.6' × 1.8', m12.0v, SB 13.5   03h17.2m −32°35'
Finder Chart 15-6   ★★

16/18" Scopes–150x: NGC 1288, a very faint galaxy, has a 1.5' circular halo that slightly brightens to a weakly concentrated nonstellar nucleus.

NGC 1292 Galaxy Type SA(s)bc? II-III
ø2.8' × 1.2', m11.9v, SB 13.1   03h18.2m −27°37'
Finder Chart 15-5   ★★★

12/14" Scopes–125x: NGC 1292 is 3' SW of a pair of 11th magnitude stars separated by 24". Its rather faint halo is elongated 1' × 0.5' N–S and has a faint stellar nucleus. A magnitude 12.5 star is 2.5' north.

16/18" Scopes–150x: NGC 1292 appears fairly faint, its 1.5' × 0.75' N–S halo slightly but broadly brightening toward the center and a faint stellar nucleus.

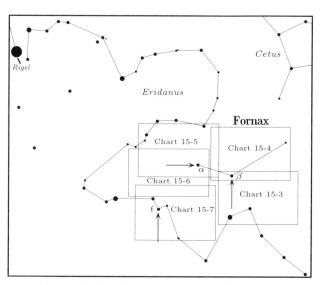

Master Finder Chart 15-2. Fornax Chart Areas
Guide stars indicated by arrows.

NGC 1302 Galaxy
Type (R)SB(r)0/a
ø4.3′ × 3.7′, m10.4v, SB 13.3
03h19.9m −26°04′
Finder Chart 15-5 ★★★

12/14″ Scopes-125x: NGC 1302 is very faint, but its round 1.5′ diameter halo suddenly brightens at its center to a stellar nucleus. An 11th magnitude star lies 2′ NNW.

16/18″ Scopes-150x: This galaxy has a very bright stellar nucleus embedded in a small, diffuse core that is surrounded by a much fainter 2.5′ halo slightly elongated NNW–SSE.

NGC 1316 Galaxy
Type (R′)SAB(s)0°: pec
ø13.5′ × 9.3′, m8.2v, SB 13.3
03h22.7m −37°12′
Finder Chart 15-7
Figures 15-5 & 15-7 ★★★★

8/10″ Scopes-100x: NGC 1316 is the brightest member of the Fornax I Galaxy cluster. It is also the strong radio source Fornax A. The galaxy is on an E–W line with 7th and 8th magnitude stars respectively 10′ and 20′ to its east. NGC 1316 has an 2.5′ × 1.5′ halo elongated NE–SW that moderately brightens to a 1′ core.

16/18″ Scopes-150x: In larger telescopes NGC 1316 has a distinct oval halo elongated 3.5′ × 2.5′ NE–SW around a dense core with a bright nonstellar nucleus. Galaxy NGC 1317 is only 6′ north.

NGC 1317 (NGC 1318) Galaxy Type (R′)SAB(rs)0/a
ø3.5′ × 3.0′, m10.8v, SB 13.2   03h22.8m −37°06′
Finder Chart 15-7, Figures 15-5 & 15-7 ★★★

8/10″ Scopes-100x: NGC 1317, 6′ north of NGC 1316 in the Fornax I Galaxy Cluster, is fairly bright with a circular 1′ diameter halo and a bright core.

16/18″ Scopes-150x: NGC 1317 is a bright, circular object 1.5′ across with a bright core and stellar nucleus.

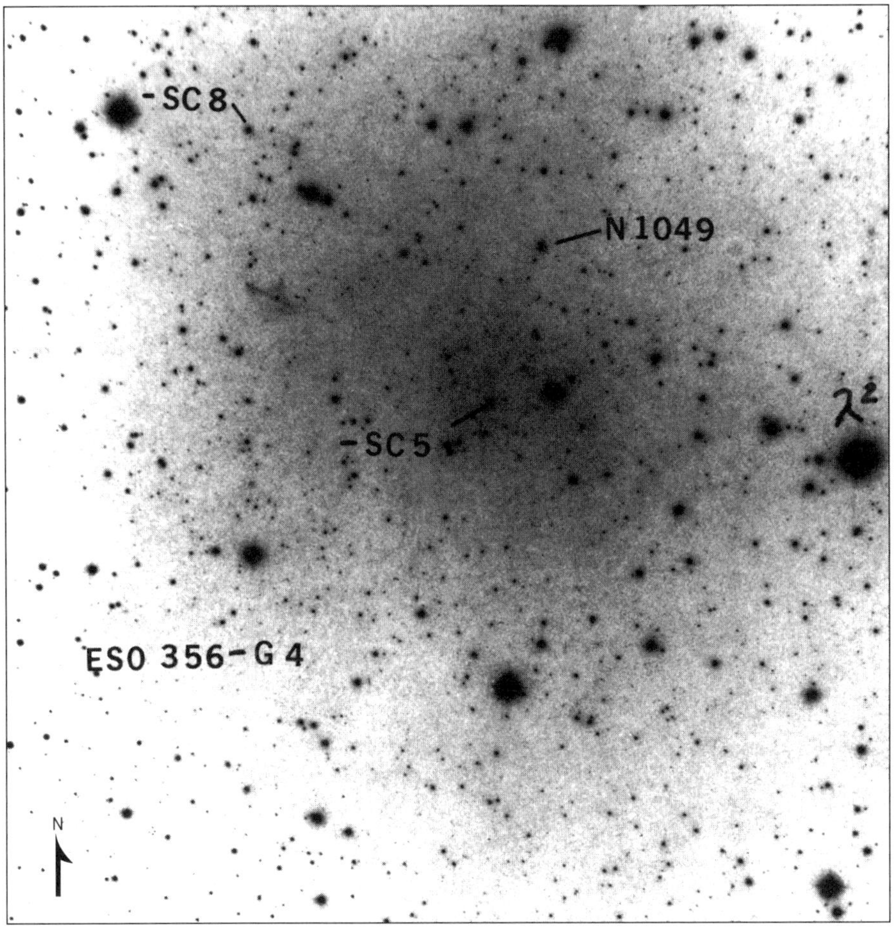

Figure 15-4. *The Fornax Dwarf Galaxy (ESO 356-G4) has exceptionally low surface brightness and is visible only as a slight brightening in the star field. Three of its globular clusters, NGC 1049, E356-SC08 (SC-8) and E356-SC05 (SC-5), appear on this 60 minute exposure on hypered 2415 film taken with an 8″, f4 Wright-Newtonian telescope by William Harris.*

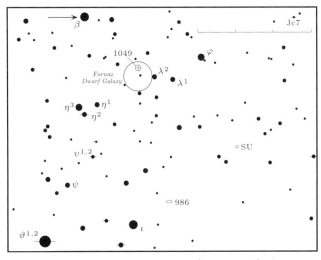

Finder Chart 15-3. β For: 02h49.1m −32°24′

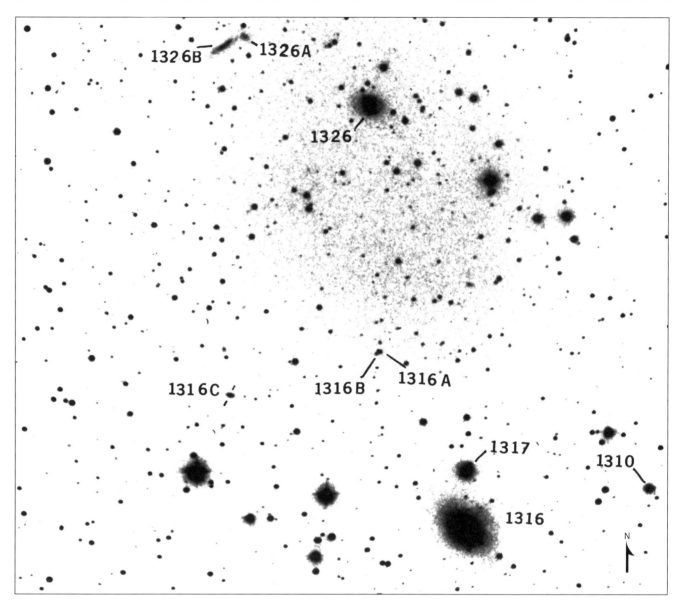

Figure 15-5. *William Harris recorded the NGC 1316 Galaxy Group within the Fornax I Galaxy Cluster for 60 minutes on hypered 2415 Kodak Tech Pan film with an 8", f4 Wright Newtonian telescope at prime focus. North is up.*

NGC 1326 Galaxy Type (R)SB(r)0+
ø4.4′ × 3.2′, m10.4v, SB 13.1   03ʰ23.9ᵐ −36°28′
Finder Chart 15-7, Figure 15-5   ★★★

16/18″ Scopes–150x: NGC 1326, a member of the Fornax I Galaxy Cluster, is 4′ SSE of an 11th magnitude star which is part of a NE–SW star-chain. NGC 1326 has a bright, circular 1.5′ halo slightly elongated ENE–WSW and a bright stellar nucleus.

NGC 1341 Galaxy Type SAB(s)b: II-III
ø1.8′ × 1.3′, m12.1v, SB 12.9   03ʰ28.0ᵐ −37°09′
Finder Chart 15-7   ★★

16/18″ Scopes–150x: NGC 1341, a member of the Fornax I Galaxy Cluster, is a very faint, circular object slightly elongated NW–SE with uniform illumination. A 13th magnitude star lies on its SE edge.

NGC 1339 Galaxy Type E3 pec
ø2.0′ × 1.4′, m11.6v, SB 12.6   03ʰ28.1ᵐ −32°17′
Finder Chart 15-6   ★★

12/14″ Scopes–125x: NGC 1339 is 6′ SE of a 25″ wide pair of 11th and 13th magnitude stars aligned in P.A. 35°. Its very faint, round, compact halo is less than 1′ across and has a stellar nucleus.

16/18″ Scopes–150x: NGC 1339 has a stellar nucleus embedded in a faint, circular 1.25′ halo with diffuse edges.

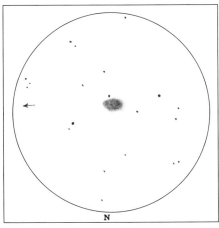
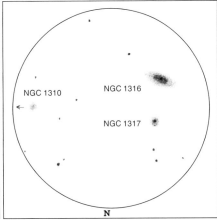
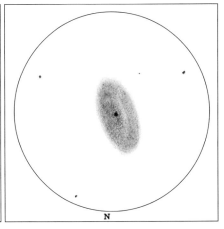

Figure 15-6. NGC 1255
13", f4.5-115x, by Tom Polakis

Figure 15-7. NGC 1310, 1316, & 1317
13", f4.5-166x, by Tom Polakis

Figure 15-8. NGC 1360
20", f4.5-175x, by Richard W. Jakiel

NGC 1344 (=1340) H257¹ Galaxy
Type (R')SAB(rs)0°
ø4.7' × 3.0', m10.2v, SB 12.9   03ʰ28.3ᵐ −31°04'
Finder Chart 15-6   ★★★

16/18" Scopes–150x: NGC 1344 is at the hypotenuse of the right triangle it forms with a 9.5 magnitude star 6' east and a 10.5 magnitude star 5.5' north. It is a fairly bright galaxy elongated 2.5' × 1.5' NNW–SSE, its well concentrated halo growing smoothly brighter to a stellar nucleus.

NGC 1351 Galaxy Type SA0–
ø3.0' × 2.0', m11.3v, SB 13.1   03ʰ30.5ᵐ −34°52'
Finder Chart 15-6   ★★★

12/14" Scopes–125x: This member of the Fornax I Galaxy Cluster is a fairly bright 1' × 0.5' NW–SE oval with a bright stellar nucleus.

16/18" Scopes–150x: NGC 1351 is moderately bright, elongated 1.25' × 0.75' NW–SE, and has, except for its prominent stellar nucleus, uniform surface brightness.

NGC 1350 Galaxy Type (R')SAB(r)ab
ø6.2' × 3.2', m10.3v, SB 13.4   03ʰ31.1ᵐ −33°38'
Finder Chart 15-6   ★★★★

16/18" Scopes–150x: NGC 1350, a Fornax I Galaxy, is 6' SW of a magnitude 6.5 star. Its small bright core contains a stellar nucleus and is within a 3.5' × 1.75' halo elongated NNE–SSW.

NGC 1360 PK220-53.1 Planetary Nebula Type 3
ø390", m9.4v, CS 10.98v   03ʰ33.3ᵐ −25°51'
Finder Chart 15-5, Figure 15-8   ★★★★★

8/10" Scopes–100x: NGC 1360, 9' NW of an 11th magnitude star, is a large, faint, circular glow with a prominent central star. It appears much brighter with a UHC filter.

12/14" Scopes–125x: NGC 1360 displays a fine oval-shaped disk elongated 7' × 4' NNE–SSW with an 11.5 magnitude star at its center. A 13th magnitude star lies on its north edge.

16/18" Scopes–150x: Very impressive–especially when viewed through an O-III filter! The disk is a very bright 7' × 5' NNE–SSW oval, and the central star is conspicuous.

20/22" Scopes–200x: Superb! The planetary's 9' × 5' oval disk is well-defined, and slightly brighter toward the magnitude 11.5 central star. A somewhat darker area can be seen in the southern portion of the disk.

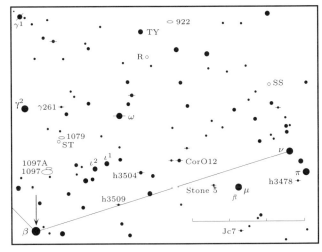

Finder Chart 15-4. β For: 02ʰ49.1ᵐ −32°24'

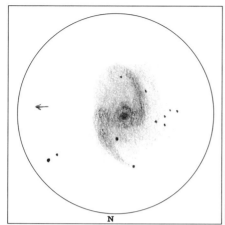
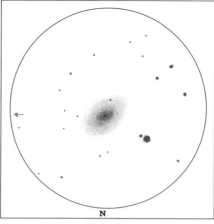
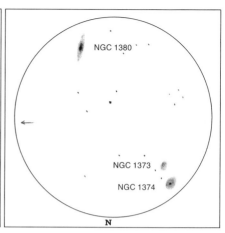

Figure 15-9. NGC 1365
17.5", 4.5–200x, by Steve Pattie

Figure 15-10. NGC 1371
20", f4.5–175x, by Richard W. Jakiel

Figure 15-11. NGC 1373, 1374, & 1380
20", f4.5–175x, by Richard W. Jakiel

NGC 1365 Galaxy Type (R')SB(s)b IC
⌀8.9' × 6.5', m9.3v, SB 13.6   03ʰ33.6ᵐ −36°08'
Finder Chart 15-7, Figures 15-9 & 15-15   ★★★★

8/10" Scopes–100x: NGC 1365 lies in the foreground of the Fornax I Galaxy Cluster and is therefore not a true member. It is faint but obvious, its prominent core surrounded by a diffuse, circular 3' diameter halo.

12/14" Scopes–125x: NGC 1365 is a fairly faint barred spiral galaxy with a diffuse 3.5' halo. The bar, projecting east and west from the galaxy's bright circular core, can be just glimpsed with averted vision.

16/18" Scopes–150x: This fairly bright galaxy has a well-defined 4' long E–W bar with a bright core at its center. From the ends of the bar spiral arms sweep north and south, gradually tapering to points. A magnitude 13.5 star is at the point of the northern arm 1.5' NW of the galaxy's center.

20/22" Scopes–175x: Impressive! This bright, large barred spiral is one of the finest of its class. The core is a bright oval about 50" × 40" with a sharp stellar nucleus. The wispy spiral arms curve north and south from the ends of the E–W bar and form a rounded Z-shaped halo. Traces of dust lanes can be glimpsed along the major axis.

NGC 1366 H857³ Galaxy Type S0–
⌀1.9' × 0.9', m11.9v, SB 12.3   03ʰ33.9ᵐ −31°12'
Finder Chart 15-6   ★★

12/14" Scopes–125x: NGC 1366, 6.5' south of a 6th magnitude star, displays a bright stellar nucleus embedded in a small 1' × 0.5' N–S halo.

16/18" Scopes–150x: The halo extends to 1.5' × 0.75' N–S but is very faint in contrast to its brilliant stellar nucleus.

NGC 1371 H262² Galaxy Type (R')SB(r)a
⌀4.9' × 3.4', m10.6v, SB 13.5   03ʰ35.0ᵐ −24°56'
Finder Chart 15-5, Figure 15-10   ★★★

8/10" Scopes–100x: NGC 1371, 4' SW of an 8th magnitude star, has a fairly faint, small, diffuse halo surrounding a bright core.

12/14" Scopes–125x: NGC 1371 has a moderately faint, oval halo elongated 2' × 1' NW–SE with an extended core containing a faint stellar nucleus.

16/18" Scopes–150x: Larger telescopes show NGC 1371 as a fairly bright, kidney bean-shaped object about 3.5' × 1.75' NW–SE with a bright oblong core and a pinpoint nucleus.

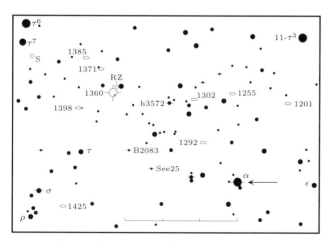

Finder Chart 15-5. α For: 03ʰ12.1ᵐ −28°59'

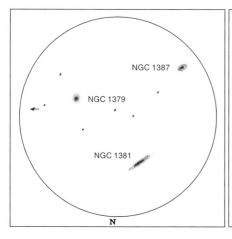
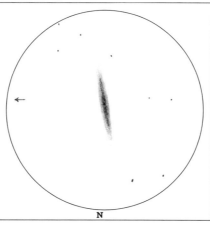
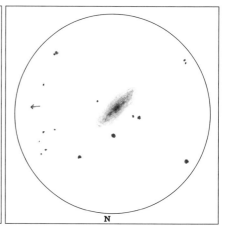

Figure 15-12. NGC 1379, 1381, & 1387
13", f4.5–166x, by Tom Polakis

Figure 15-13. NGC 1406
20", f4.5–175x, by Richard W. Jakiel

Figure 15-14. NGC 1425
20", f4.5–175x, by Richard W. Jakiel

NGC 1373 Galaxy Type S0–:
ø1.1′ × 0.9′, m13.0v, SB 12.8 $03^h35.1^m$ $-35°11′$
Finder Chart 15-7, Figures 15-11 & 15-15   ★

16/18" Scopes–150x: NGC 1373, 2′ east of a 10th magnitude star and 4.5′ NW of NGC 1374, is an extremely faint, small, circular, diffuse glow.

NGC 1375 Galaxy Type SAB0–:
ø2.2′ × 1.0′, m12.0v, SB 12.7 $03^h35.2^m$ $-35°16′$
Finder Chart 15-7, Figure 15-5   ★★

12/14" Scopes–125x: NGC 1375, just south of NGC 1374 in the Fornax I Galaxy Cluster, is much fainter and more elongated than its companion. Its 1′ × 0.25′ E–W halo is moderately brighter at center.

NGC 1374 Galaxy Type SA(rs)0–:
ø2.6′ × 2.4′, m11.0v, SB 12.8 $03^h35.3^m$ $-35°14′$
Finder Chart 15-7, Figures 15-11 & 15-15   ★★★

12/14" Scopes–125x: NGC 1374, 2.5′ north of NGC 1375 in the Fornax I Galaxy Cluster, is noticeably brighter than its companion. Its tiny circular halo, less than 1′ in diameter, has a stellar nucleus.

NGC 1379 Galaxy Type E0:
ø2.6′ × 2.5′, m11.0v, SB 12.9 $03^h36.1^m$ $-35°27′$
Finder Chart 15-7, Figures 15-12 & 15-15   ★★★

12/14" Scopes–125x: NGC 1379 is another member of the Fornax I Galaxy Cluster. Its tiny halo is only 30″ in diameter but is fairly bright and gradually brightens toward its center.

16/18" Scopes–150x: NGC 1379 has a bright, circular 1′ diameter halo with a broad diffuse core and a faint stellar nucleus. The field is well sprinkled with faint stars.

NGC 1380 Galaxy Type SB(s)0–
ø4.8′ × 2.8′, m10.0v, SB 12.7 $03^h36.5^m$ $-34°59′$
Finder Chart 15-7, Figures 15-11 & 15-15   ★★★★

12/14" Scopes–125x: NGC 1380 belongs to the Fornax I Galaxy Cluster. It has a bright 2′ × 1′ halo elongated N–S with a bright, circular core and a tiny nucleus.

16/18" Scopes–150x: NGC 1380 shows a prominent core surrounded by a much fainter 2.5′ × 1.25′ halo. A 14th magnitude star lies 1′ SW of center.

20/22" Scopes–175x: At this aperture, the halo is bright, highly elongated 4′ × 1.5′ N–S, and its interior is mottled and edges diffuse. The core is bright and slightly oval, its nucleus distinctly non-stellar. Some bright streaks are visible along the major axis.

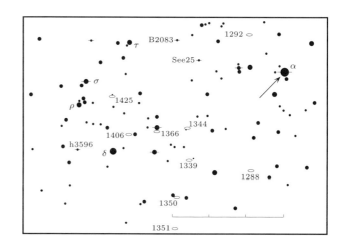

Finder Chart 15-6. α For: $03^h12.1^m$ $-28°59′$

NGC 1381 Galaxy Type S0°: sp
⌀2.6' × 1.0', m11.5v, SB-12.4   03ʰ36.6ᵐ −35°18'
Finder Chart 15-7, Figures 15-12 & 15-15   ★★★

12/14" Scopes–125x: NGC 1381 is a member of the Fornax I Galaxy Cluster. Its highly extended, well concentrated core is embedded in a fainter envelope elongated 1.5' × 0.5' NW–SE.

16/18" Scopes–150x: NGC 1381 is a fairly bright edge-on galaxy elongated 2.5' × 0.75' NW–SE with a slender, mottled core and a bright stellar nucleus. One very faint star lies 2' ESE and another 4' NW.

NGC 1386 Galaxy Type S0°
⌀3.2' × 1.2', m11.2v, SB 12.5   03ʰ36.9ᵐ −36°00'
Finder Chart 15-7, Figure 15-15 (In Eridanus) ★★★

12/14" Scopes–125x: Although located in Eridanus, NGC 1386 is a member of the Fornax I Galaxy Cluster. It is fairly bright, elongated 2' × 0.5' NNE–SSW, and has a much brighter core with a stellar nucleus.

16/18" Scopes–150x: This galaxy is moderately bright and elongated 2.5' × 1' NNE–SSW, its halo gradually brightening to a small mottled core with a stellar nucleus. A 9th magnitude star is 5' south and two faint stars about 2' NW.

NGC 1387 Galaxy Type SAB(s)0–
⌀3.1' × 2.8', m10.8v, SB 13.0   03ʰ37.0ᵐ −35°31'
Finder Chart 15-7, Figures 15-12 & 15-15   ★★★

12/14" Scopes–125x: NGC 1387, in the Fornax I Galaxy Cluster, has a fairly bright, circular 1' diameter halo gradually brightening to a faint stellar nucleus.

16/18" Scopes–150x: NGC 1387 is a fairly bright

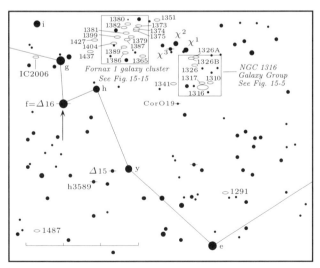

Finder Chart 15-7. f Eri: 03ʰ48.5ᵐ −37°37'

1' × 0.75' ESE–WNW oval smoothly brightening toward a small, well concentrated oval nucleus.

NGC 1382 Galaxy Type SAB(s)0–
⌀1.3' × 1.2', m12.8v, SB 13.1   03ʰ37.1ᵐ −35°12'
Finder Chart 15-7, Figure 15-15   ★

12/14" Scopes–125x: NGC 1382, a minor member of the Fornax I Galaxy Cluster, is only a faint, tiny, diffuse spot. It lies 18' west and slightly north of an 8th magnitude star.

NGC 1389 Galaxy Type SAB(s)0–
⌀2.5' × 1.5', m11.4v, SB 12.7   03ʰ37.2ᵐ −35°45'
Finder Chart 15-7, Figure 15-15 (In Eridanus) ★★★

12/14" Scopes–125x: NGC 1389 is another member of the Fornax I Galaxy Cluster located just across the border in Eridanus. This galaxy is fairly bright but small, only 45" across, with a slight NNE–SSW elongation.

16/18" Scopes–150x: This galaxy has a moderately bright halo less than 1' across slightly elongated NNE–SSW. Its broadly but slightly brighter core has a diffuse nonstellar nucleus. The galaxy is at the SW corner of the 3' × 4' parallelogram which it forms with three 10th to 12th magnitude stars.

NGC 1385 H263² Galaxy Type SB(s)d: pec
⌀3.6' × 2.4', m10.7v, SB 12.9   03ʰ37.5ᵐ −24°30'
Finder Chart 15-5   ★★★

8/10" Scopes–100x: NGC 1385, 2.5' SSW of a 10.5 magnitude star, appears fairly faint and diffuse with a circular 1.5' diameter halo.

12/14" Scopes–125x: NGC 1385 has a moderately faint halo elongated 2.5' × 1.5' N–S with a slight central brightening.

16/18" Scopes–150x: Larger instruments reveal a fairly bright halo elongated 3' × 1.75' N–S with a large, well concentrated 1.25' diameter core containing a stellar nucleus offset to the west.

NGC 1399 Galaxy Type E+1
⌀8.1' × 7.6', m8.8v, SB 13.1   03ʰ38.5ᵐ −35°27'
Finder Chart 15-7, Figure 15-15   ★★★

12/14" Scopes–125x: NGC 1399 and NGC 1404 lying 13' to its SSE form a conspicuous galaxy-pair in the Fornax I Galaxy Cluster. Both are quite obvious with bright round 1' diameter halos, but NGC 1399 is slightly larger and fainter than NGC 1404 and has a brighter core.

16/18" Scopes–150x: NGC 1399 has a bright, circular 2' diameter halo with a broad, prominent core and a

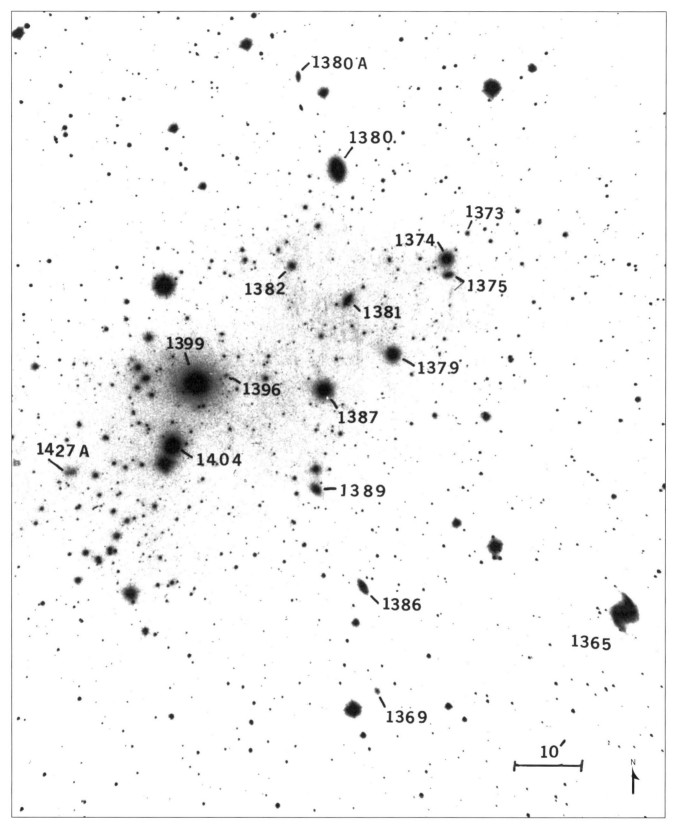

Figure 15-15. *This negative print photograph of the Fornax I Galaxy Cluster can be used as a finder chart for positive identification of each object. Some of the Fornax I galaxies lie across the border in Eridanus. William Harris made this 60 minute exposure on hypered 2415 film with an 8″, f4 Wright Newtonian reflector at prime focus.*

faint stellar nucleus. A faint star lying just north of the center gives the impression that the galaxy has a twin nucleus.

NGC 1398　Galaxy　Type (R′)SB(r)ab
⌀7.1′ × 5.2′, m9.5v, SB 13.3　　　　03ʰ38.9ᵐ　−26°20′
Finder Chart 15-5　　　　　　　　　　★★★★

8/10″ Scopes–125x: NGC 1398, 7′ east of a 9.5 magnitude star, has a fairly bright, small, round halo containing a prominent core.

12/14″ Scopes–125x: NGC 1398 is a moderately bright 1.5′ × 1′ E–W oval that gradually brightens to a diffuse core with a stellar nucleus.

16/18″ Scopes–150x: This bright galaxy has a stellar nucleus within a bright core embedded in a 3′ diameter halo fading evenly to its diffuse periphery. With averted vision at 175x the envelope looks nearly 5′ wide and slightly elongated E-W.

NGC 1404　Galaxy　Type E1:
⌀4.8′ × 3.9′, m9.7v, SB 12.7　　　　03ʰ38.9ᵐ　−35°35′
Finder Chart 15-7, Figure 15-15　　　　★★★

12/14″ Scopes–125x: NGC 1404, a Fornax I Galaxy Cluster member, is 2.5′ NNW of an 8th magnitude star. Its bright, round halo is rather small, less than 1′ in diameter, and gradually brightens to a uniformly illuminated core.

16/18″ Scopes–150x: NGC 1404 is a smaller version of NGC 1399 lying 9′ to the NNW. It is quite bright with a circular 1.5′ diameter halo and a dense core containing a very faint stellar nucleus. A very faint star is embedded in the halo 45″ SW of the galaxy's center.

NGC 1406　Galaxy　Type Sb: sp
⌀4.5′ × 0.9′, m11.6v, SB 13.0　　　　03ʰ39.4ᵐ　−31°19′
Finder Chart 15-6　　　　　　　　　　★★★★

12/14″ Scopes–125x: NGC 1406, 6.5′ east of a 9th magnitude star, is a nice but faint edge-on galaxy elongated 3′ × 0.5′ NNE–SSW, its bright center fading to tapered ends.

16/18″ Scopes–150x: NGC 1406 has a fairly faint halo elongated 3.25′ × 0.5′. The halo has tapered tips and gradually brightens to a round core containing a faint stellar nucleus.

20/22″ Scopes–175x: A fine edge-on galaxy! NGC 1406 is elongated 3.50′ × 0.75′ in P.A. 15°, its moderately concentrated core containing a stellar nucleus. The halo is mottled along the major axis with hints of an equatorial dust lane.

NGC 1425　H852² Galaxy　Type SA(rs)bc: II
⌀5.4′ × 2.2′, m10.8v, SB 13.3　　　　03ʰ42.2ᵐ　−29°54′
Finder Chart 15-6, Figure 15-14　　　　★★★★

12/14″ Scopes–125x: This galaxy is rather faint, elongated 3′ × 1.5′ NW–SE, its halo moderately brightening to an inconspicuous stellar nucleus. To the north of the galaxy is a row of four faint stars, the SE of which is a wide E–W double.

16/18″ Scopes–150x: NGC 1425 is fairly conspicuous in 17.5″ scopes. Its 4′ × 2′ halo fades suddenly at the tips; and embedded within its well concentrated core is a faint stellar nucleus.

20/22″ Scopes–175x: In large instruments NGC 1425 is a large oval elongated 5′ × 2.5′ NW-SE containing an irregular knotty core with a stellar nucleus. Along the major axis on either side of the nucleus are several condensations. The outer envelope has a mottled texture and a moderate surface brightness.

NGC 1427　Galaxy　Type SA0−:
⌀3.4′ × 2.5′, m10.9v, SB 13.1　　　　03ʰ42.3ᵐ　−35°25′
Finder Chart 15-7　　　　　　　　　　★★★

8/10″ Scopes–100x: This Fornax I Galaxy Cluster member appears faint and diffuse, its 1.5′ × 1′ halo elongated ENE–WSW.

16/18″ Scopes–150x: NGC 1427 is fairly bright, its 1.75′ × 1.25′ halo elongated ENE-WSW. The core is moderately brighter and contains a stellar nucleus. A 13th magnitude star lies 1.5′ west.

NGC 1437　Galaxy　Type (R′)SB(s)b? II-III
⌀3.0′ × 2.0′, m11.7v, SB 13.5　　　　03ʰ43.6ᵐ　−35°52′
Finder Chart 15-7　　　　*(In Eridanus)*★★

16/18″ Scopes–150x: NGC 1437, just across the border in Eridanus, is a member of the Fornax I Galaxy Cluster. It is a very faint 2′ × 1.5′ smudge elongated NNW–SSE without central brightening.

Chapter 16

Gemini, the Twins

16.1 Overview

Gemini, the Twins, is one of the twelve constellations of the Zodiac. Its brightest stars are in two long parallel rows, each representing one of the Twins. The bright stars at the NE ends of these rows mark the Twin's heads and bear their names, Castor and Pollux. In classical mythology Castor and Pollux were warlike heroes especially honored as the protectors of sailors. Their presence on board a ship in peril was believed to be marked by flames leaping from the mast, an electrostatic phenomenon related to lightening (hence its occurrence during storms) and later called "St. Elmo's fire." According to myth Pollux was immortal but Castor was not. After Castor's death Pollux begged Jupiter to let him die too and join his brother in the Netherworld. But an immortal cannot die, so the best Jupiter could do was to permit Pollux to spend alternate days in Elysium among the gods and in Hades with his beloved brother. In the sky the inseparability of the brothers is symbolized by the fact that their two stars set in the west, the direction of the Netherworld, simultaneously.

Gemini's feet are washed by the Milky Way, so the constellation contains a wide variety of celestial objects — emission and planetary nebulae, open clusters, double and variable stars. Because the Twins' heads are outside the Milky Way, the constellation also contains a surprising number of galaxies. Gemini's two finest objects are the large open cluster Messier 35 and the bright planetary nebula NGC 2392, the "Clown Face" or "Eskimo" nebula.

Gemini: JEM-in-eye
Genitive: Geminorum, JEM-i-nor-um
Abbrevation: GEM
Culmination: 9pm–Feb. 19, midnight–Jan. 5
Area: 514 square degrees
Showpieces: 15 Gem, 20 Gem, 38 Gem, $\lambda = 54$ Gem, $\delta = 55$ Gem, $\alpha = 66$ Gem, $\kappa = 77$ Gem, $\Sigma 1108$, M35 (NGC 2168), NGC 2129, NGC 2266, NGC 2392
Binocular Objects: $\alpha = 66$ Gem, M35 (NGC 2168), NGC 2129, NGC 2158, NGC 2331, NGC 2355, NGC 2392

16.2 Interesting Stars

15 Geminorum = S, h 70 Double Star   Spec. K0
m6.6, 8.0; Sep. 27.1″; P.A. 204° $06^h27.8^m$ +20°47′
Finder Chart 16-3   ★★★

4/6″ Scopes–75x: 15 Geminorum is a color contrast pair of yellowish and blue stars.

20 Geminorum = Σ924 Double Star   Spec. F8
m6.3, 6.9; Sep. 20.0″, P.A. 210° $06^h32.3^m$ +17°47′
Finder Chart 16-4   ★★★

8/10″ Scopes–100x: 20 Geminorum is a 20″ pair of nearly equally bright yellow and pale white stars.

38 Geminorum = Σ982 Double Star   Spec. F0
m4.7, 7.7; Sep. 7.1″; P.A. 145° $06^h54.6^m$ +13°11′
Finder Chart 16-4   ★★★

8/10″ Scopes–100x: 38 Geminorum is a nice yellow and pale blue pair that may be comfortably split in small telescopes.

Zeta (ζ) = 43 Geminorum Var. Star Spec. G0 Ib–G3Ib
m3.6 to 4.1 in 10.15 days $07^h04.1^m$ +20°34′
Finder Chart 16-5   *Mekbuda*   ★ ★ ★ ★

Zeta Gem is one of the brightest examples of the pulsating giant Cepheid variables. It ranges between magnitudes 3.6 to 4.1 in a period of 10.15 days. Kappa at (m3.57) and Upsilon (m4.07) may be used for visual comparison. An 8th magnitude star at 96″, and an 11th magnitude star 87″ distant, are not true companions. Zeta Geminorum is some 1,400 light years away.

Gemini, the Twins

Constellation Chart 16-1

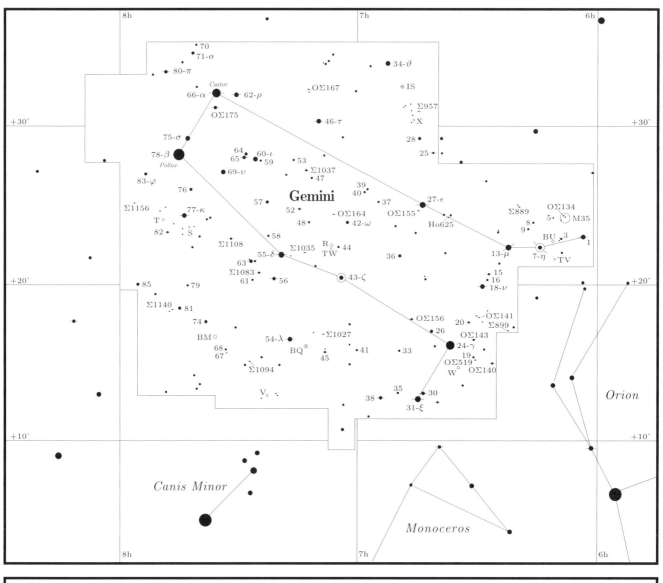

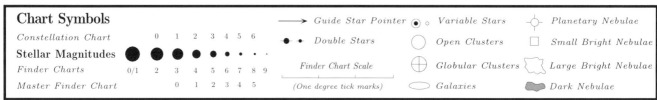

Table 16-1. Selected Variable Stars in Gemini

Name	HD No.	Type	Max.	Min.	Period (Days)	F*	Spec. Type	R.A. (2000)	Dec.	Finder Chart No. & Notes
BU Gem	42543	Lc?	5.7	7.5			M1-M2	06^{h}12.3^m	+22°54'	16-3 3/4° NW of 7-η
7-η Gem	42995	SRb(E)	3.2	3.9	232	0.05	M3	14.9	+22 30	16-3
W Gem	46595	Cδ	6.5	7.3	7.91	0.30	F5-G1	35.0	+15 20	16-4 Orange & yellow dou
X Gem	48912	M	7.5	13.6	263	0.49	M5-M6	47.1	+30 17	16-1
43-ζ Gem	52973	Cδ	3.6	4.1	10.15	0.50	F7-G3	07^{h}04.1^m	+20 34	16-5 Bright Cepheid Var.
R Gem	53791	M	6.0	14.0	369	0.36	S2-S8	07.4	+22 42	16-5
BQ Gem	55383	SRb	6.6	7.0	50		M4	13.4	+16 10	16-6 1.25° SSW of 54-λ
V Gem	57770	M	7.8	14.9	275	0.45	M4-M5	23.2	+13 06	16-6
S Gem	62045	M	8.2	14.7	293	0.42	M4-M7	43.0	+23 27	16-8 1° south of 77-κ
T Gem	63334	M	8.0	15.0	287	0.50	S1-S9	49.3	+23 44	16-8 1.25° SW of 77-κ

F* = The fraction of period taken up by the star's rise from min. to max. brightness, or the period spent in eclipse.

Table 16-2. Selected Double Stars in Gemini

Name	ADS No.	Pair	M1	M2	Sep."	P.A.°	Spec	R.A. (2000)	Dec	Finder Chart No. & Notes
OΣ134	4744		7.3	9.1	31.0	188	G0	06^{h}09.3^m	+24°26'	16-3 In NNE edge of M35
3 Gem	4751	AB	5.8	9.9	0.5	339	B1	09.7	+23 07	16-3
7-η Gem	4841		3.3	8.8	1.4	226	M	14.9	+22 30	16-3 Orange & yellow suns
Σ889	4930		7.6	9.9	21.3	237	K2	19.9	+25 01	16-3
Σ899	4991		7.3	8.3	2.3	20	A0	22.8	+17 34	16-4
OΣ140	5062	AB	6.9	9.4	2.8	118	B9	26.6	+15 31	16-4
15 Gem	5080	AB	6.6	8.0	27.1	204	K0	27.8	+20 47	16-3 Yellow & blue
18-ν Gem	5103	AB	4.2	8.7	112.5	329	B5	29.0	+20 13	16-3 White & bluish
OΣ141	5121		7.6	9.7	2.2	142	A0	30.0	+17 54	16-4
OΣ143	5146	AB	6.3	9.4	7.9	103	K0	31.2	+16 56	16-4
Fox	5146	AC		10.9	44.8	344				
OΣ519	5152		7.9	10.2	8.2	78	G5	31.3	+15 44	16-4
20 Gem	5166	AB	6.3	6.9	20.0	210	F8	32.3	+17 47	16-4 Yellow & White
Ho 625	5270	AB	6.7	11.9	13.4	194	A5	38.3	+24 27	16-5
	5270	AC		9.3	47.4	251				
25 Gem		AB	6.4	11.6	30.7	46	K0	41.3	+28 12	16-1
		AC		10.3	55.9	57				
27-ϵ Gem	5381		3.0	9.0	110.3	94	G5	43.9	+25 08	16-5 Deep & pale yellow
OΣ156	5447		6.8	7.0	0.6	*226	A0	47.4	+18 12	16-4 White pair
Σ957	5403	AB	7.3	8.8	3.5	93	A0	45.2	+30 50	16-1
OΣ155	5409		7.3	10.2	15.5	261	K2	45.4	+24 40	16-5
38 Gem	5559		4.7	7.7	7.1w	*145	F0	54.6	+13 11	16-4 White & pale blue
OΣ164	5775		7.3	10.3	13.7	50	K2	07^{h}06.2^m	+24 52	16-5
Σ1027	5816		8.3	8.4	6.9	355	K0	08.3	+16 55	16-6 Equal orange pair
Σ1035	5858		8.2	8.2	8.7	41	F5	12.0	+22 17	16-5 Yellowish pair
Σ1037	5871		7.2	7.2	1.1c	*313	F5	12.8	+27 13	16-5 Equal pair of yellow suns
OΣ167	5884		7.3	10.4	5.3	158	A5	13.5	+32 09	16-7
Σ1053	5945		7.1	9.8	13.8	310	A0	16.8	+24 32	16-8
54-λ Gem	5961		3.6	10.7	9.6	33	A2	18.1	+16 32	16-6 Bluish-white pair
55-δ Gem	5983		3.5	8.2	5.8c	*226	F0	20.1	+21 59	16-8 Yellow & reddish purple
Σ1083	6060		7.2	8.3	6.4	44	A5	25.6	+20 30	16-8 Greenish-white & yellow
Σ1094	6086		7.4	8.4	2.5	96	B8	27.4	+15 19	16-6
63 Gem	6089	AB	5.2	9.4	42.9	324	F5	27.7	+21 27	16-1
Σ1108	6160		6.5	8.3	11.5	178	G5	32.8	+22 53	16-8 Yellow & blue
66-α Gem	6175	AB	1.9	2.9	4.0w	*68	A0 A0	34.6	+31 53	16-7 AB: Blue-white duo
	6175	AC		8.8	72.5	164	M1			C: lilac
OΣ175	6185		5.8	6.4	0.4	330	K0	35.1	+30 58	16-7 Tight orange pair
77-κ Gem	6321		3.6	8.1	7.1	240	G5	44.4	+24 24	16-8 Deep yellow & white
80-π Gem	6364	AB	5.1	11.2	21.0	214	K2	47.5	+33 25	16-7 A: Orange, C: bluish
	6361	AC		10.2	91.9	314				
Σ1140	6376		7.9	9.6	6.3	273	G5	48.4	+18 20	16-1
Σ1156			7.8	10.1	18.6	159	K0	56.0	+24 40	16-8

Footnotes: *= Year 2000, a = Near apogee, c = Closing, w = Widening. Finder Chart No: All stars listed in the tables are plotted in the large Constellation Chart, but when a star appears in a Finder Chart, this number is listed. Notes: When colors are subtle, the suffix -ish is used, e.g. bluish.

Lambda (λ) = 54 Gem. (Σ1061) Double Star   Spec. A2
m3.6, 10.7; Sep. 9.6"; P.A. 33°   07ʰ18.1ᵐ +16°32'
Finder Chart 16-6   ★★★★

8/10" Scopes–100x: This unequal pair has a brilliant bluish-white primary with a faint bluish companion.

Delta (δ) = 55 Gem. (Σ1066) Double Star   Spec. F2 IV
m3.5, 8.2; Sep. 5.8"; P.A. 226°   07ʰ20.1ᵐ +21°59'
Finder Chart 16-8   *Wasat*   ★★★★

8/10" Scopes–125x: This pair has bright yellowish primary and a pale blue secondary.

12/14" Scopes–150x: Delta Geminorum is a close pair of unequally bright stars with a fine yellow and reddish-purple color contrast.

Σ1108 Double Star   Spec. G5
m6.6, 8.3; Sep.11.5"; P.A. 178°   17ʰ32.8ᵐ +22°53'
Finder Chart 16-8   ★★★★

4/6" Scopes–50x: This nice pair has a yellow primary with a tiny bluish companion.

Alpha (α) = 66 Gem. (Σ1110) Double Star Spec. A0 & A0
m1.9, 2.9; Sep. 4.0"; P.A. 68°   07ʰ34.3ᵐ +31°53'
Finder Chart 16-7   *Castor*   ★★★★

Castor, the 23rd brightest star in the sky, is a fine double for larger scopes. Castor A and B orbit one another in a period of at least four centuries. The stars' mean separation is approximately 90 A.U. or about 8 billion miles, a distance equal to the diameter of our entire Solar System. A magnitude 8.8 red dwarf 73" from the bright pair seems to be a true physical member of the system and is designated Castor C, though its orbital period must be thousands of years long. Each of the three components, A, B, and C, is a spectroscopic binary, so Castor is a complex system of six stars in three binary groupings all orbiting each other mutually. Castor C is also a variable star, an eclipsing binary (designated YY Geminorum) with the magnitude range of 9.1 to 9.6. The Castor system lies some 45 light years away.

8/10" Scopes–175x: Alpha's close A-B stars appear brilliant white while the C star is a dull lilac.

12/14" Scopes–225x: Castor is a close pair of bright blue-white stars with a wide, faint companion 73" away.

Kappa (κ) = 77 Gem. (OΣ179) Double Star   Spec. G5
m3.6, 8.1; Sep. 7.1"; P.A. 240°   07ʰ44.4ᵐ +24°24'
Finder Chart 16-8   ★★★★

8/10" Scopes–150x: Kappa Geminorum is a fine double of a brilliant orangish-yellow primary with a faint bluish companion.

Beta (β) = 78 Geminorum   Spec. K0 III
Mag: Apparent 1.14, Absolute 1.1   07ʰ45.1ᵐ +28°01'
Constellation Chart 16-1   *Pollux*   ★★★★

Unlike Caster, Pollux is a only a single star. Its spectral type is K0, implying a rather cool surface temperature of 4,500°K; hence the star's fine orange-yellow color. Pollux is about 35 light years away, so its absolute magnitude is +1.1, a luminosity 35 times that of the Sun. The star's diameter is probably 4 or 5 times that of the Sun.

Pollux, at apparent magnitude 1.14 is the 17th brightest star in the sky. Thus it is in fact brighter than the magnitude 1.7 Castor, the Alpha star of the constellation. It has been suggested that one of these stars has changed in brightness since their designations were assigned by Bayer early in the 17th century; but this is astrophysically unlikely.

16.3 Deep-Sky Objects

NGC 2129 Cr 77 Open Cluster 40★ Tr Type III 3 p,
⌀7', m6.7v, Br★ 7.36v   06ʰ01.0ᵐ +23°18'
Finder Chart 16-3, Figure 16-1   ★★★★

8/10" Scopes–100x: NGC 2129 is visible in an 8 × 50 finder as a haze around 7.4 and 8.6 magnitude stars aligned N–S. At 100x, this neat little cluster appears fairly bright and compact with five 10th to 11th magnitude, ten 12th magnitude, and a couple dozen fainter stars irregularly concentrated around the two lucidae.

12/14" Scopes–125x: This fine, bright cluster contains three dozen stars in a 7' area concentrated around two stars of magnitudes 7.4 and 8.6 aligned N–S. The cluster is divided into two irregular star chains

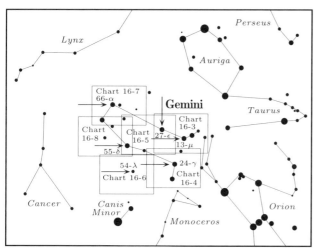

Master Finder Chart 16-2. Gemini Chart Areas
Guide stars indicated by arrows.

running E–W, making the group appear elongated in this direction.

**IC 2157   Collinder 80   Open Cl. 20★
Tr Type III 2 p, ⌀8′, m8.4v,
Br★ 11.08v   06ʰ05.0ᵐ +24°00′**
Finder Chart 16-3   ★★★

8/10″ Scopes–100x: This cluster is rather faint and small with fifteen randomly scattered stars. Three 11th magnitude stars stand out against the fainter members.

12/14″ Scopes–125x: IC 2157 has twenty faint stars in a 5′ area, its three 11th magnitude lucidae prominent against the fainter cluster members. An indistinct star-poor lane running roughly E–W divides the cluster into two irregular clumps. The southern clump has seven stars, including a 9th magnitude orange field star. The northern clump consists of a condensation of faint stars embedded in a faint glow. IC 2156 is visible in the same field 6′ NW.

Figure 16-1. *Open cluster NGC 2129 is a knot of stars around two bright stars. Lee C. Coombs made this 10 minute exposure on Kodak 103a-O Spectroscopic film at prime focus with a 10″, f5 Newtonian reflector.*

**NGC 2158   Collinder 81   Open Cluster   Tr Type II 3 r
⌀5′, m8.6v, Br★ 12.40v   06ʰ07.5ᵐ +24°06′**
Finder Chart 16-3, Figure 16-2   ★★★

NGC 2158 is 13,000 light years distant toward the rim of our Galaxy. It is thus six times farther away than its giant neighbor, M35, and one of the most remote open clusters that can be seen in small telescopes. Its high star-density and symmetrical distribution are comparable to those of the less-condensed globular clusters. If it was as close as M35, NGC 2158 would be one of the finest open clusters in the sky.

8/10″ Scopes–100x: At low power, this cluster is a milky glow in the rich star field just SW of Messier 35. At 100x, NGC 2158 is a nice but faint splash of tiny stars in a 4′ area. Several dozen threshold stars are just visible against a background haze growing considerably richer near its center.

12/14″ Scopes–125x: NGC 2158 is a very compressed, rich gathering of faint stars contrasting nicely with the bright, scattered Messier 35. Because of the crowding of its stars toward its center, it has the appearance of a loose globular. Fifty 13th magnitude and fainter stars can be resolved in a 5′ area against a background haze. A 10.5 magnitude star stands out on the SE edge. Several chains of brighter stars radiate from the cluster: one at the west edge trails SW, another trails NE from the east side toward Messier 35.

**NGC 2168   Messier 35   Open Cl. 200★   Tr Type III 2m
⌀28′, m5.1v, Br★ 8.18v   06h08.9m +24°20′**
Finder Chart 16-3, Figure 16-2   ★★★★★

Messier recorded this object in 1764, but the Swiss astronomer de Cheseaux had already mentioned a cluster at this location in a list he prepared in 1745. Its distance is estimated at about 2,200 light years.

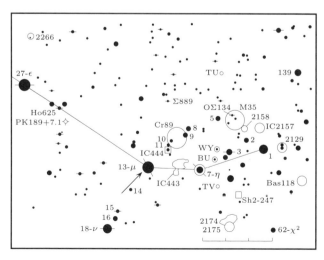

Finder Chart 16-3. 13–μ Gem: 06ʰ22.9ᵐ +22°31′

Figure 16-2. *The bright, large open cluster Messier 35 (NGC 2168) and the faint, small open cluster NGC 2158 are a fine contrast at low power. Martin C. Germano made this 30 minute exposure on hypered 2415 film with an 8″, f5 Newtonian at prime focus.*

8/10″ Scopes–50x: A magnificent cluster! M35 has seventy-five stars are resolved in a 30′ area, including more than twenty of magnitudes 7.5 to 10. The outer edge is highly irregular, the periphery of the cluster blending imperceptibly into the surrounding star field. Outliers up to half a degree from the cluster center double the star count. An E–W star-poor lane divides the cluster. The northern portion is box-shaped with a prominent curved star chain on its NW side. The southern portion is highly extended ESE–WNW. A star chain runs through the center of the cluster along the star-poor lane.

12/14″ Scopes–50x: This superb cluster has at least 150 stars of 8th magnitude and fainter in a 1.5° area. Though M35 lacks either a central condensation or a well-defined periphery, many of its stars are distributed in attractive gatherings and chains. The most prominent chain, in the northern part of the cluster, curves NE–SW, ending at the beautiful yellow and blue double OΣ134 (7.3, 9.1; 31″; 188°). Another star chain extends eastward to the magnitude 5.8 star 5 Geminorum. A third chain runs E–W along the star-poor gap that divides the cluster. The star-glittering haze of the open cluster NGC 2158 to the SW provides a fine contrast to M35's sprawl of bright stars.

IC443 SNR/Emission Nebula
ø50′ × 40′, Photo Br 2–5, Color – $06^h16.9^m$ +22°47′
Finder Chart 16-3, Figure 16-3   ★

12/14″ Scopes–100x: IC 443, a supernova remnant, requires an O-III filter to be seen as an extremely dim, irregular boomerang-shaped wisp elongated 30′ × 15′ NW–SE. Four faint stars are embedded. The nebula is brighter on its NE edge and fades southward.

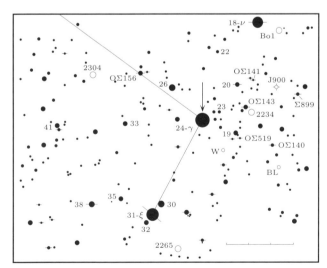

Finder Chart 16-4. 24–γ Gem: $06^h37.6^m$ +16°24′

Figure 16-3. *IC 443 is a supernova remnant visible as an extremely faint, extended wisp NE of Eta (η) = 7 Geminorum. Martin C. Germano made this 30 minute exposure on hypered 2415 film with an 8", f5 Newtonian at prime focus.*

IC 444 Reflection Nebula
ø8' × 4', Photo Br 3-5, Color 1-4 $06^h20.4^m$ +23°16'
Finder Chart 16-3   ★

12/14" Scopes–125x: This nebula is a very faint, diffuse, circular 5' diameter glow surrounding the star 12 Geminorum (m6.95v).

Collinder 89 Open Cluster 15★ Tr Type IV 2 p
ø35', m5.7p, Br★ – $06^h18.0^m$ +23°38'
Finder Chart 16-3   ★★★

12/14" Scopes–75x: Collinder 89 is a large, loose, and sparse mixture of 15 bright and faint stars. The brighter stars form a semicircle concave to the south, in which is enclosed several loose gatherings of faint stars. The WNW and ESE ends of the semicircle are marked by the bright stars 9 and 10 Geminorum (m6.25v and 6.6v), and the northern extremity of its arc by an E–W pair of 9th magnitude stars.

Bochum 1 Open Cluster 30★ Tr Type –
ø –, m7.9v, Br★ 8.42v $06^h25.5^m$ +19°46'
Finder Chart 16-4   ★★

12/14" Scopes–125x: Bochum 1 is a very faint cluster of a few stars embedded in a dim background glow. The brightest star (m8.42v) is on the cluster's west edge.

PK194+2.1 Jonckheere 900 Planetary Neb. Type 3b+2
ø>8', m11.7v, CS 16.5:v $06^h25.9^m$ +17°47'
Finder Chart 16-4, Figure 16-4   ★★★

12/14" Scopes–225x: This tiny, greenish-grey planetary nebula is a fairly faint, circular 10" smudge.

16/18" Scopes–250x: J900 has a fairly bright, greenish halo, somewhat oval-shaped 10" × 8" NE–SW, without a central star evident. A 13th magnitude star is just SSW, and more stars trail southward.

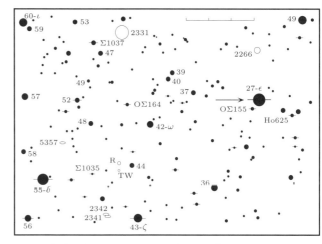

Finder Chart 16-5. 27–ϵ Gem: $06^h43.9^m$ +25°08'

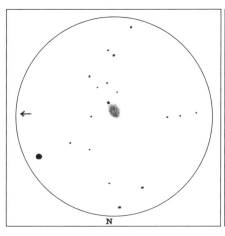

Figure 16-4. J900
16", f5–275x, by George de Lange

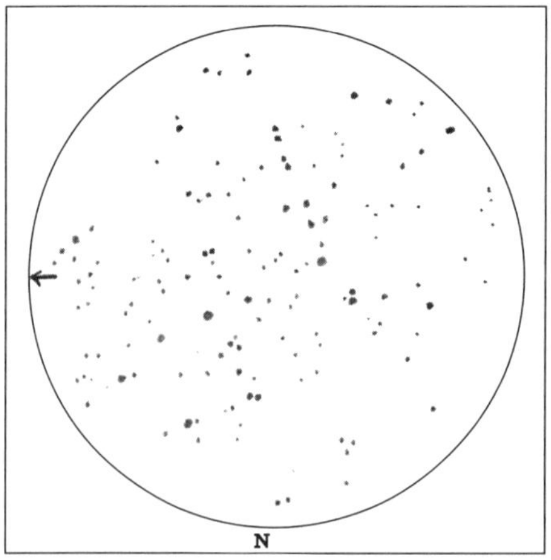

Figure 16-5. NGC 2234
12.5", f5–125x, by G. R. Kepple

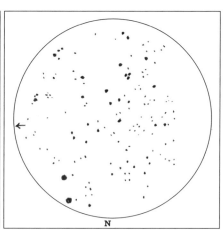

Figure 16-6. NGC 2265
12.5", f5–125x, by G. R. Kepple

NGC 2234 H9[8] Open Cluster
[⌀25′, Br★ 11.0v]   06ʰ29.3ᵐ +16°41′
Finder Chart 16-4, Figure 16-5   ★★

12/14″ Scopes–125x: Users of *Norton's Star Atlas* may be acquainted with this cluster but other atlases do not plot this supposedly nonexistent cluster. At 75x NGC 2234 is visible as a loose group of forty 11th magnitude and fainter stars spread over a 25′ area. 125x reveals sixty stars in a 10′ area, the brighter stars forming two loose concentrations: to the NW is an irregular pentagon formed by 8 or 9 stars; and to the SE is a 6′ long NNE–SSW stream of a dozen stars.

NGC 2265 Open Cluster
[⌀7′, Br★ 12.0v]   06ʰ41.5ᵐ +11°56′
Finder Chart 16-4, Figure 16-6   ★★

12/14″ Scopes–125x: NGC 2265 is another *New General Catalogue* open cluster later classed as nonexistent. At the plotted location, centered 8′ SSE of a pair of 10th magnitude stars, 150x shows a dozen 12th magnitude and fifty 13th magnitude stars irregularly scattered over a 7′ area. The "cluster," if such it be, is difficult to distinguish from the rich Milky Way star field.

NGC 2266 H21[6] Open Cluster 50★ Tr Type II 2 m
⌀6′, m9.5p, Br? 11.0p   06ʰ43.2ᵐ +26°58′
Finder Chart 16-5, Figure 16-7   ★★★★

8/10″ Scopes–100x: NGC 2266, a rich, attractive cluster, is a conspicuous and highly compressed group of 50 stars in a triangular outline. The SW apex of the triangle is marked by an 8.5 magnitude star, from which a string of stars trails NNE.

12/14″ Scopes–125x: An interesting cluster! It is fairly bright, quite rich, and highly concentrated. The cluster covers a 5′ area; but in its richest concentration alone, in the southern part, are not less than fifty 11th to 15th magnitude stars. At this power the cluster looks irregular rather than triangular, but a couple nice star chains become prominent. One begins at the magnitude 8.5 star at the SW edge of the cluster and curves eastward. A second chain curves around the cluster from the north to the SW.

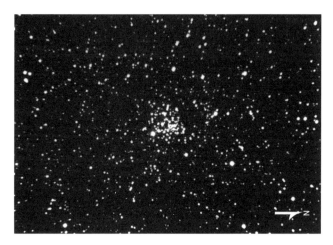

Figure 16-7. *NGC 2266 is a bright, rich, highly concentrated cluster of fifty stars. Lee C. Coombs made this 10 minute exposure on 103a-O film with a 10″, f5 Newtonian at prime focus. North is to the right.*

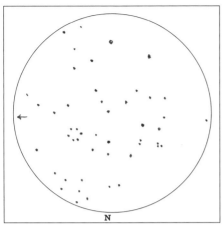
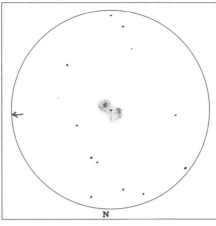
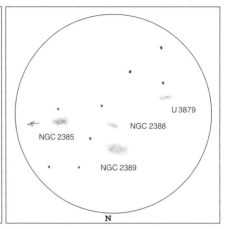

Figure 16-8. NGC 2331
13", f5.6-100x, by Steve Coe

Figure 16-9. NGC 2371 & 2372
12.5", f5-250x, by Chris Schur

Figure 16-10. NGC 2389 Group
16", f5-185x, by Bob Erdmann

NGC 2304 H2⁶ Open Cluster 30★ Tr Type II 1 p
ø5', m10.0p, Br★ –   06ʰ55.0ᵐ +18°01'
Finder Chart 16-4   ★★

8/10" Scopes–100x: This cluster is a faint mist of 20 stars scattered around half a circle. A magnitude 11.5 star lies on the NW edge, and two magnitude 12.5 stars mark the half-circle's northern arc.

12/14" Scopes–125x: NGC 2304 is a moderately small cluster of several dozen 13th magnitude and fainter stars against a background glow of unresolved haze. Its members are distributed in an irregularly semicircular area elongated 5' × 4' NE–SW. An 11.5 magnitude star stands out on the NW side. Many outlying stars spread to the east and NE of the cluster.

NGC 2331 Collinder 126 Open Cl. 30★ Tr Type IV 1 p
ø18', m8.5p, Br★ 9.0p   07ʰ07.2ᵐ +27°21'
Finder Chart 16-5, Figure 16-8   ★★★

8/10" Scopes–100x: This large, coarse cluster has 25 faint members, 15 of the 10th and 11th magnitudes. The brighter stars are in an H-shaped asterism oriented E–W. The cluster lies in a fairly rich Milky Way field.

12/14" Scopes–100x: NGC 2331 is a fairly bright, loose, irregular scattering of thirty 9th to 13th magnitude stars in a 15' area. A long zigzagging chain loops almost completely around the cluster, ending just at a ringlet on the WNW side.

NGC 2339 H769² Galaxy Type SAB(rs)bc II–III
ø2.4' × 1.8', m11.8v, SB 13.2   07ʰ08.3ᵐ +18°47'
Finder Chart 16-6   ★★

12/14" Scopes–125x: NGC 2339, 3' SW of an 11.5 magnitude star, has a moderately faint, circular 1.5' diameter halo slightly brightening to a core just discernible from the rest of the galaxy. A very faint star is at the NE edge, and a 12.5 magnitude star lies 1.75' west.

16/18" Scopes–150x: NGC 2339 is rather dim and appears slightly elongated 2' × 1.75' N–S. The envelope brightens gradually to a central core which contains a faint stellar nucleus.

NGC 2341 U3708 Galaxy Type Pec
ø0.8' × 0.8', m13.2v, SB 12.6   07ʰ09.3ᵐ +20°35'
NGC 2342 U3709 Galaxy Type S pec
ø1.3' × 1.2', m12.6v, SB 12.9   07ʰ09.4ᵐ +20°38'
Finder Chart 16-5   ★/★★

12/14" Scopes–125x: NGC 2341 and NGC 2342 are 1.5° east of 3.8 magnitude Zeta (ζ) = 43 Geminorum. NGC 2341 is a round smudge nearly as bright but much smaller than the brighter and larger NGC

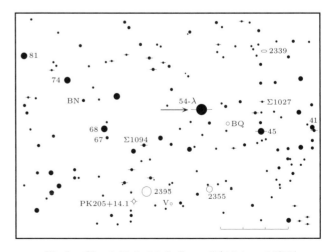

Finder Chart 16-6. 54–λ Gem: 07ʰ18.1ᵐ +16°32'

Figure 16-11. *NGC 2355 is an irregular cluster of forty faint stars. Lee C. Coombs made this 10 minute exposure on 103a-O film with a 10", f5 Newtonian at prime focus.*

2342. NGC 2341 has a tiny, very faint core; but NGC 2342 has uniform surface brightness.

16/18" Scopes–150x: NGC 2342 is the brighter and larger of a galaxy-pair with NGC 2341 lying 2.5' to its SW. NGC 2342 is fairly bright but diffuse, and has a 1.25' × 1' NE–SW elongated halo with a well concentrated center. NGC 2341 is similar in brightness but has only a 30" diameter halo with a tiny core. A 13th magnitude star lies 30" north.

NGC 2355 H6⁶ Open Cluster 40★ Tr Type II 2 p
ø9', m9.7p, Br★ 13.0p   07ʰ16.9ᵐ +13°47'
Finder Chart 16-6, Figure 16-11   ★★

8/10" Scopes–100x: NGC 2355, just SSW of an 8th magnitude star, has thirty faint members irregularly scattered in a 9' area. Its SW portion is more concentrated.

12/14" Scopes–125x: NGC 2355 is a moderately faint, irregular cluster of 40 stars in a 9' × 7' NNW–SSE area. Behind the group's twenty brighter stars its fainter members form an S-shaped diffuse background. An 11th magnitude double can be seen in the NW portion of the cluster, and a yellow 9.7 magnitude field star is to the ESE of its center.

NGC 2371-2 H316² Planetary Nebula Type 3a+6
ø55", m11.3v CS 14.80v   07ʰ25.6ᵐ +29°29'
Finder Chart 16-7, Figure 16-9   ★★★

12/14" Scopes–200x: NGC 2371-2 is an interesting object resembling a peanut, its lobes aligned NE–SW. The SW lobe is the brighter and more concentrated and has a nucleus. A faint outer shell envelopes each lobe. The central star, visible only with averted vision, lies between the lobes.

16/18" Scopes–250x: This double-lobed planetary resembles Messier 76 in Perseus. Its SW lobe is more concentrated and has a bright area on its NW edge. The slightly fainter NE lobe has a bright area on its SE edge. Both lobes are surrounded by a much fainter outer shell. The central star, located between the lobes, twinkles faintly in and out of resolution.

NGC 2395 Cr 144 Open Cluster 30★ Tr Type III 1 p
ø12', m8.0v, Br★ 9.96v   07ʰ27.1ᵐ +13°35'
Finder Chart 16-6, Figure 16-12   ★★★

8/10" Scopes–75x: This open cluster is a fairly bright scattering of forty 9th magnitude and fainter stars divided into two sections. The northern group is larger, more concentrated, and irregularly extended N–S. A detached, distorted keystone-shaped asterism lying to the SE may not belong to the cluster.

12/14" Scopes–100x: NGC 2395 is elongated 20' × 10' N–S with 45 stars. The cluster has two concentrations of stars linked by a star chain. The northern group is larger, somewhat triangular, and contains most of the stars. The southern group is loose and poor but accented by a bright triangle of one 11th and two 10th magnitude stars.

NGC 2379 Galaxy Type SA0:
ø1.0' × 1.0', m13.5v, SB 13.4   07ʰ27.4ᵐ +33°49'
Finder Chart 16-7   ★★

16/18" Scopes–150x: This faint galaxy has a round 1' diameter halo slightly brighter in the center. It is within, and near the western vertex of, a nearly equilateral triangle of 12th magnitude stars.

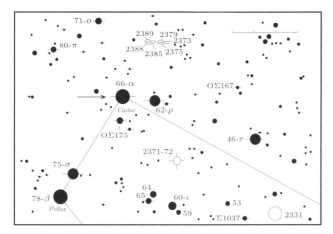

Finder Chart 16-7. 66-α Gem: 07ʰ34.6ᵐ +31°53'

Figure 16-12. *PK205+14.1, the Medusa Nebula (upper right) is an unusual crescent-shaped planetary nebula. Open cluster NGC 2395 (lower left center) is a loose scattering of stars not well detached from the star field. Martin C. Germano made this 90 minute exposure on hypered 2415 film with an 8″, f5 Newtonian at prime focus.*

NGC 2385 Galaxy Type ?
⌀0.7′ × 0.3′, m14.2v, SB 12.4   07^{h}28.4^m +33°50′

NGC 2388 Galaxy Type S?
⌀0.9′ × 0.6′, m13.7v, SB 12.9   07^{h}28.8^m +33°49′

NGC 2389 Galaxy Type SAB(rs)c
⌀1.8′ × 1.4′, m12.9v, SB 13.8   07^{h}29.1^m +33°51′
Finder Chart 16-7, Figure 16-10   ★/★/★

16/18″ Scopes–150x: NGC 2389 is the brightest member of a trio of faint galaxies 2.5° NNW of Castor. It is very faint, about 1.5′ in diameter, slightly oval ENE–WSW and has no discernible central brightening. NGC 2388, 5′ SSW of NGC 2389, is slightly dimmer and much more elongated: its 1′ × 0.5′ ENE–WSW halo is uniformly bright. NGC 2385, 18′ WSW of NGC 2389, is the faintest of the three galaxies, its dim diffuse halo elongated 0.75′ × 0.25′ E–W. UGC 3879, 18′ ESE of NGC 2385, is a faint, diffuse streak elongated 1.5′ × 0.25′ ESE–WNW.

PK205+14.1 Sh2-274 Abell 21 Planetary Nebula
⌀615″, m10.3v, CS 15.99v   07^{h}29.0^m +13°15′
Finder Chart 16-6, Figure 16-12   Medusa Nebula   ★★★

12/14″ Scopes–75x: This unusual planetary nebula is a very faint, rather large, irregular half moon-shaped object. Its NE end is brighter. Several stars are involved in the nebula.

16/18″ Scopes–100x: This faint planetary is best viewed with an O-III filter at low power. The brighter area is a huge crescent-shaped object partially filled in with fainter nebulosity. There are three or more stars in the arc but none in the central area. The arc

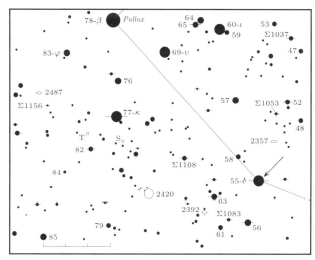

Finder Chart 16-8. 55-δ Gem: 07^{h}20.1^m +21°59′

Figure 16-13. *NGC 2420 is a rich, well-concentrated star cluster with at least fifty stars resolved in medium-sized instruments. Martin C. Germano made this 90 minute exposure on hypered 2415 Tech Pan film with an 8″, f5 Newtonian reflector at prime focus.*

is somewhat brighter along the outer rim. It is also faintly visible without filter. Open cluster NGC 2395 is half a degree NW.

NGC 2392 H45[4] PK197+17.1 Plan. Neb. Type 3b+3b
ø>15″, m9.2v, CS 10.5v $07^h29.2^m$ +20°55′
Finder Chart 16-8   ★★★★★
   Eskimo Nebula or Clown Face Nebula

4/6″ Scopes–150x: NGC 2392, forms an equilateral triangle with a yellowish 8.5 magnitude star 1.5′ North and a 12th magnitude star NW. It has a small but nice bright round, bluish disk surrounding a conspicuous central star.

8/10″ Scopes–200x: Lovely! This bright planetary displays an obvious bluish disk surrounding a bright central star. The planetary conspicuously "blinks on" when you stare at the central star until the nebula-glow disappears and then suddenly look away. 250x reveals an indistinct dark ring halfway out from center to the periphery.

16/18″ Scopes–250x: Beautiful! The Eskimo Nebula exhibits a fine luminous, bluish-colored disk divided by a concentric dark ring. The central portion is 15″ across with a 10.4 magnitude central star. The faint 45″ diameter outer shell surrounds the indistinct dark ring, which encircles the inner bright disk about halfway out from the nebula's center to its periphery.

NGC 2420 H1[6] Open Cluster 100★ Tr Type I 2 r
ø10′, m8.3v, Br★ 11.06v $07^h38.5^m$ +21°34′
Finder Chart 16-8, Figure 16-13   ★★★

8/10″ Scopes–100x: This open cluster is a fairly large cluster of three dozen 10th to 13th magnitude stars in a 10′ × 7′ N–S oval. It appears rich and well-concentrated. Its brightest star is near the west edge. The cluster lies between two 9th magnitude stars aligned N–S.

12/14″ Scopes–100x: NGC 2420 is a faint splash of stars between two approximately N–S 9th magnitude stars. Fifty 11th to 13th magnitude members over a 10′ diameter area stand out against a an unresolved background haze. At the center is an 8′ × 2′ N–S concentration; but the loosely scattered outer part of the cluster is elongated E–W. Outlying cluster members are spread further out on the east side than they do to the west.

NGC 2486 Galaxy Type Sa
ø1.7′ × 1.0′, m13.3v, SB 13.7 $07^h57.9^m$ +25°09′

NGC 2487 Galaxy Type SBb
ø2.3′ × 2.0′, m12.5v, SB 14.0 $07^h58.3^m$ +25°08′
Finder Chart 16-8   ★/★

16/18″ Scopes–150x: NGC 2486 and NGC 2487 are a close pair of faint, oval galaxies without central brightening. NGC 2487 is the larger and brighter of the two, its halo elongated about 2′ × 1′ ENE–WSW. NGC 2486, to the WNW, is elongated only about 1.25′ × 1′ ESE–WNW.

Chapter 17

Lacerta, the Lizard

17.1 Overview

Lacerta was created by the Polish astronomer Johannes Hevelius in the late 17th century. (The other constellations introduced by Hevelius still recognized today are Leo Minor, Lynx, Scutum, Sextans, and Vulpecula.) Lacerta is a small constellation of faint stars covering only 201 square degrees and wedged between Cygnus, Cepheus, Andromeda, and Pegasus. Its northern half extends into the Milky Way and contains some fine open clusters. The constellation also has several good double stars and a planetary nebula

Lacerta's most famous object is BL Lacertae, the prototype of a class of quasi-stellar object (QSO) characterized by a lack of emission lines and rapid, relatively large, light variations. BL Lacertae itself varies between the 14th and 17th magnitudes. The energy output of BL Lacertae objects and "normal" QSOs are thought to be powered by a huge black hole encircled by a vortexing accretion disk that is literally consuming the host galaxy. Because BL Lacertae is extremely remote it is not visually impressive even in large amateur instruments.

> **Lacerta:** La-SIR-ta
> **Genitive:** Lacertae, La-SIR-tee
> **Abbreviation:** Lac
> **Culmination:** 9pm–Oct. 12, midnight–Aug. 28
> **Area:** 201 square degrees
> **Best Deep-Sky Objects:** Σ2894, Σ2902, 9 Lac, OΣ475, IC 1434, NGC 7209, NGC 7243, NGC 7245
> **Binocular Objects:** NGC 7209, NGC 7243

17.2 Interesting Stars

Σ2876 Double Star   Spec. F8
m7.8, 9.3; Sep. 11.8″; P.A. 68° $22^h12.0^m$ +37°39′
Finder Chart 17-4   ★★★

4/6″ Scopes–50x: Struve 2876 is a fine double of white and blue stars.

Σ2894 Double Star   Spec. F0
m6.1, 8.3; Sep. 15.6″; P.A. 194° $22^h18.9^m$ +37°46′
Finder Chart 17-4   ★★★★

4/6″ Scopes–50x: This is a good color contrast pair of yellowish and blue stars.

Σ2902 Multiple Star   Spec. G5
AB: m7.6, 8.5; Sep. 6.4″; P.A. 89° $22^h23.6^m$ +45°21′
Finder Chart 17-3   ★★★★

4/6″ Scopes–50x: Σ2902 consists of a close 6″ double of bright stars, yellowish and white in color, with another double to the NE of magnitude 12.1 and 12.9 separated by 5″. In the same field of view are two more doubles, one close and the other wide.

8 Lacertae = Σ2922 Multiple Star   Spec. B3 & B5
AB: m5.7, 6.5; Sep. 22.4″; P.A. 186° $22^h35.9^m$ +39°38′
Finder Chart 17-4   ★★★★

8/10″ Scopes–75x: 8 Lacertae has so many components that it might be classed as a poor open cluster rather than merely a multiple star. The four closest members, all white or bluish-white, are within 82″ of each other. A fifth component lies 336″ away to the WSW. Several 13th and 14th magnitude stars are also involved.

OΣ475 Double Star   Spec. B3
m6.8, 10.8; Sep. 15.5″; P.A. 73° $22^h39.1^m$ +37°23′
Finder Chart 17-4   ★★★★

4/6″ Scopes–50x: This is a fine unequal double of a white star with a faint bluish companion. Another faint double is 10′ SW.

h975 Double Star   Spec. B9
m5.6, 9.5; Sep. 51.0″; P.A. 243° $22^h55.7^m$ +36°21′
Finder Chart 17-4   ★★★★

4/6″ Scopes–50x: Herschel 975 is a wide, unequal pair with a bright, white primary and a pale blue secondary.

Lacerta, the Lizard

Constellation Chart 17-1

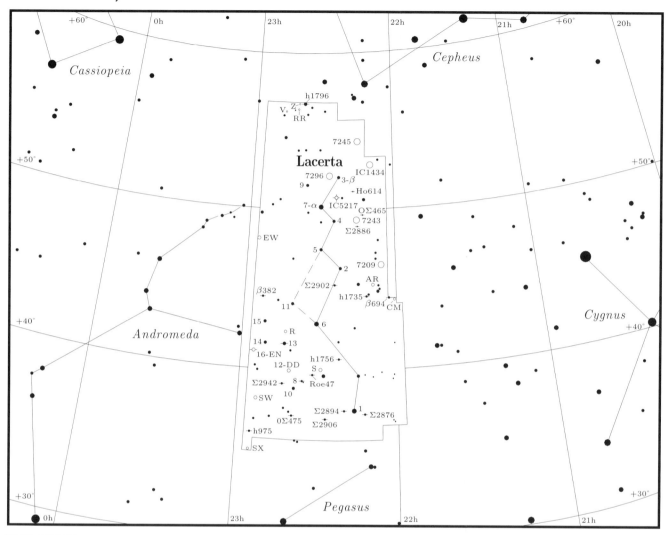

Chart Symbols

Constellation Chart		0	1	2	3	4	5	6
Stellar Magnitudes	●	●	●	●	•	·	·	
Finder Charts	0/1	2	3	4	5	6	7	8	9
Master Finder Chart		0	1	2	3	4	5	

→ Guide Star Pointer ⊙ ○ Variable Stars ⌖ Planetary Nebulae
•—• Double Stars ○ Open Clusters □ Small Bright Nebulae
Finder Chart Scale ⊕ Globular Clusters Large Bright Nebulae
(One degree tick marks) ○ Galaxies Dark Nebulae

Table 17-1. Selected Variable Stars in Lacerta

Name	HD No.	Type	Max.	Min.	Period (Days)	F*	Spec. Type	R.A. (2000)	Dec.	Finder Chart No. & Notes
CM Lac	209147	EA	8.20	9.15	1.60	0.10	A2+A8	22^{h}00.1^m	+44°33′	17-1
AR Lac	210334	EA/RS	6.11	6.77	1.98	0.17	G2+K0	08.7	+45 45	17-3
S Lac	213191	M	7.6	13.9	241	0.46	M5-M8	29.0	+40 19	17-4
Z Lac	214975	Cδ	7.94	8.85	10.88	0.43	F6-G6	40.9	+56 50	17-1
RR Lac		Cδ	8.43	9.22	6.41	0.30	F6-G2	41.4	+56 26	17-1
R Lac	215254	M	8.5	14.8	299	0.41	M5-M5.5	43.3	+42 22	17-4
V Lac	240073	Cδ	8.42	9.39	4.98	0.25	F5-G0	48.6	+56 19	17-1
SX Lac		SRd	9.0	10.0	190	0.47	K2	56.0	+35 12	17-4

F* = The fraction of period taken up by the star's rise from min. to max. brightness, or the period spent in eclipse.

Table 17-2. Selected Double Stars in Lacerta

Name	ADS No.	Pair	M1	M2	Sep."	P.A.°	Spec	R.A. (2000) Dec	Finder Chart No. & Notes
β694	15578	AB	5.6	8.1	0.8	4	A0	22^{h}02.9^m +44°39'	17-4
OΣ459	15576		7.9	10.6	10.8	196	A0	02.9 +39 34	17-4
h1735	15679	AB	6.9	8.4	27.1	110	B9	09.3 +44 51	17-3
Σ2876	15723		7.8	9.3	11.8	68	F8	12.0 +37 39	17-4 Yellowish & blue
OΣ465	15727	AB	7.3	10.8	14.8	322	F8	12.0 +50 12	17-3
Σ2886	15778		7.6	9.8	20.3	109	F8	14.7 +49 21	17-3
Ho 614	15783		7.8	10.3	4.7	175	A0	15.1 +51 29	17-3
Σ2894	15828	AB	6.1	8.3	15.6	194	F0	18.9 +37 46	17-4 Yellowish & blue
h1756	15874	AB	6.7	10.6	22.1	286	K2	21.9 +40 40	17-4
	15874	AC		13.2	22.4	325			
	15874	AD		11.3	58.0	76			
Σ2902	15900	AB	7.6	8.5	6.4	89	G5	23.6 +45 21	17-3 Yellowish & white pair
Σ2906	15942		6.5	10.1	4.3	2	B3	26.8 +37 27	17-4
Roe 47	16031	AB	5.8	9.8	43.1	158	A3	32.4 +39 47	17-4
	16031	AC		10.1	32.4	344			
	16031	AD		9.4	105.7	216			
	16031	DE		9.8	6.6	175			
8 Lac	16095	AB	5.7	6.5	22.4	186	B3 B5	35.9 +39 38	17-4 All white quadruple
	16095	AC		10.5	48.8	169			
	16095	AD		9.3	81.8	144			
h1796	16140		5.5	10.8	30.9	10	M	38.6 +56 48	17-1
OΣ475	16143		6.8	10.8	15.5	73	B3	39.1 +37 23	17-4 White & bluish
13 Lac	16227		5.1	10.5	14.6	129	K0	44.1 +41 49	17-4
Σ2942	16228	AB	6.1	8.3	2.8	280	K5	44.1 +39 28	17-4
β450	16228	AC		11.5	10.6	235			
β382	16345	AB	5.8	7.8	c0.9	*224	A3	53.7 +44 45	17-1
h1828	16345	ABxC		10.7	28.0	356			
h975	16376		5.6	9.5	51.0	243	B9	55.7 +36 21	17-4 White & pale blue
16 Lac	16381	AB	5.6	11.6	27.6	344	B3	56.4 +41 36	17-1
	16381	AC		8.7	62.8	48			

Footnotes: *= Year 2000, a = Near apogee, c = Closing, w = Widening. Finder Chart No: All stars listed in the tables are plotted in the large Constellation Chart, but when a star appears in a Finder Chart, this number is listed. Notes: When colors are subtle, the suffix *-ish* is used, e.g. *bluish*.

17.3 Deep-Sky Objects

NGC 7209 H53⁷ Open Cluster 25★ Tr Type III 1 p
ø25', m7.7v, Br★ 9.02v   22^{h}05.2^m +46°30'
Finder Chart 17-3, Figure 17-1   ★★★

8/10" Scopes–100x: NGC 7209 is a large but irregularly scattered cluster of seventy-five 10th magnitude and fainter stars spread over a 25' area framed by 8th and 9th magnitude stars. A 5th magnitude field star is to the north.

12/14" Scopes–100x: NGC 7209, set in a rich Milky Way field, consists of over a hundred members, a large number of which are 11th and 12th magnitude objects, irregularly scattered over a 25' area. It is not well concentrated, but its stars are bunched in short chains and angles, and several triples can be seen in its eastern section.

IC 1434 Cr 445 Open Cluster 40★ Tr Type II 1 p
ø7', m9.0p, Br★ 12.0p   22^{h}10.5^m +52°50'
Finder Chart 17-3, Figure 17-2   ★★★

8/10" Scopes–100x: This is a faint but attractive cluster of fifty 12th magnitude and fainter stars in a 12' area. Three brighter stars form a triangle on the south edge.

12/14" Scopes–100x: IC 1434 is a fairly faint, large, rich cluster of seventy 12.5 magnitude and fainter stars in a 12' area. Three of the brighter stars form a prominent triangle on the southern edge. A remarkable chain of 20 stars meanders NE from the northern edge.

NGC 7223 H862³ Galaxy Type SB(rs)bc
ø1.8' × 1.2', m12.2v, SB 12.9   22^{h}10.2^m +41°00'
Finder Chart 17-4   ★★★

16/18" Scopes–125x: NGC 7223, 8' east of an 8th magnitude star, appears fairly bright but is diffuse and elongated 1.5' × 1.0' N–S. Two stars are just to its south, one nearly touching the galaxy's southern edge. A knot is suspected just west of a star at the northern edge.

NGC 7243 H75⁸ Open Cluster 40★ Tr Type IV 2 p
ø21', m6.4v, Br★ 8.47v   22^{h}15.3^m +49°53'
Finder Chart 17-3, Figure 17-3   ★★★

8/10" Scopes–100x: This is a large, irregular cluster composed of five coarse groupings of stars arranged

Figure 17-1. *Open cluster NGC 7209 is a large, irregularly scattered open cluster framed by somewhat brighter stars. Martin C. Germano made this 40 minute exposure on 2415 film with an 8″, f5 Newtonian reflector.*

Figure 17-2. *Open cluster IC 1434 is an attractive cluster rich in faint stars. Lee C. Coombs made this ten minute exposure on 103a-O film at prime focus with a 10″, f5 Newtonian reflector.*

in a semicircle opening to the north. The cluster stands out well from the abundant star field, its NE and SW concentrations being the most conspicuous. The brighter stars make the cluster appear elongated ENE–WSW.

12/14″ Scopes–100x: NGC 7243 is a nice, bright cluster of seventy-five 9th magnitude and fainter stars in irregular clumpings spread over a 20′ area. The groups are arranged in an irregular half moon, its "terminator" being the line of brighter stars along the cluster's northern edge. Near the center is the double Σ2890, two 9th magnitude stars 9″ apart in P.A. 11°. Another 40 stars are scattered north and

east of the semicircle. A wide, conspicuous pair of 9th magnitude stars lies 5′ SW.

NGC 7245 H29[8] Open Cluster Tr Type II 1 p
ø6.5′, m9.2v, Br★ 12.75v $22^h15.3^m$ +54°20′

Finder Chart 17-3, Figure 17-4   ★★★

8/10″ Scopes–100x: This nice but faint cluster is a rich elongated group of forty stars, most 13th magnitude and fainter. Several 11th magnitude stars are at the west edge, and a 9th magnitude star is somewhat outside the cluster to the east.

12/14″ Scopes–100x: NGC 7245 is a concentration of forty 13th magnitude and fainter stars against a twinkling background of threshold stars in a 7′ area elongated E–W. The central area is noticeably concentrated. A 9th magnitude star stands out on the east edge, and several 11th magnitude stars are on the west and SSE sides.

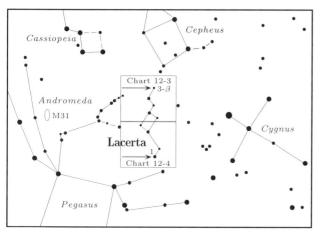

Master Finder Chart 17-2. Lacerta Chart Areas
Guide stars indicated by arrows.

NGC 7240 Galaxy Type S0–:
ø0.5′ × 0.5′, m14.2v, SB 12.5 $22^h15.4^m$ +37°17′

NGC 7242 Galaxy Type E+:
ø2.3′ × 1.7′, m12.9v, SB 14.2 $22^h15.7^m$ +37°18′

Finder Chart 17-4   ★/★★

16/18″ Scopes–125x: This pair of faint galaxies is half a degree south of 1 Lacertae (m4.13v) in a field rich in faint stars. The easternmost galaxy, NGC 7242, is very faint, diffuse, and elongated 2′ × 1.5′ NNE–SSW. NGC 7240, 3.5′ to the WSW, is a small,

Figure 17-3. *Open cluster NGC 7243 is highly irregular with several isolated clumps of stars. Lee C. Coombs made this ten minute exposure on 103a-O film at prime focus with a 10″, f5 Newtonian reflector.*

Figure 17-4. *NGC 7245 (bottom) is a rich concentration of faint stars while IC 1442 (top, right) is a box-shaped group. Lee C. Coombs made this ten minute exposure on 103a-O film with a 10″, f5 Newtonian reflector.*

faint, round 30″ patch. A third galaxy in the area, IC 1441, 4′ west of NGC 7242, is a very faint 20″ patch.

IC 1442   Open Cluster 20★   Tr Type II 2 m
ø3.5′, m9.1v, Br★ 11.43v   22ʰ16.5ᵐ +54°03′
Finder Chart 17-3, Figure 17-4   ★★

12/14″ Scopes–100x: This is a faint cluster of thirty stars contained in a 5′ wide "box" of 11th and 12th magnitude stars.

16/18″ Scopes–125x: IC 1442 is a faint, loose, diamond-shaped cluster, the tips aligned N–S and E–W, of three dozen magnitude 11.5 and fainter stars in a 7′ area. The north and south tips of the diamond are marked by magnitude 11.5 stars. A detached row of five 11th magnitude stars runs NNE from the northern tip. Outlying stars spread NW and SW from the main concentration. Open cluster NGC 7245 is located 22′ NNW.

NGC 7248   H863³   Galaxy Type S0–:
ø1.8′ × 0.9′, m12.4v, SB 12.8   22ʰ16.9ᵐ +40°30′
Finder Chart 17-4   ★★

16/18″ Scopes–125x: NGC 7248 has a faint halo, elongated 1′ × 0.5′ NW–SE, with a bright stellar nucleus. A pair of 12th and 13th magnitude stars 6″ apart lies 2.5′ WSW, and a pair of 13th magnitude stars 10″ apart is 2′ east. Galaxy NGC 7250 is located 17′ ENE.

NGC 7250   H864³   Galaxy Type Sdm?
ø1.2′ × 0.6′, m12.6v, SB 12.1   22ʰ18.3ᵐ +40°35′
Finder Chart 17-4   ★★

12/14″ Scopes–100x: NGC 7250 is a fairly faint, lens-shaped galaxy elongated 1′ × 0.5′ NNW–SSE with a sudden brightening in the core area. An 11.5 magnitude star is 1′ from the SSE tip.

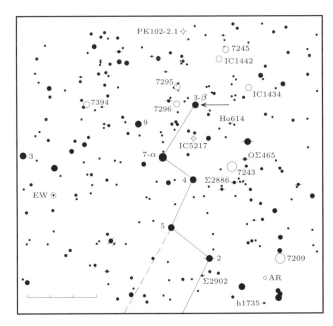

Finder Chart 17-3. 3-β Lac: 22ʰ23.5ᵐ +52°14′

NGC 7265 Galaxy Type S0–
ø2.4′ × 1.9′, m12.2v, SB 13.7   22^{h}22.4^m +36°14′
Finder Chart 17-4   ★★

16/18″ Scopes–125x: This galaxy is 15′ SSE of the magnitude 7.5 star centering a degree long NE–SW row of three bright stars, a magnitude 5 star at the NE end. NGC 7265 has a faint, circular 1.5′ diameter halo with a bright core and a stellar nucleus. It is the brightest galaxy in a group with NGC 7263–64 12′ lying NNW and NGC 7273–74–76 lying 20′ ESE.

IC 5217 Planetary Nebula Type 2
ø6″, m11.3v, CS 15.4v   22^{h}23.9^m +50°58′
Finder Chart 17-3   ★★★

12/14″ Scopes–100x: IC 5217, 2′ north of an 11.5 magnitude star, is a bright, small, bluish disk which responds well to an O-III filter.

16/18″ Scopes–125x: IC 5217 has a bright, bluish 6″ disk slightly elongated N–S with a brighter core.

NGC 7273 Galaxy Type ?
ø0.8′ × 0.5′, m13.7v, SB 12.6   22^{h}24.1^m +36°13′

NGC 7274 Galaxy Type E
ø1.5′ × 1.5′, m12.8v, SB 13.5   22^{h}24.2^m +36°08′

NGC 7276 Galaxy Type ?
ø0.9′ × 0.9′, m13.7v, SB 13.3   22^{h}24.2^m +36°06′
Finder Chart 17-4   ★/★/★

16/18″ Scopes–125x: NGC 7274, the brightest and largest of a trio of galaxies, is 6′ SSW of a 50″ pair of 11.5 magnitude stars. It has a prominent core with a stellar nucleus embedded in a faint, round 1′ diameter halo. NGC 7273, 4.5′ north of NGC 7274, has a faint, small, slightly N–S elongated halo that brightens suddenly at the core. A magnitude 11.5 star is 2′ west. NGC 7276, the southernmost of the three galaxies, is 2.5′ SSE of NGC 7274 and has a small round halo with a faint stellar nucleus. It forms an isosceles triangle with 7274 and a 12th magnitude star 2.5′ NE.

NGC 7296 Cr 452 Open Cluster 20★ Tr Type III 2 p
ø4′, m9.7p, Br★ 10.0p   22^{h}28.2^m +52°17′
Finder Chart 17-3   ★★★

12/14″ Scopes–100x: This cluster is situated between the 4.4 magnitude star Beta (β) = 3 Lacertae and a 6.5 magnitude star 15′ ENE. It is a fairly conspicuous concentration of twenty 12th to 13th magnitude stars in a 4′ area. An 11th magnitude star stands out at the western edge.

16/18″ Scopes–125x: NGC 7296 stands out well from the rich field just east of an 11th magnitude star centering a N–S row of 11th or 12th magnitude stars. The cluster is a small, fairly faint, oblong patch of three dozen 12th to 13th magnitude stars concentrated in a 4′ area.

NGC 7295 Open Cluster
[ø1′, Br★ 10.v]   22^{h}28.4^m +52°50′
Finder Chart 17-3   ★★★

12/14″ Scopes–125x: Though catalogued as nonexistent, NGC 7295 can be seen as a faint, unconcentrated 1′ diameter group of fifteen 11th to 13th magnitude stars surrounding a 10th magnitude lucida.

PK102-2.1 Abell 79 Planetary Nebula
ø>59″, m15.0v, CS 17.6v   22^{h}26.3^m +54°49′
Finder Chart 17-3   ★

16/18″ Scopes–125x: This planetary nebula is extremely faint, its 50″ diameter disk visible only through an O-III filter. It lies just east of a double of 8th and 9th magnitude stars separated by 10″.

NGC 7394 Open Cluster
[ø10′, Br★ 10.v]   22^{h}50.6^m +52°10′
Finder Chart 17-3   ★★★

12/14″ Scopes–125x: Though another allegedly "nonexistent" open cluster, NGC 7394 is a conspicuous group of nineteen 10th to 12th magnitude stars in a 10′ × 3′ area spreading NW from its southernmost bright star. Two 7th magnitude stars aligned N–S are near the cluster: one at the cluster's SE edge; the other 9′ NE. The surrounding field is not particularly rich.

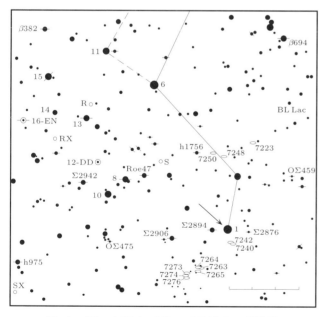

Finder Chart 17-4. 1 Lac: 22^{h}15.9^m +37°45′

Chapter 18

Lepus, the Hare

18.1 Overview

According to a late myth, Lepus had been a bird that was a changed into a hare by Ostara, the goddess of spring. Ostara permitted the hare the power to run as fast as formerly it had been able to fly. Moreover, once a year the hare was allowed to again lay eggs — the pagan origin of the custom of Easter eggs. (The word "Easter" is derived from "Ostara," the Christian commemoration of Christ's Crucifixion and Resurrection being in early Medieval times fused with the old pagan celebration of the vernal equinox dedicated to the goddess of spring.)

Lepus' brightest stars form an easily-recognized parallelogram due south of Orion. Because of its southerly declination, mid-northern observers must wait for Lepus to culminate for the best views of its objects. The astronomical glories of the Hunter overshadow the more modest offerings of the Hare. Nevertheless Lepus should not be ignored, for it contains several interesting objects, including the multicolored star-group h 3780, the globular cluster Messier 79, and the poppy red variable R Leporis, Hind's Crimson Star.

Lepus: LEE-pus
Genitive: Leporis, LEP-or-is
Abbrevation: Lep
Culmination: 9pm–Jan. 28, midnight–Dec. 14
Area: 290 square degrees
Showpieces: h3752, 13–γ Lep, R Lep, M79 (NGC 1904), NGC 1964
Binocular Objects: 13–γ Lep, R Lep, M79 (NGC 1904), NGC 2017

18.2 Interesting Stars

R Leporis Variable Star Type M   Spec. C6 IIe
m5.5 to 11.7 in 432.13 days $04^h59.6^m$ $-14°48'$
Finder Chart 18-4   Hind's Crimson Star   ★★★★

In October 1845 John R. Hind of London discovered the blood red carbon star R Leporis, a long-period pulsating red variable, now known as Hind's "Crimson star." It is one of the most vividly-hued stars in the entire sky.

R Leporis varies between magnitudes 6 and 11.5 in a period of around 430 days. Its maximum also varies in a long secondary cycle of 40 years from magnitude 5.5 down to magnitude 6.5 and back. The full 6.2 magnitude difference between R Leporis' extreme maxima and minima corresponds to an actual brightness difference of 300 times. Like other red variables in general, and carbon stars in particular, the star's color at maximum is not as intense as at minimum: at maximum R Leporis is only copper-hued, but near minimum it approaches true crimson. R Leporis, like other carbon stars, has a spectrum with strong bands of carbon compounds, which are efficient absorbers of blue light. Carbon stars are very cool, their surface temperatures being 2600°K and less.

Kappa (κ) = 4 Leporis (Σ661) Double Star   Spec. B8 III
m4.5, 7.4; Sep. 2.6"; P.A. 358° $05^h13.2^m$ $-12°56'$
Finder Chart 18-4, Figure 18-1   ★★★★

8/10" Scopes–175x: Not easily split. At 115x both components appear white.

12/14" Scopes–200x: Moderately high power shows a bright white and a blue star in contact.

S473 = ADS 3883 Double Star   Spec. B8.5 V
m6.7, 8.7; Sep. 20.6"; P.A. 305° $05^h17.6^m$ $-15°13'$
Finder Chart 18-4   ★★★★

2/3" Scopes–175x: South 473, a white and blue duo, is a nice, wide pair suitable for small instruments.

S476 = ADS 3910 Double Star   Spec. B8 & B8
m6.2, 6.4; Sep. 39.4"; P.A. 18° $05^h19.3^m$ $-18°31'$
Finder Chart 18-4   ★★★★

2/3" Scopes–175x: South 476 is a bright, wide pair of white stars also resolvable in binoculars.

Lepus, the Hare

Constellation Chart 18-1

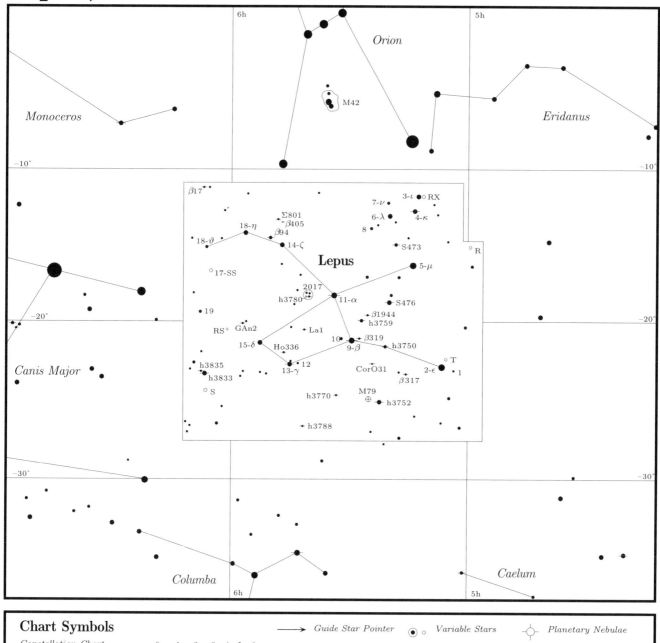

Table 18-1. Selected Variable Stars in Lepus

Name	HD No.	Type	Max.	Min.	Period (Days)	F*	Spec. Type	R.A. (2000)	Dec.	Finder Chart No. & Notes
R Lep	31996	M	5.5	11.7	432	0.55	C6	04^{h}59.6^m	−14°48′	18-4 Hind's Crimson Star
T Lep	32803	M	7.4	13.5	368	0.47	M6–M8	05^{h}04.8^m	−21 54	18-3
RX Lep	33664	Lb	5.0	7.0			M6	11.4	−11 51	18-4
RS Lep	40640	EA	9.3	10.9	1.28	0.17	A2	59.3	−20 13	18-5
S Lep	41698	SRb	7.1	8.9	90		M6	06^{h}05.8^m	−24 12	18-5

F* = The fraction of period taken up by the star's rise from min. to max. brightness, or the period spent in eclipse.

Table 18-2. Selected Double Stars in Lepus

Name	ADS No.	Pair	M1	M2	Sep."	P.A.°	Spec	R.A. (2000)	Dec	Finder Chart No. & Notes
4–κ Lep	3800		4.5	7.4	2.6	358	B8	05^{h}13.2^m	−12°56′	18-4 White pair
β317	3819	AB	7.4	10.7	8.6	12	G5	13.9	−22 59	18-3
	3819	AC		12.6	18.1	47				
S473	3883		6.7	8.7	20.6	305	B8	17.6	−15 13	18-4
S476	3910		6.2	6.4	39.4	18	B8 B8	19.3	−18 31	18-4 Wide white duo
h3750	3930		4.7	8.5	4.2	282	A0	20.4	−21 14	18-3 Yellowish-white & blue
h3752	3954	AB	5.4	6.6	3.2	97	G0 A3	21.8	−24 46	18-3 AB: Yellowish & bluish
CorO 31	3993		7.4	9.7	18.0	283	A3	23.7	−22 17	18-3
β1944			6.9	10.9	2.6	75	A0	25.0	−19 22	18-1
h3759	4034		5.8	8.6	27.1	318	F5 F5	26.0	−19 42	18-1 Unequal yellowish pair
β319	4042		7.4	10.5	3.9	231	A3	26.4	−20 43	18-3
9–β Lep	4066	AB	2.8	7.3	2.8	330	G0	28.2	−20 46	18-3
11–α Lep	4146	AB	2.6	11.1	35.8	156	F0	32.7	−17 49	18-4
h3770	4157		7.7	11.2	4.0	19	F8	33.5	−24 20	18-3
β321	4254	AB	6.4	7.9	0.8	146	B9	39.3	−17 51	18-1 = NGC 2017
h3780	4254	ABxI			89.2	102				See visual description
	4254	AC		8.5	89.2	136				
	4254	AE		8.4	76.1	7				
	4254	AF		8.1	128.8	299				
	4254	AG		9.5	59.8	49				
	4254	AH		12.4	41.8	310				
	4254	CD		9.2	1.5	357				
La1 Hd78	4260	AB	6.9	7.9	11.0	123	B8	39.7	−20 26	18-1 Fine white & purple pair
	4260	AC		11.3	32.2	83				
h3788	4281		7.8	9.6	25.9	154	F5	41.3	−26 21	18-3
13–γ Lep	4334	AB	3.7	6.3	96.3	350	F8 G5	44.5	−22 27	18-5 Vivid & pale yellow stars
Ho 336	4339		6.7	11.7	19.4	238	B3	44.8	−21 40	18-5
β405	4397		8.1	10.6	14.3	126	K2	47.9	−13 32	18-1
Σ801	4410	AxBC	7.4		27.1	326	K0	48.5	−13 22	18-1
Rst 5502	4410	BC	10.8	10.8	0.2	96				
β94	4432		5.5	8.9	2.3	173	G5	49.6	−14 29	18-1
GAn 2	4503		7.4	9.7	9.8	20	G5	54.5	−19 42	18-5
h3833	4704		5.5	10.6	44.3	72	A2	06^{h}06.5^m	−23 07	18-5
h3835			7.5	11.0	30.2	83	A0	06.9	−23 06	18-5
β17	4741		6.8	10.8	3.1	184	A2	08.4	−11 09	18-1

Footnotes: *= Year 2000, a = Near apogee, c = Closing, w = Widening. Finder Chart No: All stars listed in the tables are plotted in the large Constellation Chart, but when a star appears in a Finder Chart, this number is listed. Notes: When colors are subtle, the suffix -*ish* is used, e.g. *bluish*.

h3750 = ADS 3930 Double Star   Spec. A0
m4.7, 8.4; Sep. 4.2″; P.A. 282°   05^{h}20.4^m −21°14′
Finder Chart 18-3   ★★★★

2/3″ Scopes–100x: This close yellowish-white and blue pair was called "a most beautiful double" by William Herschel.

h3752 = ADS 3954 Double Star   Spec. G0 & A3
m5.4, 6.6; Sep. 3.2″; P.A. 97°   05^{h}21.8^m −24°46′
Finder Chart 18-3   ★★★★

2/3″ Scopes–100x: Herschel 3752, a bright close pair, has yellowish and blue-white stars. A magnitude 9 companion is about 60″ distant from the bright pair.

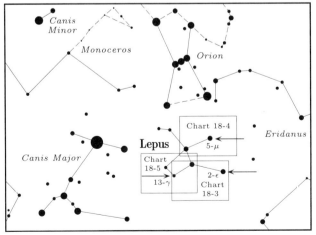

Master Finder Chart 18-2. Lepus Chart Areas
Guide stars indicated by arrows.

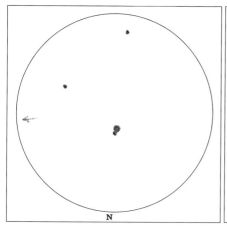

Figure 18-1. Kappa (4–κ) Leporis
12.5″, f5–225x, by G. R. Kepple

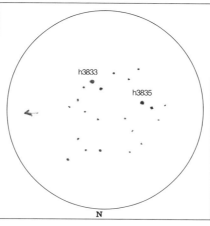

Figure 18-2. h3833 & h3835
13″, f5.6–100x, by Steve Coe

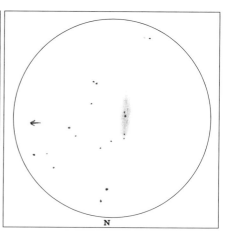

Figure 18-3. NGC 1744
13″, f4.5–115x, by Tom Polakis

h3759 = ADS 4034 Double Star   Spec. F5 & F5
m5.8, 8.6; Sep. 27.1″; P.A. 318° $05^h26.0^m$ $-19°42'$
Constellation Chart 18-1   ★★★

2/3″ Scopes–100x: Herschel 3759 is a wide, unequally bright, pair of yellowish stars.

La1 = ADS 4260 Double Star   Spec. B8
AB: m6.9, 7.9; Sep. 11.0″; P.A. 123° $05^h39.7^m$ $-20°26'$
Finder Chart 18-5   ★★★

2/3″ Scopes–100x: This is a fine white and purple duo.

Gamma (γ) = 13 Leporis Double Star Spec. F8 & G5
AB: m3.7, 6.3; Sep. 96.3″; P.A. 350° $05^h44.5^m$ $-22°27'$
Finder Chart 18-5   ★★★★

2/3″ Scopes–50x: Lovely! The primary appears bright yellowish and the companion pale orange.

8/10″ Scopes–75x: In medium size telescopes the stars appear pale yellow and vivid yellow.

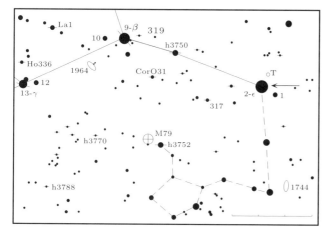

Finder Chart 18-3. 2-ε Lep: $05^h05.5^m$ $-22°22'$

18.3 Deep-Sky Objects

NGC 1744 Galaxy Type SB(s)d III-IV
⌀5.1′ × 2.5′, m11.3v, SB 13.9 $05^h00.0^m$ $-26°01'$
Finder Chart 18-3, Figure 18-3   ★★

12/14″ Scopes–125x: NGC 1744 is very faint, elongated 4′ × 1′ N–S, and slightly brighter at the center. Several 13th to 14th magnitude stars lie at the north end, and one near the center may be a stellar nucleus.

16/18″ Scopes–150x: Two faint stars superimposed at the center are surrounded by a large dim halo elongated 5 × 2′ N–S.

NGC 1784 Galaxy Type (R′)SB(r)c II
⌀4.6′ × 2.7′, m11.7v, SB 14.3 $05^h05.4^m$ $-11°52'$
Finder Chart 18-4, Figure 18-4   ★★

12/14″ Scopes–125x: NGC 1784, 1.25′ south of a 12″ wide pair of 12th and 14th magnitude stars, has a rather faint halo, elongated 2.5′ × 1.5′ E–W, with an unconcentrated core.

16/18″ Scopes–150x: NGC 1784 is fairly faint, elongated 3′ × 1.5′ E–W, and has a mottled core containing a very faint stellar nucleus. The halo's eastern edge is noticeably brighter than the western. A 12th magnitude star lies 3′ east, and another 2′ west, of the galaxy.

NGC 1832 H292² Galaxy Type SB(r)bc I-II
⌀2.1′ × 1.5′, m11.3v, SB 12.4 $05^h12.1^m$ $-15°41'$
Finder Chart 18-4, Figure 18-5   ★★★

12/14″ Scopes–125x: NGC 1832, half a degree north of the magnitude 5.5 Mu (μ) = 5 Leporis, may be readily spotted 1′ west of a 10th magnitude star and

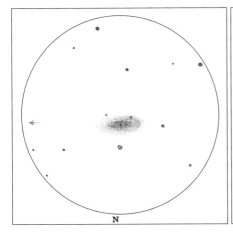
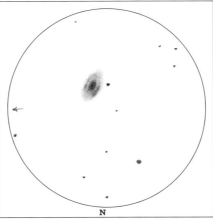
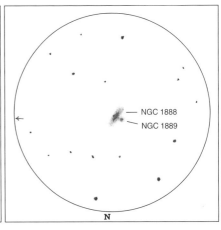

Figure 18-4 NGC 1784
16", f5–195x, by George de Lange

Figure 18-5 NGC 1832
16", f5–195x, by George de Lange

Figure 18-6. NGC 1888 & NGC 1889
12.5", f5–125x, by G. R. Kepple

5′ SW of a 9th magnitude star. The halo is fairly bright, elongated 1.75′ × 1.25′, and has a faint, circular core with a stellar nucleus.

16/18″ Scopes–150x: NGC 1832 appears fairly bright, elongated 2.75′ × 1.5′ NNW–SSE with a large inconspicuous core and a faint stellar nucleus. Bright areas are visible at the halo's NE and SW edges.

NGC 1888 Galaxy Type Sb: pec sp
ø3.0′ × 1.2′, m11.9v, SB 13.2   05ʰ22.58ᵐ −11°30′

NGC 1889 Galaxy Type SA?(r:)0+ pec
ø0.6′ × 0.4′, m13.3v, SB 11.6   05ʰ22.59ᵐ −11°29′
Finder Chart 18-4, Figure 18-6   ★★★/★

12/14″ Scopes–125x: NGC 1888 is fairly bright, elongated 2.5′ × 1′ NW–SE, and has a brighter center. NGC 1889, touching the larger galaxy's NE edge, is a faint, tiny, circular glow with some brightening at center.

NGC 1904 Messier 79 Globular Cluster Class V
ø8.7′, m7.8v,   05ʰ24.5ᵐ −24°33′
Finder Chart 18-3, Figure 18-7   ★★★★

NGC 1904 was discovered by Mechain in October 1790, and a few months later Messier added it as the 79th entry into his catalogue of imposter comets. It is located 34′ ENE of the double star h3752 (see Table 18-2). Its distance is estimated to be 41,000 light years.

12/14″ Scopes–150x: M79 is within a dipper-shaped asterism of 9th to 11th magnitude stars. It is a fine, bright globular with a large, dense core surrounded by a 5′ diameter halo. Three dozen stars may be resolved against a glowing background. A 5th magnitude star lies half a degree WSW.

16/18″ Scopes–175x: Messier 79 is a nice globular containing a bright, dense 1′ diameter core surrounded by a 7′ diameter halo of much looser outlying stars. A 12th magnitude star is visible at the north edge. Except for the dense core, the cluster is well resolved. A N–S chain of brighter stars stands out across the face. A second star chain extends 4′ south, pointing to a 9th magnitude star 10′ from the globular's center. Another 9th magnitude star lies 8′ north.

IC 418 Planetary Nebula Type 4
ø12″, m9.3v, CS 10.17v   05ʰ27.5ᵐ −12°42′
Finder Chart 18-4, Figure 18-8   ★★★

12/14″ Scopes–125x: This bright but tiny planetary nebula

Figure 18-7. *Globular cluster Messier 79 (NGC 1904) has a large core surrounded by a much looser periphery of stars. Lee C. Coombs made this 10 minute exposure on 103a-E Kodak Spectroscopic film with a 10″, f5 Newtonian.*

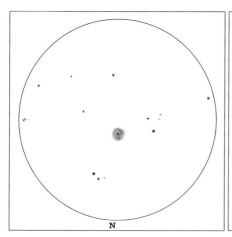
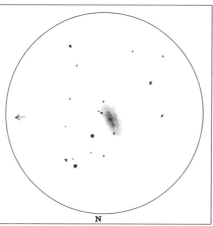
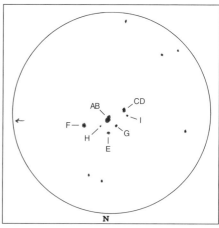

Figure 18-8. IC 418
13″, f5.6–270x, by Steve Coe

Figure 18-9. NGC 1964
16″, f5–120x, by George de Lange

Figure 18-10. NGC 2017 (h3780)
16″, f5–120x, by George de Lange

needs high power for a good view. A bluish-green 10″ diameter disk surrounds a prominent 10th magnitude central star.

16/18″ Scopes–150x: IC 418 displays a nice bluish-green 14″ × 11″ N–S oval disk with a bright yellow central star.

NGC 1964 H214⁴ Galaxy Type SA(s)bc I-II
ø5.0′ × 2.1′, m10.7v, SB 13.1   05ʰ33.4ᵐ −21°57′
Finder Chart 18-3, Figure 18-9   ★★★★

8/10″ Scopes–100x: This fairly bright galaxy has a 3′ × 1′ halo elongated NNE–SSW with a much brighter center and a stellar nucleus. A faint star is involved 45″ west of center.

16/18″ Scopes–150x: NGC 1964 is placed 1.5′ SE of a 10.5 magnitude star, at the south apex of a thin 2′ × 0.5′ acute triangle with magnitude 9.5 and 11 stars to its NW. Another 11th magnitude star lies 4.75′ SW of the galaxy, NGC 1964 has a bright stellar nucleus in a small oval core surrounded by a 4′ × 1.5′ halo elongated NNE–SSW. Although indistinct, some spiral structure can be glimpsed within the granular halo. Three stars are involved: a 13.5 magnitude star 45″ west of the galaxy's center, a 14th magnitude star on its NW edge, and a very faint star on its SW side.

NGC 2017 h3780 Open Cluster 8★ or Multiple Star
ø−, m−, Br★ 8.0v   05ʰ39.4ᵐ −17°51′
Finder Chart 18-4, Figure 18-10   ★★★★

12/14″ Scopes–125x: This star-group, 2° east of magnitude 2.6 Alpha (α) = 11 Leporis, bears both an NGC number and a double star designation. It is either a very sparse open cluster (rather like the five central

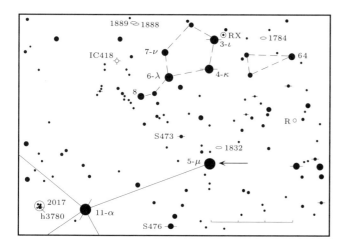
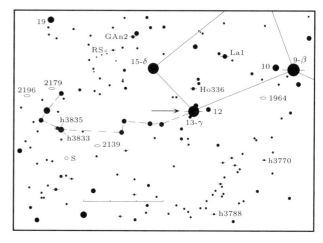

Finder Chart 18-4. 5–μ Lep: 05ʰ12.9ᵐ −16°12′

Finder Chart 18-5. 13–γ Lep: 05ʰ44.5ᵐ −22°27′

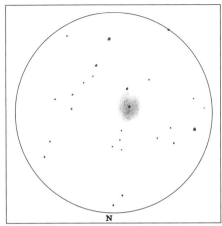 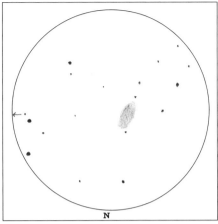 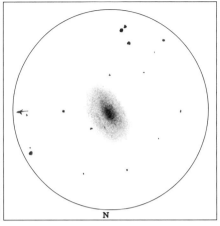

Figure 18-11. NGC 2139
13", f5.6–100x, by Steve Coe

Figure 18-12. NGC 2179
16", f5–200x, by George de Lange

Figure 18-13. NGC 2196
16", f5–200x, by George de Lange

stars of the Big Dipper) or a populous multiple star. It consists of five brighter, and four fainter, stars, four of the brighter stars in two close doubles. The primary, which has a yellowish color, is a very close pair (AB: 6.4, 7.9; 0.8"; 146°) that requires at least 250x to even look oblong. To its SE is the other close double in the group (CD: 8.5, 9.2; 1.5"; 357°) which has a nice orange color and at 250x shows as two tiny disks in contact. The westernmost star (F) in the group is bluish. At this aperture the four faint members of the "cluster" have nondescript color. Several other multiple star systems are in the surrounding star field.

NGC 2139 H264² Galaxy Type SB(s)c: pec II
⌀3.0' × 2.3', m11.4v, SB 13.3   06ʰ01.1ᵐ −23°40'

Finder Chart 18-5, Figure 18-11   ★★★

12/14" Scopes–125x: NGC 2139, 3.5' north of a 9th magnitude star, is fairly faint with a uniform, circular 1.5' diameter halo.

16/18" Scopes–150x: This galaxy's fairly bright, well defined, uniformly luminous disk is elongated 2' × 1.5' N–S and has a faint stellar nucleus. A 13th magnitude star lies 1' north.

NGC 2179 Galaxy Type (R')SB(rs)0/a?
⌀1.9' × 1.4', m12.3v, SB 13.2   06ʰ08.0ᵐ −21°45'

Finder Chart 18-5, Figure 18-12   ★★

12/14" Scopes–125x: This faint galaxy extends NNW–SSE between two 14th magnitude stars near the tips of its major axis. Its halo is an elongated 1' × 0.5' diffuse smudge. A wide blue and yellow double star shares the field of view.

16/18" Scopes–150x: NGC 2179 has a fairly faint, 1.5' × 0.75' NNW–SSE halo with little central brightening.

NGC 2196 H265² Galaxy Type (R')SA(rs)ab? I-II
⌀3.0' × 2.5', m11.1v, SB 13.1   06ʰ12.2ᵐ −21°48'

Finder Chart 18-5, Figure 18-13   ★★★

12/14" Scopes–125x: NGC 2196 has a well concentrated 1.5' × 1' NE–SW oval halo with a tiny core. The field is well sprinkled with faint stars. A 10th magnitude star lies 7' ENE.

16/18" Scopes–150x: This galaxy is larger, brighter, and easier than nearby NGC 2179. Its bright, extended core is embedded within a 2' × 1.5' NE–SW halo. A 14th magnitude star lies 1' west.

Chapter 19

Lynx, the Lynx

19.1 Overview

Lynx was created by the Polish astronomer Johannes Hevelius (A.D. 1611–1687) to fill the gap between Auriga and Ursa Major. It is a rather obscure constellation; even its brightest star Alpha Lyncis is barely third magnitude. Not only is the constellation faint, but it is devoid of star clusters and nebulae. But, don't be too quick to write it off; there are many fine double stars and galaxies in Lynx to reward the telescopist.

Lynx: Links
Genitive: Lyncis, LIN-sis
Abbrevation: Lyn
Culmination: 9pm–Mar. 5, midnight–Jan. 19
Area: 545 square degrees
Showpieces: 5 Lyn, 19 Lyn, 38 Lyn, NGC 2683
Binocular Objects: 5 Lyn, NGC 2683

19.2 Interesting Stars

5 Lyncis   Triple Star   Spec. K2
AB: m5.3, 9.8; Sep. 31.4″; P.A. 139° $06^h26.8^m$ +58°25′
Constellation Chart 19-1   ★★★★

2/3″ Scopes–35x: 5 Lyncis is a fine, wide pair of yellow and blue stars. A third star, magnitude 7.9, is 96″ distant in position angle 272°. 6 Lyncis and two other, very wide doubles make the 5 Lyncis neighborhood an unusually rich off-Milky-Way star field. One of the wide doubles, 1° SE of 5 Lyncis, consists of orange and blue stars, and the other, 40′ east, is two white stars.

19 Lyncis   Quadruple Star   Spec. B8 & A
AB: m5.6, 6.5; Sep. 14.8″; P.A. 315° $07^h22.9^m$ +55°17′
Finder Chart 19-4   ★★★★

2/3″ Scopes–35x: At low power the close A–B pair of 19 Lyncis is a fine color contrast double of yellowish-white and bluish-white stars.

12/14″ Scopes–100x: The A–B pair, both white stars, form an acute triangle with the blue magnitude D component 215″ north. The magnitude 10.9 C component is 74″ WNW of the B star.

Σ1282   Double Star   Spec. F8
m7.5, 7.5; Sep. 3.6″; P.A. 279° $08^h50.7^m$ +35°04′
Finder Chart 19-7   ★★★

2/3″ Scopes–60x: Struve 1282 is a close pair of equally bright yellowish stars. It is in the shaft of an arrow-shaped asterism that resembles Sagitta.

12/14″ Scopes–125x: Struve 1282 is composed of two lovely deep yellow suns.

Σ1333   Double Star   Spec. A5
m6.4, 6.7; Sep. 1.6″; P.A. 49° $09^h18.4^m$ +35°22′
Finder Chart 19-7   ★★★

2/3″ Scopes–100x: Quite pretty! Struve 1333 is a very close pair of white stars.

38 Lyncis   Double Star   Spec. A2
AB: m3.9, 6.6; Sep. 2.7″; P.A. 229° $09^h18.8^m$ +36°48′
Finder Chart 19-7   ★★★★

2/3″ Scopes–100x: 38 Lyncis is a beautiful white and rust colored pair that splits nicely at moderate power.

Σ1338   Double Star   Spec. F2
AB: m6.5, 6.7; Sep. 0.4″; P.A. 16° $09^h21.0^m$ +38°11′
Finder Chart 19-7   ★★★

12/14″ Scopes–250x: These two stars appear as a bright yellow oval, but are not split.

Lynx, the Lynx (Bobcat)

Constellation Chart 19-1

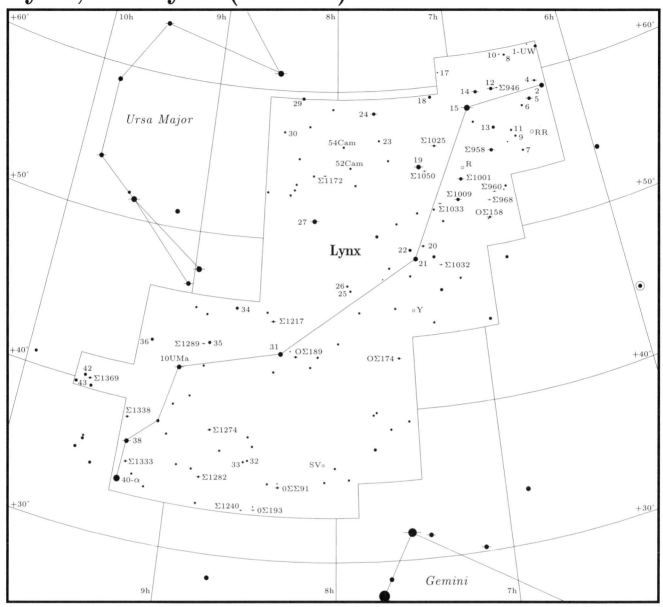

Table 19-1. Selected Variable Stars in Lynx

Name	HD No.	Type	Max.	Min.	Period (Days)	F*	Spec. Type	R.A. (2000)	Dec.	Finder Chart No. & Notes
RR Lyn	44691	EA	5.6	6.0	9.4	0.04	A3	$06^h26.4^m$	+56°17′	19-1
R Lyn	51610	M	7.2	14.5	378	0.44	S2-S6	$07^h01.3^m$	+55 20	19-1
Y Lyn	58521	SRc	7.8	10.3	110		MS	28.2	+45 59	19-5
SV Lyn	66175	SRb	8.2	9.1	70		M5	$08^h03.7^m$	+36 21	19-3

F* = The fraction of period taken up by the star's rise from min. to max. brightness, or the period spent in eclipse.

Table 19-2. Selected Double Stars in Lynx

Name	ADS No.	Pair	M1	M2	Sep"	P.A.°	Spec	R.A. (2000)	Dec	Finder Chart No. & Notes
4 Lyn	4950	AB	6.2	7.7	0.8	124	A2	06h22.1m	+59°22'	19-1
5 Lyn	5036	AB	5.3	9.8	31.4	139	K2	26.8	+58 25	19-1 AB: Yellow and blue
	5036	AC		7.9	96.0	272				
Σ946	5368		8.3	10.1	4.1	130	F5	44.8	+59 27	19-1
12 Lyn	5400	AB	5.3	5.9	1.7*	* 70	A2	46.2	+59 27	19-1 AB: Yellowish-white
	5400	AC		7.3	8.7	308				C: bluish
	5400	AD		10.6	107.0	256				
Σ958	5436	AB	6.3	6.3	4.8	257	F5 F5	48.2	+55 42	19-1 Pale yellow pair
Σ960	5462		8.0	9.9	21.9	67	F0	49.6	+53 02	19-1 Yellowish & white
Σ968	5517	AB	8.6	9.6	20.7	288	A3	52.9	+52 41	19-1 Yellowish-white & red
	5517	AC		9.6	8.3	59				
OΣ158	5531	AB	7.0	11.3	16.8	304	F5	53.4	+51 31	19-1
	5531	AC		10.7	56.0	63				
15 Lyn	5586	AB	4.8	5.9	0.9	33	G0	07h03.1m	+54 10	19-1
Σ1001	5706	AB	7.7	9.3	9.0	65				
	5706	AC		9.6	9.7	66				
	5706	BC			1.7	355				
Σ1009	5746	AB	6.9	7.0	4.1	150	A2	05.7	+52 45	19-1 Equal white pair
Σ1025	5854	AB	8.3	8.6	25.6	133	K0	12.9	+55 49	19-1 Yellowish-orange stars
Σ1032	5879	AB	7.7	11.0	26.6	110	A0	13.9	+48 30	19-1
	5879	AC		9.8	32.7	319				
Σ1033	5896		7.7	8.3	1.5	279	F0	14.8	+52 33	19-1 Close yellowish stars
Σ1050	5968	AB	8.1	8.8	19.3	20	A0	19.9	+54 55	19-4
20 Lyn	6004		7.3	7.4	15.0	254	F0 F0	22.3	+50 09	19-5 Twin yellowish-white stars
19 Lyn	6012	AB	5.6	6.5	14.8	315	B8 A	22.9	+55 17	19-4 AB: Both white
OΣ174	6191		6.5	8.1	2.0	85	F0	35.9	+43 02	19-1
Σ1172	6545		7.7	9.9	1.6	243	A0	08h04.6m	+54 45	19-4
OΣ189	6675		6.8	9.9	4.2	293	A0	14.8	+43 02	19-1
OΣΣ91		AB	7.1	8.8	92.6	220	G0	19.5	+35 03	19-1 Yellow and blue
Σ1217	6783		7.7	9.2	29.0	241	G0 K0	24.3	+44 57	19-5
OΣ193	6821		7.6	11.6	14.1	295	K0	28.1	+33 32	19-1
Σ1240	6866	AB	7.7	10.7	26.8	78	A0	33.2	+33 26	19-1
	6866	AC		10.5	51.1	246				
Σ1272	7000		8.1	9.6	20.4	343	F8	48.3	+34 36	19-7
Σ1274	7005		7.4	9.1	8.9	41	A2	49.0	+38 21	19-7 White & blue
Σ1282	7034		7.5	7.5	3.6	279	F8	50.7	+35 04	19-7 Lovely deep yellow pair
Σ1289	7075		8.2	9.0	3.9	5	F8	54.8	+43 35	19-6
Σ1333	7286		6.4	6.7	1.6	49	A5	09h18.4m	+35 22	19-7 Pretty white pair
38 Lyn	7292	AB	3.9	6.6	2.7	229	A2	18.8	+36 48	19-7 White & yellow
Σ1338	7307	AB	6.5	6.7	0.4c	* 16	F2	21.0	+38 11	19-7 Yellow unsplit oval
Σ1369	7438	AB	7.0	8.0	24.7	148	F2	35.4	+39 57	19-1 All yellowish stars
	7438	AC		8.7	117.8	325				

Footnotes: *= Year 2000, a = Near apogee, c = Closing, w = Widening. Finder Chart No: All stars listed in the tables are plotted in the large Constellation Chart, but when a star appears in a Finder Chart, this number is listed. Notes: When colors are subtle, the suffix -*ish* is used, e.g. *bluish*.

Σ1369 Triple Star   Spec. F2 & G0
AB: m7.0, 8.0; Sep. 24.7"; P.A. 148° 09h35.4m +39°57'
Constellation Chart 19-1   ★★★

2/3" Scopes–100x: Struve 1369 is an easy, beautiful pair of yellowish stars. A third yellowish component, a magnitude 8.7 star, lies 118" distant in position angle 325°.

19.3 Deep-Sky Objects

NGC 2415 Galaxy Type Im?
ø0.9' × 0.9', m12.4v, SB 12.0   07h36.9m +35°15'
Finder Chart 19-3   ★★★

16/18" Scopes–150x: NGC 2415 is 20' NW magnitude 5.6 star of 70 Geminorum on Lynx's SW boundary. Its moderately bright, round halo is less than 1' across and has a brighter center. A 9.5 magnitude star lies 2' NE.

Figure 19-1. *The remote globular cluster NGC 2419, at the east end of a line with three bright stars, is considered to be an interglactic wanderer. Chris Schur made this 25 minute exposure on 2415 film with a 14", f5 Newtonian reflector.*

NGC 2419 Globular Cluster Class 2
ø4.1′, m10.3v, Br★ 17.3v   07ʰ38.1ᵐ +38°53′
Finder Chart 19-3, Figure 19-1   ★★

NGC 2419 is known as the "Intergalactic Wanderer" because it lies 300,000 light years away from our Galaxy's center (300,000 light years from the Solar System) out beyond even the Magellanic Clouds, and has a true space velocity greater than the escape velocity from our Galaxy at its location. It lies on the side of the sky opposite the globular-cluster-rich interior of our Galaxy, and therefore no globulars are to be found anywhere near it. The brightest stars in NGC 2419 are red and yellow giants several hundred times brighter than our Sun.

8/10″ Scopes–125x: NGC 2419 is at the east end of a slightly curved, E–W row of one 9th and two 7th magnitude stars. It has a faint, circular 2′ diameter halo of unresolved stars.

12/14″ Scopes–125x: NGC 2419 has a fairly faint, round, 2′ diameter halo with the overall consistency of the typical tailless comet. A few stars are resolved around its irregular periphery. The globular's face is diffuse and only slightly brighter toward the center. A diamond-shaped asterism of four 13.5 to 14th magnitude stars boxes the cluster.

16/18″ Scopes–150x: In a 17.5″ telescope NGC 2419 is a fairly bright, somewhat mottled 2.5′ halo with a few stars resolved at the ragged edges of its periphery. The central area is slightly brighter and broadly concentrated.

NGC 2444 U4016 Galaxy Type Ring A
ø1.7′ × 1.2′, m13.2v, SB 13.8   07ʰ46.9ᵐ +39°02′

NGC 2445 U4017 Galaxy Type Ring B
ø2.3′ × 1.8′, m13.2v, SB 14.7   07ʰ46.9ᵐ +39°01′
Finder Chart 19-3   ★/★

12/14″ Scopes–125x: NGC 2444 and NGC 2445 are two nearly stellar knots separated by 1′ NW–SE. Both are surrounded by faint halos which seem to merge. NGC 2444 is the NW component.

PK164+31.1 Je 1 Planetary Nebula Type 4
ø399″, m12.1v, CS 16.8   07ʰ57.8ᵐ +53°25′
Finder Chart 19-4, Figure 19-2   ★

12/14″ Scopes–125x: This huge planetary, 5′ ENE of a magnitude 12.5 star, has such low surface brightness that it requires a UHC filter to be seen at all. Its halo is diffuse, and only irregularly brighter toward the center. Several faint stars are embedded in it.

16/18″ Scopes–150x: In an O-III filter this planetary is a very faint, diffuse 4′ disk brighter along its NW and SE edges. The SE arc is a little brighter than the NW one.

20/22″ Scopes–150x: PK164+31.1 is quite faint, but very large for a planetary nebula. It is very difficult without a filter. However, a UHC filter brings out its annular structure and it resembles the Helix Nebula in Aquarius. 175x reveals nebulous condensations along the NW and SE edges. Several faint stars are embedded in the nebula-glow.

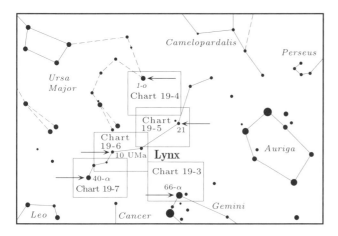

Master Finder Chart 19-2. Lynx Chart Areas
Guide stars indicated by arrows.

NGC 2474 U4114 Galaxy Type E0
ø0.6′ × 0.6′, m13.3v, SB 12.0   07ʰ57.9ᵐ +52°51′

NGC 2475 U4114 Galaxy Type E1
ø0.8′ × 0.8′, m13.1v, SB 12.4   07ʰ58.0ᵐ +52°51′
Finder Chart 19-4   ★★/★★

12/14″ Scopes–125x: NGC 2474 and NGC 2475 are a faint but striking galaxy-pair 3′ SW of a 9th magnitude star. Both have prominent cores surrounded by faint, round halos less than 1′ in diameter. The two systems are aligned E–W, their halos almost in contact. NGC 2475, the easternmost object, seems a little brighter.

NGC 2500 H709³ Galaxy Type SB(rs)d II-III
ø2.6′ × 2.6′, m11.6v, SB 13.5   08ʰ01.9ᵐ +50°44′
Finder Chart 19-5   ★★

12/14″ Scopes–125x: NGC 2500 is a diffuse, circular 2′ glow without much central brightening. The galaxy lies just south of a loose sprinkling of ten faint stars. Three more stars are to the north. A faint star nearly touches the ESE tip.

NGC 2537 H55⁴ Galaxy Type SB(s)m II-III
ø1.6′ × 1.4′, m11.7v, SB 12.4   08ʰ13.2ᵐ +46°00′
Finder Chart 19-5, Figure 19-3 Bear Paw Galaxy ★★★

12/14″ Scopes–125x: NGC 2537, 2′ NW of an 11th magnitude star, is fairly faint and round and resembles a planetary nebula. Its 1′ diameter halo has uniform surface brightness and well defined edges.

16/18″ Scopes–150x: NGC 2537 is moderately bright, its circular 1.5′ halo with a darker center and a knot along its NW edge.

20/22″ Scopes–175x: Larger instruments reveal an interesting object of fairly high surface brightness. The halo has an annular appearance with a peculiar mottled, semicircular arc running along the northern periphery all the way from the west to the NE. The knots of brightness along this arc resemble the pads of a paw-print.

NGC 2541 H710³ Galaxy Type SA(s)cd III-IV
ø7.4′ × 3.3′, m11.8v, SB 15.1   08ʰ14.7ᵐ +49°04′
Finder Chart 19-5, Figure 19-4   ★★

16/18″ Scopes–150x: NGC 2541 has a rather low surface brightness 3′ × 2′ NNW–SSE halo with an oval-shaped core and a faint nonstellar nucleus. An 11th magnitude star lies 4.5′ NNE and a 12th magnitude star 3′ SW.

Figure 19-2. *The extremely faint planetary nebula PK164+31.1 is visible in larger instruments as two large, NW–SE arcs. Martin C. Germano made this 65 minute exposure on 2415 film made with an 8″, f5 Newtonian reflector.*

NGC 2549 U4313 Galaxy Type Sa(r)0°
ø3.7′ × 1.2′, m11.2v, SB 12.7   08ʰ19.0ᵐ +57°48′
Finder Chart 19-4   ★★★

12/14″ Scopes–125x: NGC 2549, 12′ WNW of a 6th magnitude star, has a nice bright, lens-shaped halo elongated 2′ × 0.5′ N–S with pointed tips. The core, also highly extended, is well concentrated toward a stellar nucleus.

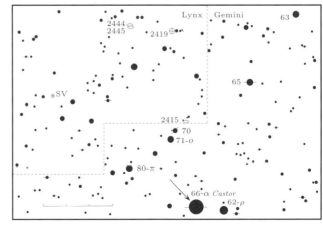

Finder Chart 19-3. 66-α Gem: 07ʰ34.6ᵐ +31°53′

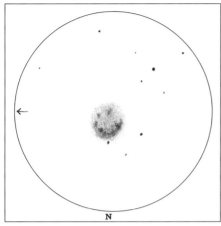

Figure 19-3. NGC 2537
20", f4.5–175x, by Richard W. Jakiel

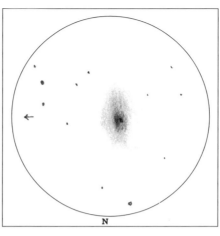

Figure 19-4. NGC 2541
12.5", f5–150x, by G. R. Kepple

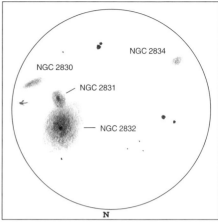

Figure 19-5. NGC 2831-32 Area
20", f4.5–315x, by Richard W. Jakiel

NGC 2552 H711³ Galaxy Type SA(s)m?
ø3.6' × 2.4', m12.1v, SB 14.3   08ʰ19.3ᵐ +50°01'
Finder Chart 19-5   ★

12/14" Scopes–125x: NGC 2552 is a very faint, amorphous glow about 2' across. A 12th magnitude star lies 3' NE.

NGC 2683 H200¹ Galaxy Type SA(rs)b II–III
ø8.4' × 2.4', m9.8v, SB 12.9   08ʰ52.7ᵐ +33°25'
Finder Chart 19-7, Figure 19-6   ★★★★

8/10" Scopes–100x: This fine, bright, edge-on galaxy has a highly elongated 7' × 1.5' NE–SW halo with a slightly brighter highly extended core. 10' south of the galaxy is a parallelogram of 10th to 11th magnitude stars, and a 12th magnitude star is 2.5' east.

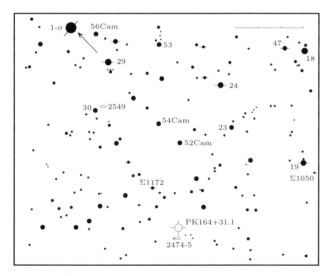

Finder Chart 19-4. 1-o UMa: 08ʰ30.3ᵐ +60°43'

12/14" Scopes–125x: This beautiful object is elongated 8' × 1.5' NE–SW with a mottled halo and hints of a dust lane that extends for nearly half the length of the major axis. The center appears moderately brighter, and the tips fade rapidly.

16/18" Scopes–150x: Impressive! This relatively high surface brightness galaxy covers a generous portion of the field of view. Its 9' × 1.5' NE–SW image is considerably mottled, and several bright knots are scattered along its major axis. A small irregular dust lane runs just NE of the center. SE of the center bulges a crescent-shaped nucleus. A 14th magnitude star lies just beyond the NE tip, and a 13th magnitude star nearly touches the halo 2.25' NNE of the center.

NGC 2712 U4708 Galaxy Type SB(r)b: I
ø3.2' × 1.6', m12.1v, SB 13.7   08ʰ59.5ᵐ +44°55'
Finder Chart 19-6   ★★

12/14" Scopes–125x: This galaxy appears fairly faint, diffuse, elongated 1.5' × 0.75' N–S, and has a faint stellar nucleus.

16/18" Scopes–150x: NGC 2712 shows a fairly bright halo elongated 2.25' × 1' N–S with a slight brightening to a diffuse core and a faint stellar nucleus.

NGC 2770B Galaxy Type ?
ø0.8' × 0.4', m14.2v, SB 12.8   09ʰ09.3ᵐ +33°08'
NGC 2770 U4806 Galaxy Type SA(s)c:
ø3.3' × 0.9', m12.2v, SB 13.3   09ʰ09.6ᵐ +33°08'
Finder Chart 19-7   ★/★★

16/18" Scopes–125x: NGC 2770, 2.5' south of a 10th magnitude star, is a moderately faint spindle elongated 2.5' × 0.75' NNW–SSE with a very faint stellar nucleus. NGC 2770B, 3′ west, may be glimpsed with averted vision as an extremely faint 20″ diameter haze.

NGC 2776 U4838 Galaxy Type SAB(rs)c I-II
ø3.0′ × 3.0′, m11.6v, SB 13.9   09ʰ12.2ᵐ +44°57′
Finder Chart 19-6   ★★

8/10″ Scopes–100x: This galaxy, 8′ NNE of a magnitude 7.5 star, is rather faint and diffuse, its round 1.5′ diameter halo slightly brighter in the center.

12/14″ Scopes–125x: NGC 2776 has a fairly faint, circular 2′ diameter halo with a broad but poorly concentrated 1′ diameter core and a faint stellar nucleus. 175x shows some mottling.

NGC 2782 H167¹ Galaxy Type SAB(rs)a pec
ø3.8′ × 2.9′, m11.6v, SB 14.1   09ʰ14.1ᵐ +40°07′
Finder Chart 19-6   ★★★

8/10″ Scopes–100x: NGC 2782, 7′ south of a 9th magnitude star, is quite obvious but rather small. Its circular 1.25′ diameter halo contains a prominent stellar nucleus.

12/14″ Scopes–125x: NGC 2782 is a fairly bright object elongated 2′ × 1.5′ NNE–SSW. Within its tiny core is a bright stellar nucleus. A 12th magnitude star lies 2.75′ south.

NGC 2793 U4894 Galaxy Type SB(s)m
ø1.2′ × 0.9′, m13.1v, SB 13.0   09ʰ16.8ᵐ +34°26′
Finder Chart 19-7   ★★

8/10″ Scopes–100x: NGC 2793, 1° west of magnitude 3.1 Alpha (α) = 40 Lyncis, is a tiny galaxy contained in a diamond-shaped asterism of 10th magnitude stars, the star at the diamond's northern vertex a 15″ NNE–SSW double. The galaxy's halo appears

Figure 19-6. *NGC 2683 is a fine bright edge-on galaxy visible in small instruments. Martin C. Germano made this 65 minute exposure on 2415 film with an 8″, f5 Newtonian.*

round and is uniform in surface brightness.

12/14″ Scopes–125x: NGC 2793 is fairly faint, round, and small, its halo less than 1′ in diameter. No core or nucleus is visible.

NGC 2798 H708² Galaxy Type SB(s)a pec
ø2.6′ × 0.8′, m12.3v, SB 13.0   09ʰ17.4ᵐ +42°00′

NGC 2799 H708² Galaxy Type SB(s)m?
ø1.9′ × 0.5′, m13.8v, SB 13.6   09ʰ17.5ᵐ +42°00′
Finder Chart 19-6   ★★/★

12/14″ Scopes–125x: The faint pair of galaxies NGC 2798 and NGC 2799 are near the Ursa Major border WSW of a wide pair of magnitude 8 and 8.5 stars. NGC 2798 is the brighter and larger of the two: its

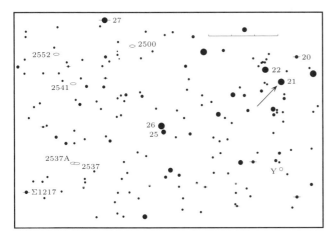

Finder Chart 19-5. 21 Lyn: 07ʰ26.7ᵐ +49°13′

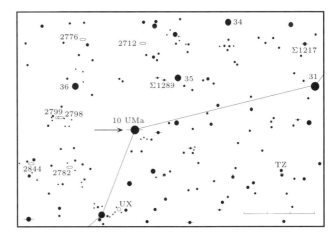

Finder Chart 19-6. 10 UMa: 09ʰ00.6ᵐ +41°47′

Figure 19-7. *NGC 2832 lies at the heart of the Abell 779 galaxy cluster. Martin C. Germano made this 60 minute exposure on hypered 2415 film with an 8", f5 Newtonian.*

1.5′ × 0.75′ NNW–SSE halo contains a bright, extended core with a stellar nucleus. Its companion, NGC 2799, 1.5′ to the ESE, is a very faint, small glow elongated 0.75′ × 0.25′ NW–SE surrounding a very faint stellar nucleus.

NGC 2831 Galaxy Type E0
⌀1.4′ × 1.4′, m12.3v, SB 12.9 $09^h19.7^m$ +33°44′

NGC 2832 U4942 Galaxy Type E+2:
⌀3.0′ × 2.1′, m11.9v, SB 13.7 $09^h19.8^m$ +33°44′
Finder Chart 19-7, Figures 19-5 & 19-7   ★/★★

8/10″ Scopes–125x: NGC 2832 is the brightest object in the Abell 779 galaxy group 1° SW of magnitude 3.1 alpha (α) = 40 Lyncis. It appears fairly faint, about 1′ in diameter, and slightly brighter toward the center. Approximately 3′ to its east and SSE are two double stars, the former a coarse pair of 11th magnitude stars and the latter magnitude 10 and 11.5 stars only 10″ apart.

12/14″ Scopes–125x: NGC 2832 has a fairly bright but small 1′ diameter halo with a weakly concentrated central area. The surrounding field is sprinkled with unidentified Abell 779 galaxy group members at the limit of detection.

16/18″ Scopes–150x: NGC 2832 is fairly bright, its 2.5′ diameter halo slightly elongated NNW–SSE and containing a faint stellar nucleus. NGC 2831 is a faint knot within the SW edge of the halo of NGC 2832. Eight more faint galaxies lie within half a degree. NGC 2830, 2′ SW of NGC 2832, is a small, faint nebulous patch. NGC 2825, 5′ west, is a very faint E–W streak. NGC 2834, 5′ SE, is a very faint, tiny oval patch. Other Abell 779 galaxies can be identified on Figure 19-7.

NGC 2844 H628³ Galaxy Type SA(r)a:
⌀1.7′ × 0.8′, m12.9v, SB 13.1 $09^h21.8^m$ +40°09′
Finder Chart 19-6   ★★

8/10″ Scopes–100x: NGC 2844, a faint, tiny galaxy, is at the southern vertex of a 6′ × 7′ triangle with two 7th and 8th magnitude stars. The halo is circular, less than 1′ across, and slightly brighter at its center.

16/18″ Scopes–150x: The halo appears moderately faint, elongated 1.25′ × 0.75′ N–S, and has a tiny core with a very faint stellar nucleus. A magnitude 12.5 star lies 2.25′ NW.

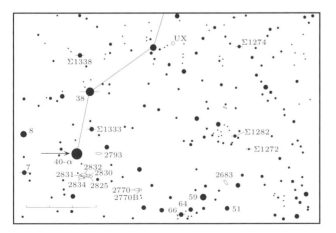

Finder Chart 19-7. 40-α Lyn: $09^h21.1^m$ +34°24′

Chapter 20

Monoceros, the Unicorn

20.1 Overview

Monoceros the Unicorn is a late addition to the celestial sphere. It probably was invented sometime late in the 16th century, but it first appears on a celestial globe constructed in 1613 by the Dutch cartographer Petrus Kaerius.

Though it is cut through by the NW–SE sweep of the winter Milky Way, Monoceros lacks bright stars and is not distinguished by any conspicuous star-patterns. It is, however, surrounded by brilliant constellations: Orion is to its west, Canis Major to its south, and Canis Minor to its north. What it lacks in bright stars Monoceros more than makes up for in clusters and nebulae. Within it are the often-photographed Rosette Nebula, Cone Nebula, and Hubble's Variable Nebula — the last one of the most intriguing nebulae in the sky for the visual observer. The finest of its many open clusters are M50, the Rosette Nebula's central cluster NGC 2244, the Christmas Tree cluster NGC 2264, and the very rich NGC 2506.

Monoceros: Mo-NOS-er-os
Genitive: Monocerotis, Mo-NOS-er-o-tis
Abbrevation: Mon
Culmination: 9pm–Feb.19, midnight–Jan. 5
Area: 482 square degrees
Showpieces: 11–β Mon, M50 (NGC 2323), NGC 2244, NGC 2261, NGC 2264, NGC 2301
Binocular Objects: Cr 91, Cr 92, Cr 96, Cr 95, Cr 97, Cr 106, Do 25, M50 (NGC 2323), NGC 2215, NGC 2232, NGC 2236, NGC 2237, NGC 2244, NGC 2251, NGC 2252, NGC 2261, NGC 2264, NGC 2286, NGC 2301, NGC 2302, NGC 2324, NGC 2343, NGC 2353, NGC 2506

20.2 Interesting Stars

Epsilon (ϵ) = 8 Monocerotis   Triple Star   Spec. A5 & A5
AB: m4.5, 6.5; 13.4″; 27° $06^h23.8^m$ +04°36′
Finder Chart 20-4   ★★★★

8/10″ Scopes–125x: This triple star is set in a grand Milky Way field. The colors of the close pair are pale yellow, but the tertiary is too faint at this aperture for its color to be discerned. A nice red star lies 5′ WNW.

12/14″ Scopes–150x: Epsilon is a lovely close pair of yellowish stars with a faint bluish-white third component (AC: 12.7; 93.7″; 254°).

Beta (β) = 11 Monocerotis   Triple Star Spec. B2 & B2
AB: m4.7, 5.2; Sep. 7.3′; P.A. 132° $06^h28.8^m$ 07°02′
BC: m5.2, 6.1; Sep. 2.8″; P.A. 106°
Finder Chart 20-3   ★★★★

Beta Monocerotis was discovered by Sir William Herschel in 1781. It is a magnificent triple of three comparably bright stars — a very unusual situation because multiples typically consist of two bright and one or more faint members. All Beta Mon stars are a striking blue-white color. The system is around 700 light years away.

8/10″ Scopes–150x: This triple star forms a nice acute triangle with all bluish-white stars.

20.3 Deep-Sky Objects

NGC 2149 Reflection Nebula
ø3′ × 2′, Photo Br 3–5; Color 1–4 $06^h03.5^m$ −09°44′
Finder Chart 20-3, Figure 20-1   ★★★

12/14″ Scopes–125x: NGC 2149, embedded in a very nice star field, is a fairly bright, grey oval elongated 2′ × 1′ N–S with a brighter center.

16/18″ Scopes–150x: This fairly obvious bean shaped reflection nebula is elongated 3′ × 1.5′ N–S. The western edge has a bright streak and therefore is more distinct than the diffuse eastern edge.

Monoceros, the Unicorn   Constellation Chart 20-1

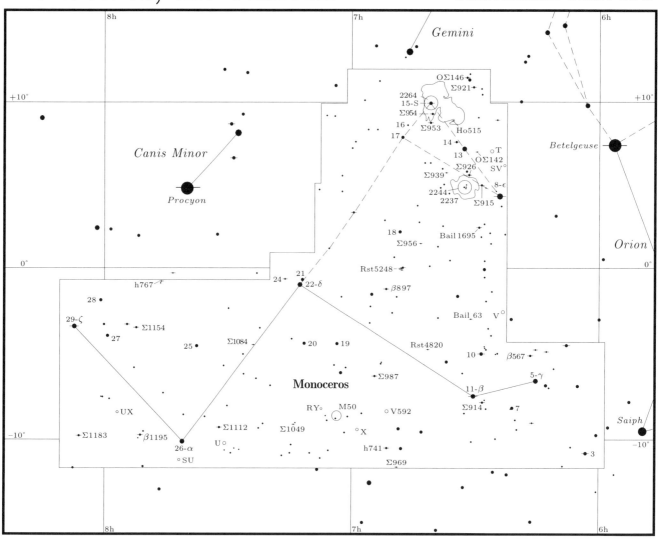

Table 20-1. Selected Variable Stars in Monoceros

Name	HD No.	Type	Max.	Min.	Period (Days)	F*	Spec. Type	R.A. (2000)	Dec.	Finder Chart No. & Notes
SV Mon	44320	Cδ	7.6	8.8	15.23	0.38	F6-G4	06ʰ21.4ᵐ	+06°28′	20-4
V Mon	44639	M	6.0	13.7	333	0.46	M5-M8	22.7	−02 12	20-5
T Mon	44990	Cδ	5.5	6.6	27.02	0.27	F7-K1	25.2	+07 05	20-4
X Mon	51478	SRa	6.9	10.0	155	0.51	M3-M4	57.2	−09 04	20-6
RY Mon		SRa	7.7	9.2	466	0.43	N3	07ʰ06.9ᵐ	−07 33	20-6
U Mon	59693	RVb	6.1	8.1	92	0.22	F8-K0	30.8	−09 47	20-7
UX Mon	65607	EA	8.0	8.9	5.90	0.17	A6-G2	59.3	−07 30	20-7

F* = The fraction of period taken up by the star's rise from min. to max. brightness, or the period spent in eclipse.

Table 20-2. Selected Double Stars in Monoceros

Name	ADS No.	Pair	M1	M2	Sep."	P.A.°	Spec	R.A. (2000)	Dec	Finder Chart No. & Notes
3 Mon	4615		5.0	8.5	1.8	354	B8	06ʰ01.8ᵐ	−10°36′	20-3
β567	4865		6.0	10.2	4.1	242	A2	15.5	−04 55	20-3
8-ε Mon	5012	AB	4.5	6.5	13.4	127	A5 A5	23.8	+04 36	20-4 Pale yellow pair
Σ914	5070		6.4	8.7	21.1	298	A0	26.7	−07 31	20-3
10 Mon		AB	5.1	9.3	77.2	256	B3	28.0	−04 46	20-3
		AC		9.3	80.7	231				
Σ915	5097	AB	7.4	8.4	5.9	40	B9	28.2	+05 16	20-4
	5097	AC		11.0	38.8	126				
11-β Mon	5107	AB	4.7	5.2	7.3	132	B2 B2	28.8	−07 02	20-3 All bluish-white triplet
	5107	AC		6.1	10.0	124	B2			
	5107	AD		12.2	25.9	56				
		BC			2.8	106				
Bail 1695			6.2	10.2	19.3	299	M	29.2	+02 39	20-5
OΣ142	5124		7.7	11.2	8.6	353	B3	29.9	+07 07	20-4
Bail 63			7.1	11.0	13.3	222	G5	30.4	−03 01	20-3
Σ921	5153		6.1	8.3	16.3	4	B0 A5	31.2	+11 15	20-4
Σ926	5162		7.1	8.5	10.9	288	A0	31.7	+05 46	20-4
OΣ146	5170		6.0	9.6	31.1	140	K0	32.4	+11 40	20-4
14 Mon	5211		6.5	10.7	10.5	209	A0	34.8	+07 34	20-4
Σ939		AB	8.3	9.6	30.1	106	B5 B8	35.9	+05 18	20-4
		AC		9.7	39.7	50				
Ho515	5262	AB	7.5	11.7	10.1	253	G5	37.5	+09 09	20-4
Rst 4820			7.5	11.2	6.0	298	K5	40.3	−04 28	20-3
15 (S) Mon	5322	AB	4.7	7.5	2.8	213	O5	41.0	+09 54	20-4 Blue giant
		AC		9.8	16.6	13				
Σ953	5328		7.2	7.7	7.1	330	F5	41.2	+08 59	20-4
Σ954	5327		7.1	9.6	12.8	153	B3	41.2	+09 28	20-4
Σ956	5364	AB	7.9	10.9	6.2	189	B2	42.7	+01 43	20-5
	5364	AC		9.0	35.8	157				
	5364	AD		10.9	56.0	180				
Σ969	5463		7.8	10.8	7.2	317	B9	48.0	−11 06	20-3
Rst 5248			7.8	11.5	7.0	143	G0	48.2	+00 18	20-5
β897	5505		5.8	11.3	5.9	30	F2	50.8	−00 32	20-5
h741	5516		7.4	10.9	16.1	226	K5	51.2	−10 05	20-6
Σ987	5557		7.1	7.2	1.2	170	A3	54.1	−05 51	20-6
Σ1049	5904		7.9	9.7	3.6	40	A0	07ʰ13.6ᵐ	−08 55	20-6
Σ1084	6047		7.1	9.6	14.1	286	K0	24.0	−03 59	20-6
Σ1112	6158		6.1	8.8	23.4	112	F5	32.1	−08 53	20-7
h767			7.8	9.5	21.6	164	A0	45.3	−00 26	20-1
β1195	6412		7.3	7.6	0.3	91	A0	51.3	−09 24	20-7
Σ1154	6421		7.0	9.2	2.5	354	A5	52.1	−03 03	20-1
Σ1183	6588	AB	6.0	8.7	30.9	327	A0 A0	08ʰ06.5ᵐ	−09 15	20-7
29-ζ	6717	AB	4.3	10.0	32.0	105	G0	08.6	−02 59	20-1
	6717	AC		7.8	66.5	245				

Footnotes: *= Year 2000, a = Near apogee, c = Closing, w = Widening. Finder Chart No: All stars listed in the tables are plotted in the large Constellation Chart, but when a star appears in a Finder Chart, this number is listed. Notes: When colors are subtle, the suffix *-ish* is used, e.g. *bluish*.

NGC 2170 Reflection Nebula
⌀2′ × 2′, Photo Br 1-5, Color 1-4   06ʰ07.5ᵐ −06°24′
Finder Chart 20-3   ★★

12/14" Scopes-125x: This reflection nebula is a very faint, diffuse circular glow surrounding a 9.5 magnitude star.

16/18" Scopes-150x: NGC 2170 is a fairly obvious circular nebulosity around a 9.5 magnitude star. 5′ east is vdB 69, a similar but smaller glow surrounding a 9th magnitude star.

NGC 2182 H38[4] Reflection Nebula
⌀2.5′ × 2.5′, Photo Br 1-5, Color 1-4 06ʰ09.5ᵐ −06°20′
Finder Chart 20-3   ★★

16/18" Scopes150x: NGC 2182 is between its fellow reflection nebulae NGC 2183-85 and vdB 69 and, like vdB 69, is a circular glow around a white 9th magnitude central star.

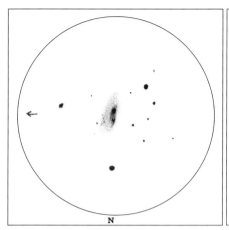
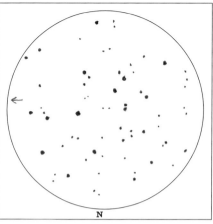
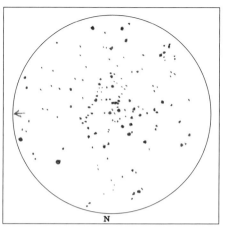

Figure 20-1. NGC 2149
16″, f5–200x, by Alister Ling

Figure 20-2 NGC 2215
12.5″, f5–100x, by G. R. Kepple

Figure 20-3 NGC 2225-26
12.5″, f5–125x, by G. R. Kepple

NGC 2183 Reflection Nebula
ø1′ × 1′, Photo Br 3-5, Color 1-4   06ʰ10.8ᵐ −06°13′

NGC 2185 H20⁴ Reflection Nebula Type vF 3 B
ø2.5′ × 2.5′, Photo Br 3-5, Color 1-4 06ʰ11.1ᵐ −06°13′
Finder Chart 20-3   ★★/★★

16/18″ Scopes150x: NGC 2183 is a wispy patch of haze involved with three faint stars forming a long triangle. Just to its east is the faint diffuse glow of NGC 2185. The O-III filter has little effect on reflection nebulae but most often the Deep-Sky filter or other broad band filters helps increase contrast between these objects and the sky.

NGC 2215 Cr90 Open Cluster 40★ Tr Type II 2 p
ø10′, m8.4v, Br★ 10.52v   06ʰ21.0ᵐ −07°17′
Finder Chart 20-3, Figure 20-2   ★★★

8/10″ Scopes 75x: NGC 2215, 35′ NNE of magnitude 5.3 star 7 Monocerotis, is a loose assemblage of about twenty 11th magnitude and fainter stars in a 10′ area.

12/14″ Scopes–100x: NGC 2215 is a large, fairly obvious group of ten 11th and 12th magnitude and fifteen fainter stars scattered around the cluster's magnitude 10.5 lucidae.

16/18″ Scopes–125x: NGC 2215 consists of thirty 11th to 13th magnitude stars scattered in a 10′ area around the cluster's magnitude 10.5 lucida. The group is arranged in a very rough "V," the SE leg of which is the most star-rich. Several short rows of stars are aligned N–S. Two of the cluster's brighter stars are conspicuous on its east and west edges.

Collinder 91 Open Cluster 20★ Tr Type IV 2 p
ø17′, m6.4p   06ʰ21.7ᵐ +02°22′
Finder Chart 20-5   ★★★

12/14″ Scopes–100x: Collinder 91 is a bright, sparse, diamond-shaped group in a 15′ area. A 6th magnitude star anchors the SW edge; and a trio consisting of a 7th magnitude star flanked N–S by 9th magnitude stars marks the center. Close attention is needed to distinguish the cluster from the surrounding star field.

Collinder 92 Open Cluster
ø11′, m8.6v   06ʰ22.9ᵐ +05°07′
Finder Chart 20-4   ★

12/14″ Scopes–100x: Collinder 92 is large, faint, and loose, a poor cluster weakly concentrated in a 10′ area. Several other nearby star aggregations are better cluster candidates.

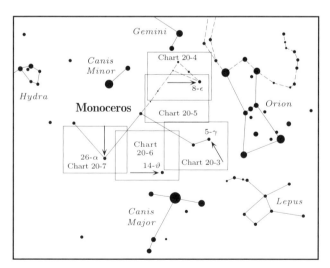

Master Finder Chart 20-2. Monoceros Chart Areas
Guide stars indicated by arrows.

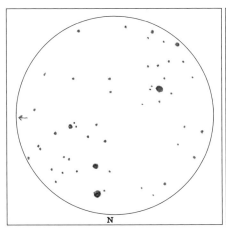

Figure 20-4 NGC 2232
12.5", f5–125x, by G. R. Kepple

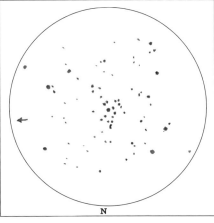

Figure 20-5 NGC 2236
13", f5.6–165x, by Steve Coe

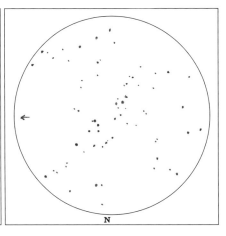

Figure 20-6. NGC 2251
12.5", f5–100x, by G. R. Kepple

Dolidze 22   Open Cluster   10★   Tr Type IV 1 p
⌀18'   06ʰ23.3ᵐ +04°39'
Finder Chart 20-4   ★

12/14" Scopes–100x: Dolidze 22 has little resemblance to a cluster. It is a large, loose E–W patch of very faint stars, its east edge marked by the nice bright double 8 Monocerotis (4.5, 6.5; 13.4"; 127°). Two more doubles, faint, wide pairs, are west of B Mon.

NGC 2219   Open Cluster
[⌀8', Br★ 6.5v]   06ʰ23.4ᵐ −04°41'
Finder Chart 20-3   ★★

12/14" Scopes–100x: NGC 2219 is catalogued as a nonexistent cluster. However, what is seen at the specified location is an irregular group of 11th to 14th magnitude stars scattered around a bright E–W pair of magnitude 6.5 and 7 stars. The brighter, western star is surrounded by a 5' halo of 13th–14th magnitude stars, richest to the north and east of the star. North and south of the eastern, magnitude 7, star are concentrations of 11th–12th magnitude stars, the southern concentration triangular, the 7th magnitude star and a 9th magnitude star to the SSW marking two of the triangle's vertices. 10th and 11th magnitude stars to the north of this triangle form it into an arrow-shaped asterism similar to Sagitta. An E–W star chain concave to the south crosses the arrow formation north of the 7th magnitude star.

NGC 2225   Open Cluster
[⌀7', Br★ 12.v]   06ʰ26.6ᵐ −09°39'
NGC 2226   Open Cluster
[⌀7', Br★ 12.v]   06ʰ26.7ᵐ −09°39'
Finder Chart 20-3, Figure 20-3   ★★★

12/14" Scopes–100x: The NGC 2225 + NGC 2226 open cluster pair have been classed "nonexistent." However they stand out well from the surrounding star field and even show different features: the western, NGC 2225, aggregation has in general fainter stars and is more scattered, whereas the eastern, NGC 2226, group is smaller but of brighter members. Together the two groups (if they really are separate) occupy a 7' area and have fifty 11th–12th magnitude stars. A deep notch cuts into the cluster(s) from the north; and a starless gap is in the south. Intersecting E–W

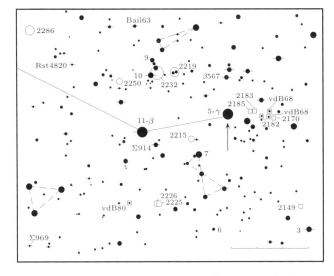

Finder Chart 20-3. 5–γ Mon: 06ʰ14.8ᵐ −06°16'

Figure 20-7. *This rich region of Monoceros includes four reflection nebulae and an open cluster. Reflection nebula IC2169 is visible near top center with open cluster Collinder 95 to its left. From left to right near bottom center are RNs IC 446, NGC 2245, and IC 2247. Martin C. Germano made this 90 minute exposure on 2415 Kodak Tech Pan film with an 8", f5 Newtonian reflector.*

and N–S star chains are in the western and SW parts of this "double" cluster.

NGC 2232 Open Cluster 20★ Tr Type IV 3 p
ø29', m3.9v, Br★ 5.03v $06^h26.6^m$ $-04°45'$
Finder Chart 20-3, Figure 20-4   ★★★

12/14" Scopes–100x: NGC 2232 is a very large, bright, loose cluster divided into two irregular groups aligned NNW–SSE. The SSE group is the more concentrated, fanning southward from the magnitude 5.0 star 10 Monocerotis in a triangular or V-shaped scattering. The NNW group contains a curved NNE–SSW row of 7th magnitude stars flanked by 6th magnitude stars, the northern star of the row magnitude 6.5 star 9 Monocerotis.

NGC 2236 H5[7] Open Cluster 50★ Tr Type III 2 p
ø7', m8.5v, Br★ 10.97v $06^h29.7^m$ $+06°50'$
Finder Chart 20-4, Figure 20-5   ★★

8/10" Scopes–100x: This faint cluster is visible as a soft, round glow in which several stars may be resolved — one 11th magnitude star near the group's center, and five 12th magnitude stars, three south and two east of the center. With averted vision several 13th magnitude stars may also be discerned.

12/14" Scopes–125x: NGC 2236 contains three 11th magnitude and thirty 12th to 13.5 magnitude stars in a 7' area. The three 11th magnitude members are in a thin SE pointing triangle, the densest concentration of cluster stars around the triangle's SE tip.

Collinder 96 Open Cluster 15★ Tr Type IV 2 p
ø7', m7.3v, Br★ 8.75v $06^h30.3^m$ $+02°52'$
Finder Chart 20-5   ★★

12/14" Scopes–100x: Collinder 96 is 16' NE of the 6th magnitude double Bail 1695 (6.2, 10.2; 19.3"; 299°). Several 9th magnitude field stars lead the way from the double to the cluster, a fairly obvious group of fifteen stars in a semicircle. Two 9th magnitude stars stand out from the fainter cluster members. 14' east is the double OΣ 144 (7.7, 12.4; 22.5"; 145°).

Collinder 95 Open Cluster 10★ Tr Type IV 2 n
ø19' $06^h30.5^m$ $+09°56'$
Finder Chart 20-4, Figure 20-7   ★★★

12/14" Scopes–100x: Collinder 95 is a large, very loose cluster. Its brightest members form a "Y" upon the background of much fainter stars. At the south edge is an X-shaped asterism resembling the arrangement of five dots on a die.

Figure 20-8. *The extremely faint Rosette Nebula, NGC 2337-39, has several open clusters embedded in it: NGC 2244 is at the upper right side of the Rosette's "cavity," NGC 2239 is to the lower left of center within the "cavity," NGC 2252 is the fish hook to the lower right of the nebula, and Cr 104 is the 20′ N-S streamer of stars at the extreme right center. This 50 minute exposure was made through a 25A filter on hypered Kodak 2415 Tech Pan film with a 400mm telephoto lens at f4.0 by John Ebersole, M.D.*

IC 446 (IC 2167)   Reflection Nebula
ø5′ × 4′, Photo Br 3-5, Color 1-4 $06^h31.0^m$ +10°27′
Finder Chart 20-4, Figure 20-7   ★★

16/18″ Scope–150x: IC 446 is an easy circular glow surrounding a 12th magnitude star. The nebula lies between a pair of N–S 10th magnitude stars. Dark nebula LDN 1607 lies to the south.

IC 2169   Reflection Nebula
ø25′ × 20′, Photo Br 3-5, Color 1-4 $06^h31.2^m$ +09°54′
Finder Chart 20-4, Figure 20-7   ★★

16/18″ Scopes–75x: IC 2169 is a faint, diffuse haze fanning south from a 20″ NE–SW pair of 9th and 10th magnitude stars. Half a dozen fainter stars are also embedded in the nebula.

Collinder 97   Open Cluster 15★ Tr Type IV 3 p
ø21′, m5.4p $06^h31.3^m$ +05°55′
Finder Chart 20-4   ★★★

12/14″ Scopes–100x: The large, loose cluster Collinder 97 is framed by a triangle of magnitude 6.5 stars, the triangle's western vertex the short-range variable AX Monocerotis. Fifteen 9th magnitude and fainter stars are loosely scattered over a 20′ area. On the southern edge is a N–S star stream. 125x shows very faint concentrations of stars within the cluster.

NGC 2237-39   Emission Nebula   Rosette Nebula
ø80′ × 60′, Photo Br 1-5, Color 3-4 $06^h32.3^m$ +05°03′

NGC 2244   H2⁷   Open Cluster   Type pB 1 R
ø23′, m4.8v, Br★ 5.84v $06^h32.4^m$ +04°52′
Finder Chart 20-4, Figure 20-8   ★/★★★★

The Rosette Nebula is one of the most overlooked of the larger emission nebulae. Despite its notoriously low surface brightness, the nebula is easy in 10×50 binoculars on dark clear nights; and even without a filter the Rosette's wreathlike annularity is visible in wide-angle eyepieces on large telescopes.

William Herschel discovered the nebula and assigned a different number for each portion that he saw as independently visible. Hence the brightest segments of the Rosette have the NGC numbers 2237, 2238, 2239, and 2246, though the whole object is customarily designated NGC 2237. The apparent size of the Rosette is astonishing: it is over 1° in diameter and therefore covers four times as much sky as the Moon!

The Rosette Nebula is estimated to be 4,900 light-years away. Hence its true diameter is 90 light years. The central hole" is about 30 light years across. Within this "hole" is the nebula's involved open cluster, NGC 2244, whose super-hot O-type stars provide the ultraviolet radiation which fluoresces the nebula-gas.

The fine rosy color of the Rosette on photographs comes from its hydrogen gas: after a photon of ultraviolet radiation from the central star cluster knocks an electron off an atom of hydrogen, that electron soon recombines with another electronless hydrogen ion; and when it does, it gives back all the energy it gained from the original photon — not all at once, but in increments as it cascades down the new hydrogen atom's orbital shells. The electron's drop from the third to the second lowest orbital shell in the hydrogen atom emits a photon of radiation of wavelength 6562Ångstroms, which is the red end of the visible spectrum. The Rosette Nebula's gas also contains a fair amount of oxygen atoms which have lost two electrons to ultraviolet photons. When electrons recombine with doubly-ionized oxygen ions, they emit radiation at wavelengths 4959 and 5007 Å in the blue-green region of the spectrum. In the Rosette this blue-green O III radiation is completely swamped by the red H II emission; but in such dense emission nebulae as M42 in Orion, the blue-green radiation is more prominent.

NGC 2244, because it contains hot O-type stars, cannot be very old. Calculations of how long it would take the radiation pressure from the O stars in NGC 2244 to clear out the "hole" in the Rosette Nebula suggest that the stars began shining less than half a million years ago. Star-formation might still be occurring in the Rosette Complex: "close-up" photos of the Nebula show silhouetted against it thin, dark nebulous ropes of dust

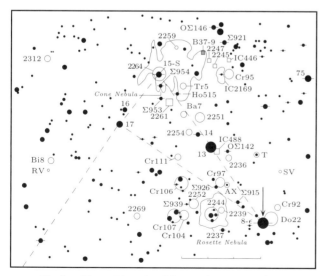

Finder Chart 20-4. 8–ε Mon: $06^h23.8^m$ +04°36′

and tiny opaque "globules," the latter very likely contracting on their way to protostars. The initial "push" to get dust clouds in the Rosette Complex to begin gravitational contraction into globules, and ultimately into protostars, could come either from the radiation pressure upon such clouds from the O stars of NGC 2244 or from the outward expansion of the relatively hot Rosette Nebula itself into surrounding cool material.

8/10″ Scopes–50x: NGC 2244 is a very bright, 25′ long cluster of three dozen 6th to 12th magnitude stars embedded in traces of nebula-glow.

12/14″ Scopes–60x: A 38mm giant Erfle eyepiece shows several dark lanes snaking across the faint, annular glow of nebulosity surrounding the bright open cluster NGC 2244. The fifteen brightest cluster stars fall along two NW–SE parallel rows, but some fifty stars can be seen in a 25′ area. The bright star at the SE side of the cluster, 12 Monocerotis (m5.8), has a fine yellow color that contrasts well with the O stars of the group, but it is in fact a foreground object and not involved with the Rosette Complex. An O-III or a UHC filter will enhance the nebula considerably and bring out many dark lanes, glowing streaks, and swirls of nebulosity. Two dark globules in the western and northern sections are especially noticeable.

NGC 2250 Open Cluster 10★ Tr Type IV 2 p
ø7′, m8.9p, Br★ 12.0p $06^h32.8^m$ −05°02′
Finder Chart 20-3   ★★

12/14″ Scopes–125x: NGC 2250, located between two 7th magnitude stars 22′ to its NE and SW, is a coarse cluster of eight 12th to 14th magnitude stars in an 8′ E–W area.

NGC 2245 Reflection Nebula
ø5′ × 4′, Photo Br 1-5, Color 2-4 $06^h32.7^m$ +10°10′
Finder Chart 20-4, Figure 20-7   ★★★

16/18″ Scopes–100x: NGC 2245, just 2′ SW of an 8th magnitude star, is an excellent object under a clear dark sky! It is a bright 1.5′ diameter comet-shaped nebulosity that fans to the SW of an 11th magnitude white star which looks like a comet's nucleus. Reflection nebula NGC 2247 lies 12′ NNE.

NGC 2247 Reflection Nebula
ø6′ × 6′, Photo Br 1-5, Color 2-4 $06^h33.2^m$ +10°20′
Finder Chart 20-4, Figure 20-7   ★★

16/18″ Scopes–100x: NGC 2247 is much fainter than reflection nebula NGC 2245 to its SSW. It is a faint, circular glow surrounding an 11th magnitude white star. A magnitude 8.5 star is 5′ NW and a 9th magnitude star lies to the south.

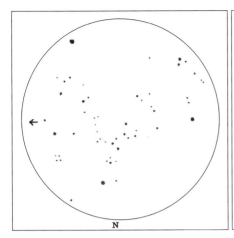
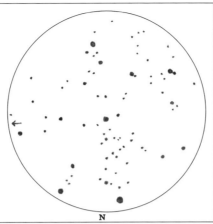
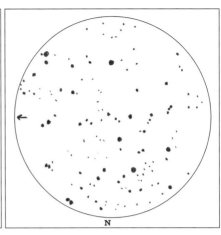

Figure 20-9. NGC 2252
12.5", f5–100x, by G. R. Kepple

Figure 20-10. Collinder 106
12.5", f5–80x, by G. R. Kepple

Figure 20-11. NGC 2260
12.5", f5–125x, by G. R. Kepple

NGC 2251 Cr101 Open Cluster 30★ Tr Type III 2 p
ø10', m7.3v, Br★ 9.10v   06ʰ34.7ᵐ +08°22'
Finder Chart 20-4, Figure 20-6   ★★★

8/10" Scopes–100x: This irregular cluster lies 3/4° north of magnitude 6.5 14 Monocerotis. It is a fairly bright, large, elongated formation of several dozen stars in a 10' × 4' NW–SE area.

12/14" Scopes–100x: NGC 2251 is a bright, large, attractive 12' long NW–SE stream of stars. On its western side is a 9th magnitude star with a faint companion 10" away. Its southern wing contains some nebulous haze. A few reddish stars are sprinkled among its 40 members.

NGC 2252 Open Cluster 30★ Tr Type IV 2 p
ø20', m7.7v, Br★ 9.0p   06ʰ35.0ᵐ +05°23'
Finder Chart 20-4, Figures 20-8 & 20-9   ★★★

12/14" Scopes–75x: NGC 2252 lies 3.4° NE of open cluster NGC 2244 just outside the edge of the Rosette Nebula. It is quite large, irregular, and loose, its three dozen stars in a wishbone with the prongs diverging toward the SE and SW from the N–S stem. The cluster has eight 10th to 11th magnitude stars, but the majority of its members are 11.5 to 12.5. Near the cluster center is a 10" wide N–S pair of 10th magnitude stars.

NGC 2254 Open Cluster 50★ Tr Type I 2 p
ø4', m9.1v, Br★ 11.85v   06ʰ36.0ᵐ +07°40'
Finder Chart 20-4   ★★★

12/14" Scopes–75x: NGC 2254, 17' NE of 14 Monocerotis (m6.4), is an irregular, 2' diameter group of a dozen faint stars in two rows extending eastward and southward. A small N–S arc of stars concave to the east is visible. The brightest stars are near the north edge.

Collinder 104 Open Cluster 15★ Tr Type IV 1 p n
ø21', m9.6p   06ʰ36.5ᵐ +04°49'

Collinder 107 Open Cluster 15★ Tr Type IV 3 p
ø35', m5.1p, Br★ 7.14v   06ʰ37.7ᵐ +04°44'
Finder Chart 20-4, Figure 20-8   ★★/★★★

12/14" Scopes–75x: Collinder 107 is a bright, large, irregular and very poorly concentrated cluster whose brightest members form a 35' long acute triangle pointing NW. At the NW vertex of the triangle is a magnitude 6.2 star; and the SE short side of the triangle has a magnitude 7.1 star at the SW end and an 8th magnitude star at the NE end. Separate concentrations of fainter stars are around the NW vertex and SE short side of the acute triangle. Collinder 104, just west of Collinder 107, is a conspicuous 20' long N–S chain of relatively faint stars. East of the southern end of the chain is a detached trio of magnitude 7.5 to 8 stars.

Basel 7 Open Cluster 15★
ø–, m8.5v, Br★ 10.35v   06ʰ36.6ᵐ +08°21'
Finder Chart 20-4   ★★

12/14" Scopes–100x: Basel 7 is an irregular, mildly concentrated, 8' group of twenty 11th to 13th magnitude stars. It is more concentrated in the northern section but the brightest star, about 10.5 magnitude, lies at the southern edge. Just NE of this star is a tiny starless gap.

Figure 20-12. *The very faint emisssion nebula Sharpless 2-282 is visible through an O-III filter as a hazy streak. Open cluster NGC 2262 is at the upper left, and Collinder 110 is to the far right. Martin C. Germano made this 80 minute exposure on hypered Kodak 2415 Tech Pan film with an 8", F5 Newtonian reflector at prime focus. North is to the right.*

Collinder 106 Open Cluster 20★ Tr Type III 3 p
ø45', m4.6p $06^h37.1^m$ +05°57'
Finder Chart 20-4, Figure 20-10   ★★★

12/14" Scopes–100x: Collinder 106, a nice cluster, is a bright, loose, irregular 45' wide group of 70 stars of varying magnitudes most concentrated toward the north. The two northernmost stars of the cluster are 6th and 7th magnitude while the two southernmost are 7–7.5 magnitude. The star at the cluster's center is a magnitude 7.5 object. These five stars together form a large "X." An interesting star chain is south of the cluster.

Trumpler 5 Open Cluster 150★ Tr Type II 3 r n
ø8', m10.9p, Br★ 17.0p $06^h37.6^m$ +09°26'
Finder Chart 20-4, Figure 20-14   ★

16/18" Scopes–100x: Trumpler 5 is a rich concentration of extremely faint stars. However, even at this aperture, and under the best of skies, the cluster does not resolve but can be seen only as a 20' wide nebulous patch.

Sharpless 2–282 Emission Nebula
ø25' × 10', Photo Br 3-5, Color 3-4 $06^h38.0^m$ +01°31'
Finder Chart 20-5, Figure 20-12   ★

16/18" Scopes–75x: Sharpless 2-282, just SW of a 6th magnitude star, is visible with an O-III filter as a very faint, diffuse haze elongated N–S.

NGC 2260 H48[8] Open Cluster
[ø20', Br★ 8.0v] $06^h38.1^m$ −01°28'
Finder Chart 20-5, Figure 20-11   ★★★

12/14" Scopes–100x: NGC 2260 is an officially "nonexistent" cluster nevertheless very much in existence! It is bright and large with 80 to 90 members irregularly spread over a 20' diameter area between two 8th magnitude stars. The brighter cluster stars are in two NE–SW concentrations. The larger NE group is a pentagon with an 8th magnitude star on its northern corner and a wide double just north of its center. The SW group is elongated E–W: an 8th magnitude star is on its SW edge, but it is star-poor in the center. Each group, in addition to its 8th magnitude lucida, has a dozen 9th to 11th magnitude, and a dozen 12th magnitude, stars.

NGC 2262 H37[7] Open Cluster 35★ Tr Type I 2 p
ø3.5', m11.3p   06ʰ38.4ᵐ +01°11'
Finder Chart 20-5, Figures 20-12 & 20-16   ★★★

12/14" Scopes-100x: NGC 2262 is a rich, circular condensation of three dozen 12th magnitude and fainter stars in a 3.5' diameter area. The cluster, though only mildly compressed, appears rich relative to the surrounding field. Nevertheless, the surrounding field itself is, especially to the NE of the cluster, richly sprinkled with 10th to 11th magnitude stars. Emission nebula Sh2-282 is visible as a very faint haze in the star field to the NW.

Collinder 110 Open Cluster 70★ Tr Type III 1
ø12', m10.5p   06ʰ38.4ᵐ +02°01'
Finder Chart 20-5, Figure 20-12   ★★

12/14" Scopes-100x: Collinder 110 is large, irregular, and moderately faint but stands out fairly well as an enrichment in the surrounding star field. It has seventy 11th to 13th magnitude stars in a 12' area, its members gathered into rows and irregular groups separated by star-poor gaps.

NGC 2259 Cr108 Open Cluster 25★ Tr Type II 2 p n
ø4.5', m10.8p, Br★ 14.0p   06ʰ38.6ᵐ +10°53'
Finder Chart 20-4   ★

12/14" Scopes-125x: NGC 2259, 14' east of a magnitude 6.5 star (with a 9th magnitude star between the two), is a small, rich patch of a dozen faint stars resolved against a granular background haze. Near its north edge is a 7" double of magnitude 11.5 and 12.5 stars.

16/18" Scopes-150x: At this aperture fifteen NGC 2259 stars are easily resolved and an equal number near resolution in the background haze. The cluster is compact, irregularly round, and about 4' across. It is embedded in the northern part of the same nebulous complex to which the Cone Nebula belongs, and on dark, transparent nights patches of dark and bright nebulosity can be glimpsed in the vicinity of the cluster.

Collinder 111 Open Cluster or Asterism?
ø3.2', m7.0p   06ʰ38.7ᵐ +06°54'
Finder Chart 20-4   ★★

12/14" Scopes-100x: The obscure cluster Collinder 111 is a small, sparse collection of a dozen faint stars positioned just SW of an 8th magnitude star.

Figure 20-13. *NGC 2261, Hubble's Variable Nebula, is a bright comet-shaped nebulosity. Lee C. Coombs made this 10 minute photo on 103a-E film with a 10", f5 Newtonian at prime focus.*

NGC 2261 H2[4] Emission & Reflection Nebula
ø3.5' × 1.5', Photo Br 1-5, Color 2-4 06ʰ39.2ᵐ +08°44'
Finder Chart 20-4, Figure 20-13   ★★★

Hubble's Variable Nebula

Hubble's Variable Nebula is a peculiar emission and reflection nebula enveloping the erratic variable R Monocerotis. The nebula was first observed by Sir William Herschel in 1783, and the variability of its associated star discovered by Schmidt in 1861. The variability of the nebula itself was found by Edwin Hubble in 1916 from a series of photographs. Hubble's Variable Nebula was the first object photographed by the 200" Hale Telescope at Mt. Palomar in 1949.

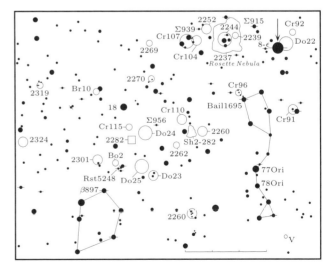

Finder Chart 20-5. 8-ε Mon: 06ʰ23.8ᵐ +04°36'

Figure 20-14. *This wide field photograph of the NGC 2264 complex shows a wealth of nebulous detail and several star clusters and nebulae. The Cone Nebula is the little dark notch piercing the top of the small bright patch near the center. Trumpler 5 is the bright concentration of stars embedded in nebulosity at left center. Chris Schur made this 20 minute exposure through a 92 Wratten filter on Kodak 2415 film with an 8", f1.5 Schmidt Camera.*

The variations in brightness, shape, and detail of NGC 2261 seem to be caused by shadows cast through it by dense dust clouds drifting near its illuminating star, R Monocerotis. In the eyepiece the nebula has the appearance of a comet, with the star in the "head" near — but not on — the southern edge and the nebula fanning outward to the north as the "tail." In larger telescopes some blue might be discerned in the nebula — reflected light from R, which would itself look bluish if it was not embedded in dust.

Hubble's Variable Nebula is believed to be an outlying member of the huge NGC 2264 nebulous complex centered a degree away. The entire complex is about 3,000 light years distant. NGC 2261 consequently is about 3 light years long and 1.5 light years wide, and R Mon is as luminous as about 80 suns.

4/6" Scopes–100x: NGC 2261 is readily apparent as a small, diffuse, cometary patch with R Monocerotis at the southern tip and the nebulosity fanning northward.

8/10" Scopes–125x: Hubble's Variable Nebula is a fairly bright comet-shaped nebulosity with R Monocerotis at the southern tip acting as the comet's nucleus. The glowing nebulosity spreads 1.5' northward, curving west.

12/14" Scopes–150x: This emission and reflection nebula is an interesting wedge-shaped glow fanning north from R Monocerotis. The nebula is quite bright around the star, fades slightly, brightens again about 1' north of the star, and fades again quite abruptly at its northern edge. 200x shows dark markings within the nebulosity. The features in the nebulosity change unpredictably and without regard to the light variations of R.

NGC 2264 H5[8] Emission Nebula
ø35' × 15', Photo Br 1-5, Color 3-4   06^{h}41.1^m +09°53'

NGC 2264 Cr112 Open Cluster 40★ Tr Type IV 3 p
nø20', m3.9v, Br★ 4.62v   06^{h}41.1^m +09°53'
Finder Charts 20-4 & 20-5, Figs. 20-14 & 20-15   ★/★★★★
Christmas Tree Cluster

NGC 2264 is a large, bright cluster easily visible in finder scopes and binoculars. When viewed in normal inverting astronomical telescopes, with south up, its shape resembles a Christmas tree. The cluster spans half a degree and contains 20 bright and another hundred or so fainter stars. At the tree's base lies the brightest star of the cluster, the 5th magnitude variable S (15) Monocerotis, a visual double with an 8.5 magnitude companion 3″ away in P.A. 213°.

NGC 2264 is embedded in an extensive but tenuous nebulosity which may be glimpsed with large scopes under clear, dark skies. Careful inspection of the area with a nebula filter at low power can reveal many faint wispy streaks. To the south, beyond the tip of the Christmas tree, lies the famous "Cone Nebula," a beautiful object on photographs but very difficult to detect visually.

The cluster spans some 20 light years and lies about 3,000 light years away. It therefore is the foreground of the Rosette Nebula complex to the south, but nearly twice as far from us as the nebulae and supergiant blue stars in the Belt and Sword of Orion.

10 × 50 Binoculars: NGC 2264 is a bright, large triangular cluster easily visible in binoculars. The cluster spans half a degree and has twenty bright, and perhaps sixty faint, stars. The Christmas tree appears right side up in normal inverting astronomical telescopes, but is upside-down in binoculars. Under exceptionally good observing conditions NGC 2264 is visible to the naked eye as a bright Milky Way spot.

8/10″ Scopes–75x: This grand cluster is a naked eye object and conspicuous in an 8 × 50 finder or binoculars as a small triangular star group — the Christmas Tree. The top of the tree, pointing south, is decorated by a 6th magnitude star. The tree is outlined by magnitude 8–9.5 stars, and the bright blue-white 15 Monocerotis marks its trunk. Some three dozen cluster members can be counted. Faint nebulosity can be glimpsed just SW of 15 Mon. The Cone Nebula lies just south of the magnitude 6 star at the Tree's tip, but is only visible as a star-poor space.

12/14″ Scopes–100x: NGC 2264 is a bright, large, loose cluster of 40 stars concentrated around 15 Monocerotis (m4.6), with another sixty stars in the area 30′ south to the magnitude 6.5 star at the top of the Christmas Tree. The cluster is faintly illuminated with nebulosity which can be brought to life

Figure 20-15. *NGC 2264, the Christmas Tree Cluster is decorated by bright stars with 15-S Monocerotis marking the trunk and the Cone Nebula trimming the top of the tree. Martin C. Germano made this 45 minute exposure on 2415 film with an 8″, f5 Newtonian reflector.*

with a UHC or O-III filter. Just SW of 15 Mon is a bright patch of nebulosity, from which a fainter haze can be followed out for a couple degrees north and east and half a degree south to the star at the top of the Tree. The Cone Nebula is just visible as a dark notch south of this star.

Dolidze 23 Open Cluster 20★ Tr Type IV 2 p
⌀12′   **06ʰ43.2ᵐ 00°00′**
Finder Chart 20-5   ★★★

12/14″ Scopes–100x: Dolidze 23 is a fairly bright, loose, sparse, 12′ diameter cluster of several dozen stars surrounding a bright trio, two of which are 8th magnitude objects. The highest concentration of cluster stars lies west of the bright trio.

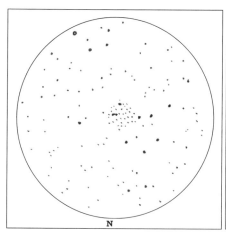

Figure 20-16. NGC 2262
13", f5.6–165x, by Steve Coe

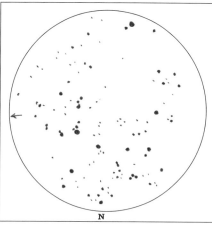

Figure 20-17. NGC 2270
12.5", f5–125x, by G.R. Kepple

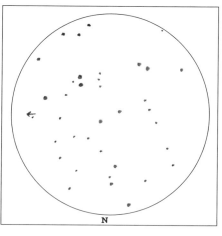

Figure 20-18. NGC 2286
12.5", f5–100x, by G.R. Kepple

NGC 2269 Open Cluster 12★ Tr Type II 2 p
ø4′, m10.0:v, Br★ 11.61v   06ʰ43.9ᵐ +04°34′
Finder Chart 20-4   ★★

12/14″ Scopes–75x: At low powers NGC 2269 is a faint patch in a nice star field. At 125x the cluster resolves into a couple dozen 13th magnitude and fainter stars in a 4′ × 2′ NW–SE area. A slightly brighter star stands out near the center of the group. The cluster is encircled by 12th magnitude stars.

NGC 2270 H36⁷ Open Cluster
[ø12′ × 5′, Br★ 9.0v]   06ʰ43.9ᵐ +03°26′
Finder Chart 20-5, Figure 20-17   ★★★

12/14″ Scopes–100x: Though catalogued as "nonexistent," NGC 2270 is identifiable as a loose and irregular aggregation of fifty stars in a NE–SW 12′ × 5′ area. The cluster has many small star-knots and doubles, but no central compression. A starchain begins at a 10th magnitude star at the cluster's east edge, runs ENE–WSW through the magnitude 9 star at the cluster's center, and then loops around the group's SW quadrant, returning to the 9th magnitude star. An 11th magnitude member lies near the cluster's northern edge.

Dolidze 24 Open Cluster 40★ Tr Type III 1 p
ø18′   06ʰ44.2ᵐ +01°36′
Finder Chart 20-5   ★★

12/14″ Scopes–75x: Dolidze 24 is faint and loose with thirty stars in an 18′ area. 9th magnitude stars are on both its east and west edges, with fifteen 11th and 12th magnitude and many more faint stars scattered between them. The cluster is densest to the east.

125x reveals several dozen additional threshold stars.

Dolidze 25 Open Cluster 50★ Tr Type IV 2 p n
ø23′, m7.6v, Br★ 8.87v   06ʰ45.1ᵐ +00°18′
Finder Chart 20-5   ★★★

12/14″ Scopes–75x: Dolidze 25 is visually unique. It consists of two wishbone-shaped groups joined by a pair of bright stars at their stems. The eastern wishbone is larger, longer, and seems to be centered in the cluster's "official" position. Its fork has an extension to the west formed by a long faint stream of stars.

Collinder 115 Open Cluster 50★ Tr Type III 2 p
ø7′, m9.2p   06ʰ46.5ᵐ +01°46′
Finder Chart 20-5   ★★

12/14″ Scopes–75x: Collinder 115 is a fairly faint cluster of forty 12th magnitude and fainter stars in a 20′ × 10′ NE–SW. An 11th magnitude star stands out on the north side. At low powers the cluster is merely a granular patch; but 175x resolves fifty stars.

NGC 2282 IC 2172 Reflection Nebula
ø3′ × 3′, Photo Br 3-5, Color 2-4   06ʰ46.9ᵐ +01°19′
Finder Chart 20-5   ★★

12/14″ Scopes–75x: NGC 2282 is a very faint 2′ patch of nebulosity associated with a 10th magnitude star. The nebula is uneven in brightness and fans NW from the star.

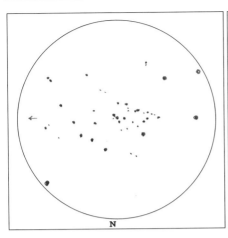

Figure 20-19. NGC 2302
12.5", f5–100x, by G.R. Kepple

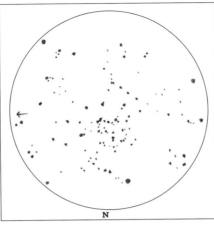

Figure 20-20. NGC 2309
12.5", f5–150x, by G.R. Kepple

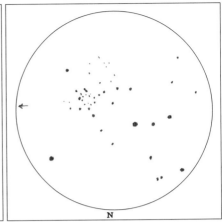

Figure 20-21. NGC 2311
12.5", f5–70x, by G.R. Kepple

NGC 2286 Cr117 Open Cluster 50★ Tr Type IV 3 m
⌀14′, m7.5v, Br★ 9.71v   06ʰ47.6ᵐ −03°10′
Finder Chart 20-6, Figure 20-18   ★★★

12/14" Scopes–75x: NGC 2286 is a loose, irregular group of thirty 12th magnitude and fainter stars in a 10′ area with no central concentration. Outlying stars within a 25′ area bring the star count to 50. Two wide pairs of magnitude 9.5–10 stars stand out in the southern portion. The SW of these two star-pairs forms a thin triangle with another star of similar magnitude.

Bochum 2 Open Cluster 10★
⌀1.5′, m9.7v, Br★ 10.86v   06ʰ48.9ᵐ +00°23′
Finder Chart 20-5   ★★

12/14" Scopes–75x: Bochum 2 is a faint, obscure cluster 10′ east of a triangle of one 7th and three 8.5 magnitude stars. It consists of one 10.8 and ten 13th to 14.5 magnitude stars in a 3′ area. A trio of 11th magnitude stars spaced 30″ apart lies 2′ SE.

NGC 2299 Open Cluster
[⌀6′, Br★ 9.0v]   06ʰ51.1ᵐ −07°00′
Finder Chart 20-6   ★★

12/14" Scopes–125x: NGC 2299 is another cluster classified as "nonexistent" because it tends to be lost in its rich Milky Way star field. It includes some three dozen 12th to 14th magnitude stars in a triangle extending 6′ NNW from a 9th magnitude star. In the field to the south and SW are other stellar concentrations. A star stream terminating with an 11th magnitude star extends 10′ SW.

NGC 2301 Open Cluster 80★ Tr Type I 3 m
⌀12′, m6.0v, Br★ 8.01v   06ʰ51.8ᵐ +00°28′
Finder Chart 20-5, Figure 20-22   ★★★★

2/3" Scopes–75x: Beautiful! NGC 2301 is a prominent N–S string of bright stars ending in a bright knot. North of the string is a second star-knot, and a host of faint stars spreads eastward.

8/10" Scopes–75x: This nice cluster is large and rich with sixty stars, most between 9.5 and 11th magnitude, in a 10′ area. A stem consisting of two 8.5 and two 9th magnitude stars extends southward.

12/14" Scopes–75x: NGC 2301 is one of the more im-

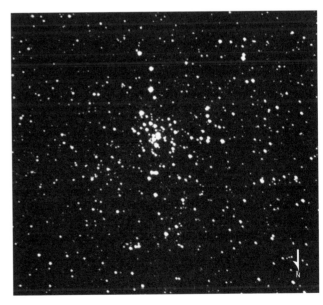

Figure 20-22 *NGC 2301 is a bright, large, irregular cluster with many star chains. Lee C. Coombs made this five minute exposure on 2415 film with a 10", f5 Newtonian reflector at prime focus.*

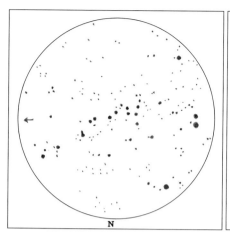
Figure 20-23. NGC 2319
12.5″, f5–125x, by G. R. Kepple

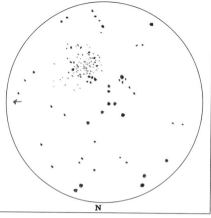
Figure 20-24. NGC 2324
12.5″, f5–70x, by G. R. Kepple

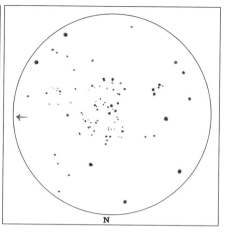
Figure 20-25. NGC 2335
12.5″, f5–100x, by G. R. Kepple

pressive clusters in Monoceros. It is bright and large, its sixty stars distributed in a distorted, highly elongated N–S "Y"-formation. The stem of the "Y" lies toward the south, splitting in two as it extends northward. The most concentrated portion of the cluster runs N–S. The NE prong of the "Y" extends out of the field of view into a large sparse group of bright stars. Three colorful stars form a N–S row at the cluster's center: The two southernmost forming a beautiful wide gold and blue pair, the other displaying a nice reddish tint.

NGC 2302 H39⁸ Open Cluster 30★ Tr Type II 2 p
⌀2.5′, m8.9v, Br★ 10.24v   06ʰ51.9ᵐ −07°04′
Finder Chart 20-6, Figure 20-19   ★★★

12/14″ Scopes–75x: NGC 2302, 3′ SE of a 7th magnitude star, is conspicuous enough to be spotted at low power without difficulty. At 125x, it is loose and irregular, its 20 stars forming a nearly empty circlet except for an obvious "Y" with its stem pointing SW. A 9th magnitude star at the east edge forms a triangle with two more stars of magnitudes 9 and 9.5 farther east. A nice magnitude 8.5 double lies 15′ south of the cluster and a pretty orange star is visible to the SW.

Biurakan 10 Open Cluster 20★ Tr Type I 3 p
⌀4′, m10.4v, Br★ 10.69v   06ʰ52.2ᵐ +02°56′
Finder Chart 20-5   ★★

12/14″ Scopes–75x: Biurakan 10 is a small, faint group of one 11th magnitude and a dozen 13th to 14th magnitude stars in a 2′ area. The brightest star is part of a "V" pointing west. The cluster lies 7′ SE of a 6th magnitude star which forms an equilateral triangle with a 7th magnitude star 22′ south and a 7.5 magnitude star 25′ SE.

NGC 2306 Open Cluster or Asterism?
⌀20′   06ʰ53.4ᵐ −07°19′
Finder Chart 20-6   ★★★

12/14″ Scopes–75x: NGC 2306 is a large 20′ group of 25 to 30 faint stars lying within a triangle of three 8th magnitude stars. A very close 10th magnitude double is on the south edge, and a 12th magnitude double lies at the center.

NGC 2309 H18⁶ Open Cluster 40★ Tr Type II 2 m
⌀3′, m10.5p, Br★ 13.0p   06ʰ56.2ᵐ −07°12′
Finder Chart 20-6, Figure 20-20   ★★★

12/14″ Scopes–75x: NGC 2309 is a faint, small cluster 4′ SSW of a 9th magnitude star in an attractive star

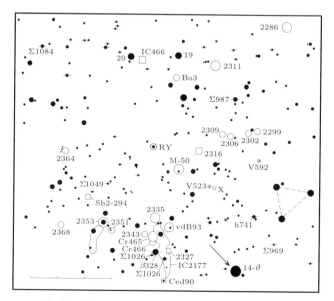
Finder Chart 20-6. 14–ϑ CMa: 06ʰ54.2ᵐ −12°02′

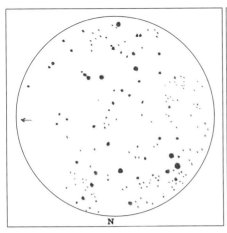

Figure 20-26. NGC 2351
12.5″, f5-125x, by G. R. Kepple

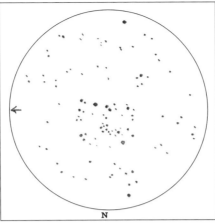

Figure 20-27. NGC 2353
8″, f5-100x, by A. J. Crayon

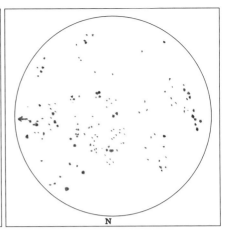

Figure 20-28. NGC 2364
12.5″, f5-125x, by G. R. Kepple

field. At low power it appears as just a field concentration. 125x, however, resolves fifteen 12th to 13th magnitude stars and an equal number of fainter stars in a 3′ area. An 11th magnitude star with a 10″ companion stands out just south of the cluster's center. Several more doubles, all with 12th to 12.5 magnitude stars, are north of center and at the NNE edge.

16/18″ Scopes-150x: NGC 2309, 1.75° NW of open cluster Messier 50, is a small, fairly compressed group of forty 12th to 14th magnitude stars in a 4′ area. The center is marked by an 11th magnitude star with a 12th magnitude companion 10″ away to the ENE. Several other pairs are also visible.

NGC 2311 Cr123 Open Cluster 50★ Tr Type III 2 p
ø6′, m9.6p, Br★ 12.0p   06ʰ57.8ᵐ −04°35′
Finder Chart 20-6, Figure 20-21   ★★★

12/14″ Scopes-75x: NGC 2311 is a fairly rich cluster 10′ WSW of a row of one 9th and two 8th magnitude stars. The cluster contains fifteen 11th to 12th magnitude stars in a 7′ area against an unresolved background haze. 175x, however, reveals several dozen stars to the 14th magnitude.

NGC 2312 Open Cluster
[ø6′ × 4′, Br★ 13.v]   07ʰ04.2ᵐ +01°03′
Finder Chart 20-4   ★

12/14″ Scopes-125x: NGC 2312 is officially listed as "nonexistent" because it is almost overwhelmed by the rich Milky Way field around it. It contains three dozen 13th to 14th magnitude stars in a 6′ × 4′ NNW–SSE triangular area partially set off from the surrounding star field by a few small, nearly starless voids around its perimeter. 16′ NW is a 1.5′ wide E–W pair of 9th magnitude stars.

Biurakan 8 Open Cluster 70★ Tr Type II 2 n
ø5′, Br★ 14.0p   06ʰ58.1ᵐ +06°26′
Finder Chart 20-4   ★★★

12/14″ Scopes-75x: Biurakan 8, 7′ SSE of an 8th magnitude star, contains fifteen 14th magnitude stars in a 5′ area against a background haze. 225x resolves 30 stars. The cluster is flanked by 12th magnitude stars, two 4′ and 5.5′ to the cluster's east, and two 4′ and 6.5′ to its west.

NGC 2316 Emission & Reflection Nebula
ø4′ × 3′, Photo Br 2-5, Color 2-4   06ʰ59.7ᵐ −07°46′
Finder Chart 20-6   ★★★

12/14″ Scopes-100x: NGC 2316 is a faint, small, oval haze around a very dim star that can be split at high power. The nebula is located just north of three faint stars.

NGC 2319 Open Cluster
ø15′, Br★ 8.v   07ʰ01.1ᵐ +03°04′
Finder Chart 20-5, Figure 20-23   ★★★

12/14″ Scopes-75x: NGC 2319 is another officially "nonexistent" cluster, but we had no trouble finding it. Herschel described the group as a "linear cluster of stars from 11th to 13th magnitude forming a bent line nearly 15′ long, terminated on the following side by a magnitude 8 star..." The line angles E–W and then curves WNW and consists of sixteen 8th to 12th magnitude stars. Twenty 13th magnitude stars are haphazardly sprinkled around. A second 8th magnitude star is NNW of the one that anchors the cluster-line's east end. The group contains many paired stars.

Figure 20-29. *Messier 50 (NGC 2323) is the finest cluster in the constellation of Monoceros. It has 150 stars spread over 25 minutes of arc. Martin C. Germano made this 30 minute exposure on hypered Kodak 2415 Tech Pan film with an 8", f5 Newtonian reflector at prime focus.*

NGC 2323 Messier 50 Open Cl. 80★ Tr Type II 3m
ø16′, m5.9v, Br★ 7.85v   07ʰ03.2ᵐ −08°20′
Finder Chart 20-6, Figure 20-29   ★★★★★

The finest cluster in Monoceros, Messier 50, is visible in binoculars as a bright, large, irregularly round concentration of stars. It seems to have been first seen by G.D. Cassini some time before 1711, but Charles Messier rediscovered it on April 5, 1772 while observing the comet of that year. M50 is about 2,900 light years away, 14 light years in diameter, and is as luminous as 6,400 suns.

8/10″ Scopes–50x: Low power shows ten 8 to 9.5 magnitude and fifty 10th to 12th magnitude stars in a 20′ area. The central concentration is shaped like a blunt arrowhead pointing north.

12/14″ Scopes–75x: Messier 50 is a large, splendid assemblage of bright stars in a rich star field. At least 150 stars from 8th to 14th magnitude are visible in a 25′ area. In the center is a rich, heart-shaped concentration offset by a starless void to its north. The cluster's magnitude 7.8 reddish-orange lucida is conspicuous in its southern part. From the eastern edge a stream of stars stretches SE.

Bochum 3 Open Cluster 25★
ø4′, m9.9v, Br★ 11.19v   07ʰ03.4ᵐ −05°04′
Finder Chart 20-6   ★★★

12/14″ Scopes–75x: Bochum 3, 20′ NNW of a 6th magnitude star, contains a dozen 12th to 14th magnitude stars in a 4′ area with a row of four 11th to 11.5 magnitude stars crossing through its center. 225x resolves another dozen members.

NGC 2324 H38⁷ Open Cluster 70★ Tr Type II 2 r
ø7′, m8.4v, Br★ 10.35v   07ʰ04.2ᵐ +01°03′
Finder Chart 20-5, Figure 20-24   ★★★

12/14″ Scopes–75x: NGC 2324 is a large, faint, granular patch in a fine field of bright stars. A "Y" of seven magnitude 9 to 9.5 stars lies just NE of the cluster. 100x resolves five 10th to 11th magnitude and fifty 12.5 magnitude and fainter stars in a circular 8′ area. 22′ north is the wide, bright double, OΣΣ 82 (6,7, 7.4; 90″; 318°).

vdB 93 Gum 1 Emission & Reflection Nebula
ø20′ × 20′, Photo Br 1-5, Color 2-4   07ʰ04.3ᵐ −10°28′
Finder Chart 20-6, Figures 20-30 & 20-31   ★★

16/18″ Scopes–75x: van den Bergh 93 is a very faint, irregular nebula surrounding a bright 6th magnitude star that interferes with observations. It is

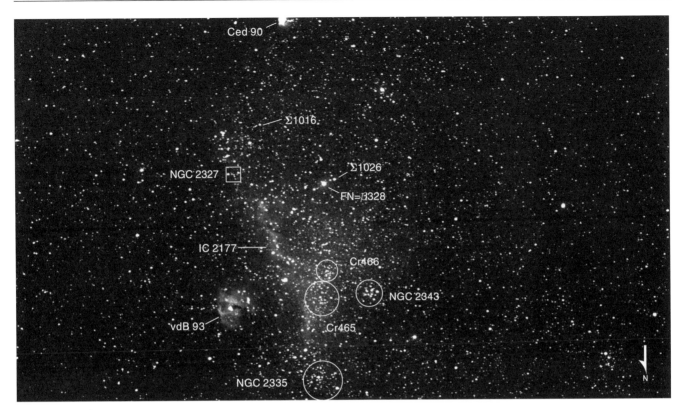

Figure 20-30. *Emission nebula IC 2177 (at center) is a faint diffuse streamer resembling the spread wings of an eagle. Open cluster NGC 2335 marks the northern edge (bottom center) and the bright but small emission & reflection nebula Cederblad 90 (top center) lies at the southern edge in Canis Major. vdB93 is the nebula to the west (left) of the eagle's wings and open cluster NGC 2343 is to the east (right). Open cluster NGC 2353 lies at the lower right corner. John Ebersole, M.D. took this 40 minute photograph at sea level through a 25A filter with a 300mm telephoto lens on hypered Kodak 2415 Tech Pan film.*

visible through an O-III filter as a circular haze with a dark notch penetrating from the nebula's NE edge almost to its center. To the east of vdB 93 are the open clusters NGC 2335, NGC 2343, Collinder 465, and Collinder 466, rich knots in a fine star field laced with faint nebulae.

IC 2177   Gum 2   Emission Nebula
ø120′ × 40′, Photo Br 3-5, Color 3-4 $07^h05.3^m$ −10°38′
Finder Chart 20-6, Figures 20-30 & 20-31
★★

16/18″ Scopes–75x: A UHC filter shows IC 2177 to be a very faint, much elongated streamer of pale nebulosity running 2° N–S. Its northern edge is marked by the open cluster NGC 2335 and its southern edge, across the border in Canis Major, is the concentrated emission + reflection knot Cederblad 90.

NGC 2335   Cr127   Open Cluster 35★   Tr Type III 3 m n
ø12′, m7.2v, Br★ 9.50v $07^h06.6^m$ −10°05′
Finder Chart 20-6, Figures 20-25 & 20-30   ★★

12/14″ Scopes–75x: NGC 2335 lies 5′ WSW of a 7th magnitude star. It is a loose, irregular sprinkling of several dozen 11.5 to 13th magnitude stars spread over a 12′ area. It would be more conspicuous if the surrounding star field were not so star-crowded. It is more concentrated in its NE section, wherein some of its stars form a rather conspicuous parallelogram.

Collinder 465   Open Cluster 35★   Tr Type IV 2 p
ø9′, m10.1p $07^h07.2^m$ −10°37′
Finder Chart 20-6, Figure 20-31   ★★

12/14″ Scopes–75x: Collinder 465, 12′ west of NGC 2343, is not as obvious as its NGC neighbor. It is a loose, irregular scattering of about thirty 11th magnitude and fainter stars in a 9′ area with slightly richer concentrations toward its north and south edges. Open cluster Collinder 466 is a somewhat more compressed cluster immediately south of Cr 465.

Collinder 466   Open Cluster 25★   Tr Type III 2 p n
ø4′, m11.1p $07^h07.3^m$ −10°49′
Finder Chart 20-6, Figure 20-31   ★★★

12/14″ Scopes–75x: Collinder 466 is smaller but much more concentrated than Collinder 465 to the north. Several dozen 11th magnitude and fainter stars are

Figure 20-31. *This close up of vdB 93 (lower center) and the northern part of IC 2177 was made by Martin C. Germano on hypered 2415 film for 75 minutes with an 8", f5 Newtonian. Open cluster NGC 2343 is at right. Collinder 465 & 466 share the rich N–S star-stream midway between NGC 2343 and IC 2177.*

visible in a 4′ × 3′ N–S area. The southern portion is more concentrated.

NGC 2343 H33[8] Open Cluster 20★ Tr Type III 3 m n
⌀6′, m6.7v, Br★ 8.39v   07ʰ08.3ᵐ −10°39′
Finder Chart 20-6, Figures 20-30 & 20-31   ★★★

12/14″ Scopes–75x: NGC 2343 is immediately apparent at low power as a rectangular concentration of relatively bright stars in a field rich in faint stars. Ten 9th to 11th and about twenty 12th-13th magnitude cluster members can be seen in a 9′ area. The cluster's lucida, a magnitude 8.5 orange star with an 11th magnitude companion 10″ away, lies in the eastern part of the group.

NGC 2351 Open Cluster
[⌀10′, Br★ 11.v]   07ʰ13.5ᵐ −10°29′
Finder Chart 20-6, Figure 20-26   ★★★

12/14″ Scopes–75x: The exact location of the nonexistent cluster NGC 2351 seems to be in doubt. However, at the position given by the Webb Society, west of a wide pair of 9th and 9.5 magnitude stars aligned NNE–SSW, we found forty stars, including a dozen of magnitudes 11 and 12, scattered over a 10′ area. If a large, loose concentration to the SW of the Webb position is part of the cluster, it increases the group's size to 20′ and its star population to eighty. 125x brings out several starless voids between the two concentrations.

NGC 2353 H34[8] Open Cluster 30★ Tr Type II 2 p
⌀20′, m7.1v, Br★ 9.19v   07ʰ14.6ᵐ −10°18′
Finder Chart 20-6, Figures 20-27 & 20-30   ★★★

4/6″ Scopes–50x: NGC 2353 is an irregular gathering of 30 stars spreading NE from a 6th magnitude blue-white star. To the cluster's NW and SE are 6th magnitude orange stars.

8/10″ Scopes–75x: NGC 2353 has three dozen 10th magnitude and fainter stars irregularly distributed in a 10′ area. The magnitude 6.0 cluster lucida is on its SW edge. A conspicuous double of 9th magnitude stars 20″ apart is 2′ NE of the lucida.

12/14″ Scopes–100x: In large telescopes NGC 2353 is a bright, loose aggregation of two dozen 10th to 11th magnitude, and another two dozen fainter, stars in a 10′ area. The 6th magnitude lucida and a double of 9th magnitude stars are conspicuous near the southern edge. At this magnification the numerous

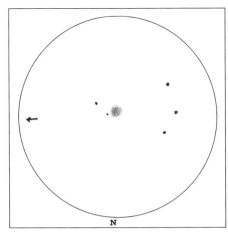
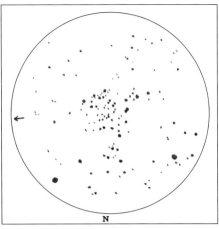
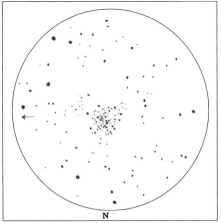

Figure 20-32. PK221+5.1
16", f5–300x, by Alister Ling

Figure 20-33. Melotte 72
10", f7–115x, by Dr. Leonard Scarr

Figure 20-34. NGC 2506
13", f5.6–100x, by Steve Coe

10th and 11th magnitude stars sprinkled north of the main group do not look like cluster members.

NGC 2364 Open Cluster
[ø20', Br★ 10.5v]   07ʰ20.8ᵐ −07°34'
Finder Chart 20-6, Figure 20-28   ★★

12/14" Scopes–125x: NGC 2364 is another Monoceros cluster so crowded by its rich star field that it has come to be classed as "nonexistent." However, Herschel saw "two small, pretty close groups of pretty large stars in the Milky Way, rather a remarkable cluster." What appears at the catalogue position is a large irregularly scattered group with two concentrations. The eastern concentration is a 7' long horseshoe open to the SW. The western concentration is a 3' clump of stars attached to the SW end of the horseshoe. In both concentrations are a total of fifteen 10th to 12th magnitude and seventy fainter stars. 6' to the east is an isolated group of twenty stars highly elongated 4' × 1' NNE–SSW, its brighter stars along its eastern flank. Because of this star stream's complete isolation, it is doubtful that it was one of the two groups Herschel saw. Herschel probably would have assigned a separate NGC number to any aggregation this far away. The larger group with its two areas of concentrations is most likely his NGC 2364.

NGC 2368 Cr138 Open Cluster 15★ Tr Type IV 2 p
ø5', m11.8p   07ʰ21.0ᵐ −10°23'
Finder Chart 20-6   ★★

12/14" Scopes–75x: NGC 2368 is a faint, fairly rich cluster of fifteen 12th magnitude and fainter stars in a 4' area. It is divided by a NNW–SSE starless band into two triangular groups, the western of which is the more populous and includes the cluster's 11th magnitude lucida.

PK221+5.1 Minkowski 3-3 Planetary Nebula Type 4
ø13", m14.8v   07ʰ26.6ᵐ −05°22'
Finder Chart 20-7, Figure 20-32   ★

16/18" Scopes–75x: In an O-III filter this planetary has a nice, fairly bright, round 13" disk with well defined edges.

Melotte 72 Cr 156 Open Cluster 40★ Tr Type II 1 p
ø9', m10.1p   07ʰ38.4ᵐ −10°41'
Finder Chart 20-7, Figure 20-33   ★★★

8/10" Scopes–75x: Melotte 72, just SE of a reddish 7th magnitude star, is a faint but nice cluster of three dozen 12th magnitude and fainter stars filling a 4' wide triangular area.

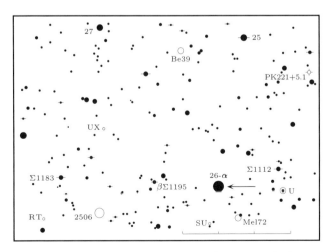

Finder Chart 20-7. 26–α Mon: 07ʰ41.3ᵐ −09°34'

12/14" Scopes–75x: This cluster is a fairly rich, chevron-shaped group of 12th magnitude and fainter stars covering a 5' area.

NGC 2506   H37[6]   Open Cluster 150★   Tr Type I 2 r
ø6', m7.6v, Br★ 10.76v   08ʰ00.2ᵐ −10°47'
Finder Chart 20-7, Figure 20-34   ★★★

8/10" Scopes–100x: This cluster is quite rich and concentrated, with 75 magnitude 12 and fainter stars in a tight 7' area. Half a dozen 11th magnitude stars stand out against the fainter members.

12/14" Scopes–125x: NGC 2506 is a nice but faint 8' wide triangular arrangement of a hundred mostly 12th to 14th magnitude stars, against which the cluster's half dozen magnitude 11–11.5 stars stand out quite well. The brighter stars form wide pairs on the east and west edges of the group. A relatively star-empty notch juts into the south side. Numerous magnitude 9.5 to 11 stars are sprinkled around the cluster, but they give the impression of being foreground objects.

Chapter 21

Orion, the Hunter

21.1 Overview

Orion the Hunter is a magnificent winter constellation formed of a distinctive pattern of bright stars. Orion is nearly as well known as the Big Dipper asterism in Ursa Major. The Hunter's Belt is a conspicuous NW–SE line of 2nd magnitude stars cutting the celestial equator and culminating during the mid evenings in January. From west to east the Belt stars' names are Mintaka, Alnilam, and Alnitak, all from the Arabic meaning "Belt" or "Girdle." The Hunter's right shoulder is the 1st magnitude star Betelgeuse (Alpha Orionis) and his left shoulder the 2nd magnitude Bellatrix (Gamma Orionis). The right knee is at Rigel (Beta Orionis) and the left knee is Saiph (Kappa Orionis). Hanging from the belt is a N–S line that includes the Great Nebula M42 and Iota Orionis. Above the line connecting the shoulder stars is a triangle of one 3rd magnitude and two 4th magnitude stars that marks the Hunter's head. Before him he holds a lion skin (the two Omicrons and six Pis) to take the charge of Taurus the Bull; and over his head he holds his club (Mu, Nu, Xi, and the two Chis).

Orion's stars are good guides to the other stars and constellations of winter. His Belt points SE toward Sirius in Canis Major and NW toward the Hyades Star Cluster and the 1st magnitude star Aldebaran in the face of Taurus the Bull. The meanders of the celestial River Eridanus begin near Rigel with Lambda and Beta Eridani. Betelgeuse is at the NW vertex of the Winter Triangle of 1st magnitude stars, Procyon being the NE vertex and Sirius the south vertex of the triangle.

Because of the distinctive pattern of its stars, Orion has throughout history been envisioned as some important person or deity. The medieval Arabians called him Al Jabbar, "The Giant." Likewise, the medieval Hebrews knew the star-group as Gibbor, "Giant," and identified him with Nimrod, whom Genesis calls "the mighty hunter before the Lord." However, the Babylonians and Sumerians had called our Orion Sipazianna, "Steadfast Shepherd of Heaven," his sheep being the stars in general. (The planets were "Wandering Sheep.")

In Greek and Roman mythology Orion was the son

Orion: Oh-RYE-an
Genitive: Orionis, OR-e-oh-nis
Abbrevation: Ori
Culmination: 9 pm–Jan. 27, midnight–Dec. 13
Area: 1,231 square degrees
Showpieces: 41-ϑ^1 Ori, 44-ι Ori, M42 (NGC 1976), M43 (NGC 1982), M78 (NGC 2068), NGC 1980, NGC 1981
Binocular Objects: 34-δ Ori, 41-ϑ^1–43-ϑ^2, Ori, 43-ϑ^2 Ori, Cr70 (Belt Stars), M42 (NGC 1976), M43 (NGC 1982), M78 (NGC 2068), NGC 1662, NGC 1980, NGC 1981, NGC 2024, NGC 2112, NGC 2169, NGC 2175, NGC 2194

of Neptune/Poseidon, god of the Sea, and of the sea-nymph Euryale. He was a handsome man of gigantic size and strength, so tall that he could walk through deep water without wetting his head. He had no fear of any animal, and once he even threatened to exterminate all the animals of the Earth. When Gaia, the Goddess of the Earth, heard of this threat, she became furious and sent a scorpion to kill him. The scorpion stung Orion, mortally wounding him, but the legendary hero Aeschulapis/Ophiuchus, founder of medicine, gave him an antidote for the poison which saved his life. Thus in the sky Orion sets as Scorpius rises, but Ophiuchus stands upon the Scorpion, trampling it underfoot.

Orion offers an outstanding variety of telescopic and binocular attractions. Its showpiece *par excellence* is the Great Orion Nebula, considered by many to be the finest diffuse nebula in the sky and one of the most beautiful objects visible through a telescope. However, many other nebulae can be seen in the vast gas-and-dust complex which covers most of the constellation. Moreover, Orion contains several outstanding double, triple, and multiple stars, most of which are a beautiful bluish-white because they are young stars that have been only recently born in the Orion Complex.

Orion, the Hunter

Constellation Chart 21-1

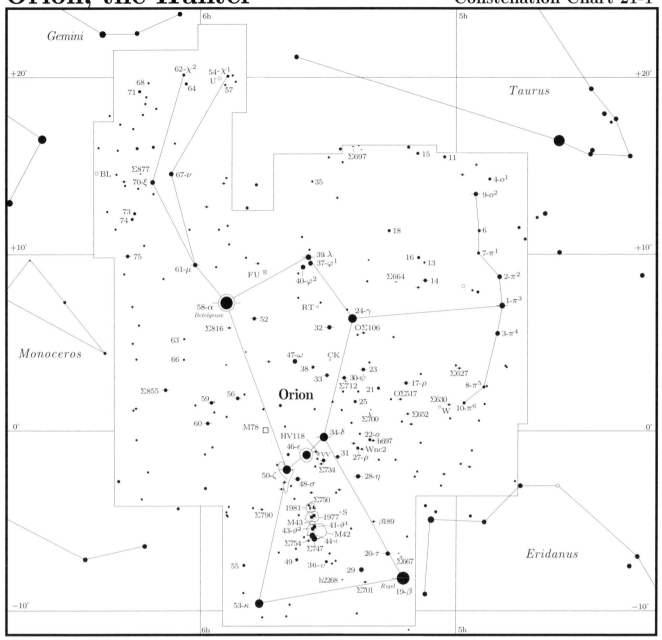

Chart Symbols

Constellation Chart — Stellar Magnitudes: 0 1 2 3 4 5 6	→ Guide Star Pointer ● ○ Variable Stars	⌀ Planetary Nebulae
Finder Charts: 0/1 2 3 4 5 6 7 8 9	●—● Double Stars ○ Open Clusters	▫ Small Bright Nebulae
Master Finder Chart: 0 1 2 3 4 5	Finder Chart Scale (One degree tick marks) ⊕ Globular Clusters	Large Bright Nebulae
	⬭ Galaxies	Dark Nebulae

Table 21–1. Selected Variable Stars in Orion

Name	HD No.	Type	Max.	Min.	Period (Days)	F*	Spec. Type	R.A. (2000)	Dec.	Finder Chart No. & Notes
S Ori	36090	M	7.5	13.5	419	0.48	M6.5–M8	05^{h}29.0^m	−04°42′	21-5
CK Ori	36217	SR	5.9	7.1	120		K2	30.3	+04 12	21-1
VV Ori	36695	EB	5.1	5.5	1.48		B1	33.5	−01 09	21-6
58-α Ori	39801	SRc	0.4	1.3	2110		M1–M2	55.2	+07 24	21-7
U Ori	39816	M	4.8	12.6	372	0.38	M6.5	55.8	+20 10	21-9

F* = The fraction of period taken up by the star's rise from min. to max. brightness, or the period spent in eclipse.

Table 21-2 Selected Double Stars in Orion

Name	ADS No.	Pair	M1	M2	Sep."	P.A.°	Spec	R.A. (2000)	Dec	Finder Chart No. & Notes	
Σ627	3597		6.6	7.0	21.3	260	A0 A0	05ʰ00.6ᵐ	+03°37′	21-1	Fine white pair
Σ630	3623	AB	6.5	7.7	14.2	50	B8	02.0	+01 37	21-1	White, purplish
	3623	BC		10.8	0.5	25					
A2630	3623	AD		9.5	130.3	99					
14 Ori	3711		5.8	6.5	0.8	*322	F0	07.9	+08 30	21-3	
Σ652	3764		6.1	7.6	1.7	183	F5	11.8	+01 02	21-1	Yellowish & bluish
17–ϱ Ori	3797	AB	4.5	8.3	7.0	64	K0	13.3	+02 52	21-1	Orangish & blue
OΣ517	3799	AB	6.9	7.1	0.5	237	A5 G0	13.5	+01 58	21-1	
	3799	ABxC		12.9	6.8	137					
19–β Ori	3823	AB	0.1	6.8	9.5	202	B8	14.5	−08 12	21-5	Rigel. Blue-white, blue
Σ667	3825		7.1	8.6	4.2	313	K2	14.7	−07 04	21-5	
Σ664	3827		7.7	8.2	4.7	170	F0	15.2	+08 26	21-3	
20–τ Ori	3877	AB	3.6	13.6	35.2	250	B5	17.6	−06 51	21-5	
h2259	3877	AD		10.8	36.0	60					
β188	3877	BC		11.9	3.7	51					
β189	3926		6.4	11.1	4.4	285	B9	20.4	−05 22	21-5	Nice! White & blue
h697	3941	AB	5.7	11.4	32.7	60	B3	21.5	−00 25	21-6	
	3941	AC		10.9	37.6	110					
OΣ106	3949		7.2	10.7	9.3	42	F5	22.1	+05 24	21-4	
Σ700	3968		7.7	7.9	4.7	5	A0	23.1	+01 03	21-6	
Σ697	3969		6.9	8.4	26.0	285	B8 A	23.5	+16 02	21-3	
Σ701	3978		6.0	7.8	5.9	142	A0	23.3	−08 25	21-5	White & bluish
Wnc 2	3991	AxBC	6.1	7.8	2.7	161	F5	23.9	−00 52	21-6	Yellowish-white pair
28–η Ori	4002	AB	3.6	5.0	1.5	80	B1	24.5	−02 24	21-6	Two white stars
H V 167	4002	AC		9.4	115.1	51					
Σ712	4033		7.5	9.5	3.1	63	B9	26.5	+02 58	21-1	
h2268	4071		7.0	9.6	26.0	300	G5	28.7	−08 23	21-5	
31 Ori	4091		4.7	9.9	12.7	87	K5	29.7	−01 06	21-6	Orangish & light blue
32 Ori	4115		4.5	5.8	1.1	41	B3	30.8	+05 57	21-4	
33 Ori	4123	AB	5.8	7.1	1.8	27	B3	31.2	+03 18	21-1	White & pale blue
34–δ Ori	4134	AB	2.0	13.7	32.8	227	B0	32.0	−00 18	21-6	Bluish-white primary,
Σ I 14	4134	AC		6.3	52.6	359	B0				pale blue secondary
Σ734	4150	AB	6.5	8.2	1.8	354	B3	33.1	−01 42	21-6	White & bluish stars
H V 119	4150	AC		8.2	29.4	243					
β1049	4150			9.3	0.7	295					
H V 118	4159		6.2	9.8	27.5	264	B3	34.1	−01 02	21-6	
Σ747	4182		4.8	5.7	35.7	223	B1 B1	35.0	−06 00	21-5	Both stars white
39–λ Ori	4179	AB	3.5	5.6	4.4	43	O8 B0	35.1	+09 56	21-4	Close white pair
	4179	AC		11.1	28.6	184					Blue
	4179	AD		11.1	78.3	271					Blue
41–ϑ¹ Ori	4186	AB	6.7	7.9	8.8	31	B0 B2	35.3	−05 23	21-5	Trapezium
	4186	AC		5.1	12.8	132	B0 O6				
	4186	AD		6.7	21.5	96					
	4186	AE		11.1	4.1	351					
	4186	CF		11.5	4.0	122					
43–ϑ² Ori	4188	AB	5.2	6.5	52.5	92	B0 B0	35.4	−05 25	21-5	
	4188	AC		8.1	128.7	97	B5				
44–ι Ori	4193	AB	2.8	7.3	11.3	141	O9 B7	35.4	−05 55	21-5	White, bluish-white
Σ750	4192		6.5	8.5	4.2	60	B5	35.5	−04 22	21-5	White & blue
Σ754	4212		5.7	8.9	5.2	287	B3	36.6	−06 04	21-5	
48–σ Ori	4241	ABxC	3.8	10.3	11.4	238	B0	38.7	−02 36	21-6	Bluish-white primary
	4241	ABxD		7.5	12.9	84					bluish companions
	4241	ABxE		6.5	42.6	61					Σ761 in field
50–ζ Ori	4263	AB	1.9	4.0	c2.3	*165	B0 B0	40.8	−01 57	21-6	Brillant white pair
	4263	AC		9.9	57.6	10					
Σ790	4361		6.4	8.7	6.9	89	K0	46.0	−04 16	21-6	Yellow & deep blue
52 Ori	4390		6.1	6.1	1.6	210	A3	48.0	+06 27	21-4	Equal white pair
Σ816	4499		6.8	9.3	4.3	289	B9	54.9	+05 52	21-7	White & pale blue
59 Ori	4555		6.1	9.7	36.7	204	A5	58.4	+01 50	21-6	Both stars bluish
60 Ori			5.2	11.8	19.1	30	A0	58.8	+00 33	21-6	
Σ855	4749		6.0	7.0	29.3	114	A0	06ʰ09.0ᵐ	+02 30	21-7	White, bluish
Σ877	4840		7.5	8.0	5.6	263	B9	14.7	+14 35	21-8	

Footnotes: *= Year 2000, a = Near apogee, c = Closing, w = Widening. Finder Chart No: All stars listed in the tables are plotted in the large Constellation Chart, but when a star appears in a Finder Chart, this number is listed. Notes: When colors are subtle, the suffix -ish is used, e.g. bluish.

21.2 Interesting Stars

Beta (β) = 19 Orionis (Σ668) Double Star   Spec. B8, B5
m0.1, 6.8; Sep. 9.5″; P.A. 202° $05^h14.5^m$ $-08°12'$
Finder Chart 21-5   Rigel   ★★★★

At magnitude 0.14 Rigel is the seventh brightest star in the sky. It is a supergiant of spectral type B8 and therefore has a fine bluish-white color in binoculars and small telescopes. Rigel's companion is a bluish B5 star 9″ away to the SSW. The pair show no orbital motion but no doubt are physically related because they have the same radial velocity. The Rigel system is estimated to be 900 light years away. Thus the absolute magnitudes of the two stars are -7.1 and -0.3, luminosities of 58,000 and 110 Suns, respectively. If the supergiant component of Rigel was at the same distance from us as Sirius, it would have an apparent magnitude of -10.0, about as bright as the quarter Moon.

8/10″ Scopes–150x: Exquisite! Rigel shines like a brilliant blue-white diamond with a delicate blue companion.

Eta (η) = 28 Orionis (Dawes 5) Triple Star   Spec. B1 IV
AB: m3.6v, 5.0; Sep. 1.5″; P.A. 80° $05^h24.5^m$ $-02°24'$
Finder Chart 21-6   ★★★★

8/10″ Scopes–250x: Beautiful! Eta's close AB pair appear as two white disks in contact.

Delta (δ) = 34 Orionis (E I 14) Double Star   Spec. O9.5 II
AC: m2.0v, 6.9; Sep. 52.6″; P.A. 359° $05^h32.0^m$ $-00°18'$
Finder Chart 21-6   Mintaka   ★★★★

8/10″ Scopes–100x: Delta has a bluish-white primary with a pale blue secondary.

Lambda (λ) = 39 Orionis Quad. Star   Spec. O8 III, B0 V
AB: m3.5, 5.6; Sep. 4.4″; P.A. 43° $05^h35.1^m$ $+09°56'$
Finder Chart 21-4   ★★★★

8/10″ Scopes–100x: The close AB stars of Lambda appear white, while its distant companions located 28″ and 78″ away look bluish. Lambda Orionis marks the head of Orion and forms an interesting triangle with Phi-1 (φ^1) and Phi-2 (φ^2) Orionis, both 4th magnitude stars.

Theta-One (ϑ^1) = 41 Orionis Multiple Star
CD: 5.1, 6.7; Sep. 13.4″; P.A. 241°
AB: 6.7, 7.9; Sep. 8.8″; P.A. 031° $05^h35.3^m$ $-05°23'$
Finder Chart 21-5   The Trapezium   ★★★★★

Theta-one Orionis lies in the very heart of the Great Orion Nebula M42 and is one of the best known multiple stars in the entire sky. Its four 5th to 8th magnitude components form a quadrangle called the Trapezium. They are lettered in order of right ascension rather than in the customary sequence of brightness; hence the brightest Trapezium star is in fact Theta-one C. Several 11th magnitude stars in and around the Trapezium are undoubtedly physically involved with it. One of these is 4″ ESE of C, and another 4″ north of A. Moreover, within 5′ of the Trapezium are more than 400 very faint stars, extremely young objects recently condensed from the giant cloud of gas and dust within which the Orion Nebula and the Trapezium are embedded. The Trapezium cluster is estimated to be less than 100,000 years old.

8/10″ Scopes–125x: The surrounding nebula probably affects the color perception of these stars: the C component looks off-white, the B component bluish, and A and D yellowish.

Iota (ι) = 44 Orionis Triple Star   Spec. O9 III, B7 III p
AB: m2.8, 7.3; Sep. 11.3″; P.A. 141° $05^h35.4^m$ $-05°55'$
Finder Chart 21-5   ★★★★★

8/10″ Scopes–100x: Iota is a fine triple star for small telescopes with a white primary and a pale blue companion 11″ away. An 11th magnitude reddish companion is located 50″ away in P.A. 103°.

Sigma (σ) = 48 Orionis Multiple Star Spec. O9.5V+B3, A2
ABxC: m4.0, 10.3; Sep. 11.4″; P.A. 238° $05^h38.7^m$ $-02°36'$
Finder Chart 21-6   ★★★★

8/10″ Scopes–100x: Sigma is a quadruple with a brilliant white primary and three bluish companions, the two brightest magnitude 7.5 and 6.5 stars, not physically involved with it.

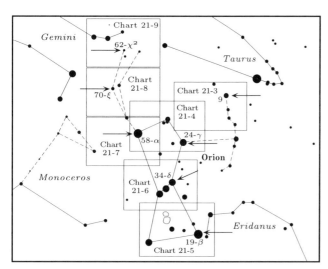

Master Finder Chart 21-2. Orion Chart Areas
Guide stars indicated by arrows.

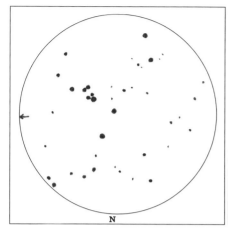 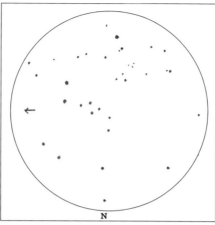 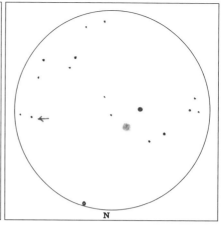

Figure 21-1. NGC 1662
12.5″, f5–75x, by G. R. Kepple

Figure 21-2. NGC 1663
12.5″, f5–75x, by G. R. Kepple

Figure 21-3. Jonckheere 320
17.5″, f4.5–250x, by Steve Coe

Zeta (ζ) = 50 Orionis Triple Star   Spec. O9.5 Ib, B0 III
AB: m1.9, 4.0; Sep. 2.3″; P.A. 165°   05^{h}40.8^m −01°57′
Finder Chart 21-6   *Alnitak*   ★★★★

8/10″ Scopes–100x: Zeta Orionis lies in a bright field of several emission and reflection nebulae. 15′ east of Zeta is the large patch of NGC 2024, an emission nebula bisected by a N–S dust rift. Extending south from Zeta is the 1° long reef of IC 434, a faint and difficult glow upon which is silhouetted the famous but even more difficult Horsehead Nebula. The Zeta double is quite close but a rather easy split, its components both bluish-white.

Alpha (α) = 58 Orionis Variable Star   Spec. M2 Iab
m0.40 to 1.3   05^{h}55.2^m +07°24′
Constellation Chart 21-1   *Betelgeuse*   ★★★★

Though Betelgeuse is a red supergiant star, it looks orange through a telescope and even in binoculars. Like most other red giants and supergiants, it is a pulsating variable, its observed extremes being magnitudes 0.4 and 1.3. However, its variations are not even as imperfectly periodic as those of the standard Mira-type long-period variables: Betelgeuse seems to be completely irregular, though some observers have suspected a six year cycle in its pulsations. Betelgeuse is around 590 light years distant, and has an absolute magnitude of about −5.5, a luminosity of 13,000 Suns. The stars surface temperature is 3,100°K. At its extreme it is probably a billion miles in diameter: if it were at the center of our Solar System, its surface would be almost out to the orbit of Jupiter! However in all that vast volume is only about twenty solar masses of material. Hence the average density of the star is less than one ten-thousandth the density of the atmosphere at the Earth's surface.

21.3 Deep-Sky Objects

NGC 1662 Cr55 Open Cluster 35★   Tr Type I 2
pø20.0′, m6.4v, Br★ 8.34v   04^{h}48.5^m +10°56′
Finder Chart 21-3, Figure 21-1   ★★★

12/14″ Scopes–75x: NGC 1662 is a loose but obvious group of several dozen 9th magnitude and fainter stars in a 20′ area. A dozen of the brighter stars lie in a NW–SE chain, its NW terminus marked by a wide 9th and 10th magnitude pair.

NGC 1663 Open Cluster
ø −   04^{h}48.6^m +13°10′
Finder Chart 21-3, Figure 21-2   ★★★

12/14″ Scopes–75x: NGC 1663, 14′ NNE of an 8th magnitude star, is a moderately faint but rich concentration of fifteen 11th to 12th magnitude and fifteen 13th magnitude and fainter stars spread over an 8′ area.

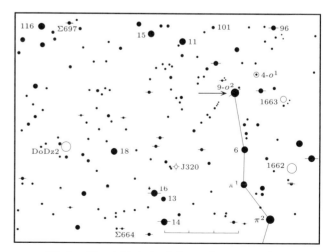

Finder Chart 21-3. 9–o² Ori: 04^{h}56.4^m +13°31′

Figure 21-4. *NGC 1788 is in interesting reflection nebula involved with a pair of 10th magnitude stars. Martin C. Germano made this 55 minute exposue on 2415 film with an 8", f5 Newtonian.*

The northern portion contains the highest concentration of stars while the two brightest stars are somewhat detached to the SSW.

Jonckheere 320 PK190-17.1 Planetary Neb. Type 2+4
ø7.0″, m12.0v, CS 14.31v $05^h05.6^m$ +10°42′
Finder Chart 21-3, Figure 21-3   ★★

12/14″ Scopes–250x: Jonckheere 320, 4′ NNW of a 9th magnitude star, appears stellar at medium power. 250x is needed to distinguish its diffuse, greyish 7″ disk.

16/18″ Scopes–300x: Larger instruments reveal a greenish 7″ halo with a slightly brighter center in which is embedded the faint 14th magnitude central star.

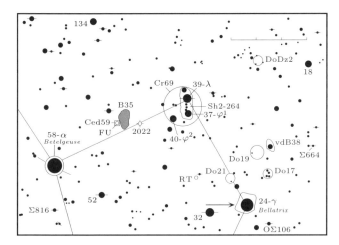

Finder Chart 21-4. 24–γ Ori: $05^h25.1^m$ +06°21′

NGC 1788 Reflection Nebula
ø5′ x 3′, Photo Br 1–5, Color 1–4 $05^h06.9^m$ −03°21′
Finder Chart 21-5, Figure 21-4   ★★★

12/14″ Scopes–125x: This interesting reflection nebula is contained by a keystone of one 10th and three 9th magnitude stars. The brightest area surrounds two 10th magnitude stars aligned approximately E–W, and a wide N–S pair of 12th magnitude stars between them. The nebula spreads NE of its central stars, its faint wispy filaments reaching as far as the NE star in the keystone. Overall, the nebula is elongated E–W. A dark void lies between the keystone's SE and SW stars.

16/18″ Scopes–150x: NGC 1788 is a fine, bright reflection nebula. Its brightest section surrounds two 10th magnitude stars. Faint, wispy extensions spread east, NE and SW. The most conspicuous nebula-glow is a 4′ × 2′ N–S banana-shaped area, concave to the east, just west of one of the 10th magnitude stars. Overall, the nebula measures 8′ × 5′ E–W.

Dolidze 17 Open Cluster or Asterism? Tr Type IV 2 p
ø12′ $05^h22.4^m$ +07°07′
Finder Chart 21-4, Figure 21-13   ★★★

12/14″ Scopes–100x: Dolidze 17 is a fairly bright, loose, irregular cluster of 20 stars without a central compression. Toward the SSW side of the cluster is a triangle of three 8th magnitude stars, and toward the NNE a parallelogram of four 8th–9th and two 10th magnitude stars. Another thirteen 11th to 13th magnitude cluster members are scattered around these two asterisms.

Dolidze 19 Open Cluster 12★ Tr Type IV 1 p
ø23′ $05^h23.7^m$ +08°11′
Finder Chart 21-4   ★★★

12/14″ Scopes–75x: Dolidze 19 is even less of an open cluster than Dolidze 17. It contains two dozen 10th to 13th magnitude stars in a very loose, irregular 23′ area not well detached from the surrounding star field. The ten brighter stars traverse the center of the cluster from east to west. Just east of the center is a prominent triangle of stars.

Dolidze-Dzimselejsvili 2 Open Cluster 12★ Tr Type III 2 p
ø12′ $05^h23.9^m$ +11°28′
Finder Chart 21-3   ★★

12/14″ Scopes–75x: This cluster is a small, moderately compressed group of eight 9th to 11th and a dozen 12th–13th magnitude stars in a 12′ × 8′ N–S area. It is just detached from the well populated star field to its east.

Figure 21-5. *Three reflection nebulae are east of Mintaka, Delta (34-δ) Orionis: IC 423 is the ear-shaped nebula at the upper left; IC 424 is the small hazy patch with two glowing eyes below and to the right of IC 423; and IC 426 is the hazy irregular patch at the far right. Martin C. Germano made this 85 minute exposure on hypered 2415 film with an 8″, f5 Newtonian.*

Dolidze 21 Open Cluster Tr Type IV 2 p
ø12.0′   05ʰ27.4ᵐ +07°04′
Finder Chart 21-4   ★

12/14″ Scopes–100x: Dolidze 21, 12′ NE of a 5th magnitude star, is a very loose scattering of a dozen 12th magnitude and fainter stars in a 12′ area. It lies just NE of a group of brighter field stars resembling a bird in flight with outstretched wings and tail. A 5th magnitude star marks the SW wing tip.

IC 423 Reflection Nebula
ø6′ × 4′, Photo Br 2–5, Color 1–4   05ʰ33.4ᵐ −00°37′
IC 424 Reflection Nebula
ø2′ × 2′, Photo Br 2–5, Color 1–4   05ʰ33.6ᵐ −00°25′
Finder Chart 21-6, Figure 21-5   ★/★

12/14″ Scopes–100x: IC 423 is a very low surface brightness haze elongated 5′ × 3′ NNW–SSE and pointed in the general direction of Delta (δ) = 34 Orionis. IC 424, 10′ NNE of IC 423, is just detectable as a diffuse featureless glow immediately NE of a 9th magnitude star. Two 12th magnitude stars embedded in its southern portion stare like a pair of sinister eyes.

Collinder 70 Belt Stars 100★
ø150′   05ʰ35.0ᵐ −01:°
Finder Chart 21-6   ★★★★

8 × 50 Binoculars: The three 2nd magnitude stars of Orion's Belt form a conspicuous 3° NW–SE line. Among them is a rich sprinkling of 5th to 8th magnitude, and fainter, stars which collectively are catalogued as the open cluster Collinder 70. This is

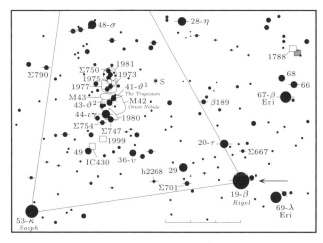

Finder Chart 21-5. 19–β Ori: 05ʰ14.5ᵐ −08°12′

Figure 21-6. *The emission-reflection nebula complex NGC 1973-75-77 is quite easily visible but often neglected because Messier 42 to its south steals the spotlight. Open cluster NGC 1981 is the loose grouping of stars at center. Martin C. Germano made an 85 minute exposure on hypered 2415 Kodak Tech Pan film with an 8″, f5 Newtonian at prime focus.*

a magnificent field for binoculars, its star-richness spread over by the blue-white glare of the three Belt stars.

Collinder 69   Open Cluster 20★   Tr Type II 3 p n
ø65.0′, m2.8p   **05ʰ35.1ᵐ +09°56′**
Finder Chart 21-4   ★★★★

8 × 50 Binoculars: The Head of Orion is marked by the small triangle of 3rd magnitude Lambda (λ) = 39 Orionis to the north and two 4th magnitude stars Phi-1 (φ) and Phi-2 Orionis to the SW and SE. These stars, and fainter ones around them, are catalogued as the open cluster Collinder 69. It is an ideal binocular group. In addition to the bright triangle, Cr 69 has an attractive N–S line of three 6th–7th magnitude stars between Lambda and Phi-1. The cluster is listed in *Sky Catalogue 2000* but is not plotted on any star map of which we are aware. The very faint emission nebula Sharpless 2–264 surrounds the cluster.

8/10″ Scopes–60x: An O-III filter and averted vision reveal a very faint glow around most of the brighter stars.

NGC 1973-75-77   Emission & Reflection Nebula
ø5′ × 5′, Photo Br 1–5, Color 1–4   05ʰ35.1ᵐ −04°44′
Finder Chart 21-5, Figures 21-6 & 21-7   ★★★

12/14″ Scopes–75x: NGC 1973-75-77 is a fairly bright nebulous complex between Messier 42 on the south and open cluster NGC 1981 on the north. Its proximity to the Great Orion Nebula leaves it outclassed and generally neglected; however, it is an interesting object that deserves attention. These three nebulae are merely the illuminated parts of the same large cloud of gas and dust. NGC 1977 is the largest of the three; but it is the most difficult to see because of the brilliance of its magnitude 4.7 illuminating star 42 Orionis. It is a large, diffuse semicircular glow concave to the south. NGC 1973, immediately north of NGC 1977, is a fairly obvious, diffuse 5′ × 3′ knot, elongated E–W. NGC 1975 is a much smaller blotch lying just NE of NGC 1973. The area is awash with very faint nebulous swirls.

NGC 1981   Cr73   Open Cluster 20★   Tr Type III 2 p n
ø25.0′, m4.6v   **05ʰ35.2ᵐ −04°26′**
Finder Chart 21-5, Figures 21-6 & 21-7   ★★★★

8/10″ Scopes–50x: NGC 1981 is one degree north of the Great Orion Nebula, the NGC 1973-75-77 complex between them. This bright, coarse cluster contains forty stars in a 25′ E–W area. Its most conspicuous feature is two nearly parallel, roughly N–S, star rows concave east and SE. The eastern row is of three 6th magnitude stars, the western of one 9th, one 8th, and two 7th magnitude stars (south to north). The western row is accompanied by more fainter stars than the eastern. Outlying cluster members are scattered in ragged rows to the north and east.

12/14″ Scopes–75x: NGC 1981 is a bright, very loose, irregular cluster of fifty stars in a 30′ area. Three 6th magnitude stars aligned N–S are separated from the larger, richer concentration of fainter stars to the west. The western portion is dominated by a "Y" of four 7th to 9th magnitude stars. Three dozen, mostly 11th and 12th magnitude stars are loosely scattered about the "Y."

NGC 1980 Emission Neb. & Open Cl. 37★
ø14′ × 14′, Photo Br 1–5, Color 1–4
05ʰ35.4ᵐ −05°54′
Finder Chart 21-5, Figure 21-7   ★★★★

12/14″ Scopes–75x: NGC 1980 is located just south of the Great Orion Nebula in a very fine Milky Way field. It is a bright cluster of thirty stars embedded in a 20′ large, NNE–SSW elongated nebula. The cluster is dominated by the colorful triple Iota (ι) = 44 Orionis, which has a bluish-white primary and secondary, and a red third star (see Table 21-2). Iota is embedded in a patch of nebulosity about 5′ in diameter. Another patch of nebulosity nearly the same size is 8′ SW and contains the double star Σ747 (4.8, 5.7; 357″; 223°). Further west is the double Σ745 (8.4, 9.3; 29″; 347°). All these objects, along with Messier 42 and Messier 43, may be seen in a one degree field — a most impressive sight indeed!

NGC 1976 Messier 42 Emission & Reflection Nebula
ø65′ × 60′, Photo Br 1–5, Color 3–4 05ʰ35.4ᵐ −05°27′
Finder Chart 21-5, Figures 21-7 & 21-8   ★★★★★
The Great Orion Nebula

This magnificent object is perhaps the favorite target for deep-sky observers, beginners and advanced amateurs alike. Under dark skies Messier 42 is visible to the naked eye as a diffuse patch around Theta (ϑ) Orionis, the middle star in the Sword of Orion. Theta Orionis is an extremely wide (135″) binocular double,

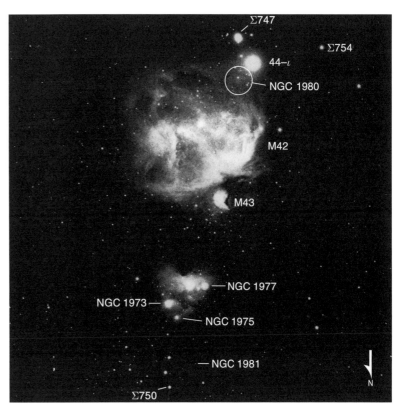

Figure 21-7. *The Sword of Orion includes the splendid nebulae Messier 42, and Messier 43, the nebulous complex NGC 1973-75-77, and the open cluster NGC 1981. Gary Capps made this 20 minute exposure on hypered 2415 film with a 12″, f5 Newtonian at prime focus.*

the western component of which, Theta-one (ϑ¹) Orionis, is the famous Trapezium multiple star. M42 is the finest example of a diffuse emission nebula visible to mid-northern hemisphere observers. It covers one square degree, four times the size of the full moon.

The Orion Nebula was not mentioned in any known ancient or medieval records, and Galileo never noted seeing it. Nicholas Peiresc first called attention to it between 1610 and 1611, and his report was followed with observations by Cysatus (1618) and the first detailed drawings, which included the Trapezium region, by Huygens in 1656. Messier recorded the nebula as the 42nd and 43rd objects in his pseudo-comet catalogue on March 4, 1769. Both William and John Herschel viewed the region in 1774 and 1825, respectively, and the latter's description of the nebula as "the breaking up of a mackerel sky" is still cited today. Draper took the first successful photograph of the nebula in 1880 with an 11-inch refractor.

An emission nebula is a cloud of gas glowing by basic fluorescence. The first thing necessary is the presence of a hot, luminous star emitting lots of high-energy ultraviolet radiation. The ultraviolet photons from the star knock the electrons off the neutral hydrogen atoms (HI)

Figure 21-8. *The bright, beautiful nebulous complex and stellar nursery of Messier 42 and 43 are, without a doubt, the most observed objects in the winter sky. Martin C. Germano made this 30 minute exposure on hypered 2415 Kodak Tech Pan film with an 8″, f5 Newtonian at prime focus.*

hydrogen — an "HI region" — that covers the entire constellation of Orion, a cloud that can be traced by its low energy 21 cm hydrogen radio emission. The Nebula also lies on the Solar System side of a dense, cool cloud of molecular hydrogen, H_2, a cloud that can be traced by the shortwave radio emission of its carbon monoxide, CO, molecules. This molecular cloud, in which are embedded all the nebulae in and around the Sword of Orion, is the raw material from which the stars of the Sword have condensed in the astronomically recent past.

The Trapezium multiple star, Theta-one (ϑ^1) Orionis, and the cluster of several hundred faint stars around it are the most recent products of the Orion Molecular Cloud. Stars like the "O6" lucida of the Trapezium are only a few hundred thousand years old. Moreover, many of the faint stars around the Trapezium prove to be over-luminous for their spectral type, which implies they are radiating energy from gravitational contraction and have not yet settled down onto the stable main sequence. The Trapezium Cluster also has an abundance of nebular variables, erratic stars in which energy derived from the last stages of the star's gravitational contraction is competing with the energy derived from the first stages of the nucleosynthesis of helium from hydrogen in its core. Only 100″ NW of the Trapezium are a group of strong emitters of infrared radiation, the "Kleinmann-Low Sources" and the "Becklin-Neugebauer Object." They are thought to be the still-dust-enshrouded stars of a newly-formed open cluster: the dust in which they are embedded absorbs their ultraviolet and visible light and radiates it as infrared light.

The Orion Nebula is estimated to be around 1,600 light years distant. Its 1° apparent size therefore corresponds to a true size of 30 light years. The nebula's integrated apparent magnitude of 5 (its brightness if all its light was collected into a starlike point) implies an integrated absolute magnitude of −4.5, a luminosity of 5,200 suns.

The Orion Nebula can be viewed through a telescope even from the brightest cities! The nebula is a soft, greenish patch of light extending outward into the nighttime sky with tenuous filaments of radiance reminiscent of fine-textured cirrus clouds. Because it is fairly large, M42-43 will fill a good portion of the field of a standard 2000mm focal length 8″ SCT. Use low power for perspective of the entire nebulous complex, then study it with higher magnification.

in the nebula's gas, thus creating free electrons and hydrogen ions (HII — basically a free proton). When the electrons and the hydrogen ions recombine, the former give back the energy they gained from the ultraviolet photons, emitting various wavelengths of lower-energy radiation as they cascade back down the orbital shells of the hydrogen atom. An especially efficient transition is from the third down to the second lowest energy shell, during which the electron emits a photon of red light of wavelength 6562 Å, the "hydrogen-Alpha line." This radiation is the reason so many emission nebulae — the Rosette, the Lagoon, and M16 in Serpens, to name three — have such a conspicuous red color in photographs.

Interstellar gas clouds are primarily of hydrogen, but they also contain helium and trace amounts of carbon, nitrogen, oxygen, sulfur, neon, chlorine, and other elements. When the gas' temperature and density are favorable, emission lines from the fluorescence of one of these other elements can compete with the intensity of the Hα line. The greenish appearance of the Orion Nebula is radiation from doubly ionized oxygen at wavelengths 4959 and 5007 Å.

The Orion Nebula is just one small illuminated area — one "HII region" — in the vast cloud of neutral

4/6″ Scopes–50x: Words cannot do justice to this magnificent gem! Even at low power M42 fills the entire field of view. Its shape is rather like that of an eagle in flight. The brightest part of the nebulosity, the bird's body, is outlined on the north side by a prominent dark bay that points in the direction of the eagle's flight. To either side curve out and back (south) the bird's wings. The head and bill are separated from the rest of the body by a dark lane, the eagle's eye marked by a 10th magnitude star. The brightest part of the nebula is noticeably light green. Exquisite detail is visible throughout the entire complex. The Trapezium is located in the central portion, and Theta-2 (ϑ^2) Orionis is embedded near the leading edge of the SE wing.

8/10″ Scopes–75x: On a good night, two winglike arcs of nebulosity curve from the northern edge all the way around to the south where they merge in a diffuse mass of swirls mixed with dark streaks. The entire complex covers a one degree area. The blazing stars of the Trapezium and the bright nebula-glow within which they are embedded are almost surrounded by a dark gap of obscuration. Messier 43 is a bright nebulous arc to the north detached from the main body of M42 by a dark streak.

12/14″ Scopes–100x: This splendid nebula is impressive at all powers! Start with low power for perspective then continue observing with increased magnification until only a portion is visible–you will be amazed at the wealth of detail. The densest portion is adorned by the by the four bright and two fainter star-gems of the Trapezium. The northern boundary is marked by the large dark rift that separates M43 from M42. A large bay of obscuration, popularly called the "fish's mouth" because of its profile, extends from the Trapezium to a less dense dark rift. The bay is bordered on its SW by bright nebulosity and on its NE by a large, faint patch of glow with embedded stars. Two broad, bright nebulous arcs curve from the dark bay out and back toward the south. The western of these arcs has a pinkish or rose tint. The colors in the nebula's fainter streaks are much more subtle. The nebulosity south of the Trapezium is extremely bright and markedly mottled — the effect that John Herschel called "the breaking up of a mackerel sky." In a UHC or O-III filter the east and west regions of M42 become more extensive and reveal delicate, diffuse lacy streaks mixed with dark areas however the filters kill the nebula's delicate hues. The southern expanse of M42, between the two arcing wings, gradually fades southward.

NGC 1982   Messier 43   Emission & Reflection Nebula
ø20′ × 15′, Photo Br 1–5, Color 3–4 05^{h}35.6^m −05°16′
Finder Chart 21-5, Figure 21-7 ★★★★

8/10″ Scopes–75x: Messier 43 is a large bright nebula surrounding a magnitude 7.5 star just north of M42. Its brightest part is a comma concave to the east and bordered on the NE by a dark streak. Faint arcs of nebulosity spread from the bright comma toward the NW, east, and south, those spreading south abruptly terminated by the dark lane of obscuration that cuts M43 off from M42.

12/14″ Scopes–100x: Although low power encompasses more field, higher magnification is needed to see all the detail visible in M43. The nebula is bordered to the south by a dark lane that cuts it off from M42. Its brightest section is a 15′ long N–S crescent, concave to the east, around a magnitude 7.5 star. A generous sprinkling of magnitude 9.5 to 11 stars surround it, especially on the north and NW. A dark bay running NE–SW nearly bisects it. Faint nebulosity may be traced westward and eastward. At 150x, the main portion is a complex structure of dark areas and mottled streaks to the NE of the 7.5 magnitude central star.

NGC 1990   Emission & Reflection Nebula
ø50′ × 50′, Photo Br 4–5, Color 1–4 05^{h}36.2^m −01°12′
Finder Chart 21-6 ★★★

12/14″ Scopes–100x: NGC 1990 is an extremely faint glow centered upon Epsilon (ϵ) = 46 Orionis. The nebula could not be confirmed visually, however, there are traces of nebulosity in the area as is the case with most of the constellation. According to Brian Skiff this object may not even exist.

NGC 1999   Emission & Reflection Nebula
ø2′ × 2′, Photo Br 1–5, Color 1–4   05^{h}36.5^m −06°42′
Finder Chart 21-5, Figure 21-9 ★★★

8/10″ Scopes–75x: NGC 1999, 1° south of the Orion Nebula, is a fairly bright, circular nebula around a magnitude 9.3 star. It resembles a planetary nebula with a bright central star.

12/14″ Scopes–100x: NGC 1999 is a bright, 2′ diameter, circular nebula with a 9th magnitude star embedded noticeably east of its center. 200x reveals a slightly darker center.

IC 426   Reflection Nebula
ø7′ × 7′ Photo Br 2–5, Color 1–4   05^{h}36.8^m −01°15′
Finder Chart 21-6, Figure 21-5 ★★★

12/14″ Scopes–100x: IC 426 is a very faint, diffuse, triangular patch of haze spreading ENE from a 9th

Figure 21-9. *NGC 1999 (lower left) is a fairly bright, circular nebula resembling a planetary with a bright central star. IC 430 (to the lower left of bright star at upper right) is very faint and difficult to see. Martin C. Germano made this 50 minute exposure on hypered 2415 film with an 8", f5 Newtonian at prime focus.*

NGC 2024 Emission Nebula
ø30'×30', Ph Br 2–5, Col 3–4   05ʰ40.7ᵐ 02°27'
Finder Chart 21-6, Figure 21-10   ★★★

12/14" Scopes–75x: NGC 2024 lies immediately east of Zeta (ζ) = 50 Orionis, a magnitude 1.8 star that seriously interferes with observation of the nebula. Nevertheless NGC 2024 can be seen in 10×50 binoculars as an asymmetry toward the NE in the glow around Zeta; and the dark N–S lane through the nebula is visible at 70x in 6" telescopes if Zeta is outside the field of view. In moderately large telescopes equipped with an O-III filter, NGC 2024 covers half a degree wide area with wispy streaks bisected by the broad N–S dark band. The overall shape of the nebula suggests a maple leaf and has three distinct brighter zones. The eastern zone is the largest: near its south end a tributary dark streak branches from the main N–S rift and extends east into it. The western bright zone, the nearest to Zeta, is virtually detached from the other two by the N–S rift. The northern bright zone is tenuously attached to the other two bright areas by tendrils of faint nebulosity.

16/18" Scopes–125x: NGC 2024 is a very large emission nebula full of striking detail when skies are good but nearly invisible when any atmospheric haze is present. Zeta Orionis must be placed out of the field of view; and O-III and UHC filters much enhance the details that can be observed. NGC 2024 covers a 30' area with patches of nebulosity rifted and stretched by subtle dark bands and threads. A broad N–S dust lane nearly bisects the nebula: the lane is very dense toward the south but becomes more transparent to the north. A secondary dark lane runs NE from near the south end of the main rift into the eastern, and brighter, wing of the nebula. At least a dozen magnitude 10 to 12 stars are embedded in NGC 2024. The field is full of delicate wisps and festoons, and visually interesting structural detail.

magnitude star. A thin dark E–W streak along the south side of the nebula separates it from a stream of stars lying further south. Many interesting double stars and stellar groupings are in the area.

IC 430 Reflection Nebula
ø11' × 11', Photo Br 2–5, Color –   05ʰ38.5ᵐ −07°05'
Finder Chart 21-5, Figure 21-9   ★

12/14" Scopes–75x: IC 430, 5' NW of 49 Orionis, is a very faint, fan-shaped nebula elongated 5' × 3' NW–SE. It is difficult to see because of the glare of 4.8 magnitude 49 Orionis: an excellent night and averted vision are needed.

IC 431 Reflection Nebula
ø5' × 3', Photo Br 1–5, Color 1–4   05ʰ40.3ᵐ −01°27'
IC 432 Reflection Nebula
ø8' × 4', Photo Br 1–5, Color 1–4   05ʰ40.9ᵐ −01°29'
Finder Chart 21-6, Figure 21-10   ★★/★★★

12/14" Scopes–100x: Both of these nebulae are faint bluish glows around bright stars. IC 431 consists of two detached areas surrounding 8th and 9th magnitude stars. IC 432 is the brighter and larger of the two nebulae: it spreads SE from its 7th magnitude illuminating star.

Barnard 33 Dark Nebula The Horsehead Nebula
ø6' × 4', Opacity 4   05ʰ40.9ᵐ −02°28'

IC 434 Emission Nebula
ø60' × 10', Photo Br 1–5, Color 3–4 05ʰ41.0ᵐ −02°24'
Finder Chart 21-6, Figure 21-10   ★/★★

12/14" Scopes–75x: B33, the Horsehead Nebula, is a small, dark nebula silhouetted against the very faint glow of IC 434, a narrow reef of very tenuous nebulosity extending a degree south of Zeta (ζ) = 50

Figure 21-10. *The highlights of the Zeta Orionis region are emission nebula NGC 2024 to the star's east (lower right) and emission nebula IC 434 extending south from the star (top center). Upon the middle of the IC 434 reef is silhouetted the elusive Horsehead Nebula. Other emission or emission + reflection nebulae near Zeta Orionis are NGC 2023 and IC 435 (near center and far right) and IC 432 (at bottom). Martin C. Germano made this 45 minute exposure on hypered 2415 film with an 8″, f5 Newtonian at prime focus.*

Figure 21-11. *NGC 2022 is a fairly obvious annular planetary nebula with a slightly darker center. Martin C. Germano made this 20 minute exposure on hypered 2415 Tech Pan film with an 8", f5 Newtonian at prime focus.*

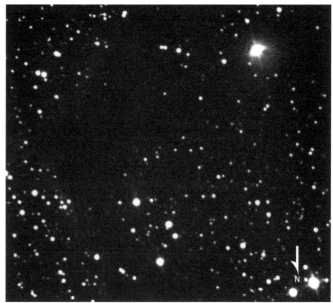

Figure 21-12. *Dark nebula Barnard 35 is a conspicuous starless area east of diffuse nebula Cederblad 59. Martin C. Germano made this 70 minute exposure on hypered 2415 film with an 8", f5 Newtonian at prime focus.*

Orionis. The two objects are exceedingly difficult to detect and require excellent sky transparency. O-III and UHC filters help enhance the contrast between the Horsehead and its IC 434 background, but B33 is also one of the few objects which responds well to a Hydrogen-Beta filter.

16/18" Scopes–125x: The Horsehead Nebula is so difficult that even with instruments of this aperture you may not see it even when looking directly at its position. However, with averted vision under ideal observing conditions the orientation of the horsehead figure — facing north toward Zeta Orionis — may just be discerned. The IC 434 reef upon which it is silhouetted is more defined on its eastern edge and becomes thinner and fainter further south.

NGC 2023 Emission & Reflection Nebula
ø10′ × 10′, Photo Br 1–5, Color 1–4 05ʰ41.6ᵐ −02°16′
Finder Chart 21-6, Figure 21-10 ★★★

12/14" Scopes–100x: This is a fairly bright, diffuse nebulosity around an 8th magnitude star located just SW of the nebula's center. The nebula-glow is brightest in the center and spreads NE toward a 12th magnitude star.

16/18" Scopes–125x: NGC 2023 is an interesting nebula involved with an 8th magnitude star. The nebula spreads NE of the bright star toward one of the 12th magnitude. Dark streaks are visible on the nebula's NE and west sides.

NGC 2022 PK196-10.1 Planetary Nebula Type 4+2
ø18″, m11.9v, CS 14.9v 05ʰ42.1ᵐ +09°05′
Finder Chart 21-7, Figure 21-11 ★★★

8/10" Scopes–175x: NGC 2022, though rather faint, is a tiny, clearly recognizable greyish disk of uniform surface brightness.

12/14" Scopes–200x: This planetary has a fairly obvious disk slightly elongated 20″ × 18″ NNE–SSW, slightly greenish in tint, and slightly granular in texture.

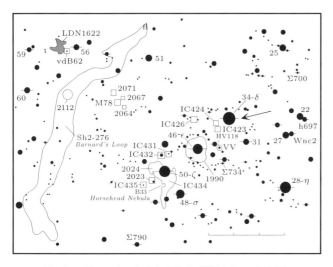

Finder Chart 21-6. 34–δ Ori: 05ʰ32.0ᵐ −00°17′

Orion 273

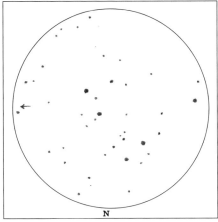

Figure 21-13. Dolidze 17
12.5", f5–100x, by G. R. Kepple

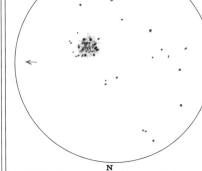

Figure 21-14. NGC 2141
12.5", f5–75x, by G. R. Kepple

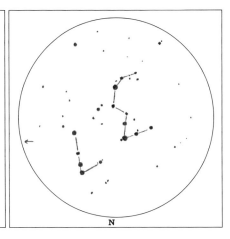

Figure 21-15. NGC 2169
8", f6–100x, by A. J. Crayon

16/18" Scopes–225x: NGC 2022 lies 10′ SE of two 8th magnitude stars which point to it in the finder. Half a dozen 11th magnitude stars are scattered to the west between the nebula and the 8th magnitude stars. The disk is fairly bright, slightly elongated 25″ × 20″ NNE–SSW and distinctly annular with a slightly darker center. The planetary marks the tip of the handle in a little dipper formation of stars extending to the NNE.

IC 435   Reflection Nebula
ø4′ × 3′, Photo Br 1–5, Color –   05ʰ43.0ᵐ   −02°19′
Finder Chart 21-6, Figure 21-10   ★★

12/14" Scopes–100x: IC 435 is a faint, circular bluish glow surrounding an 8.3 magnitude star.

NGC 2039   Open Cluster
[ø20′ × 12′, Br★ 8.v]   05ʰ44.1ᵐ +08°38′

NGC 2063   H2⁸   Open Cluster
[ø3′ × 2′, Br★ 13.v]   05ʰ46.8ᵐ +08°48′
Finder Chart 21-7   ★★/★★

12/14" Scopes–100x: Both NGC 2039 and NGC 2063 are classified as "nonexistent" by the RNGC, but the problem seems to have been in the original observations of the stargroups in this area: W. Herschel described a "small group of faint stars" but J. Herschel observed a much larger stellar aggregation. Possibly W. Herschel's small group is embedded in J. Herschel's large group. What can be seen in the area is a small concentration of ten faint stars, the brightest magnitude 12.5, arranged in a 7′ tall "V," its prongs extending north and NW. This small clump is at the southern edge of a large scattered group of thirty 8th to 13th magnitude stars elongated 20′ × 12′ NW–SE. The latter might be J. Herschel's large cluster. Its northern edge is marked by an 8th magnitude star to the NW from which a row of six magnitude 12.5–14 stars extend east to one of the 9th magnitude.

If the coordinates are correct, NGC 2063 located 40′ ENE seems too far away to be a part of NGC 2039. We observed a small knot of ten 13th to 14th magnitude stars elongated 3′ × 2′ NW–SE with the brightest star being 13th magnitude at the SE edge. Just south of this star is a curved line of six 13th magnitude stars concave to the north. We feel that the position of the larger cluster may be correct, but the location of NGC 2063 is still very much in doubt.

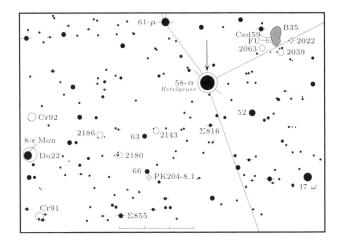

Finder Chart 21-7. 58-α Ori: 05ʰ55.2ᵐ +07°24′

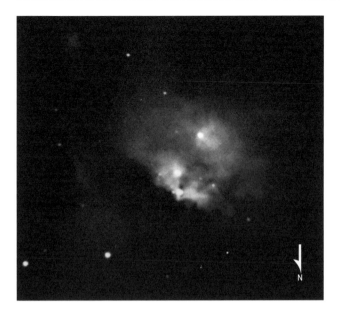

Figure 21-16. *Like a pair of eyes, two 10th magnitude stars peer out of the ghostly glow of Messier 78. NGC 2064 and NGC 2067 (upper and lower left) are visible as very faint hazes. Dr. Harvey Freed made this 50 minute exposure on hypered 2415 film with a 10", f6 Newtonian at prime focus.*

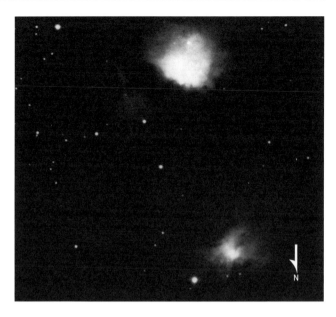

Figure 21-17. *Messier 78 (NGC 2068) is the large nebulous patch at the top and NGC 2071 the smaller nebula to the lower right. This 45 minute photo was taken by Martin C. Germano on hypered Kodak 2415 Tech Pan film with an 8", f5 Newtonian reflector.*

Cederblad 59 Emission & Reflection Nebula
⌀3' × 2', Photo Br 1–5, Color 3–4   05^{h}45.3^m +09°04'

Barnard 35 Dark Nebula
⌀20' × 10', Opacity 5   05^{h}45.5^m +09°03'
Finder Chart 21-7, Figure 21-12   ★★/★★

16/18" Scopes–100x: Cederblad 59 is a small, faint haze around the 12th magnitude variable star FU Orionis. The nebula is excited by the star and varies in size and brightness in concert with the star's light changes. Cederblad 59 is just east of the nearly starless 20' × 10' NW–SE dust cloud Barnard 35. Three faint stars are immediately east of B35's northern extension.

NGC 2064 Emission & Reflection Nebula
⌀1.5' × 1', Photo Br 2–5, Color 2–4   05^{h}46.3^m 00°00'
Finder Chart 21-6, Figure 21-16   ★

12/14" Scopes–100x: NGC 2064, 3' SW of Messier 78, is a small, very faint, diffuse nebulosity elongated N–S.

NGC 2067 Emission & Reflection Nebula
⌀8' × 3', Photo Br 2–5, Color 2–4   05^{h}46.5^m +00°06'
Finder Chart 21-6, Figure 21-16   ★

12/14" Scopes–100x: NGC 2067, just WNW of Messier 78, is a very faint 6' × 3' NNE–SSW haze. The nebula is broadest to the NNE and narrows into a thin taper toward the SSW. A very transparent night is needed for a good view of this nebula because its glow is not enhanced by filters.

NGC 2068 Messier 78 Emission & Reflection Nebula
⌀8' × 6', Photo Br 1–5, Color 2–4   05^{h}46.7^m +00°03'
Finder Chart 21-6, Figure 21-16   ★★★★

8/10" Scopes–100x: Messier 78 is a bright fan-shaped nebulosity with two 10th magnitude stars embedded in it. The nebula is sharply defined along the northern edge, but fades southward to a diffuse edge.

12/14" Scopes–125x: Two 10th magnitude stars shine like eyes out of a white, fan-shaped nebula that suggests a sheet draped over the head of a ghost. The brightest area is around the "eyes" along the northern edge. The nebula fans out to the south like a comet tail, the two 10th magnitude stars resembling a split comet nucleus. At high power on a good night some mottling might be glimpsed near the stars and along the east edge. NGC 2067, 4' NW of M78, is a very faint diffuse haze.

NGC 2071 Reflection Nebula
⌀7' × 5'   05^{h}47.2^m −00°18'
Finder Chart 21-6, Figure 21-17   ★★

12/14" Scopes–125x: NGC 2071, 15' NNE of Messier 78, is a faint glow surrounding a double star with a 10th magnitude primary and a faint secondary 15" SSW.

16/18" Scopes–150x: NGC 2071 is a faint 4' diameter haze surrounding a 10th magnitude star with a faint companion. The nebula fades from its center more quickly toward the south than in other directions.

Figure 21-18. *Open cluster NGC 2112 is a rich cluster of three dozen faint stars. Lee C. Coombs made this 15 minute exposure on 103a-O Kodak Spectroscopic film with a 10", f5 Newtonian reflector at prime focus.*

Figure 21-19. *LDN 1622 is a conspicuous dark patch NE of the faint reflection nebula vdB62 and the bright star 56 Orionis. Martin C. Germano made this 90 minute exposure on hypered 2415 film with an 8", f5 Newtonian reflector.*

Berkeley 21 Open Cluster 40★ Tr Type II 3 m
ø3.5′, m11.1v, Br★ 14.80v   05ʰ51.7ᵐ +21°47′
Finder Chart 21-9   ★

16/18″ Scopes–100x: Berkeley 21 is an extremely faint granular patch that can be glimpsed only with averted vision.

NGC 2112 Cr76 Open Cluster 50★ Tr Type II 3 m n
ø11.0′, m8.4v, Br★ 10.04v   05ʰ53.9ᵐ +00°24′
Finder Chart 21-6, Figure 21-18   ★★

12/14″ Scopes–75x: NGC 2112 is a fairly faint, moderately compressed open cluster of faint stars just SE of a 10th magnitude foreground field star probably not associated with the cluster. At low power, only ten 12th magnitude stars stand out against a rich haze spanning a 10′ diameter.

16/18″ Scopes–75x: NGC 2112 is a faint but rich patch of three dozen 13th magnitude and fainter stars in a 12′ × 6′ E–W area. A single bright star lies on the NW edge. A few 12th magnitude stars stand out near the center; and a faint star chain extends westward to a 10th magnitude star.

van den Bergh 62 Reflection Nebula
ø15′   05ʰ53.5ᵐ +01°42′

LDN 1622 Dark Nebula
ø7′ × 5′, Opacity 6   05ʰ54.6ᵐ +02°00′
Finder Chart 21-6, Figure 21-19   ★/★★

12/14″ Scopes–125x: LDN 1622 is a conspicuous dark patch 25′ NE of 56 Orionis (m4.80). It is irregularly shaped, and has a prominent, narrow extension to the NNE. The SW area is broader but less visible, though a row of four 11th–13th magnitude stars outline its western edge. The reflection nebula vdB 62, beyond the SW edge of the dark cloud, is a very faint haze surrounding and spreading 2′ SW from a magnitude 9.5 star. Averted vision helps, but, because this is a reflection nebula, filters do not.

Basel 11B Open Cluster 12★ Tr Type II 2 m
ø10.0′, m8.9v, Br★ 11.48v   05ʰ58.2ᵐ +21°58′
Finder Chart 21-9   ★★★

12/14″ Scopes–75x: Basel 11B is a loose and irregular cluster of two dozen stars in a 10′ × 5′ NE–SW area. It has a dozen 11th to 12th magnitude and a dozen 13th magnitude and fainter members in a zigzagging streamer. To its NNW is a triangle of 9th magnitude stars; and just SE is a large spread-wing asterism of three 11th–12th magnitude, two 10th magnitude, and several 8th magnitude stars.

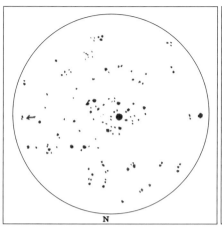

Figure 21-20. NGC 2180
12.5″, f5–125x, by G. R. Kepple

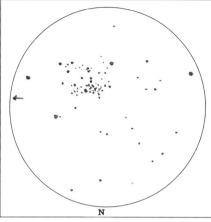

Figure 21-21. NGC 2186
12.5″, f5–125x, by G. R. Kepple

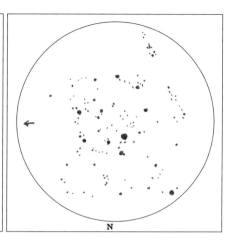

Figure 21-22. NGC 2202
12.5″, f5–150x, by G. R. Kepple

NGC 2143 Open Cluster
[ø20′, Br★ 12.v]   06ʰ03.0ᵐ +05°43′
Finder Chart 21-7   ★★★

12/14″ Scopes–100x: NGC 2143, another of the RNGC "nonexistent" clusters, is a highly irregular scattering of eighty 12th to 14th magnitude stars in a 20′ diameter triangular area. The brighter stars are widely separated into small clumps, three of the clumps defining the northern, SW, and SE edges. The cluster does not stand out well from the star field. A 9th magnitude star lies near the eastern edge.

NGC 2141 Cr79 Open Cluster 100★ Tr Type II 3 r
ø10′, m9.4v, Br? 13.33v   06ʰ03.1ᵐ +10°26′
Finder Chart 21-8, Figure 21-14   ★★

12/14″ Scopes–125x: NGC 2141, less than a degree NNW of Mu (μ) = 61 Orionis, is a small concentration of twenty 13th magnitude stars embedded in a dim, granular haze 10′ across. It is just NW of the open end of a large V-shaped asterism of 11th and 12th magnitude stars.

16/18″ Scopes–150x: In larger instruments, NGC 2141 appears fairly rich with a dozen stars plainly visible and another three dozen intermittently resolved in a 10′ area. A close double of 10th and 12th magnitude stars lies at its south edge.

NGC 2169 Cr38 Open Cluster 30★ Tr Type I 3 p n
ø6′, m5.9v, Br★ 6.94v   06ʰ08.4ᵐ +13°57′
Finder Chart 21-8, Figure 21-15   ★★★

8/10″ Scopes–100x: This neat cluster spells the number "37" (however, the numerals do not appear upright due to their field orientation). The cluster has seventeen 7.5 to 12th magnitude stars in a 5′ × 7′ area. Nine stars form the "3" on the SE, and five stars form the "7" on the NW. A few more stars not involved in the asterisms are sprinkled about.

12/14″ Scopes–125x: NGC 2169 is a fairly bright cluster of two loose groups. The NW group is a N–S row of five stars, the two northernmost reddish. A star east of the north end of the row makes it form a figure "7." The SE group has nine stars zigzagging to form a crude figure "3." The easternmost star in this group is reddish, contrasting nicely with the bluish tint of other stars in the group. The cluster's lucida is the double Σ848 (8.3, 9.0′; 2.5″; 108°), which is in the "7." 9th magnitude stars are 28″ and 43″ south of Σ848 and are considered to be part of a multiple system with its close pair. Fainter members of this multiple are a magnitude 12.0 star 14.5″ away in P.A. 204° and a magnitude 12.5 star 15″ away in P.A. 295°. All these stars may not be orbiting the

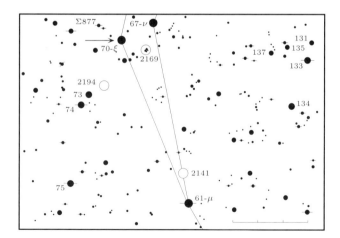

Finder Chart 21-8. 70–ξ Ori: 06ʰ11.9ᵐ +14°13′

Figure 21-23. *Emission nebula NGC 2174 is a large glow embedded in open star cluster NGC 2175. NGC 2175S is the small, dense knot of stars at right. Martin C. Germano made this 60 minute exposure on hypered 2415 film with an 8", f5 Newtonian at prime focus.*

close Σ848 pair; but they are no doubt NGC 2169 members and therefore at least indirectly related to Σ848. At 175x the count of cluster members increases to around three dozen.

Cederblad 62 Reflection Nebula
ø3′ × 2′, Photo Br 2–5, Color 1–5 $06^h07.8^m$ +18°40′
Finder Chart 21-9   ★★

12/14″ Scopes–125x: Cederblad 62, 3′ west of an 8th magnitude star, is a highly extended N–S streak around a 13th magnitude star. A 20″ pair of 10th magnitude stars lies 8′ SW.

NGC 2180 Open Cluster
[ø10′, Br★ 9.v] $06^h09.6^m$ +04°43′
Finder Chart 21-7, Figure 21-20   ★★★

12/14″ Scopes–100x: Though catalogued as "nonexistent" in the RNGC, NGC 2180 is quite apparent as a small, faint group of thirty 12th to 14th magnitude stars in a 10′ area loosely distributed around and mostly west of a 9th magnitude star. Half a dozen of the stars form an E–W line across the northern edge. A 4′ concentration of 14th magnitude stars is 10′ SW of the 9th magnitude star.

NGC 2174 Emission Nebula
ø40′ × 30′, Photo Br 1–5, Color 3–4 $06^h09.7^m$ +20°30′

NGC 2175 Cr84 Open Cluster 20★ Tr Type IV 3 p n
ø18′, m6.8v, Br★ 7.55v $06^h09.8^m$ +20°19′
Finder Chart 21-9, Figure 21-23   ★★/★★★

12/14″ Scopes–75x: Emission nebula NGC 2174 is a faint 20′ diameter glow surrounding a 7.5 magnitude star, its brightest area just north of the star. A faint

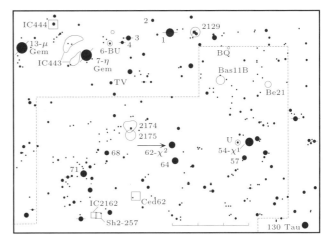

Finder Chart 21-9. 62–χ² Ori: $06^h03.9^m$ +20°08.3′

Figure 21-24. *NGC 2194 is a a rich, concentrated cluster of 12th magnitude and fainter stars. Lee C. Coombs made this 10 minute exposure on 103a-O Kodak Spectroscopic film with a 10", f5 Newtonian at prime focus. South is up.*

haze is also around a knot of 20 stars known as open cluster NGC 2175S lying 17' ENE of the 7.5 magnitude star. The catalogue position of NGC 2175, some 10' south of the magnitude 7.5 star, seems incorrect, for there is nothing resembling a cluster south of the star, though a star-dense triangle is just NNE of the star. Martin C. Germano's photo shows that the NGC 2175 open cluster is in fact superimposed upon the whole nebula rather than isolated in its southern part.

NGC 2186   Cr85   Open Cluster 30★   Tr Type II 2
ø4.0', m8.7v, Br★ 9.82v   06ʰ12.2ᵐ +05°27'
Finder Chart 21-7, Figure 21-21   ★★★

12/14" Scopes–125x: NGC 2186 is a small, unconcentrated cluster of eighteen 9.5 to 12th magnitude stars in a 4' area. The brighter stars, which include two of magnitude 9.5 and two of magnitude 10, form an arrowhead pointing east. The arrowhead also contains most of the fainter stars, but some spill out toward the NE. Outlying stars to the north and east add fifteen more members to the cluster.

16/18" Scopes–150x: This loose cluster has thirty stars in a 5' area. Two magnitude 9.5 stars are on its east and west edges. Just west of the easternmost 9.5 magnitude star is a semicircle of five faint stars. The cluster has three magnitude 10–10.5, sixteen magnitude 11–12, and a sprinkling of fainter stars. Near the center is a 10" wide magnitude 11 and 12 double. A magnitude 12.5 pair is on the southern edge of the cluster. The densest concentration in the cluster is just west of its center.

NGC 2194   Cr87   Open Cluster 80★   Tr Type III 1 r
ø8', m8.5v, Br★ 12.07v   06ʰ13.8ᵐ +12°48'
Finder Chart 21-8, Figure 21-24   ★★★

12/14" Scopes–75x: NGC 2194, a faint but rich cluster, lies half a degree NW of 73 Orionis (m5.3). It is highly concentrated with eight 12th magnitude, and fifteen 13th to 14th magnitude, stars resolved against a granular background 8' across. The brighter stars are near the NE and SW edges.

16/18" Scopes–150x: This open cluster is attractive but faint, best with larger apertures. It is a rich, compact cluster of fifty stars concentrated in an irregular 10' area and embedded in an unresolved background glow. Eight of the brighter stars form a rectangle within which is the majority of the fainter members. Faint outlying stars fall along several star chains, the most prominent extending SW. This cluster resembles NGC 2158 in Gemini. A 2' knot of 20 faint stars lies 15' ENE.

NGC 2202   Open Cluster
ø7', Br★ 12.v   06ʰ16.9ᵐ +05°59'
Finder Chart 21-7, Figure 21-22   ★★★

12/14" Scopes–125x: NGC 2202 is another cluster classed as "nonexistent" in the RNGC. We observed sixteen 12th to 13th magnitude, and at least twice as many 14th magnitude, stars sprinkled over a 7' area around, but mostly west of, a 9th magnitude star.

Chapter 22

Pegasus, the Winged Horse

22.1 Overview

The Pegasus was the winged horse of Greek mythology, though the image of a winged horse originated in Babylonia. When Perseus cut off the head of Medusa, some of the snake-haired monster's blood fell into the sea and up from the foaming compound flew the Pegasus. Perseus mounted the mighty steed and was riding it when he rescued the princess Andromeda from the clutches of Cetus the Sea Monster. Pegasus later became the bearer of Zeus' thunder and lightening.

Only the front half of Pegasus, upside down, appears in the sky. The Horse's body is marked by the four stars of the Great Square, a conspicuous asterism that makes the constellation easy to identify. The NE star of the Great Square, Alpheratz, is shared by Pegasus with Andromeda, and thus in older star charts was designated both Alpha Andromedae and Delta Pegasi. However, when the modern constellation boundaries were set the star was officially assigned to Andromeda — which is why on modern charts Pegasus has no star designated Delta.

Pegasus contains 1,121 square degrees and is the seventh largest of the eighty-eight constellations. Nevertheless, mostly because it is off the Milky Way, it is rather poor in objects. Its best offering is also its lone Messier entry, the globular cluster M15. Pegasus, like most off-Milky Way constellations, has quite a few galaxies; but, apart from the nearly edge-on spiral NGC 7331, they are faint and small and require dark skies and moderate-to-large aperture to make much of a visual impression.

Pegasus: PEG-a-sus
Genitive: Pegasi, PEG-a-si
Abbrevation: Peg
Culmination: 9 pm–Oct. 16, midnight–Sept. 1
Area: 1,121 square degrees
Showpieces: M15 (NGC 7078), NGC 7331, NGC 7332, NGC 7479, NGC 7814
Binocular Objects: 1 Peg, 3 Peg, 53-β Peg, M15 (NGC 7078), NGC 7217, NGC 7331, NGC 7332, NGC 7814

22.2 Interesting Stars

AG Pegasi Variable Star   **Spec. WN6+M1-3**
m6.0 to 9.4 in 830 days $21^h51.0^m$ $+12°38'$
Finder Chart 22-3   ★★★★

AG Pegasi is a Z Andromedae star, a type of variable which couples an otherwise normal cool M-type red giant with a compact blue star (sometimes a subdwarf such as are found at the center of many planetary nebulae) which is tearing material off the bloated envelope of its companion. The variations of AG Peg are possibly a combination of normal long-period variable pulsations in the M-star and explosive activity on the hot surface of the compact star. Such systems, because of the matter exchange, are also known as symbiotic variables.

Σ2841 Double Star   **Spec. K0**
m6.4, 7.9; Sep. 22.3″; P.A. 110° $21^h54.3^m$ $+19°43'$
Finder Chart 22-4   ★★★★

12/14″ Scopes–100x: Struve 2841 is a fine, wide double for small telescopes and binoculars. The components are yellow and greenish.

Σ2848 Double Star   **Spec. A2**
m7.2, 7.5; Sep. 10.7″; P.A. 56° $21^h58.0^m$ $+05°56'$
Finder Chart 22-3   ★★★★

8/10″ Scopes–100x: Struve 2848 is an attractive double of pale yellowish stars.

Pi-1 (π^1) = 27 Pegasi Multiple Star   **Spec. K0**
AB: m5.6, 11.9; Sep. 27.7″; P.A. 323° $22^h09.2^m$ $+33°10'$
Finder Chart 22-5   ★★★★

2/3″ Scopes–50x: Pi-1 Pegasi is a beautiful, deep yellow

Pegasus, the Winged Horse Constellation Chart 22-1

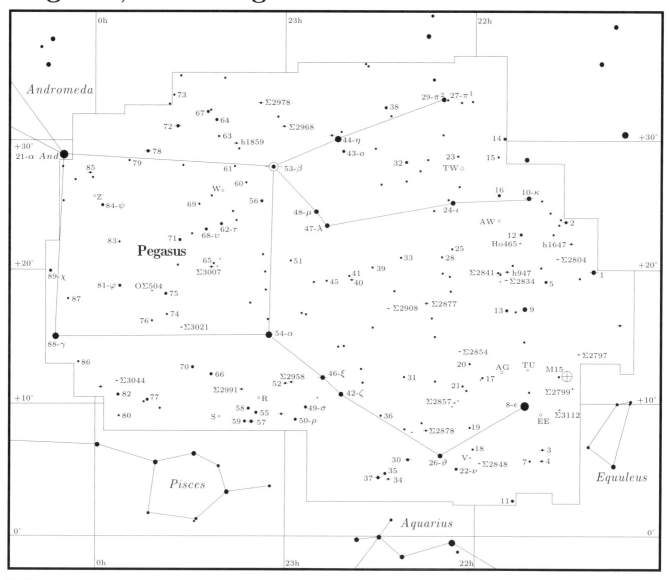

Table 22-1. Selected Variable Stars in Pegasus

Name	HD No.	Type	Max.	Min.	Period (Days)	F*	Spec. Type	R.A. (2000)	Dec.	Finder Chart No. & Notes
EE Peg	206155	EA	6.9	7.6	2.63	0.10	A3+F4	21^{h}40.0^m	+09°11'	22-3
AG Peg	207757	Z And	6.0	9.4	830		WN6+M1-3	51.0	+12 38	22-3
AW Peg	207956	EA	7.8	9.2	10.62	0.10	A3-F5	52.3	+24 01	22-4
V Peg	209127	M	7.0	15.0	302	0.44	M3-M6	22^{h}01.0^m	+06 07	22-3
R Peg	218292	M	6.9	13.8	378	0.44	M6-M9	06.6	+10 33	22-8
S Peg	220033	M	7.1	13.8	319	0.47	M5-M8	20.6	+08 55	22-8
Z Peg	224709	M	7.7	13.6	325	0.50	M6-M8	00^{h}00.1^m	+25 53	22-11

F* = The fraction of period taken up by the star's rise from min. to max. brightness, or the period spent in eclipse.

Table 22-2. Selected Double Stars in Pegasus

Name	ADS No.	Pair	M1	M2	Sep."	P.A.°	Spec	R.A. (2000)	Dec	Finder Chart No. & Notes
1 Peg	14909	AB	4.1	8.2	36.3	311	K0	21ʰ22.1ᵐ	+19°48′	22-4
Σ2797	14977		7.3	8.8	3.3	217	A2	26.7	+13 41	22-1
Σ2799	15007	AB	7.5	7.5	1.6	276	F2	28.9	+11 05	22-3
	15007	AC		9.3	136.2	336				
h1647		AB	6.0	8.7	41.0	177	M3	29.0	+22 11	22-4
		AC		10.4	41.2	126				
Σ2804	15076	AB	7.6	8.3	3.1	350	F5	33.0	+20 43	22-4
Σ3112	15092		7.9	9.7	7.0	238	G0	34.5	+09 30	22-3
3 Peg	15147	AB	6.0	8.3	39.2	349	A0	37.7	+06 37	22-3
4 Peg	15157		5.7	11.7	26.5	336	F0	38.5	+05 46	22-3
10-κ Peg	15281	AC	4.7	10.6	13.8	292	F5	44.6	+25 39	22-1 Yellow primary
Ho 465	15311	AB	7.1	9.1	42.6	246	A2	46.5	+22 10	22-4
	15311	BC		10.9	4.7	79				
h947	15383	AB	5.8	9.1	19.4	95	B9	51.6	+19 50	22-4
	15383	AC		11.1	23.8	320				
Σ2834	15386	AB	6.9	10.2	4.3	295	F2	51.7	+19 18	22-4
	15386	AC		12.5	26.0	121				
Σ2841	15431	AB	6.4	7.9	22.3	110	K0	54.3	+19 43	22-4 Yellow & greenish
Σ2848	15493		7.2	7.5	10.7	56	A2	58.0	+05 56	22-3 Pale yellow pair
Σ2854	15596		7.6	7.9	2.0	83	F5	22ʰ04.4ᵐ	+13 39	22-1
Σ2857	15630		7.4	9.1	19.8	113	A2	06.2	+10 06	22-3
	15767	AC		9.7	66.0	122				
Σ2877	15763	AB	6.5	9.7	16.0	13	K2	14.3	+17 11	22-1 Deep yellow, bluish
Σ2878	15767	AB	6.8	8.3	1.5	122	A0	14.5	+07 59	22-1
34 Peg	15935	AB	5.8	12.3	3.5	224	G0	26.6	+04 24	22-1 Yellow primary
	15935	AC		12.8	103.3	272				
Σ2908	15967		7.7	9.4	9.0	116	K0	28.2	+17 16	22-1
37 Peg	15988		5.8	7.1	c 0.7	*118	F5	30.0	+04 26	22-1
46-ξ Peg	16261	AB	4.2	12.2	11.5	100	F5	46.7	+12 10	22-8 Yellow primary
	16261	AC		11.0	145.0	15				
Σ2958	16389		6.6	8.9	3.8	13	A3	56.9	+11 51	22-8
52 Peg	16428		6.1	7.4	c 0.7	*330	F0	59.2	+11 44	22-8 Yellow pair
Σ2968	16443		6.6	9.1	3.4	90	A0	23ʰ00.7ᵐ	+31 05	22-6
53-β Peg	16483	AB	2.4	11.6	108.5	211	M	03.8	+28 05	22-6 Scheat: Reddish
	16483	AC		9.4	253.1	98				
Σ2978	16519		6.3	7.5	8.4	145	A2	07.5	+32 50	22-1
57 Peg	16550		5.1	9.7	32.6	198	M	09.5	+08 41	22-8 Orange & blue
Σ2991	16603		5.9	9.2	33.6	359	K0	13.4	+11 04	22-8
h1859			6.4	9.9	35.2	121	F5	14.4	+29 46	22-1
Σ3007	16713	AB	6.6	9.6	5.9	91	G0	22.8	+20 34	22-9 Yellow & ashy
Σ3021	16812		7.9	9.1	8.6	308	F8	31.4	+16 13	22-1
72 Peg	16836		5.7	5.8	0.5	* 97	K2	34.0	+31 20	22-11
OΣ504	16942		7.4	10.2	7.7	175	K0	42.6	+18 40	22-9
78 Peg	16957		5.0	8.1	1.0	235	K0	44.0	+29 22	22-11 Deep yellow primary
Σ3044	17079		7.3	7.9	19.0	282	F0 F0	53.0	+11 55	22-10

Footnotes: *= Year 2000, a = Near apogee, c = Closing, w = Widening. Finder Chart No: All stars listed in the tables are plotted in the large Constellation Chart, but when a star appears in a Finder Chart, this number is listed. Notes: When colors are subtle, the suffix -ish is used, e.g. *bluish*.

star with four distant companions of magnitudes 10.1, 10.6, 11.9, and 12.5. Pi-1 forms a nice binocular pair with the yellow Pi-2 Pegasi (m4.3) 7′ east.

Σ2877 Double Star   **Spec. K2**
m6.5, 9.7; Sep. 16.0″; P.A. 13°   22ʰ14.3ᵐ +17°11′
Constellation Chart 22-1   ★★★★

12/14″ Scopes–100x: Struve 2877 is a fine double of deep yellow and bluish stars.

Eta (η) = 44 Pegasi Multiple Star   **Spec. G0**
AxBC: m2.9, 9.9; Sep. 90.4″; P.A. 339°   22ʰ43.0ᵐ
+30°13′
Finder Chart 22-5   ★★★★

12/14″ Scopes–100x: Eta Pegasi is a binocular double of yellow and blue stars, though a telescope is necessary to see the color of the secondary.

Figure 22-1. *Messier 15 is a fine, bright globular with a blazing core and many star chains radiating outward like a splash of paint. Evered Kreimer made this 30 minute exposure on Tri-X film with a 12.5", f7 Newtonian reflector.*

Figure 22-2. *NGC 7094 is a large, faint planetary nebula visible as a round, uniform glow with a 13.7 magnitude center star. Martin C. Germano made this 80 minute exposure on hypered 2415 film with a 14.25", f5 Newtonian reflector.*

57 Pegasi Double Star   Spec. M
m5.1, 9.7; Sep. 32.6"; P.A. 198° $23^h09.5^m$ +08°41'
Finder Chart 22-8   ★★★★

8/10" Scopes–100x: 57 Pegasi is a beautiful pair with an orange primary and a blue companion.

Σ3007 Double Star   Spec. G0
AB: m6.6, 9.6; Sep. 5.9"; P.A. 91° $23^h22.8^m$ +20°34'
Finder Chart 22-9   ★★★★

8/10" Scopes–100x: Struve 3007 has a deep yellow primary with a grayish-white companion.

22.3 Deep-Sky Objects

NGC 7078 Messier 15 Globular Cluster Class IV
⌀12.3', m6.0v $21^h30.0^m$ +12°10'
Finder Chart 22-3, Figure 22-1   ★★★★★

This impressive globular was found by Maraldi in September, 1746 while he searched for the de Chessaux Comet. Messier rediscovered the object in 1764. Messier 15 lies 30,600 light years away and has a diameter of about 130 light years.

- *8/10" Scopes–100x:* This globular lies 5' SSW of a 7.5 magnitude star which is at the northern vertex of a triangle of bright stars containing the object. It has an intense, compact core surrounded by a halo of faint stars. At 150x a few stars can be resolved across the core. Star streamers radiate from the core in all directions, though the western part of the halo is less star-rich.

- *12/14" Scopes–125x:* Messier 15 is a fine, bright globular with a blazing core that thins rapidly 5' out from the center. Many arms and chains radiate from the core-like splashes of whitewash, reaching to about 13' from the cluster center. 175x resolves the core's 13th magnitude stars and reveals a dark blotch SW of the center.

NGC 7094 PK66-28.1 Planetary Nebula Type 4
⌀95", m13.4v, CS 13.73v $21^h36.9^m$ +12°47'
Finder Chart 22-3, Figure 22-2   ★★

12/14" Scopes–250x: NGC 7094, located 1.5° ENE of globular cluster Messier 15, is a very faint, round,

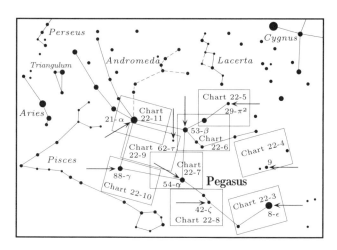

Master Finder Chart 22-2. Pegasus Chart Areas
Guide stars indicated by arrows.

uniform glow surrounding a 13.7 magnitude central star. A fainter star lies within the halo near the NE edge. A row of three 13.5 magnitude stars lies to the ESE.

NGC 7137   H261²   Galaxy   Type SAB(rs)c III
⌀1.5′ × 1.2′, m12.4v, SB 12.9   21ʰ48.2ᵐ +22°10′

Finder Chart 22-4   ★★

8/10″ Scopes–100x: This galaxy is surrounded by a one degree circlet of four 7th and two 8th magnitude stars, a convenient bull's eye at which to aim a viewfinder. It has a very faint, round 1′ diameter halo of uniform surface brightness. Half a degree NE is an interesting configuration of stars consisting of a pair and a triplet 3′ apart.

12/14″ Scopes–125x: NGC 7137 has a rather faint, round 1.5′ diameter halo with a faint stellar nucleus. A 14th magnitude star lies at the NW edge.

NGC 7177   H247²   Galaxy   Type SAB(r)b I-III
⌀3.0′ × 1.9′, m11.2v, SB 12.9   22ʰ00.7ᵐ +17°44′

Finder Chart 22-4   ★★★

8/10″ Scopes–100x: This small, faint galaxy is a nearly round hazy glow, slightly brighter in the center. At 150x it is irregularly round with suggestions of uneven surface texture and brightness. A curved row of three magnitude 10.5 to 12 stars lies to the west.

12/14″ Scopes–125x: NGC 7177 has a fairly bright, diffuse, uniform, oval halo elongated 2′ × 1′ E–W. Its faint, nearly stellar nucleus appears about as bright as a 13th magnitude star lying 1′ SSW.

NGC 7217   H207²   Galaxy   Type (R)SA(r)ab II-III
⌀3.5′ × 3.0′, m10.1v, SB 12.5   22ʰ07.9ᵐ +31°22′

Finder Chart 22-5   ★★★

8/10″ Scopes–100x: NGC 7217 has a moderately bright 2′ diameter halo that gradually brightens towards its center. At 150x a very faint stellar nucleus is suggested. The galaxy is in a fairly rich star field with a 10.5 magnitude star 3′ SE.

12/14″ Scopes–125x: The halo is moderately concentrated, about 2.5′ in diameter, and gradually brightens to a prominent nonstellar nucleus. At 175x, the halo shows a slight granular texture.

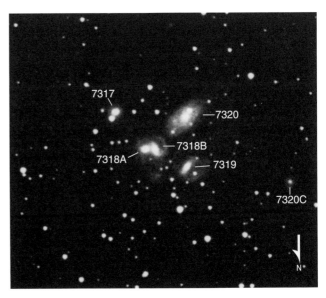

Figure 22-3. *Stephan's Quintet is a group of very faint galaxies challenging even for large instruments. Martin C. Germano made this 90 minute exposure on hypered 2415 film with a 14.25″, f5 Newtonian.*

NGC 7317   Galaxy   Type E2   Stephan's Quintet
⌀2.5′ × 0.5′, m13.6v, SB 11.9   22ʰ35.9ᵐ +33°57′

NGC 7318A   Galaxy   Type E2 pec
⌀1.7′ × 1.2′, m13.4v, SB 13.4   22ʰ36.0ᵐ +33°58′

NGC 7318B   Galaxy   Type E2 pec
⌀0.9′ × 0.9, m13.4v, SB13.0   22ʰ36.0ᵐ +33°58′

NGC 7319   Galaxy   Type SB(S)bc pec
⌀1.5′ × 1.1′, m13.1v, SB 13.5   22ʰ36.1ᵐ +33°59′

NGC 7320   Galaxy   Type SA(s)d
⌀1.7′ × 0.9′, m12.6v, SB 12.9   22ʰ36.1ᵐ +33°57′

Finder Chart 22-5, Figure 22-3   ★/★/★/★/★

12/14″ Scopes–125x: These five faint galaxies, all within a 3.5′ area, comprise Stephan's Quintet. NGC 7320, on the southeastern edge of the group, the most

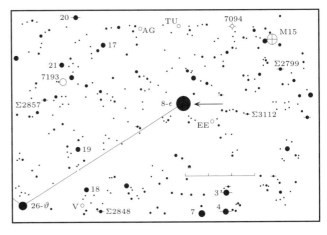

Finder Chart 22-3. 8–ε Peg: 21ʰ44.1ᵐ +09°53′

Figure 22-4. *NGC 7331, a nearly edge-on spiral, is visible in small telescopes but becomes really impressive through larger instruments. Four small companion galaxies are labeled in the photo above. Martin C. Germano made this 90 minute exposure on hypered 2415 film with a 14.25", f5 Newtonian. North is to the right.*

conspicuous member, nevertheless is only a very faint, small halo of even surface brightness, elongated 1' × 0.5' NW–SE. The other four galaxies, even with the assistance of averted vision, are extremely faint and undetailed smudges.

16/18" Scopes–150x: NGC 7320, the southeastern galaxy of Stephen's Quintet, is the group's brightest and most easily-observed member. Its very faint, diffuse halo is elongated 1.25' × 0.75' NW–SE and slightly brighter in the center. The westernmost galaxy of the quintet, NGC 7317, could be mistaken for a double with the 13th magnitude star just to its NW edge for it appears nearly stellar, being only a tiny 25" spot with a faint core. A couple minutes NW of NGC 7320 is the close interacting galaxy-pair NGC 7318A–B, two faint concentrations sharing a common halo: NGC 7318B is the brighter and larger of the two faint blobs, and boasts a faint stellar nucleus; NGC 7318A is smaller, circular, and has a more concentrated halo. The NGC 7318A–B pair spans a 1' × 0.75' E–W area. To its ENE, and almost due north of NGC 7320, is the faintest Stephen's Quintet galaxy, NGC 7319, an extremely faint 45" × 25" NW–SE slash.

NGC 7331 H531² Galaxy Type SA(s)b I-II
ø10.5' × 3.7', m9.5v, SB 13.3   22ʰ37.1ᵐ +34°25'
Finder Chart 22-5, Figure 22-4   ★★★★★

8/10" Scopes–100x: NGC 7331 is a fine bright galaxy highly elongated 6' × 1.5' N–S. It has a large, extended, well concentrated core containing a stellar nucleus. A wide pair of 13th magnitude stars is on the NNW edge.

12/14" Scopes–125x: The halo is bright, elongated 8' × 2' NNW–SSE, and has a large, bright, oval core

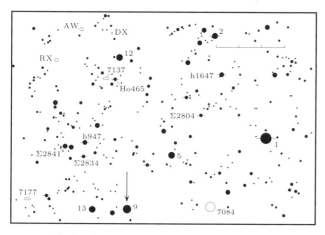

Finder Chart 22-4. 9 Peg: 21ʰ44.5ᵐ +17°21'

containing a stellar nucleus. Four extremely faint companion galaxies are to the east of NGC 7331 and may be glimpsed with averted vision under good skies. Use Figure 22-4 to confirm your sightings.

16/18" Scopes–150x: This bright, large spiral is impressive in larger instruments. The halo is elongated 10' × 3' N–S and has a prominent 1.5' × 1' oval core with a stellar nucleus. The halo is speckled near the core, and a subtle dust lane on the west side can be glimpsed with averted vision. Four dim, tiny companion galaxies are just visible to the east. The easiest, NGC 7335, located 3.5' NE of the center of NGC 7331, is a featureless 45" × 25" blip. NGC 7336, just 2' NNE of NGC 7335, is even smaller and fainter, being an extremely faint circular spot seen only with averted vision. NGC 7337, 5' SE of the big galaxy, is so tiny that it appears to be a double with a faint star 10" to its SE. The farthest east of these companion galaxies, NGC 7340, is only a very faint, tiny spot. Two 10th magnitude stars 7' east of NGC 7331 point SSE at NGC 7340.

NGC 7332 H233² Galaxy Type S0 pec sp
ø3.7' × 1.0', m11.1v, SB 13.4   22^{h}37.4^m +23°48'

Finder Chart 22-6, Figure 22-5   ★★★★

8/10" Scopes–100x: This galaxy forms a fine 5' pair with NGC 7339 to its east, the two situated between a 30' wide N–S pair of magnitude 6.5 stars. NGC 7332 has a diffuse halo elongated 1.5' × 0.5' NNW–SSE with a broad, prominent core and a bright stellar nucleus. A 10th magnitude star 2.5' SSE is part of a chain that ends just SW of NGC 7339.

12/14" Scopes–125x: NGC 7332 is a fine, bright edge-on galaxy elongated 2.5' × 0.75' NNW–SSE with an extended core and a prominent nonstellar nucleus. 175x shows tapered ends 3' in length. Companion galaxy NGC 7339 is a faint carbon copy aligned so that it points at NGC 7332.

NGC 7339 Galaxy Type SAB(s)bc:?
ø2.6' × 0.8', m12.2v, SB 12.9   22^{h}37.8^m +23°47'

Finder Chart 22-6, Figure 22-5   ★★★

8/10" Scopes–100x: NGC 7339 is a fainter version of NGC 7332 lying 5' west. It is about the same size as its companion, 1.5' × 0.5' (its E–W elongation pointing at its mate), but its halo is much more diffuse than that of NGC 7332 and shows no central brightening.

12/14" Scopes–125x: NGC 7339 has a 2' × 0.75' halo elongated E–W with a slight but broad central concentration. Its tips are not as tapered and are more diffuse than those of its companion, NGC 7332.

Figure 22-5. *NGC 7332 (left) and NGC 7339 are a fine pair of edge-on galaxies. Alexander Brownlee made this 90 minute exposure on hypered 2415 film with a 16", f5 Newtonian reflector.*

NGC 7448 H251² Galaxy Type SA(rs)bc II-III
ø2.5' × 1.0', m12.6v, SB 12.6   23^{h}00.1^m +15°59'

Finder Chart 22-7   ★★★

8/10" Scopes–100x: NGC 7448 is situated between 10.5 magnitude stars 2.5' east and 4.5' west. It has a fairly faint halo elongated 1.75' × 1' N–S with a uniform surface brightness.

12/14" Scopes–125x: NGC 7448 appears fairly bright, elongated 2.25' × 1' NNW–SSE, and slightly brighter in the center. Two 13th magnitude stars about 15" apart are 2' NW.

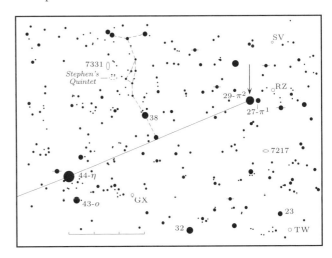

Finder Chart 22-5. 29–π² Peg: 22^{h}09.9^m +33°11'

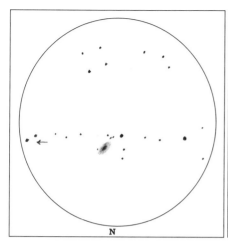
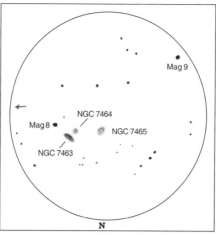
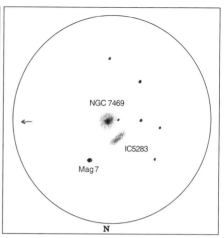

Figure 22-6. NGC 7457
8", f6–80x, by A. J. Crayon

Figure 22-7. NGC 7463, 7464, & 7465
12.5", f5–125x, by G. R. Kepple

Figure 22-8. NGC 7469 & IC 5283
16", f5–125x, by Bob Erdmann

NGC 7457 H212² Galaxy Type SA(RS)0
⌀4.4' × 2.5', m11.2v, SB 13.6   23ʰ01.0ᵐ +30°09'
Finder Chart 22-6, Figure 22-6   ★★★

8/10" Scopes–100x: This galaxy is just west of an open "V" of stars and north of a wide 12th magnitude double. The halo is fairly bright, elongated 1' × 0.5' NW–SE, and has a well concentrated core.

12/14" Scopes–125x: NGC 7457 displays a bright oval halo elongated 1.5' × 1' NW–SE with a brighter center and a nonstellar nucleus. The galaxy is just north of an ENE–WSW chain of a dozen stars.

NGC 7454 H249² Galaxy Type E4
⌀1.8' × 1.5', m11.8v, SB 12.7   23ʰ01.1ᵐ +16°23'
Finder Chart 22-7   ★★★

8/10" Scopes–100x: NGC 7454, located 4' west of a 9.5 magnitude star, is a very faint, small, circular glow with a slight central brightening. An 8th magnitude star is 18' SE.

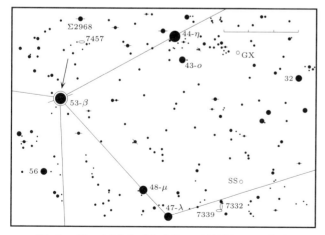

Finder Chart 22-6. 53-β Peg: 23ʰ03.7ᵐ +28°05'

12/14" Scopes–125x: NGC 7454 appears fairly faint, elongated 1.5' × 1' NW–SE, and moderately brighter in its center.

NGC 7463 Galaxy Type SABb: pec
⌀2.9' × 0.6', m13.2v, SB 13.7   23ʰ01.9ᵐ +15°59'

NGC 7464 Galaxy Type E1 pec:
⌀0.5' × 0.5', m13.3v, SB 11.6   23ʰ01.9ᵐ +15°58'

NGC 7465 Galaxy Type (R')SB(s)0°:
⌀1.2' × 0.7', m12.6v, SB 12.2   23ʰ02.0ᵐ +15°58'
Finder Chart 22-7, Figure 22-7   ★/★/★★

12/14" Scopes–125x: This trio of faint galaxies is 2' ENE of an 8th magnitude star and a degree NW of Alpha (α) = 54 Pegasi. NGC 7465, at the east end of the group, is the brightest and most conspicuous of the three; NGC 7464 is the smallest, faintest, and most difficult. NGC 7463, at the NW corner of the group, is the largest but very thin, its uniformly illuminated halo elongated 2.5' × 0.75' ENE–WSW. NGC 7464, just SE of NGC 7463 and the nearest of the group to the 8th magnitude field star, is only a very faint, round 30" glow without central brightening. NGC 7465 has a 1' diameter halo that is considerably brighter toward its center. To the south of the galaxies is a slightly curved E–W line of three magnitude 10.5 stars pointing SE toward a 10' distant 9th magnitude field star.

NGC 7469 H230³ Galaxy Type (R)SAB(rs)a
⌀1.5' × 1.0', m12.3v, SB 12.6   23ʰ03.3ᵐ +08°52'

IC 5283 Galaxy Type SA(r)cd pec?
⌀0.7' × 0.4', m13.8v, SB 12.2   23ʰ03.3ᵐ +08°54'
Finder Chart 22-8, Figure 22-8   ★★/★

12/14" Scopes–125x: This close pair of faint galaxies is

Figure 22-9. *NGC 7479 is a bright SBc galaxy with an S-shaped spiral pattern and a conspicuous central bar. Martin C. Germano made this 90 minute exposure on hypered 2415 film with a 14.25″, f5 Newtonian reflector.*

near the south vertex of a triangle consisting of one 7th and two 8th magnitude stars. NGC 7469 has a bright nucleus surrounded by a much fainter, diffuse halo 45″ in diameter. IC 5283, located 1.5′ NE of NGC 7469, appears very faint and is elongated 1′ × 0.25′ NW–SE without any central brightening.

16/18″ Scopes–150x: These two galaxies lie 5′ NW of a 7th magnitude star in a field of relatively bright stars. NGC 7469 has a faint halo elongated 1.25′ × 1′ WNW–ESE surrounding a bright stellar nucleus. A 13th magnitude star touches the east edge. IC 5283, 1.5′ NE is a faint, diffuse 2′ × 0.5′ NW–SE streak.

NGC 7479 H55¹ Galaxy Type SB(s)c I-II
⌀4.0′ × 3.1′, m10.8v, SB 13.4   23ʰ04.9ᵐ +12°19′
Finder Chart 22-8, Figure 22-9   ★★★★

8/10″ Scopes–100x: NGC 7479, located 3′ north of a 10th magnitude star, has a faint, diffuse halo elongated 3′ × 2′ N–S and moderately brighter at its center.

12/14″ Scopes–125x: NGC 7479 is a fairly bright galaxy elongated 4′ × 3′ N–S within which is a prominent bar containing a bright center. With averted vision, very faint arms may be seen curving from either end of the bar.

16/18″ Scopes–150x: In large telescopes the N–S bar of NGC 7479 is fairly bright, 4′ in length, and has faint extensions from either end. The southern extension curves westward and nearly wraps around a 14th magnitude star. The northern extension stops just short of a 13th magnitude star, curves eastward, then fades suddenly.

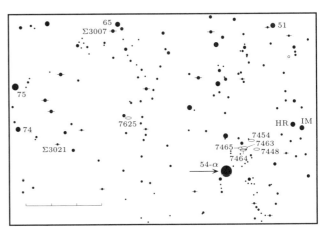

Finder Chart 22-7. 54–α Peg: 23ʰ04.7ᵐ +15°12′

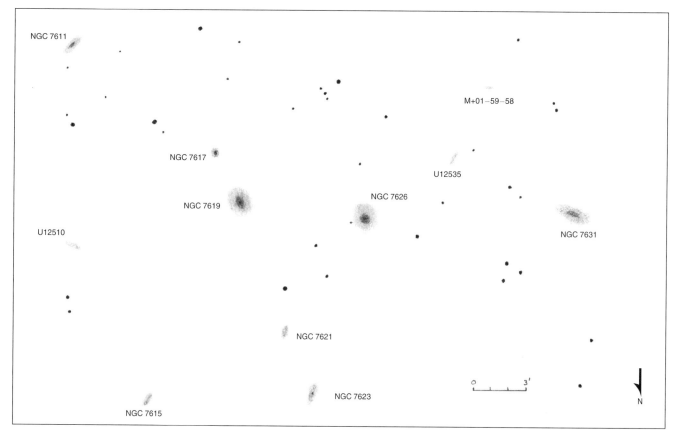

Figure 22-10. *The Pegasus I Galaxy Cluster sketched by G. R. Kepple at 150x with a 17.5, f4.5 Newtonian reflector. South is up.*

NGC 7619 H439² Galaxy Type E
ø2.8′ × 2.5′, m11.1v, SB 13.0   23ʰ20.2ᵐ +08°12′
Finder Chart 22-8, Figure 22-10   ★★★

8/10″ Scopes–100x: NGC 7619 is near the center of the 250 million light year distant Pegasus I Galaxy Cluster. It forms an equilateral triangle with the galaxy NGC 7626 lying 7′ east and a 10th magnitude star to its north. NGC 7619 is the brightest Pegasus I member: it has a fairly bright but diffuse 1′ diameter halo around a stellar nucleus.

12/14″ Scopes–125x: NGC 7619 displays a bright stellar nucleus surrounded by a uniform halo elongated 1.25′ × 1′ NE–SW.

NGC 7623 Galaxy Type SA0°:
ø1.6′ × 1.0′, m12.8v, SB 13.2   23ʰ20.5ᵐ +08°24′
Finder Chart 22-8, Figure 22-10   ★★

12/14″ Scopes–125x: NGC 7623, located 6′ north of a 10th magnitude star, is the third brightest member of the Pegasus I Galaxy Cluster. Its fairly obvious 30″ diameter halo is slightly elongated N–S and surrounds a stellar nucleus.

NGC 7625 H250² Galaxy Type SA(rs)a pec
ø1.4′ × 1.4′, m12.1v, SB 12.7   23ʰ20.5ᵐ +17°14′
Finder Chart 22-7   ★★

8/10″ Scopes–100x: NGC 7625 is 7′ WSW of a 6.5 magnitude star and has a bright but small 1′ diameter halo surrounding a tiny core.

12/14″ Scopes–125x: This galaxy has a bright, circular 1.5′ diameter halo around a small core containing a faint stellar nucleus.

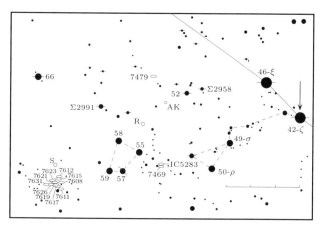

Finder Chart 22-8. 42–ζ Peg: 22ʰ41.5ᵐ +10°50′

Figure 22-11. *PK104-29.1 is a very faint, large planetary nebula requiring larger instruments for a good view. Martin C. Germano made this 90 minute exposure on hypered 2415 Kodak Tech Pan film with a 8", f5 Newtonian.*

Figure 22-12. *In large telescopes NGC 7741 can be seen to have a prominent central bar surrounded by very faint spiral arms. Alexander Brownlee made this 90 minute exposure on hypered 2415 film with a 16", f5 Newtonian.*

NGC 7626 H440² Galaxy Type E pec:
ø2.4′ × 1.9′, m11.1v, SB 12.6   23ʰ20.7ᵐ +08°13′
Finder Chart 22-8, Figure 22-10   ★★★

8/10″ Scopes–100x: NGC 7626, the second brightest member of the Pegasus I Galaxy Cluster, is nearly the twin of NGC 7619, 7′ to the west, but appears slightly fainter and has a less prominent stellar nucleus.

12/14″ Scopes–125x: NGC 7626 has a 1′ diameter halo slightly elongated NNE-SSW with a faint stellar nucleus. A 14th magnitude star lies less than 1′ west. 8′ ENE is a 1′ long triangle of one 11.5 and two 12th magnitude stars.

NGC 7631 Galaxy Type SA(r)b:
ø1.6′ × 0.7′, m13.1v, SB 13.1   23ʰ21.4ᵐ +08°13′
Finder Chart 22-8, Figure 22-10   ★

12/14″ Scopes–125x: NGC 7631 is the easternmost member of the Pegasus I Galaxy Cluster. This much-tilted spiral galaxy appears as a very faint uniform glow elongated 1.25′ × 0.5′ ENE–WSW.

NGC 7673 Galaxy Type (R′)SAc? pec
ø1.6′ × 1.5′, m12.8v, SB 13.6   23ʰ27.7ᵐ +23°35′
Finder Chart 22-9, Figure 22-13   ★★

12/14″ Scopes–125x: NGC 7673, located 3/4° ENE of Nu (ν) = 68 Pegasi, has a prominent but tiny core with a stellar nucleus surrounded by a much fainter halo elongated 1′ × 0.5′ N–S. A few minutes east of NGC 7673 are two stars of magnitudes 8.0 and 8.5 that point toward but just slightly north of the galaxy. 3′ south of these stars is galaxy NGC 7677, which has a tiny core surrounded by a very faint diffuse 30″ diameter halo. Two 12th magnitude stars lie 1′ NW of NGC 7677, the northernmost with a 13th magnitude companion to its SW.

NGC 7678 H226² Galaxy Type SAB(rs)c I-II
ø2.3′ × 1.7′, m11.8v, SB 13.2   23ʰ28.5ᵐ +22°25′
Finder Chart 22-9, Figure 22-14   ★★★

12/14″ Scopes–125x: NGC 7678 is within, and very near the southern vertex, of a triangle of 11th magnitude stars. It is a faint, diffuse, oval glow elongated 1.5′ × 1′ NE–SW with a small, faint core.

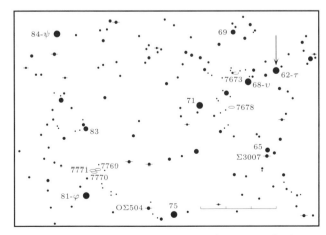

Finder Chart 22-9. 62–τ Peg: 23ʰ20.6ᵐ +23°44′

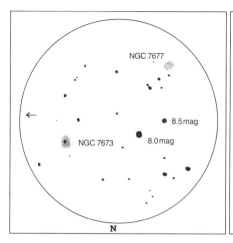

Figure 22-13. NGC 7673 & NGC 7677
12.5″, f5-175x, by G. R. Kepple

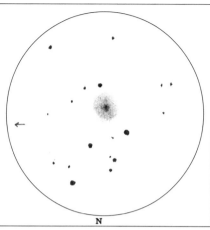

Figure 22-14. NGC 7678
13″, f5.6-200x, by Steve Coe

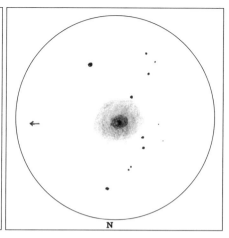

Figure 22-15. NGC 7743
17.5″, f4.5-225x, by G. R. Kepple

PK104-29.1 Jones 1 Planetary Nebula Type 3b
ø332″, m12.1v, CS 16.13v   23^{h}35.9^m +30°28′
Finder Chart 22-11, Figure 22-11   ★★

16/18″ Scopes–125x: Jones 1, 18′ SW of a 6th magnitude star, is very faint but very large. Its 5′ diameter disk is somewhat brighter on the northern and southern edges, like polar caps. At least a dozen stars are scattered across the nebula's glow, including four in an L-shaped asterism at the planetary's center.

NGC 7741 H208² Galaxy Type SB(s)cd II-II
ø4.0′ × 2.7′, m11.3v, SB 13.7   23^{h}43.9^m +26°05′
Finder Chart 22-11, Figure 22-12   ★★★

12/14″ Scopes–125x: This very faint galaxy is just SSE of a 25″ wide NNW–SSE pair of 9th and 12th magnitude stars. Four faint stars are scattered around this double. The galaxy's halo is diffuse, seems elongated E–W, and is slightly brighter at its center. A 10.5 magnitude star lies 3′ ESE.

16/18″ Scopes–150x: In larger telescopes, the halo of NGC 7741 is faint, appears elongated 3′ × 2′ E–W, and contains a central bar extended ESE–WNW. Averted vision brings out the very faint spiral arms. At 175x the outer halo enveloping the spiral arms is elongated N–S. This is the type of object for which a clear, steady atmosphere is absolutely indispensable: with poorer seeing the galaxy is merely a round circular glow without inner detail.

NGC 7742 H255² Galaxy Type SA(r)b
ø2.0′ × 1.8′, m11.6v, SB 12.9   23^{h}44.3^m +10°46′
Finder Chart 22-10   ★★★

8/10″ Scopes–100x: NGC 7742, half a degree NNE of 77 Pegasi (m5.1), is a faint, round 1′ diameter glow 1′ west of an 11th magnitude star. The center brightens moderately.

12/14″ Scopes–125x: NGC 7742 has a fairly bright, round 1.5′ diameter halo that gradually brightens to a stellar nucleus. A 13th magnitude star is 2.5′ SE.

NGC 7743 H256² Galaxy Type (R)SB(s)0+
ø2.3′ × 1.9′, m11.5v, SB 12.9   23^{h}44.4^m +09°56′
Finder Chart 22-10, Figure 22-15   ★★★

12/14″ Scopes–125x: NGC 7743, located 2.75′ NE of a 10th magnitude star, has a fairly bright, round 1′ halo brightening to a broad core. A 12th magnitude star lies 1′ SSE.

16/18″ Scopes–125x: NGC 7743 has a prominent core with a stellar nucleus surrounded by a 1.5′ diameter halo which fades abruptly at its periphery. The field is rather well populated with bright stars, the majority to the north of the galaxy.

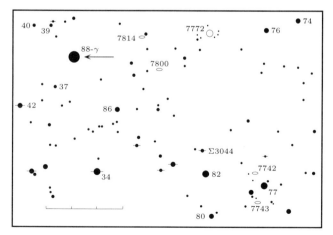

Finder Chart 22-10. 88–γ Peg: 00^{h}13.2^m +15°11′

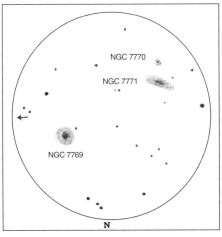
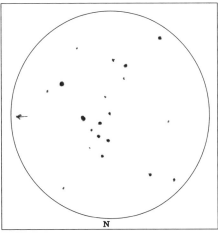
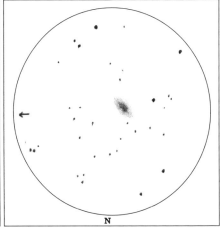

Figure 22-16. NGC 7769, 7770, & 7771
17.5″, f4.5–225x, by G. R. Kepple

Figure 22-17. NGC 7772
12.5″, f5–200x, by G. R. Kepple

Figure 22-18. NGC 7800
17.5″, f4.5–225x, by G. R. Kepple

NGC 7769 H230² Galaxy Type (R)SA(rs)b II
ø1.6′ × 1.6′, m12.0v, SB 12.8   23ʰ51.1ᵐ +20°09′

NGC 7770 Galaxy Type S0/a?
ø0.9′ × 0.8′, m13.8v, SB 13.3   23ʰ51.4ᵐ +20°06′

NGC 7771 Galaxy Type SB(s)a
ø2.3′ × 1.1′, m12.2v, SB 13.1   23ʰ51.4ᵐ +20°07′

Finder Chart 22-9, Figure 22-16   ★★★/★/★★

16/18″ Scopes–150x: NGC 7769 is the brightest and westernmost member of a nice trio of galaxies with NGC 7770 and NGC 7771. It has a fairly bright, circular 1′ diameter halo around a slightly brighter core with a very faint stellar nucleus. A 14th magnitude star lies just beyond the galaxy's SE edge. NGC 7771, located 5.5′ to the ESE of NGC 7769, appears just slightly fainter and is elongated: its 1.5′ × 0.75′ ENE–WSW halo contains a faint extended 1′ × 0.25′ core. NGC 7770, just south of NGC 7771, is the smallest and faintest of the three galaxies: merely a very faint, round smudge surrounding a stellar nucleus.

NGC 7772 Open Cluster 10★
ø5.0′   23ʰ51.8ᵐ +16°15′

Finder Chart 22-10, Figure 22-17   ★★

8/10″ Scopes–100x: The star-poor open cluster NGC 7772 is in a sparse field NE of three 9th magnitude stars. It contains a dozen magnitude 13–14 members in a 5′ area. Its two brightest stars are in its SW sector and point NNE at a triangular knot of faint stars.

NGC 7800 Galaxy Type Im?
ø2.1′ × 1.5′, m12.6v, SB 13.7   23ʰ59.6ᵐ +14°49′

Finder Chart 22-10, Figure 22-18   ★★

16/18″ Scopes–150x: NGC 7800 is moderately faint, elongated 2′ × 1.5′ NE–SW, and has a slightly brighter center. To its north is a curved row of three 10th to 11th magnitude stars, and to its south a triangle of fainter stars. A 13th magnitude star is 1.5′ SE. Just NNE is a "T" of 14th magnitude stars, the closest nearly touching the galaxy's edge.

NGC 7814 H240² Galaxy Type SA(s)ab: sp
ø6.0′ × 2.5′, m10.6v, SB 13.4   00ʰ03.3ᵐ +16°09′

Finder Chart 22-10, Figure 22-19   ★★★★

8/10″ Scopes–100x: NGC 7814, a nice, bright target for small telescopes, is 8′ SE of a 7th magnitude star. It is slightly elongated about 2.75′ × 2′ NW–SE and slightly brighter in its center. A wide pair of 9th magnitude stars lies 8′ SW.

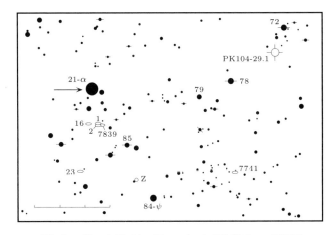

Finder Chart 22-11. 21–α And: 00ʰ08.2ᵐ +29°05′

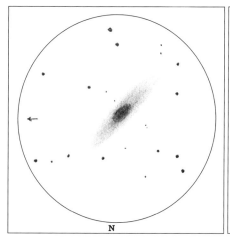

Figure 22-19. NGC 7814
17.5″, f4.5–225x, by G. R. Kepple

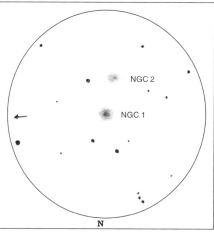

Figure 22-20. NGC 1 & NGC 2
17.5″, f4.5–225x, by G. R. Kepple

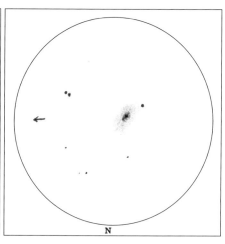

Figure 22-21. NGC 23
17.5″, f4.5–225x, by G. R. Kepple

12/14″ Scopes–125x: The halo is bright, elongated 3′ × 2′ NW–SE, and contains a broad but well-condensed core.

16/18″ Scopes–150x: NGC 7814 is a nice surprise after all the other faint galaxies in Pegasus. Medium power shows a bright halo elongated 5′ × 2′ NW–SE with a 1′ long oval.

NGC 7839 Galaxy Type ?
ø –   00ʰ07.0ᵐ +27°38′

NGC 1 Galaxy Type SA(s)b: II-III
ø1.6′ × 1.1′, m12.9v, SB 13.4   00ʰ07.3ᵐ +27°43′

NGC 2 Galaxy Type Sab
ø1.0′ × 0.6′, m14.2v, SB 13.5   00ʰ07.3ᵐ +27°41′
Finder Chart 22-11, Figure 22-20   ★★/★/★

12/14″ Scopes–125x: This trio of faint galaxies is 1/2° east of a 5th magnitude star. NGC 1, the northernmost of the three, is very faint, though its round 1′ diameter halo is of uniform surface brightness. NGC 2, just south and very slightly east of NGC 1, is only a diffuse, round, nearly stellar glow. NGC 7839, 4′ west of NGC 1, has a bright center and is much more obvious than NGC 2.

16/18″ Scopes–150x: NGC 1 is just south of the easternmost in a triangle of 10th magnitude stars. It is faint but obvious, elongated 1.25′ × 1.0′ E–W, and slightly brighter in the center. NGC 2, 2′ south, is a very faint, round glow. NGC 7839, 4′ west, is much brighter with a prominent core.

NGC 16 H15⁴ Galaxy Type SAB0–
ø1.8′ × 1.0′, m12.0v, SB 12.5   00ʰ09.1ᵐ +27°44′
Finder Chart 22-11   ★★

8/10″ Scopes–100x: This galaxy's 1′ halo is slightly elongated N–S and slightly brighter in the center.

12/14″ Scopes–125x: NGC 16 has a fairly well concentrated halo elongated 1.5′ × 0.75′ NNE–SSW and moderately brighter in the center. A wide pair of 11th and 12th magnitude stars point at the galaxy from the west.

NGC 23 H147³ Galaxy Type SB(s)a
ø2.0′ × 1.5′, m12.0v, SB 13.1   00ʰ09.9ᵐ +25°55′
Finder Chart 22-11, Figure 22-21   ★★

12/14″ Scopes–125x: This galaxy is an obvious but diffuse circular glow with a bright, nearly stellar nucleus. Immediately SE of the nucleus is a star that is nearly one magnitude fainter than the nucleus. A wide double of 13th magnitude stars lies 1.5′ WSW. Galaxy NGC 26 is 10′ SE.

16/18″ Scopes–150x: NGC 23 has a fairly bright halo about 1′ in diameter somewhat elongated NNW–SSE. The stellar nucleus appears brighter than the 12.5 magnitude star 20″ to its SE. The galaxy is 7′ SE of a 9th magnitude star flanked by two 10th magnitude stars.

Chapter 23

Perseus, the Hero

23.1 Overview

Perseus was the slayer of Medusa, the snake-haired Gorgon, the mere sight of which would turn any mortal to stone. She seemed to have the perfect defense against attack: how can you approach, let alone slay, what you don't even dare look at? Perseus, however, got around this problem by looking not directly at *her*, but only her *reflection* in his brass shield. In the sky Perseus is shown holding the Medusa's Head, marked by the asterism of Beta (β), Pi (π), Rho (ρ), and Omega (ω) Persei, in one hand and his Scimitar, marked by the famous Perseus Double Cluster, over his head in the other. Though the variable star Beta Persei, Algol, was one of the Medusa's eyes in the original ancient Greek conception of the constellation Perseus, no evidence has survived that Greek astronomers had noticed the star's light changes.

Immediately after his exploit with the Medusa, as he was riding the winged horse Pegasus which had leaped up from the Gorgon's blood, Perseus came upon the predicament of Andromeda, chained to the rocks by the sea as a sacrifice to Cetus the Sea Monster. She was thus exposed because the sea god Poseidon was angry with the claim of her mother Cassiopeia to be more beautiful than the sea nymphs. All the figures in this story — Perseus, the Medusa, Pegasus, Cetus, Andromeda, Cassiopeia, and Cepheus (Andromeda's father) — are enconstellated in the same region of the heavens.

Perseus lies in the Milky Way and therefore is rich in such typical Milky Way objects as open clusters and diffuse nebulae. It does, however, also have quite a few interesting galaxies, which is rather unusual for a Milky Way group. Perseus' two Messier objects, the open cluster M34 and the planetary nebula M76, though fine sights, are far from the most outstanding things in the constellation. That honor goes to the justifiably famous Perseus Double Cluster, NGC 869 & NGC 884, which was known to the ancient Greeks but rather inexplicably omitted by Messier from his catalogue. What the Double Cluster is to telescopes the Alpha Persei Moving Group, one of the nearest open clusters to the Solar System, is to binoculars. Because it was included neither in the

---

Perseus: PUR-see-us
Genitive: Persei, PUR-see-eye
Abbrevation: PER
Culmination: 9 pm–Dec. 22, midnight–Nov. 7
Area: 615 square degrees
Showpieces: 13-ϑ Per, 15-η Per, 26-β Per (Algol), M34 (NGC 1039), M76 (NGC 650-1), NGC 869 & NGC 884 (Double Cluster)
Binocular Objects: 26-β Per(Algol), 57 Per, IC 348, Mel 20, M34 (NGC 1039), NGC 744, NGC 869 & NGC 884 (Double Cluster), NGC 957, NGC 1023, NGC 1245, NGC 1342, NGC 1444, NGC 1528, NGC 1545, NGC 1582, Tr 2

Messier list nor the NGC, the Alpha Persei group is often neglected by observers.

23.2 Interesting Stars

Σ268 Double Star Spec. A2
m6.8, 8.1; Sep. 2.7″; P.A. 129° $02^h29.4^m$ +55°32′
Finder Chart 23-3   ★★★★

4/6″ Scopes–100x: Struve 268 is a close pair of whitish and bluish stars. Struve 270 lies 10′ east.

Σ270 Double Star Spec. F5
m6.8, 8.1; Sep. 2.7″; P.A. 129° $02^h30.8^m$ +55°33′
Finder Chart 23-3   ★★★★

4/6″ Scopes–100x: Struve 270 is a wide pair of yellowish-white and bluish stars. Struve 268 lies 10′ west.

Theta (ϑ) = 13 Persei (Σ 296) Double Star   Spec. F8
AB: m4.1, 9.9; Sep. 20.0″; P.A. 305° $02^h44.2^m$ +49°14′
Constellation Chart 23-1   ★★★★★

4/6″ Scopes–100x: Theta Persei has a bright yellow primary with a blue companion. Struve 304, a wide pair of white and blue stars lies 40′ east.

Perseus, the Hero

Constellation Chart 23-1

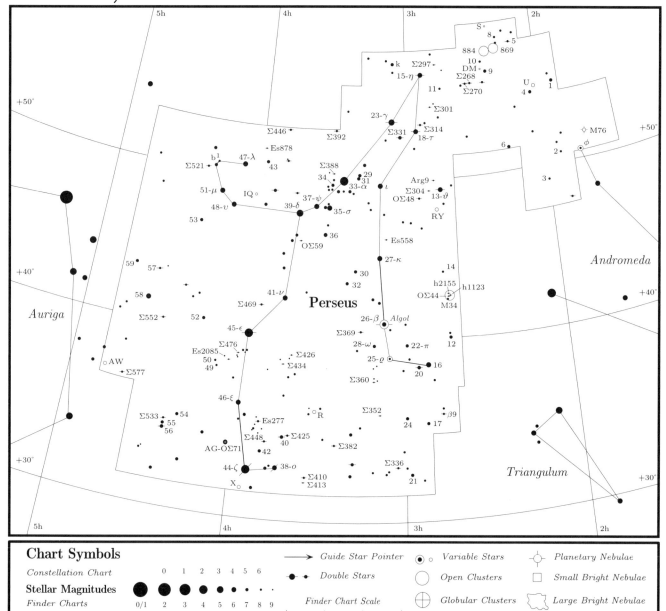

Table 23-1 Selected Variable Stars in Perseus

Name	HD No.	Type	Max.	Min.	Period (Days)	F*	Spec. Type	R.A. (2000)	Dec.	Finder Chart No. & Notes
U Per	12025	M	7.4	12.3	321	0.46	M5–M7	$01^h59.6^m$	+54°49′	23-3
S Per	14528	Src	7.9	11.5			M3–M6	$02^h22.9^m$	+58 35	23-3
DM Per	14871	DM	7.7	8.4	2.27	0.16	B6	26.0	+56 06	23-3
RY Per	17034	EA	8.5	10.7	6.86	0.12	B8+G8	45.7	+48 09	23-1
25–ϱ Per	19058	SRb	3.3	4.0	50		M4	$03^h05.2^m$	+48 50	23-4
26–β Per	19356	EA(x)	2.12	3.40	2.86	0.14	B8+G5	08.2	+40 57	23-4 Algol
X Per	24534	gC(x)	6.07	7.00			09.5	55.4	+31 03	23-1
IQ Per	24909	EA	7.72	8.27	1.74	0.12	B8+A0	59.7	+48 09	23-5
AG Per	25833	EA	6.71	7.00	2.02	0.12	B4+B5	$04^h06.9^m$	+33 27	23-7
AW Per	30282	Cd	7.04	7.85	6.46	0.25	F6–G0	47.8	+36 43	23-1

F* = The fraction of period taken up by the star's rise from min. to max. brightness, or the period spent in eclipse.

Table 23-2 Selected Double Stars in Perseus

Name	ADS No.	Pair	M1	M2	Sep."	P.A.°	Spec	R.A. (2000)	Dec	Finder Chart No. & Notes
Σ268	1878		6.8	8.1	2.7	129	A2	02^{h}29.4^m	+55°32'	23-3 Whitish & bluish
Σ270	1901	AB	7.4	9.2	21.2	303	F5	30.8	+55 33	23-3 Yellowish & bluish
	1901	AC		10.6	38.3	338				
	1901	AD		12.5	42.4	270				
h1123	2048		8.0	8.0	20.0	248	A0 A0	42.0	+42 47	23-4 In M34. White pair
OΣ44	2052		8.4	9.1	1.4	55	B9	42.2	+42 42	23-4 In M34. Both stars white
h2155	2060		8.0	10.4	17.1	321	B9	42.7	+42 48	23-4 In M34. White & bluish
13–ϑ Per	2081	AB	4.1	9.9	w20.0	*305	F8	44.2	+49 14	23-1 Fine gold & blue pair
Σ297	2094	AB	8.5	8.8	15.8	277	A0	45.4	+56 34	23-3 A-B pair white
	2094	AC		10.6	28.4					
β9	2117	AB	6.4	8.5	1.7	188	F2	47.1	+35 33	23-1
Arg 9	2112		8.8	8.8	3.1	148	F8	47.1	+50 07	23-1 Yellowish-white pair
Σ301	2115		7.8	8.8	8.1	16	A0	47.6	+53 57	23-3 Whitish & bluish
Σ304	2139		7.6	9.1	25.2	289	B9	48.8	+49 11	23-1 White & blue
15–η Per	2157	AB	3.8	8.5	28.3	300	K0	50.7	+55 54	23-3 Orange & blue
	2157	AC		9.8	66.6	268				
	2157	CD		10.3	5.2	114				
Σ314	2185	ABxC	6.5	7.1	1.6	308	B9	52.9	+53 00	23-3
OΣ48	2192		6.5	10.6	6.7	317	K0	53.4	+48 34	23-5
20 Per	2200	AB	5.6	6.7	0.2	*0	F0	53.7	+38 20	23-4 Yellowish & white
	2200	ABxC		10.1	14.1	237				
18–τ Per	2202	AB	4.0	10.6	51.7	106	G0	54.3	+52 46	23-3 Yellowish & white
	2202	BC		11.7	3.5	87				
Σ331	2270		5.3	6.7	12.1	85	B5	03^{h}00.9^m	+52 21	23-5 White pair
Σ336	2286		6.9	8.4	8.4	8	G5	01.5	+32 25	23-6 Yellowish & pale blue
Es558	2341		7.8	9.7	8.4	358	B9	06.7	+45 45	23-1
Σ352	2364		7.7	9.8	3.5	2	A0	08.8	+35 28	23-6
Σ360	2390		8.1	8.3	w2.6	*125	G0	12.1	+37 12	23-6
Σ369	2443		6.7	8.0	3.5	28	A0	17.2	+40 29	23-4 White & blue
Σ382	2514	AB	5.6	9.1	4.4	153	A0	24.5	+33 32	23-6
Σ388	2548		8.1	9.1	2.8	210	F0	28.7	+50 26	23-5 Yellowish pair
Σ392	2566		7.4	9.6	25.8	347	K0	30.3	+52 54	23-5 Yellowish-orange & bluish
Σ410	2622		6.6	10.6	5.4	208	F0	35.0	+32 01	23-6
Σ413	2625		8.6	8.6	2.5	128	F0	35.4	+33 41	23-6
Σ425	2668		7.6	7.6	1.8	76	F5	40.1	+34 07	23-6 Equal yellow stars
OΣ59	2669		8.1	8.4	2.7	352	G5	40.7	+46 02	23-1
Σ426	2677	AB	7.8	9.3	19.8	341	A3	40.8	+39 07	23-7
40 Per	2699		5.0	9.5	20.0	238	B2	42.4	+33 58	23-6
Σ434	2717		7.8	8.6	31.0	85	K5	44.0	+38 22	23-7 Deep yellow & blue
38–o Per	2726		3.8	8.3	1.0	37	B1	44.3	+32 17	23-7
Σ448	2772		6.7	9.2	3.2	16	B3	47.9	+33 36	23-7
Σ446	2783	AB	6.9	9.1	8.6	253	B0	49.5	+52 39	23-8 In NGC 1444
	2783	AC		12.0	12.1	39				
	2783	AD		10.3	66.5	336				
	2783	DE		10.7	2.6	232				
Es277	2794	AB	6.8	9.8	30.2	142	F0	50.2	+34 49	23-7
44–ζ Per	2843	AB	2.9	9.5	12.9	208	B1	54.1	+31 53	23-7
Σ469	2884		6.8	10.3	9.1	148	A2	57.3	+41 53	23-9
45–ε Per	2888	AB	2.9	8.1	8.8	10	B1	57.9	+40 01	23-1 Blue-white, blue
Es878	2896		7.7	9.9	12.3	224	K2	58.9	+51 30	23-8
Σ476	2932		7.9	9.1	21.9	287	K2	04^{h}01.6^m	+38 40	23-7
Es2085	2956		7.8	9.9	4.4	266	A0	03.8	+37 58	23-7
OΣ71(AG)	2990	AB	6.9	8.9	0.8	220	B3	06.9	+33 27	23-7
Σ521	3141		7.5	9.6	2.1	257	G0	21.8	+50 02	23-8
Σ533	3185	AB	7.2	8.7	19.6	61	B9	24.4	+34 19	23-7
56 Per	3188		5.9	8.7	4.2	22	F5	24.6	+33 58	23-7 Lovely yellow pair
Σ552	3273		7.0	7.2	9.0	114	B8	31.4	+40 01	23-9
57 Per		AD	6.1	6.8	116.2	198	F0 F0	33.4	+43 04	23-9
Σ577	3390		8.6	8.6	c1.0	*4	F8	42.3	+37 30	23-1

Footnotes: *= Year 2000, a = Near apogee, c = Closing, w = Widening. Finder Chart No: All stars listed in the tables are plotted in the large Constellation Chart, but when a star appears in a Finder Chart, this number is listed. Notes: When colors are subtle, the suffix -ish is used, e.g. *bluish*.

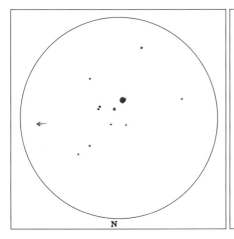

Figure 23-1. Eta Persei 15-η
12.5", f5-150x, by G. R. Kepple

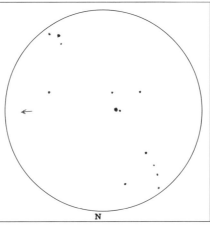

Figure 23-2. Σ331
4", f10-40x, by Mark T. Stauffer

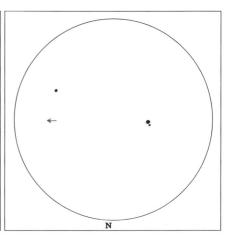

Figure 23-3. Σ369
4", f10-65x, by Mark T. Stauffer

Eta (η) = 15 Persei (Σ 307) Double Star   Spec. K0
AB: m3.8, 8.5; Sep. 28.3"; P.A. 300° 02ʰ50.7ᵐ +55°54'
Finder Chart 23-3, Figure 23-1   Miram   ★★★★★

4/6" Scopes–100x: Eta Persei is a gorgeous color-contrast double of golden and blue stars. It is reminiscent of the famous Albireo (Beta Cygni). 66" almost due west of the primary is a close double of magnitude 10 and 10.5 stars that seems to be part of the system. A dozen faint stars form a small cluster around the Eta Persei multiple.

Σ331 Double Star   Spec. B5
m5.3, 6.7; Sep. 12.1"; P.A. 85°   03ʰ00.9ᵐ +52°21'
Finder Chart 23-5, Figure 23-2   ★★★★

4/6" Scopes–100x: Struve 331 is a fine white and bluish-white duo.

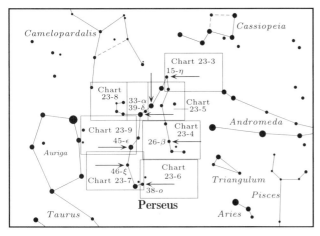

Master Finder Chart 23-2. Perseus Chart Areas
Guide stars indicated by arrows.

Beta (β) = 26 Persei Eclipsing Binary Spec. B8 + G8
m2.12 to 3.40 in 2.87 days   03ʰ08.2ᵐ +40°57'
Finder Chart 23-4   Algol   ★★★★★

Algol, the "Demon Star," is the most famous of all the eclipsing variable stars. It is a fine object for amateurs with or without optical aid. The star goes through a 10 hour eclipse every 2.86739 days — that is, every 2 days, 20 hours, 48 minutes, and 56 seconds. The star's five-hour fall from magnitude 2.1 to magnitude 3.4 frequently can be observed in the course of a single night. Sometimes the whole ten-hour eclipse — from maximum to minimum and back to maximum — will occur in one night.

John Goodriche timed Algol's period accurately in 1782 and suggested that a fainter companion was at least partially eclipsing the brighter star. The primary is a white B8 main sequence star some 100 times more luminous than the Sun, its companion a late G or early K type star, only two or three times brighter than the Sun. The components are quite close, only six million miles center to center. From our line of sight the eclipse is not total: the secondary covers only about 79% of the primary's disk. Midway through the period a slight secondary minimum occurs as the brighter star passes in front of its fainter companion. A third companion exists, and a fourth component is suspected. The Algol system is about 95 light years away.

Σ369 Double Star   Spec. A0
m6.7, 8.0; Sep. 3.5"; P.A. 28°   03ʰ17.2ᵐ +40°29'
Finder Chart 23-4, Figure 23-3   ★★★★

4/6" Scopes–100x: Struve 369 is a fine double of yellowish-white and pale blue stars.

Σ392   Double Star   Spec. K0
m7.4, 9.6; Sep. 25.8″; P.A. 347°
03ʰ30.3ᵐ +52°54′
Finder Chart 23-5   ★★★★

4/6″ Scopes–100x: Struve 392, easily split in small telescopes, is a wide double of yellow and pale blue stars.

Zeta (ζ) = 44 Persei Multiple Star   Spec. B1
AB: m2.9, 9.5; Sep.12.9″; P.A. 208°
03ʰ54.1ᵐ+31°53′
Finder Chart 23-7   ★★★★

Zeta is a bright, blue-white star with three faint companions: a 9.5 magnitude star 12.9″ away, an 11th magnitude star 33″ away, and a 9.5 magnitude star 94″ distant. The closest of the three shares the primary's proper motion.

Zeta is the brightest member of the Zeta Persei Moving Group, an association of young, hot O and B stars moving outward from their common point of formation. The group's rate of expansion coupled with its size suggests that its stars were born only about one million years ago. Other members of the group include Omicron, Xi, 40, and the variables X and AG Persei. The Zeta Persei association is around 1,300 light years distant; so the absolute magnitude of Zeta is −6.2, a luminosity of 25,000 Suns.

56 Persei   Multiple Star   Spec. F5
m5.9, 8.7; Sep. 4.2″; P.A. 22°   04ʰ24.6ᵐ +33°58′
Finder Chart 23-7   ★★★★

4/6″ Scopes–100x: 56 Persei has a fine golden primary and a yellowish secondary.

23.3 Deep-Sky Objects

NGC 650-1 Messier 76   Planetary Nebula Type 3+6
⌀65″, m10.1v, CS 15.87v   01ʰ42.4ᵐ +51°34′
Finder Chart 23-3, Figure 23-4   ★★★★★

The Little Dumbbell

M76 was discovered by P. Mechain in September, 1780, and Messier rediscovered it some six weeks later. It is called the "Barbell" or the "Little Dumbbell" Nebula because of its resemblance to Messier 27 in Vulpecula. William Herschel assigned it two numbers, and J.L.E. Dreyer agreed by giving its separate lobes different NGC numbers: 650 and 651.

4/6″ Scopes–100x: The Little Dumbbell Nebula is 14′ west of a 7th magnitude stars and appears as a small

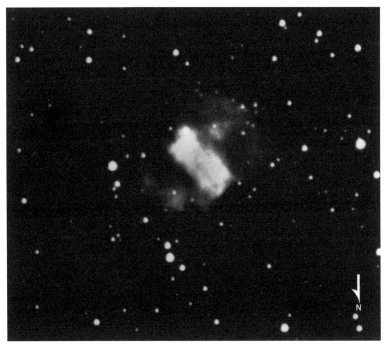

Figure 23-4. *Messier 76 (NGC 650-1), The Little Dumbbell, is an interesting peanut-shaped planetary nebula with two connecting nodules. Martin C. Germano made this 50 minute exposure on hypered Kodak 2415 Tech Pan film with a 14.5″, f5 Newtonian reflector at prime focus.*

peanut-shaped nebulosity elongated NE-SW. Averted vision shows two the distinct nodules.

12/14″ Scopes–125x: M76 lies within a small right triangle composed of a magnitude 9.5 to the east of the nebula and two magnitude 10–10.5 stars to its west and NW. It appears as a pair of bright, small NE-SW glowing disks in contact. The SW nodule is the brighter: it protrudes slightly beyond the

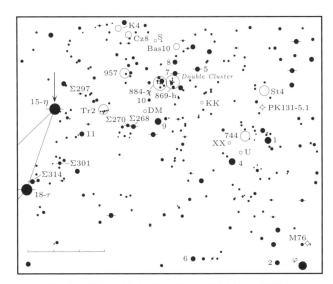

Finder Chart 23-3. 15-η Per: 02ʰ50.7ᵐ +55°54′

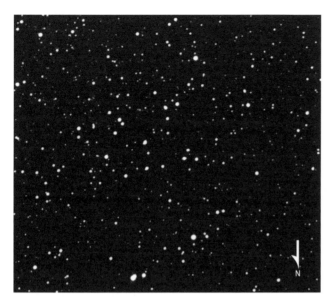

Figure 23-5. *Stock 4 is a large, loose, irregular cluster that does not stand out well from the surrounding star field. Martin C. Germano made this 37 minute exposure on 2415 film with a 14.5", f5 Newtonian reflector at prime focus.*

Figure 23-6. *NGC 744 is an irregularly scattered open cluster of forty stars in clumpings and short chains. Lee C. Coombs made this 10 minute exposure on 103a-0 film with a 10", f5 Newtonian reflector at prime focus.*

triangle's southern side and contains a 13th magnitude star near one edge. The fainter NE lobe has an extended halo at the threshold of visibility.

16/18" Scopes–150x: The Little Dumbbell is a remarkable object with or without a filter. Most planetary nebulae have circular disks; but Messier 76 exhibits two bright nodules with a fairly bright connecting bridge, the whole surrounded by a much fainter, diffuse outer shell. The outer halo, even in a large telescope, requires very good viewing conditions to be seen without a filter; but an O-III filter enhances the view significantly, showing several dark lanes on the east and west sides of the central area and a faint but distinct outer envelope.

Stock 4   Open Cluster 15★   Tr Type III 1 p
⌀20′, m –, Br★ 11.0p $01^h52.8^m$ +57°04′
Finder Chart 23-3, Figure 23-5   ★★★

12/14" Scopes–125x: This cluster is moderately faint, loose, and irregular, containing only 25 stars in a 20′ area. Most of the group's members are 11th magnitude and fainter, but three 9th magnitude stars are near its north, WSW, and SSE edges.

16/18" Scopes–150x: Stock 4 is loose and scattered, composed of fifty 11th magnitude and fainter stars in a 20′ area. The stars lie along irregular chains that meander completely around the cluster, enclosing a peanut-shaped N–S starless void.

PK131-5.1   BV-3   Planetary Nebula
⌀30.0″, m14.2v $01^h53.7^m$ +56°25′
Finder Chart 23-3, Figure 23-11   ★★★

12/14" Scopes–125x: PK131-5.1 appears very faint, small, and round. A star 40″ east is at the eastern vertex of a 10′ long star-triangle.

16/18" Scopes–150x: This planetary nebula has a round, 30″ disk easily visible with direct vision.

NGC 744   Open Cluster 20★   Tr Type IV 2 p
⌀11″, m7.9v, Br★ 10.44v $01^h58.4^m$ +55°29′
Finder Chart 23-3, Figure 23-6   ★★★

8/10" Scopes–100x: NGC 744, located 7′ SSW of an 8th magnitude star, is a coarse cluster of twenty 10th magnitude and fainter stars scattered over a 7′ area.

12/14" Scopes–125x: NGC 744 is a loose, irregular scattering of forty 10th magnitude and fainter stars in numerous clumps and short star chains. In the northern section the brighter members lie along a curved chain concave to the north. On the NW edge is a triangle of wide double stars. Outside the SE edge, separated from the main group by a starless gap, is a 14″ wide pair of magnitude 11.5 stars.

Basel 10   Open Cluster 12★   Tr Type II 1 p
⌀2′, m9.9v, Br★ 10.23v $02^h18.8^m$ +58°19′
Finder Chart 23-3, Figure 23-12   ★★★

12/14" Scopes–125x: Basel 10 is a faint, neat little group a degree north of the Double Cluster. At low

Figure 23-7. *The Double Cluster of NGC 869 and NGC 884 (left to right) is a splendid sight at low power in small telescopes. NGC 884 is 300 light years further away than its companion. Lee C. Coombs made this 10 minute exposure on Kodak 103a-0 Spectroscopic film with a 10", f5 Newtonian reflector.*

power averted vision is needed to detect the half dozen fainter stars sprinkled around the cluster's two 10.5–11 lucidae. 250x reveals two rows of stars aligned NE-SW. The northern row is longer and contains a knot of four stars. The brighter stars near the center, distributed in a triangle, all are wide doubles.

NGC 869 H33⁶ Open Cluster 200★ Tr Type I 3 r
⌀29′, m5.3v, Br★ 6.55v   02ʰ19.0ᵐ +57°09′

NGC 884 H34⁶ Open Cluster 115★ Tr Type II 2 p
⌀29′, m6.1v, Br★ 8.05v   02ʰ22.4ᵐ +57°07′

Finder Chart 23-3, Figure 23-7   ★★★★★/★★★★★

The Double Cluster

The Double Cluster was known as a nebulous "star" to the ancient Greeks and the Babylonians before them. In the classical constellation figure of Perseus it marked the Scimitar with which the Hero decapitated the Medusa. In Babylonia the Double Cluster was the NW end of a celestial Scimitar that occupied most of the stars of the later Greek Perseus. On older star atlases NGC 869 is sometimes designated "h" and NGC 884 "χ." For some reason Messier did not include the Double Cluster in his famous catalogue, though he certainly was aware of its existence.

The two components of the Double Cluster are in fact rather different from one another. The brightest stars of NGC 869 are blue early-B giants and supergiants, whereas the brightest stars of NGC 884 are mostly white late-B/early-A, or red M-type, supergiants. This implies that NGC 869 is significantly younger than NGC 884, 6 million as opposed to 14 million years old. The two groups also seem to be at slightly different distances: recent estimates place NGC 869 7,200 light years away and NGC 884 7,500 light years distant, thought this difference is less than the uncertainties in the estimates. However, this is not a chance alignment of clusters, for NGC 869 and NGC 884 are the cluster-cores of the vast Perseus OB1 association of young supergiant and blue stars which surrounds them. Therefore the two groups were formed (a few million years apart) from the same huge interstellar cloud of gas and dust. Even binoculars show that NGC 869 and NGC 884 are not well detached from the

Figure 23-8. *NGC 957 is a fine, moderately concentrated open cluster of forty stars. Lee C. Coombs made this 10 minute exposure on 103a-0 film with a 10″, f5 Newtonian reflector at prime focus.*

Figure 23-9 *NGC 1023 is a bright SB0 galaxy with a large, well concentrated core. Harvey Freed, D.D.S. made this 50 minute exposure on 2415 film with a 10″, f6.3 Newtonian reflector at prime focus.*

surrounding rich star field of Perseus OB1, but blend imperceptibly into it. Stars from the association cover an approximately 8° × 6° area of the Milky Way, and include as distant members the blue supergiants 5 and 9 Persei and the red M-type supergiant variables XX and KK Perseus. The true size of the Perseus OB1 is about 1040 × 780 light years. The densest parts of NGC 869 and NGC 884 are around 30′ across, which implies true cluster diameters in excess of 60 light years. It is from the Double Cluster and its surrounding Perseus OB1 association that the Perseus Spiral Arm, the next spiral feature out from our Orion-Cygnus arm, has been named.

4/6″ Scopes–125x: These two splendid clusters are just half a degree apart in the same rich Milky Way star field and require low power, wide field oculars to be seen together. Giant binoculars also provide a spectacular view. Both groups are extremely large, very bright, and very rich. NGC 869 is the more compressed of the two. A relatively starless NW–SE lane separates the main concentration of NGC 869 from a semidetached segment to its NW.

8/10″ Scopes–125x: An exquisite sight! NGC 869 has several hundred 8th to 14th magnitude stars, plus two 6.5 magnitude stars embedded in its main concentration. A dark lane separates this concentration from a much looser and fainter group to the NW. NGC 884 is elongated E–W and contains 150 stars ranging from magnitude 6.5 to magnitude 14. It is larger but not as bright nor as concentrated as its neighbor. NGC 884's brighter stars are concentrated in its SW portion, but the majority of its members are loosely scattered to the north and NE. Near the NE edge is an arc of stars concave to the south.

12/14″ Scopes–125x: These two clusters are a stunning sight in low power, wide field oculars. Each cluster is half a degree in diameter, but outliers merge into the rich star-field that encompasses both groups. NGC 869 is the more compressed of the two, and has over 200 predominantly white and bluish-white members. A relatively star-poor gap cutting NE–SW through the cluster separates a looser concentration to the NW from the denser southeastern core of bright stars. NGC 884 to the east has 175 mostly white and bluish-white stars. However three red supergiants provide a nice color contrast. A dark star-poor gap protrudes from the cluster's SSE edge in toward the center. Another reddish star lies between the two clusters.

Czernik 8 Open Cluster 10★ Tr Type II 3 m
⌀7″, m9.7v, Br★ 9.87v   02^{h}33.0^m +58°44′
Finder Chart 23-3, Figure 23-13   ★★★

12/14″ Scopes–125x: Czernik 8, centered 8′ north of an 8th magnitude star, has eighteen 10th to 13th magnitude members in a roughly trapezoidal 7′ × 5′ area elongated N–S.

NGC 957 Open Cluster 30★ Tr Type III 2 p
⌀11″, m7.6v, Br★ 9.49v   02^{h}33.6^m +57°32′
Finder Chart 23-3, Figure 23-8   ★★★

4/6″ Scopes–50x: This fine cluster appears moderately faint, small, rich and elongated E–W with several dozen stars visible.

8/10″ Scopes–100x: NGC 957 is a fairly concentrated cluster of thirty 11th to 12th magnitude stars spread E–W over a 10′ × 5′ area. The two brightest members mark the group's SW and SSE corners, the latter star a fine, wide double with white and blue components.

12/14″ Scopes–100x: Forty stars are moderately concentrated in a 12′ × 5′ ENE–WSW area. At the SW corner is a magnitude 9.5 star and at the SE corner a 23″ wide double of magnitude 8 and 10 components. The fainter cluster members, magnitude 11–14 objects are spread north from these two bright stars. The cluster tapers west to a point directed at a 7th magnitude field star 7′ away.

Trumpler 2 Open Cluster 20★ Tr Type III 2p
ø20′, m5.9v, Br★ 7.38v   02ʰ37.3ᵐ +55°59′
Finder Chart 23-3, Figure 23-14   ★★★

8/10″ Scopes–75x: Trumpler 2 is a large open cluster spread E–W over a 20′ area. Its brighter stars lie along a jagged E–W chain that crosses the cluster center: the chain includes two 8th magnitude stars, one near the cluster center and the other on its western edge, and seven magnitude 9–9.5 stars, two in a close double near the center. Another two dozen 11th to 14th magnitude stars are sprinkled around on either side of the chain.

12/14″ Scopes–100x: Trumpler 2 is a bright, large, irregular group of forty stars in a 20′ × 10′ E–W area. Its brightest members, three 8th magnitude and six 9–9.5 magnitude stars, form a zigzagging E–W chain. Near the cluster center is a color-contrast pair of a blue 8th magnitude and a gold 9th magnitude star. A faint star chain twists south to a 9th magnitude star on the cluster's southern edge. 6′ north of the cluster is a detached 8th magnitude star.

NGC 1003 H240² Galaxy Type SA(s)cd III
ø6.4′ × 2.8′, m11.4p, SB 14.4   02ʰ39.3ᵐ +40°52′
Finder Chart 23-4   ★★★

12/14″ Scopes–125x: NGC 1003, located 2′ NE of a 10.5 magnitude star, appears moderately faint, elongated 3′ × 1′ E–W, and slightly brighter at its center.

16/18″ Scopes–150x: NGC 1003 has a fairly bright 4′ × 1.25′ E–W halo gradually brighter toward its center. A very faint knot is near the NW edge and a 13th magnitude star on the NE edge.

Figure 23-10. *Messier 34 (NGC 1039) is a large, attractive open cluster of bright stars for small telescopes. Lee C. Coombs made this 10 minute exposure on 103a-0 film with a 10″, f5 Newtonian reflector.*

NGC 1023 H156¹ Galaxy Type SB(rs)0–
ø8.6′ × 4.2′, m9.3v, SB 13.1   02ʰ40.4ᵐ +39°04′
Finder Chart 23-4, Figure 23-9   ★★★

4/6″ Scopes–100x: This galaxy is bright and easy in small instruments, appearing highly elongated E–W with a diffuse edge and a sudden brightening at center.

8/10″ Scopes–100x: The halo is a bright lens elongated 5′ × 1.5′ E–W and has a well concentrated core. Several stars of the 10th and 11th magnitude are 4′ to 6′ SW.

12/14″ Scopes–125x: NGC 1023 has a bright halo elongated 7′ × 2′ E–W with a prominent core and a stellar nucleus. 200x reveals a faint, diffuse tuft on the eastern tip. A very faint star is embedded 1′ west of the core.

NGC 1039 Messier 34 Open Cluster 60★ Tr Type II3m
ø35′, m5.2v, Br★ 7.33v   02ʰ42.0ᵐ +42°47′
Finder Chart 23-4, Figure 23-10   ★★★★

Charles Messier discovered this cluster in August, 1764. The group is thought to be approximately 180 million years old and about 1,500 light years away.

4/6″ Scopes–100x: This cluster, attractive at low power, has twenty bright stars in a 10′ area surrounded by twice as many fainter outlying members. The brighter stars encircle the cluster core in a 30′ diameter ring.

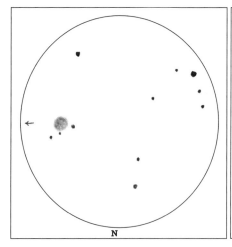
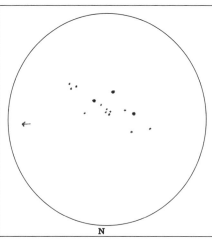
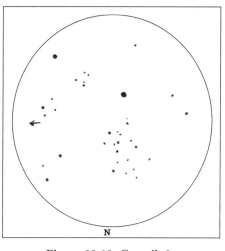

Figure 23-11. PK131-5.1
17.5″, f4.5–200x, by Dr. Jack Marling

Figure 23-12. Basel 10
12.5″, f5.6–225x, by Alister Ling

Figure 23-13. Czernik 8
12.5″, f5–100x, by G. R. Kepple

8/10″ Scopes–100x: Messier 34 is a bright, very large, loose, irregular cluster of sixty 8th to 12th magnitude stars distributed in small groups and short chains.

12/14″ Scopes–125x: This is a grand cluster of 80 stars, including seven of the 8th magnitude and two dozen of the 9th–10th magnitude. The magnitude 9–10 stars are in an obvious E–W concentration. The cluster's magnitude 7.3 blue-white lucida is on its SSE edge. The faintest cluster members are scattered about in clumps and short chains. M34 has an abundance of wide pairs. Near the center is h1123, a 20″ wide double of white magnitude 8.5 stars. Near the south edge is the much tighter OΣ44, whose magnitude 8.4 and 9.1 components are only 1.4″ apart.

NGC 1058 H633² Galaxy Type SA(rs)c III
ø3.5′ × 3.4′, m11.2v, SB 13.7   02ʰ43.5ᵐ +37°21′
Finder Chart 23-4   ★★★

12/14″ Scopes–125x: NGC 1058, situated within a circlet of 11th to 13th magnitude stars, is a fairly faint, diffuse 2′ diameter glow without central brightening. A very faint star or knot is on the NW edge.

PK144-15.1 Abell 4 Planetary Nebula Type 3b
ø22″, m14.4v, CS 19.3v   02ʰ45.4ᵐ +42°33′
Finder Chart 23-4, Figure 23-15   ★★★

16/18″ Scopes–125x: This planetary nebula is 1.5′ WNW of an 8th magnitude star, and is a very faint, round 20″ disk slightly brighter in its center. Though the planetary can be seen without it, the O-III filter considerably enhances the nebula's contrast with its background.

NGC 1160 Galaxy Type Scd:
ø1.4′ × 0.7′, m12.8v, SB 12.6v   03ʰ01.2ᵐ +44°58′
Finder Chart 23-4   ★★★

16/18″ Scopes–150x: NGC 1160 is the smaller and fainter of a galaxy-pair with NGC 1161 lying 3′ south. It has a diffuse 1′ × 0.5′ NE–SW oval halo slightly brighter in the center. A tiny N–S knot of 13th to 14th magnitude stars lies to the north.

NGC 1161 Galaxy Type S0
ø2.8′ × 2.1′, m11.0v, SB 12.8   03ʰ01.2ᵐ +44°55′
Finder Chart 23-4   ★★★

16/18″ Scopes–150x: NGC 1161 is the larger, brighter, more conspicuous of a galaxy-pair with NGC 1160 3′ north. Its 1.5′ × 0.75′ NNE–SSW halo contains a round core within which is embedded a faint stellar nucleus. Just SW is a wide pair of magnitude 9.5 and 10 stars, and two 11th magnitude stars are 1.5′ and 2.75′ east.

NGC 1167 Galaxy Type SA0–
ø3.3′ × 2.4′, m12.4v, SB 14.5   03ʰ01.7ᵐ +35°12′
Finder Chart 23-6   ★★★

12/14″ Scopes–125x: NGC 1167, three-quarters degree east of magnitude 4.9 star 24 Persei, is a rather faint with a circular 2′ diameter halo considerably brighter in its center.

NGC 1169 H620² Galaxy Type SAB(r)b I-II
ø4.8′ × 2.8′, m11.2v, SB 13.9   03ʰ03.6ᵐ +46°23′
Finder Chart 23-5   ★★★

12/14″ Scopes–125x: NGC 1169, located 1.5′ west of a 12.5 magnitude star, has a very faint, circular 1.5′

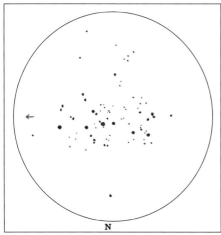
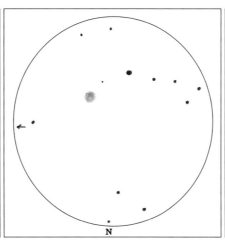
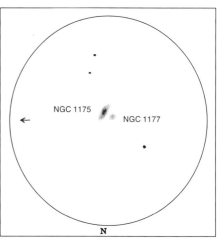

Figure 23-14. Trumpler 2
10″, f4–35x, by Alister Ling

Figure 23-15. PK144-15.1
17.5″, f4.5–150x, by Dr. Jack Marling

Figure 23-16. NGC 1175
12.5″, f5.6–180x, by Alister Ling

diameter halo with a prominent core and a stellar nucleus. 175x reveals a 13.5 magnitude star embedded just SW of the nucleus. A very faint companion lies just NNE of the 12.5 magnitude star.

NGC 1171 U2510 Galaxy Type Scd:
ø2.9′ × 1.3′, m12.3v, SB 13.6   03ʰ04.0ᵐ +43°24′
Finder Chart 23-4   ★★★

12/14″ Scopes–125x: NGC 1171, located 3′ NE of a 12th magnitude star, appears very faint and elongated 2′ × 0.5′ NW–SE. A large prominent core extends along the major axis. A 13.5 magnitude star lies 1.75′ NW of the galaxy's center.

NGC 1175 H607² Galaxy Type SA(r)0+
ø2.1′ × 0.7′, m12.8v, SB 13.1   03ʰ04.5ᵐ +42°20′

NGC 1177 IC 281 Galaxy Type S?
ø0.5′ × 0.5′, m14.5v, SB 12.8   03ʰ04.6ᵐ +42°22′
Finder Chart 23-4, Figure 23-16   ★★/★

12/14″ Scopes–125x: This pair of galaxies is located 7′ SE of an 8th magnitude star. NGC 1175 is faint, elongated 1′ × 0.25′ NNW–SSE and has a very extended core. NGC 1177, 1.5′ NE of its companion, is an extremely faint, tiny diffuse smudge of indefinite dimensions.

NGC 1186 Galaxy Type SB(r)bc:
ø3.7′ × 1.5′, m11.4v, SB 13.1   03ʰ05.5ᵐ +42°50′
Finder Chart 23-4   ★★★

16/18″ Scopes–150x: NGC 1186, in a field well-populated with faint stars, is a very faint, diffuse glow elongated about 1.5′ × 0.75′ NW–SE with a slightly brighter center. A 13th magnitude star embedded in the galaxy's halo just SW of its of center mimics a stellar nucleus. A pair of magnitude 13.5 and 14 stars is just SE, one of them touching the edge of the galaxy's halo.

NGC 1193 Open Cluster 40★ Tr Type II 3 m
ø1.5′, m12.6p, Br★ 14.0p   03ʰ05.8ᵐ +44°23′
Finder Chart 23-4   ★

16/18″ Scopes–150x: NGC 1193, 5′ SE of a wide pair of 8th magnitude stars, is a very faint concentration of a dozen 13th magnitude and fainter stars in a 3′ area. A 12th magnitude star stands out on the WNW edge, and a 7″ wide pair of 14th magnitude stars is just south of the cluster's center. Half a dozen threshold stars can be resolved at 200x.

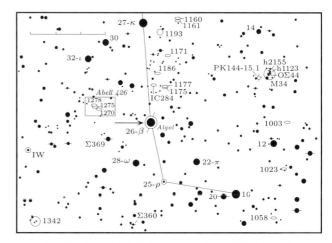

Finder Chart 23-4. 26–β Per: 03ʰ08.2ᵐ +40°57′

Chapter 23

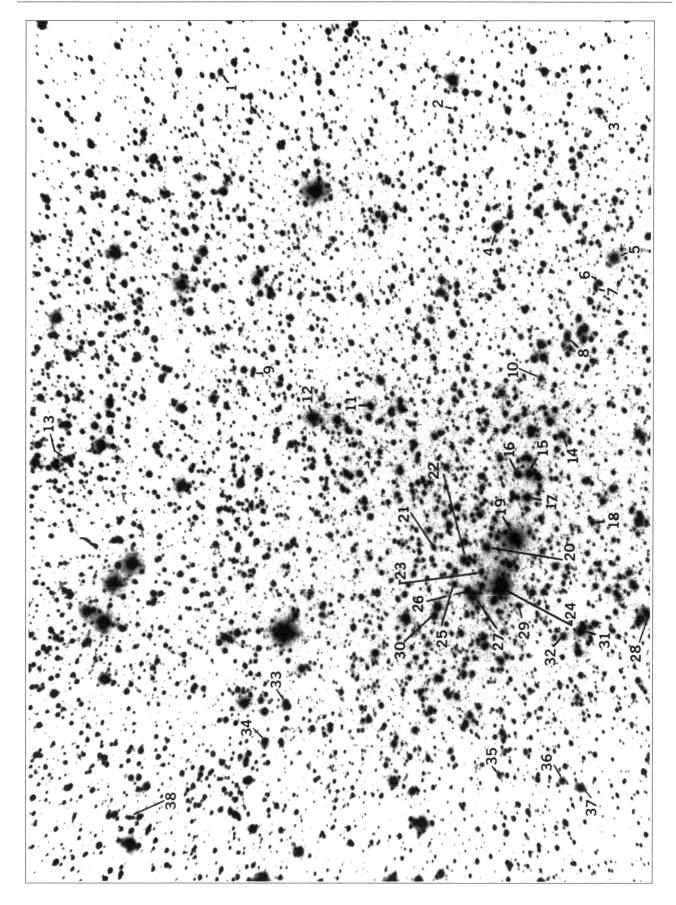

Table 23-3. Perseus I Galaxy Cluster (Abell 426)

ID#	Number	Type	Size	Mag	SB	R. A.	Dec.	Description
1.	U2608	Galaxy	1.0' × 0.9'	14.0v		03ʰ15.0ᵐ	+41°02'	Fairly faint, small, slightly elongated
2.	Z540.067			15.3p		15.4	+41 37	Ex faint, very small, round 9m★ 3' W
3.	NGC 1250	S0°: sp	2.4' × 1.0'	12.8v	13.6	15.4	+41 21	Faint, small, elongated 159°, br cntr, SN
4.	NGC 1257	Sa	1.3' × 0.2'	13.7v	12.1	16.4	+41 31	Very faint, small, very el 68°, br cntr
5.	IC 310		1.3' × 1.3'	14.3v		16.7	+41 20	Fairly faint, fy small, round, br cntr, f SN
6.	NGC 1259			14.7v		17.0	+41 21	Ex faint, very small, m15.5★ at W edge
7.	U2626		1.2' × 0.2'	15.7p		17.0	+41 21	Ex faint, ex small, two ★ close NE
8.	NGC 1260	S0/a: sp	1.3' × 0.6'	13.3v	12.9	17.5	+41 24	Fairly faint, fy small, oval E–W, sl br cntr
9.	U2639		1.2' × 0.2'	15.6p		17.8	+41 58	Ex faint, small, elongated streak E–W
10.	NGC 1264		1.1' × 1.0'	14.6v	14.6	17.9	+41 26	Very faint, small, round, low SB
11.	IC 312		1.2' × 0.6'	14.9p		18.1	+41 45	Faint, small, oval 125°, br cntr, 13m★ 1' E
12.	NGC 1265	E+	1.8' × 1.6'	12.1v	13.1	18.3	+41 52	Very faint, small, round 11m★ on W side
13.	U2654		1.6' × 0.6'	14.6p		18.7	+42 18	
14.	Z540.087			15.3p		18.4	+41 25	Faint, small, sl el N–S, 14m★ 1' N
15.	NGC 1267	E+	1.1' × 0.9'	13.1v	12.9	18.7	+41 28	Faint, very small, round, f SN, arc of 4★ S
16.	NGC 1268	SAB(rs)b:	1.0' × 0.8'	13.4v	13.0	18.8	+41 29	Ex faint, small, round, faint SN
17.	NGC 1270	E:	0.9' × 0.7'	13.1v	12.4	19.0	+41 28	Faint, very small, sl el N–S, sl br cntr
18.	NGC 1271	SB0?		15.4p		19.2	+41 21	Very faint, small, sl el, br cntr, faint SN
19.	NGC 1272	E+	2.2' × 2.0'	11.8v	13.2	19.3	+41 29	Fairly bright, small, round sl br cntr
20.	NGC 1273	SA(r)0°?	1.1' × 1.1'	13.2		19.4	+41 32	Fairly faint, very small, round, br cntr
21.	U2665			15.5p		19.4	+41 38	
22.	M+07-07-061		0.2' × 0.2'	15.0p		19.6	+41 35	
23.	NGC 1274	E3	0.5' × 0.4'	14.1v		19.7	+41 33	Faint, very small, sl br cntr, sl elongated
24.	NGC 1275	Pec	3.2' × 2.3'	11.9v	13.9	19.8	+41 31	Fy bright, fy small, oval E–W, sud br cntr
25.	NGC 1276					19.9	+41 33	
26.	NGC 1277	S0+: pec	0.7' × 0.2'	13.5v	11.2	19.8	+41 34	Faint, very small, oval E–W, sud br cntr
27.	NGC 1278	E pec:	1.4' × 1.1'	12.4v	12.7	19.9	+41 33	Fairly faint, fy small, oval, sud br cntr
28.	U2673		1.6' × 1.4'	15.6p		20.0	+41 15	Very faint, small, round, sl br cntr
29.	NGC 1279			16.6v		20.1	+41 28	Very faint, very small, sl elongated N–S
30.	NGC 1281	E5	0.6' × 0.4'	13.3v	11.6	20.1	+41 38	Faint, small, el ENE–WSW, 10m★ 1.5' W
31.	NGC 1282	E:	1.4' × 1.1'	12.9v	13.2	20.2	+41 22	Faint, fairly small, oval NE-SW, br cntr
32.	NGC 1283	E1:	0.9' × 0.6'	13.6v	12.8	20.3	+41 24	Very faint, small, sl el, two 14m★ 1' N
33.	IC 313		1.3' × 0.9'	15.1p		21.0	+41 54	Very faint, small, round, 14.5★ SE edge
34.	IC 316		1.5' × 0.8'	15.0p		21.4	+41 56	Very faint, small, oval N–S, sl br cntr
35.	Z540.115			15.6p		21.5	+41 31	Very faint, v small, v elongated NW–SE
36.	NGC 1293	E0	1.0' × 1.0'	13.4v	13.2	21.6	+41 24	Very faint, small, round, faint SN
37.	NGC 1294	SA0-?	1.5' × 1.3'	13.2v	13.8	21.7	+41 22	Faint, small, round, suddenly br cntr
38.	U2696		1.0' × 0.1'	15.7p		22.0	+42 11	

Description Abbreviations: br = brighter, cntr = center, el = elongated, ex = extremely, f = faint, fy = fairly, r = round, sl = slightly, ★ = star, sud = suddenly, SN = stellar nucleus, SB = surface brightness, v = very.

Figure 23-17. *(Opposite page) The Perseus I Galaxy Cluster is a challange even for larger telescopes. It is about 300 million light years from us. William Harris made this 60 minute exposure on hypered 2415 Kodak Tech Pan film with an 8", f4 Wright Newtonian reflector at prime focus. North is up.*

Figure 23-18. *NGC 1220 is a small knot of faint stars forming a triangle with two bright stars. Lee C. Coombs made this 10 minute exposure on 103a-0 film with a 10″, f5 Newtonian.*

Figure 23-19. *NGC 1342 is a loose, irregular scattering around an E–W chain of brighter stars. Lee C. Coombs made this 10 minute exposure on 103a-0 film with a 10″, f5 Newtonian.*

IC 284   Galaxy   Type SAdm
ø4.6′ × 2.3′, m12.5v, SB 14.0   03ʰ06.2ᵐ +42°23′
Finder Chart 23-4   ★★★

16/18″ Scopes–150x: IC 284 is 15′ SSW of an 8th magnitude star and 2.5′ SE of a 17″ wide NNE–SSW pair of 11th magnitude stars. A large, faint central concentration is surrounded by an extremely faint 2.5′ × 1.25′ NNE–SSW halo. 13th magnitude stars touch the edge of the halo NE and SW of the galaxy's center.

NGC 1220   Open Cluster   15★   Tr Type II 2 p
ø1.6′, m11.8p, Br★ 13.0p   03ʰ11.7ᵐ +53°20′
Finder Chart 23-5, Figure 23-18   ★

8/10″ Scopes–125x: NGC 1220, a faint, small cluster at the northern corner of an 8′ × 12′ triangle with a 7th and an 8th magnitude star, has ten 13th magnitude members in a small 1.5′ × 1′ N–S knot.

12/14″ Scopes–150x: This cluster is a faint, triangular formation of about fifteen 13th magnitude and fainter stars pointing north. Two 13th magnitude stars are detached to the south and west.

King 5   Open Cluster   40★   Tr Type I 2 m
ø7′, m –, Br★ 13.0v   03ʰ14.5ᵐ +52°43′
Finder Chart 23-5   ★★★

8/10″ Scopes–125x: King 5 is a condensation of faint stars just SW of a 10th magnitude star. Half a dozen 13th magnitude stars resolve against a background haze. A second 10th magnitude star lies 5′ north, and an 8th magnitude star is 7′ ENE of the cluster.

12/14″ Scopes–125x: The fairly faint group King 5 contains several dozen 13th to 14th magnitude stars in a 6′ equilateral triangle. Some outlying stars are north of the triangle.

NGC 1245   H25⁶   Open Cluster   200★   Tr Type III 1 r
ø10′, m8.4v, Br★ 11.16v   03ʰ14.7ᵐ +47°15′
Finder Chart 23-5, Figure 23-20   ★★★

4/6″ Scopes–35x: Small telescopes show NGC 1245 as a very faint, round misty patch framed by a thin triangle of 8th to 9.5 magnitude stars pointing NE. The cluster lies north of the brightest star of the triangle.

8/10″ Scopes–100x: This rich cluster has fifty faint stars in a 10′ diameter area.

12/14″ Scopes–125x: NGC 1245 is a fine, rich cluster of seventy-five 12.5 to 14th magnitude stars in a circular 10′ area. Numerous short arcs and star chains form a broad halo around an irregular central void. Three 12th magnitude stars stand out among the 13th magnitude stars along the southern edge.

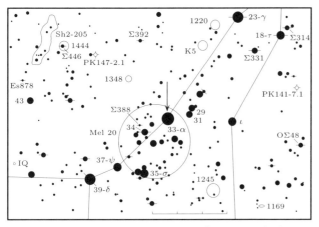

Finder Chart 23-5. 33–α Per: 03ʰ24.3ᵐ +49°52′

Figure 23-20. *NGC 1245 is a fine, rich cluster of seventy-five faint stars concentrated in a 10' area. Martin C. Germano made this 40 minute exposure on Kodak 2415 Tech Pan film with a 14.5", f5 Newtonian reflector at prime focus.*

NGC 1275 Galaxy Type Pec
ø3.2' × 2.3', m11.9v, SB 13.9   03ʰ19.8ᵐ +41°31'
Finder Chart 23-4, Figure 23-17   ★★★

NGC 1275 is the strong extragalactic radio source Perseus A = 3C84. It is either an exploding galaxy, or two galaxies in collision or in the process of merger. NGC 1275 is the brightest member of the Perseus I Galaxy Cluster (Abell 426), which is roughly 300 million light years away and almost 10 million light years in diameter.

12/14" Scopes–125x: NGC 1275, located near the center of the Perseus I Cluster, has a fairly bright nucleus surrounded by a small, faint, circular halo.

16/18" Scopes–150x: NGC 1275 shows an irregularly bright core with a stellar nucleus embedded in a faint 2' × 1.5' E–W oval halo. A faint star is 1' NE of center.

Melotte 20 Cr 39/40 Open Cl. 50★ Tr Type III 3 m
ø185', m1.2v   03ʰ22.+49°
Finder Chart 23-5 Alpha Persei Moving Group   ★★★★

Melotte 20, the Alpha Persei Moving Group, is the fine field of stars around, and extending SE from, the group's lucida, Alpha (α) Persei. It includes Psi (ψ), 29, 31, and 34 but not Sigma (σ) Persei. Delta (δ) and Epsilon (ε) Persei are probable outlying members, for they share the group's proper motion toward the SE. The Alpha Persei group's distance is some 540 light years so the true length of its core (from 29 to Psi) is some 33 light years. Over one hundred members have been identified. The cluster core is best observed with 10–20x binoculars or RFTs at low powers.

10x50 Binoculars: Forty stars are visible in a 3° area with many attractive star chains and wide pairs.

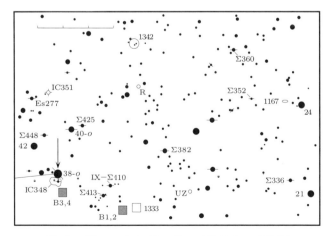

Finder Chart 23-6. 38–*o* Per: 03ʰ44.3ᵐ +32°17'

Figure 23-21. *NGC 1333 is a glowing patch of nebulosity surrounding an 11th magnitude star. The dark nebulae Barnard 1 and 2 are the conspicuous starless voids to its north and south. Martin C. Germano made this 90 minute exposure on 2415 Kodak Tech Pan film with an 8", f5 Newtonian reflector at prime focus.*

NGC 1342 H88⁸ Open Cluster 40★ Tr Type III 3 p
ø14′, m6.7v, Br★ 8.75v   03ʰ31.6ᵐ +37°20′
Finder Chart 23-6, Figure 23-19   ★★★

4/6″ Scopes–35x: Small telescopes show a fairly bright, large, loose gathering of stars. Several of the brighter cluster members are toward its NNE edge.

8/10″ Scopes–100x: This cluster is a bright, large, scattered group of fifty 9th magnitude and fainter stars covering a 15′ area.

12/14″ Scopes–125x: NGC 1342 is a very loose, irregular scattering of stars with no central compression. Sixty stars are visible in a 17′ area. A number of the group's 9th–10th magnitude members lie along a zigzagging E–W star chain that makes the cluster look elongated E–W. The majority of its faint stars are spread north of this chain. On the NNE edge, is a 3.25′ × 2.5′ triangle of 8th, 9th, and 10th magnitude stars is nearly detached from the main body of the cluster by a starless void.

NGC 1333 Reflection Nebula
ø6′ × 3′, Photo Br 3–5, Color 3–4   03ʰ29.3ᵐ +31°25′

Barnard 1, 2, 202-6 Dark Nebula
ø160′ × 70′, Opacity 5   03ʰ32.1ᵐ +31°10′
Finder Chart 23-6, Figure 23-21   ★★★/★★★

8/10″ Scopes–100x: Reflection nebula NGC 1333 is a fairly conspicuous 8′ × 6′ NE–SW oval glow. The nebula spreads SW from a bright star at its NE tip.

12/14″ Scopes–125x: NGC 1333 is a 9′ × 7′ NE–SW oval. A magnitude 10.8 star is embedded in its glow near the NE edge. Along the southern edge is a noticeably brighter patch. North and south of NGC 1333 are the distinct, irregular starless opacities of the dust clouds Barnard 1 and 2.

NGC 1348 H84⁸ Open Cluster 30★ Tr Type III 2m
ø5′, m –, Br★ 8.53v   03ʰ33.8ᵐ +51°26′
Finder Chart 23-5, Figure 23-28   ★★★

12/14″ Scopes–125x: NGC 1348 is a rectangular group of 17 stars in a 7′ area. It has only two brighter stars among its faint population, but the cluster stands out well from the surrounding star field at low power.

PK147-2.1   Minkowski 1-4   Planetary Nebula
ø4″, m13.6v   03ʰ41.7ᵐ +52°17′
Finder Chart 23-5, Figure 23-29   ★★★

16/18″ Scopes–275x: This tiny planetary nebula, located 7′ west of an 8th magnitude star, is easily spotted by blinking with an O-III filter at high power. It has a round 5″ disk of uniform surface brightness but no visible central star. A 3″ wide E–W pair of magnitude 12–12.5 stars lies between the planetary and the 8th magnitude star.

Barnard 3, 4   LDN 1470   Dark Nebula
ø100′ × 70′, Opacity 5   03ʰ44.0ᵐ +31°47′

IC 348   Open Cluster   20★   Tr Type IV 2 p n
ø7′, m7.3, Br★ 8.53v   03ʰ44.5ᵐ +32°17′

IC 348   Reflection Nebula
ø10′ × 10′, Photo Br 2–5, Color 2–4 03ʰ44.5ᵐ +32°17′
Finder Chart 23-6, Figure 23-22   ★★★/★★/★

12/14″ Scopes–125x: IC 348, 7′ south of magnitude 3.8 star Omicron (*o*) = 38 Persei, is a faint 5′ glow surrounding a magnitude 8.5 star.

16/18″ Scopes–150x: IC 348 is a faint 7′ diameter patch of nebulosity with an involved cluster of twenty magnitude 8.5 to 13 stars. The nebula-glow is more conspicuous around a magnitude 8.5 star near the NE edge and around several magnitude 9.5–10 stars on its south and SE edges. Because this is a reflection nebula, the Deep-Sky filter should be used rather than the O-III filter to improve its contrast with the sky background. The glare from magnitude 3.8 Omicron (*o*) = 38 Persei 7′ NNW also interferes. To the south and SW are the irregular dark E–W patches of the dust clouds Barnard 3 and 4, remnants of the interstellar material from which were formed Omicron (*o*), Zeta (ζ), and the other members of the Zeta Persei Association.

IC 351   PK159–15.1   Planetary Nebula   Type 2a
ø7″, m12.0v, CS 15.8   03ʰ47.5ᵐ +35°03′
Finder Chart 23-7, Figure 23-30   ★★★

8/10″ Scopes–200x: IC 351 is 3.5′ NW of a diminutive triangle of magnitude 10, 12, and 13 stars. This tiny planetary has a fairly faint, circular disk of even surface brightness. The 10th magnitude star in the triangle has a 14th magnitude companion to its SSW.

16/18″ Scopes–250x: IC 351 is quite bright, elongated 8″ × 6″ NE–SW, and has a slight bluish-green tint.

NGC 1444   Open Cluster   Tr Type IV 1 p
ø4′, m6.6v, Br★ 6.77v   03ʰ49.4ᵐ +52°40′
Finder Chart 23-5   ★★★

8/10″ Scopes–100x: This small cluster has ten stars

Figure 23-22. *IC 348 is a faint patch of nebulosity with an involved open cluster of twenty stars. Martin C. Germano made this 85 minute exposure on 2415 film with an 8″, f5 Newtonian reflector at prime focus.*

scattered around a bright multiple. A 7th magnitude star is 12′ SE.

12/14″ Scopes–125x: NGC 1444 is an interesting, compact group of twenty stars, most 12th to 14th magnitude, surrounding the multiple Struve 446 (AB: 6.9, 9.1; 8.6″; 253°). A third companion (AC: 6.9, 12.0; 12.1″; 39°) lies to the NE, while two 10th magnitude components are located much further NW of this triple is a bright double (DE: 10.3, 10.7; 2.6″; 232°). From this double and the bright Σ446 triple the cluster's stars straggle 8′ NNE. The cluster is widest, about 3′ across, at its northern end.

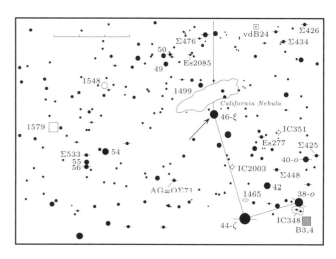

Finder Chart 23-7. 46–ξ Per: 03ʰ58.9ᵐ +35°47′

Figure 23-23. *NGC 1499, the California Nebula, is an extremely faint patch of nebulosity almost three degrees long. Martin C. Germano made this 80 minute exposure on 2415 film with an 8″, f5 Newtonian reflector at prime focus.*

vdB24 Reflection Nebula
⌀5′ × 3′, Photo Br 3–5, Color 2–4 $03^h49.6^m$ +38°59′
Finder Chart 23-7   ★★★

16/18″ Scopes–150x: This very faint reflection nebula has the appearance of a comet, for it fans to the south of its magnitude 8.8 illuminating star like a comet tail fanning out from a comet nucleus.

NGC 1465 Galaxy Type S0/a
⌀1.7′ × 0.5′, m13.7v, SB 13.4 $03^h53.6^m$ +32°28′
Finder Chart 23-7   ★★★

12/14″ Scopes–125x: NGC 1465, located 35′ north of 2.8 magnitude star Zeta (ζ) = 44 Persei, is an extremely faint, elongated smudge 10′ ENE of a 7th magnitude star.

16/18″ Scopes–150x: NGC 1465 is a very dim spindle elongated 1′ × 0.25′ NNW–SSE.

IC 2003 PK161-14.1 Planetary Nebula Type 2
⌀7″, m12.5v, CS 15.3v $03^h56.4^m$ +33°52′
Finder Chart 23-7, Figure 23-31   ★★★

8/10″ Scopes–200x: At low power IC 2003 appears moderately bright but stellar; 200x is needed to show a definite disk. This planetary is situated in a group of ten 11th to 14th magnitude stars, including a wide pair of 13th magnitude stars 3.5′ NNW, aligned so that they point toward it.

12/14″ Scopes–250x: IC 2003 shows a bright 8″ diameter disk slightly brighter in the core but without a visible central star. A 13.5 magnitude star is visible 15″ SW.

NGC 1499 Emission Nebula
⌀160′ × 40′, Photo Br 1–5, Color 3–4 $04^h00.7^m$ +36°37′
Finder Chart 23-7, Figure 23-23   ★
California Nebula

16 × 80 Binoculars: In 10×50 binoculars the California Nebula is an extremely faint rectangular patch north of its illuminating star, Xi (ξ) = 46 Persei. In 16×80 glasses with a Hydrogen-Beta filter, the rectangular patch is sharper and better defined along its northern edge.

8/10″ Scopes–40x: NGC 1499 is nearly three degrees long and beautiful in photographs but is very disappointing and difficult to detect visually. Using a nebula filter, it is only a slight brightening of the sky.

12/14″ Scopes–60x: NGC 1499 is an exceptionally large, extremely faint 3° × 1° arc elongated generally E–W and concave to the south. The O-III filter does little

Figure 23-24. *NGC 1491 is a fairly conspicuous fan-shaped emission nebula. Martin C. Germano made this 45 minute exposure on 2415 film with an 8″, f5 Newtonian reflector at prime focus.*

Figure 23-25. *NGC 1513 is a moderately faint but rich concentration of 60 stars. Lee C. Coombs made this 10 minute exposure on 103a-O film with an 10″, f5 Newtonian reflector at prime focus.*

to enhance the view. Several very low surface brightness streaks are just visible.

NGC 1491 H258¹ Emission Nebula
ø25′ × 25′, Photo Br 1–5, Color 3–4   04ʰ03.4ᵐ +51°19′
Finder Chart 23-8, Figure 23-24   ★★★

8/10″ Scopes–100x: This emission nebula is just west of an 11th magnitude star. It is a diffuse, fairly bright, 6′ fan-shaped glow.

12/14″ Scopes–125x: NGC 1491 is a bright, triangular nebulosity elongated NNE–SSW. It is bright and concentrated at the SSW apex, becoming dimmer and diffuse as it fans NE. The nebula responds well to both the UHC and O-III filters. An 11th magnitude star is embedded just inside the east edge. Four wide pairs of 13th to 14th magnitude stars are sprinkled around the northern edge of the nebula, and a fifth pair is to its SW.

NGC 1496 Open Cluster 10★ Tr Type II 1 p
ø6′, m9.6p, Br★ 12.0p   04ʰ04.4ᵐ +52°37′
Finder Chart 23-8   ★★★

12/14″ Scopes–125x: NGC 1496, located 3.5′ NE of a 10th magnitude star, has eighteen faint stars in a 3′ area. The cluster's brighter members form a 2′ long C-shaped asterism open to the ESE, a 12th magnitude star, the second brightest in the group, on its NE edge. The cluster contains at least five faint doubles, four in the "C."

NGC 1513 H60⁷ Open Cluster 50★ Tr Type II 1m
ø9′, m8.4v, Br★ 11.19v   04ʰ10.0ᵐ +49°31′
Finder Chart 23-8, Figure 23-25   ★★★

4/6″ Scopes–50x: This cluster is a faint, rich, compressed patch of faint stars distributed in a figure "9."

8/10″ Scopes–100x: NGC 1513, SW of a 45″ wide double of magnitude 10 and 11.5 stars, contains thirty 12th to 14th magnitude stars in a 10′ × 8′ NW–SE area. The oval outline of the cluster is broken on its NE edge by a starless gap that extends nearly to the group's center.

12/14″ Scopes–125x: NGC 1513 is a moderately faint but rich concentration of 60 stars in a distorted

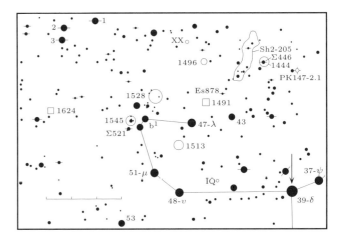

Finder Chart 23-8. 39-δ Per: 03ʰ42.9ᵐ +47°47′

Figure 23-26. *NGC 1528 is a bright, rich, star cluster with long star chains. Martin C. Germano made this 30 minute exposure on Kodak 2415 Tech Pan film with an 8", f5 Newtonian reflector.*

Figure 23-27. *NGC 1545 is an irregular cluster of faint stars scattered around a triangle of brighter stars. Lee C. Coombs made this 10 minute exposure on 103a-O film with a 10", f5 Newtonian reflector at prime focus.*

dumbbell-shape, pinched from the east and west sides. The northern group is elongated 9′ × 4′ E–W with a 10–11.5 magnitude pair on its east edge. The southern group is richer, elongated 10′ × 6′ ESE–WNW, and has an irregular circlet on the SE edge.

NGC 1528   H61⁷   Open Cluster 40★   Tr Type II 2m
ø23′, m6.4v, Br★ 8.75v   04ʰ15.4ᵐ +51°14′
Finder Chart 23-8, Figure 23-26   ★★★

4/6″ Scopes–50x: NGC 1528 is fairly bright, large and rich, its stars mildly concentrated toward its center. A rather wide, relatively starless lane runs NW–SE through the cluster. Several short arcing star chains extend in various directions from the main group.

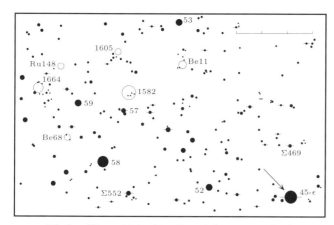

Finder Chart 23-9. 45–ϵ Per: 03ʰ57.9ᵐ +40°01′

8/10″ Scopes–100x: This bright, large, triangular group has sixty stars in a 24′ area. In its western portion the cluster's brighter stars, 9th to 10th magnitude objects, form an irregular circlet.

12/14″ Scopes–125x: This fine cluster is bright, rich, and irregularly concentrated, its stars falling along chains and streams aligned generally NW–SE and separated by conspicuous starless voids. It contains seventy-five 9th to 14th magnitude members. The cluster's overall shape is triangular, the vertices — which fall to the north, west and SE — marked by fairly bright stars.

Berkeley 11   Open Cluster 35★   Tr Type II 3 m
ø5.4′, m10.4v, Br★ 11.75v   04ʰ20.6ᵐ +44°55′
Finder Chart 23-9   ★★★

12/14″ Scopes–125x: At low power Berkeley 11 is only a faint blotch nestled in a star chain just west of a wide pair of 9.5 magnitude stars. At higher powers a dozen very faint stars, the brightest magnitude 11.75, can be resolved in a 6′ diameter haze. Five stars, the brightest another magnitude 11.75 object, straggle in a chain out of the cluster toward the NW.

NGC 1545   Open Cluster 20★   Tr Type II 2 p
ø18′, m6.2v, Br★ 7.13v   04ʰ20.9ᵐ +50°15′
Finder Chart 23-8, Figure 23-27   ★★★

4/6″ Scopes–35x: Small telescopes show a scattering of faint stars surrounding a triangle of bright stars. The three bright stars make the cluster's catalogued

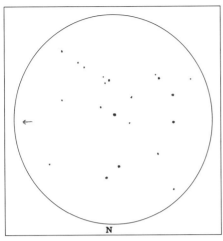

Figure 23-28. NGC 1348
12.5″, f5.6–250x, by Alister Ling

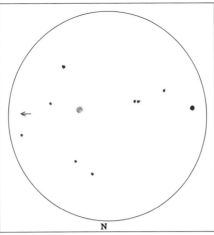

Figure 23-29. PK147-2.1
17.5″, f4.5–280x, by Dr. Jack Marling

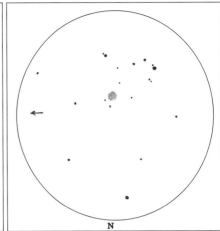

Figure 23-30. IC 351
17.5″, f4.5–250x, by Dr. Jack Marling

integrated magnitude of 6.2 higher than it should be, for these stars are probably foreground objects.

8/10″ Scopes–100x: This cluster forms the NE corner of a half degree long parallelogram with a 7th and two 5th magnitude stars. Thirty faint stars in an 18′ area are scattered in and around a triangle of 8.5–9.5 magnitude stars.

12/14″ Scopes–125x: NGC 1545 is an interesting cluster of 45 stars loosely concentrated over a 25′ area. Just west of its center is a 2.5′ × 1.25′ triangle, pointing west, of one 8th and two 9th magnitude stars of contrasting colors: the eastern star is orange, and the northern star is yellow. 7.5′ north of this triangle, near the cluster's northern edge, is the double Σ519 (7.9, 9.5; 18.3″; 346°), the primary of which is orange.

NGC 1548 Open Cluster Tr Type IV 1 p
⌀30′, m –, Br★ –   04^{h}21.0^m +36°56′
Finder Chart 23-7   ★★★

12/14″ Scopes–125x: NGC 1548 is an irregular scattering of several dozen 11th to 13th magnitude stars in a 4′ area. On the north edge of the group is a wide pair of magnitude 11 and 12 stars. Just south is an E–W stream of fifteen stars. On the southern edge is a 12th magnitude star with an E–W concentration of ten stars immediately to its north.

NGC 1579 H217[1] Reflection Nebula
⌀12′ × 8′, Photo Br 1–5, Color 1–4   04^{h}30.2^m +35°16′
Finder Chart 23-7, Figure 23-32   ★★★

12/14″ Scopes–100x: NGC 1579, just 12′ east of a 7th magnitude star, is a fairly bright reflection nebula. The northern position is a brighter N–S oval, and a

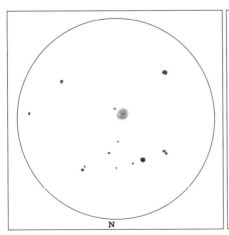

Figure 23-31. IC 2003
17.5″, f4.5–250x, by Dr. Jack Marling

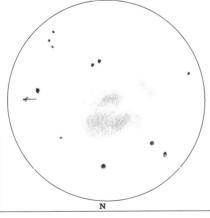

Figure 23-32. NGC 1579
20″, f4.5–175x, by Richard W. Jakiel

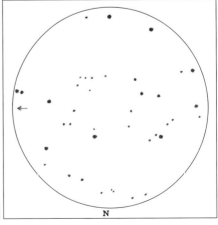

Figure 23-33. NGC 1582
12.5″, f5–130x, by G. R. Kepple

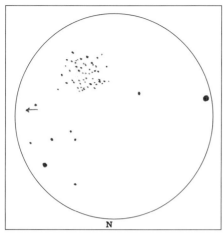
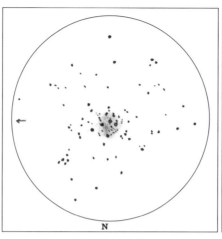
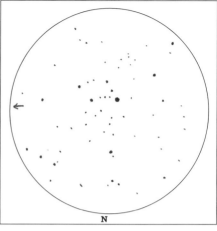

Figure 23-34. NGC 1605
12.5", f5–130x, by G. R. Kepple

Figure 23-35. NGC 1624
13", f4.5–165x, by Tom Polakis

Figure 23-36. Ruprecht 148
12.5", f5–160x, by G. R. Kepple

fainter, nearly detached, glow fans westward.

20/22" Scopes–125x: NGC 1579 is a bright, diffuse patch elongated 5′ × 3′ N–S. On the east side is a dense oval core, and on the west and SW some detached filamentary segments. At least eight faint stars are embedded in the nebula-glow.

NGC 1582 Open Cluster 20★ Tr Type IV 2 p
ø37′, m7.0p, Br★ 9.0p   04ʰ32.0ᵐ +43°51′
Finder Chart 23-9, Figure 23-33   ★★★

12/14" Scopes–125x: NGC 1582 is a bright, irregularly scattered cluster of thirty-five 9th to 13th magnitude stars in a 35′ × 17′ E–W area. It has four 9th and four 10th magnitude members. Its stars are grouped in short rows and in small curved chains. A longer chain of faint stars is detached to the north.

NGC 1605 Open Cluster 40★ Tr Type III 1 m
ø5′, m10.7v, Br★ 12.52v   04ʰ35.0ᵐ +45°15′
Finder Chart 23-9, Figure 23-34   ★★★

12/14" Scopes–125x: NGC 1605, located 7′ west of an 8th magnitude star, is a hazy 5′ group of faint stars. 165x is required to resolve 18 cluster members, the three brightest being magnitude 12.5. The cluster is in a void between two loose groups of brighter field stars.

NGC 1624 H49⁵ Open Cluster 12★ Tr Type I 2 p n
ø1.9′, m10.4v, Br★ 11.77v   04ʰ40.4ᵐ +50°27′

NGC 1624 H49⁵ Emission Nebula
ø5′ × 5′, Photo Br 1–5, Color 3–4   04ʰ40.5ᵐ +50°27′
Finder Chart 23-8, Figure 23-35   ★★★

12/14" Scopes–125x: NGC 1624 consists of a dozen 12th magnitude and fainter stars embedded in a compact 3′ wide diffuse nebula. The brightest involved star is in the center of this small complex at the angle of an L-shaped asterism of four stars. Along the eastern edge is a short star chain concave to the west, and at the western edge a short row of three stars.

Berkeley 68 Open Cluster 60★ Tr Type IV 2 p
ø10′, m9.8v, Br★ 13.68v   04ʰ44.5ᵐ +42°04′
Finder Chart 23-9   ★★★

12/14" Scopes–125x: Berkeley 68 is in a Milky Way field sprinkled with bright stars. It is a hazy, poorly resolved, 16′ long E–W patch of mostly magnitude 13.5 and fainter stars. Its six brightest stars form a 7′ × 5′ isosceles triangle pointing south. A magnitude 9.5 star is at the NE corner of the triangle and another at the midpoint of its NE–SW side.

Ruprecht 148 Open Cluster 15★ Tr Type IV 2 p
ø8′, m9.5v, Br★ 9.91v   04ʰ46.5ᵐ +44°44′
Finder Chart 23-9, Figure 23-36   ★★★

12/14" Scopes–125x: The rather loose, faint Ruprecht 148 contains some three dozen 12th magnitude and fainter stars in an irregularly extended 8′ long ESE–WNW area. The 10 magnitude lucida marks the NE corner of a parallelogram of brighter cluster members.

Chapter 24

Pisces, the Fishes

24.1 Overview

Pisces, one of the twelve constellations of the Zodiac, was figured by the ancient Greeks as two Fish, one south of the Great Square of Pegasus, facing west and the other between Pegasus and Aries facing north. The two Fish were bound by cords around their tails, the cord from the Western Fish running east and the cord from the Northern Fish running south, the two meeting at what the Greek writer Aratos (270 B.C.) called a "great and beautiful star... the knot of tails." Today the Knot Star is generally assumed to have been our Alpha Piscium. But this is not a "great and beautiful star," just a 4th magnitude object. The Greek Knot Star was almost certainly our Omicron Ceti, the variable Mira, toward which the two cords from the tails of the Fishes arc, and which when at its 2nd magnitude high maxima, is indeed "great and beautiful."

The Greeks inherited Pisces from the ancient Babylonians, to whose god of water and wisdom, Ea, fish were sacred. (The Goat-fish of our Capricornius and the streaming Water-jar of our Aquarius were also sacred icons of Ea.) A remembrance of the Babylonian origin of the constellation Pisces appears in a Greek myth about the goddess of love Venus and her son Cupid. The two of them were being pursued by the fire-breathing monster Typhoon and to escape throw themselves into the Euphrates River — which of course flowed past the walls of Babylon — and were transformed into two fish. Venus herself was a Greek inheritance from Babylonia, where she had been called Ishtar and was already identified with the planet.

Pisces contains the vernal equinox, the point where the Sun moves across the celestial equator into the northern hemisphere each year on the first day of spring. However, because of precession, two thousand years ago the vernal equinox was in Aries, hence its (today inappropriate) name "the First Point of Aries." A couple thousand years hence the vernal equinox shall be in Aquarius.

Pisces is well off the Milky Way and therefore, though it is fairly large, is not rich in objects. It does contain some fine double stars, several of which are beautiful color contrast pairs, and one Messier object, the face-on spiral galaxy M74. Most of its other galaxies — and, like other off-Milky Way constellations, Pisces does not lack galaxies — are rather small, faint, and visually unimpressive even in large telescopes.

Pisces: PIE-seez
Genitive: Piscium, PISH-ee-um
Abbrevation: Psc
Culmination: 9 pm–Nov.11, midnight–Sept. 27
Area: 889 square degrees
Showpieces: 35 Psc, 42 Psc, 51 Psc, 55 Psc, 65 Psc, 74–ψ Psc, 113–α Psc, M74 (NGC 628) NGC 7541, NGC 128, NGC 488
Binocular Objects: 74–ψ Psc, 77 Psc, M74 (NGC 628)

24.2 Interesting Stars

35 Piscium Double Star Spec. F0
m6.0, 7.6; Sep. 11.6″, P.A. 148° $00^h15.0^m$ +08°49′
Constellation Chart 24-1   ★★★★

4/6″ Scopes–75x: 35 Piscium is a nice double of light yellow and blue stars.

38 Piscium Multiple Star Spec. F5
ABxC: m7.9, 7.8; Sep. 4.3″, P.A. 236° $00^h17.4^m$ +08°53′
Constellation Chart 24-1   ★★★★

4/6″ Scopes–75x: 38 Piscium is an attractive pair of yellow stars in a field sprinkled with faint stars. A 12th magnitude third companion lies 63″ SE. The A component is also an unresolvable double.

42 Piscium Double Star Spec. K0
m6.2, 10.1; Sep. 28.5″, P.A. 324° $00^h22.4^m$ +13°29′
Constellation Chart 24-1   ★★★★

4/6″ Scopes–75x: 42 Piscium is an easily separated double of an orange primary with a blue secondary.

Pisces, the Fishes

Constellation Chart 24-1

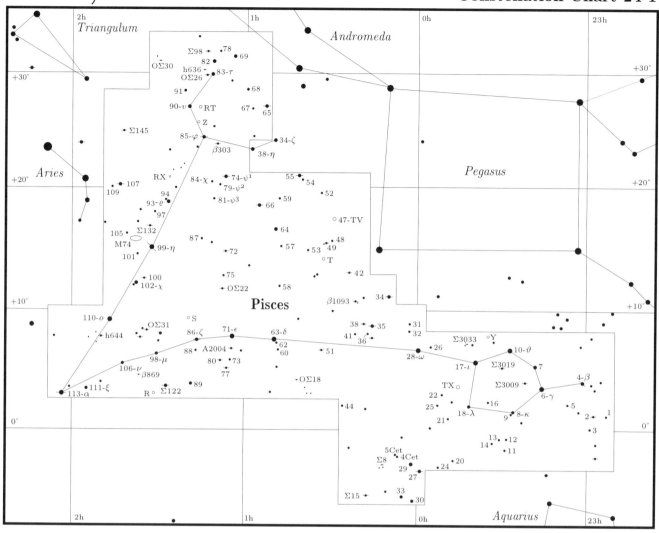

Table 24-1. Selected Variable Stars in Pisces

Name	HD No.	Type	Max.	Min.	Period (Days)	F*	Spec. Type	R.A. (2000)	Dec.	Finder Chart No. & Notes	
Y Psc	221700	EA	9.0	12.0	3.76	0.10	A3+K0	$23^h34.4^m$	+07°55'	24-4	
TX Psc	223075	Lb	6.9	7.7			N0 (C6)	46.4	+03 29	24-4	Fine red color
TV Psc	2411	SR	4.6	5.4	70		M3	$00^h28.0^m$	+17 54	24-1	
T Psc	–	SR	9.2	12.3	260		M5	32.0	+14 36	24-1	
RT Psc	7307	SRb	8.2	10.4	70		M	$01^h13.8$	+27 08	24-6	
Z Psc	7561	SRb	8.8	10.1	144		C7,3	16.1	+25 46	24-6	Reddish
S Psc	7773	M	8.2	15.3	405	0.42	M5-M7	17.6	+08 56	24-8	
RX Psc	–	M	8.8	14.6:	280	0.44	M1	25.6	+21 23	24-1	
R Psc	9203	M	7.1	14.8	344	0.44	M3-M6	30.6	+02 53	24-8	

F* = The fraction of period taken up by the star's rise from min. to max. brightness, or the period spent in eclipse.

Table 24-2. Selected Double Stars in Pisces

Name	ADS No.	Pair	M1	M2	Sep."	P.A.°	Spec	R.A. (2000)	Dec	Finder Chart No. & Notes
Σ3009	16730		6.8	8.8	7.0	230	K2	23^{h}24.3^m	+03°43'	24-4 Orange & whitish
Σ3019	16803		7.6	8.6	10.7	185	A3	30.7	+05 15	24-4
Σ3033	16958		8.8	8.8	3.3	6	F2	43.9	+07 15	24-4
27 Psc	17137		4.9	10.2	1.3	292	K0	58.7	−03 33	24-1
34 Psc	122		5.5	9.4	7.7	160	B8	00^{h}10.0^m	+11 09	24-1
Σ8	144		7.8	9.4	7.8	292	F8	11.6	−03 05	24-1
35 Psc	191		6.0	7.6	11.6	148	F0	15.0	+08 49	24-1 Yellowish & blue
Σ15	205		7.8	10.3	4.7	198	G5	15.9	−05 36	24-1
38 Psc	238	ABxC	7.9	7.8	4.3	236	F5	17.4	+08 53	24-1 Attractive yellow pair
β1093	287		7.0	7.9	0.7	103	A0	20.9	+10 59	24-1
42 Psc	303		6.2	10.1	28.5	324	K0	22.4	+13 29	24-1 Orange & blue
49 Psc	420		6.9	10.7	20.9	103	A0	30.8	+16 02	24-1
51 Psc	449		5.7	9.5	27.5	83	A0	32.4	+06 57	24-5 Blue-white & greenish
55 Psc	558		5.4	8.7	6.5	194	K0	39.9	+21 26	24-1 Yellowish-orange & blue
OΣ18	588	AB	7.8	9.4	w1.6	*213	F8	42.4	+04 10	24-5
	588	AC		12.1	42.8	270				
65 Psc	683		6.3	6.3	4.4	297	F0	49.9	+27 43	24-6 Fine yellowish pair
66 Psc	746	AB	6.2	6.9	w0.5	*199	A0	54.6	+19 11	24-1
A2004	874		7.0	9.9	1.4	245	A0	01^{h}03.8^m	+06 46	24-5
74–ψ Psc	899	AB	5.6	5.8	30.0	159	A2 A0	05.6	+21 28	24-1 Matched bluish-white
77 Psc	903	AB	6.8	7.6	33.0	83	F2 F2	05.8	+04 55	24-5 Wide yellowish pair
OΣ22	920		7.2	10.3	8.6	197	F0	07.1	+11 33	24-1
β303	955		7.3	7.5	0.7	290	F0	09.7	+23 48	24-1 Close yellowish pair
Σ98	988		7.0	8.0	19.5	248	A0	12.9	+32 05	24-6
OΣ26	990	AB	6.2	10.0	10.8	258	K0	13.0	+30 04	24-6
	990	AC		12.0	113.9	342				
85–φ Psc	995	AB	4.7	10.1	7.8	227	K0	13.7	+24 35	24-1
86–ζ Psc	996	AB	5.6	6.5	23.0	63	A5 F8	13.7	+07 35	24-1 White & yellowish
h636	1000		7.5	10.0	20.4	288	A0	14.3	+30 33	24-6
OΣ30	1134	AB	8.2	11.8	4.4	238	F8	25.6	+31 33	24-6
	1134	AC		8.1	56.8	105	G0			
Σ122	1148		6.6	8.6	6.0	328	B8	26.9	+03 32	24-8
Σ132	1202	AB	6.9	9.9	43.4	348	G5	32.1	+16 57	24-1
	1202	AC		10.9	68.7	229				
	1202	AD		10.0	133.0	113				
	1202	Dd		10.6	5.5	288				
OΣ31	1214		6.6	10.7	4.1	78	K0	33.3	+08 13	24-8
100 Psc	1238	AB	7.3	8.4	15.5	77	A3	34.8	+12 34	24-7 Bluish primary
β869	1257		7.9	11.6	5.1	197	K0	36.3	+04 19	24-8
Σ145	1326	AB	6.2	10.8	10.5	31	F5	41.3	+25 45	24-1
	1326	AC		11.0	82.4	338				
107 Psc		AB	5.2	11.6	19.0	248	G5	42.5	+20 16	24-1
h644	1435	AB	7.3	11.7	17.5	278	K0	48.7	+07 41	24-8
113–α Psc	1615		4.2	5.1	c1.8	*272	A2	02^{h}02.0^m	+02 46	24-8 Close bluish-white pair

Footnotes: *= Year 2000, a = Near apogee, c = Closing, w = Widening. Finder Chart No: All stars listed in the tables are plotted in the large Constellation Chart, but when a star appears in a Finder Chart, this number is listed. Notes: When colors are subtle, the suffix *-ish* is used, e.g. *bluish*.

51 Piscium Double Star　　　　　　　　　　**Spec. A0**
m5.7, 9.5; Sep. 27.5", P.A. 83°　　　　　00^{h}32.4^m +06°57'
Finder Chart 24-5　　　　　　　　　　　　★★★★

4/6" Scopes–75x: 51 Piscium is a gorgeous blue-white and greenish pair.

55 Piscium Double Star　　　　　　　　　　**Spec. K0**
m5.4, 8.7; Sep. 6.5", P.A. 194°　　　　　00^{h}39.9^m +21°26'
Constellation Chart 24-1　　　　　　　　★★★★★

4/6" Scopes–75x: 55 Piscium is a beautiful double with a vivid color contrast of yellowish-orange and blue stars.

Wolf 28 White Dwarf Star　　　　　　　　**Spec. DF/DG**
m12.3　　　　　　　　　　　　　　　　00^{h}49.1^m +05°25'
Finder Chart 24-5　　*Van Maanen's Star*　★★

8/10" Scopes–100x: Wolf 28 is of special interest because it is one of the few white dwarf stars visible in amateur telescopes. White dwarfs are the end product of the evolution of solar-type stars. At the point when the core of a red giant of one or two solar masses runs out of helium to nucleosynthesize into carbon, it violently contracts, the sudden burst of energy thus released puffing off the star's outer layers. The outer layers expand into a planetary

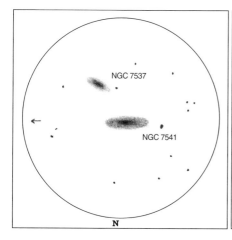

Figure 24-1. NGC 7537 & NGC 7541
13", f5.6–100x, by Steve Coe

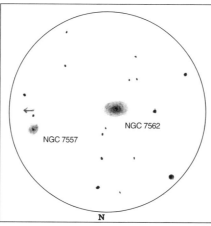

Figure 24-2. NGC 7557 & NGC 7562
12.5", f5–200x, by G. R. Kepple

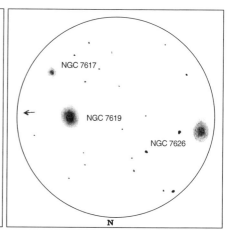

Figure 24-3. NGC 7617, 7619, & 7626
12.5", f5–200x, by G. R. Kepple

nebula; but the core, with no energy production to counteract the pull of gravity, continues contracting until the electrons in its now extremely dense gas are as packed together as the laws of quantum mechanics allow them to be. The core at that point stops contracting and is said to be "degenerate." It is now a true white dwarf, shining only by residual thermal energy. White dwarfs are thus very dense: van Maanen's Star has the mass of the Sun in a sphere about the size of the Earth, and consequently is a million times as dense as water. Because of their extremely small radiating surfaces, it takes billions of years for a white dwarf to cool from a true white DA dwarf to a yellowish DF or DG star like van Maanen's. This object is a close neighbor to the Solar System, only 13.8 light years away. Its very large annual proper motion of 2.98" in P.A. 155° led to its discovery.

65 Piscium Double Star Spec. F0
m6.3, 6.3; Sep. 4.4", P.A. 297° $00^h49.9^m$ +27°43′
Finder Chart 24-6 ★★★★

4/6" Scopes–100x: 65 Piscium is a fine yellowish starpair.

Psi-one (ψ^1) = 74 Piscium Double Star Spec. A2 & A0
m5.6, 5.8; Sep. 30.0", P.A. 159° $01^h05.6^m$ +21°28′
Constellation Chart 24-1 ★★★★

4/6" Scopes–50x: Psi-one Piscium is a fine, matched pair of bright, bluish-white stars.

77 Piscium Double Star Spec. F2 & F2
m6.8, 7.6; Sep. 33.0", P.A. 83° $01^h05.8^m$ +04°55′
Finder Chart 24-5 ★★★★

4/6" Scopes–50x: 77 Piscium is a nice, wide pair of yellowish stars easily split in small telescopes.

Zeta (ζ) = 86 Piscium Double Star Spec. A5 & F8
m5.6, 6.5; Sep. 23.0", P.A. 63° $01^h13.7^m$ +07°35′
Constellation Chart 24-1 ★★★★

4/6" Scopes–100x: Zeta has a white primary and a yellowish secondary.

Alpha (α) = 113 Piscium Double Star Spec. A2
m4.2, 5.1; Sep. 1.8", P.A. 272° $02^h02.0^m$ +02°46′
Finder Chart 24-8 *Alrisha* ★★★★

12/14" Scopes–200x: Alpha Piscium is a very close pair of bright bluish-white stars that are rather difficult to split. At 200x, it appears as two disks in contact.

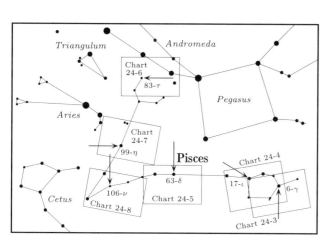

Master Finder Chart 24-2. Pisces Chart Areas
Guide stars indicated by arrows.

24.3 Deep-Sky Objects

NGC 7428   U12262   Galaxy   Type (R:)SAB(r)a: pec
ø2.3′ × 1.2′, m12.5v, SB 13.5 $22^h57.3^m$ $-01°02′$
Finder Chart 24-3   ★★

12/14″ Scopes–100x: NGC 7428, located 26′ NNW of a 6th magnitude star, is a rather dim, small, circular object with a bright center.

16/18″ Scopes–150x: This galaxy has a small, bright core embedded in a moderately faint 2′ × 1.25′ NNE–SSW halo. A 12th magnitude star lies 2.25′ NNE.

NGC 7537   H429²   Galaxy   Type SAbc:
ø1.9′ × 0.5′, m13.2v, SB 13.0 $23^h14.6^m$ $+04°30′$
Finder Chart 24-3, Figure 24-1   ★★

12/14″ Scopes–125x: NGC 7537 is much smaller and fainter than its companion NGC 7541 3′ to its NE. It has a faint 1′ × 0.25′ ENE–WSW halo pointing toward the companion. The center is slightly brighter with a very faint stellar nucleus.

16/18″ Scopes–150x: NGC 7537, a companion to the larger brighter NGC 7541 lying 3′ NE, has a well concentrated 1.5′ × 0.5′ NE–SW halo with a small, bright core. 13th magnitude stars are 1.25′ east and 1.75′ west.

NGC 7541   H430²   Galaxy   Type SB(rs)bc: pec II
ø3.1′ × 1.0′, m11.7v, SB 12.8 $23^h14.7^m$ $+04°32′$
Finder Chart 24-3, Figure 24-1   ★★★★

8/10″ Scopes–125x: This galaxy is the brighter and larger of a fine pair with NGC 7537 lying 3′ SW. NGC 7541 has a bright, extended core surrounded by a moderately bright halo elongated 2.5′ × 1′ ESE–WNW. A 12th magnitude star lies 2′ from the galaxy's center beyond the eastern tip.

16/18″ Scopes–150x: NGC 7541 has a bright 3′ × 1′ ESE–WNW halo with a large, extended, irregularly illuminated core and a very faint stellar nucleus. The center is somewhat mottled and slightly brighter along the southern flank. A small equilateral triangle of 13.5 magnitude stars lies 3.75′ ESE.

NGC 7562   H467²   Galaxy   Type E2-3
ø2.1′ × 1.6′, m11.6v, SB 12.7 $23^h16.0^m$ $+06°41′$
Finder Chart 24-4, Figure 24-2   ★★★

8/10″ Scopes–100x: This galaxy has a fairly bright, circular 1.25′ halo moderately brighter toward its center.

16/18″ Scopes–150x: NGC 7562 displays a bright core with a stellar nucleus embedded in a 2′ × 1.5′ E–W halo. 1.5′ NW is a 14th magnitude star and 2′ east a 13th magnitude star. 5′ west is NGC 7557, a very faint 30″ glow surrounding a faint stellar nucleus.

NGC 7611   U12509   Galaxy   Type SB0+:
ø1.2′ × 0.6′, m12.5v, SB 12.0 $23^h19.6^m$ $+08°04′$
Finder Chart 24-4   ★★★

8/10″ Scopes–125x: NGC 7611, a member of the Pegasus I Galaxy Cluster, is 6′ NW of a 7th magnitude yellowish-orange star. It is a fairly faint object, elongated 1′ × 0.25′ NW–SE, and has a faint stellar nucleus.

12/14″ Scopes–150x: NGC 7611, located 12′ SW of NGC 7619 in Pegasus, has a fairly bright halo elongated 1.25′ × 0.5′ NW–SE with a poorly concentrated core and a prominent stellar nucleus. A very wide pair of 13.5 magnitude stars lies 1.5′ NNW.

NGC 7617   Galaxy   Type SA0°:
ø0.8′ × 0.6′, m13.8v, SB 12.8 $23^h20.2^m$ $+08°10′$

NGC 7619   H439²   Galaxy   Type E
ø2.8′ × 2.5′, m11.1v, SB 13.0 $23^h20.2^m$ $+08°12′$

NGC 7626   H440²   Galaxy   Type E pec:
ø2.4′ × 1.9′, m11.1v, SB 12.6 $23^h20.7^m$ $+08°13′$
Finder Chart 24-4, Figure 24-3   ★★/★★★/★★★

12/14″ Scopes–125x: NGC 7619 and NGC 7626 are fairly bright twin elliptical galaxies 7′ E–W apart in the Pegasus I Galaxy Group. They are technically in Pegasus, just north of the Pegasus/Pisces border. NGC 7619, the westernmost of the pair, is elongated 1.5′ × 1.25′ NE–SW and has a slightly oval 50″ × 40″ core. NGC 7626 has a 1.5′ diameter halo slightly elongated NNE–SSW with a round 50″ core. A 13.5 magnitude star lies 1′ west. NGC 7617, a small, faint companion of NGC 7619 3′ lying to its SSW in Pisces

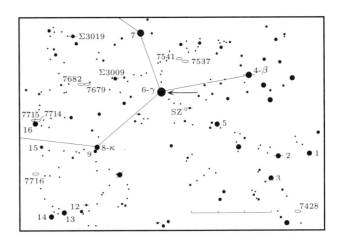

Finder Chart 24-3. 6–γ Psc: $23^h17.1^m$ $+03°17$

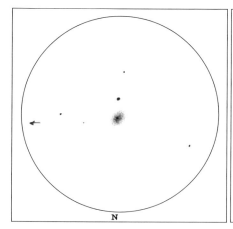

Figure 24-4. NGC 7716
8″, f10 SCT–100x, by Stan Howerton

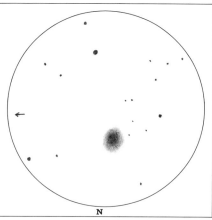

Figure 24-5. NGC 7750
13″ f5.6–100x, by Steve Coe

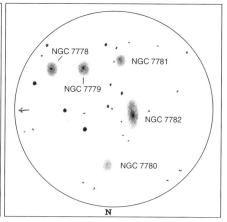

Figure 24-6. NGC 7782 galaxy group
13″ f5.6–100x, by Steve Coe

proper, has a prominent stellar nucleus surrounded by a faint, tiny halo elongated 25″ × 12″ NNE–SSW.

NGC 7679 U12618 Galaxy Type (R':)Sb? pec
ø1.7′ × 1.1′, m12.9v, SB 13.5   23^{h}28.8^m +03°31′
Finder Chart 24-4   ★★

8/10″ Scopes–100x: NGC 7679, located 5.5′ SE of a 10th magnitude star, is faint, elongated 1′ × 0.75′ E–W, and has a very faint stellar nucleus. A magnitude 11.5 magnitude star is 2′ WNW and a 12th magnitude star 2′ NNE. Galaxy NGC 7682 lies 4.5′ ENE.

16/18″ Scopes–150x: NGC 7679 has a tiny, bright core with a stellar nucleus surrounded by a much fainter diffuse halo elongated 1.5′ × 1′ E–W. At 200x, a star 5.5′ NW is seen as a very close double with 10th magnitude components in contact aligned NE–SW.

NGC 7682 U12622 Galaxy Type SB(r)a: pec
ø1.0′ × 0.8′, m13.2v, SB 12.8   23^{h}29.1^m +03°32′
Finder Chart 24-3   ★★

12/14″ Scopes–125x: NGC 7682, located 4.5′ ENE of NGC 7679, is very faint, small, and round.

16/18″ Scopes–150x: NGC 7682 has a fairly faint, circular 1′ diameter halo slightly elongated N–S. It is slightly brighter toward the center where a faint stellar nucleus is just visible.

NGC 7714 U12699 Galaxy Type SB:(s)b? pec
ø1.6′ × 1.2′, m12.5v, SB 13.0   23^{h}36.2^m +02°09′
Finder Chart 24-3   ★

12/14″ Scopes–125x: NGC 7714, a very faint object, lies just 4′ NW of the magnitude 5.7 star 16 Piscium, the glare of which makes the galaxy difficult to see. Its halo is diffuse, about 1′ in diameter, and slightly brighter in the center. A 12th magnitude star is 1′ SW. Immediately to its east is its close companion galaxy NGC 7715, a very faint streak elongated 1.5′ × 0.5′.

NGC 7716 U12702 Galaxy Type SA:(r)b II
ø2.1′ × 1.7′, m12.1v, SB 13.3   23^{h}36.5^m +00°18′
Finder Chart 24-3, Figure 24-4   ★★

12/14″ Scopes–125x: NGC 7716, located 2′ north of a 9.5 magnitude star, has a very faint 1′ diameter halo slightly brighter toward its center, where a faint stellar nucleus is just visible.

16/18″ Scopes–150x: NGC 7716 appears fairly faint and has a circular 2′ diameter halo with a weakly concentrated core containing a stellar nucleus.

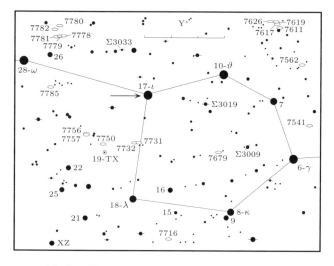

Finder Chart 24-4. 17–ι Psc: 23^{h}39.9^m +05°38′

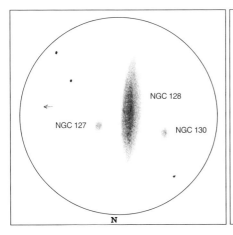
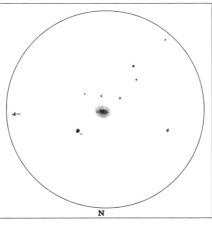
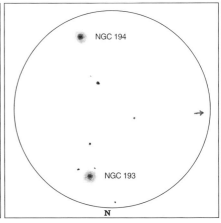

Figure 24-7. NGC 127, 128, 130
20" f4.5–315x, by Richard W. Jakiel

Figure 24-8. NGC 182
8", f10 SCT–100x, by Stan Howerton

Figure 24-9. NGC 193 & NGC 194
8", f10 SCT–100x, by Stan Howerton

NGC 7731   U12737   Galaxy (R:)Sa: pec
⌀1.4′ × 0.9′, m13.5v, SB 13.6   23^{h}41.6^m +03°45′
Finder Chart 24-4   ★★

12/14″ Scopes–125x: NGC 7731, located 1.25′ west of an 11.5 magnitude star, appears faint, elongated 1′ × 0.5′ NE–SW, and slightly brighter at its center. Its companion galaxy, NGC 7732, 1.5′ to its SE, is only a faint smudge elongated 1′ × 0.25′ E–W.

NGC 7750   H427³   Galaxy Type (R′)SB(r)b: II
⌀1.6′ × 0.7′, m12.9v, SB 12.9   23^{h}46.7^m +03°47′
Finder Chart 24-4, Figure 24-5   ★★

12/14″ Scopes–125x: NGC 7750 lies 20′ NNE of the beautiful variable 19 = TX Piscium (mag 6.9 to 7.7), a poppy red carbon star. Its halo is fairly faint, elongated 1.5′ × 0.75′ N–S, and slightly brighter at the center. A 12th magnitude star is 3′ ESE and an 11th magnitude star 6′ SE.

NGC 7757   U12788   Galaxy Type SA(rs)c
⌀2.3′ × 2.0′, m12.7v, SB 14.2   23^{h}48.8^m +04°10′
Finder Chart 24-4   ★★

12/14″ Scopes–125x: NGC 7757, located 13′ east of a wide pair of 8th magnitude stars, is a very faint, diffuse, round 1.5′ glow. To the north, two magnitude 12.5 and 13 stars 1′ apart form a line with the galaxy.

16/18″ Scopes–150x: NGC 7757 has a faint, diffuse, round 2′ diameter halo of even surface brightness. Averted vision brings out some granulation. A 12.5 magnitude star lies 3′ west.

NGC 7778   Galaxy Type E
⌀1.1′ × 1.0′, m12.7v, SB 12.7   23^{h}53.3^m +07°52′
Finder Chart 24-4, Figure 24-6   ★★

16/18″ Scopes–150x: NGC 7778 is the westernmost of a group of five galaxies, of which the brightest, NGC 7782, is 10′ NE. Just 2′ west is NGC 7779. NGC 7778 is smaller and fainter than its close companion, with a circular 1′ diameter halo and a stellar nucleus.

NGC 7779   H232³   Galaxy Type (R′)SA0/a:
⌀1.3′ × 0.9′, m12.7v, SB 12.7   23^{h}53.4^m +07°52′
Finder Chart 24-4, Figure 24-6   ★★

16/18″ Scopes–150x: NGC 7779 is a close pair with NGC 7778 lying 2′ east. It is the second brightest in a group of five galaxies. NGC 7779 is as bright but more conspicuous than its companion, its slightly larger halo elongated 1.25′ × 1′ N–S. The nucleus is not as bright as that of NGC 7778. NGC 7881, 5′

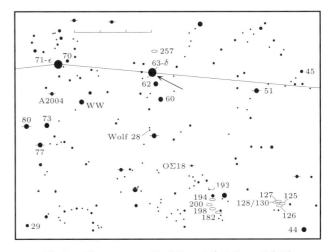

Finder Chart 24-5. 63-δ Psc: 00^{h}48.7^m +07°35′

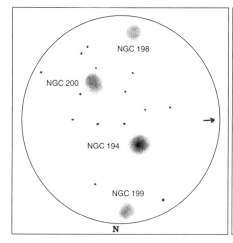

Figure 24-10. NGC 194, 198, 199, & 200
11", f10 SCT–100x, by L. J. Kemble

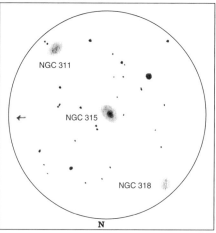

Figure 24-11. NGC 311, 315, & 318
12.5", f5–200x, by G. R. Kepple

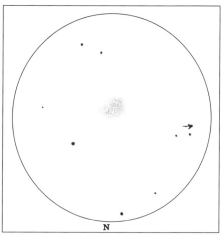

Figure 24-12. Andromeda II
11", f10 SCT–100x, by L. J. Kemble

ESE, is a very faint, round 1′ diameter smudge. NGC 7782, the brightest member of the galaxy group, is 8′ NE.

NGC 7782 H233³ Galaxy Type SA(s)b I-II
⌀1.9′ × 1.2′, m12.2v, SB 13.0   23ʰ53.9ᵐ +07°58′
Finder Chart 24-4, Figure 24-6   ★★

16/18″ Scopes–150x: NGC 7782 is the brightest in a group of five galaxies that includes NGC 7780 6′ NNW and the E–W row of NGC 7778-79-81 to the SW. Both NGC 7780 and NGC 7781 are extremely faint, diffuse, round smudges seen only with averted vision. NGC 7782 has a faint 2′ × 0.75′ N–S halo that gradually brightens to an inconspicuous stellar nucleus. A 15th magnitude star lies 1′ east.

NGC 7785 H468² Galaxy Type E5-6
⌀1.8′ × 1.2′, m11.6v, SB 12.3   23ʰ55.3ᵐ +05°55′
Finder Chart 24-4   ★★

8/10″ Scopes–125x: NGC 7785 lies at the NW vertex of a 3′ × 2.5′ triangle with two 10th magnitude stars. A 9th magnitude star lies 4′ WNW. NGC 7785 has a fairly obvious halo elongated 1.25′ × 0.5′ NW–SE with a stellar nucleus.

12/14″ Scopes–150x: NGC 7785 has a fairly bright, 1.5′ × 0.75′ NW–SE halo with a small, well concentrated core and a nonstellar nucleus.

NGC 125 H869³ Galaxy Type (R:)SA0+ pec
⌀1.6′ × 1.5′, m12.1v, SB 12.9   00ʰ28.8ᵐ +02°50′
Finder Chart 24-5   ★★

12/14″ Scopes–125x: NGC 125 is a faint, small, round companion of NGC 128, which is 6.5′ to its WSW. To the south is a 20″ NNW–SSE double of 12th magnitude stars.

16/18″ Scopes–150x: NGC 125 has a moderately concentrated core with a stellar nucleus surrounded by a much fainter, circular 1′ diameter halo.

NGC 128 H854² Galaxy Type S0°: pec sp
⌀3.0′ × 1.0′, m11.8v, SB 12.8   00ʰ29.2ᵐ +02°51′
Finder Chart 24-5, Figure 24-7   ★★★★

8/10″ Scopes–100x: NGC 128 is the brightest in a group of five galaxies, of which only NGC 128 and NGC 125, 6.5′ to its WSW, are visible in 8″ telescopes. NGC 128 is fairly bright, elongated 1.75′ × 0.5′ N–S, and very slightly brighter through the center. NGC 125 is a very faint, round smudge.

16/18″ Scopes–150x: NGC 128 is a moderately bright lenticular galaxy elongated 2.5′ × 0.75′ N–S with an extended, mottled core and a nonstellar nucleus. It is flanked by two faint, circular, nearly identical companions, NGC 130 lying 1.5′ NE and NGC 127 located 2′ NW.

NGC 182 H870³ Galaxy Type SAB(rs:)ab pec
⌀2.1′ × 1.7′, m12.4v, SB 13.7   00ʰ38.2ᵐ +02°44′
Finder Chart 24-5, Figure 24-8   ★★

8/10″ Scopes–100x: NGC 182, located 4′ SE of a 7.5 magnitude blue star, is just visible as a very faint, round 1.5′ glow slightly brighter at the center.

16/18″ Scopes–150x: NGC 182 has a well concentrated core with a faint stellar nucleus surrounded by a 2′ diameter halo. It is the brightest and southernmost in a group of a dozen galaxies. A 13th magnitude star lies 1.75′ SE.

Pisces 323

NGC 193 H595³ Galaxy Type SB?(s?)0°
⌀1.6′ × 1.5′, m12.2v, SB 13.0   00ʰ39.3ᵐ +03°20′
Finder Chart 24-5, Figure 24-9   ★★

8/10″ Scopes-100x: NGC 193, located 3′ ESE of a 10th magnitude star, is a faint, small, circular glow, elongated 1.25′ × 1′ NE–SW with a slightly brighter center. The view is similar in 12″ scopes.

16/18″ Scopes-150x: NGC 193 has a fairly faint circular 1.5′ halo with a well concentrated core. 13th magnitude stars are on the western edge and near the nucleus.

NGC 194 H856² Galaxy Type E1-2
⌀1.6′ × 1.6′, m12.2v, SB 13.0   00ʰ39.3ᵐ +03°02′
Finder Chart 24-5, Figures 24-9 & 24-10   ★★★

8/10″ Scopes-100x: NGC 194 lies 5′ south of a 7.5 magnitude star in the NGC 182 galaxy group. It has a faint stellar nucleus embedded in a diffuse, round 1.25′ diameter halo. Galaxies NGC 198 and NGC 200 lie to the south and SSE, respectively, and require averted vision.

16/18″ Scopes-150x: A fairly faint, 2′ diameter halo surrounds a small, bright core within which is embedded a stellar nucleus. Galaxy NGC 200 lies 10′ SSE.

NGC 198 H857² Galaxy Type SA(rs:)c II
⌀1.2′ × 1.2′, m13.2v, SB 13.4   00ʰ39.4ᵐ +02°48′
Finder Chart 24-5, Figure 24-10   ★★

8/10″ Scopes-100x: NGC 198 lies about 14′ almost due south of NGC 194. It is very faint, small, round, diffuse and requires averted vision to be seen. It is smaller but more conspicuous than NGC 200 lying 6′ NNE.

16/18″ Scopes-150x: NGC 198 is fairly faint, its round 1′ diameter halo containing a slightly brighter center.

NGC 200 U420 Galaxy Type SB(s)c pec I
⌀1.8′ × 1.1′, m12.6v, SB 13.2   00ʰ39.6ᵐ +02°53′
Finder Chart 24-5, Figure 24-10   ★★

8/10″ Scopes-125x: NGC 200, though larger than NGC 198 lying 6′ to its SSW, is fainter and like that galaxy requires averted vision. The halo seems elongated N–S.

12/14″ Scopes-125x: NGC 200 appears very faint, rather small, and round with a slightly brighter center. NGC 199, NGC 194, NGC 200, and NGC 198 form a N–S zigzagging line of galaxies.

16/18″ Scopes-150x: The 1.5′ × 0.75′ NNW–SSE halo is faint, diffuse and slightly brighter in its center. NGC 198 lies 6′ SSW.

Figure 24-13. *NGC 266 lies just north of an 8th magnitude star which interferes with observation. In medium-aperture telescopes the galaxy shows a bright core with a stellar nucleus surrounded by a faint, circular halo. Alexander Brownlee made this 60 minute exposure on 103a-O film with a 16″ Newtonian reflector.*

NGC 257 H863² Galaxy Type Scd:
⌀2.0′ × 1.5′, m12.6v, SB 13.7   00ʰ48.1ᵐ +08°19′
Finder Chart 24-5   ★★

12/14″ Scopes-125x: NGC 257, 3/4° north of magnitude 4.4 Delta (δ) = 63 Piscium, has a faint 1.5′ × 1′ E–W halo gradually brighter toward its center.

16/18″ Scopes-150x: NGC 257 is a fairly faint galaxy with a 2′ × 1.5′ E–W oval halo containing a broad well concentrated core.

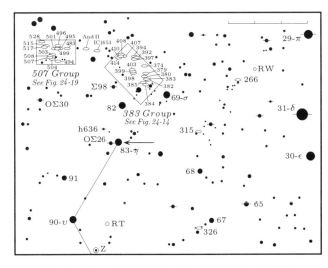

Finder Chart 24-6. 83-τ Psc: 01ʰ11.6ᵐ +30°05′

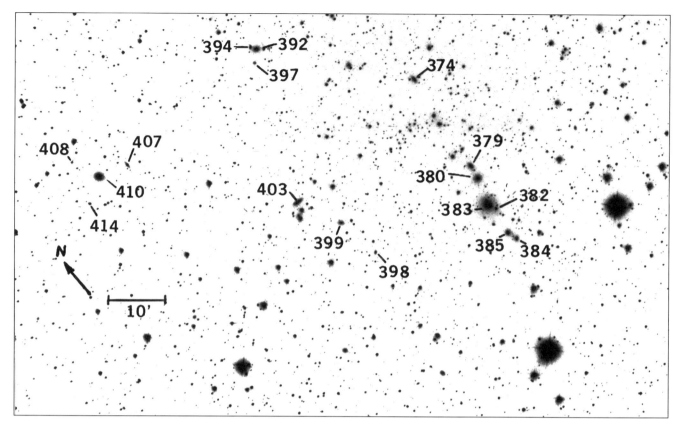

Figure 24-14. *This photograph can be used to find the members of the NGC 383 galaxy group. William Harris recorded this 60 minute exposure on hypered 2415 film with an 8", f4 Wright Newtonian. North is at upper left corner.*

NGC 266 H153³ Galaxy Type SB(rs)ab
ø3.2′ × 3.0′, m11.6v, SB 13.9   00ʰ49.8ᵐ +32°16′
Finder Chart 24-6, Figure 24-13   ★★

8/10″ Scopes–100x: NGC 266, located 4′ north of an 8th magnitude star, is a faint, small galaxy with a diffuse, circular halo considerably brighter in the center.

16/18″ Scopes–150x: NGC 266 has a fairly conspicuous 2.5′ diameter halo slightly elongated E–W containing a bright core with a stellar nucleus. A string of 13.5 magnitude stars is to the NNW.

NGC 315 H210² Galaxy Type E+
ø3.0′ × 2.6′, m11.2v, SB 13.2   00ʰ57.8ᵐ +30°21′
Finder Chart 24-6, Figure 24-11   ★★

12/14″ Scopes–125x: NGC 315 lies 3.5′ NNW of an 8th magnitude star at the center of a NE–SW row of three galaxies. It has a moderately faint 2.5′ × 2′ NE–SW halo with a bright core and a stellar nucleus. NGC 311, located 5.5′ SW, is nearly the same size but has only a slight central brightening. NGC 318, located 5.5′ NE of NGC 315, is much fainter and smaller than its two companions, its halo only 1′ in diameter.

NGC 326 U601 Galaxy Type E+ pec
ø1.1′ × 1.1′, m13.2v, SB 13.3   00ʰ58.4ᵐ +26°52′
Finder Chart 24-6   ★★

16/18″ Scopes–150x: NGC 326 lies near the center of an isosceles triangle of three bright stars, a magnitude 6.5 star at the south vertex 5′ from the galaxy, a magnitude 9 star at the NNE vertex, and a neat, close double at the NW vertex. The galaxy has a faint, round 1.5′ diameter halo with a slight brightening at center.

NGC 374 Galaxy Type S0/a
ø1.2′ × 0.5′, m13.4v, SB 12.7   01ʰ07.1ᵐ +32°46′
Finder Chart 24-6, Figure 24-14   ★★

16/18″ Scopes–150x: NGC 374, a member of the NGC 383 galaxy group, is very faint, elongated 1.25′ × 0.75′ N–S, and considerably brighter at its center. It is flanked by 15th magnitude stars 1′ NE and 1′ WSW.

NGC 379 H215² Galaxy Type S0
⌀1.5′ × 0.8′, m12.9v, SB 12.9   01ʰ07.3ᵐ +32°31′

NGC 380 H216² Galaxy Type E2
⌀1.3′ × 1.1′, m12.5v, SB 12.8   01ʰ07.3ᵐ +32°29′
Finder Chart 24-6, Figure 24-14   ★★/★★

16/18″ Scopes–150x: NGC 379 and NGC 380, members of the NGC 383 galaxy group, are a close, 2′ N–S pair. NGC 379, the northernmost of the two, has a fairly faint, diffuse halo elongated 1′ × 0.75′ N–S. NGC 380 has a conspicuous core with a stellar nucleus embedded in a fairly faint, circular 1.25′ diameter halo.

NGC 382 U688 Galaxy Type E:
⌀0.3′ × 0.3′, m13.2v, SB 10.5   01ʰ07.4ᵐ +32°24′

NGC 383 H217² Galaxy Type SA0–:
⌀2.0′ × 1.7′, m12.4v, SB 13.6   01ʰ07.4ᵐ +32°25′
Finder Chart 24-6, Figure 24-14   ★/★★

12/14″ Scopes–125x: NGC 383 is the brightest member of a well-populated galaxy group spread over a 1.5° area. It is fairly bright and its round 1.5′ diameter halo contains a much brighter center. Just 30″ to its SSW is its companion galaxy, NGC 382, a very faint, tiny, round spot. The immediate vicinity is well-provided with galaxy-pairs: roughly 5′ north is the NGC 379+380 pair, and 5′ south is the faint, tiny NGC 384+385.

16/18″ Scopes–150x: In large telescopes NGC 383 has a large bright core embedded in a circular 2′ diameter halo. NGC 382, 30″ SSW, remains at these apertures merely a very faint, nearly stellar, spot.

NGC 384 U686 Galaxy Type E3
⌀1.1′ × 0.8′, m13.1v, SB 12.8   01ʰ07.4ᵐ +32°18′

NGC 385 U687 Galaxy Type SA0–:
⌀1.3′ × 1.0′, m13.0v, SB 13.1   01ʰ07.4ᵐ +32°19′
Finder Chart 24-6, Figure 24-14   ★/★

16/18″ Scopes–150x: NGC 384 and NGC 385 are a nearly identical N–S pair of galaxies 4′–6′ south of NGC 383. Both galaxies are very faint, small, and circular, and have prominent cores. NGC 385, slightly larger and brighter than NGC 384, has a 1.25′ diameter halo elongated N–S.

NGC 392 Galaxy Type S0–:
⌀1.2′ × 0.9′, m12.7v, SB 12.6   01ʰ08.4ᵐ +33°07′

NGC 394 Galaxy Type ?
⌀0.5′ × 0.2′, m14.0v, SB 11.3   01ʰ08.4ᵐ +33°08′
Finder Chart 24-6, Figure 24-14   ★★/★

16/18″ Scopes–150x: NGC 392 has a bright core with a sharp stellar nucleus embedded in a circular 1′ diameter halo. A 13th magnitude star is 1′ SW. 1′ NE is NGC 392's close companion NGC 394, a tiny 20″ diameter spot with a stellar nucleus. NGC 397 lies 3′ SE. These three galaxies are north of the main body of the NGC 383 galaxy group

NGC 397 Galaxy Type ?
⌀0.7′ × 0.5′, m14.5v, SB 13.2   01ʰ08.4ᵐ +33°06′
Finder Chart 24-6, Figure 24-14   ★

16/18″ Scopes–150x: NGC 397, located 3′ SE of NGC 392, is an extremely dim, nonstellar spot.

NGC 398 Galaxy Type ?
⌀0.3′ × 0.2′, m14.6v, SB 11.4   01ʰ08.8ᵐ +32°30′
Finder Chart 24-6, Figure 24-14   ★

16/18″ Scopes–150x: NGC 398, a minor member of the NGC 383 galaxy group due east of the cluster's core, is barely visible as an extremely faint, very tiny round spot.

NGC 399 Galaxy Type SBa:
⌀1.1′ × 0.9′, m13.6v, SB 13.4   01ʰ09.0ᵐ +32°37′
Finder Chart 24-6, Figure 24-14   ★

16/18″ Scopes–150x: NGC 399, an NGC 383 galaxy group member, has a faint core with a stellar nucleus enveloped by a circular 1′ halo.

NGC 403 U717 Galaxy Type S0/a:
⌀1.8′ × 0.7′, m12.5v, SB 12.6   01ʰ09.2ᵐ +32°45′
Finder Chart 24-6, Figure 24-14   ★★

12/14″ Scopes–125x: This NGC 383 galaxy group member is a faint 1.5′ × 0.5′ E–W streak. To its immediate south is a triangle of 11th–12th magnitude stars.

16/18″ Scopes–150x: NGC 403 displays a bright core with a small nucleus embedded in a faint 2′ × 0.75′ E–W halo.

NGC 407 Galaxy Type S0/a: sp
⌀1.8′ × 0.3′, m13.4v, SB 12.6   01ʰ10.6ᵐ +33°07′
Finder Chart 24-6, Figure 24-14   ★★

12/14″ Scopes–125x: The faint NGC 407, located 3.25′ WSW of NGC 410, appears elongated 1.25′ × 0.5′ N–S.

16/18″ Scopes–150x: NGC 407 has a prominent core embedded in a highly extended 2′ × 0.5′ N–S halo. A 13th magnitude star lies 2′ WSW.

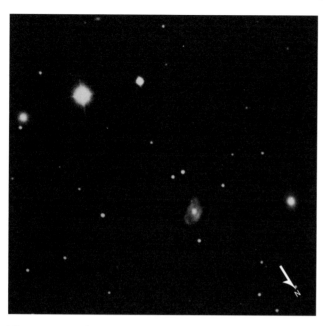

Figure 24-15. *Galaxies NGC 467-470-474 (left to right) are a faint trio for larger instruments. South is up in this 60 minute exposure on 103a-O film with a 16" Newtonian by Alexander Brownlee.*

NGC 408 Galaxy Type ?
ø –, m –   01ʰ10.9ᵐ +33°06′
Finder Chart 24-6, Figure 24-14   ★★

16/18" Scopes–150x: NGC 408, an extremely small, faint galaxy 3′ NNE of NGC 410, is a nearly stellar spot.

NGC 410 H220² Galaxy Type E+
ø2.3′ × 1.8′, m11.5v, SB 12.9   01ʰ11.0ᵐ +33°09′
Finder Chart 24-6, Figure 24-14   ★★

12/14" Scopes–125x: This galaxy, at the extreme northeastern end of the NGC 383 galaxy group, is a very faint, small round smear of light.

16/18" Scopes–150x: NGC 410 is a fairly obvious galaxy with a stellar nucleus, a large bright core, and a 2′ × 1.5′ NE–SW oval halo. It is accompanied by three very faint, tiny companions: NGC 407 is 5′ west, NGC 408 only 2.75′ NNE, and NGC 414 about 5′ SE.

NGC 414 Galaxy Type ?
ø0.7′ × 0.4′, m13.5v, SB 12.0   01ʰ11.3ᵐ +33°06′
Finder Chart 24-6, Figure 24-14   ★

12/14" Scopes–125x: This galaxy is very faint, small, and round.

16/18" Scopes–150x: NGC 414 appears faint, elongated 0.50′ × 0.25′ NE–SW, and slightly brighter at the center.

Andromeda II Galaxy Type E?
ø –, m13.5p   01ʰ16.4ᵐ +33°27′
Finder Chart 24-6, Figure 24-12   ★

16/18" Scopes–150x: Andromeda II, an extremely faint galaxy, is a dwarf-spheroidal member of our Local Galaxy Group lying about 2.4 million light years away, nearly the same as Messier 31. It has a diffuse, circular 2′ diameter halo.

NGC 467 H108¹ Galaxy Type S0– (or E0)
ø2.5′ × 2.5′, m11.8v, SB 13.7   01ʰ19.2ᵐ +03°18′
Finder Chart 24-8, Figure 24-15 & 24-16   ★★

8/10" Scopes–100x: NGC 467, located 3′ NW of an 8th magnitude star at the western end of a short galaxy-chain with NGC 470 and NGC 474, has a moderately faint, 1′ diameter halo with a slightly brighter center.

12/14" Scopes–125x: NGC 467 is fairly faint, its 1.5′ diameter halo moderately brightening to a mottled core. A 13th magnitude star lies 2′ SW and a 14th magnitude star 1.5′ west.

NGC 470 H250³ Galaxy Type SA(r)b pec II
ø3.0′ × 1.8′, m11.8v, SB 13.5   01ʰ19.7ᵐ +03°25′
Finder Chart 24-8, Figures 24-15 & 24-16   ★★

8/10" Scopes–100x: NGC 470 is in the middle of a NE–SW line with NGC 474 lying 5.5′ east and NGC 467 lying 7′ SW. It has a faint, diffuse 2′ × 1′ NNW–SSE halo slightly brighter in the center.

12/14" Scopes–125x: NGC 470 has a fairly faint halo elongated 2.5′ × 1.5′ NNW–SSE with a moderately concentrated core. 13th magnitude stars are 1.5′ north, 2.25′ SE, and (two stars) 2.25′ south.

NGC 473 U859 Galaxy Type SAB(r)0/a:
ø2.6′ × 1.7′, m12.5v, SB 14.0   01ʰ19.9ᵐ +16°33′
Finder Chart 24-7, Figure 24-17   ★★

8/10" Scopes–100x: NGC 473, located 5′ NW of a 10th magnitude orange star, has a faint 1′ × 0.5′ NNW–SSE halo slightly brighter in the center.

12/14" Scopes–125x: NGC 473 is fairly faint, elongated 1.5′ × 1′ NNW–SSE, and has a faint stellar nucleus. Two very faint stars are at its eastern edge.

NGC 474 H251³ Galaxy Type (R′)SAB0° pec
ø10.0′ × 9.2′, m11.5v, SB 16.3   01ʰ020.1ᵐ +03°25′
Finder Chart 24-8, Figures 24-15 & 24-16   ★★

8/10" Scopes–100x: NGC 474, ENE of galaxies NGC 467 and NGC 470, has a prominent nonstellar nucleus embedded in a much fainter, circular 1.5′ diameter

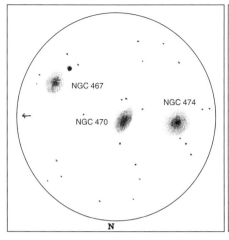

Figure 24-16. NGC 467, 470, & 474
12.5", f5–200x, by G. R. Kepple

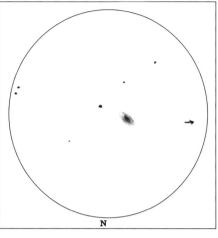

Figure 24-17. NGC 473
8", f10 SCT–100x, by Stan Howerton

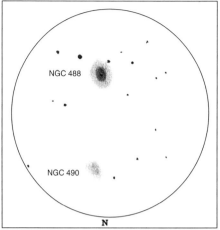

Figure 24-18. NGC 488 & NGC 490
13", f5.6–100x, by Steve Coe

halo. It appears much smaller than its catalog size: apparently the outer halo is extremely faint and not visible in medium-size telescopes.

12/14" Scopes–125x: NGC 474 appears nearly the same size as but somewhat brighter than NGC 470 lying 5′ ENE. A very faint 2.5′ diameter halo surrounds the galaxy's bright but tiny nucleus.

NGC 488 H252³ Galaxy Type SA(r)b I
⌀5.5′ × 4.0′, m10.3v, SB 13.5   01ʰ21.8ᵐ +05°15′
Finder Chart 24-8, Figure 24-18   ★★★★

8/10" Scopes–100x: This galaxy is easily found 9′ west of an 8.5 magnitude bluish star and just north of an 11.5 magnitude star, the latter at the east end of an ENE–WSW row with two other magnitude 11.5 stars. NGC 488 is fairly bright, elongated 2′ × 1.5′ N–S, and has a well concentrated core.

12/14" Scopes–125x: NGC 488 has a bright, large 3.5′ × 2.5′ N–S halo with diffuse edges and a bright core containing a stellar nucleus. Four very faint galaxies are to the NE, but only NGC 490, lying 8′ NE, is visible with direct vision as a diffuse 2′ × 1.5′ NE–SW oval.

NGC 483 U906 Galaxy Type S?
⌀0.7′ × 0.7′, m13.1v, SB 12.2   01ʰ22.0ᵐ +33°32′
Finder Chart 24-6, Figure 24-19   ★★

16/18" Scopes–150x: NGC 483, the westernmost member of the NGC 507 galaxy group, is just west of a small triangle of 10.5 to 12th magnitude stars. It has a faint, round 30″ diameter halo slightly brighter in the center.

NGC 495 U920 Galaxy Type (R′)SB(s)0/a pec:
⌀1.2′ × 0.7′, m12.9v, SB 12.6   01ʰ22.9ᵐ +33°28′
Finder Chart 24-6, Figure 24-19   ★★

16/18" Scopes–150x: NGC 495, a member of the NGC 507 galaxy group, is at the SW corner of a galaxy-triangle with NGC 496 lying 4′ to its NNE and the brighter NGC 499 is 3′ to its east. It has a faint 1.5′ × 1′ NNW–SSE halo that is slightly brighter in the center. A triangle of 12th magnitude stars surrounds it.

NGC 494 U919 Galaxy Type Sab
⌀1.9′ × 0.7′, m12.8v, SB 13.0   01ʰ23.0ᵐ +33°11′
Finder Chart 24-6, Figure 24-19   ★★

16/18" Scopes–150x: NGC 494, a member of the NGC 507 galaxy group, lies 7′ south of a 7th magnitude star. Its halo is faint, elongated 2′ × 0.75′ E–W, and has a prominent core. A 14th magnitude star touches its southern edge; and 2′ SW is a curved row of three 13th magnitude stars.

NGC 499 IC 1686 Galaxy Type S0–
⌀1.7′ × 1.3′, m12.1v, SB 12.8   01ʰ23.2ᵐ +33°28′
Finder Chart 24-6, Figure 24-19   ★★

12/14" Scopes–125x: NGC 499, one of the most conspicuous members of the NGC 507 galaxy group, is 10′ NNW of a 7th magnitude star. It is the brightest and largest galaxy in a subgroup with NGC 495 lying 4′ NNW. It has a bright core embedded in a 2′ × 1.5′ ENE–WSW halo.

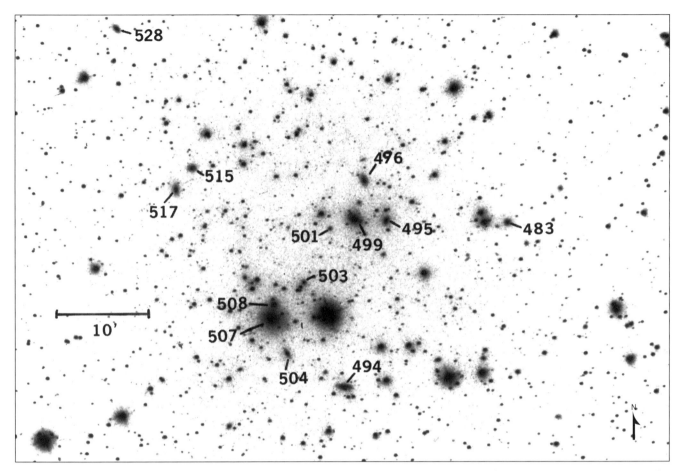

Figure 24-19. *This photograph of the NGC 507 galaxy cluster makes a helpful finder chart. As a starting point, setting circles may be set on the bright star (m7.8) below center at 01h23.1m +33°17. William Harris recorded this 60 minute exposure on hypered 2415 Kodak Tech Pan film with an 8", f4 Wright Newtonian.*

NGC 496 Galaxy Type Sbc
ø1.6′ × 0.8′, m13.3v, SB 13.4   01ʰ23.3ᵐ +33°33′
Finder Chart 24-6, Figure 24-19   ★

16/18″ Scopes–150x: NGC 496, an NGC 507 galaxy group member, is the faintest in a subgroup with NGCs 495 and 496. Its very faint, diffuse halo is elongated 1.25′ × 0.5′ NNE–SSW and not perceptibly brighter in its center.

NGC 501 Galaxy Type –
ø –, m14.5v, SB –   01ʰ23.4ᵐ +33°27′
Finder Chart 24-6, Figure 24-19   ★

16/18″ Scopes–150x: NGC 501, a minor member of the NGC 507 galaxy group, is 9′ north of a 7th magnitude star and 3.5′ ESE of NGC 499, and is a very faint, nearly stellar spot.

NGC 504 U935 Galaxy Type S0
ø1.6′ × 0.4′, m13.0v, SB 12.4   01ʰ23.5ᵐ +33°13′
Finder Chart 24-6, Figure 24-19   ★★

16/18″ Scopes–150x: NGC 504, a member of the NGC 507 galaxy group, has a tiny, bright core embedded in a faint 1.25′ × 0.5′ NE–SW halo.

NGC 503 Galaxy Type E?
ø –, m14.1v, SB –   01ʰ23.5ᵐ +33°21′
Finder Chart 24-6, Figure 24-19   ★

16/18″ Scopes–150x: NGC 503, another faint member of the NGC 507 galaxy group, is visible only as a very faint 30″ diameter spot. It can be found near, and just outside, the apex of a tiny, NW pointing, triangle of magnitude 13–14 stars.

NGC 507 U938 Galaxy Type SA(r)0°
ø4.1′ × 4.1′, m11.2v, SB 14.1   01ʰ23.7ᵐ +33°15′
Finder Chart 24-6, Figure 24-19   ★★

16/18″ Scopes–150x: NGC 507, the largest galaxy in the NGC 507 galaxy group, is 6′ east of a pretty blue and gold double star. It has a large prominent core surrounded by a diffuse 1.5′ diameter halo. NGC 508 lies 1.5′ north.

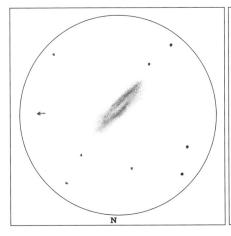

Figure 24-20. NGC 520
20″, f4.5–175x, by Richard W. Jakiel

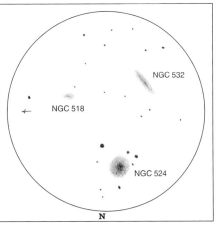

Figure 24-21. NGC 518, 524, & 532
13″, f5.6–100x, by Steve Coe

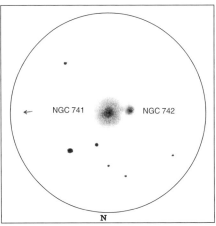

Figure 24-22. NGC 741 & NGC 742
12.5″, f5–250x, by G. R. Kepple

NGC 508 U939 Galaxy Type E0:
⌀1.3′ × 1.3′, m13.1v, SB 13.5 $01^h23.7^m$ +33°17′
Finder Chart 24-6, Figure 24-19   ★

16/18″ Scopes–150x: NGC 508, a very faint member of the NGC 507 galaxy group, is located 1′ north of NGC 507. It is a very small, round 30″ diameter object just 30″ SW of a 13th magnitude star.

NGC 514 H252² Galaxy Type SAB(rs)c II
⌀3.9′ × 2.9′, m11.7v, SB 14.1 $01^h24.1^m$ +12°55′
Finder Chart 24-7   ★★

8/10″ Scopes–100x: NGC 514, located 3′ WNW of a 9.5 magnitude star, is a faint, diffuse, round 1.5′ diameter object of uniform surface brightness. Two 13th magnitude stars lie 2′ east, and a 12.5 magnitude star is 3.5′ SW.

12/14″ Scopes–125x: The halo appears faint, circular, and about 2′ across, and slightly brighter in the center.

16/18″ Scopes–150x: NGC 514 shows a weakly concentrated core containing an inconspicuous nucleus surrounded by a faint halo elongated 3′ × 2′ E–W. Two wide pairs of very faint stars are visible 1′ south and 2′ east.

NGC 515 U956 Galaxy Type S0
⌀1.3′ × 1.0′, m13.0v, SB 13.2 $01^h24.6^m$ +33°29′
Finder Chart 24-6, Figure 24-19   ★

16/18″ Scopes–150x: NGC 515 is a galaxy-pair with NGC 517 lying 3′ to its SSE. It has a very faint 1.25′ × 1′ NW–SE halo with a faint stellar nucleus.

NGC 520 H253³ Galaxy Type Pec
⌀4.6′ × 1.9′, m11.4v, 13.6 $01^h24.6^m$ +03°48′
Finder Chart 24-8, Figure 24-20   ★★★

12/14″ Scopes–125x: NGC 520 is a pair of irregular galaxies that appear as one fairly bright object elongated 2.5′ × 0.75′ NW–SE. Two parallel bright streaks are separated by a dark lane through the center.

16/18″ Scopes–150x: These two close irregular galaxies appear as one object with an elongated, polygonal-shape with wispy plumes at both ends. The bright 3.5′ × 1′ NW–SE halo is bisected into two unequal parts by an irregular dust lane through the object's center. The NW section of the hole is the larger and contains several bright spots. Magnitude 13.5 stars lie 2′ and 3.75′ out from the center beyond the NW tip.

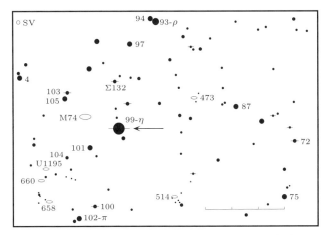

Finder Chart 24-7. 99–η Psc: 01h31.5m +15°21′

Figure 24-23. *Messier 74 (NGC 628) has a large, bright core but its faint spiral structure requires larger instruments to be seen. Harvey Freed, D.D.S. made this 58 minute exposure on hypered 2415 film with a 10″, f6.3 Newtonian reflector.*

NGC 517 U960 Galaxy Type S0
ø2.1′ × 1.1′, m12.4v, SB 13.2   01ʰ24.7ᵐ +33°27′
Finder Chart 24-6, Figure 24-19   ★★

16/18″ Scopes–150x: NGC 517, on the NE side of the NGC 507 galaxy group, is a galaxy-pair with NGC 515 located 3′ NNW. It is about half a magnitude brighter and much more elongated than its companion. The halo measures 2′ × 1′ NNE–SSW and is well concentrated at its center. An extremely faint star is on the NNW edge.

NGC 524 H151³ Galaxy Type SA(rs)0+
ø3.5′ × 3.5′, m10.2v, SB 12.8   01ʰ24.8ᵐ +09°32′
Finder Chart 24-8, Figure 24-21   ★★★★

8/10″ Scopes–100x: The bright galaxy NGC 524 is within an equilateral triangle with 2′ sides composed of 11th and 12th magnitude stars. Its round 1.5′ diameter halo has a tiny but bright core.

12/14″ Scopes–125x: NGC 524 is the brightest member of a group of ten galaxies. It is at the northern vertex of a triangle with NGC 518, located 18′ to its SW, and NGC 532, the same distance SSE. It has a large, bright core containing a nonstellar nucleus and embedded in a round 2.5′ diameter halo that fades rapidly out from the core. NGC 518 is only a faint 1′ × 0.5′ E–W smudge.

NGC 532 H556³ Galaxy Type Sab? sp
ø2.6′ × 0.8′, m12.9v, SB 13.6   01ʰ25.3ᵐ +09°16′
Finder Chart 24-8, Figure 24-21   ★★

12/14″ Scopes–125x: NGC 532, located 18′ SSE of NGC 524, is very faint, highly elongated 1.5′ × 0.5′ NNE–SSW, and very little brighter in its center.

16/18″ Scopes–150x: NGC 532 appears fairly faint, elongated 2.5′ × 0.75′ NNE–SSW, and gradually brighter toward its center. A 13th magnitude star is 2.5′ south.

NGC 528 U988 Galaxy Type S0
ø1.8′ × 1.1′, m12.5v, SB 13.1   01ʰ25.5ᵐ +33°41′
Finder Chart 24-6, Figure 24-19   ★

16/18″ Scopes–150x: NGC 528, on the far NE edge of the NGC 507 galaxy group, appears very faint, elongated about 1.75′ × 1′ NNE–SSW, and slightly brighter at its center.

Figure 24-24. *NGC 660 (top left) has a large, faint, diffuse halo and a mottled core. UGC 1195 (bottom right) is elongated in the same direction but much fainter and smaller. Martin C. Germano made this 80 minute exposure on hypered 2415 film with an 8″, f5 Newtonian reflector.*

NGC 628 Messier 74 Galaxy Type SA(s)c I
⌀11.0′ × 11.0′, m9.4v, SB 14.4   01ʰ36.7ᵐ +15°47′
Finder Chart 24-7, Figure 24-23   ★★★★

This faint face-on galaxy was discovered by Mechain in September 1780 and confirmed by Messier a month later. Because the halo with the spiral arms has low surface brightness, the core is the galaxy's most conspicuous portion. Hence in small telescopes M74 has the appearance of an unresolved globular cluster. Sir John Herschel recorded it as such in his observations, and the object was still incorrectly classified as a globular in Dreyer's *New General Catalogue*. M74 is similar in structure to Messier 33 but over fifteen times farther, lying 40 million light years distant. It is estimated to be nearly the same size as our own Milky Way Galaxy.

8/10″ Scopes–100x: Messier 74 is a fairly faint, large, circular object. Its broad, well-condensed core is surrounded by a mottled, uneven 8′ diameter halo. Foreground field stars are numerous just east and west of the galaxy.

12/14″ Scopes–125x: This bright galaxy is large, irregularly round, and has a large, bright core. 175x reveals a stellar nucleus in a granular core upon which several stars are superimposed. Faint spiral arms wind out from the core counterclockwise into the halo.

16/18″ Scopes–150x: In larger instruments, the 10′ diameter halo is irregular with many bright knots and dark spaces visible between the two counterclockwise spiral arms. The southern arm curves further from the core than the northern arm, which uncurls eastward from the NW edge of the core. The core is bright, about 2.5′ across, and has several bright

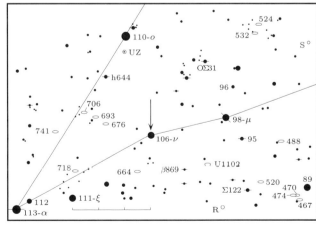

Finder Chart 24-8. 106-ν Psc: 01ʰ41.4ᵐ +05°29′

knots and several foreground stars embedded in it. The halo reaches east and west just beyond two rows of 12th and 13th magnitude stars.

NGC 658   U1192   Galaxy   Type SB II–III
ø3.2′ × 1.6′, m12.5v, SB 14.1 $01^h42.2^m$ +12°35′
Finder Chart 24-7   ★

16/18″ Scopes–150x: NGC 658, located between a 9th magnitude star 2′ NNE and a 12th magnitude star 3′ SW, is a very faint, 2′ × 1′ NNE–SSW oval glow.

UGC 1195   Galaxy   Type Im:
ø3.2′ × 1.0′, m12.9v, SB 14.1 $01^h42.4^m$ +13°58′
Finder Chart 24-7, Figure 24-24   ★★

16/18″ Scopes–150x: UGC 1195 has a well concentrated core embedded in a faint, diffuse, 2′ × 0.75′ NE–SW oval halo. A 13th magnitude star is just beyond the NE tip.

NGC 660   H253²   Galaxy   Type SB(s)a pec
ø9.2′ × 4.2′, m11.2v, SB 15.0 $01^h43.0^m$ +13°38′
Finder Chart 24-7, Figure 24-24   ★★★

8/10″ Scopes–100x: NGC 660, located 9′ SE of an 8th magnitude star and 4.5′ NE of a wide 10th magnitude double has a moderately faint, diffuse halo elongated 6′ × 2′ NE–SW and slightly brighter at its center.

16/18″ Scopes–150x: NGC 660 has a large, diffuse halo elongated 8′ × 4′ NE–SW, within which is a large, faint, mottled core, uneven in brightness because of faint streaks. 1.25′ ESE is a 16″ wide double of magnitude 13 and 14 stars. UGC 1195, located 20′ NNW, is a faint 2.5′ × 0.75′ NE–SW streak with a suggestion of a core. A magnitude 13.5 star touches the galaxy's NE tip.

NGC 664   U1210   Galaxy   Type Sb: II
ø1.6′ × 1.4′, m12.8v, SB 13.5 $01^h43.8^m$ +04°14′
Finder Chart 24-8   ★★

16/18″ Scopes–150x: NGC 664 is a faint, diffuse object with a round 1.25′ diameter halo.

NGC 676   H42⁴   Galaxy   Type S0/a: sp
ø4.6′ × 1.7′, m9.6v, SB 11.7 $01^h49.0^m$ +05°54′
Finder Chart 24-8   ★★★

8/10″ Scopes–125x: NGC 676 is a rather faint, diffuse galaxy but has a 10th magnitude star superimposed on its core. The tips of its 3′ × 1′ N–S halo are tapered.

16/18″ Scopes–150x: This lenticular galaxy has a moderately faint halo elongated 3.5′ × 1′ N–S with tapered ends and brighter north and east of its center. The 10th magnitude star embedded exactly at the center makes it impossible to determine if the galaxy has a nucleus of its own. At 175x faint stars can be seen just east of the center, and 1.5′ SSW of the center on the edge of the halo.

NGC 693   H859²   Galaxy   Type S0/a?
ø2.9′ × 1.5′, m12.4v, SB 13.9 $01^h50.5^m$ +06°09′
Finder Chart 24-8   ★★

16/18″ Scopes–150x: NGC 693 is a fairly conspicuous lenticular galaxy elongated 2′ × 1′ ESE–WNW with a much brighter center. An 11th magnitude star is 1.25′ ENE and an 8th magnitude star 10′ SSE. NGC 706 lies 22′ NE while NGC 676 is 26′ SW.

NGC 706   H596²   Galaxy   Type Sbc?
ø1.9′ × 1.6′, m12.5v, SB 13.6 $01^h51.8^m$ +06°18′
Finder Chart 24-8   ★★

16/18″ Scopes–150x: NGC 706 is moderately faint with a circular 1.5′ diameter halo that contains a very faint stellar nucleus visible during moments of good seeing. A 12.5 magnitude star is 1′ north. NGC 693 is 22′ SW.

NGC 718   H270²   Galaxy   Type SAB(s)a
ø2.4′ × 2.0′, m11.7v, SB 13.3 $01^h53.2^m$ +04°12′
Finder Chart 24-8   ★★★

8/10″ Scopes–100x: This galaxy has a fairly faint, round, 1′ diameter halo containing a prominent stellar nucleus.

16/18″ Scopes–150x: NGC 718 displays a well concentrated core containing a bright stellar nucleus displaced north of center. The 2′ diameter halo is fairly faint and slightly oval-shaped NE–SW.

NGC 741   H271²   Galaxy   Type E0:
ø3.0′ × 3.0′, m11.2v, SB 13.4 $01^h56.4^m$ +05°38′
Finder Chart 24-8, Figure 24-22   ★★

8/10″ Scopes–100x: NGC 741, located 2.5′ SE of an 11th magnitude star, has a rather faint, round 1′ diameter halo with a stellar nucleus.

16/18″ Scopes–150x: NGC 741 is fairly faint. Its 1.5′ diameter halo has diffuse edges and a tiny core containing a faint stellar nucleus. A 13th magnitude star is 45″ east of center. Two very faint companion galaxies, stellar in appearance, are 1′ east and 1.5′ NNW.

Chapter 25

Piscis Austrinus, the Southern Fish

25.1 Overview

Piscis Austrinus, the Southern Fish, has been depicted as drinking the streams falling from the water-jar of Aquarius. This is another constellation which the Greeks inherited from the Babylonians, who knew it simply as The Fish. It was sacred to the Mesopotamian god of the Fresh Waters and of Wisdom, named both Ea ("Temple of Water") and Enki ("Prince of the Place" — that is, his sacred city Eridu, which lay on the margins of the fish-crowded marshes).

Despite its antiquity, Piscis Austrinus has no bright stars but its 1st magnitude bluish-white lucida, Fomalhaut. This star-name means "Mouth of the Fish" but is no older than the Medieval Arabian astronomers, who titled stars simply after their places in the Greek constellation figures. Piscis Austrinus is rather small, its 245 square degrees making it 60th in area among the 88 constellations. It is far from the Milky Way and therefore not rich in objects other than galaxies — and the constellation's far southern position in the sky makes its faint galaxies a challenge for mid-northern latitude observers. Clear, transparent nights are absolutely essential for observing in Piscis Austrinus.

Piscis Austrinus: PIE-sis Os-TRY-nus
Genitive: Piscis Austrini, PIE-sis Os-TRY-ni
Abbrevation: PsA
Culmination: 9 pm–Oct. 9, midnight–Aug. 25
Area: 245 square degrees
Best Deep-Sky Objects: 12-η PsA, 17-τ PsA, Dunlop 241, H N 119, NGC 7174-76, NGC 7221, NGC 7314, IC 5271

25.2 Interesting Stars

Eta (η) = 12 Piscis Austrini (β276) Double Star   Spec. B8
m5.8, 6.8; Sep. 1.7"; P.A. 115° $22^h00.8^m$ $-28°27'$
Finder Chart 25-3   ★★★★

8/10'" Scopes–175x: Eta Piscis Austrini appears as two bluish-white disks nearly in contact.

Beta (β) = 17 Piscis Austrini Double Star   Spec. A0
m4.4, 7.9; Sep. 30.3"; P.A. 172° $22^h32.5^m$ $-32°21'$
Finder Chart 25-4   ★★★★

4/6" Scopes–75x: Beta Piscis Austrini has a yellowish-white primary and a bluish secondary.

Dunlop 241 Double Star   Spec. K0 K0
m5.8, 7.6; Sep. 89.5"; P.A. 31° $22^h36.6^m$ $-31°40'$
Finder Chart 25-4   ★★★★

4/6" Scopes–75x: Dunlop 241 is a nice, wide pair of orange stars.

H VI 119 (h5356) Double Star   Spec. G0 F5
AB: m6.3, 7.3; Sep. 86.6"; P.A. 160° $22^h39.7^m$ $-28°20'$
AC: m6.3, 8.1; Sep. 3.2"; P.A. 67°
Finder Chart 25-4   ★★★★

4/6" Scopes–200x: This W. Herschel discovery consists of a deep yellow primary with a close yellow companion (the "C" component) and a much more distant light bluish star (the "B" component).

25.3 Deep-Sky Objects

IC 5131 Galaxy Type SAB0°:
ø1.6' × 1.4', m12.4v, SB 13.0 $21^h47.4^m$ $-34°53'$
Finder Chart 25-3   ★★

8/10" Scopes–100x: IC 5131, located 15' south of a 9th magnitude star, is a very faint 1' diameter smudge slightly brighter at its center. A 13th magnitude star lies 1.25' WSW, and galaxy NGC 7130 is 12' SE.

Piscis Austrinus, the Southern Fish Constellation Chart 25-1

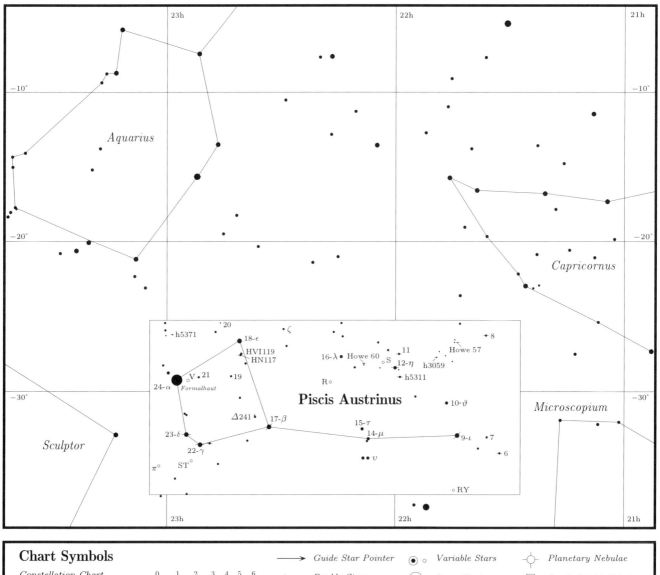

Table 25-1. Selected Variable Stars in Piscis Austrinus

Name	HD No.	Type	Max.	Min.	Period (Days)	F*	Spec. Type	R.A. (2000)	Dec.	Finder Chart No. & Notes
RY PsA		M	8.8	>14.0	224		Me	21^{h}47.5^m	−36 12	25-3
S PsA	209400	M	8.0	14.5	224	0.42	M3-M5	22^{h}03.8^m	−28 03	25-3
R PsA	211493	M	8.5	14.7	292	0.38	M4-M5	18.0	−29 36	25-3
ST PsA		M	9.5	>13.0	179		Me	54.3	−34 23	25-4
V PsA	216692	SRb	9.3	10.5	148		Mb	55.3	−29 37	25-4

F* = The fraction of period taken up by the star's rise from min. to max. brightness, or the period spent in eclipse.

Table 25-2. Selected Double Stars in Piscis Austrinus

Name	ADS No.	Pair	M1	M2	Sep."	P.A.°	Spec	R.A. (2000) Dec	Finder Chart No. & Notes
Howe 57	15257		8.3	9.6	1.6	297	F0	21^{h}44.0^m −26 31	25-3
9–ι PsA			4.3	11.3	20.0	290	A0	44.9 −33 03	25-3
h3059	15365		7.4	10.5	25.2	253	B9	50.8 −27 56	25-3
h5311		AB	7.1	10.6	40.6	292	K0	59.5 −29 03	25-3
		AC		10.9	48.6	227			
11 PsA	15509		7.5	10.5	11.7	36	F0	59.6 −27 38	25-3	
12–η PsA	15536		5.8	6.8	1.7	115	B8	22^{h}00.8^m −28 27	25-3	Bluish-white pair
Howe 60	15647		8.1	9.5	3.1	143	G0	07.7 −28 04	25-3	
17–β PsA			4.4	7.9	30.3	172	A0	31.5 −32 21	25-3	Yellowish & bluish
Δ241			5.8	7.6	89.5	31	K0 K0	36.6 −31 40	25-4	Nice orange pair
H VI 119	16149	AB	6.3	7.3	86.6	160	G0 F5	39.7 −28 20	25-4	Deep yellow & bluish
H N 117	16149	BC		8.1	3.2	67				C: pale yellow
22–γ PsA			4.5	8.0	4.2	262	A0	52.5 −32 53	25-4	Pale & deep yellow
23–δ PsA			4.2	9.2	5.0	244	G3	55.9 −32 32	25-4	Yellow primary
h5371	16400		7.4	8.9	8.9	343	G0	57.8 −26 06	25-4	

Footnotes: *= Year 2000, a = Near apogee, c = Closing, w = Widening. Finder Chart No: All stars listed in the tables are plotted in the large Constellation Chart, but when a star appears in a Finder Chart, this number is listed. Notes: When colors are subtle, the suffix -*ish* is used, e.g. *bluish*.

NGC 7130 IC 5135 Galaxy Type SA pec
ø1.7′ × 1.5′, m12.0v, SB 12.9   21^{h}48.3^m −34°57′
Finder Chart 25-3, Figure 25-1   ★★

8/10″ Scopes–100x: This galaxy, nearly the twin of IC 5131 located 12′ to its NW, is a very faint, round glow about 1′ in diameter. Galaxy NGC 7135 lies 17′ east near a triangle of 10th magnitude stars.

12/14″ Scopes–125x: NGC 7130 has a fairly faint 1′ diameter halo that is slightly brighter in its center, at which is a faint stellar nucleus.

NGC 7135 Galaxy Type S0°: pec
ø3.0′ × 2.1′, m11.3v, SB 13.1   21^{h}49.8^m −34°53′
Finder Chart 25-3, Figure 25-1   ★★

8/10″ Scopes–100x: This galaxy is 4′ SE of a triangle of 10th and 11th magnitude stars. It has a faint 2′ × 0.75′ NNE–SSW halo that is moderately brighter at its center.

12/14″ Scopes–125x: NGC 7135 contains a stellar nucleus surrounded by a weakly concentrated core embedded in a much fainter 2.25′ × 1′ NNE–SSW halo. A very faint star or knot is just inside the halo's west edge.

NGC 7154 Galaxy Type SB(s)d: II-III
ø2.4′ × 1.7′, m12.4v, SB 13.8   21^{h}55.4^m −34°49′
Finder Chart 25-3   ★★

12/14″ Scopes–125x: NGC 7154, lying 10′ south of a wide pair of 8th and 9th magnitude stars, is a very faint, diffuse, circular glow of uniform surface brightness.

16/18″ Scopes–150x: NGC 7154 shows a mottled halo with several knots at the NNW and SSW edges. A 13th magnitude star is 2.5′ north and an 11.5 magnitude star is 8′ NW.

NGC 7163 Galaxy Type (R′)SB(rs)b pec II
ø2.0′ × 0.9′, m12.7v, SB 13.2   21^{h}59.3^m −31°53′
Finder Chart 25-3   ★★

16/18″ Scopes–150x: NGC 7163 is 7′ west of a 9th magnitude star and 35′ west of the north end of the NGC 7176 galaxy group. It has a very faint, diffuse halo elongated 1.5′ × 1′ ESE–WNW and slightly brighter at its center.

NGC 7172 Galaxy Type Sa: pec sp
ø2.8′ × 1.4′, m11.8v, SB 13.1   22^{h}02.0^m −31°52′
Finder Chart 25-3, Figure 25-2   ★★★

12/14″ Scopes–125x: NGC 7172 is the northernmost member of the NGC 7176 galaxy group. It is a fairly bright but diffuse 2′ × 1′ E–W glow. 175x reveals the

Master Finder Chart 25-2. Piscis Austrinus Chart Areas
Guide stars indicated by arrows.

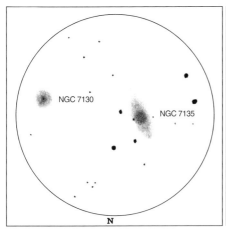

Figure 25-1. NGC 7130 & NGC 7135
13", f5.6–100x, by Steve Coe

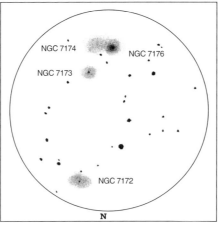

Figure 25-2. NGC 7172, 7173, 7174, & 7176
13", f5.6–100x, by Steve Coe

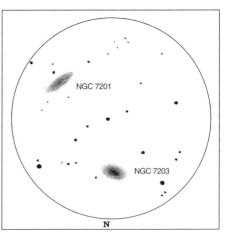

Figure 25-3. NGC 7201 & NGC 7203
16", f5–200x, by G. R. Kepple

very faint stellar nucleus, and a 15th magnitude star near the southern flank. A 10.5 magnitude star lies 2.5' SE, and a 13th magnitude star 1.5' SW.

NGC 7173   Galaxy   Type E1: pec
ø4.4' × 2.6', m11.1v, SB 13.6 $22^h02.06^m$ $-31°58'$
Finder Chart 25-3, Figure 25-2   ★★★

12/14" Scopes–125x: NGC 7173, only 1.5' NW of interacting galaxies NGC 7174-76, is a conspicuous 1' diameter spot with a faint stellar nucleus. The surface brightness fades smoothly to diffuse edges.

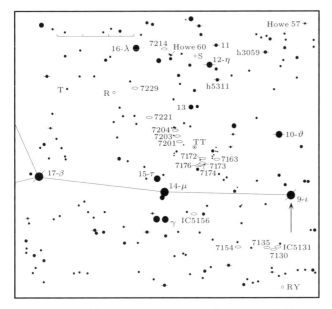

Finder Chart 25-3. 9-ι PsA: $21^h44.9^m$ $-33°02'$

NGC 7174   Galaxy   Type S0/a? pec
ø3.4' × 2.2', m11.3v, SB 13.3 $22^h02.11^m$ $-31°59'$

NGC 7176   Galaxy   Type S0–? pec
ø3.5' × 2.2', m11.1v, SB 13.2 $22^h02.14^m$ $-31°59'$
Finder Chart 25-3, Figure 25-2   ★★★

12/14" Scopes–125x: This interacting galaxy-pair is at the southern end of a chain of galaxies that includes NGC 7173 just 1.5' NW and NGC 7172 lying 7' north. At 100x NGC 7176 and NGC 7174 appear as one; but 175x reveals them to be two galaxies in contact. NGC 7176, the eastern component, is the brighter and larger galaxy, and has a sharp stellar nucleus. NGC 7174 is significantly fainter, slightly smaller, and lacks a nucleus.

16/18" Scopes–150x: At first glance these two galaxies appear as a single object elongated 2' × 1' E–W. 175x reveals two halos in contact, each with a faint core. The easternmost object, NGC 7176, is much the brighter and contains a bright stellar nucleus. Two threshold stars lie immediately north of the galaxy-pair, one on the edge of their halo and the other 30" from the center of NGC 7176.

IC 5156   Galaxy   Type Sab: pec
ø2.3' × 0.8', m12.1v, SB 12.6 $22^h03.3^m$ $-33°50'$
Finder Chart 25-3   ★★★

12/14" Scopes–125x: IC 5156 can be easily found by aiming the viewfinder between a 1/2° wide E–W pair of 7th and 8th magnitude stars. The galaxy has a faint 1' diameter halo slightly elongated N–S with a slight central brightening.

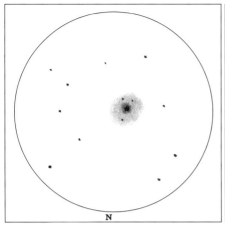
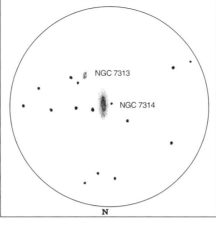
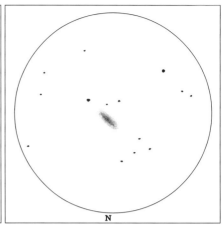

Figure 25-4. NGC 7214
13", f5.6–165x, by Steve Coe

Figure 25-5. NGC 7313 & NGC 7314
13", f4.5–100x, by Tom Polakis

Figure 25-6. IC 5269
13", f5.6–165x, by Steve Coe

NGC 7201 Galaxy Type SAa:
ø1.7′ × 0.6′, m12.7v, SB 12.7   22ʰ06.5ᵐ −31°16′
Finder Chart 25-3, Figure 25-3   ★★★

12/14" Scopes–125x: NGC 7201 is the southernmost in a NNE–SSW row of galaxies that includes NGC 7203 located 6′ NNE and NGC 7201 lying 13′ NNE. It is faint, small, and round.

16/18" Scopes–150x: NGC 7201 has a faint 1.5′ × 0.5′ NW–SE halo that slightly brightens at its center. One 13th magnitude star is on the halo's edge 25″ from the galaxy's center and another is beyond the SE tip 1.5′ from the center.

NGC 7203 Galaxy Type (R′)SB(r)0/a
ø1.7′ × 1.0′, m12.6v, SB 13.0   22ʰ06.7ᵐ −31°10′
Finder Chart 25-3, Figure 25-3   ★★★

12/14" Scopes–125x: NGC 7203 is the middle member of a row of faint galaxies with NGC 7201 located 6′ SSW and NGC 7704 7′ NNE. It is faint, diffuse, and has a round 1′ diameter halo.

16/18" Scopes–150x: NGC 7203 has a fairly faint 1.5′ × 1′ ENE–WSW halo with a well concentrated center. The galaxy lies between 9.5 magnitude stars 2.75′ to the ENE and 5′ WSW. A 14th magnitude star is 45″ west of its center.

NGC 7204 Galaxy Type Pec
ø1.3′ × 0.6′, m13.3v, SB 13.3   22ʰ06.9ᵐ −31°03′
Finder Chart 25-3   ★★★

12/14" Scopes–125x: NGC 7204 is the faintest, and most northerly of a NNE–SSW line of galaxies with NGCs 7201 and 7203. It lies south of a group of relatively bright stars, including one of the 8th magnitude 7′ to its NNE. The galaxy is a very faint, small, diffuse, irregular smudge.

16/18" Scopes–150x: NGC 7204 is a diffuse 1′ × 0.25′ E–W streak that on its NW end touches a faint star. A 12th magnitude star lies 6′ east.

NGC 7214 Galaxy Type SB(s)bc: pec
ø2.3′ × 1.3′, m11.7v, SB 12.7   22ʰ09.1ᵐ −27°49′
Finder Chart 25-3, Figure 25-4   ★★★

8/10" Scopes–100x: NGC 7214 resembles a globular cluster, and early observers mistook it for one. It has a faint, round 1.75′ diameter halo.

12/14" Scopes–125x: NGC 7214 appears fairly faint, though its circular 2′ diameter halo contains a prominent nonstellar nucleus. Several stars are superimposed on the halo, and a 12th magnitude star lies to the south. Four fainter companion galaxies lie nearby.

NGC 7221 Galaxy Type SB(rs)bc: pec
ø1.9′ × 1.4′, m11.2v, SB 13.1   22ʰ11.3ᵐ −30°37′
Finder Chart 25-3   ★★★

8/10" Scopes–100x: NGC 7221 is quite faint, small, and round with a slightly brighter center. A 10th magnitude star lies 4.5′ NW.

16/18" Scopes–150x: In large telescopes the halo of NGC 7221 can be traced with averted vision out to a diameter of 2′. It has a faint ring structure and a small core. A knot or star is 25″ NNW of the core's center. A 12th magnitude star touches the halo's NNE edge.

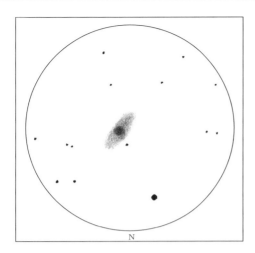

Figure 25-7. IC 5271
13", f5.6–100x, by Steve Coe

NGC 7229   Galaxy   Type SB(s)c I-II
ø1.5′ × 1.5′, m12.6v, SB 13.3   22ʰ14.0ᵐ −29°23′
Finder Chart 25-3   ★★★

12/14″ Scopes–125x: NGC 7229, located 7′ west of a 9th magnitude star, has a faint round 1.5′ diameter halo that is slightly brighter in its center. 5.5′ SW is a 20″ wide pair of 13th magnitude stars.

16/18″ Scopes–150x: NGC 7229 has a circular 2′ diameter halo with a faint stellar nucleus. Two 14th magnitude stars lie 1.5′ ESE and 1.25′ SW of center.

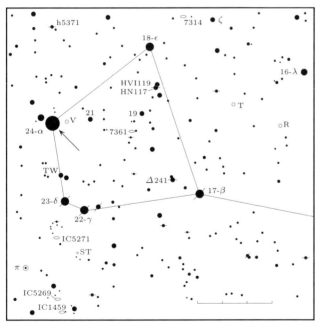

Finder Chart 25-4. 24–α PsA: 22ʰ57.7ᵐ −29°37′

NGC 7314   Galaxy   Type SAB(rs)c: II
ø4.2′ × 1.7′, m10.9v, SB 12.9   22ʰ35.8ᵐ −26°03′
Finder Chart 25-4, Figure 25-5   ★★★

8/10″ Scopes–100x: NGC 7314 lies between an 8th magnitude star 14′ to its east and a 9th magnitude star 9′ west. Although it is the brightest galaxy in the constellation, NGC 7314 is somewhat disappointing: its 3′ × 1.5′ halo is fairly faint and diffuse and otherwise undistinguished. A 12th magnitude star lies 2′ west.

12/14″ Scopes–125x: NGC 7314 has a fairly bright 4′ × 2′ N–S halo with a broad, well-concentrated, mottled core lacking a stellar nucleus. A 14th magnitude star lies near the eastern edge. 5′ to the SW of NGC 7314, near a NE–SW pair of magnitude 10.5 and 12 stars, is the galaxy NGC 7313, a faint 30″ × 20″ NNW–SSE smudge.

NGC 7361   Galaxy   Type Sb III
ø4.0′ × 0.9′, m12.2v, SB 13.4   22ʰ42.3ᵐ −30°03′
Finder Chart 25-4   ★★★

12/14″ Scopes–125x: NGC 7361, located 6′ east of an 8th magnitude star, is a faint, diffuse 2.5′ × 1′ N–S streak. A faint star is 2.5′ from the southern tip.

16/18″ Scopes–150x: NGC 7361 has a 3′ × 1′ N-S spindle-shaped halo. The inner region is slightly brighter and has a granular or mottled texture.

IC 5269   Galaxy   Type SB(s)0+:
ø1.8′ × 0.9′, m12.4v, SB 12.8   22ʰ57.7ᵐ −36°02′
Finder Chart 25-4, Figure 25-6   ★★★

12/14″ Scopes–125x: IC 5269 has a fairly faint spindle-shaped halo elongated 1.25′ × 0.25′ NE–SW with some brightening through center.

16/18″ Scopes–150x: IC 5269 shows a bright core embedded in a 1.5′ × 0.5′ NE–SW halo. To the NE of the galaxy is a 4′ long arc of 14th magnitude stars convex to the SE.

IC 5271   Galaxy   Type Sb:
ø2.9′ × 1.0′, m11.7v, SB 12.7   22ʰ58.0ᵐ −33°45′
Finder Chart 25-4, Figure 25-7   ★★★

12/14″ Scopes–125x: IC 5271, located 10′ NNE of an 8th magnitude star, has a fairly faint 2′ × 1′ NW–SE halo with a broad, circular 1′ core. A 12.5 magnitude star lies 2′ north of the galaxy's center.

Chapter 26

Puppis, the Ship's Stern

26.1 Overview

Puppis was once part of the ancient Greek constellation Argo Navis, the Ship. Due to its vast extent, Argo was divided in the 1750s by Lacaille into Puppis, the Ship's Stern; Pyxis, the Compass; Vela, the Sails; and Carina, the Keel. Puppis alone still covers a respectable 673 square degrees of sky.

In Greek mythology the ship Argo was the vessel commanded by Jason and sailed by his fifty Argonauts from Greece through the Bosporus end of the Black Sea in search of the Golden Fleece. When they had successfully returned with the Fleece, Athena commemorated the event by placing the Argo in the heavens. A number of the celestial constellations have Argonautic connections: the Ram with the Golden Fleece was of course Aries; the ship's surgeon was Aesculapius, enconstellated as Ophiuchus; Hercules and the Twins, Castor and Pollux, in the sky as Gemini, were on the voyage; and even Jason's tutor, the Centaur Chiron, was honored in the stars of Centaurus. Apparently some early Greek mythographer made a conscious attempt to relate existing star-groups to the Argonautic legend. The voyage itself probably has historic origins in the late 2nd millennium Mycenean explorations of the Black Sea in search of gold or trade routes.

Puppis is east and southeast of Canis Major and through it runs an exceptionally star-rich stretch of the Milky Way. It is star-rich because in this direction through the plane of our Galaxy dust clouds are rather scattered and we can see long distances out toward the Galactic rim. Because this is a dust and gas-poor direction, Puppis contains relatively few of the emission and reflection nebulae that are embedded in clouds of interstellar matter. However, because the view is so long and so clear, Puppis is exceptionally rich in open clusters. Indeed, the *Sky Catalogue 2000* lists 73 open clusters in Puppis. The variety of Puppis open clusters is truly remarkable: several are very large and very loose, good binocular objects; some are relatively large and very rich, glorious sights in small telescopes; and a number are faint and distant and require medium to

---

Puppis: PUP-is
Genitive: Puppis, PUP-is
Abbreviation: Pup
Culmination: 9 pm–Feb. 22, midnight–Jan. 8
Area: 673 square degrees
Showpieces: M46 (NGC 2437), M47 (NGC 2422), M93 (NGC 2447), NGC 2438, NGC 2440, NGC 2452, NGC 2477, NGC 2539
Binocular Objects: Cr135, Mel 71, M46 (NGC 2437), M47 (NGC 2422), M93 (NGC 2447), NGC 2423, NGC 2437, NGC 2438, NGC 2439, NGC 2447, NGC 2451, NGC 2467, NGC 2477, NGC 2482, NGC 2483, NGC 2527, NGC 2539, NGC 2546, NGC 2567, NGC 2579

---

large telescopes to be seen at best advantage. Three of the Puppis open clusters have Messier numbers, M46, M47, and M93, and a fourth, NGC 2477, undoubtedly would have made it into the Messier catalogue but for its far southern declination.

26.2 Interesting Stars

R65 = AB, Δ30 = AC Triple Star   Spec. F2 & F2
AC: m6.0, 9.0; Sep. 12.4″; P.A. 314° $06^h29.8^m$ $-50°14'$
Constellation Chart 26-1   ★★★★

4/6″ Scopes–50x: Δ30 = Dunlop 30, the wide A–C pair of this triple, are two pale yellow stars. At 150x the B component, only 0.8″ from the magnitude 6.0 A star and with it catalogued as the double R65, shows up as a slight NW–SE elongation in the A component.

Y Puppis = Δ31 Double Star   Spec. G7 & A0
m5.0, 8.3; Sep. 13.0″; P.A. 321° $06^h38.6^m$ $-48°13'$
Constellation Chart 26-1   ★★★★

2/3″ Scopes–50x: Y Puppis is a pleasing pair with a yellow primary and a white secondary.

Puppis, the Ship's Stern

Constellation Chart 26-1

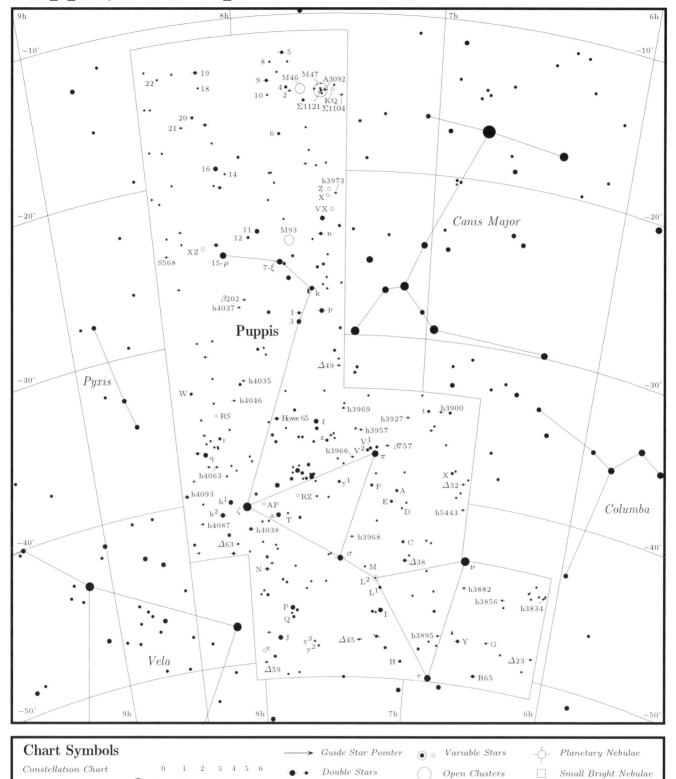

Table 26-1. Selected Variable Stars in Puppis

Name	HD No.	Type	Max.	Min.	Period (Days)	F*	Spec. Type	R.A. (2000)	Dec.	Finder Chart No. & Notes
L² Pup		SRb	2.6	6.2	140	0.40	M5	07ʰ13.5ᵐ	−44°39′	26-1
VX Pup		Cd	7.73	8.51	3.01	0.53	F5-F8	32.6	−21 56	26-5
X Pup		Cd	7.82	9.24	25.96	0.14	F6-G2	32.8	−20 55	26-5
AP Pup		Cd	7.11	7.78	5.08	0.51	F5-F8	57.8	−40 07	26-7
V Pup		EB	4.7	5.2	1.45	.	B1+B3	58.2	−49 15	26-1
RS Pup		Cd	6.53	7.62	41.38	0.24	F9-G7	08ʰ13.1ᵐ	−34 35	26-7
XZ Pup		EA	8.0	10.9	2.19	0.21	A0	13.5	−23 57	26-5

F* = The fraction of period taken up by the star's rise from min. to max. brightness, or the period spent in eclipse.

Table 26-2. Selected Double Stars in Puppis

Name	ADS No.	Pair	M1	M2	Sep."	P.A.°	Spec	R.A. (2000)	Dec	Finder Chart No. & Notes
h3834		AB	5.9	9.4	4.8	220	F5	06ʰ04.7ᵐ	−45°05′	26-1 Yellow & orange pair
		AC		6.2	196.7	321	F8			
Δ23			7.2	7.4	2.6	*124	G5	04.8	−48 28	26-1
h3856			6.7	9.7	34.4	4	K0	22.9	−45 38	26-1
R65		AB	6.0	6.1	0.8	*267	F2	29.8	−50 14	26-1
Δ30		AC		9.0	12.4	314	F2			Yellow & reddish
Y Pup			5.0	8.3	13.0	321	G7 A0	38.6	−48 13	26-1 Orangish & white
h3882		AC	7.7	9.9	18.1	331	B8	38.7	−45 04	26-1
h5443			6.1	10.5	15.4	107	B5	41.2	−40 21	26-3
Δ32			6.5	8.0	8.0	277	A3	42.3	−38 24	26-3
h3895			6.9	11.0	26.2	64	G5	46.7	−47 48	26-1 Orange & blue
Δ38		AB	5.6	7.2	20.5	122	G0 G0	07ʰ04.0ᵐ	−43 36	26-1 AB: Yellow pair
		AC		8.1	184.8	334	K2			C: Orange
h3927		AB	6.5	7.6	3.1	149	F0	05.0	−34 47	26-3
β757			6.0	8.5	2.5	66	B5	12.4	−36 33	26-3 Bluish-white
Δ45			7.1	8.3	22.7	157	A A	21.4	−48 32	26-1
h3957			7.5	8.8	7.7	194	F8	22.3	−35 55	26-3
h3966			6.9	7.0	7.0	322	A3	24.7	−37 17	26-3
h3969			7.0	7.7	17.4	226	F8 F8	27.0	−34 19	26-3
Δ49			6.5	7.2	8.9	53	B3	28.9	−31 51	26-1
σ Pup			3.3	9.4	22.3	74	M0 G5	29.2	−43 18	26-1
Σ1104	6126	AB	6.4	7.5	2.1	358	F8	29.4	−15 00	26-4 Fine yellow pair
	6126	AC		10.7	20.7	188				
	6126	AD		11.2	42.4	8				
h3973	6159		8.3	9.3	9.0	38	B8	31.9	−20 56	26-5
n Pup	6190		5.8	5.9	9.6	114	F4 F5	34.3	−23 28	26-5
P Pup	6205	AB	4.6	9.3	38.4	156		35.4	−28 22	26-1
	6205	BC		10.0	42.2	130				
A3092	6208	AB	5.7	12.3	5.2	176	B5	36.1	−14 30	26-4 (In M47) A: blue-white
Σ1120	6208	ABxC		9.6	19.6	36				C: blue
Σ1121	6216	AB	7.9	7.9	7.4	305	B9 B9	36.6	−14 29	26-4 (In M47) White pair
k Pup	6255	AB	4.5	4.7	9.9	318	B5 B8	38.8	−26 48	26-5 Matched white double
2 Pup	6348	AB	6.1	6.8	16.8	339	A0 A0	45.5	−14 41	26-4 White duo
5 Pup	6381		5.6	7.7	2.2	5	F5	47.9	−12 12	26-4
Howe 65			5.1	8.1	3.0	274	F3	52.3	−34 42	26-7
Δ59			6.5	6.5	16.4	47	B3 B3	59.2	−49 59	26-1
β202	6535	AB	6.6	9.3	7.6	161	B9	08ʰ01.9ᵐ	−27 13	26-6 Fine white pair
h4037	6544	AB	7.1	8.8	7.1	244	G5	02.6	−27 33	26-6
h4038			5.5	8.5	27.0	346	A0	02.7	−41 19	26-1 White & reddish
h4035			5.8	9.5	35.0	134	K2	03.1	−32 28	26-6
h4046			6.0	8.4	22.1	88	G5 A	05.7	−33 34	26-7 Golden & white
Δ63			6.6	7.7	5.5	81	A0	09.8	−42 38	26-1 Pretty white stars
h4063			7.5	9.6	17.7	350	B8	15.5	−37 22	26-7
h4087		AB	7.7	8.0	1.5	278	F5	22.1	−41 00	26-7
h4093		AxBC	6.7	7.9	8.1	124	A0 A0	26.3	−39 04	26-7 Yellowish-white pair

Footnotes: *= Year 2000, a = Near apogee, c = Closing, w = Widening. Finder Chart No: All stars listed in the tables are plotted in the large Constellation Chart, but when a star appears in a Finder Chart, this number is listed. Notes: When colors are subtle, the suffix -ish is used, e.g. bluish.

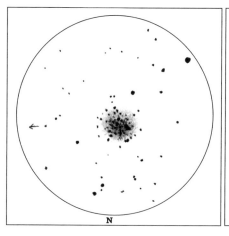

Figure 26-1. NGC 2298
17.5", f4.5–200x, by G. R. Kepple

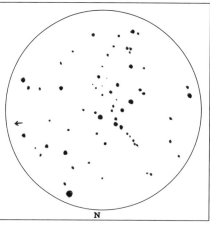

Figure 26-2. Trumpler 7
12.5", f5–200x, by G. R. Kepple

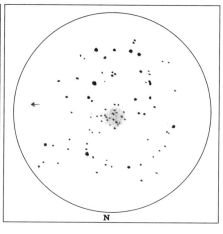

Figure 26-3. NGC 2401
13", f5.6–165x, by Steve Coe

Δ38   Triple Star   Spec. G0 & G0
AB: m5.6, 7.2; Sep. 20.5"; P.A. 122°   07^{h}04.0^m −43°36′
Constellation Chart 26-1   ★★★★

2/3" Scopes–50x: Dunlop 38 is an attractive triple star for small telescopes. The AB stars are yellow and the C component, 185" to the NNW, is orange.

Sigma (σ) Puppis = Δ51   Double Star   Spec. M0 & G5
m3.3, 9.4; Sep. 22.3"; P.A. 74°   07^{h}29.2^m −43°18′
Constellation Chart 26-1   ★★★★

2/3" Scopes–50x: Sigma Puppis, though an unequally bright pair, is a fine double of deep orange and yellow stars.

Master Finder Chart 26-2. Puppis Chart Areas
Guide stars indicated by arrows.

Σ1104   Quadruple Star   Spec. F8
AB: m6.4, 7.5; Sep. 2.1"; P.A. 358°   07^{h}29.4^m −15°00′
Finder Chart 26-4   ★★★★

4/6" Scopes–150x: Struve 1104 AB is a beautiful but close pair of golden yellow stars that require good seeing conditions for a clean split. For the positions of the C and D components, see Table 26-2.

k Puppis = Δ535   Double Star   Spec. B5 & B8
AB: m4.5, 4.7; Sep. 9.9"; P.A. 318°   07^{h}38.8^m −26°48′
Finder Chart 26-5   ★★★★

2/3" Scopes–50x: k Puppis is a remarkably beautiful pair of equally bright blue-white stars.

h4038   Double Star   Spec. A0
m5.5, 8.5; Sep. 27.0"; P.A. 346°   08^{h}02.7^m −41°19′
Constellation Chart 26-1   ★★★★

2/3" Scopes–50x: J. Herschel 4038 consists of a bright pale yellowish primary and a reddish secondary.

h4046   Double Star   Spec. G5 & A
m6.0, 8.4; Sep. 22.1"; P.A. 88°   08^{h}05.7^m −33°34′
Finder Chart 26-7   ★★★★

2/3" Scopes–50x: Herschel 4046, in the middle of the faint star-rich Puppis Milky Way, consists of a bright golden primary with a white attendant. A deep red star is just north of the companion, and a close, faint double lies just west of the primary.

26.3 Deep-Sky Objects

NGC 2298 Globular Cluster Class VI
ø6.8', m9.4v, Br★ 13.4v   06^{h}49.0^m −36°00'
Finder Chart 26-3, Figure 26-1   ★★★

12/14" Scopes–150x: The globular cluster NGC 2298, situated in a rich Milky Way field, has a moderately faint, 3' diameter halo with a broad, mildly concentrated, core. Half a dozen very faint stars can be resolved around the periphery. An 11th magnitude star lies 3' SSE.

16/18" Scopes–175x: NGC 2298 has a 4' diameter halo growing broadly brighter toward its center. The halo has a granular texture, and its periphery, around which a dozen very faint stars can be resolved, is irregular. A double consisting of magnitude 12.5 and 14 components 9" apart lies on the halo's NNE edge.

NGC 2310 H152³ Galaxy Type S0
ø4.0' × 0.8', m11.7v, SB 12.8   06^{h}54.0^m −40°52'
Finder Chart 26-3   ★★★

12/14" Scopes–125x: NGC 2310, located 10' west of an 8th magnitude star, has a faint lenticular 3.0' × 0.5' NE–SW halo with some brightening at its center. A 12.5 magnitude star lies 2' ESE.

Collinder 135 Open Cluster Tr Type IV 2 p
ø50', m2.1v, Br★ 2.71v   07^{h}17.0^m −36°50'
Finder Chart 26-3, Figure 26-4   ★★★

12/14" Scopes–60x: Collinder 135 is a very large but very loose and poorly concentrated open cluster of some thirty stars of a wide variety of magnitudes in a 50' area. Its lucida is the double Pi (π) Puppis (2.7, 8.0; 69.2"; 213°) which couples a bright golden primary with a faint bluish companion. Three bright bluish stars are along the northern edge of the cluster, the eastern two a wide double of magnitude 4.5 stars. A chain of 8th magnitude stars extends SE.

Trumpler 7 Open Cluster 30★ Tr Type II 3 p n
ø5', m7.9v, Br★ 9.13v   07^{h}27.3^m −24°02'
Finder Chart 26-5, Figure 26-2   ★★★

12/14" Scopes–125x: Trumpler 7 is a faint cluster of thirty 9th magnitude and fainter stars irregularly scattered over a 5' area. At the NE edge is a 2.5' long NW–SE string of eight stars. At its SE end is a wide NW–SE pair of 10th magnitude stars. At its NW end is a 10th magnitude star with two magnitude 12.5 companions. West of this end of the chain is the cluster's 9th magnitude lucida. Another 9th magnitude star is 5' further NW from the end of the chain.

Figure 26-4. *Spanning a degree of sky, Collinder 135 is a bright, loose cluster without central compression. Martin C. Germano made this 10 minute exposure on hypered 2415 film with an 8", f5 Newtonian reflector.*

NGC 2396 H36⁸ Open Cluster 30★ Tr Type IV 1 m
ø10', m7.4p, Br★ 11.0p   07^{h}28.1^m −11°44'
Finder Chart 26-4   ★★★

12/14" Scopes–100x: NGC 2396 is a moderately faint open cluster on the Puppis/Canis Major border toward Puppis' extreme NW corner. A single 9th magnitude star is on the cluster's western edge. It contains thirty 11th to 13th magnitude members spaced evenly over a 10' area. An orange star is near the cluster center. NGC 2396 is 7' south of the multiple star β332 which consists of extremely close (0.7") 6th and 8th magnitude stars accompanied at 20" and 23" by magnitude 8.5 and 9.5 companions.

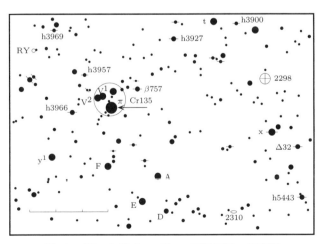

Finder Chart 26-3. π Pup: 07^{h}17.1^m −37°06'

Figure 26-5. *van den Bergh 97 (right side of dark lane) is an extremely faint, small slice of nebula separated from Sharpless 2-302 (left side of dark lane) by a narrow dark lane. Open clusters Bochum 4 and Bochum 5 lie along the nebula's west (left) side but are difficult to separate from the star field. Martin C. Germano made this 80 minute exposure on hypered 2415 film with an 8″, f5 Newtonian reflector.*

Czernik 29 Open Cluster 40★ Tr Type II 2 m
⌀3.8′, m10.3, Br★ 12.22v   07ʰ28.3ᵐ −15°24′

Haffner 10 Open Cluster 40★ Tr Type III 2 m
⌀4.2′, m11.5:v, Br★ 12.68v   07ʰ28.6ᵐ −15°23′
Finder Chart 26-4   ★/★

12/14″ Scopes–125x: Cz 29 and Haf 10 are two adjoining condensations of faint stars near the Puppis/Canis Major frontier SW of the M46 + M47 cluster pair. Cz29 is the larger of the two, with eighteen 12th magnitude and fainter members spread over a 5′ area. At 125x a starless lane can be seen running ESE–WNW through the cluster's center. A 9th magnitude star lies west of the cluster. Haf 10, on the NE edge of Cz 29, is a 2′ NE–SW string of ten faint stars. To its east is a 3.5′ × 2′ triangle of 10th magnitude stars.

NGC 2401 H65⁷ Open Cluster 20★ Tr Type I 1 p
⌀2′, m12.6p, Br★ −   07ʰ29.4ᵐ −13°58′
Finder Chart 26-4, Figure 26-3   ★★

12/14″ Scopes–150x: NGC 2401, located 7′ west of a 7th magnitude star in a rich star field, is a hazy, 2′ diameter clump of seven faint stars embedded in a background haze.

16/18″ Scopes–175x: At low power, NGC 2401 appears only as a nebulous patch NNE of two 10th magnitude stars. 175x resolves fifteen 13.5 magnitude and fainter cluster members spread over a 2′ × 1.5′ E–W area. A circlet of 13th magnitude stars is at the eastern edge, and a 15″ wide pair of 12th and 13th magnitude stars stands out 2′ NNW of the cluster center.

Bochum 5 Open Cluster 12★ Tr Type I 2 p
⌀11′, m −, Br★ 7.78v   07ʰ30.9ᵐ −17°04′
Finder Chart 26-4, Figure 26-5   ★★

12/14″ Scopes–125x: This cluster is situated on the west edge of the faint nebulous complex Sharpless 2-302 + van den Bergh 97. One orange and six white stars

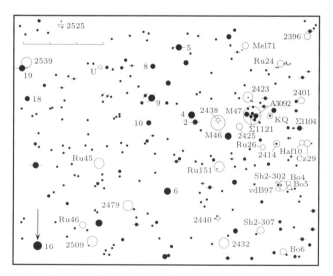

Finder Chart 26-4. 16 Pup: 08ʰ09.0ᵐ −19°15′

Puppis 345

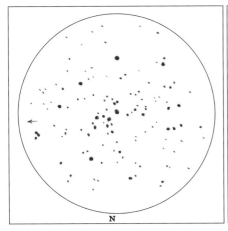

Figure 26-6. NGC 2414
17″ f4.5–200x, by G. R. Kepple

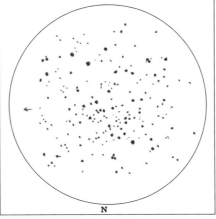

Figure 26-7. NGC 2421
12.5″, f5–175x, by G. R. Kepple

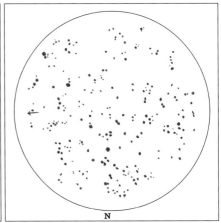

Figure 26-8. NGC 2423
12.5″, f5–150x, by G. R. Kepple

form a horseshoe asterism. Bochum 5 is more obvious than Bochum 4 lying 5′ north.

16/18″ Scopes–150x: Bochum 5 is a conspicuous "C" of nine stars, the magnitude 7.8 lucida at the SE tip. South of this crescent is a very faint patch of stars located near the position plotted for Bochum 5 in *Uranometria 2000.0*.

Bochum 4 Open Cluster 30★ Tr Type II 2 m n
ø23′, m7.3p, Br★ 8.05p $07^h31.0^m$ $-16°57′$
Finder Chart 26-4, Figure 26-5   ★★

12/14″ Scopes–125x: Bochum 4, just north of Bochum 5, is an irregular, scattered assembly of fifteen magnitude 7.8 and fainter stars.

16/18″ Scopes–150x: Bochum 4 is an irregular, uncompressed cluster of about twenty 7.8 magnitude and fainter stars in a 5′ area. It has nine 9th–10th magnitude stars, two each on the northern and western edges, and two pairs near the SE edge. Three fainter cluster members are near the eastern edge. The extremely faint nebula Sharpless 2-302 is just east.

Sh2-302 Gum 6 Emission Nebula
ø20′ × 20′, Photo Br 4-5, Color 3-4 $07^h31.6^m$ $-16°58′$

vdB97 Reflection Nebula
ø2′ × 2′, Photo Br 3-5, Color 1-4 $07^h32.5^m$ $-16°54′$
Finder Chart 26-4, Figure 26-5   ★/★

16/18″ Scopes–100x: Sharpless 2-302 is extremely faint and diffuse. A slight haze can be seen through O-III filters just NW of an 8th magnitude star at the position plotted for the nebula in *Uranometria 2000.0*.

20/22″ Scopes–125x: The nebulosity of Sharpless 2-302 is a 10′ diameter, very faint, low contrast haze west of two 8th magnitude stars. van den Bergh 97 is a

thin roughly N–S wing of the main complex divided from it by a narrow N–S dark lane.

Ruprecht 24 Open Cluster 15★ Tr Type IV 1 p
ø2′, m –, Br★ 11.0p $07^h31.9^m$ $-12°45′$
Finder Chart 26-4   ★★★

12/14″ Scopes–125x: Ruprecht 24 is a moderately rich group of thirty 11th magnitude and fainter stars in a 12′ area. A thin triangle of 11th magnitude stars pointing west stands out against the fainter members.

Bochum 6 Open Cluster 40★ Tr Type I 2 m n
ø10′, m9.9v, Br★ 10.65v $07^h32.0^m$ $-19°26′$
Finder Chart 26-5   ★★★

12/14″ Scopes–125x: Bochum 6 is between a 6th magnitude star 17′ east and a 6.5 magnitude star 7′ WSW. Fifteen 11th magnitude and fainter members are spread over a 5′ area upon some background haze. A magnitude 10.5 star is on the south edge.

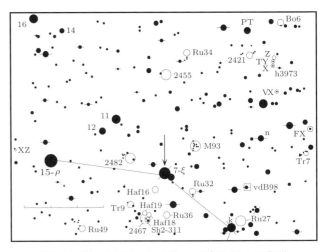

Finder Chart 26-5. 7-ξ Pup: $07^h49.3^m$ $-24°52′$

Figure 26-9. *Messier 47 (left of center) is a bright, very large, moderately rich but irregularly scattered group centered around a clump of six bright stars. NGC 2423 (at right) is a large unconcentrated group that appears more like an enrichment of the Milky Way than a true cluster. NGC 2425 (top left) is a small knot of faint stars. Martin C. Germano made this 30 minute exposure on hypered 2415 Kodak Tech Pan film with an 8", f5 Newtonian at prime focus.*

NGC 2414 H37[8] Open Cluster 35★ Tr Type I 2 m
ø4', m7.9p, Br★ 8.21v   07ʰ33.3ᵐ −15°27'
Finder Chart 26-4, Figure 26-6   ★★★

12/14" Scopes–125x: NGC 2414 is a faint but conspicuous cluster of fifteen 12th magnitude and fainter stars scattered irregularly around the magnitude 8.2 lucida. Three short star chains stand out against a background haze.

16/18" Scopes–150x: NGC 2414 is a rich 2' knot of several dozen 12th magnitude and fainter stars clustered around the 8th magnitude lucida. Ten of the brighter stars form a "U" open to the north just NW of the 8th magnitude star; and just to its south is a 6" NW–SE double of 12th magnitude stars. 3' SW of the cluster center is a NW–SE row of three widely spaced 10th magnitude stars.

Ruprecht 26 Open Cluster Tr Type III 1 p
ø4', m –   07ʰ34.9ᵐ −15°32'
Finder Chart 26-4   ★★

12/14" Scopes–125x: Ruprecht 26, located 10' SW of an 8th magnitude star, is a mildly compressed gathering of several dozen faint stars in a 5' area. Just beyond the eastern edge is a 9th magnitude field star; and a loose group of 10th magnitude stars is NNE.

Sh2-307 Gum 7 Emission Nebula
ø4' × 4', Photo Br★ 3-5, Color 4-4   07ʰ35.5ᵐ −18°46'
Finder Chart 26-4   ★

16/18" Scopes–100x: Sharpless 2-307 is an extremely faint emission nebula 8' NW of an 8th magnitude star. It can be seen with the assistance of an O-III filter as a hazy circular 3' glow.

NGC 2421 H67[7] Open Cluster 70★ Tr Type I 2 m
ø10', m8.3v, Br★ 10.45v   07ʰ36.3ᵐ −20°37'
Finder Chart 26-5, Figure 26-7   ★★

4/6" Scopes–75x: NGC 2421 is 10' SE of a wide pair of 9.5 magnitude stars. It is a rich, compressed cluster of three dozen faint stars in a triangular area, its brightest members along its SE and SW edges.

12/14" Scopes–100x: NGC 2421 is a fairly rich cluster of forty 11th to 13th magnitude stars in an 8' wide triangle. Magnitude 10.5 stars mark the triangle's SE and SW tips, and at the cluster center is an 18" double of magnitude 11 and 11.5 stars.

van den Bergh 98 Reflection Nebula
ø10′ × 10′, Photo Br 3-5, Color 1-4
07ʰ36.4ᵐ −25°20′
Finder Chart 26-5   ★

12/14″ Scopes–100x: vdB 98 is a very faint E–W glow around a 7th magnitude star. With a UHC filter, the nebula appears somewhat brighter north of the star.

NGC 2422 Messier 47 Open Cluster 30★
Tr Type I 2 m, ø29′, m4.4v, Br★ 5.68v
07ʰ36.6ᵐ −14°30′
Finder Chart 26-4, Figure 26-9   ★★★★★

M47 is an extremely large, very bright open cluster visible to the unaided eye as a hazy Milky Way spot and partially resolvable even in 10×50 binoculars. Its brightest member is a magnitude 5.7 blue-white B2 star; and all its other bright members — one of magnitude 6.2 and two each of magnitudes 6.5 and 7 — are blue-white B-type objects contrast their color with the splendid orange of the variable KQ Puppis 40′ due west of the cluster. M47 is about 1,700 light years distant, around 15 light years in diameter, and only 25–30 million years old.

4/6″ Scopes–50x: M47 contains fifty stars of a wide range of magnitudes in a half degree area. The cluster is best at low powers.

8/10″ Scopes–60x: This bright, large cluster is highly irregular, containing some 75 members scattered in knots and chains separated by several starless voids. The brightest star, located near the west edge, is a nice double (Σ1120: 5.7, 9.6; 19.6″; 36°). Another very nice double is at the very center of the group (Σ1121: 7, 7.5; 7.4″; 304°).

12/14″ Scopes–75x: Nice! Messier 47 is a bright, very large, moderately rich but irregularly scattered group centered upon a clump of six bright stars. The brighter stars are aligned E–W, the magnitude 5.7 lucida at the west end, a magnitude 6.2 star at the east end, and several magnitude 6.5–7 stars in a knot at the center. Four 9th magnitude stars are part of a star chain running NNE–SSW just west of the central concentration. Several other star chains may be sorted out, including a loose one that winds northward in the direction of open cluster NGC 2423.

NGC 2423 H28⁷ Open Cluster 40★ Tr Type II 2 m
ø19′, m6.7v, Br★ 9.02v   07ʰ37.1ᵐ −13°52′
Finder Chart 26-4, Figures 26-8 & 26-9   ★★★

8/10″ Scopes–50x: NGC 2423 is so large and unconcentrated that it appears more like an enrichment of the (already rich) Milky Way star field around it than a true autonomous cluster. Low powers show it at its best. It contains some forty 11th to 12th magnitude stars loosely scattered around the magnitude 9.0 cluster lucida. Its brighter members form a 3′ triangle with the 9th magnitude star at the northern vertex.

12/14″ Scopes–75x: NGC 2423 is a large, fairly rich cluster of sixty 11th magnitude and fainter stars in a 15′ area centered upon the magnitude 9.0 lucida. At low powers the stars seem uniformly distributed; but 100x reveals them to be in star-rich streaks separated by star-poor voids. The most conspicuous concentration fans SE from the magnitude 9.0 star at the cluster center, but another group of fifteen faint stars extends SE.

Melotte 71 Open Cluster 80★ Tr Type II 2 r
ø9′, m7.1v, Br★ 10.18v   07ʰ37.5ᵐ −12°04′
Finder Chart 26-4, Figure 26-10   ★★★★

8/10″ Scopes–75x: Excellent! At low power Melotte 71 resembles a very loose globular cluster. At least fifty 11th magnitude and fainter stars resolve against a hazy background, the brighter stars lying along the east and west sides and the fainter members concentrated at the center. At the SW edge is a 3′ × 1.25′ isosceles triangle pointing south. The triangle's NE apex is marked by a 9″ wide ESE–WNW pair of 10.5 magnitude stars.

Figure 26-10. *Melotte 71 is a very rich cluster of faint stars that resembles a loose globular cluster. Lee C. Coombs made this 10 minute exposure on 103a-O film with a 10″, f5 Newtonian reflector.*

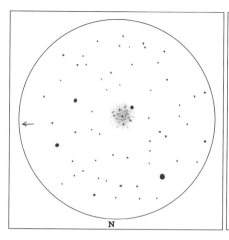
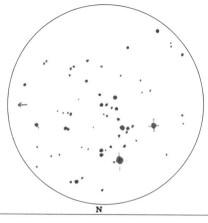
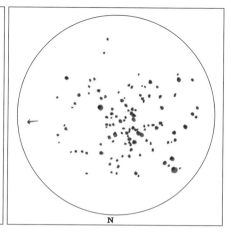

Figure 26-11. Bochum 15
13" f5.6–100x, by Steve Coe

Figure 26-12. NGC 2439
17" f4.5–100x, by G. R. Kepple

Figure 26-13. NGC 2432
13" f5.6–100x, by Steve Coe

Ruprecht 27 Open Cluster 30★ Tr Type II 2 m
ø18', m –, Br★ 12.0p   07ʰ37.5ᵐ –26°36'
Finder Chart 26-5   ★★

12/14" Scopes–125x: Ruprecht 27 is an inconspicuous cluster just north of a loose scattering of bright stars that includes the beautiful 4th magnitude double Kappa (κ) Puppis. It has fifteen 12th magnitude and fainter stars in an irregular 18' × 9' area. Near the south end is an E–W chain of magnitude 9.5 stars.

NGC 2425 H87⁸ Open Cluster 30★ Tr Type II 1 m
ø3.3', m –, Br★ 14.0p   07ʰ38.3ᵐ –14°52'
Finder Chart 26-4, Figure 26-9   ★

12/14" Scopes–125x: This faint cluster is between and just south of Messier 46 and Messier 47. It is an inconspicuous knot of fifteen very faint stars irregularly scattered over a 5' area elongated E–W.

16/18" Scopes–150x: NGC 2425 is a faint cluster of several dozen 13th to 15th magnitude stars spread over a 5' × 2' ESE–WNW area.

Bochum 15 Open Cluster 12★ Tr Type IV 2 p n
ø3', m6.3v, Br★ 7.65v   07ʰ40.1ᵐ –33°33'
Finder Chart 26-7, Figure 26-11   ★★

12/14" Scopes–125x: Bochum 15 is 7' SW of a 22' wide equilateral triangle of 8th magnitude stars. It contains half a dozen 13th magnitude stars in a 4' area superimposed upon a hazy background. A 9th magnitude star is just east of the cluster.

Haffner 13 Open Cluster 15★ Tr Type III 3 p
ø14', m –, Br★ 8.0p   07ʰ40.5ᵐ –30°07'
Finder Chart 26-6   ★★

12/14" Scopes–100x: Haffner 13 is an irregular grouping of fifteen 7th to 13th magnitude stars in a 15' area. The brighter members are aligned NE–SW with an 8th magnitude star at the NE edge and a 7th magnitude star at the SW edge. A wide 8th magnitude pair is near the cluster's center, and a Y-shaped asterism is on its SE edge.

NGC 2439 Cr 158 Open Cluster 80★ Tr Type II 3 m
ø10', m6.9v, Br★ 8.90v   07ʰ40.8ᵐ –31°39'
Finder Chart 26-6, Figure 26-12   ★★★

4/6" Scopes–50x: This cluster is a faint, moderately large, rich, irregular group of 30 stars scattered to the SW of a 2.5' × 1.75' triangle composed of the reddish-orange magnitude 6.6 variable star R Puppis, thought to be a true physical member, and two 9th magnitude stars.

12/14" Scopes–100x: NGC 2439 is a fairly conspicuous cluster of fifty-five 9th magnitude and fainter stars spanning a 10' area. On the north edge is the bright reddish R Puppis. The majority of the cluster members are in a star stream trailing SSW from R. Many wide doubles are visible.

NGC 2432 H36⁶ Open Cluster 50★ Tr Type II 1 m
ø7', m10.2p, Br★ 8.21v   07ʰ40.9ᵐ –19°05'
Finder Chart 26-4, Figure 26-13   ★★

4/6" Scopes–50x: NGC 2432 is a faint, tiny N–S splash of a dozen stars.

12/14" Scopes–100x: NGC 2432 is a moderately faint, highly N–S elongated group of thirty 12th to 13th magnitude stars. The brighter stars lie along a jagged chain which splits at the northern end, a secondary prong diverging NE. The fainter cluster stars, including some doubles, are scattered around the chain.

Figure 26-14. *Messier 46 (NGC 2437) is a moderately bright, rich cluster with a hundred stars visible. The disk of planetary nebula NGC 2438, a foreground object, is a fine contrast to the rich stellar background. Lee C. Coombs made this ten minute recording on 103a-E Spectroscopic film with a 10", 5 Newtonian reflector at prime focus.*

Ruprecht 151 Open Cluster 30★ Tr Type III 2 m
ø14', m –, Br★ 12.0p $07^h41.3^m$ $-16°15'$
Finder Chart 26-4   ★★

12/14" Scopes–125x: Ruprecht 151, centered 15' SW of a 7th magnitude star, is a fairly faint assembly of a dozen 12th magnitude and fainter stars spread irregularly over a 12' area.

16/18" Scopes–150x: Ruprecht 151 is a large, loose, gathering of three dozen 12th to 14th magnitude stars in a 15' long rectangular area. A magnitude 9.5 star is at the SE edge. An abundance of bright stars is visible in a low power field, especially to the west of the cluster.

NGC 2437 Messier 46 Open Cl. 100★ Tr Type II 2r
ø27', m6.1v, Br★ 8.68v $07^h41.8^m$ $-14°49'$

NGC 2438 H39[4] PK231+4.2 Planetary Neb. Type 4+2
ø66", m11.0v, CS 17.5:v $07^h41.8^m$ $-14°44'$
Finder Chart 26-4, Figure 26-14   ★★★★★/★★★★

M46, discovered by Messier in 1771, is in the star- and cluster-rich northern Puppis Milky Way just 1.5° east of its companion open cluster M47. Though M46, like M47, is very large and bright, the two clusters are anything but twins: M47 is beautiful because of its half dozen bright blue-white stars, and M46 is beautiful because of the crowded richness of its uniformly faint stars. Even in 10×50 binoculars can the striking difference between the two clusters be appreciated. M46 is also notable for the presence of a planetary nebula, NGC 2438, among its stars near the cluster's NE edge. However, the planetary is a foreground object perhaps 3,300 light years away to the cluster's 5,000. The true diameter of M46 is nearly 40 light years and its true luminosity around 9,000 suns. It is estimated to be 300 million years old, ten times the age of the neighboring blue-star rich M47.

4/6" Scopes–50x: M46, though fainter than M47, is more impressive. It is large and very rich, its 75 stars evenly distributed and nearly uniform in brightness. At low power the planetary nebula NGC 2438 is a fuzzy "star;" at 75x is shows a definite disk.

8/10" Scopes–75x: M46 has eighty 11th to 14th magnitude stars in a 25' × 20' E–W area. In general the distribution of stars is uniform and three to four times more dense than the surrounding Milky Way star field, but the northern portion of the cluster appears slightly richer than its southern portion, and just south of the cluster center is a nearly starless gap. Planetary nebula NGC 2438 is just NW

Figure 26-15. *Planetary nebula NGC 2440 has a conspicuous oval disk. Martin C. Germano made this 10 minute exposure on 2415 film with an 8", f5 Newtonian reflector.*

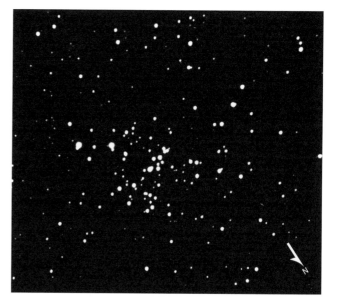

Figure 26-16. *M93 (NGC 2447) is a bright, large, irregular open cluster with more than a hundred stars. Don Walton made this 20 minute exposure on hypered 2415 film with an 11", f10 SCT.*

of an 11th magnitude star in the NE quadrant of the cluster. It lies just where the denser portion of the cluster begins to thin. At 200x the planetary's bright 60" disk shows up well, and a 13th magnitude star can be seen near its center.

12/14" Scopes–75x: M46 is moderately bright and very rich, its hundred 11th magnitude and fainter members evenly distributed over a 25' area. The cluster's magnitude 8.7 lucida is in its western portion. The northern area is noticeably richer and more compressed, and has a greater expanse of outlying stars to its east and west. The southern portion is an irregular loop around a starless gap. The planetary nebula NGC 2438 is embedded near the northern edge in the richest area of stars, four stars converging near it. The planetary's bright disk, a nice contrast to its star-rich setting, is 65" in diameter and subtly annular. Its outer ring fades noticeably at the NW edge. A 13th magnitude star with a very faint companion is 15" SW of the planetary's center, but the actual magnitude 17.7 illuminating central star is not visible in amateur instruments.

NGC 2440 H64⁴ PK234+2.1 Planetary Neb. Type 5+3
ø14/32", m9.4v, CS 18.9v   07ʰ41.9ᵐ −18°13'
Finder Chart 26-4, Figure 26-15   ★★★★

8/10" Scopes–200x: This planetary nebula is a distinct oval disk 3' SW of a 9th magnitude star and 7' NW of a triple of magnitude 9.5, 10.3, and 11.1 components. It is elongated 20" × 15" NE–SW.

12/14" Scopes–225x: NGC 2440 has a bright greenish oval disk elongated 35" × 25" NE–SW and much brighter in the center.

16/18" Scopes–250x: In larger telescopes, the disk appears as an irregular 50" × 30" NE–SW oval. With averted vision a very faint star may be seen near the disk's center; but this is not the 19th magnitude central star. Several slightly brighter spots are NW and SE of center. The disk is surrounded by six faint stars, the three brighter all to its south. At high power the triple star to the SW resolves into a quadruple. The 9th magnitude star to the NE is a close double (ADS 6309 AB: 9.4, 9.9; 1.8"; 291°).

Ruprecht 30 Open Cluster 30★ Tr Type IV 2 m
ø4', m − , Br ★ 11.p   07ʰ42.2ᵐ −31°28'
Finder Chart 26-6   ★★★

12/14" Scopes–125x: Ruprecht 30, located 18' NE of open cluster NGC 2439, is a gathering of twenty 11th magnitude and fainter stars in a 4' area.

Ruprecht 31 Open Cluster 15★ Tr Type I 3 p
ø2', m −, Br ★ 11.p   07ʰ42.7ᵐ −35°35'
Finder Chart 26-7   ★★

12/14" Scopes–125x: Ruprecht 31 is a faint, highly concentrated knot of fifteen 11th magnitude and fainter stars in a 2' area. Its two brightest members are on the cluster's NW and SE edges.

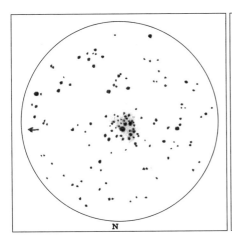
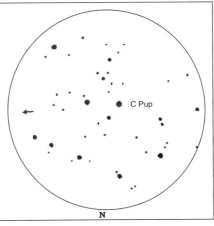
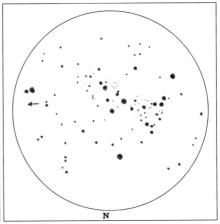

Figure 26-17. Haffner 15
13″, f5.6–100x, by Steve Coe

Figure 26-18. NGC 2451
12.5″, f5–75x, by G. R. Kepple

Figure 26-19. Ruprecht 34
12.5″, f5–150x, by G. R. Kepple

NGC 2447 Messier 93 Open Cluster 80★ Tr Type I3r
ø22′, m6.2:v, Br★ 8.20v   07ʰ44.6ᵐ −23°52′
Finder Chart 26-5, Figure 26-16   ★★★★★

Discovered by Messier in March 1781, M93 is a luminous wedge-shaped star cluster that is as bright but less populous than M46. It lies some 3,400 light years away, so its true size is about 20 light years and its true luminosity 4,000 suns.

8/10″ Scopes–60x: M93 is 1.25° NW of the magnitude 3.34 golden yellow Xi (ξ) = 7 Puppis. It has fifty 8th magnitude and fainter members in a 20′ area. The stars are concentrated in a SW-pointing wedge, some brighter cluster members scattered west and east of the wedge tip. The two brightest M93 stars, east of the wedge tip, both have widely separated companions.

12/14″ Scopes–75x: Messier 93 is a magnificent sight! It is a bright, large, rich, irregular cluster of 100 stars in a 22′ × 10′ E–W area. A conspicuous starless gap at the center is flanked to the east by a star-dense wedge pointing SW. Some irregular star chains extend east and west from the wedge tip. The two brightest cluster members are a wide E–W pair due east of the wedge tip.

Ruprecht 32 Open Cluster 30★ Tr Type III 2 p n
ø5′, m8.4v, Br ★ 9.55v   07ʰ45.0ᵐ −25°31′
Finder Chart 26-5   ★★★

12/14″ Scopes–125x: Ruprecht 32, just north of a wide N–S pair of 9.5 magnitude stars, is a fairly obvious cluster of twenty 11th magnitude and fainter stars in an uncompressed, irregular group. The NNW portion contains a 6′ × 2′ NNW–SSE patch of the brighter cluster stars. A spray of faint outlying members spreading SSE from the most concentrated area extends the overall dimensions of the group to 8′ × 4′. A 6.5 magnitude star lies 18′ west.

Haffner 15 Open Cluster 35★ Tr Type II 2 m
ø3.5′, m9.4v, Br ★ 10.45v   07ʰ45.3ᵐ −32°47′
Finder Chart 26-6, Figure 26-17   ★★★

12/14″ Scopes–125x: Haffner 15 is a fairly conspicuous cluster of a dozen stars resolved against a 3′ diameter granular background haze.

NGC 2451 Open Cluster 40★ Tr Type II 2 m
ø45′, m2.8v, Br★ 3.60v   07ʰ45.4ᵐ −37°58′
Finder Chart 26-7, Figure 26-18   ★★★★

4/6″ Scopes–40x: NGC 2451 is a very bright, coarse, irregular cluster of thirty 6th to 11th magnitude stars surrounding magnitude 3.6 star c Puppis. Near the center, two bright stars form a triangle with c Puppis.

8/10″ Scopes–50x: NGC 2451 is a colorful scattering of yellow and blue stars around the bright orange-red c Puppis. It has thirteen 6th to 9th magnitude and forty fainter stars in a 45′ × 30′ E–W area. The three brightest members form a triangle near the center. Along the northern edge is an ENE–WSW star chain of one 7.5 magnitude, three 9th magnitude, and seven fainter stars. The cluster's NE and SW edges are marked by 6th magnitude stars. NGC 2451, at a distance of about 710 light years, is one of the nearest open clusters. c Puppis is a K-type luminous giant as bright as 1,600 suns.

Ruprecht 34 Open Cluster 35★ Tr Type III 2 p
ø4′, m 9.5:v, Br ★ –   07ʰ45.9ᵐ −20°23′
Finder Chart 26-5, Figure 26-19   ★★★

12/14″ Scopes–125x: Ruprecht 34 is at the north end of

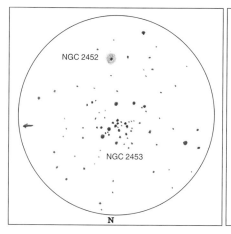
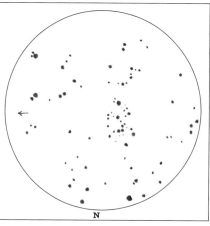
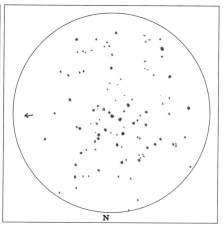

Figure 26-20. NGC 2452 & NGC 2453
13″, f5.6–135x, by Steve Coe

Figure 26-21. Ruprecht 36
12.5″, f5–200x, by G. R. Kepple

Figure 26-22. NGC 2455
12.5″, f5–200x, by G. R. Kepple

an arc of three magnitude 9.5 stars curving NW from a 7th magnitude star. It contains thirty magnitude 11–14 stars in a fairly well concentrated 7′ triangular area, the majority of its members in the eastern part of the triangle. A 10th magnitude star is 5′ NNW of the cluster center.

NGC 2452 PK243-1.1 Planetary Nebula Type 4+3
ø>19″, m12.0v, CS 16.11v $07^h47.4^m$ $-27°20′$
Finder Chart 26-6, Figure 26-20   ★★★★

12/14″ Scopes–225x: This planetary nebula lies 6′ south of open cluster NGC 2453. Moderately high power shows a bright, greenish 20″ disk.

16/18″ Scopes–250x: The fairly bright NGC 2452 has a distinctly annular disk elongated 30″ × 25″ N–S. The northern and southern edges are slightly brighter. A 13th magnitude star is 1′ SW, and a 14th magnitude star almost touches the northern edge. The central star is not visible.

NGC 2453 Cr162 Open Cluster 30★ Tr Type I 2 m
ø5′, m8.3v, Br★ 9.46v $07^h47.8^m$ $-27°14′$
Finder Chart 26-6, Figure 26-20   ★★★

12/14″ Scopes–125x: This cluster has fifteen 12th magnitude and fainter stars in a 5′ area. The central concentration of ten stars is elongated N–S.

16/18″ Scopes–150x: NGC 2453 contains twenty stars, most 12th magnitude and fainter, in a 6′ area. On the NW edge is a 2.5′ long NW–SE row of one 11th and two 10th magnitude stars. On the SE edge is a very dense knot of faint cluster members. 10th magnitude field stars are 3′ south, 4′ SE, and 4.5′ east of the cluster.

Ruprecht 36 Open Cluster 30★ Tr Type IV 1 m
ø4′, m9.6:v, Br★ 10.28v $07^h48.5^m$ $-26°18′$
Finder Chart 26-6, Figure 26-21   ★★

12/14″ Scopes–225x: Ruprecht 36, located 1/3° south of the magnitude 4.5 Omicron (o) Puppis, is a faint cluster of twenty 12th magnitude and fainter stars running north from the 10th magnitude lucida. It covers a 3′ × 2′ area, some of its brighter members in a box-shaped asterism on its northern edge.

NGC 2455 Cr163 Open Cluster 50★ Tr Type III 2 m
ø7′, m10.2p, Br★ – $07^h49.0^m$ $-21°18′$
Finder Chart 26-5, Figure 26-22   ★★

4/6″ Scopes–75x: NGC 2455, located just west of an 8th magnitude star, consists of ten 12th magnitude and fainter stars against a hazy backdrop.

8/10″ Scopes–100x: Twenty 12th to 14th magnitude stars in an 8′ area resolve against a rich, granular background.

12/14″ Scopes–125x: Several dozen 12th to 14th magnitude stars are visible in an irregular, uncompressed 8′ diameter area. Nearly all the stars are distributed along NW–SE chains, the northwesternmost chain being the longest. At 175x another half dozen cluster members reach the threshold of resolution. An E–W chain of twenty 11th to 12th magnitude field stars is 10′ NE of the cluster.

Haffner 16 Open Cluster 30★ Tr Type I 2 m
ø5′, m10.0:v, Br★ 11.62v $07^h50.3^m$ $-25°27′$
Finder Chart 26-5, Figure 26-23   ★★

12/14″ Scopes–225x: Haffner 16 is 35′ SSE of Xi (ξ) = 7 Puppis and 40′ NE of Omicron (o) Puppis. It contains twenty 11.5 magnitude and fainter stars in a moderately well compressed 5′ area. The 3′ diameter

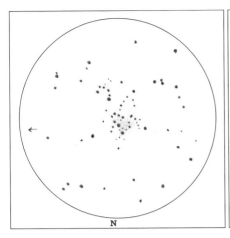
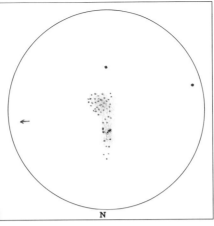
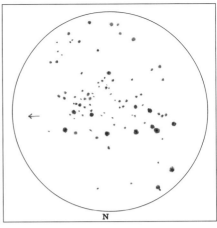

Figure 26-23. Haffner 16
13", f5.6–135x, by Steve Coe

Figure 26-24. NGC 2477
6", f8–80x, by Andrew D. Fraser

Figure 26-25. Haffner 18
12.5", f5–300x, by G. R. Kepple

northern section is the richest part of the cluster, but an arm extending 2' SW gives the group an overall NE–SW elongation. An E–W row of four magnitude 12.5 stars, two of them in the center of the row, is isolated 3' SE of the cluster.

NGC 2477 Open Cluster 160★ Tr Type I 2 r
⌀27', m5.8v, Br★ 9.81v   07ʰ52.3ᵐ −38°33'
Finder Chart 26-7, Figure 26-24   ★★★★

4/6" Scopes–50x: NGC 2477, just north of magnitude 4.5 b Puppis, is a rich NE–SW teardrop-shaped aggregation of at least one hundred stars. If farther north, NGC 2477 would be a sight to exceed M46; but its far southerly declination means that it cannot be seen to full advantage by most U.S. and European observers even on the clearest of nights. Its brighter stars are in its northern extension; but the richest star-concentration is in its southern lobe.

8/10" Scopes–75x: This bright cluster contains several hundred 10th magnitude and fainter stars in a 25' area. The southern section, which has many star-clumps and star-chains, is clearly its richest part. On the northern side is a 35" wide NNW–SSE double of magnitude 9.5 and 10.5 stars.

12/14" Scopes–75x: NGC 2477 is a grand, rich assemblage of at least 250 stars. Most are rather faint, the cluster lucida being only 9.8 magnitude. The richest area, in the southern part of the cluster, is 15' across, but overall the group is extended 30' × 20' NE–SW. A string of brighter stars exits the cluster on its NE side. Starless streaks set off star-knots and tiny groupings. NGC 2477 is about 4,200 light years away, 33 light years in diameter, and as luminous as 14,500 suns. It is a rather evolved open cluster, its age being 1.3 billion years, four times that of M46.

Haffner 18 Open Cluster 15★ Tr Type I 3 p n
⌀1', m9.3v, Br★ −   07ʰ52.5ᵐ −26°22'
Finder Chart 26-6, Figures 26-25 & 26-29   ★★★

12/14" Scopes–125x: Haffner 18 is a well defined 4' × 3' triangular concentration of thirty 10th to 13th magnitude stars just NE of the bright NGC 2467 emission nebula. The cluster is most concentrated along the SW side of the triangle, its SE and NW sides being more straggling and its center almost empty. Haffner 18 looks like it is part of the NGC 2467 complex; but in reality it is in the background of the 15,000 light year distant NGC 2467 and might be as much as 22,500 light years away. If so, it marks one of the extreme outer spiral arms of our Galaxy.

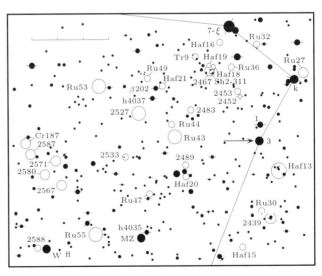

Finder Chart 26-6. 3 Pup: 07ʰ43.8ᵐ −28°57'

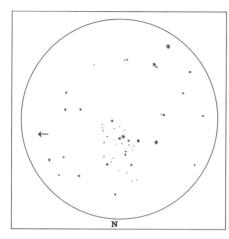
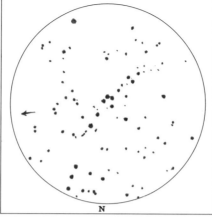
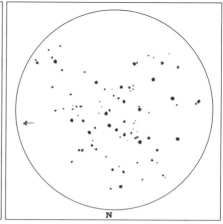

Figure 26-26. Haffner 19
12.5″, f5–300x, by G. R. Kepple

Figure 26-27. NGC 2482
12.5″, f5–150x, by G. R. Kepple

Figure 26-28. NGC 2479
12.5″, f5–125x, by G. R. Kepple

NGC 2467 Sh2-311 Gum 9 Emission Nebula
⌀16′ × 12′, Photo Br 1–5, Color 3–4
07ʰ52.5ᵐ −26°24′

NGC 2467 H22⁴ Open Cluster 50★ Tr Type I 3 m n
⌀14′, m7.1p, Br★ −   07ʰ52.6ᵐ −26°23′
Finder Chart 26-6, Figure 26-29   ★★/★★★

4/6″ Scopes–75x: NGC 2467 is a small but rather bright patch of haze embedded in a rich star field, the most concentrated part of which forms a "Y." The nebula

Figure 26-29. *The open cluster + emission nebula NGC 2467 is a scattered field condensation of stars in and around a rather bright haze. A conspicuous 4′ long NW–SE line of stars NE of the nebula (to its lower right in the photo) is the most condensed part of the open cluster Haffner 18. Haffner 19 is a small star-knot north (beneath) of Haffner 18. Lee C. Coombs took this ten minute exposure on 103a-E film with a 10″, f5 Newtonian.*

is more prominent on its west side around a magnitude 8.1 O7 star, probably one of the hot stars powering the nebula's fluorescence. The surface brightness of the central part of the nebula is so high that it is easily seen in 10×50 binoculars as a small spot. However, with averted vision in medium size telescopes the nebula appears elongated E–W, stretching across the arms of the Y-shaped asterism.

12/14″ Scopes–100x: In larger telescopes, and with an O-III filter, NGC 2467 is a conspicuous 4′ × 3′ oval glow in a rich star field. In the SW part of the nebula is a magnitude 8.1 O7 star and just north of its center a magnitude 9.2 O6 star, both objects no doubt involved in its illumination. A total of about fifteen stars can be counted in the nebula. Just north of the magnitude 9.2 star is an E–W dark lane that separates the brighter part of the nebula from the fainter northern section. Another dark lane intrudes into the nebula from the SE almost to its center. The brightest part of the nebula is along its southern rim. The star field is richer north of the nebula, and in it are the two open clusters Haffner 18 (4′ NE of the magnitude 9.2 star) and Haffner 19 (4′ north of Haffner 18). These two clusters are, however, background objects not involved with the NGC 2467 open cluster and its surrounding association, Puppis OB2.

Haffner 19 Open Cluster 30★ Tr Type I 3 p n
⌀1.8′, m9.4v, Br★ −   07ʰ52.7ᵐ −26°15′
Finder Chart 26-6, Figures 26-26 & 26-29   ★★

12/14″ Scopes–125x: Haffner 19 is 4′ north of Haffner 18 and 10′ NNE of the center of open cluster and emission nebula NGC 2467. It is a tiny, hazy concentration of a dozen stars in a 2′ area centered upon a prominent 4.4″ pair of 10.8 and 12.5 magnitude

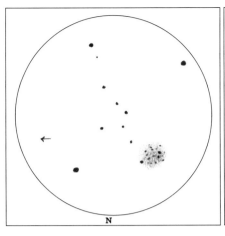
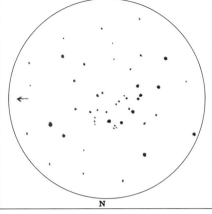
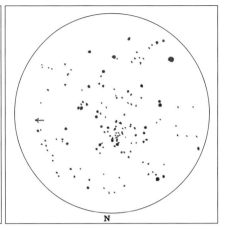

Figure 26-30. Trumpler 9
12.5", f5–100x, by G. R. Kepple

Figure 26-31. NGC 2489
13", f5.6–135x, by Steve Coe

Figure 26-32. NGC 2509
12.5", f4.5–175x, by G. R. Kepple

stars. The cluster's brighter stars are in a 1.5′ long E–W line, around the western end of which is the haze of faint unresolved members.

NGC 2482   H10⁷   Open Cluster 40★   Tr Type IV 1 m
ø12′, m7.3v, Br★ 10.04v   07ʰ54.9ᵐ −24°18′
Finder Chart 26-5, Figure 26-27   ★★★

4/6″ Scopes–75x: NGC 2482 is a quite rich cluster of thirty 10th to 12th magnitude stars irregularly distributed in a 12′ area. Concentrations on the SE and NW sides of the cluster are connected by a rich star stream through the cluster center. A starless void is just south of the cluster center.

12/14″ Scopes–100x: NGC 2482 is a cluster rich in 10th to 11.5 magnitude stars with fainter members abundantly sprinkled throughout. A total of sixty-five stars are irregularly distributed in a 15′ area, the densest concentration being on the cluster's east side. A thick star chain runs NNW–SSE through the cluster center, and a second chain exits the cluster toward the west. In the field to the NE are one 8th and three 9th magnitude stars; and an 8th magnitude star is on the cluster's west edge.

NGC 2479   H58⁷   Open Cluster 45★   Tr Type III 1 m
ø7′, m9.6p, Br★ –   07ʰ55.1ᵐ −17°43′
Finder Chart 26-4, Figure 26-28   ★★★

4/6″ Scopes–75x: The faint cluster NGC 2479 lies 12′ NE of a 20″ wide NE–SW double of 9th magnitude stars. It is a large, rich concentration of nearly uniformly faint stars.

12/14″ Scopes–100x: NGC 2479 is a rich, dense concentration of fifty 12th magnitude and fainter stars in a somewhat triangular 7′ area. It is mildly concentrated toward its center, but a starless patch in-

trudes into the cluster from the ESE. A string of stars flairs eastward from the cluster's NE edge.

Trumpler 9   Harvard 2   Open Cluster 20★   Tr Type II 2p
ø5′, m8.7:v, Br★ 10.09v   07ʰ55.3ᵐ −25°56′
Finder Chart 26-6, Figure 26-30   ★★★

12/14″ Scopes–100x: Trumpler 9 is a hazy concentration of faint stars 3′ east of a N–S string of 8.5 to 9th magnitude stars, at the north end of which is a wide magnitude 9.5 double. The cluster has thirty 10th to 13th magnitude members in an irregular 5′ area somewhat elongated NW–SE. SW of the group is a N–S string of 7th magnitude stars.

NGC 2483   Open Cluster 30★   Tr Type III 2 m
ø10′, m7.6v, Br★ 9.26v   07ʰ55.9ᵐ −27°56′
Finder Chart 26-6   ★★★

12/14″ Scopes–100x: NGC 2483 is a coarse, irregular splattering of thirty 9.3 to 13th magnitude stars in a 10′ area. Several star strings can be discerned, and the group's east edge is marked by two magnitude 9.5 members.

NGC 2489   H23⁷   Open Cluster 45★   Tr Type I 2 m
ø6′, m7.9v, Br★ 11.11v   07ʰ56.2ᵐ −30°04′
Finder Chart 26-6, Figure 26-31   ★★★

4/6″ Scopes–50x: NGC 2489 is 13′ north of an E–W arc of three magnitude 4.5 to 6.5 stars, the red star PX Puppis at the arc's center. It is a faint, fairly rich 4′ cluster, concentrated on the west and more scattered to the east.

12/14″ Scopes–100x: NGC 2489 is a circular cluster of fifty 11th to 13th magnitude stars in a 6′ area. It is noticeably denser in the central 3.5′, where the brighter stars are distributed in a 3′ × 1.25′ oval. The outer regions are about half as concentrated as the center.

Haffner 20 Open Cluster 20★ Tr Type II 1 p
⌀2.2′, m11.0v, Br★ 13.13v   07ʰ56.3ᵐ −30°24′
Finder Chart 26-6   ★★

12/14″ Scopes–100x: Haffner 20 is 5′ south of the 6th magnitude reddish star PX Puppis. It is a faint 2′ knot of 13th to 15th magnitude stars. Half a dozen stars may be resolved against a hazy background.

Ruprecht 43 Open Cluster 25★ Tr Type IV 1 p
⌀13′, m –, Br★ 12.0p   07ʰ54.9ᵐ −24°18′
Finder Chart 26-6   ★★

12/14″ Scopes–100x: Ruprecht 43 is a fairly faint cluster of a dozen stars resolved against a granular background. A large triangle of magnitude 9.5 stars is superimposed upon the faint stars and the background haze.

Ruprecht 44 Open Cluster 40★ Tr Type IV 2 m
⌀5′, m7.2v, Br★ –   07ʰ59.0ᵐ −28°35′
Finder Chart 26-6   ★★★

12/14″ Scopes–125x: Ruprecht 44 is a rich cluster of fifty 9.5 and fainter stars gathered in a 5′ area. The fainter stars are concentrated at the center and the brighter stars are around the edges, including an arc of six 9.5 to 10.5 magnitude stars along the northern boundary and a lone 9.5 magnitude star at the southern edge.

Ruprecht 45 Open Cluster 35★ Tr Type IV 1 m
⌀11′, m –, Br★ 13.0p   07ʰ59.6ᵐ −16°18′
Finder Chart 26-4   ★★

12/14″ Scopes–100x: Ruprecht 45 is a gathering of three dozen faint stars in an 11′ area. Eight 13th to 14th magnitude stars are resolved around a bright star near the cluster center, with a granular background of partially resolved members behind them. On the east side of the group is a pentagon of stars, and on the west side is a diamond-shaped asterism.

NGC 2509 H1⁸ Open Cluster 70★ Tr Type I 1 m
⌀8′, m9.3p, Br★ –   08ʰ00.7ᵐ −19°04′
Finder Chart 26-4, Figure 26-32   ★★★

12/14″ Scopes–100x: NGC 2509 is a moderately rich and somewhat compressed group of forty 10th to 13th magnitude stars in an 8′ area. The distribution of stars in the cluster suggests a diamond ring, the gemstone represented by a rich 2′ concentration of faint stars on the northern edge connected by an irregular loop to a magnitude 8.5 star to the SSE.

16/18″ Scopes–125x: NGC 2509 is a moderately bright, rich concentration of fifty stars in an 8′ area. Eight 10th to 12th magnitude stars form a ring south of a 2′ concentration of fainter stars. Near the NE edge of the cluster is a 25″ wide double of magnitude 10 and 12 stars, and on the east edge is a 13″ magnitude 12 and 13 pair. A tiny clump of 15 stars is on the NNW side.

Haffner 21 Open Cluster 20★ Tr Type I 1 p
⌀2.2′, m10.3v, Br★ 12.14v   08ʰ01.2ᵐ −27°10′
Finder Chart 26-6   ★★

12/14″ Scopes–100x: Haffner 21 is nestled in a star-rich area 10′ west of the double β202 (6.6, 9.3; 7.6; 161°). This cluster is a small knot of fifteen 10.3 magnitude and fainter stars in a 1′ area.

Ruprecht 46 Open Cluster 15★ Tr Type I 2 p
⌀2′, m9.1v, Br★ 9.43v   08ʰ02.1ᵐ −19°28′
Finder Chart 26-4, Figure 26-33   ★★

12/14″ Scopes–100x: The small cluster Ruprecht 46 has a dozen 13th magnitude and fainter stars in a 2′ × 1′ NE–SW area. A magnitude 9.4 star lies at the center of a Y-shaped asterism, the stem of which points NE. A few faint stars are scattered along the east edge of the group, and two are detached to the ENE.

Ruprecht 47 Open Cluster 20★ Tr Type II 1 p
⌀5′, m9.6:v, Br★ 10.90v   08ʰ02.3ᵐ −31°06′
Finder Chart 26-6   ★★★

12/14″ Scopes–100x: Ruprecht 47, in a rich Milky Way field just east of an 8th magnitude star, is a rich, well compressed concentration of twenty 11th to 13th magnitude stars in a 5′ area.

Ruprecht 49 Open Cluster 10★ Tr Type II 1 p
⌀2.5′, m9.6:v, Br★ 10.11v   08ʰ03.1ᵐ −26°47′
Finder Chart 26-6, Figure 26-34   ★★

12/14″ Scopes–125x: Ruprecht 49 is a small 2.5′ concentration of a dozen 11th to 13th magnitude stars. A 10th magnitude star near the northern edge is partially encircled by an arc of four 13th magnitude stars swinging from its NW to its SSE side. A hazy splash of very faint stars extends east from center. Another 10th magnitude star is detached 4′ SE.

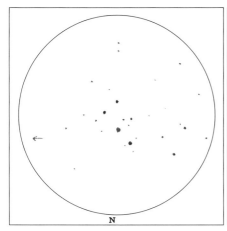

Figure 26-33. Ruprecht 46
12.5", f5–250x, by G. R. Kepple

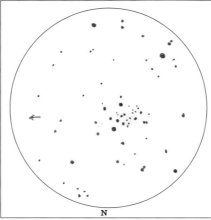

Figure 26-34. Ruprecht 49
12.5", f5–250x, by G. R. Kepple

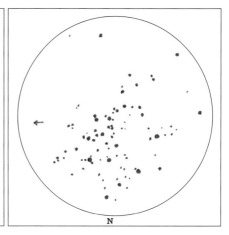

Figure 26-35. NGC 2527
12.5", f5–100x, by G. R. Kepple

NGC 2527 H30⁸ Open Cluster 40★ Tr Type II 2 m
⌀16′, m6.5v, Br★ 8.59v   08ʰ05.3ᵐ −28°10′
Finder Chart 26-6, Figure 26-35   ★★★

4/6" Scopes–50x: NGC 2527 is bright and large, its 25 stars mildly compressed in a 20′ area.

8/10" Scopes–75x: NGC 2527, located in a rich Milky Way field, is a loose, unconcentrated cluster of three dozen 9th to 13th magnitude stars in a 12′ area with outliers spread over a 25′ area. A magnitude 7.5 star lies on the ESE edge.

12/14" Scopes–75x: NGC 2527 is a fairly bright, large, loose assemblage of fifty stars in a 22′ area. The central 10′ contains a dozen 9th to 10th magnitude members. South of the center is a 5′ long hook-shaped asterism of sixteen stars. A compact group of a dozen 11th to 12th magnitude stars is along the cluster's eastern edge. The surrounding field is rather well populated with bright Milky Way stars.

NGC 2525 A135 Galaxy Type SB(rs)cd II
⌀3.2′ — 2.0′, m11.6v, SB 13.5   08ʰ05.6ᵐ −11°26′
Finder Chart 26-4, Figure 26-36   ★★★

12/14" Scopes–125x: NGC 2525 lies between N–S 9th magnitude stars. The halo is moderately faint, elongated 3′ × 1.5′ ENE–WSW, and slightly brighter at its center.

16/18" Scopes–150x: This galaxy has a fairly bright but diffuse 3′ × 2′ ENE–WSW halo containing a broad, mildly concentrated core. A 13th magnitude star lies on the ESE edge, and a wide 14th magnitude pair points toward the galaxy's center from its WSW side.

NGC 2533 Cr175 Open Cluster Tr Type II 2 m
⌀10′, m7.6v, Br★ 8.99v   08ʰ07.0ᵐ −29°54′
Finder Chart 26-6   ★★★

4/6" Scopes–50x: In small telescopes NGC 2533 is a misty patch surrounding a 9th magnitude star.

8/10" Scopes–100x: Fifteen 12th to 14th magnitude stars are visible in a 3′ area surrounding a 9th magnitude star.

12/14" Scopes–125x: NGC 2523 is a faint, somewhat triangular cluster of twenty 9th to 14th magnitude stars. An 11th magnitude star is at the triangle's SW vertex, with the cluster's 9th magnitude lucida located just to its NE.

NGC 2539 H11⁷ Open Cluster 50★ Tr Type III 2 m
⌀21′, m6.5v, Br★ 9.15v   08ʰ10.7ᵐ −12°50′
Finder Chart 26-4, Figure 26-37   ★★★★

4/6" Scopes–50x: This cluster is a bright, large, rich, well resolved mass of 80 to 100 magnitude 9.2 and fainter stars in a 25′ area. 19 Puppis (m4.7) lies at the SE edge, but is not a true physical cluster member.

12/14" Scopes–75x: NGC 2539 is a rich, irregular expanse of 125 stars with three concentrations of stars spreading 30′ NW from the magnitude 4.7 star 19 Puppis. Most of its members seem to lie in arcs around three relatively star-poor gaps. The cluster lucida is a magnitude 9.2 orange-red giant.

Ruprecht 53 Open Cluster 40★ Tr Type IV 3 m
⌀18′, m , Br★ 10.0p   08ʰ10.8ᵐ −27°01′
Finder Chart 26-6   ★★

12/14" Scopes–75x: Ruprecht 53 is not well detached from the surrounding star field, its three dozen faint stars spread loosely and irregularly over an 18′ area.

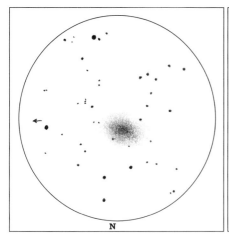
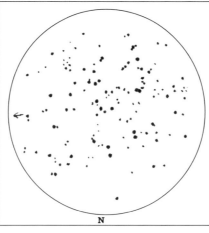
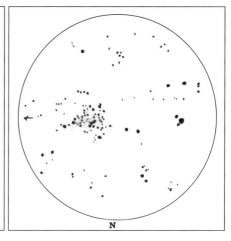

Figure 26-36. NGC 2525
12.5″, f5–250x, by G. R. Kepple

Figure 26-37. NGC 2539
12.5″, f5–125x, by G. R. Kepple

Figure 26-38. NGC 2568 (Pismis 1)
12.5″, f5–150x, by G. R. Kepple

The brighter cluster stars, four of them magnitude 9.5 objects, are somewhat concentrated toward its center. A wide pair of magnitude 9.5 stars is near the group's NNE edge. SW of the cluster is a group of four bright field stars.

Ruprecht 55   Open Cluster 12★   Tr Type III 1 p
ø17′, m7.8v, Br★ 8.56v $08^h12.3^m$ $-32°36′$
Finder Chart 26-6   ★★

12/14″ Scopes–75x: Ruprecht 55 is a sparse cluster of a dozen faint stars spread over a 17′ area. A lone 8.5 magnitude star stands out NW of the group's center.

NGC 2546   Open Cluster 40★   Tr Type III 2 m
ø40′, m6.3v, Br★ 8.22v $08^h12.4^m$ $-37°38′$
Finder Chart 26-7   ★★★

8/10″ Scopes–50x: NGC 2546 is an exceptionally large cluster of fifty 8th magnitude and fainter stars in a 50′ area. A 6th magnitude deep yellow star is on the southern edge, and a magnitude 6.4 blue-white star on the NNW edge, the fainter cluster members sprinkled between them. The magnitude 6.4 star seems to be a true physical member of the group.

12/14″ Scopes–75x: This is a bright, extremely large, irregularly scattered cluster without central compression. 75 stars are in a NNW–SSE 50′ × 25′ area, the brighter cluster members in a stream between the two 6th magnitude stars on the south and NNW edges. The NNW portion of the cluster is richer.

Ruprecht 56   Open Cluster 40★   Tr Type III 2 m
ø42′, m –, Br★ 9.0p $08^h12.6^m$ $-40°28′$
Finder Chart 26-7   ★★★

12/14″ Scopes–100x: Ruprecht 56 contains forty stars irregularly broadcast SW of the 4th magnitude h^2 Puppis. In its center one 8th and a dozen 9–9.5 magnitude stars form a distinct 15′ × 7′ N–S group. Outliers are scattered to 45′.

NGC 2568   Pismis 1   Open Cluster 30★   Tr Type II 2 m
ø4.6′, m10.7v, Br★ 12.76v $08^h18.0^m$ $-37°06′$
Finder Chart 26-6, Figure 26-38   ★★★

12/14″ Scopes–75x: NGC 2568, located half a degree south of the magnitude 4.5 g Puppis, contains three dozen magnitude 12.8 and fainter members in a well concentrated triangular area open to the NNW. The SW side is a highly concentrated string of 13th–14th magnitude stars. A 13th magnitude star marks the NE vertex. In the center of the triangle is a conspicuous knot of stars. A bright 10th magnitude field star is 7′ east of the cluster.

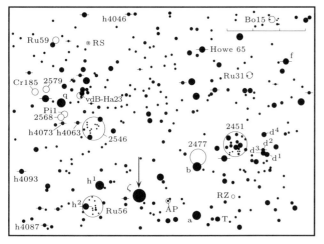

Finder Chart 26-7. ζ Pup: $08^h03.6^m$ $-40°01′$

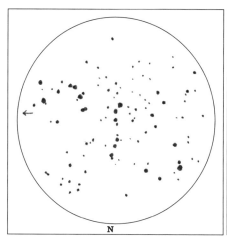
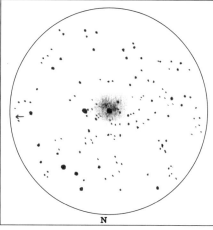
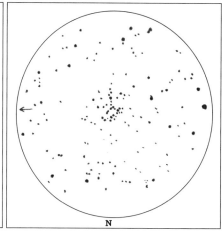

Figure 26-39. NGC 2567
12.5″, f5-250x, by G. R. Kepple

Figure 26-40. NGC 2579
13″, f5.6-200x, by Steve Coe

Figure 26-41. NGC 2588
12.5″, f5-200x, by G. R. Kepple

NGC 2567 H64⁷ Open Cluster 40★ Tr Type II 2 m
ø10′, m7.4v, Br★ 10.10v   08ʰ18.6ᵐ −30°38′
Finder Chart 26-6, Figure 26-39   ★★★

12/14″ Scopes-100x: NGC 2567 is a moderately rich but uncompressed cluster of fifty 10th to 14th magnitude stars sprinkled NE of a 9th magnitude star. A N–S row of six stars runs through the cluster's center; and on its west side is a V-shaped asterism, the star at its tip a 12″ wide double of 11th magnitude stars.

NGC 2571 H39⁷ Open Cluster 30★ Tr Type III 2 m
ø30′, m7.0v, Br★ 8.82v   08ʰ18.9ᵐ −29°44′
Finder Chart 26-6   ★★★

4/6″ Scopes-50x: This cluster is bright, moderately large, and fairly compressed. A bright double star is superimposed against a background of glittering stars. A large E–W dark lane bisects the cluster.

12/14″ Scopes-75x: NGC 2571 is a large, loose cluster of some twenty-five stars, most 12th magnitude and fainter, in a 15′ × 10′ NW–SE area. A wide 9th magnitude pair dominates the cluster's center, the southernmost of the two stars having a 12th magnitude companion 15″ to its SE.

Ruprecht 59 Open Cluster 20★ Tr Type III 1 p
ø5′, m9.0v, Br★ 10.17v   08ʰ19.1ᵐ −34°27′
Chart 26-7   ★★★

12/14″ Scopes-75x: Ruprecht 59, located 8′ NNW of a 6th magnitude star, is an unconcentrated scattering of twenty 10th magnitude and fainter stars in a 5′ area.

NGC 2579 Gum 11 Emission & Reflection Nebula
ø3′, Photo Br 3-5, Color 2-4   08ʰ21.0ᵐ −36°15′

NGC 2579 Open Cluster 20★ Tr Type IV 2 p
ø8′, m7.5v, Br★ 9.51v   08ʰ21.1ᵐ −36°11′
Finder Chart 26-7, Figure 26-40   ★★/★★★

12/14″ Scopes-125x: The emission nebula NGC 2579, centered 20′ north of a magnitude 4.5 star, is a fairly bright 3′ × 2′ E–W oval glow. It is brighter along its eastern side. Near the nebula's center is its magnitude 9.5 illuminating star. 10th magnitude stars are 2′ west and 6.5′ NW of the central star. Half a dozen fainter stars are scattered around the nebula's periphery. The involved cluster is not especially concentrated, with only twenty 11th to 13th magnitude stars scattered over a 10′ area.

NGC 2580 Open Cluster 50★ Tr Type II 2 m
ø7′, m9.7v, Br★ –   08ʰ21.6ᵐ −30°19′
Finder Chart 26-6   ★★

12/14″ Scopes-125x: This inconspicuous cluster does not stand out well from the rich Milky Way field. It contains only 18 magnitude 11 and fainter stars in an 8′ area.

16/18″ Scopes-150x: NGC 2580 has several dozen 11th magnitude and fainter stars in a 7′ area. The brighter stars form an "X" against the fainter cluster members.

NGC 2588 Open Cluster 20★ Tr Type II 1 p
ø2′, m11.8p, Br★ –   08ʰ23.2ᵐ 32°59′
Finder Chart 26-6, Figure 26-41   ★★

12/14″ Scopes-125x: NGC 2588, located 23′ east of the 5th magnitude w Puppis and 10′ SSW of a 7th magnitude star, is a faint 2′ concentration of twenty

13th magnitude and fainter stars. Eight 14th magnitude stars arcing along the northern edge partially encircle the faint knot of stars at the cluster center.

Collinder 185   Open Cluster 35★   Tr Type III 2 p
⌀8′, m7.8v, Br★ 10.08 $08^h22.5^m$ $-36°10′$
Finder Chart 26-7   ★★

12/14″ Scopes–125x: Collinder 185 is 15′ east of the emission nebula/open cluster NGC 2579 and 20′ NE of a 4.5 magnitude star. It is a rather faint, unconcentrated scattering of eighteen 10th magnitude and fainter stars in a 9′ area.

NGC 2587   Open Cluster 40★   Tr Type III 2 m
⌀9′, m9.2p, Br★ – $08^h23.5^m$ $-29°30′$
Finder Chart 26-6   ★★★

12/14″ Scopes–125x: NGC 2587 appears west of a 9th magnitude field star as a gathering of a dozen faint stars against a background haze. An array of ten stars lies to the east, and an 11.5 magnitude star is just west.

16/18″ Scopes–150x: This cluster is a moderately faint but compact group of thirty faint stars in a 7′ × 5′ N–S triangular area. 9.5 magnitude stars are at the triangle's eastern and northern vertices. The southern part of the cluster is the most concentrated, and a row of five brighter stars are along the very southern edge.

Collinder 187   Open Cluster 20★   Tr Type IV 1 p
⌀7′, m9.6p, Br★ – $08^h24.2^m$ $-29°09′$
Finder Chart 26-6   ★★★

12/14″ Scopes–125x: Collinder 187, located 10′ SSE of a 6.5 magnitude star, is an irregular, unconcentrated spray of twenty 10th magnitude and fainter stars in a 7′ area.

Chapter 27

Pyxis, the Marier's Compass

27.1 Overview

Pyxis, the Compass, was originally in the huge ancient constellation Argo Navis, the Ship of the Argonauts. Its stars had been part of the Ship's mast, and often on old star charts were labelled Malus, the Mast. But in the mid-17th century Lacaille renamed these stars Pyxis Nautica, the Mariner's compass. This is of course an anachronism, for the Argonauts had sailed by the stars. The Greek-letter designations we have for the brighter stars in the main divisions of Argo, Carina the Keel, Puppis the Stern, and Vela the Sails, we also owe to Lacaille. However, he assigned no Greek letters to the faint stars of Pyxis: the Greek letters of Pyxis are of more recent vintage.

27.2 Interesting Stars

h4166 Triple Star   Spec A0 & A
AxBC: m6.7, 8.6; Sep. 13.7″; P.A. 153°   09^{h}03.0^m −33°36′
Finder Chart 27-4   ★★★★

2/3″ Scopes–75x: This triple is in an attractive, well populated, star field. The AB stars are a fine pale yellowish pair for small telescopes, but the BC pair remains unresolved.

27.3 Deep-Sky Objects

NGC 2613 H266^2 Galaxy Type Sb II
ø7.6′ × 1.9′, m10.5v, SB 13.2   08^{h}33.4^m −22°58′
Finder Chart 27-3, Figure 27-1   ★★★★

8/10″ Scopes–100x: This galaxy lies 10′ SE of an 8th magnitude star, a wide magnitude 9.5 and 11 double between it and the star. It appears fairly bright, elongated 4′ × 1′ ESE–WNW, and has a brighter center.

Pyxis: PIK-sis
Genitive: Pyxidis, PIK-i-dis
Abbrevation: Pyx
Culmination: 9 pm–Mar. 21, midnight–Feb. 4
Area: 221 square degrees
Best Deep-Sky Objects: NGC 2613, NGC 2627, NGC 2818, NGC 2818A
Binocular Objects: NGC 2627, NGC 2818

12/14″ Scopes–125x: NGC 2613 has a moderately bright 5′ × 2′ ESE–WNW halo containing a well concentrated oval core with a very faint stellar nucleus. A 12″ pair of magnitude 12 and 13 stars are at the NW tip, and a tiny triangle of 13th magnitude stars is just off the NE edge of the halo, the SE two of which are only 10″ apart. South of the galaxy is a NW–SE arc of three 13th magnitude stars.

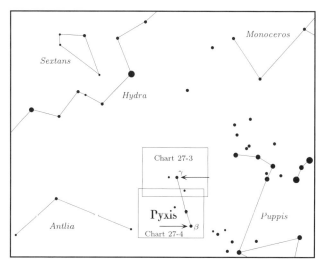

Master Finder Chart 27-2. Pyxis Chart Areas
Guide stars indicated by arrows.

Pyxis, the Mariner's Compass Constellation Chart 27-1

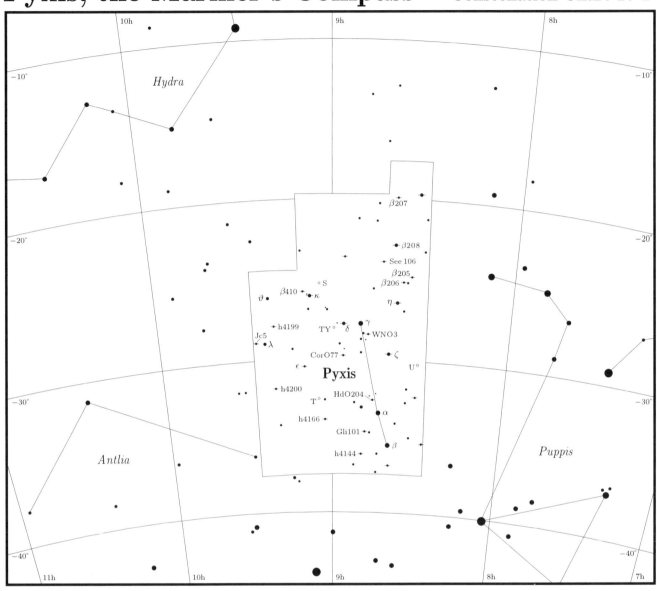

Table 27-1. Selected Variable Stars in Pyxis

Name	HD No.	Type	Max.	Min.	Period (Days)	F*	Spec. Type	R.A. (2000)	Dec.	Finder Chart No. & Notes
U Pyx	72085	SR	8.6	9.4	345		K5	08^{h}29.9^m	−30°19′	27-4
TY Pyx	77137	E/RS	6.87	7.47	3.20		G5+G5	59.7	−27 49	27-3
T Pyx		Nr	6.3	14.0	7000		P	09^{h}04.7^m	−32 23	27-4
S Pyx	78000	M	8.0	14.2	206	0.45	M3-M5	05.1	−25 05	27-3

F* = The fraction of period taken up by the star's rise from min. to max. brightness, or the period spent in eclipse.

Table 27-2. Selected Double Stars in Pyxis

Name	ADS No.	Pair	M1	M2	Sep."	P.A.°	Spec	R.A. (2000) Dec	Finder Chart No. & Notes
β205	6871	AB	6.9	7.0	0.6	*337	A5	08^{h}33.1^m −24°36'	27-3
β206	6887		8.2	8.4	1.8	279	G0	35.4 −25 07	27-3
β207	6903		6.4	9.4	4.3	103	K5	38.7 −19 44	27-1
ζ Pyx	6923		4.9	9.1	52.4	61	G4 G0	39.7 −29 34	27-3 h4120
See 106	6974	AB	6.8	11.5	17.9	238	K0	44.7 −23 47	27-3
		BC		12.0	3.5	333			
HdO 204		AB	6.9	12.0	10.0	310	F2	45.1 −32 15	27-4
		AC		12.0	13.0	275			
Gli 101		AB	6.7	10.6	51.8	131	K5 A2	47.4 −34 36	27-4
WNO 3			8.2	8.6	25.6	286	F5 F5	47.6 −28 22	27-3
h4144			7.0	9.2	2.4	315	B9	50.4 −35 56	27-4
CorO 77			8.4	9.3	10.2	200	F5	57.1 −29 51	27-4
h4166		AxBC	6.7		13.7	153	A0 A	09^{h}03.0^m −33 36	27-4 Pale yellowish pair
Rst 3619		BC	8.6	11.8	0.8	77			
κ Pyx	7202		4.6	9.8	2.1	263	M0	08.0 −25 52	27-3
β410	7220		7.4	8.9	1.7	15.8	A0	09.8 −25 48	27-3
ε Pyx		AxBC	5.6	10.5	17.8	147	A3 A3	09.9 −30 22	27-4
h4199		AB	8.2	9.5	11.7	111	A0 A0	20.0 −27 46	27-3
h4200			7.3	7.9	3.1	73	A0	20.7 −31 46	27-4
Jc 5	7379		6.5	7.2	0.6	264	B8	26.7 −28 47	27-3

Footnotes: *= Year 2000, a = Near apogee, c = Closing, w = Widening. Finder Chart No: All stars listed in the tables are plotted in the large Constellation Chart, but when a star appears in a Finder Chart, this number is listed. Notes: When colors are subtle, the suffix -*ish* is used, e.g. *bluish*.

NGC 2627 H63^7 Open Cluster 60★ Tr Type III 2 m ⌀11', m8.4v, Br★ 11.p   08^{h}37.3^m −29°57'

Finder Chart 27-3, Figure 27-2   ★★★

8/10" Scopes–100x: NGC 2627 is 40' SW of the 5th magnitude star Zeta Pyxidis. It is a moderately rich cluster of three dozen 11th to 13th magnitude stars in a 9' × 5' E–W area. The center is well concentrated and has a faint background haze.

12/14" Scopes–125x: NGC 2627 contains sixty 11th to 14th magnitude stars in a 10' × 5' area, most concentrated E–W through the cluster's center. An 8th magnitude star lies 5' SW, and two more 8th magnitude stars lie to the east, one near the cluster's edge. Two wide doubles are in the cluster, one on its WNW side, the other on the ENE side.

NGC 2635 Cr190 Open Cluster 15★ Tr Type I 3 p ⌀3', m11.2v, Br★ 12.45v   08^{h}38.5^m −34°46'

Finder Chart 27-4   ★★

8/10" Scopes–125x: This small, faint cluster is 40' NNW of 4th magnitude Beta (β) Pyxidis. Ten stars are embedded in a granular background haze.

12/14" Scopes–150x: Twenty 12.5 magnitude and fainter stars resolve in a 3' area.

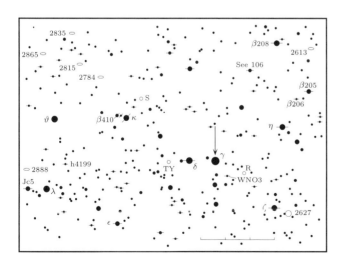

Finder Chart 27-3. γ Pyx: 08^{h}50.5^m −27°43'

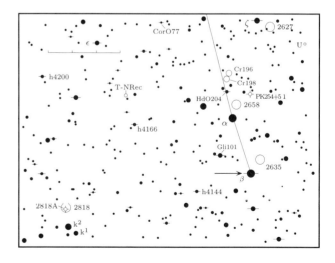

Finder Chart 27-4. β Pyx: 08^{h}40.1^m −35°19'

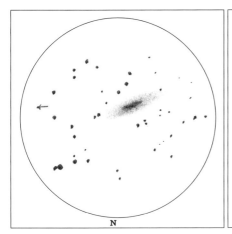
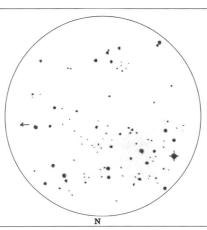
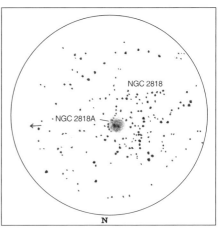

Figure 27-1. NGC 2613
13″, f4.5–165x, by Tom Polakis

Figure 27-2. NGC 2627
12.5″, f5–250x, by G. R. Kepple

Figure 27-3. NGC 2818 & NGC 2818A
20″, f4.5–175x, by Richard W. Jakiel

PK254+5.1 Min 3-6 Planetary Nebula Type 2a
ø9″, m11.0v, CS 13.9v   08ʰ40.7ᵐ −32°23′
Finder Chart 27-4   ★★★

16/18″ Scopes–225x: This planetary nebula has a rather bright 9″ × 7″ N–S oval disk. It responds well to an O-III filter.

NGC 2658 Cr195 Open Cluster 80★ Tr Type II 2 m
ø12′, m9.2p, Br★ 12.p   08ʰ43.4ᵐ −32°39′
Finder Chart 27-4   ★★

8/10″ Scopes–100x: NGC 2658, located 40′ north of Alpha (α) Pyxidis, has twenty 12th magnitude and fainter stars in a 5′ area embedded in an unresolved background haze.

12/14″ Scopes–125x: NGC 2658 is a loose, irregular scattering of thirty 12th magnitude and fainter stars in a 10′ area. A tiny starless void is near the center, and a 10th magnitude star on the SSW edge.

NGC 2818 Cr206 Open Cluster 40★ Tr Type II 2 m n
ø9′, m8.2v, Br★ 11.32v   09ʰ16.0ᵐ −36°37′
NGC 2818A PK261+8.1 Planetary Nebula Type 3 b
ø38″, m11.6v, CS ?   09ʰ16.0ᵐ −36°38′
Finder Chart 27-4, Figure 27-3   ★★★/★★★

8/10″ Scopes–100x: NGC 2818 is a rather faint cluster with forty 11.5 magnitude and fainter stars in a 7′ area. The planetary nebula NGC 2818A is a fairly bright 35″ disk embedded near the cluster's western side.

12/14″ Scopes–125x: NGC 2818 has seventy 11.5 magnitude and fainter stars in an irregularly concentrated 8′ area. Most of its members are concentrated in a 4′ central area, including the cluster's magnitude 11.3 lucida and a conspicuous jagged N–S chain of faint stars. The planetary has a bright, 40″ diameter greenish disk without a visible central star, but a faint star is embedded in the nebula's south side. At 250x a faint streak can be seen running E–W through the nebula center.

NGC 2888 Galaxy Type SA0°:
ø1.9′ × 1.4′, m12.3v, SB 13.2   09ʰ26.3ᵐ −28°02′
Finder Chart 27-3   ★★

12/14″ Scopes–125x: NGC 2888 is a faint, tiny, diffuse galaxy somewhat elongated NNW–SSE.

16/18″ Scopes–200x: This galaxy has a moderately faint 45″ × 30″ NNW–SSE halo of uniform surface brightness in which is embedded a stellar nucleus.

Chapter 28

Sculptor, the Sculptor

28.1 Overview

Sculptor was invented around 1760 by the French cleric and observer of the southern skies Nicolas Louis de Lacaille. His original name for his new constellation was l'Atelier du Sculptor, the Sculptor's Workshop, but the title was later shortened for convenience to Sculptor. Lacaille was simply foisting a name that pleased his fancy upon the heavens, for the poor scattering of magnitude 4.5 and fainter stars south of the ancient Greek constellation of Cetus does not resemble anything in particular.

Sculptor's main claim to fame is the south galactic pole, located just north of Alpha Sculptoris. This marks one of the two directions in which points the axis of rotation of our Milky Way Galaxy. (The north galactic pole is 180° away in the constellation of Coma Berenices.) Because toward the galactic poles we look in a direction perpendicular to the dust-laden spiral arms of our Galaxy out into intergalactic space, Sculptor is a constellation rich in galaxies. It contains the Sculptor System, a dwarf spheroidal member of our Local Galaxy Group a mere 260,000 light years away from the Milky Way but of far too low surface brightness to be suitable for amateur telescopes. Better for observing are the members of the Sculptor Galaxy Group, a handful of loose-structured spirals only 8 million light years from our Local Group and very possibly the closest galaxy cluster to our own.

28.2 Interesting Stars

Kappa-1 (κ^1) Sculptoris Double Star   Spec. F2 & F2
m6.1, 6.2; Sep. 1.4″; P.A. 265°   00^{h}09.3^m −27°59′
Finder Chart 28-5   ★★★★
8/10″ Scopes–150x: Kappa-1 is a tight pair of equally bright yellow stars.

Sculptor: SKULP-tor
Genitive: Sculptoris, SKULP-tor-is
Abbrevation: Scl
Culmination: 9 pm–Nov. 10, midnight–Sept. 26
Area: 475 square degrees
Showpieces: NGC 24, NGC 55, NGC 134, NGC 253, NGC 288, NGC 613, NGC 7793
Binocular Objects: Blanco 1, NGC 24, NGC 55, NGC 134, NGC 253, NGC 288, NGC 300, NGC 613, NGC 7793

Tau (τ) Sculptoris Double Star   Spec. F0
m6.0, 7.1; Sep. 2.2″; P.A. 338°   01^{h}36.1^m −29°54′
Finder Chart 28-6   ★★★★
8/10″ Scopes–150x: Tau Sculptoris is a close pair of bright yellow stars.

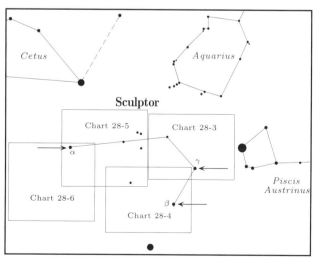

Master Finder Chart 28-1. Sculptor Chart Areas
Guide stars indicated by arrows.

365

Sculptor, the Sculptor

Constellation Chart 28-2

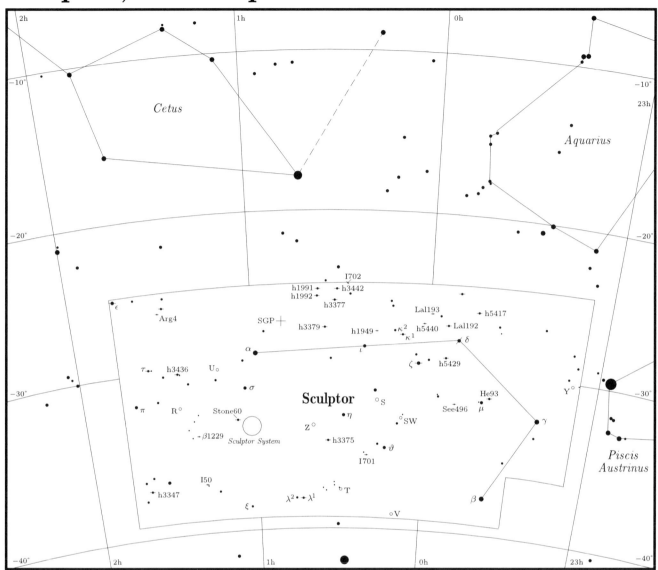

Table 28-1. Selected Variable Stars in Sculptor

Name	HD No.	Type	Max.	Min.	Period (Days)	F*	Spec. Type	R.A. (2000)	Dec.	Finder Chart No. & Notes
Y Scl	218541	SRb	8.7	10.3			M6	$23^h09.1^m$	$-30°08'$	28-3
SW Scl	–	SR	9.4	11.7	144		M1–M4	$00^h06.2^m$	$-32\ 49$	28-4
V Scl	409	M	8.7	15.0	296	0.48	M4–M6	08.6	$-39\ 14$	28-4
S Scl	1115	M	5.5	13.6	365	0.48	M–M8	15.4	$-32\ 03$	28-5
T Scl	2585	M	8.5	13.5	201	0.49	M3	29.2	$-37\ 55$	28-1
U Scl	–	M	8.3	15.2	333	0.39	M5	$01^h11.6^m$	$-30\ 07$	28-6
R Scl	8879	SRb	9.1	12.8	370		C6	27.0	$-32\ 33$	28-6

F* = The fraction of period taken up by the star's rise from min. to max. brightness, or the period spent in eclipse.

Table 28-2. Selected Double Stars in Sculptor

Name	ADS No.	Pair	M1	M2	Sep."	P.A.°	Spec	R.A. (2000)	Dec	Finder Chart No. & Notes
Howe 93	–		6.5	9.8	5.5	251	K0	23^{h}37.1^m	−31°52′	28-3 Orange primary
h5417	16963		6.3	9.0	8.5	320	F5	44.5	−26 15	28-3
δ Scl	17021	AB	4.5	11.5	3.9	243	A0	48.9	−28 08	28-3
(h3216)	17021	AC		9.3	74.3	297	G			
See 496	–		7.8	12.5	16.8	192	F0	51.5	−32 26	28-3
h5429	–		7.5	10.6	28.8	226	K5	53.7	−29 24	28-3
Lal 192	17090		6.9	7.5	6.6	270	A2	54.4	−27 03	28-3
Lal 193	17150		8.2	8.5	10.6	170	F0 F0	59.5	−26 31	28-3
h5440	17177		8.6	8.9	3.3	288	F8	00^{h}02.0^m	−27 09	28-3
κ^1 Scl	111	AB	6.1	6.2	1.4	265	F2 F2	09.3	−27 59	28-5 Equal yellow pair
h1949	–		8.6	8.9	74.6	322	K2 K5	18.5	−27 57	28-5
I 701	–		8.3	8.5	0.6	58	K0	20.2	−35 54	28-4
I 702	370		8.1	9.9	1.3	82	F8	27.9	−24 51	28-5
h3442	456		6.9	11.0	21.9	196	K0	32.7	−25 22	28-5
h3377	466		7.5	9.7	20.1	60	K0 M5	33.6	−26 06	28-5
h3375	–		6.6	8.4	5.3	168	G0	33.7	−35 00	28-1
h3379	514		7.9	11.5	14.6	231	F2	36.7	−27 25	28-5
h1991	–		6.6	9.7	46.9	95	K0	38.8	−25 06	28-5
h1992	–		7.8	8.4	45.4	246	A3 K	38.9	−25 36	28-5
λ^1 Scl	–		6.7	7.0	0.7	3	A0	42.7	−38 28	28-1
Stone 60	–		6.6	10.6	8.6	219	G5	01^{h}04.5^m	−33 32	28-6
I 50	–		7.7	9.7	70.9	247	K0 F8	17.4	−37 16	28-6
β1229	–		8.6	8.8	1.0	286	F0	19.3	−34 30	28-6
h3436	–		6.9	9.7	9.8	126	K0	27.1	−30 14	28-6
Arg 4	–		8.2	8.9	17.9	72	A3	32.3	−26 33	28-1
τ Scl	–		6.0	7.1	1.1	122	F0	36.1	−29 54	28-6 Yellow stars
h3347	–		7.2	9.0	20.2	276	K0 K0	39.7	−37 28	28-1
ε Scl	1394	AB	5.4	8.6	4.7	34	F0	45.6	−25 03	28-1 Unequal yellow pair

Footnotes: *= Year 2000, a = Near apogee, c = Closing, w = Widening. Finder Chart No: All stars listed in the tables are plotted in the large Constellation Chart, but when a star appears in a Finder Chart, this number is listed. Notes: When colors are subtle, the suffix *-ish* is used, e.g. *bluish*.

Epsilon (ε) Sculptoris Double Star   Spec. F0
m5.4, 8.6; Sep. 4.7″; P.A. 34°   01^{h}45.6^m −25°03′
Constellation Chart 28-2   ★★★★

8/10″ Scopes–150x: Epsilon Sculptoris presents a pair of unequally bright yellow stars.

28.3 Deep-Sky Objects

NGC 7507 Galaxy Type E0
ø3.1′ × 3.0′, m10.6v, SB 12.9   23^{h}12.1^m −28°32′
Finder Chart 28-3   ★★★

16/18″ Scopes–150x: NGC 7507, located 6′ NW of a 9.5 magnitude star, is fairly bright but rather small with only a 1′ diameter halo. The core is small and bright, and contains a stellar nucleus. An 11.5 magnitude star 3′ NW of the galaxy is part of an irregular pentagon of stars of different magnitude surrounding it. Galaxy NGC 7513 is 18′ NE.

NGC 7513 Galaxy Type (R′)SB(s)b pec
ø2.9′ × 1.7′, m11.9v, SB 13.5   23^{h}13.2^m −28°22′
Finder Chart 28-3   ★★★

16/18″ Scopes–150x: NGC 7513 is 5′ SE of a 9th magnitude star and 3′ NNW of a wide pair of 8th and 9th magnitude stars. Its moderately faint halo is elongated 1.5′ × 1′ ENE–WSW and contains a bright core slightly extended E–W. Galaxy NGC 7507 lies 18′ SW.

IC 5332 Galaxy Type SAB(s)cd II
ø6.0′ × 5.8′, m10.3v, SB 14.0   23^{h}34.5^m −36°06′
Finder Chart 28-4, Figure 28-1   ★★

12/14″ Scopes–125x: This very faint galaxy is located 15′ NE of a 20″ wide magnitude 7.0 and 11.7 double.

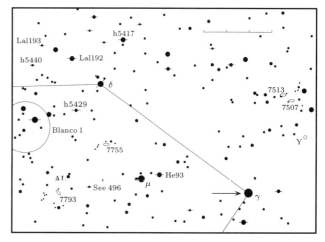

Finder Chart 28-3. γ Scl: 23^{h}18.8^m −32°32′

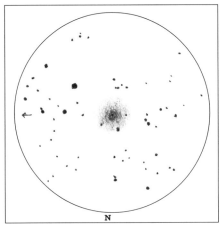 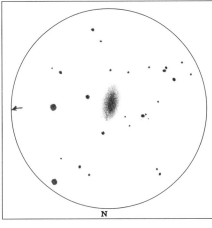 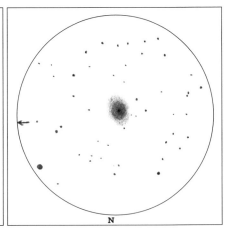

Figure 28-1. IC 5332
12.5", f5–200x, by G. R. Kepple

Figure 28-2. NGC 7713
12.5", f5–200x, by G. R. Kepple

Figure 28-3. NGC 7755
12.5", f5–200x, by G. R. Kepple

It is a diffuse 1.5' diameter glow.

16/18" Scopes–150x: IC 5332 has a faint 2.5' diameter halo with a faint 1' core that contains a stellar nucleus. Centered 6' WSW of the galaxy is a 4' diameter circlet of 9th to 12th magnitude stars; and 4' ENE is a tight semicircle, open to the north, of faint stars.

NGC 7713 Galaxy Type SB(s)d: IV
ø4.7' × 2.0', m11.1v, SB 13.4 $23^h36.5^m$ $-37°56'$
Finder Chart 28-4, Figure 28-2   ★★★

12/14" Scopes–125x: NGC 7713, located 40' ESE of 4th magnitude Beta (β) Sculptoris, has a faint 3' × 1.5' N–S halo with a slightly brighter center. 9th magnitude stars 4' west and 8' WNW form a triangle with the galaxy. A 12.5 magnitude star is 2.5' from the galaxy's center off its northern tip, and an 11th magnitude star lies 1.5' from the center just WSW of the halo.

NGC 7755 Galaxy Type SB(rs)bc: II
ø3.6' × 3.0', m11.4v, SB 13.8 $23^h47.9^m$ $-30°31'$
Finder Chart 28-3, Figure 28-3   ★★★

8/10" Scopes–100x: NGC 7755, located 10' SE of an 8th magnitude star is very faint, elongated 2' × 0.75' NE–SW, and has a well concentrated core.

12/14" Scopes–125x: NGC 7755 has a bright core containing a stellar nucleus surrounded by a faint 3' × 1' NE–SW halo.

NGC 7793 Galaxy Type SA(s)d III–IV
ø10.5' × 6.2', m9.2v, SB 13.6 $23^h57.8^m$ $-32°35'$
Finder Chart 28-3, Figure 28-4   ★★★★

NGC 7793 is one of the smaller members of the 8 million light year distant Sculptor Galaxy Group. The true size and the absolute magnitude that correspond to its apparent size of 10.5' × 6.2' and apparent magnitude of 9.2 are 24,500 × 14,500 light years and –17.8, rather small compared to the 100,000 light years and –20.5 of our own Milky Way Galaxy.

12/14" Scopes–125x: This galaxy, despite its low surface brightness, is immediately obvious. The halo is elongated 6.5' × 3.5' E–W and has a small, faint core. A 12th magnitude star lies 2' north.

16/18" Scopes–150x: NGC 7793 has an irregularly bright 8' × 4' E–W halo that gradually brightens to a mottled core containing an indistinct nucleus. A close triplet of 10th to 11th magnitude stars lies 8' north.

Figure 28-4. *In larger telescopes, NGC 7793 shows a large, irregularly bright halo with a mottled core. Martin C. Germano made this 65 minute exposure on 2415 film with a 14.5", f5 Newtonian reflector.*

Figure 28-5. *NGC 24 is a highly elongated streak with a well concentrated extended core. Harvey Freed D.D.S. made this 60 minute exposure on hypered 2415 Kodak Tech Pan film with a 10″, f6 Newtonian.*

Figure 28-6. *NGC 55 is a large, thin galaxy with an unevenly illuminated, mottled halo spanning a length of half a degree. Martin C. Germano made this 60 minute exposure on hypered 2415 film with an 8″, f5 Newtonian.*

Blanco 1 Open Cluster 30★ Tr Type III 2 m
ø90′, m4.5v, Br★ 5.02v   00ʰ04.3ᵐ −29°56′
Finder Chart 28-3   Zeta (ζ) Sculptoris Cluster   ★★★

15 ×65 Binoculars: Blanco 1 is an extremely large cluster of at least sixty irregularly scattered stars over a 1.5° area. The brighter members lie in a NW–SE chain across the group's center, magnitude 5.02 Zeta (ζ) Sculptoris anchoring the chain's NW end. A 6.4 magnitude star marks the northern edge, and a 7th magnitude star is situated at the center. The group contains five magnitude 8–8.5 and several dozen magnitude 9–9.5 stars.

NGC 24 Galaxy Type Sbc: II-III
ø6.3′ × 1.3′, m11.3v, SB 13.4   00ʰ09.9ᵐ −24°58′
Finder Chart 28-5, Figure 28-5   ★★★★

12/14″ Scopes–125x: This nice but faint galaxy is a 4′ × 0.75′ NE–SW streak with a well concentrated core. A 12th magnitude star is near the NE tip.

16/18″ Scopes–150x: NGC 24 has a 5′ × 1′ NE–SW halo that elongates to thin tips. It is broadly brighter along the major axis with a bulging center and a core that is either off-center to the NE, or seems to be off-center because the halo's SW extension is longer.

NGC 55 Galaxy Type SB(s)m III
ø30.0′ × 6.3′, m8.1v, SB 13.6   00ʰ14.9ᵐ −39°11′
Finder Chart 28-4, Figure 28-6   ★★★★★

NGC 55 is the second brightest member of the 20° wide Sculptor Galaxy Group that includes NGCs 55, 253, 300, and 7793 in Sculptor and NGCs 45 and 247 in Cetus. The true size of NGC 55 is at least 70,000 light years and its absolute magnitude is −18.9, a luminosity of three billion suns.

12/14″ Scopes–125x: Splendid! NGC 55 is a highly elongated 20′ × 4′ ESE–WNW streak with a large extended core 5′ in length and half as wide. The envelope shows some mottled, unevenly bright areas.

16/18″ Scopes–150x: NGC 55 is a superb sight. Its extremely long, thin 30′ × 4′ ESE–WNW halo contains a large 5′ × 3′ oval core. The halo is uneven in brightness and fainter in the ESE half. The core is off-center to the NNW. Several small, brighter patches are along the galaxy's major axis east of the

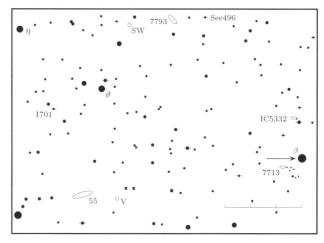

Finder Chart 28-4. β Scl: 23ʰ32.9ᵐ −37°49′

Figure 28-7. *NGC 253 is phenomenal in larger instruments, its huge, unevenly bright, mottled envelope spanning half a degree. The core is a broad, well concentrated oval with a dark lane passing along its northern edge. Martin C. Germano made this 65 minute exposure on hypered 2415 Kodak Tech Pan film with a 14.5″, f5 Newtonian.*

core, and a bright knot is in the halo just SE of the core. Three fairly bright stars lie along the southern edge of the galaxy, and a few faint foreground stars are superimposed upon the halo.

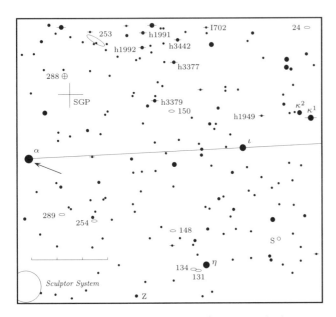

Finder Chart 28-5. α Scl: $00^h58.6^m$ $-29°21'$

NGC 131 Galaxy Type Sa:
ø1.8′ × 0.6′, m12.8v, SB 12.7 $00^h29.6^m$ $-33°16'$
Finder Chart 28-5, Figure 28-8 ★★

12/14″ Scopes–125x: NGC 131 is a faint companion of the beautiful galaxy NGC 134 lying 9′ west of it. NGC 131 has a faint 1′ × 0.5′ NE–SW oval halo containing a tiny core. A 13th magnitude star lies off the NE tip 2′ from the galaxy's center.

NGC 134 Galaxy Type SA(s)bc? II
ø8.5′ × 1.9′, m10.4v, SB 13.3 $00^h30.4^m$ $-33°15'$
Finder Chart 28-5, Figure 28-8 ★★★★★

12/14″ Scopes–125x: NGC 134 is a magnificent edge-on galaxy half a degree SE of Eta (η) Sculptoris. The 6′ × 1′ NE–SW halo contains a small, oval core. The SW edge is sharper, and a star is just outside the halo on this side.

16/18″ Scopes–150x: Impressive! NGC 134 is a fine 7′ × 1.25′ NE–SW nearly edge-on spiral visible as a luminous streak with a lengthwise dark lane. The dark lane may be seen with direct vision but averted vision does help. The core is a broad oval, the dark lane passing on its NW side. A 13th magnitude star lies on the edge of the halo just NW of the galaxy's

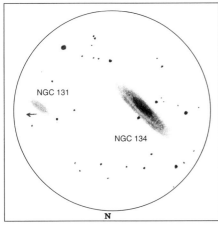

Figure 28-8. NGC 131 & NGC 134
17.5", f4.5–200x, by G. R. Kepple

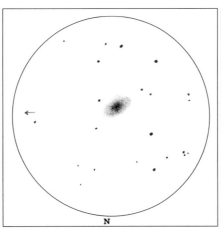

Figure 28-9. NGC 150
17.5", f4.5–200x, by G. R. Kepple

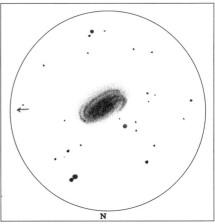

Figure 28-10. NGC 613
17.5", f4.5–200x, by G. R. Kepple

center, and a star of similar brightness is outside the halo 1.5′ SE of the center. Companion galaxy NGC 131 lies 9′ west.

NGC 148 Galaxy Type S0°: sp
ø1.9′ × 0.8′, m12.1v, SB 12.4   00ʰ34.3ᵐ −31°47′
Finder Chart 28-5   ★★★

12/14″ Scopes–125x: This galaxy is 6′ SW of a 9.5 magnitude star and west of a 3′ triangle of 12th magnitude stars. It is fairly bright but small, its 1′ × 0.5′ E–W elongated halo containing a stellar nucleus.

16/18″ Scopes–150x: NGC 148 is a bright, small, oval elongated 1.5′ × 1′ E–W containing a well concentrated center with a stellar nucleus.

NGC 150 Galaxy Type SA(s)b? I-II
ø3.4′ × 1.6′, m11.3v, SB 13.0   00ʰ34.3ᵐ −27°48′
Finder Chart 28-5, Figure 28-9   ★★★

12/14″ Scopes–125x: This moderately faint and elongated galaxy is 2.5′ × 1.5′ ESE–WNW and slightly brighter in the core.

16/18″ Scopes–150x: NGC 150 is fairly bright, elongated 3′ × 2′ ESE–WNW, and has a broad, unevenly concentrated core with a faint stellar nucleus. A 13.5 magnitude star is 1.25′ WSW of the center, and a wide pair of 13th magnitude stars lies 3′ SE.

NGC 253 Galaxy Type SAB(s)c: II
ø30.0′ × 6.9′, m7.6v, SB 13.2   00ʰ47.6ᵐ −25°17′
Finder Chart 28-5, Figure 28-7   ★★★★★
The Sculptor Galaxy

NGC 253 is the brightest member of the 8 million light year distant Sculptor Galaxy Group. Its true size is at least 70,000 light years, comparable to the diameter of our Milky Way Galaxy, and its absolute magnitude of −19.4 is a luminosity of some 4.8 billion suns, just one magnitude less than that of the Milky Way.

12/14″ Scopes–125x: Stunning! This galaxy has an incredibly long halo with a mottled texture throughout most of its 25′ × 5′ NE–SW extent. The core is well concentrated and highly extended but only moderately brighter. Two 9th magnitude stars are located SSW of the galaxy's center, and four faint stars are strung along the western side of the SW tip.

16/18″ Scopes–150x: NGC 253 is awesome in larger instruments. Its huge envelope is elongated 30′ × 5′ NE–SW and gradually concentrates to a highly extended core. A dark dust lane is visible NW of the core, and at least a dozen faint stars are embedded in the halo.

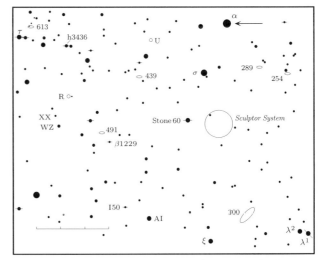

Finder Chart 28-6. α Scl: 00ʰ58.6ᵐ −29°21′

Figure 28-11. *NGC 288 is one of the lesser concentrated globular clusters with a loose periphery and a well resolved core. It is located 1.5° SE of the huge edge-on galaxy NGC 253. William Harris made this 30 minute exposure on hypered 2415 film with an 8", f/4 Wright Newtonian.*

NGC 254 Galaxy Type (R)SB(r)0+?
ø2.6' × 1.5', m11.6v, SB 12.9   00ʰ47.5ᵐ −31°25'
Finder Chart 28-6   ★★★

12/14" Scopes–125x: NGC 254, located 4.5' south of a 7th magnitude star, has a bright stellar nucleus surrounded by a 1.5' × 0.25' NW–SE halo.

16/18" Scopes–150x: NGC 254 has a moderately bright 1.75' × 0.5' NW–SE halo containing a tiny core and a conspicuous stellar nucleus.

NGC 289 Galaxy Type (R')SB(rs)b? pec I-II
ø3.7' × 2.7', m10.6v, SB 12.9   00ʰ52.7ᵐ −31°12'
Finder Chart 28-6   ★★★

12/14" Scopes–125x: This fairly bright galaxy has a 2' × 1.5' NNW–SSE oval halo with a brighter center.

16/18" Scopes–150x: NGC 289 is a moderately bright galaxy elongated 2.5' × 1.75' NW–SE with a well concentrated, broad central brightening to an inconspicuous stellar nucleus. The central area has a mottled texture. A 13th magnitude star lies 1.25' NNW.

NGC 288 H20⁶ Globular Cluster Class X
ø13.8', m8.1v, Br★ 12.6   00ʰ52.8ᵐ −26°35'
Finder Chart 28-5, Figure 28-11   ★★★★

12/14" Scopes–125x: This globular cluster is 1.8° SSE of galaxy NGC 253 and 15' south of an 8th magnitude star. It is partially enclosed by a semicircular star chain open on the SW side. The large, dense core is embedded in a much less concentrated halo 7' in diameter and irregular in outline. Outlying members spread noticeably further to the south and SW.

16/18" Scopes–150x: The loosely-structured Class X globular NGC 288 has a 9' diameter but star-poor halo around a dense 3' diameter core. Three dozen stars can be resolved in the core against a background haze. The globular's periphery is irregular in profile, outlying members extending particularly far on its SW side.

NGC 300 Galaxy Type SA(s)cd III-IV
ø20.0' × 13.0', m8.1v, SB 14.0   00ʰ54.9ᵐ −37°41'
Finder Chart 28-6, Figure 28-12   ★★★

The true diameter of NGC 300 is at least 32,500 light years and its absolute magnitude is the same as NGC 55's, −18.9. In reality, however, NGC 55 is probably significantly brighter then NGC 300: the two systems have the same apparent magnitude, 8.1, only because we see NGC 300 nearly face-on and NGC 55 nearly edge-on.

12/14" Scopes–125x: NGC 300, a member of the Sculptor Galaxy Group, has a 14' × 10' ESE–WNW oval halo with a poorly concentrated core. Superimposed upon the galaxy is a 5' × 3.5' NW–SE triangle of magnitude 9.5–10 stars the glare of which interferes with the galaxy's subtle glow. The galaxy's core is between the northern and southern stars of the triangle, and the galaxy's halo extends well beyond the triangle's NW side.

16/18" Scopes–150x: NGC 300 is a large, moderately faint galaxy elongated 16' × 10' ESE–WNW with a well concentrated, broad brightening to an inconspicuous stellar nucleus. The central area has a mottled texture. The low surface brightness of the halo is overwhelmed by a 9.5 magnitude star 2.5' SW of center that is part of a triangle with two more 9.5–10th magnitude stars superimposed upon the galaxy.

NGC 439 Galaxy Type SAB(rs)0−:
ø3.2' × 2.1', m11.4v, SB 13.3   01ʰ13.8ᵐ −31°45'
Finder Chart 28-6   ★★★

12/14" Scopes–125x: NGC 439, located 10' NE of an 8th magnitude star, has a bright, tiny core and a stellar nucleus embedded in a faint, 1' diameter halo. Its faint companion galaxy NGC 441, just 2.5' to its south, is a diffuse glow only about half its size. Two 13th magnitude stars 2.25' SW and SE of NGC 439 form a rectangle with the two galaxies.

Figure 28-12. *NGC 300 is a fairly bright but diffuse galaxy upon which is superimposed a triangle of 9.5–10th magnitude foreground stars the glare of which interferes with the fragile galaxy-glow. Martin C. Germano made this 75 minute exposure on hypered 2415 film with a 14.5″, f5 Newtonian.*

NGC 491 Galaxy Type SB(s)b? IC
⌀1.3′ × 0.9′, m12.5v, SB 12.5   01ʰ21.4ᵐ −34°03′
Finder Chart 28-6   ★★

8/10″ Scopes–100x: This galaxy is an extremely faint, small, round object on the threshold of visibility.

16/18″ Scopes–150x: This fairly faint galaxy lies just beyond the eastern tip of an equilateral triangle of 11th to 12th magnitude stars. It has a 1′ wide slightly E–W oval halo that contains a tiny bright nucleus.

NGC 613 Galaxy Type SAB(rs)bc II
⌀5.2′ × 2.6′, m10.0v, SB 12.7   01ʰ34.3ᵐ −29°25′
Finder Chart 28-6, Figure 28-10   ★★★★

8/10″ Scopes–100x: NGC 613, located just 2′ SW of a 9th magnitude star, has a bright core surrounded by a faint 4′ × 2′ NW–SE halo. A conspicuous pair of 10th and 11th magnitude stars lies 7.5′ NNW.

16/18″ Scopes–150x: NGC 613 is a fairly bright galaxy with a 4.5′ × 2.5′ NW–SE halo. The core is uneven in brightness and has a bright, nonstellar nucleus. A very faint, indistinct spiral structure is just visible with averted vision. An arm wraps around the east end and fades as it curves northward.

Chapter 29

Taurus, the Bull

29.1 Overview

Taurus the Bull is another of the many constellations which the ancient Greeks inherited from the Babylonians. In Babylonian astromythology, the celestial Bull pulled a celestial Plow (figured in the stars of our Triangulum + Gamma Andromedae) guided by a celestial Plowman (our Aries) around the heavens, the Plow's furrow being the ecliptic. In the Babylonian epic of Gilgamesh, the Bull of Heaven was sent by the volatile goddess of love and war, Ishtar (the prototype of the Greek Aphrodite and Roman Venus), to inflict drought upon the land of Mesopotamia: the Bull was killed by Gilgamesh and his half-man/half-bull sidekick Enkidu. In classical mythology Taurus was associated with the Bull who abducted Europa, the beautiful daughter of Arenor, King of Sidon. Europa was attracted to a fine white bull in her father's herd. Having stroked him and adorned him with flowers, she decided to climb on the gentle bull's back for a ride. However, the bull was actually Jupiter in disguise who had planned from the start to abduct her. As soon as she was seated, the bull ran off, straight into the sea with his beautiful rider, swimming all the way to Crete. Europa became the mother of Minos, who was destined to become King of Crete and, in turn, the father of Ariadne.

Taurus is one of the most attractive and distinctive of the constellations, and one of the few which really resembles what it has been named for: the Bull's horns extend NE to the stars Beta and Zeta Tauri, and its face is the conspicuous "V" formed by the Hyades Star Cluster, the ruddy 1st magnitude star Aldebaran marking one of the Bull's flaming eyes. The stream of the winter Milky Way cuts NW–SE behind the tips of the Bull's horns; hence the constellation is rich in open clusters and diffuse nebulae. It contains the two finest unaided-eye open clusters, the Hyades and the Pleiades, and the brightest, most easily-observed, supernova remnant, M1, the famous Crab Nebula.

Taurus: TORE-us
Genitive: Tauri, TORE-i
Abbrevation: Tau
Culmination: 9 pm–Jan. 14, midnight–Nov. 30
Area: 797 square degrees
Showpieces: Mel 25 (Hyades), M1 (NGC 1952), M45 (Pleiades), NGC 1514,
Binocular Objects: DoDz 3, DoDz 4, Mel 25 (Hyades), M1 (NGC 1952), M45 (Pleiades), NGC 1647, NGC 1746, NGC 1807, NGC 1817

29.2 Interesting Stars

Σ422 Double Star   Spec. G0
m5.9, 8.8; Sep. 6.6″; P.A. 265° $03^h36.8^m$ +00°35′
Constellation Chart 29-1   ★★★★

4/6″ Scopes–100x: Struve 422 is an attractive pair of yellow and bluish-white stars. The bright yellow star 10 Tauri lies 11′ south.

30 Tauri (Σ452) Double Star   Spec. B3
m5.1, 10.2; Sep. 9.0″; P.A. 59° $03^h48.3^m$ +11°09′
Constellation Chart 29-1   ★★★★

4/6″ Scopes–100x: 30 Tauri is an unequally bright but easy pair of beautiful bluish-white and reddish stars lying in a well populated star field. Observers have also reported the primary as green or pale yellow and the companion as purple. However, most stars with a B3 spectrum like that of 30 Tauri A appear bluish-white.

Lambda (λ) = 35 Tauri Variable Star   Spec. B3+A4
m3.3 to 3.8 in 3.95 days $04^h00.7^m$ +12°29′
Constellation Chart 29-1   ★★★★

Lambda Tauri is one of the brighter Algol-type eclipsing binary variable stars. Its light changes were noticed by J. Baxendell in England in 1848. The eclipses

Taurus, the Bull

Constellation Chart 29-1

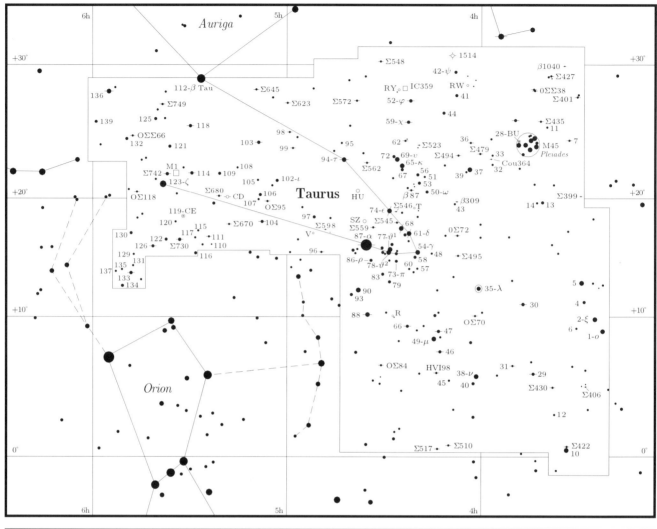

Chart Symbols

Constellation Chart	0 1 2 3 4 5 6
Stellar Magnitudes	
Finder Charts	0/1 2 3 4 5 6 7 8 9
Master Finder Chart	0 1 2 3 4 5

→ Guide Star Pointer
• — • Double Stars
Finder Chart Scale (One degree tick marks)
⊙ ○ Variable Stars
○ Open Clusters
⊕ Globular Clusters
◯ Galaxies
⟐ Planetary Nebulae
□ Small Bright Nebulae
Large Bright Nebulae
Dark Nebulae

Table 29-1. Selected Variable Stars in Taurus

Name	HD No.	Type	Max.	Min.	Period (Days)	F*	Spec. Type	R.A. (2000)	Dec.	Finder Chart No. & Notes
28-BU Tau	23862	C	4.7	5.5			B8	03h49.2m	+24°08′	29-4 In Pleiades
35-λ Tau	25204	EA	3.3	3.8	3.95	0.15	B3+A4	04h00.7m	+12 29	29-1
RW Tau	25487	EA	7.9	11.4	2.76	0.14	B8+K0	03.9	+28 08	29-3
T Tau	284419	InT	8.4	13.5			dGe-K1e	22.0	+28 27	29-6
R Tau	28309	M	7.6	14.7	323	0.41	M5-M9	28.3	+10 10	29-1
SZ Tau	29260	Cδ	6.3	6.7	3.14	0.45	F5-F9	37.2	+18 33	29-7
HU Tau	29365	EA	4.7	6.7	2.05	0.16	A8	38.3	+20 41	29-7
V Tau	30868	M	8.5	14.6	169	0.45	M0-M4	52.0	+17 32	29-7
CD Tau	34335	EA	7.2	7.9	3.43	0.08	F7+F7	05h17.5m	+20 08	29-8 = Σ674
CE Tau	36389	SRc	6.1	6.5	165		M2	32.2	+18 36	29-1

F* = The fraction of period taken up by the star's rise from min. to max. brightness, or the period spent in eclipse.

Table 29-2. Selected Double Stars in Taurus

Name	ADS No.	Pair	M1	M2	Sep."	P.A.°	Spec	R.A. (2000)	Dec	Finder Chart No. & Notes
Σ399		AB	8.1	9.6	20.2	147	G5	03ʰ30.3ᵐ	+20°07′	29-1
Σ406	2580		8.1	10.1	9.3	124	F0	30.8	+05 09	29-1
Σ401	2582		6.4	6.9	11.3	270	A0	31.3	+27 34	29-1
7 Tau	2616	AB	6.7	6.7	w0.7	*359	A2	34.3	+24 28	29-3
	2616	ABxC		10.0	22.4	56				
β1040	2633		7.8	11.5	3.6	337	A0	36.2	+29 59	29-3
Σ422	2644		5.9	8.8	6.6	265	G0	36.8	+00 35	29-1 Yellow & bluish-white
Σ430	2681	AB	6.8	9.8	26.3	56	G5	40.5	+05 08	29-1
	2681	AC		10.5	37.1	301				
Σ427	2679		7.3	8.1	6.8	208	A0	40.6	+28 46	29-3
Σ435	2708		7.3	8.8	13.0	2	F5	43.1	+25 41	29-3
OΣΣ28	2735	AB	6.7	7.0	126.7	43	F0 G0	44.6	+27 54	29-3
30 Tau	2778		5.1	10.2	9.0	59	B3	48.3	+11 09	29-1 Bluish-white & reddish
	2795			9.0	10.2	236				
Cou 364			8.0	9.7	2.0	150	K0	57.8	+22 55	29-3
Σ479	2926		7.0	7.9	7.3	128	B9	04ʰ00.9ᵐ	+23 12	29-3
OΣ70	2938		5.7	11.7	12.0	227	B8	01.8	+10 00	29-1
Σ495	2999		6.0	8.8	3.8	221	F0	07.7	+15 10	29-1 Bluish-white pair
OΣ72	3006		6.0	9.1	4.4	326	K0	08.0	+17 20	29-1
β309	3010		7.9	11.2	5.7	278	A2	08.3	+19 44	29-1
Σ494	3019		7.6	7.6	5.2	187	A3	08.9	+23 06	29-3
Σ510	3054		6.9	9.9	11.0	300	G5	12.2	+00 44	29-1
47 Tau	3072	AB	4.9	7.4	1.1	351	G5	13.9	+09 16	29-1 Deep yellow suns
		AC		11.8	29.8	226				
H VI 98	3085	AB	6.3	7.0	65.5	315	G0 G0	15.5	+06 11	29-1
	3085	AC		10.0	214.5	47	M2			
	3085	CD		10.0	52.7	139				
Σ517	3095		7.4	9.1	3.4	12	A0	16.0	+00 27	29-1
Σ523	3131		7.7	9.7	10.3	163	A0	19.8	+23 44	29-3
52-φ Tau	3137		5.0	8.4	52.1	250	K0	20.4	+27 21	29-3 Deep yellow & blue
β87	3158		6.0	9.1	1.9	170	K5	22.4	+20 49	29-6 Orange & white
59-χ Tau	3161		5.5	7.6	19.4	24	B9	22.6	+25 38	29-3 White & blue
62 Tau	3179	AB	6.2	8.6	28.9	290	B8	24.0	+24 18	29-3
65-κ & 67		AB	4.4	5.4	339.5	173	A3 F0	25.4	+22 18	29-6 Bright pale yellow pair
Σ548	3243		6.5	8.5	14.4	35	F5	28.9	+30 22	29-1
OΣ84	3279		7.3	8.2	9.5	254	G5	31.1	+06 47	29-1
Σ562	3311		6.8	10.5	2.0	274	F2	34.8	+22 42	29-5
88 Tau	3317		4.3	8.4	69.7	299	A3	35.7	+10 10	29-1
Σ572	3353		7.3	7.3	4.0	194	F0	38.5	+26 56	29-5 Pale yellow double
94-τ Tau		AC	4.3	8.6	62.8	213	B5 A0	42.2	+22 57	29-5
Σ598	3460		8.1	9.6	9.4	318	F2	48.6	+17 49	29-6,7
Σ623	3587		6.8	8.3	20.5	205	B9	59.7	+27 20	29-5
OΣ95	3672		6.9	7.5	1.1	311	A2	05ʰ05.5ᵐ	+19 48	29-7
Σ645	3730	AxBC	6.1	9.0	11.8	27	A3	09.8	+28 02	29-5 White & bluish
Σ670	3854		7.7	8.2	2.5	164	B3	16.7	+18 26	29-1 Both stars bluish
Σ674	3866		6.9	9.9	10.2	148	F5	17.5	+20 08	29-8 A = CD Tau
Σ680	3894		6.1	10.0	9.0	204	K0	19.2	+20 08	29-8 Deep yellow & bluish
111 Tau			5.0	8.8	85.7	271	G0 K0	24.4	+17 23	29-1
114 Tau	4048	AB	4.9	11.0	37.8	348	B3	27.6	+21 56	29-8
		AC		10.5	58.8	194				
		AD		11.7	74.2	280				
118 Tau	4068	AB	5.8	6.6	4.8	204	A0	29.3	+25 09	29-9 White & yellowish pair
		AC		11.6	141.3	99				
Σ730	4131		6.0	6.5	9.6	141	B9	32.2	+17 03	29-1
Σ742	4200		7.2	7.8	3.9	270	F8	36.4	+22 00	29-8 0.5° west of M1
Σ749	4208	AB	6.4	6.5	1.0	333	B8	37.1	+26 55	29-9
	4208	CD	10.4	10.9	4.4	290				
133 Tau	4381	AB	5.3	12.3	17.8	298	B5	47.7	+13 54	29-1
	4381	AC		11.6	24.9	181				
OΣΣ66			7.2	8.0	94.2	166	K2 K2	47.9	+24 41	29-9 Wide reddish-orange dbl.
OΣ118	4392	AB	6.1	7.6	0.6	318	B9	48.4	+20 52	29-8
OΣΣ67	4392	ABxC		8.6	75.5	161	A			

Footnotes: *= Year 2000, a = Near apogee, c = Closing, w = Widening. Finder Chart No: All stars listed in the tables are plotted in the large Constellation Chart, but when a star appears in a Finder Chart, this number is listed. Notes: When colors are subtle, the suffix -ish is used, e.g. bluish.

are partial, approximately 40% of the hot, bright B3 star being covered by the cooler A4 star at mid-eclipse. The variations can be followed with the unaided eye by using the magnitude 3.7 stars Gamma (γ) Tauri 6° NE and Xi (ξ) Tauri 9° SW for comparison. The Lambda Tauri system is about 400 light years away.

RW Tauri Variable Star   Spec. B8+K0
m7.9 to 11.4 in 2.76 days $04^h03.9^m$ +28°08′
Finder Chart 29-3   ★★★★

RW Tauri, 1° NW of 41 Tauri, is an eclipsing binary system in which a B8 primary star is totally eclipsed every 2.76 days by a K-type subgiant companion. The entire eclipse takes 9 hours, totality lasting 84 minutes. The depth of the eclipse, 4.49 magnitudes photographically and 3.50 magnitudes visually, is one of the greatest known for an eclipsing variable. This system is estimated to be 1,370 light years distant.

Σ495 Double Star   Spec. F0
m6.0, 8.8; Sep. 3.8″; P.A. 221° $04^h07.7^m$ +15°10′
Constellation Chart 29-1   ★★★★

4/6″ Scopes–100x: Struve 495 is a long period binary consisting of an elegant pair of yellow stars that can be just resolved in small telescopes.

47 Tauri = β547 Double Star   Spec. G5
AB: m4.9, 7.4; Sep. 1.1″; P.A. 351° $04^h13.9^m$ +09°16′
Constellation Chart 29-1   ★★★★

8/10″ Scopes–200x: 47 Tauri is a very close double which 200x shows as two deep yellow suns in contact. A 12th magnitude component lies 30″ SW.

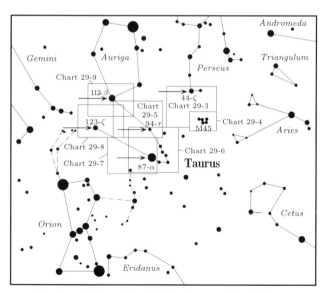

Master Finder Chart 29-2. Taurus Chart Areas
Guide stars indicated by arrows.

Phi (φ) = 52 Tauri (S, h 40) Double Star   Spec. K0
m5.0, 8.4; Sep. 52.1″; P.A. 250° $04^h20.4^m$ +27°21′
Finder Chart 29-3   ★★★★

2/3″ Scopes–25x: Phi Tauri, a pretty deep yellow and blue pair, is easily separated in small telescopes and binoculars.

T Tauri Variable Star   Spec. dGe–K1e
m8.4 to 13.5 with erratic period $04^h22.0^m$ +19°32′
Finder Chart 29-6   ★★★★

T Tauri, located 1.8° west of Epsilon Tauri, the star at the northern tip of the "V" of the Hyades, is a faint, erratic variable associated with the peculiar emission/reflection nebula NGC 1555, named Hind's Variable Nebula. The star varies unpredictably between magnitudes 8.5 and 13.5, its spectrum also changing idiosyncratically and peculiar because of its bright emission lines and strong lithium absorption lines. T Tauri is thought to be a very young star just emerged from its natal dust cloud, erratic because of the conflict between the last stages of gravitational contraction toward the stable main sequence and the first stages of hydrogen burning in its core, one of the initial products of which would be lithium. T Tauri is the prototype of the solar-mass, F, G, and K spectrum, nebular variables which are found in great abundance in such star-formation regions as M42 in Orion and M8 in Sagittarius. (Another type of nebular variable are the more massive, B-spectrum RW Aurigae stars.) NGC 1555 probably is a remnant of the dust cloud in which T Tauri had been born: eventually radiation and solar winds from T Tau should disperse the nebula. T Tauri and NGC 1555 lie within the relatively nearby (450–500 light year distant) Taurus Dark Cloud Complex, in which are many other, fainter, nebular variables and recently-formed solar mass stars.

β87 Double Star   Spec. K5
m4.9, 7.4; Sep. 1.1″; P.A. 351° $04^h22.4^m$ +20°49′
Finder Chart 29-6   ★★★★

8/10″ Scopes–200x: Burnham 87 is a very close pair consisting of an orange primary with a companion that can be seen only as a white spike to the WSW.

29.3 Deep-Sky Objects

Messier 45 Mel 22 Open Cluster 100★ Tr Type I 3 r
ø 110′, m1.2v, Br★ 2.87v $03^h47.0^m$ +24°07′
Finder Chart 29-3, Figure 29-1   Pleiades   ★★★★★

The Pleiades is one of the most beautiful objects in the entire sky no matter how you look at it — with the unaided eye, through binoculars, in large telescopes, or on a photograph. Indeed, it is every bit as beautiful in

Figure 29-1. *The Pleiades (Messier 45) is a huge, brilliant open cluster that can be described as blue sapphires wrapped in swirls of cirrus nebulosity. Martin C. Germano made this 45 minute exposure on hypered 2415 Kodak Tech Pan film with an 8", f5 Newtonian reflector at prime focus. North is to the upper right in this 6x enlargement.*

10×50 binoculars as it is in the most painstaking color photo: the photograph might bring out the delicate cirrus streaks of blue-white reflection nebulae that are around most of the stars of the cluster; but no photograph can recapture the scintillating silver-blue color the stars have in the eyepiece, a truly "celestial" blue beyond the range of the earthbound chemicals of photographic emulsions.

The ancient Mesopotamians knew that the Pleiades were something special: they named it simply "The Constellation," and it is the first star group in the official ancient Babylonian constellation canon, the earliest extant copy of which dates from around 2000 B.C. Its lesser Mesopotamian name was "The Seven Stars," and it was represented in a distinctive seven-star pattern in Assyrian art. This is something of a puzzle: for only six stars are visible in the Pleiades under poor observing conditions — and if conditions are better at least nine are easily seen. The Greeks tried to account for the difference between the number of Pleiades you can see and the number they had inherited from the Assyrians with legends about a "lost Pleiad." Probably the Assyrians and Babylonians had seen seven stars in the Pleiades simply because of the importance of the number seven in their religion and mythology.

The nine brightest stars of the Pleiades are concentrated in a field just over one degree in diameter and only a pair of binoculars are needed to enjoy this dazzling cluster. For telescopic viewing, a wide-angle, very low power ocular is needed to view the entire group. On

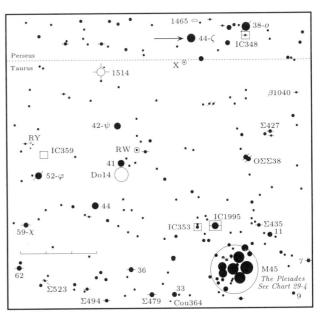

Finder Chart 29-3. 44–ζ Per: $03^h54.1^m$ +31°53′

Table 29-3. Double Stars in the Pleiades

Name	ADS No.	Pair	M1	M2	Sep."	P.A.°	Spec	R.A. (2000)	Dec	Chart No. & Notes
19 Tau			4.4	8.1	68.9	329	B6	03h45.2m	+24 28	29-4 Taygeta: White & violet
21 & 22 Tau			5.8	6.4	168.0		B8 A0	45.9	+24 33	29-4 Asterope: Binocular pair
β536	2755	AB	8.3	9.3	0.2	56	A6	46.1	+24 12	29-4
	2755	AC		8.0	39.1	306				
	2755	CD		12.0	18.1	8				
β537			8.5	10.5	0.9	175	A2	47.0	+24 49	29-4
Σ449	2766		8.5	11.0	6.8	330	A0	47.4	+24 40	29-4
Σ450	2767		7.3	9.3	6.1	265	B9	47.4	+23 55	29-4
25-η Tau		AB	2.9	8.0	117.2	289	B7 A0	47.5	+24 06	29-4 Alcyone
		AC		8.0	180.8	312	A0			
		AD		8.6	190.5	295	G0			
		BC			85.6	344				
		BD			74.7	306				
27 & 28-BU			3.6	5.1	300.0	180	B8 B8	49.2	+24 03	29-4 Atlas & Pleione
OΣΣ40			6.6	8.1	87.1	308	B9 A0	49.4	+24 23	29-4
OΣ64	2795	AB	9.0	12.5	9.0	190		49.9	+23 57	29-4
	2795	AC	6.9	9.8	3.3	238	B9	50.0	+23 51	29-4
				9.0	10.2	236				

Footnotes: *= Year 2000, a = Near apogee, c = Closing, w = Widening. Finder Chart No: All stars listed in the tables are plotted in the large Constellation Chart, but when a star appears in a Finder Chart, this number is listed. Notes: When colors are subtle, the suffix -*ish* is used, e.g. *bluish*.

clear, dark nights, when the moon is absent, swirls of nebulosity are noticeable around a couple of the brighter stars, especially Merope.

The Pleiades, 410 light years away, is the fourth most distant open cluster from the Solar System. The brightest stars of the Pleiades are all blue-white mid-B giants and subgiants with absolute magnitudes of about –1.5 to –2.5, luminosities of 330 and 830 suns. The brightest member, the B7 IIIe Eta (η) = 25 Tauri, has an apparent magnitude of 2.87 and an absolute magnitude of –2.7. The total luminosity of the cluster as a whole is 4,800 suns. The true diameter of the dipper-shaped asterism formed by the cluster's brightest members is about 7 light years, but outliers are as much as 20 light years from the cluster center and some 300 (mostly faint) stars are confirmed Pleiades members. The cluster is rich in binaries. (See Table 29–3 and Chart 29–4.) The dust which so aesthetically reflects the blue light of the stars of the Pleiades "Dipper Bowl" is not, in fact, related to the cluster: the Pleiades is some 70 million years old and has long past moved out of the interstellar cloud of gas and dust in which it was born. At present the Pleiades happens to be passing through the nearer fringe of the Taurus Dark Cloud Complex. It is no accident that on color photos the Pleiades nebulosity looks bluer than the stars themselves: blue light is scattered better than red light, hence the already blue light of the Pleiades stars is enhanced by the dust of the reflecting nebula.

4/6" Scopes–25x: What can you say about the Pleiades — they are magnificent! Six stars are visible with the naked eye from urban locations even under a full moon, but as many as 18 stars can be discerned on a good night from a dark sky site. The brighter members stand out conspicuously against the fainter stars, forming a dipper asterism. The cluster contains many attractive doubles. (See Table 29–3.) A curious chain of 7th and 8th magnitude stars begins at the double Σ450 due south of Alcyone and due east of Merope and heads first south and then SE. At low power on a good night two or three of the bright Pleiades in the "dipper bowl" can be enshrouded in faint nebulosity. Though this is a reflection nebula, an O-III filter seems to help — though it does extinguish the outer, fainter portions of the nebulae.

8/10" Scopes–35x: The Pleiades is a brilliant and beautiful naked eye star cluster with the immense apparent diameter of 110', nearly that of four full moons! The

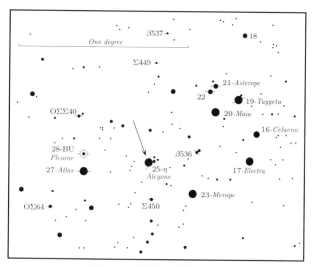

Pleiades Finder Chart 29-4.
Alcyone Eta (25-η) Tauri: 03h47.5m +24°06'

Figure 29-2. *NGC 1514 is a bright planetary nebula with an unevenly illuminated disk. Martin C. Germano made this two hour exposure on hypered 2415 film with an 8", f5 Newtonian reflector.*

Figure 29-3. *NGC 1554-55, Hind's Variable Nebula, is an extremely faint reflection nebula lying west of T Tauri. Martin C. Germano made this 65 minute exposure on hypered 2415 film with an 8", f5 Newtonian reflector.*

easternmost star, magnitude 3.6 Atlas at the end of the dipper handle, is a wide double with magnitude 5.0 Pleione 5′ to its north. Just west of magnitude 2.9 Alcyone, the brightest Pleiad, is a tiny 1′ triangle of one magnitude 6.5 and two magnitude 8.5 stars. In the middle of the dipper bowl is a wide (39″) double of magnitude 8 and 8.3 stars each of which is a close double itself. Due north of Maia at the north corner of the dipper is the very wide binocular double of magnitudes 5.8 and 6.4 21 and 22 Tauri, which cannot be resolved with the unaided eye and therefore share the name Asterope. Of the reflection nebulae around the dipper stars, that in which Merope, at the south corner of the dipper, is embedded, is by far the easiest and brightest to observe and therefore has an NGC number, 1435. The Merope nebula can be glimpsed in 10×50 binoculars under excellent observing conditions. SE of Merope a curved row of eight magnitude 9–9.5 stars trails out from the center of the cluster. Behind all these more conspicuous features of the Pleiades is a rich embellishment of at least a hundred faint group members. And everywhere you look the field is awash in the icy-blue glow of the bright blue giants of the cluster.

Dolidze 14 Open Cluster 18★ Tr Type III 2 p
ø12′, m–   04ʰ06.6ᵐ +27°26′

Finder Chart 29-3, Figure 29-8   ★

16/18″ Scopes–150x: Dolidze 41 is a very faint, easily overlooked, cluster centered 10′ south of 41 Tauri. It consists of fifteen 13th magnitude and fainter members scattered over a 12′ area south of an E–W pair of magnitude 9.5 stars.

NGC 1514 H69⁴ PK165-15.1 Planetary Neb. Type 3+2
ø>114″, m10.9p, CS 9.40v   04ʰ09.2ᵐ +30°47′

Finder Chart 29-3, Figure 29-2   ★

8/10″ Scopes–150x: NGC 1514 lies between a 16′ wide N–S pair of magnitude 8.5 stars. It has a bright 1.5′ diameter disk with an unusually bright magnitude 9.5 central star.

12/14″ Scopes–175x: NGC 1514 is a fine planetary nebula with a large 2′ disk. The disk is noticeably

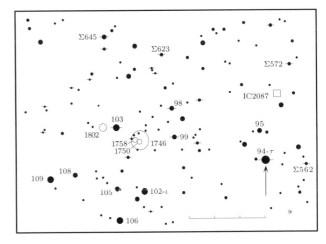

Finder Chart 29-5. 94–τ Tauri: 04ʰ42.2ᵐ +22°57′

Figure 29-4. *NGC 1647 is a large, loose cluster with fifty stars visible in small telescopes. Lee C. Coombs made this 10 minute exposure on 103a-O film with a 10", f5 Newtonian reflector. South is up.*

fainter on its NE and SW edges, which gives the planetary something of a "dumbbell" shape like that of M27 — though the effect is not nearly so pronounced. The magnitude 9.4 central star can be used to produce a "blinking" of the nebula itself: stare at the star until the halo disappears, then look suddenly

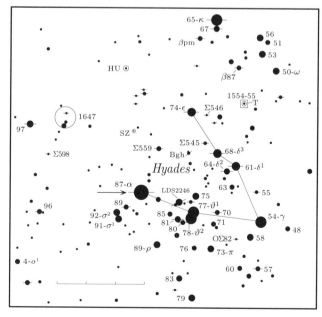

Hyades Finder Chart 29-6.
Aldebaran Alpha (87–α) Tauri: 04ʰ35.9ᵐ +16°31′

away from the star to make the nebula "reappear." A 13th magnitude star lies just outside the nebula's northern edge.

NGC 1554-55   Reflection & Emission Nebula
ø Var 1-7′, Photo Br 2-5, Color 3-4   04ʰ21.8ᵐ +19°32′
Finder Chart 29-6, Figure 29-3   ★
Hind's Variable Nebula

16/18″ Scopes–150x: Hind's Variable Nebula is a reflection/emission nebula associated with and just 40″ west of the nebular variable T Tauri (magnitudes 8.5 to 13.5). As the name suggests, the nebula changes in size, brightness, and shape — but not in concert with T Tauri's variations. The nebula is a 30″ N–S arc somewhat dimmer in the center — hence its two NGC numbers. It was once fairly easy to see but is now a challenge even for larger instruments. Under the best sky conditions, NGC 1554-55 can be seen as a small, extremely faint glow just west of T Tauri.

NGC 1647   H8⁸   Open Cluster   200★   Tr Type II 2r
ø45′ m6.4v,Br? 8.61v   04ʰ46.0ᵐ +19°04′
Finder Chart 29-6, Figure 29-4   ★★★

8/10″ Scopes–150x: This cluster is a loose, irregularly scattered group of fifty stars spread over a 45′ area. It includes a dozen magnitude 9.5–10 stars, but the majority of its members are 11th to 12th magnitude objects. Its stars are widely spaced in pairs and triplets. The zone of highest star-concentration is an E–W band through the cluster's center that is separated from peripheral concentrations by relatively star-poor gaps. Just north of the cluster center is a 30″ wide pair of 9th magnitude stars. 10′ NNW of this pair is a parallelogram, three of its corners marked by brighter cluster stars. A star chain runs NNW from the parallelogram, and a broad, loose star stream extends west from the cluster. Outside the northern edge of the cluster is a magnitude 7.5 star, and on its southern edge is a yellow 6th magnitude field star.

Hyades   Cr 50   Mel 25   Open Cluster 40★   Tr Type II 3m
ø330′, m0.5v, Br★ 3.40v   04ʰ27:ᵐ +16:°
Finder Chart 29-6   ★★★★

The Hyades, 150 light years distant, is the nearest open cluster to the Solar System after the Ursa Major Moving Group. Its core is the conspicuous 4.6° long "V" which forms the face of the celestial Bull. The bright reddish-orange 1st magnitude Aldebaran, on the top of the SE leg of the "V" in the Bull's eye, is not a true cluster member: it is in fact only 68 light years distant. The Hyades, like the Pleiades, are ideal binocular objects. In fact the Hyades are at their very best at 10x in 50mm binoculars or 60–90mm telescopes: such instruments do not magnify the appearance of "clusterness" out of the Hyades, but have sufficient light-gathering power to reveal the fine color contrast of the group's four orange giants, Gamma (γ), Delta (δ), Epsilon (ϵ), and Theta-one (θ^1) Tauri, with the white of its other bright stars. The true size of the "V" asterism is some 12 light years, but the actual extent of the richest part of the cluster is 10.4° = 27 light years from Kappa (κ) Tauri on the north to 90 Tau on the south, and many members are even more distant from the cluster's "V" core. An abundance of good doubles are in the cluster. (See Table 29-4 and Chart 29-6.)

NGC 1746 Mel 28 Open Cl 20★ Tr Type III 2 p
⌀42', m6.1v, Br★ 8.0p   05^{h}03.6^m +23°49'
Finder Chart 29-5, Figure 29-5   ★★★

8/10" Scopes–50x: NGC 1746 is an exceptionally large, loose and irregularly scattered group of 75 stars spread over a 40' area without central compression. The cluster has three concentrations, two with their own NGC designations: NGC 1750, in the western section of NGC 1746, is the richest of the two, with three dozen 10th magnitude and fainter stars in a 7' × 5' NNE–SSW area; and NGC 1758, in the eastern part of NGC 1746, has thirty 12th magnitude and fainter stars in a 5' area, its more prominent members in a semicircle open to the north. The western area is not as concentrated but encompasses most of the brighter stars, which lie in short rows, small clumps, and doubles. Low power provides the best view.

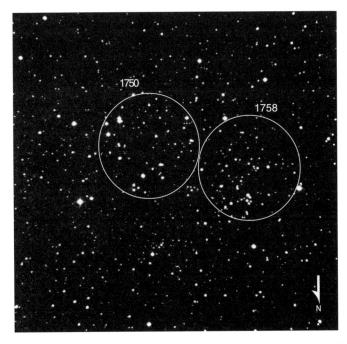

Figure 29-5. *NGC 1746 is a large, loose, irregularly concentrated cluster. Two of its subgroups have been assigned their own NGC numbers, as shown above. Lee C. Coombs made this 10 minute exposure on 103a-O film with a 10", f5 Newtonian reflector.*

Table 29-4. Double Stars in the Hyades

Name	ADS No.	Pair	M1	M2	Sep."	P.A.°	Spec	R.A. (2000)	Dec	Chart No. & Notes
57 Tau			5.6	13.5	34.6	359	F0	04^{h}20.0^m	+14°02'	29-6
OΣ82	3169	AB	7.3	8.5	1.4	*345	G0	22.7	+15 03	29-6
61-δ^1 Tau			3.8	12.5	106.6	341	K0	22.9	+17 33	29-6
64-δ^2 Tau			4.8	13.3	137.3	246	A7	24.1	+17 27	29-6
65-κ & 67 Tau		AB	4.4	5.4	339.5	17.3	A7 A7	25.4	+22 18	29-6 Pale yellow stars
		Aa		11.9	135.5	266				
		Bb		12.2	106.9	209				
68-δ^3 Tau	3206	AB	4.2	7.5	1.4	333	A2	25.5	+17 56	29-6
(H VI 101)	3206	AC		8.7	77.2	233				
70 Tau		AB	6.3	12.4	124.4	343	F8	25.6	+15 56	29-6
Σ546	3224		7.9	9.7	6.8	184	G0	27.0	+19 08	29-6
Σ545	3226		6.9	8.7	18.9	58	A0	27.1	+18 12	29-6
74-ϵ Tau			3.5	10.5	181.6	268	K0	28.6	+19 11	29-6
78-θ^2 & 77-θ^1			3.4	3.8	337.4	346	A7 K0	28.7	+15 52	29-6 Pale & deep yellow
Bgh			7.1	8.7	109.1	9	F8 G8	29.5	+17 52	29-6 Yellowish & yellow stars
80 Tau	3264		5.7	8.0	1.8	18	F0	30.1	+15 38	29-6
83 Tau			5.4	11.2	111.9	105	F0	30.6	+13 43	29-6
81 Tau			5.5	9.4	161.8	339	A5 K0	30.6	+15 42	29-6 Yellowish & orangish
LDS 2246			4.8	6.7	250.0	131		30.6	+16 12	29-6
Σ559	3297		6.9	7.0	3.1	277	B8	33.5	+18 01	29-6 Equal bluish pair
87-α Tau	3321	AB	0.9	13.4	30.4	110	K5	35.9	+16 31	29-6 Aldebaran. Bright orange
(Σ II 2)	3321	AC		11.1	121.7	34				
	3321	AF		13.4	271.4	124				
	3321	CD		13.5	1.7	274				

Footnotes: * = Year 2000, a = Near apogee, c = Closing, w = Widening. Finder Chart No: All stars listed in the tables are plotted in the large Constellation Chart, but when a star appears in a Finder Chart, this number is listed. Notes: When colors are subtle, the suffix *-ish* is used, e.g. *bluish*.

Figure 29-6. *NGC 1807 & NGC 1817 (left to right) make an interesting contrast of size and richness. Martin C. Germano made this 30 minute exposure on hypered 2415 Kodak Tech Pan film with an 8″, f5 Newtonian reflector.*

NGC 1802 H41⁸ Open Cluster [65★]
[ø13′ × 10′, m–, Br★ 9.5v]   05ʰ10.2ᵐ +24°06′
Finder Chart 29-5   ★★★

12/14″ Scopes–100x: NGC 1802, though officially classified as "nonexistent," can be seen as a coarse, irregularly scattered group of 65 stars covering a 13′ × 10′ NE–SW area. It has a dozen 9.5 to 12th magnitude stars, the majority in a "V" with a 10.5 magnitude star at its vertex. The V's southern prong, which is oriented E–W, is well-defined; its western prong extends northward into a moderate concentration of faint stars. The region between the two prongs is sparse but becomes more concentrated toward a 9.5 magnitude star lying beyond the V's opening.

NGC 1807 Mel 29 Open Cluster 20★ Tr Type II 2 p
ø17′, m7.0v, Br★ 8.60v   05ʰ10.7ᵐ +16°32′
Finder Chart 29-7, Figure 29-6   ★★★

8/10″ Scopes–50x: NGC 1807 is half a degree NNW of a 4.5 magnitude star and centered 20′ WSW of open cluster NGC 1817. At low power it appears loose and irregular with fair central compression. Thirty 9th magnitude and fainter stars are visible in a 12′ area, the brighter members in a N–S stream. 9th magnitude stars are near the north and south edges of the stream, the other prominent cluster stars scattered to the west. Near the center is a triangle of a dozen 9th–10th magnitude stars, its NW vertex marked by a 10″ wide pair of 10th and 11th magnitude stars.

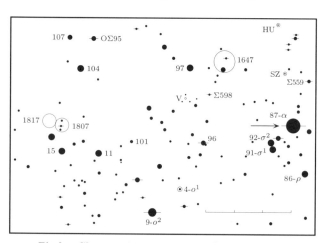

Finder Chart 29-7. 87–α Tau: 04ʰ35.9ᵐ +16°31′

NGC 1817 Cr6 Open Cluster 60★ Tr Type IV 2 r
ø15′, m7.7v, Br★ 11.17v   05ʰ12.1ᵐ +16°42′
Finder Chart 29-7, Figure 29-6   ★★★

8/10″ Scopes–50x: NGC 1817, located just NE of NGC 1807 in the same low power field, consists of a

Figure 29-7. *Messier 1, the Crab Nebula, is the debris cloud of a supernova recorded by the Chinese in 1054 A.D. Martin C. Germano took this 50 minute exposure on hypered 2415 film with a 14.5", f5 Newtonian reflector.*

chain of 9.5–10th magnitude stars running N–S with 40 faint stars spreading eastward from the chain.

12/14" Scopes–75x: NGC 1817 is a loose scattering of 75 stars in a 15' area. Its brighter members, magnitude 9.5–10 objects, lie in a NNW–SSE star chain with a field nearly devoid of stars on its west and the fainter cluster members to the east. At low power NGC 1817 and NGC 1807 to its east mimic — or parody — the Perseus Double Cluster.

Dolidze-Dzimselejsvili 3 Open Cluster 10★ Tr Type IV 2 p
ø14', m –, Br★ –   05ʰ33.7ᵐ +26°29'
Finder Chart 29-9, Figure 29-9   ★★★

8/10" Scopes–75x: This cluster, centered 40' NW of DoDz4, is fairly bright but loose and scattered. Three 9th magnitude stars near its center connect a pentagon of fainter stars at the cluster's southern edge to a wedge-shaped pattern at its northern edge. The group is not especially impressive; but it is detached fairly well from the surrounding field by an adjacent sparse area. Beyond this sparse area, however, the field is well populated with fairly bright groups of stars.

NGC 1952 Messier 1 Supernova Remnant
ø6' × 4', Photo Br 1-5, Color 3-4   05ʰ34.5ᵐ +22°01'
Finder Chart 29-8, Figure 29-7   ★★★

Messier 1 is the famous "Crab Nebula," an expanding cloud of gas from the explosion of a supernova observed by the Chinese in 1054 A.D. The supernova was so brilliant that it could be seen in the daytime for months, and was visible at night for more than a year. The Crab is the most conspicuous example of a supernova remnant.

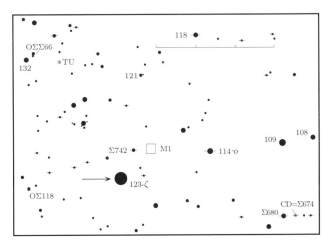

Finder Chart 29-8. 123–ζ Tau: 05ʰ37.6ᵐ +21°09'

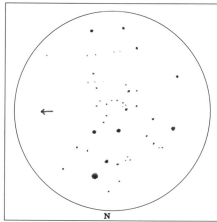
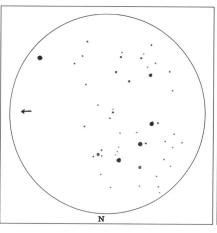
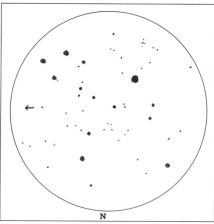

Figure 29-8. Dolidze 14
12.5″, f5–100x, by G. R. Kepple

Figure 29-9. DoDz 3
12.5″, f5–100x, by G. R. Kepple

Figure 29-10 DoDz 4
12.5″, f5–75x, by G. R. Kepple

Photographs taken decades apart show that the Crab is visibly expanding. The measured rate of expansion is about 1000 miles per second — very respectable, but probably much less than the original expansion velocity nine centuries ago. This is nevertheless virtually proof positive that the Crab Nebula is indeed the debris cloud of the 1054 supernova.

In 1968 a pulsar was detected in the center of the Crab Nebula. The Crab Pulsar, NP0532, is a 16th magnitude object that is in fact a rapidly rotating neutron star only a few miles in diameter which emits a pulse of radiation in radio, X-ray, and optical wavelengths every 0.033 second. It is the collapsed core of the star that went supernova in 1054.

The Crab Nebula gives a pleasing view at all powers and takes higher magnifications quite well. With six to eight-inch telescopes medium power, about 125x, gives the best view. A nebula filter helps some but doesn't seem to enhance the object as much as it does regular planetary or emission nebulae. The central star, or pulsar in this case, is not visible.

8/10″ Scopes–125x: The diffuse, ghostly grayish-white glow of the Crab Nebula is a nice contrast to the starry richness of the surrounding star field. It is a dim 5′ × 4′ patch elongated NW–SE.

12/14″ Scopes–150x: Messier 1 is a fairly bright, diffuse 6′ × 4′ NW–SE glow with irregular patches. It is nearly uniform in brightness with no central brightening.

16/18″ Scopes–175x: This supernova remnant is a 6′ × 4′ NW–SE oval with a highly irregular periphery. The brighter central area is a NE–SW oval. On excellent nights uneven surface brightness, and even hints of filamentary detail, can be glimpsed with averted vision. The SE edge is much fainter because of a dark bay intruding from the east. The NW edge has an indistinct notch that, with the SE bay, gives the nebula a vaguely pinched appearance.

Dolidze-Dzimselejsvili 4 Open Cluster 15★ Tr Type IV 2 p
ø28′, m –, Br★ 6.49v   05ʰ35.9ᵐ +25°57′

Finder Chart 29-9, Figure 29-10   ★★★

12/14″ Scopes–75x: This cluster is large, bright, loose, and irregular. It has several dozen stars in a 28′ area, fifteen of the brighter stars in a 32′ × 5′ ENE–WSW steam with a parallelogram formed by two 9th and two 7.5 magnitude stars at its WSW edge. The other stars in the stream are 11th to 12.5 magnitude objects. A 6.5 magnitude star, detached from the stream, is near the SE edge of the cluster and forms a 20′ isosceles triangle with two 7.5 magnitude stars beyond the cluster. Very faint stars are sprinkled north and south of the central stream.

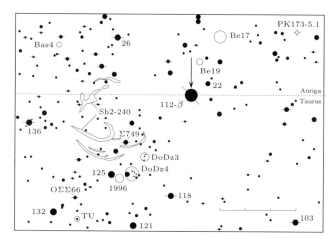

Finder Chart 29-9. 112–β Tau: 05ʰ26.3ᵐ +28°36′

NGC 1996 H42⁸ Open Cluster [100★]
[⌀15′, m–, Br★ 9.5v]   05ʰ38.2ᵐ +25°49′
Finder Chart 29-9   ★★★

12/14″ Scopes–100x: NGC 1996 was classified as "nonexistent" in the Revised NGC. Nevertheless it is easily found SW of magnitude 5.2 125 Tauri. It has a hundred 10th to 14th magnitude stars in a 15′ area. In the western part of the cluster is an 8′ triangle of one 10th and two 11th magnitude stars (the 10th magnitude star at the northern vertex) with twenty 10th to 11th and thirty 12th to 14th magnitude stars in and around it. Two star streams extend eastward from the cluster center, passing north and south of 125 Tauri, the southern stream broader but with fainter stars. A sparse streak of stars is sprinkled between the two streams.

Chapter 30

Triangulum, the Triangle

30.1 Overview

Triangulum (in ancient Greek Deltoton) the Triangle is listed among the constellations in the earliest surviving complete classical astronomical work, the *Phainomena* of Aratos, written about 270 B.C. The Greek celestial Triangle was, however, an equilateral figure, comprising our Alpha (α), Beta (β), and 12 Trianguli, rather than the present scalene figure of Alpha (α), Beta (β), and Gamma (γ). It got reduced to its present size in the late 17th century when Hevelius made a new constellation, which he called Triangulum Minor, from Iota, 10, and 12 Trianguli. Triangulum lies in an area of the sky with several distinct small star patterns which are either recognized asterisms or, like Triangulum itself, full-scale constellations: to its NE is the Head of the Medusa (Beta (β), Pi (π), Rho (ρ) and Omega (o) Persei); to its south is the Head of the Ram (Alpha (α), Beta (β), and Gamma (γ) Arietis); and to its SE is the obsolete constellation Musca Borealis, the Northern Fly (33, 35, 39, and 41 Arietis).

Although it contains no exceptionally bright stars, Triangulum is easily located because of its distinctive pattern. It is the 11th smallest of the 88 constellations, occupying only 132 square degrees. Its showpiece object is the very large, but low surface brightness, face-on spiral galaxy M33, the third largest member of the Local Galaxy Group. The constellation also contains quite a few fainter galaxies, and a handful of good color contrast double stars.

30.2 Interesting Stars

Epsilon (ϵ) = 3 Trianguli Double Star   **Spec. A2**
m5.4, 11.4; Sep. 3.9″; P.A. 118° $02^h03.0^m$ +33°17′
Finder Chart 30-3   ★★★★

8/10″ Scopes–125x: Epsilon Trianguli is an unequally bright double with a bright blue-white primary and a much fainter white secondary.

Triangulum: Tri-ANG-you-lum
Genitive: Trianguli, Tri-ANG-you-lie
Abbrevation: Tri
Culmination: 9 pm–Dec. 7, midnight–Oct. 23
Area: 132 square degrees
Best Deep-Sky Objects: Iota Tri (6–ι), M33 (NGC 598), NGC 669, NGC 670, NGC 672, NGC 684, NGC 784, NGC 925, NGC 949
Binocular Objects: M33 (NGC 598), NGC 604, Collinder 21

Iota (ι) = 6 Trianguli Double Star   **Spec. G0**
m5.3, 6.9; Sep. 3.9″; P.A. 71° $02^h12.4^m$ +30°18′
Finder Chart 30-3   ★★★★

8/10″ Scopes–125x: Iota Trianguli is a fine yellow and blue color contrast double.

Σ239 Double Star   **Spec. F5**
m7.0, 8.0; Sep. 13.8″; P.A. 211° $02^h17.4^m$ +28°45′
Constellation Chart 30-1   ★★★★

4/6″ Scopes–75x: Struve 239 is a nice color contrast pair of yellowish and light blue stars.

30.3 Deep-Sky Objects

NGC 579 Galaxy Type Scd:
⌀1.6′ × 1.5′, m13.3v, SB 14.1 $01^h31.7^m$ +33°38′

NGC 582 Galaxy Type SB?
⌀2.0′ × 0.6′, m13.2v, SB 13.2 $01^h31.9^m$ +33°30′
Finder Chart 30-3, Figure 30-2   ★/★

16/18″ Scopes–150x: These two galaxies form an 8′ NNW–SSE pair. NGC 579, the northernmost galaxy, has a faint, diffuse 1.25′ diameter halo with an inconspicuous core. NGC 582, the southern galaxy, is a 2′ × 0.5′ NE–SW spindle with tapered ends. Its core is moderately brighter and extended along the halo's major axis.

Triangulum, the Triangle

Constellation Chart 30-1

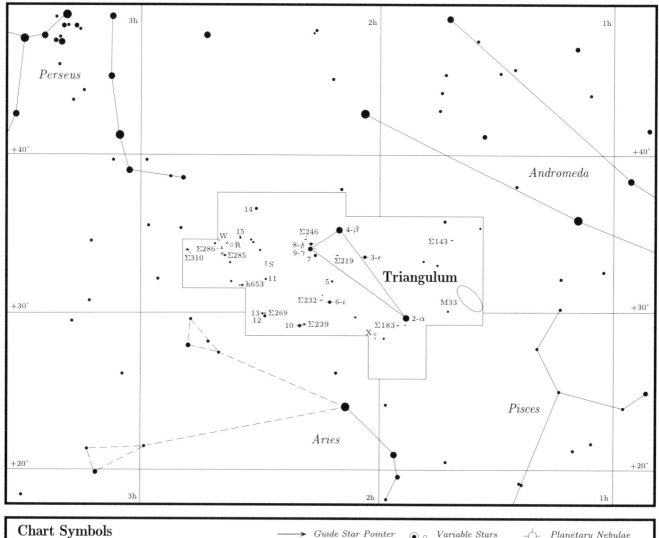

Table 30-1. Selected Variable Stars in Triangulum

Name	HD No.	Type	Max.	Min.	Period (Days)	F*	Spec. Type	R.A. (2000)	Dec.	Finder Chart No. & Notes
X Tri	12211	EA	8.9	11.89	0.97	0.22	A3+G3	02^{h}00.6^m	+27°53′	30-3
S Tri		M	8.7	>12.4	247	0.39	M2	27.3	+32 44	30-4
R Tri	16210	M	5.4	12.6	266	0.44	M4	37.0	+34 16	30-4
W Tri	16682	SRc	8.5	9.7	108		M5	41.5	+34 31	30-1

F* = The fraction of period taken up by the star's rise from min. to max. brightness, or the period spent in eclipse.

Table 30-2. Selected Double Stars in Triangulum

Name	ADS No.	Pair	M1	M2	Sep."	P.A.°	Spec	R.A. (2000) Dec	Finder Chart No. & Notes
Σ143		AB	8.1	9.4	37.8	319	G0	01^{h}40.4^m +34°21′	30-1
Σ183	1522	ABxC	7.7	8.7	5.6	164	F2	55.1 +28 48	30-3
3–ε Tri	1621		5.4	11.4	3.9	118	A2	02^{h}03.0^m +33 17	30-3 Blue-white & white
Σ219	1681		8.2	9.0	11.6	183	A0	10.3 +33 22	30-3
6–ι Tri	1697		5.3	6.9	3.9	71	G0	12.4 +30 18	30-3 Yellow & blue
Σ232	1723		8.0	8.0	6.6	246	B8	14.7 +30 24	30-3 Equal white pair
Σ239	1752		7.0	8.0	13.8	211	F5	17.4 +28 45	30-3 Yellowish & pale blue
Σ246	1764		8.2	9.4	9.2	123	F8	18.6 +34 30	30-4
10 Tri	1770		5.0	11.0	57.1	205	A2	19.0 +28 39	30-3
Σ269	1868		7.9	10.2	1.7	343	G0	28.2 +29 52	30-4 Pale yellow & bluish
h653	1947		7.4	11.1	23.0	42	K2	33.6 +31 24	30-4
15 Tri			5.7	6.9	141.3	17	M, A5	35.8 +34 41	30-4
Σ285	2004		7.5	8.2	1.7	167	K0	38.8 +33 25	30-4 Deep yellow & whitish
Σ286	2020		7.9	10.2	2.9	254	G5	39.8 +33 57	30-4 Yellow & bluish
Σ310	2150		7.3	10.5	2.6	91	A2	49.4 +33 56	30-1 White & bluish

Footnotes: *= Year 2000, a = Near apogee, c = Closing, w = Widening. Finder Chart No: All stars listed in the tables are plotted in the large Constellation Chart, but when a star appears in a Finder Chart, this number is listed. Notes: When colors are subtle, the suffix -*ish* is used, e.g. *bluish*.

NGC 598 Messier 33 Galaxy Type SA(s)cd II-III
⌀67.0′ × 41.5′, m5.7v, SB 14.2   01^{h}33.9^m +30°39′
Finder Chart 30-3, Figure 30-1   ★★

Pinwheel Galaxy

Messier 33 is the third largest member of the Local Galaxy Group after the Andromeda Galaxy and our own Milky Way. It is also the next nearest spiral galaxy after Messier 31, lying 2.4 million light years away. M33, M31, and the satellites of the latter (which include M32 and NGC 205) form one subunit of our Local Group, and our Milky Way Galaxy and its two Magellanic satellites form a second subunit of the Group.

M33 is a loosely-wound Scd spiral with a tiny nucleus and practically no central hub to speak of. M31, on the other hand, is a more compactly-wound Sb galaxy with a huge central hub and a large, bright nucleus. Thus M33 is a much lower surface brightness object than M31, particularly in its interior, and has no bright central disk to call attention to it as you scan for it at low power. M31 and M33 are more than half a million light years apart. The absolute magnitude of M33 is –18.6, a luminosity of 2.4 billion suns, which is only a little greater than that of the Large Magellanic Cloud. Both systems are also unusually blue in overall color (no doubt due to vigorous star formation in the astronomically recent past, hence to large populations of young blue stars) and both are around 50,000 light years in diameter.

M33 was discovered in August 1764 by Charles Messier, who described it as a "whitish light of almost even brightness." Its spiral structure was detected in the mid 19th century by Lord Rosse with his 6 foot reflector at Birr Castle in Parsontown, Ireland.

Because of its large size and low surface brightness M33 is a challenging object for small telescopes. The problem is that the observer, expecting to see something much smaller and brighter, fails to spot its dim glow, which covers an area larger than that of the full Moon. Sky conditions are a critical factor; the least bit of haze can render it quite invisible. But when viewed near the zenith under clear, dark skies it is easy in binoculars and with low power in small telescopes. Instruments of 16″ and larger will reveal some of the galaxy's HII regions (See Figure 30-1).

4/6″ Scopes–25x: Small telescopes show a very faint, oval glow elongated nearly N–S with diffuse edges. It is within a NNE-pointing isosceles triangle of 7.5 magnitude stars. Several triangles of magnitude 9 to 9.5 stars are 12′ NE and 10′ SE of the galaxy's center.

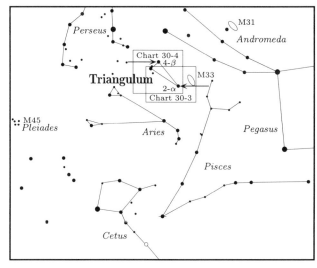

Master Finder Chart 30-2. Triangulum Chart Areas
Guide stars indicated by arrows.

Figure 30-1. *Messier 33 (NGC598) is a low surface brightness object, but large Dobsonian telescopes under clear, dark skies are able to detect spiral structure and many of the H-II regions labelled above. Martin C. Germano made this 90 minute exposure on hypered Kodak 2415 Tech Pan film with a 14.5", f5 Newtonian reflector at Mt. Pinos, California.*

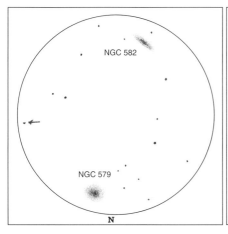

Figure 30-2. NGC 579 & NGC 582
20″, f4.5–175x, by Richard W. Jakiel

Figure 30-3. NGC 669
17.5″, f4.5–250x, by G. R. Kepple

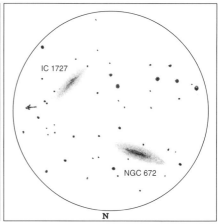

Figure 30-4. NGC 672 & IC 1727
17.5″, f4.5–175x, by G. R. Kepple

8/10″ Scopes–35x: Messier 33 is a fairly conspicuous 50′ × 30′ NNE–SSW glow gradually brighter toward its center. The periphery is diffuse and indefinite.

12/14″ Scopes–50x: Although rather faint, the 60′ × 30′ NNE–SSW halo is distinctly uneven in brightness and shows some mottling toward its center. The core is broad and diffuse. At least half a dozen 13th magnitude and fainter stars are visible against the dim halo. Averted vision reveals hints of spiral structure and dark lanes.

16/18″ Scopes–75x: From a good observing site, Messier 33 is a fairly bright, extremely large glowing object with a much brighter center. A beautiful reverse-S spiral pattern is superimposed upon the halo's background glow. The entire face shimmers with a mottled texture. Several HII regions are visible, including NGC 604 at the NNE edge. At higher powers several of the galaxy's brighter globular clusters may be seen as stellar points.

20/24″ Scopes–75x: In large aperture telescopes one may certainly see why this object was named the Pinwheel Galaxy. This is the type of deep-sky object that responds to increased aperture. If you are accustomed to smaller instruments, your first view of Messier 33 through a 20″ or larger instrument will border on shock! A fine spiral arm pattern is traceable with direct vision over a 65′ × 35′ halo, and over two dozen HII regions are visible. In the center is a small, round core with a sharp stellar nucleus. The halo is mottled and very uneven in brightness, and the spiral pattern is brighter and broader south of the nucleus. Many faint stars are embedded in the halo. The HII regions identified on Martin C. Germano's photograph (see Figure 30-1) are visible under good sky conditions.

NGC 661 H610² Galaxy Type E+:
ø2.0′ × 1.6′, m12.2v, SB 13.3   01ʰ44.2ᵐ +28°42′
Finder Chart 30-3   ★★

12/14″ Scopes–125x: This galaxy is a faint, circular 1′ diameter glow with a stellar nucleus.

16/18″ Scopes–150x: NGC 661 has a prominent stellar nucleus embedded in a slightly ENE–WSW elongated 1.25′ diameter halo.

NGC 669 U1248 Galaxy Type Sab
ø3.0′ × 0.7′, m12.3v, SB 13.0   01ʰ47.2ᵐ +35°33′
Finder Chart 30-4, Figure 30-3   ★★★

12/14″ Scopes–125x: NGC 669, located 2′ south of a row of 11th magnitude stars, is a fine 2′ × 0.25′ NE–SW spindle slightly brighter along its major axis. A 13th magnitude star lies 1.5′ south of the galaxy's center.

16/18″ Scopes–150x: Larger instruments show a nice, edge-on galaxy with a well concentrated 2.5′ × 0.5′ NE–SW halo containing a thin, faint 45″ long core.

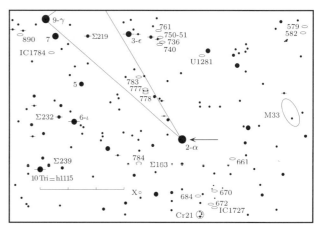

Finder Chart 30-3. 2–α Tri: 01ʰ53.1ᵐ +29°35′

Figure 30-5. *IC 1727 (lower left) and NGC 672 (upper right) form a nice galaxy pair. NGC 672 is the brighter of the two, and has a bar-like core region surrounded by a faint outer halo. IC 1727 is faint and diffuse. Martin C. Germano made this 65 minute exposure on hypered Kodak 2415 Tech Pan film with a 14.5", f5 Newtonian reflector.*

NGC 670 H611² Galaxy Type SA0
ø2.2' × 0.9', m12.7v, SB 13.3   01ʰ47.4ᵐ +27°53'
Finder Chart 30-3   ★★★

12/14" Scopes–125x: This galaxy has a thin, well concentrated 1' × 0.25" NNW–SSE halo with a tiny core extending along its major axis. It lies at the NNW vertex of a 2.5' × 1.5' isosceles triangle with 12th and 13th magnitude stars at the other corners.

16/18" Scopes–150x: NGC 670 is a fairly bright 1.25' × 0.5' NNW–SSE lens-shaped galaxy with an elongated core and a faint stellar nucleus.

IC 1727 U1249 Galaxy Type SB(s)m III-IV
ø7.3' × 2.8', m11.5v, SB 14.6   01ʰ47.5ᵐ +27°20'
Finder Chart 30-3, Figures 30-4 & 30-5   ★★

12/14" Scopes–125x: IC 1727, located 4' north of a 10.5 magnitude star, is a faint, diffuse galaxy elongated 3' × 1.5' NW–SE with only slight central brightening. A couple of faint spots are visible SE of its center. The much brighter galaxy NGC 672 is 8' NE.

NGC 672 H157¹ Galaxy Type SB(s)cd III
ø6.6' × 2.6', m10.9v, SB 13.8   01ʰ47.9ᵐ +27°26'
Finder Chart 30-3, Figures 30-4 & 30-5   ★★★★

4/6" Scopes–50x: This conspicuous galaxy has a fairly large E–W oval halo of uniform surface brightness but diffuse edges.

12/14" Scopes–125x: NGC 672 is a fairly bright galaxy elongated 4' × 1' ENE–WSW and of even surface brightness. A 14th magnitude star is 45" south, and 12.5 magnitude stars 2' WNW and 3.25' east, of the galaxy's center. Its WSW tip points toward IC 1727 lying 8' away.

16/18" Scopes–150x: This fine, bright galaxy has a 4.5' × 1.5' ENE–WSW halo containing a 1.5' × 0.5' bar-shaped core. The halo is generally diffuse except for some mottling around the core, and it is somewhat more tapered toward the east.

Collinder 21 Open Cluster 20★ Tr Type IV 2 p
ø7', m7.3v   01ʰ50.2ᵐ +27°05'
Finder Chart 30-3, Figure 30-6   ★★★

12/14" Scopes–125x: At low power Collinder 21 is an obvious 6' wide "C" of bright stars. 125x reveals a dozen 8th to 10th magnitude stars, nine of them in

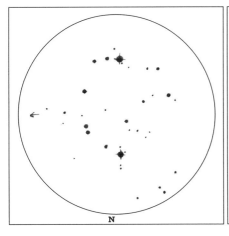

Figure 30-6. Collinder 21
12.5", f5–200x, by G. R. Kepple

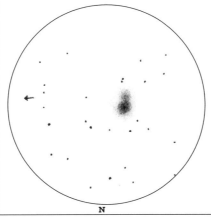

Figure 30-7. NGC 750-1
20", f4.5–175x, by Richard W. Jakiel

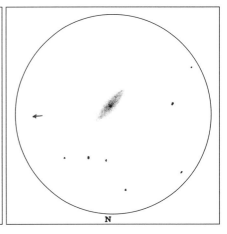

Figure 30-8. NGC 761
20", f4.5–175x, by Richard W. Jakiel

a nice quarter moon asterism with the terminator facing east and the two brightest stars at its north and south edges. An 18″ wide pair of 10.5 magnitude stars is on the WNW edge of the semicircle. Galaxy IC 1731, lying 4′ north and just west of a 13.5 magnitude star, is a faint, diffuse 1.5′ diameter glow surrounding a faint, tiny core.

NGC 684   H612²   Galaxy   Type Sb II-III
ø3.1′ × 0.8′, m12.3v, SB 13.2   01ʰ50.2ᵐ +27°39′
Finder Chart 30-3   ★★★

12/14″ Scopes–125x: NGC 684, located 2′ west of a 12th magnitude star, is faint, small, elongated 1.5′ × 0.25′ E–W, and has some brightening at its center. 6′ to the SW is a short NE–SW row of two 12th magnitude stars flanking one of the 13th magnitude.

16/18″ Scopes–150x: NGC 684 is a nice, fairly bright, lenticular galaxy elongated 3′ × 0.5′ E–W with a pronounced central bulge and a small, bright extended core containing a stellar nucleus. There are hints of a dust lane.

NGC 735   H176³   Galaxy   Type Sb I
ø1.6′ × 0.7′, m13.3v, SB 13.2   01ʰ56.6ᵐ +34°11′
Finder Chart 30-4   ★★

16/18″ Scopes–150x: NGC 735, located 1.5′ NE of a 10th magnitude star, has a diffuse, uniformly faint, 1′ × 0.25′ NW–SE halo. A 13th magnitude star is embedded near the halo's NW tip.

NGC 736   U1414   Galaxy   Type E+:
ø1.8′ × 1.6′, m12.2v, SB 13.1   01ʰ56.7ᵐ +33°03′
Finder Chart 30-4   ★★★

16/18″ Scopes–150x: NGC 736, located 6′ NNE of a 9th magnitude star, has a well concentrated, circular 1′ halo with a small bright core. A 13th magnitude star 30″ from the galaxy's center on its NNE edge was mistakenly designated NGC 737.

NGC 740   U1421   Galaxy   Type SBb?
ø1.6′ × 0.3′, m14.0v, SB 13.0   01ʰ56.9ᵐ +33°01′
Finder Chart 30-4   ★★★★

16/18″ Scopes–150x: NGC 740, located 1.5′ NW of a 10th magnitude star, is a very faint, thin edge-on galaxy elongated 1.5′ × 0.25′ and slightly brighter along its major axis. Averted vision reveals three very faint companion galaxies to the NW.

NGC 750   H222²   Galaxy   Type E pec
ø1.7′ × 1.3′, m11.9v, SB 12.6   01ʰ57.5ᵐ +33°13′

NGC 751   U1431   Galaxy   Type E pec
ø1.4′ × 1.4′, m12.8v, SB 13.4   01ʰ57.6ᵐ +33°12′
Finder Chart 30-4, Figure 30-7   ★★/★

12/14″ Scopes–125x: The interacting pair of galaxies NGC 750 and NGC 751 appear as twin nuclei in a common halo elongated 1.25′ × 0.75′ N–S. NGC 750, the northern nucleus, is the brighter and sharper of the two. 4.5′ to the SW is a close 1.5″ wide pair of 9th magnitude stars.

16/18″ Scopes–150x: NGC 750 and NGC 751 are a pair of NNW–SSE galactic nuclei that share the same 1.5′ × 1′ N–S galactic halo. The NNW nucleus, NGC 750, is moderately bright, small, and round with a stellar nucleus. NGC 751 is a little fainter, more broadly concentrated, and lacks a sharp nucleus.

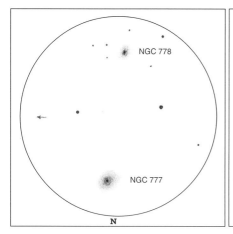 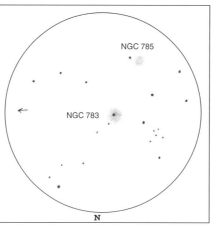 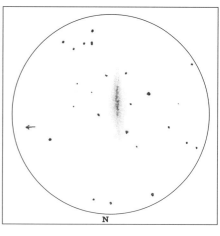

Figure 30-9. NGC 777 & NGC 778
20″, f4.5–175x, by Richard W. Jakiel

Figure 30-10. NGC 783 & NGC 785
13″, f5.6–175x, by Steve Coe

Figure 30-11. NGC 784
20″, f4.5–175x, by Richard W. Jakiel

NGC 761 Galaxy Type SBa:
⌀1.6′ × 0.5′, m13.5v, SB 13.1   01ʰ57.8ᵐ +33°23′
Finder Chart 30-4, Figure 30-8   ★

12/14″ Scopes–100x: NGC 761 is an extremely faint, diffuse 1′ × 0.5′ NW–SE glow with a stellar nucleus. 1′ to the NE is a 1′ triangle of 13th magnitude stars. The interacting galaxies NGC 750-51 lie 11′ SSW.

NGC 777 H223² Galaxy Type E1
⌀2.8′ × 2.3′, m11.4v, SB 13.3   02ʰ00.2ᵐ +31°26′
NGC 778 Galaxy Type S0:
⌀1.1′ × 0.5′, m13.2v, SB 12.4   02ʰ00.3ᵐ +31°17′
Finder Chart 30-4, Figure 30-9   ★★★/★★

12/14″ Scopes–125x: These galaxies form a nice but faint 8′ wide N–S pair. The two galaxies mark the north and south corners of a parallelogram with magnitude 8.5 and 9 stars at the east and west corners. NGC 777 has a faint but easily noticed 1.5′ diameter halo slightly elongated NW–SE with a small core containing a stellar nucleus. NGC 778 has a bright nonstellar nucleus embedded in a much fainter halo that is less than 30″ in diameter.

NGC 783 U1497 Galaxy Type Sc
⌀1.6′ × 1.3′, m12.1v, SB 12.8   02ʰ01.1ᵐ +31°52′
Finder Chart 30-4, Figure 30-10   ★★★

12/14″ Scopes–125x: NGC 783, located 10′ east of an 8th magnitude star, is a fairly faint 1′ diameter glow. A 13th magnitude star is near the WNW edge and a 12.5 magnitude star 1′ SE.

16/18″ Scopes–150x: NGC 783 has a faint stellar nucleus embedded in a 1.25′ diameter halo that fades smoothly to its edges. Companion galaxy NGC 785, lying 8′ SE, is a very faint 30″ spot with a stellar nucleus.

NGC 784 U1501 Galaxy Type SBdm: sp
⌀6.2′ × 1.7′, m11.7v, SB 14.1   02ʰ01.3ᵐ +28°50′
Finder Chart 30-3, Figure 30-11   ★★★

8/10″ Scopes–100x: This fairly faint galaxy is a 5′ × 0.5′ N–S spindle that is slightly brighter along its major axis. The halo's northern portion protrudes into a right triangle of 11th to 13th magnitude stars.

12/14″ Scopes–125x: NGC 784 is a moderately bright galaxy with a 6′ × 0.75′ N–S halo mottled throughout and containing a slightly brighter, highly extended core. A 13th magnitude star is just east of the northern tip, and a wide pair of 11th magnitude stars are SSW of the southern tip.

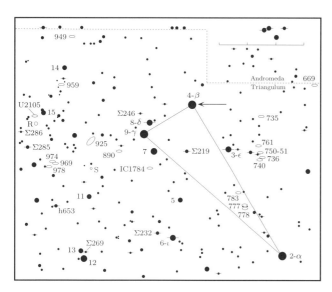

Finder Chart 30-4. 4–β Tri: 02ʰ09.5ᵐ +34°59′

Figure 30-12. *NGC 925 has a fairly bright extended core embedded in a tenuous star-studded halo. Martin C. Germano made this 90 minute exposure on hypered Kodak 2415 Tech Pan film with a 14.5", f5 Newtonian reflector.*

IC 1784   U1744   Galaxy   Type SA(rs)bc pec
ø2.1' × 1.1', m13.1v, SB 13.9 $02^h16.2^m$ +32°39'
Finder Chart 30-4   ★★

16/18" Scopes–150x: IC 1784, a fairly faint galaxy, has a 1.25' × 0.75' teardrop-shaped halo, the tapered end pointing east. The halo is diffuse and without central brightening. 2.5' to the NE is galaxy NGC 1785 an extremely faint pip elongated 30"×15" NW–SE.

NGC 890   H225²   Galaxy   Type SAB(r)0-?
ø2.9' × 2.4', m11.2v, SB 13.2 $02^h22.0^m$ +33°16'
Finder Chart 30-4, Figure 30-13   ★★★

8/10" Scopes–125x: This galaxy has a faint 1' × 0.75' ENE–WSW halo with a bright center. A faint double star lies to the NW of the galaxy.

12/14" Scopes–150x: NGC 890 shows a small, circular core with a stellar nucleus embedded in a well concentrated 1.5' × 1' ENE–WSW oval halo. Three faint stars lie nearby to the west.

NGC 925   H177³   Galaxy   Type SAB(s)d II-III
ø12.0 × 7.4', m10.1v, SB 14.8 $02^h27.3^m$ +33°35'
Finder Chart 30-4, Figure 30-12   ★★★

8/10" Scopes–100x: This galaxy has a large, extended ESE–WNW core embedded in a very tenuous halo sprinkled with foreground stars. An E–W row of five stars runs along the southern edge. Wide pairs of 12.5 magnitude stars lie on the northern and western edges.

12/14" Scopes–125x: NGC 925 has a 2' × 0.5' ESE–WNW bar-like core embedded in a much fainter, irregularly bright 6' × 4' halo. The halo extends almost to the 10th magnitude star 3.5' south of the galaxy's center. An E–W row of stars lies within the southern edge of the halo.

16/18" Scopes–150x: In larger instruments and with averted vision, NGC 925 displays faint counter-clockwise spiral arms. The central area sparkles with mottling in and around a 2' × 1' core. The halo containing the spiral pattern is extremely tenuous, but can be traced, again with averted vision, out to 8' × 5' ESE–WNW. It is studded with an abundance of foreground Milky Way stars.

NGC 949   H154¹   Galaxy   Type SA(rs)b:? III-IV
ø3.3' × 2.1', m11.8v, SB 13.7 $02^h30.8^m$ +37°08'
Finder Chart 30-4, Figure 30-14   ★★★

8/10" Scopes–100x: This galaxy is fairly faint, elongated 2' × 1' NNW–SSE, and has a much brighter, nearly stellar core.

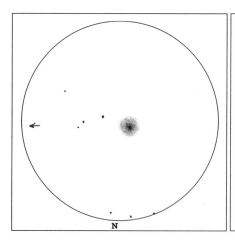

Figure 30-13. NGC 890
8", f10–170x, by Stan Howerton

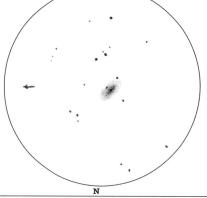

Figure 30-14. NGC 949
8", f4.5–150x, by Dennis E. Hoverter

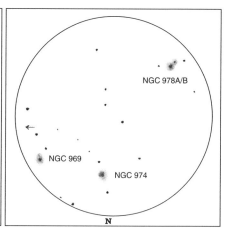

Figure 30-15. NGC 978A/B Area
12.5", f5–250x, by G. R. Kepple

16/18" Scopes–150x: NGC 949 has a nice, fairly bright 2.5' × 0.75' NNW–SSE halo elongated to tapered ends. The central area is lens-shaped 1' × 0.5', contains an inconspicuous stellar nucleus, and is irregularly illuminated, its texture decidedly granular. A 14th magnitude star is embedded on the SW edge of the halo, another faint star lies 30" south, and a wide pair of 13th magnitude stars is 2' south.

NGC 959 U2002 Galaxy Type Sdm:
⌀2.4' × 1.4', m12.4v, SB 13.5 $02^h32.3^m$ +35°30'
Finder Chart 30-4   ★★★

12/14" Scopes–125x: NGC 959, located 4' south of an 11th magnitude star, is a faint 1.5' × 1' ENE–WSW oval of uniform brightness.

16/18" Scopes–150x: NGC 959 has a well concentrated halo elongated 1.75' × 1' ENE–WSW and slightly brighter at its center.

UGC 2105 Galaxy Type (R')SB(s)a: III
⌀1.5' × 1.2', m13.5v, SB 14.0 $02^h37.7^m$ +34°26'
Finder Chart 30-4   ★★

16/18" Scopes–150x: UGC 2105, located 2' NW of an 11.5 magnitude star, appears fairly faint. The 1' diameter halo is slightly elongated NE–SW and moderately brighter at the center.

NGC 969 Galaxy Type S0
⌀2.0' × 1.5', m12.3v, SB 13.3. $02^h34.1^m$ +32°58'
Finder Chart 30-4, Figure 30-15   ★★

12/14" Scopes–125x: NGC 969 has a faint 1' × 0.5' N–S halo containing a small prominent nucleus. A 13th magnitude star lies 45" SSE. NGC 969 is the westernmost of a trio of galaxies along with NGC 974 and NGC 978.

NGC 974 Galaxy Type SAB(rs)b:
⌀3.7' × 3.2', m12.7v, SB 15.2 $02^h34.4^m$ +32°59'
Finder Chart 30-4, Figure 30-15   ★★

12/14" Scopes–100x: NGC 974 has a small, well concentrated core surrounded by a very diffuse, circular 1.5' halo. Faint stars are on the halo's NNE and SSW edges. Galaxy NGC 969 lies 4' west.

NGC 978A-B Galaxy Type S0-:
⌀2.9' × 2.4', m11.2v, SB 13.2 $02^h34.8^m$ +32°52'
Finder Chart 30-4, Figure 30-15   ★★★

12/14" Scopes–100x: NGC 978A-B is the most conspicuous of a trio of faint galaxies with NGC 969 and NGC 974 to its NW. It appears as distinct twin cores with a common faint, 1' diameter halo.

20/22" Scopes–175x: This is a faint double galaxy rather like NGC 750-751. The NW, "A," component is the larger, and its 45" × 30" oval halo has a dense core. The "B" component on the SE has only a 15" diameter halo and a sharp stellar nucleus. A 13th magnitude star lies 1' SE and a 14th magnitude star is the same distance west.

Appendix A

The Local Galaxy Group

Table A-1. Local Galaxy Group

Number	Other No.	Type	Size (')	Mag.	SB	Dist.	Con.	R.A. (2000) Dec	Page & Chart No.
UA444	WLM	IB(s)m V	11.5 × 4.0	10.6	14.6	960	Cet	00ʰ02.0ᵐ −15°28″	Page 1-155, Chart 12-3
IC 10		IBm?	7.3 × 6.4	11.3	15.3	870	Cas	20.4 +59 18	Not reviewed
NGC 147	U326	E5 pec	15.0 × 9.4	9.5	14.7	610	Cas	33.2 +48 30	Page 1-113, Chart 10-4
And III		E?	4.5 × 4.0				And	35.4 +36 31	Not reviewed
NGC 185	U396	dE3 pec	14.5 × 12.5	9.2	14.7	570	Cas	39.0 +48 20	Page 1-114, Chart 10-4
NGC 205	M110	E5 pec	19.5 × 12.5	8.1	13.9	760	And	40.4 +41 41	Page 1-17, Chart 2-4
NGC 221	M32	cE2	11.0 × 7.3	8.1	12.7	760	And	42.7 +40 52	Page 1-17, Chart 2-4
NGC 224	M31	SA(s)b I-II	185. × 75.	3.4	13.6	760	And	42.7 +41 16	Page 1-17, Chart 2-4
And I		dE3 pec	3.5: × 2.5:	13.2	16.2		And	45.7 +38 00	Not reviewed
NGC 292	SMC	Im:	350. × 350.	2.2	14.8	60	Tuc	52.6 −72 48	Southern object
Scl Dwarf	E351-G30	dE2	28 × 23	8.8	16.7	110	Scl	59.9 −33 42	Not reviewed
LGS 3		I?					Psc	01ʰ03.8ᵐ +21 53	Not reviewed
IC 1613	DDO 8	IB(s)m V	20.0 × 15.5	9.2	15.5	730	Cet	04.8 +02 07	Page 1-159
And II		E?	3.5 × 3.5	>15.0		730	Psc	16.4 +33 27	Page 1-326, Chart 24-6
NGC 598	M33	SA(s)cd II-III	67.0 × 41.5	5.7	14.2	1100	Tri	33.9 +30 39	Page 1-391, Chart 31-3
For Dwarf	E356-G04	dE2	12.0 × 10.2	8.1	13.2	130	For	02ʰ39.9ᵐ −34 34	Page 1-191, Chart 15-3
UGC-A86						630	Cam	03ʰ59.9ᵐ +67 08	Not reviewed
LMC		SB(s)m II-IV	650. × 550.	0.4	14.1	54	Dor	05ʰ23.6ᵐ −69 45	Southern Object
Car Dwarf	E206-G220	dE3	35. × 26.	20.9		85	Car	06ʰ41.6ᵐ −50 58	Southern Object
Leo III (A)	U5364	IBm V	5.0 × 3.2	12.6	15.5	1600	Leo	09ʰ59.4ᵐ +30 45	Not reviewed
Leo I	U5470	dE3	12.0 × 9.3	10.2	15.1	230	Leo	10ʰ08.4ᵐ +12 18	Not reviewed
Sex Dwarf		dE3	90. × 65.	11.	20.	85	Sex	13.2 −01 37	Not reviewed
Leo II (B)	U6253	dE0 pec	15.0 × 12.5	12.0	17.5	230	Leo	11ʰ13.5ᵐ +22 10	Not reviewed
GR8	U8091	Im V	1.0 × 0.8	14.4	14.0	1000	Vir	12ʰ58.7ᵐ +14 13	Not reviewed
UMi Dwarf	U9749	dE4	41.0 × 26.0	10.9	18.3	67	UMi	15ʰ08.8ᵐ +67 12	Not reviewed
Dra Dwarf	U10822	dE0 pec	51.0 × 31.0	9.9	17.7	75	Dra	17ʰ20.2ᵐ +57 55	Not reviewed
Sgr Dwarf	E594-G04	IB(s)m V	3.4 × 2.2	12.8	14.8		Sgr	19ʰ30.0ᵐ −17 41	Not reviewed
NGC 6822	Barnard's	IB(s)m IV-V	19.1 × 14.9	8.8	14.8	620	Sgr	44.9 −14 48	Page 2-331, Chart 56-7
Aqr Dwarf	DDO 210	Im V	2.3 × 1.2	13.9	14.9		Aqr	20ʰ46.9ᵐ −12 51	Not reviewed
IC 5152		IAB(s)m IV	4.9 × 3.0	10.6	13.4	1500	Ind	22ʰ02.9ᵐ −51 17	Southern Object
Tuc Dwarf		dE5	5.0 × 2.5	15.0:	17.5:	290	Tuc	41.7 −64 25	Southern Object
Peg Dwarf	U12613	Im V	4.6 × 2.8	12.6	15.2	1700	Peg	23ʰ28.6ᵐ +14 45	Not reviewed

Our own Milky Way Galaxy is a member of a galaxy group called the Local Group, its center lying somewhere between our galaxy and M31, the Andromeda Galaxy. Lying within a radius of 1.5 megaparsecs around this center is a cluster of over thirty galaxies. Finding each of these members is an interesting as well as a challenging project for the visual observer. The observer will see a wide variety of galactic structure even though all but three galaxies are elliptical of irregular in nature. Table A-1 lists the objects belonging to the Local Group. We have listed the page and chart numbers for the objects which were included in our sky survey. Many of these galaxies have extremely low surface brightness and will provide a challenge, however, most of them are within reach of large amateur telescopes. Of course, the observer must have a dark, transparent sky and be fully dark-adapted to detect the fainter objects. The majority of the Local Group galaxies may be found in the evening autumn sky.

Appendix B

Meteor Showers

Meteors, popularly called shooting stars throughout the ages, are unexpected streaks of light produced when a particle from space enters Earth's upper atmosphere and is vaporized from the intense heat produced by its sudden plunge into the atmosphere at heights from 80 to 112km (50-70 miles). The combined speed of the Earth and meteor can be as high as 70km per second. Meteoroids are small particles, usually about the size of a grain of sand, travelling around the Sun. Larger meteoroids, perhaps the size of a pea, will produce a very bright meteor called a fireball or bolide, if it is the size of a walnut or larger it may survive its fiery plunge to Earth and reach the ground - when it does it is called a meteorite.

Major meteor showers are produced at the same time every year when the Earth passes through an area where a comet has shed particles during its perihelion passage around the Sun. The more reliable showers are shown in the table below. ZHR stands for zenithal hourly rate, the number of meteros a single observer may see per hour when the radiant is directly overhead in a dark sky location. Less meteors may be seen when the radiant is not overhead. The actual number seen is determined by the sky's condition (moonlight and light pollution being the two worst offenders). The night of maximum usually produces the highest rate of meteors, especially after midnight when the Earth is advancing toward the meteors as it travels in its orbit. Meteors seen earlier have to catch up with the Earth and meet it more slowly. Meteors belonging to the shower may be traced back to the radiant point; those that come from a different direction are call sporadic meteors. Five to ten sporadic meteors may be seen per hour at any time under clear, dark skies.

Table B-1. Major Meteor Showers

Shower	Max	ZHR at max	Duration above 1/4 peak	Limits	Radiant R.A. & Dec.	
Quadrantids	Jan 3	85	0.8 days	Jan 1-5	$15^h 30^m$	$+50°$
Lyrids	Apr 22	15	6 days	Apr 19-24	18 16	$+34$
Eta Aquarids	May 4	30	14 days	May 1-8	22 27	00
Delta Aquarids	Jul 28	20	12 days	Jul 15 - Aug 15	22 38	-16
Perseids	Aug 11	100	4 days	Jul 25 - Aug 18	03 06	$+58$
Orionids	Oct 21	20	8 days	Oct 16 - 27	06 22	$+16$
S Taurids	Nov 3	15	30 days	Oct 10 - Dec 5	03 24	$+14$
N Taurids	Nov 13	15	30 days	Oct 10 - Dec 5	03 55	$+23$
Leonids	Nov 17	*15	4 days	Nov 14-20	10 11	$+22$
Geminids	Dec 14	95	3 days	Dec 7-15	07 30	$+33$
Ursids	Dec 22	20	2 days	Dec 17-24	14 28	$+75$

*Leonids produce a narrow storm of many thousands per hour every 33 years (1933, 1966, 1999).

Appendix C

The Editors

George Robert Kepple
"Bob" houses a 17.5″, f4.5 and a 12.5″, f5 Newtonian in his homemade 10 × 15 ft roll-off roof observatory. He is a retired Lathe and Grinder operator. His first telescope, purchased when he was in only the 4th grade, was a 3″ refractor which he used for 15 years. His forty-five years of observing experience was helpful in editing the deep-sky descriptions.

Glen W. Sanner
A pharmacist by day and an amateur astronomer by night, Glen uses an 18.5″, f5 Newtonian housed in a 12-ft diameter homemade domed observatory. He has been an amateur astronomer for over 20 years and his first telescope was a 5″, f10 SCT.

The Contributors

Martin C. Germano – Astrophotographer
Equipment: 14.5″, f5 Newtonian, 8″, f5 Newtonian
Contributed over 500 photographs

Lee C. Coombs, Ph. D. – Astrophotographer
Equipment: 10″, f5 Newtonian
Contributed over 100 photographs

Steve Coe – Observer
Equipment: 13″, f5.6 Newtonian
Contributed over 3000 observations

Steve Gottlieb – Observer
Equipment: 17.5″, f4.5 Newtonian, 13″, f4.5 Newtonian
Contributed over 3000 observations

A. J. Crayon – Observer
Equipment: 8″, f6 Newtonian
Contributed over 2000 observations

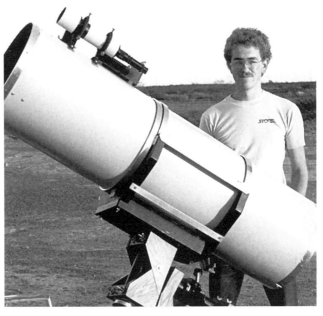

Tom Polakis – Observer
Equipment: 13″, f4.5 Newtonian
Contributed over 2000 observations

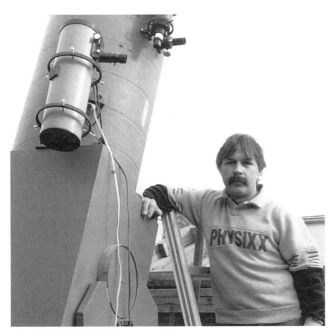

Jim Lucyk – Observer
Equipment: 17.5″, f4.5 Newtonian
Contributed over 1500 observations

Mark T. Stauffer – Observer
Equipment: 3.5″, f11, 4″, f10 Refractors
Contributed over 2000 observations

Bob Erdmann – Observer
Equipment: 16″, f5 Newtonian
Contributed over 1000 observations

George de Lange – Observer
Equipment: 13″, f4.5 Newtonian
Contributed over 1000 observations

Chris Schur – Astrophotographer
Equipment: 16″, f5 Newtonian, 12″, Schmidt Camera
Contributed over 50 photographs & 500 observations

Thomas Jager – Observer
Equipment: 12″, f5 Newtonian, 8″, f6 Newtonian
Contributed over 500 observations

Rick Dilsizian – Astrophotographer & Writer
Equipment: 17.5″, f5 Newtonian, 8″ Schmidt Camera
Observer's Guide magazine associate editor & photos

John S. Ebersole, M.D. – Astrophotographer
Equipment: 165mm, 6 × 7 equatorialy mounted camera
Contributed over 50 photographs

Jack Kramer – Observer
Equipment: 10″, f5 Newtonian, 98mm Refractor
Contributed over 500 observations

Michael Downs – Observer
Equipment: 16″, f4.5 Newtonian, 12″, f5 Newtonian
Contributed over 1000 observations

John R. Stansfield – Observer
Equipment: 2.4″, f15 Refractor
Contributed over 500 observations

Dr. Leonard Scarr – Observer
Equipment: 10″, f5 Newtonian
Contributed over 500 observations

Kirk Conyers – Observer
Equipment: 10″, f4.5 Newtonian, 4″, f12 SCT
Contributed over 500 observations

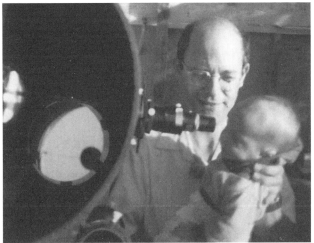

Dr. Jack Marling – Observer
Equipment: 17.5″, f5 Newtonian
Contributed over 500 observations

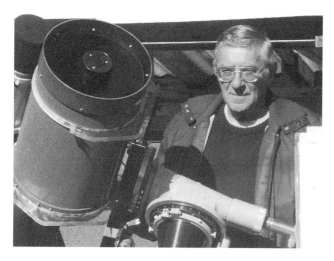

Fr. Lucian J. Kemble, OFM – Observer
Equipment: 11″, f10 SCT
Contributed over 200 observations

David M. Hasenauer – Observer
Equipment: 17.5″, f4.5 Newtonian
Contributed over 500 observations

Tom Osypowski – Observer
Equipment: 22″, f5 Newtonian
Contributed over 500 observations

Larry McHenry – Observer
Equipment: 13″, f4.5 Newtonian (Not shown)
Contributed over 250 observations

Ed & Tyler Faits – Observers
Equipment: 13″, f4.5 Newtonian, 3″, f10 Refractor
Contributed over 500 observations

Dave Selinger – Observer
Equipment: 10″, f4.5 Newtonian
Contributed over 250 observations

Contributors

Murray Cragin – Observer
Equipment: 24″, f4 Newtonian
Contributed over 500 observations

Stan Howerton – Observer
Equipment: 8″, f10 SCT
Contributed over 500 observations

Juhani Salmi – Astrophotographer
Equipment: 12″, f6.4 Newtonian
Contributed over 50 photographs

Richard W. Jakiel – Observer
Equipment: 20″, f4.5 Newtonian
Contributed over 500 observations

Dennis E. Hooverter
Observer
Equipment: 8″, f4.5 Newtonian
Contributed over 500 observations

Steve Pattie
Observer
Equipment: 17.5″, f4.5 Refl
Contributed 200 observations

Bennie Negy, Jr
Observer & Astrophotographer
Equipment: 8″, f1.5 Schmidt Camera
Contributed over 100 observations &photos

K. Alexander Brownlee
Astrophotographer
Equipment: 16″, f5 Newtonian
Contributed over 100 photographs

Tony & Daphne Hallas
Astrophotographers
12″ f7 Newtonian

Contributors

Herbert Arold
Observer
90mm, f11 Mak

Greg Bargerstock
Observer
8″ f4.5 Newtonian

Michael J. Benyo
Observer
13″ f4.5 Newtonian

Gary Capps
Astrophotographer
12″ f5 Newtonian

Andrew D. Fraser
Observer
6″ f8 Newtonian

Harvey Freed D.D.S.
Astrophotographer
10″ f6 Newtonian

Dan Gordon
Astrophotographer
70mm f5.6 refractor

William Harris
Astrophotographer
8″ f1.5 Newtonian

Evered Kreimer
Astrophotographer
12″ f6 Newtonian

Alister Ling
Observer
12.5″, f5.6 Newtonian

Barbara Lux
Observer
6″, f8 Newtonian

J. C. Mirtle
Astrophotograher
8″ f6 Newtonian

W. L. Robinson
Observer
10″ f5 Newtonian

Norman G. Smith
Observer
10″ f10 SCT

Ernest Underhay
Observer
10″ f4.5 Newtonian and 60mm, f15 Refractor

Don Walton
Astrophotographer
10″ f10 SCT

Barbara Wilson
Observer
20″ f4 Newtonian

Bibliography

Books

Cragin, M., Lucyk, J., Rappaport, B., The *Deep Sky Field Guide* to The Uranometria 2000.0, Willmann-Bell, Inc., 1993.

Eicher, D.J., *Deep-Sky Observing With Small Telescopes*, Enslow, 1989.

Eicher, D.J., *The Universe from Your Backyard*, Cambridge University Press, 1988.

Hirshfeld, A., Sinnott, R.W., *Sky Catalogue 2000.0 Vol. 1, Stars to Magnitude 8.0*, Sky Publishing Corp, 1982.

Hirshfeld, A., Sinnott, R.W., *Sky Catalogue 2000.0 Vol. 2, Double Stars, Variable Stars and Nonstellar Objects, Sky,* Publishing Corp, 1985.

Hynes, S.J., *Planetary Nebulae: A Practical Guide and Handbook for Amateur Astronomers*, Willmann-Bell, Inc., 1991.

Jones, K.G., *Messier's Nebulae and Star Clusters*, 2nd Edition, Published by Cambridge University Press, 1991.

Lovi, G., Tirion, W., *Men Monsters and the Modern Universe*, Willmann-Bell, Inc., 1989.

Luginbuhl and Skiff, *Observing Handbook and Catalog of Deep-Sky Objects,* Published by Cambridge University Press, 1990.

Sinnott, R.W. Ed., *NGC 2000.0: The Complete New General Catalog and Index Catalouges of Nebulae and Star Clusters by J.L.E. Dreyer,* Published by Sky Publishing Corp, 1988.

Staal, J.D.W., *The New Patterns in the Sky,* The McDonald and Woodward Publishing Co., 1988.

Tirion, W., *Bright Star Atlas 2000.0,* Published by Willmann-Bell, Inc., 1990.

Tirion, W., *Sky Atlas 2000.0,* Sky Publishing Corp, 1981.

Tirion, W., Rappaport, B., Lovi, G., *Uranometria 2000.0* Vol. 1, The Northern Sky to −6° Willmann-Bell, Inc., 1987.

Tirion, W., Rappaport, B., Lovi, G., *Uranometria 2000.0* Vol. 2, The Southern Sky to +6° Willmann-Bell, Inc., 1988.

Vickers, J.C., *Deep Space CCD Atlas: North,* Published by John C. Vickeres dba: Graphic Traffic Co., 1994.

Vickers, J.C., Wassilieff, A., *Deep Space CCD Atlas: South,* Published by John C. Vickers dba: Graphic Traffic Co., 1994.

Monographs

Webb Society Observing Section Reports, The Webb Society

Kepple, G.R., *Astro Cards*

Periodicals

Webb Society Quarterly Journal, The Webb Society

The Observer's Guide, Vol. 1 thru 32, by G.R. Kepple and G.W. Sanner

The Deep-Sky Observer, The Webb Society

Computer Databases and Software

Digitized Sky Survey, Space Telescope Science Institute

MegaStar, ELB Software, by Emil Bonanno

Voyager II, Carina Software

Sky Gallery with the Hubble Guide Star Catalog, Carina Software

Index

This is a cummulative Index for *The Night Sky Observer's Guide*, Volumes 1 and 2. The number (1 or 2) preceding a dash indicates volume number.

A

Abbreviations, 1-10
Abell 4, 1-302
Abell 21, 1-213
Abell 30, 1-82
Abell 33, 2-211
Abell 50, 2-184
Abell 53, 2-15
Abell 55, 2-17-18
Abell 61, 2-136
Abell 62, 2-20
Abell 67, 1-144
Abell 70, 2-24
Abell 71, 2-150
Abell 72, 2-168
Abell 75, 1-138
Abell 79, 1-220
Abell 82, 1-110
Abell 84, 1-110
Abell 262, 1-20
Abell 347, 1-23
Abell 426, 1-305, 1-307
Abell 779, 1-236
Abell 1060, 2-216
Abell 1367, 2-246-248
Abell 1656, 2-101
Abell 1890, 2-33
Abell 2065, 2-111, 2-113
Abell 2151, 2-192-195
Abell 2197, 2-195-196
Abell 2199, 2-195-196
Acamar, 1-175
Achernar, 1-175
Achilles, 2-125
ADS 3883, 1-221
ADS 3910, 1-221
ADS 3930, 1-223
ADS 3954, 1-223
ADS 4034, 1-224
ADS 4260, 1-224
ADS 6309, 1-350
ADS 7565, 2-393
ADS 7566, 2-393
ADS 9060, 2-467
ADS 13374, 2-142
ADS 17081, 1-110
Aesculapius, 1-259, 2-287, 2-311, 2-365, 2-373
AGC 11, 2-307
Aitken 1774, 2-125
Al Giedi, 2-61
Al Jabbar, 1-259
Al Kurhah, 1-134
Albireo, 1-79, 1-88, 1-296, 2-83, 2-133, 2-136
Alcmene, 2-189
Alcor, 2-385, 2-387, 2-389
Alcyone, 1-380-381
Alderamin, 1-131
Alfirk, 1-131
Algieba, 2-225, 2-228
Algol, 1-293-294, 1-296
Allen, R. H., 2-415
Almach, 1-14, 1-88
Almagest, 2-328, 2-333
Alnilam, 1-259
Alnitak, 1-259, 1-263
Alpha Persei Moving Group, 1-293, 1-307
Alphard, 2-203, 2-379
Alpheca, 2-111
Alphoratz, 1-279
Alrisha, 1-318
Alshain, 2-9
Altair, 2-9, 2-22, 2-305
Alvan Clark 11, 2-375
Ammon Ra, 1-39
Amphitrite, 2-163
Amphytrion, 2-189
Andromeda Galaxy, 1-11, 1-17-18, 1-391, 2-53, 2-193, 2-426, 2-487
Andromeda II, 1-322, 1-326
Andromeda, 1-11, 1-13-14, 1-16-18, 1-105, 1-131, 1-151, 1-215, 1-279, 1-293, 1-322, 1-326, 2-53
Antares, 2-287, 2-299, 2-333, 2-336, 2-338, 2-340
Antlia, 2-1-2, 2-4
Anubis, 1-85
Aphrodite, 2-225
Apollo, 2-117, 2-125, 2-287, 2-305
Aquarius, 1-25, 1-27-28, 1-31, 1-232, 1-315, 1-333
Aquila, 2-9-10, 2-12, 2-15, 2-17, 2-21, 2-133, 2-305, 2-311
Ara, 2-265, 2-338
Aratos, 1-81, 1-315, 1-389, 2-189, 2-415
Arcturus, 2-25, 2-31, 2-388
Arenor, 1-375
Arg 28, 2-267
Argo Navis, 1-339, 1-361
Argo, 1-171
Ariadne, 1-375, 2-111
Aries, 1-39-40, 1-42-43, 1-315, 1-339, 1-375
Arold, Herbert, 1-405
Arp, Halton, 2-444
Aselli, 1-81
Astartre, 2-423
Asterion, 2-41
Asterope, 1-380-381
Astraea, 2-257, 2-423
Athena, 1-339
Auriga, 1-47, 1-51, 1-59, 1-61, 1-229, 1-378

B

Bacchus, 2-105, 2-125
Bail 1695, 1-239
Bargerstock, Greg, 1-405
Barkhatova 1, 2-152
Barnard 1, 1-308
Barnard 2, 1-308
Barnard 3, 1 300
Barnard 4, 1-309
Barnard 26-28, 1-51-52
Barnard 29, 1-52
Barnard 33, 1-270, 1-272
Barnard 34, 1-60
Barnard 35, 1-272, 1-274
Barnard 42, 2-287
Barnard 47, 2-293
Barnard 48, 2-342-343
Barnard 50, 2-343
Barnard 51, 2-293
Barnard 53, 2-343
Barnard 55, 2-344
Barnard 56, 2-344
Barnard 57, 2-295
Barnard 59, 2-295, 2-298
Barnard 60, 2-295
Barnard 61, 2-294, 2-297
Barnard 62, 2-294, 2-297
Barnard 63, 2-296-297
Barnard 64, 2-296-298
Barnard 72, 2-287, 2-298-299
Barnard 78, 2-298-299
Barnard 79, 2-302
Barnard 86, 2-311, 2-317
Barnard 87, 2-319
Barnard 90, 2-322
Barnard 92, 2-323-324
Barnard 93, 2-323
Barnard 100, 2-353, 2-358
Barnard 101, 2-353, 2-358
Barnard 103, 2-353, 2-358, 2-360
Barnard 104, 2-353, 2-360-361
Barnard 110, 2-353, 2-361
Barnard 111, 2-353, 2-361
Barnard 113, 2-353, 2-361
Barnard 127, 2-13-14
Barnard 129, 2-13-14
Barnard 130, 2-13-14
Barnard 132, 2-9, 2-14
Barnard 133, 2-9, 2-15-16
Barnard 134, 2-15-16
Barnard 135, 2-9, 2-16
Barnard 136, 2-9, 2-16
Barnard 139, 2-18
Barnard 142, 2-9, 2-21
Barnard 143, 2-9, 2-21
Barnard 145, 2-140-141, 2-133
Barnard 150, 1-136-137
Barnard 152, 1-137
Barnard 157, 2-160
Barnard 164, 2-133, 2-161
Barnard 168, 2-133, 2-161
Barnard 169, 1-141
Barnard 170, 1-141
Barnard 171, 1-141-142
Barnard 173, 1-141-142
Barnard 174, 1-141
Barnard 228, 2-265, 2-270
Barnard 246, 2-295
Barnard 276, 2-302
Barnard 283, 2-350
Barnard 312, 2-353, 2-357-358
Barnard 343, 2-133, 2-144
Barnard 352, 2-133
Barnard 361, 2-133
Barnard 364, 2-160
Barnard's Galaxy, 2-331
Barnard's Star, 2-228, 2-290, 2-388
Bartsch, Jakob, 1-63
Basel 1, 2-359, 2-361
Basel 4, 1-60
Basel 6, 2-133, 2-139, 2-142
Basel 7, 1-245
Basel 10, 1-298, 1-302
Basel 11A, 1-96
Basel 11B, 1-275
Baxendell, J., 1-375
Bayer, Johannes, 1-151, 1-171, 2-75, 2-281, 2-328, 2-385
Bear Paw Galaxy, 1-233
Becklin-Neugebauer Object, 1-268
Beehive, 1-77, 1-81
Bellatrix, 1-259
Benyo, Michael J., 1-405
Berkeley 4, 1-116
Berkeley 10, 1-66-67
Berkeley 11, 1-312
Berkeley 17, 1-55
Berkeley 18, 1-55-56
Berkeley 19, 1-56
Berkeley 21, 1-275
Berkeley 50, 2-143
Berkeley 57, 1-145
Berkeley 58, 1-112
Berkeley 59, 1-148
Berkeley 62, 1-118
Berkeley 65, 1-126
Berkeley 68, 1-314
Berkeley 70, 1-56
Berkeley 86, 2-133, 2-139, 2-148
Berkeley 87, 2-146, 2-148
Berkeley 94, 1-143
Berkeley 97, 1-143
Bernes 149, 2-337
Bernes 157, 2-108
Bernoulli, 2-178
Betelgeuse, 1-85, 1-101, 1-131, 1-259, 1-263, 2-336
Big Dipper, 1-85, 1-91, 1-227, 1-259, 2-25, 2-111, 2-169, 2-385, 2-415, 2-423
Biurakan 1, 2-139, 2-142
Biurakan 2, 2-133, 2-139, 2-143
Biurakan 8, 1-253
Biurakan 10, 1-252

Black Pillar, 2-378
Black-Eye Galaxy, 2-99-100
Blanco 1, 1-365, 1-369
Blinking Planetary, 2-138, 2-155
Blue Flash Nebula, 2-166
Blue Snowball, 1-24
Bochum 1, 1-209
Bochum 2, 1-251
Bochum 3, 1-254
Bochum 4, 1-344-345
Bochum 5, 1-344-345
Bochum 6, 1-345
Bochum 15, 1-348
Bode, J. E., 2-99, 2-103, 2-199, 2-392
Bond, 2-389
Bootes, 2-25, 2-27-28, 2-41, 2-111, 2-388, 2-413-414
Bowl of the Pipe Nebula, 2-298-299, 2-301
Box Nebula, 2-297
Boyle, Robert, 2-1
Brahe, Tycho, 1-1
Bridal Veil Nebula, 2-150-153
Brisbane 14, 2-105, 2-107-108
Brocchi's Cluster, 2-477
Brownlee, Alexander, 1-168-169, 1-285, 1-289, 1-323, 1-326, 1-404, 2-49, 2-51-52, 2-59, 2-230, 2-234, 2-237, 2-240, 2-249, 2-445, 2-447-449, 2-453, 2-472
Brso 6, 2-69
Brso 12, 2-335
Brso 14, 2-103
Bubble Nebula, 1-108
Bug Nebula, 2-344
Burnham 28, 2-119
Burnham 87, 1-378
Burnham 126, 2-290
Burnham 271, 2-64
Burnham 441, 2-477
Burnham 442, 2-147
Burnham 800, 2-83
Burnham 920, 2-117
Burnham 1110, 2-70
Butterfly Cluster, 2-348-349

C

California Nebula, 1-310
Callimachus, 2-415
Callisto, 2-25, 2-385
Camelopardalis, 1-63, 1-65-66, 2-392
Campbell's Hydrogen Star, 2-136-137
Cancer I Galaxy Group, 1-80
Canes Venatici I Galaxy Cloud, 2-41, 2-51
Canes Venatici II Galaxy Cloud, 2-41
Canes Venatici, 2-25, 2-41, 2-44, 2-49, 2-51-99, 2-385, 2-423
Canis Major, 1-85-86, 1-88, 1-237, 1-255, 1-259, 1-339, 1-343-344
Canis Minor, 1-85, 1-101, 1-103, 1-237
Canopus, 1-1, 1-175
Capella, 1-47, 1-56
Capo 13, 2-69
Capo 62, 2-267
Capo 70, 2-335
Capps, Gary, 1-267, 1-405
Capricornus, 1-25, 1-315, 2-61-62, 2-64, 2-281
Carina, 1-339, 1-361
Cassini, G. D., 1-254
Cassiopeia, 1-11, 1-105, 1-107-108, 1-112, 1-121, 1-123, 1-131, 1-134, 1-293
Castor, 1-203, 1-206, 1-213, 1-339, 2-133
Cat's Eye Nebula, 2-169, 2-183
Ce 2, 2-487
Cederblad 59, 1-272, 1-274
Cederblad 62, 1-277
Cederblad 90, 1-93, 1-255
Cederblad 214, 1-148
Centaurus Galaxy Group, 2-222, 2-465
Centaurus, 1-339, 2-41, 2-67, 2-69-70, 2-72-76, 2-105, 2-222, 2-287, 2-311, 2-333, 2-373, 2-423, 2-451, 2-465
Cepheus, 1-105, 1-131, 1-133, 1-215, 1-293, 2-201
Ceres, 2-423
Cetus, 1-11, 1-151-152, 1-154, 1-279, 1-293, 1-365, 1-369, 2-426
CGCG 074-058, 2-29
CGCG 074-62, 2-29
CGCG 097-90, 2-247
Chara, 2-41
Chiron, 1-339, 2-67, 2-287, 2-311
Christmas Tree Cluster, 1-237, 1-248-249
Cirrus Nebula, 2-150-153
Clark, Alvan, 2-375
Clown Face Nebula, 1-214
Coalsack, 2-133
Coathanger, 2-475, 2-477-479
Cocoon Galaxy, 2-49
Cocoon Nebula, 2-161
Coe, Steve, 1-18, 1-28, 1-32-33, 1-44, 1-53, 1-66, 1-71, 1-75, 1-80, 1-83, 1-89-90, 1-94, 1-144, 1-156, 1-161-162, 1-178-179, 1-182, 1-185, 1-211, 1-224, 1-226-227, 1-241, 1-250, 1-257, 1-263, 1-290, 1-318, 1-320, 1-327, 1-329, 1-336-338, 1-342, 1-348, 1-351-353, 1-355, 1-359, 1-396, 1-400, 2-18, 2-32, 2-37-39, 2-45, 2-54, 2-88, 2-91, 2-93, 2-119-122, 2-127, 2-129, 2-181, 2-182, 2-210, 2-260, 2-262-263, 2-284, 2-339, 2-370, 2-390, 2-397, 2-403, 2-408
Collinder 3, 1-113
Collinder 6, 1-384
Collinder 15, 1-122
Collinder 21, 1-389, 1-394-395
Collinder 33, 1-128
Collinder 34, 1-128
Collinder 38, 1-276
Collinder 55, 1-263
Collinder 62, 1-55
Collinder 69, 1-266
Collinder 70, 1-259, 1-265
Collinder 73, 1-266
Collinder 76, 1-275
Collinder 77, 1-206
Collinder 79, 1-276
Collinder 80, 1-207
Collinder 81, 1-207
Collinder 84, 1-277
Collinder 85, 1-278
Collinder 87, 1-278
Collinder 89, 1-209
Collinder 90, 1-240
Collinder 91, 1-237, 1-240
Collinder 92, 1-237, 1-240
Collinder 95, 1-237, 1-242
Collinder 96, 1-237, 1-242
Collinder 97, 1-237, 1-243
Collinder 101, 1-245
Collinder 104, 1-245
Collinder 106, 1-237
Collinder 107, 1-245
Collinder 108, 1-247
Collinder 110, 1-247
Collinder 111, 1-247
Collinder 112, 1-248
Collinder 115, 1-250
Collinder 117, 1-251
Collinder 121, 1-92
Collinder 123, 1-253
Collinder 127, 1-255
Collinder 132, 1-95
Collinder 135, 1-343, 1-339
Collinder 138, 1-257
Collinder 140, 1-98
Collinder 144, 1-212
Collinder 158, 1-348
Collinder 162, 1-352
Collinder 163, 1-352
Collinder 175, 1-357
Collinder 185, 1-360
Collinder 187, 1-360
Collinder 190, 1-363
Collinder 195, 1-364
Collinder 206, 1-364
Collinder 301, 2-338
Collinder 308, 2-340
Collinder 309, 2-340
Collinder 315, 2-340
Collinder 316, 2-341-342
Collinder 317, 2-342
Collinder 319, 2-343
Collinder 322, 2-343
Collinder 323, 2-343
Collinder 324, 2-344
Collinder 325, 2-345
Collinder 326, 2-345
Collinder 333, 2-347
Collinder 335, 2-347
Collinder 340, 2-348
Collinder 342, 2-349
Collinder 344, 2-349
Collinder 352, 2-350
Collinder 379, 2-356
Collinder 384, 2-349
Collinder 387, 2-360
Collinder 390, 2-361
Collinder 392, 2-12
Collinder 399, 2-475, 2-477-478
Collinder 401, 2-21-22
Collinder 402, 2-137
Collinder 403, 2-138
Collinder 413, 2-141
Collinder 415, 2-143
Collinder 419, 2-133, 2-139, 2-147
Collinder 421, 2-146, 2-148
Collinder 428, 2-133, 2-146, 2-156
Collinder 445, 1-217
Collinder 452, 1-220
Collinder 464, 1-63, 1-71
Collinder 465, 1-255-256
Collinder 466, 1-255-256
Columba, 1-171-172, 1-175
Coma Berenices Galaxy Cluster, 2-426
Coma Berenices Star Cluster, 2-90, 2-225, 2-423
Coma Berenices, 1-365, 2-41, 2-81-82, 2-84, 2-90, 2-225, 2-385, 2-423, 2-426, 2-429-430, 2-437
Coma I Galaxy Cloud, 2-97
Coma-Virgo Galaxy Cluster, 2-81, 2-85, 2-89, 2-95-96, 2-385, 2-388, 2-423, 2-433, 2-437, 2-440, 2-444-445, 2-448-452, 2-454
Cone Nebula, 1-237, 1-247-249
Conon, 2-81
Coombs, Lee C., 1-30, 1-43, 1-51, 1-53, 1-55, 1-59-61, 1-70, 1-91, 1-96, 1-109, 1-112-114, 1-116, 1-120-124, 1-129, 1-135, 1-141, 1-143, 1-146, 1-149, 1-154, 1-168-169, 1-187, 1-207, 1-210, 1-212, 1-218--219, 1-225, 1-247, 1-251, 1-275, 1-278, 1-298-301, 1-306, 1-311-312, 1-347, 1-349, 1-354, 1-382-383, 2-12, 2-66, 2-96, 2-103, 2-155, 2-166, 2-199, 2-207, 2-278, 2-304, 2-308, 2-315-316, 2-323-324, 2-327, 2-330, 2-360, 2-362, 2-403, 2-405, 2-409, 2-429, 2-450, 2-467, 2-478-480, 2-484
Copeland's Septet, 2-245-246
Copernicus, 2-225
Cor Caroli, 2-41, 2-43
Cor Scorpionis, 2-336
Corax, 2-117
Coro 157, 2-69
Coro 167, 2-69
Coro 170, 2-259
Coro 178, 2-267
Coro 179, 2-267
Coro 193, 2-267
Coro 271, 2-267
Corona Australis, 2-105-106
Corona Borealis, 2-105, 2-111-112, 2-426
Corona Borealis Galaxy Cluster, 2-426
Corvus, 2-117-118, 2-120, 2-125
Crab Nebula, 1-375, 1-385-386
Crater, 2-117, 2-125-126
Crayon, A.J., 1-94, 1-98, 1-156, 1-162, 1-163, 1-253, 1-273, 1-286, 2-84, 2-339
Crescent Nebula, 2-143-144
Cromwell, Oliver, 2-41
Cronos, 2-67
Crux, 2-67, 2-185, 2-305, 2-333
Cupid, 1-315, 2-305
Cyclops, 2-305
Cygnus, 1-131, 1-215, 2-9, 2-133-134, 2-136, 2-144-145, 2-151, 2-153, 2-158, 2-373, 2-475
Cygnus Star Cloud, 2-133, 2-136, 2-144-145
Cynosura, 2-415
Cysatus, 1-267
Czernik 8, 1-300, 1-302

Czernik 13, 1-127
Czernik 19, 1-52
Czernik 29, 1-344
Czernik 43, 1-108-109

D

Dabih, 2-61
Dark adaptation, 1-3, 1-138
Dawes limit, 1-8
de Cheseaux, 1-207, 2-308, 2-318, 2-325, 2-327, 2-338
de Chessaux comet, 1-282
de Lacaille, Nicolas Louis, 1-191, 1-365, 2-1, 2-222, 2-281, 2-327
de Lange, George, 1-173, 1-174, 1-210, 1-225, 1-226, 1-227, 2-91, 2-93, 2-100, 2-139, 2-187
de Vaucouleurs, 1-xlviii, 1-xlv
De 0, 2-487
De 2, 2-487
De 3, 2-487
De 4, 2-487
De 5, 2-487
Delphinus, 1-139, 2-163-164, 2-185, 2-340, 2-475
Deltoton, 1-389
Deneb, 2-9, 2-133, 2-155
Denebola, 2-423
Dilsizian, Rick, 2-197, 2-277
Dobson, John, 1-6
Dog Star, 1-85, 1-101
Dolidze 3, 2-144
Dolidze 9, 2-146, 2-149
Dolidze 13, 1-116
Dolidze 14, 1-381, 1-386
Dolidze 15, 1-52
Dolidze 16, 1-54
Dolidze 17, 1-264, 1-273
Dolidze 18, 1-56-57
Dolidze 19, 1-264
Dolidze 20, 1-58
Dolidze 21, 1-265
Dolidze 22, 1-241
Dolidze 23, 1-249
Dolidze 24, 1-250
Dolidze 25, 1-250
Dolidze 26, 1-101, 1-103-104
Dolidze 28, 2-355
Dolidze 29, 2-357
Dolidze 30, 2-359
Dolidze 31, 2-359
Dolidze 32, 2-359-360
Dolidze 36, 2-139-140
Dolidze 39, 2-145, 2-147
Dolidze 40, 2-147
Dolidze 41, 2-147
Dolidze 42, 2-147
Dolidze 44, 2-146, 2-149-150
Dolidze 45, 2-156
Dolidze-Dzimselejsvili 1, 1-46
Dolidze-Dzimselejsvili 2, 1-264
Dolidze-Dzimselejsvili 3, 1-385
Dolidze-Dzimselejsvili 4, 1-386
Dolidze-Dzimselejsvili 5, 2-195
Dolidze-Dzimselejsvili 6, 2-198
Dolidze-Dzimselejsvili 7, 2-199-200
Dolidze-Dzimselejsvili 8, 2-200
Dolidze-Dzimselejsvili 9, 2-201
Dolidze-Dzimselejsvili 10, 2-141
Dolidze-Dzimselejsvili 11, 2-152
Double Cluster, 1-124, 1-127, 1-242, 1-293, 1-298-300, 1-385
Double-Double, 2-61, 2-271, 2-273-274, 2-335-336
Draco, 2-169-170, 2-172, 2-183, 2-385, 2-392, 2-415
Dreyer, John, 1-1, 1-3, 1-297, 1-331
Duck Nebula, 1-96
Dumbbell Nebula, 1-297, 2-315-316, 2-475, 2-482
Dunlop 30, 1-339
Dunlop 31, 1-339
Dunlop 38, 1-342
Dunlop 51, 1-342

Dunlop 177, 2-268
Dunlop 199, 2-337
Dunlop 236, 2-281
Dunlop 241, 1-333
Dunlop 508, 1-173
Dunlop 535, 1-173
Dunlop 536, 2-340
Dunlop 557, 2-350

E

E206-G220, 2-487
E351-G30, 2-487
E356-G04, 2-487
E356-SC05, 1-193, 1-195
E356-SC08, 1-193, 1-195
E594-G04, 2-487
Eagle Nebula, 2-378
Ebersole, John, 1-139, 1-243, 1-255
Egg Nebula, 2-155
Ellipsoidal Variables, 1-xxvi
Eltanin, 2-169
Engagement Ring, 2-417
Enki, 1-333, 2-61
Enlil, 2-25, 2-385
Equuleus, 2-185, 2-305
Erdmann, Bob, 1-134, 1-136, 1-144, 1-161, 1-211, 1-286, 2-172-173, 2-175-176
Erichthonius, 1-47
Eridanus, 1-25, 1-175, 1-177-178, 1-187, 1-200-202, 1-259, 2-426
Eskimo Nebula, 1-68, 1-203, 1-214
ESO 322-84, 2-72
ESO 356-G4, 1-191, 1-195
ESO 384-37, 2-79
ESO 446-59, 2-224
ESO 462-31, 2-284
ESO 514-3, 2-264
ESO 580-34, 2-261
Euergetes, 2-81
Europa, 1-375
Euryale, 1-259
Eurytheos, 2-189

F

Fabricius, David, 1-151
Faits, Ed, 1-405
Farnese Globe, 1-151, 2-25, 2-265, 2-423
Filamentary Nebula, 2-150, 2-153
Fish's Mouth, 2-326
Flaming Star Nebula, 1-47, 1-54
Flamsteed, 2-318, 2-353
Flamsteed, John, 2-318, 2-353
Fomalhaut, 1-333
Footprint Nebula, 2-137
Fornax, 1-191, 1-193-202, 2-426
Fornax Dwarf Galaxy, 1-191, 1-193, 1-195
Fornax I Galaxy Cluster, 1-191, 1-195-202
Fraser, Andrew D., 1-18, 1-182, 1-186-189, 1-353, 1-405
Freed, Harvey, 1-37, 1-180, 1-274, 1-300, 1-330, 1-369, 1-405, 2-48, 2-396, 2-451
Frolov 1, 1-111

G

Gaia, 1-259
Galileo, 1-1, 1-81, 1-267
Gemini, 1-68, 1-77, 1-203, 1-205-206, 1-278, 1-339
Geminos, 2-105
Gemma, 2-111
Germano, Martin C., 1-31, 1-34, 1-52, 1-54, 1-56, 1-58, 1-60, 1-67-70, 1-72-73, 1-82, 1-93, 1-97, 1-111, 1-115, 1-118, 1-122, 1-125-129, 1-137-138, 1-140-145, 1-147, 1-155, 1-157-159, 1-166-167, 1-181, 1-208-209, 1-213-214, 1-218, 1-233, 1-235-236, 1-242, 1-246, 1-249, 1-254, 1-256, 1-264-266, 1-268, 1-70-272, 1-274-275, 1-277-278, 1-282-284, 1-287, 1-289, 1-297-298, 1-307-312, 1-331, 1-343-344, 1-346, 1-350, 1-368-370, 1-373, 1-379, 1-381, 1-384-385, 1-392-394, 1-397,
2-12-13, 2-15-20, 2-23-24, 2-31, 2-46-51, 2-55, 2-57, 2-59, 2-75, 2-85-86, 2-89-90, 2-96-97, 2-99, 2-103, 2-108, 2-155, 2-166, 2-174, 2-177-178, 2-180, 2-182, 2-184, 2-192, 2-215, 2-229, 2-233-234, 2-236, 2-239, 2-241-243, 2-246, 2-261, 2-293-296, 2-299, 2-301, 2-308, 2-315-318, 2-320, 2-322-325, 2-331, 2-337, 2-342-347, 2-349-350, 2-356-358, 2-360, 2-368, 2-394-396, 2-399, 2-407, 2-413, 2-419, 2-433, 2-441, 2-444, 2-449, 2-468, 2-470, 2-479-483
Ghost of Jupiter, 2-215
Gibbor, 1-259
Gilgamesh, 2-189
Gingerich, Owen, 2-96, 2-178
Gomeisa, 1-101
Goodricke, John, 1-134, 1-296
Great Cygnus Loop 2-151
Great Hercules Cluster, 2-196-198
Great Orion Nebula, 1-267-269
Great Rift, 2-9, 2-311, 2-353, 2-360, 2-365, 2-373, 2-375, 2-377, 2-475
Great Sagittarius Star Cloud, 2-314, 2-317, 2-322, 2-324, 2-328
Great Square of Pegasus, 1-11
Groombridge 8, 2-487
Groombridge 34, 1-11
Groombridge 1830, 2-388
Gum 1, 1-254
Gum 2, 1-93, 1-255
Gum 3, 1-93
Gum 4, 1-96
Gum 5, 1-95
Gum 6, 1-345
Gum 7, 1-346
Gum 9, 1-354
Gum 11, 1-359
Gum 56, 2-342

H

Haffner 4, 1-94
Haffner 6, 1-97
Haffner 8, 1-98
Haffner 10, 1-344
Haffner 13, 1-348
Haffner 15, 1-351
Haffner 16, 1-352-353
Haffner 18, 1-353-354
Haffner 19, 1-354
Haffner 20, 1-355
Haffner 21, 1-356
Haffner 23, 1-94
Hallas, Tony and Daphne, 2-93, 2-120, 2-222, 2-377, 2-393
Halley, Edmond, 2-41, 2-75, 2-196
Hamal, 1-39
Harris, William, 1-108, 1-117, 1-195-196, 1-201, 1-305, 1-324, 1-328, 1-372, 1-405
Harvard 1, 1-129-130
Harvard 2, 1-355
Harvard 12, 2-342
Harvard 14, 2-347
Harvard 15, 2-301
Harvard 16, 2-347
Harvard 17, 2-349
Harvard 18, 2-351
Harvard 19, 2-376
Harvard 20, 2-308-309
Harvard 21, 1-110
Harvard Observatory, 2-271
Hasenauer, David M., 1-405
Hdo 124, 2-205
Hdo 127, 2-227
Hdo 230, 2-267
Hdo 232, 2-267
Hdo 240, 2-267
Hdo 242, 2-267
Hdo 244, 2-267
Hdo 252, 2-267
Hdo 283, 2-313
Helix Galaxy, 2-390
Helix Nebula, 1-25, 1-34, 1-232
Helle, 1-39

Hera, 2-385
Heracles, 2-189
Hercules, 1-66, 1-77, 1-339, 2-56, 2-67, 2-111,
    2-125, 2-169, 2-189, 2-191-193, 2-195-197,
    2-199, 2-203, 2-225, 2-305, 2-328, 2-365,
    2-368-369, 2-426
Hercules Cluster, 2-189, 2-192-193, 2-196-197,
    2-199, 2-365, 2-368-369, 2-426
Herschel 84, 2-305
Herschel 599, 2-315
Herschel 840, 2-125
Herschel 975, 1-215
Herschel 2682, 2-418
Herschel 3506, 1-191
Herschel 3555, 1-191
Herschel 3750, 1-223
Herschel 3752, 1-223
Herschel 3759, 1-224
Herschel 3857, 1-171
Herschel 4038, 1-342
Herschel 4046, 1-342
Herschel 4166, 1-361
Herschel 4423, 2-67
Herschel 4690, 2-265
Herschel 4728, 2-265
Herschel 4753, 2-248
Herschel 4788, 2-268
Herschel 5003, 2-314
Herschel 5014, 2-105
Herschel 5224, 2-281
Herschel 5356, 1-333
Herschel, Caroline, 1-1, 1-111
Herschel Classifications, 1-1
Herschel, John, 1-1, 1-10, 1-267, 1-269, 1-273,
    1-331, 1-342
Herschel, William, 1-4, 1-10, 1-30, 1-105,
    1-223, 1-237, 1-243, 1-247, 1-273, 1-297,
    1-333, 2-103, 2-144, 2-151, 2-228, 2-259,
    2-309, 2-315-316, 2-376, 2-388, 2-417, 2-451
Herschel's Garnet Star, 1-131, 1-133, 1-138
Hertzprung, Ejnar, 1-xiv
Hesiod, 2-415
Hesperides, 2-169
Hevelius, Johannes, 1-1, 1-215, 1-229, 1-389,
    2-41, 2-249, 2-328, 2-353, 2-379,
    2-410, 2-475
Hind, John R., 1-221
Hind's Crimson Star, 1-221-222
Hind's Variable Nebula, 1-378, 1-381-382
Hipparchos, 1-1, 2-185
Hodierna, 2-348
Holmberg 342a, 2-44
Holmberg IX, 2-392
Homer, 2-385, 2-415
Horsehead Nebula, 1-263, 1-270-272
Houston, Walter Scott, 1-69
Hoverter, Dennis E., 1-178, 1-188, 1-398, 2-37,
    2-408
Howerton, Stan, 1-320, 1-321, 1-327, 1-398
Hubble, Edwin, 1-247
Hubble's Variable Nebula, 1-237, 1-247-248,
    2-108
Huygens, 1-267
Hyades, 1-82, 1-134, 1-259, 1-375, 1-378,
    1-382-383, 2-90
Hydra, 1-77, 2-67, 2-73, 2-117, 2-125, 2-203,
    2-205-206, 2-216-217, 2-379, 2-423, 2-465
Hydra I Galaxy Cluster, 2-216-217

I

IC 10, 2-487
IC 59, 1-118
IC 63, 1-118
IC 166, 1-124
IC 184, 1-163
IC 281, 1-303
IC 284, 1-306
IC 289, 1-129
IC 310, 1-305
IC 312, 1-305
IC 313, 1-305
IC 316, 1-305
IC 342, 1-68
IC 348, 1-293, 1-309

IC 351, 1-309, 1-313
IC 361, 1-69-70
IC 405, 1-49, 1-54
IC 410, 1-55-56
IC 417, 1-57-58
IC 418, 1-225-226
IC 423, 1-265
IC 424, 1-265
IC 426, 1-269
IC 430, 1-270
IC 431, 1-270
IC 432, 1-270-271
IC 434, 1-263, 1-270
IC 435, 1-271, 1-273
IC 443, 1-208-209
IC 444, 1-209
IC 446, 1-242-243
IC 467, 1-72
IC 468, 1-97
IC 520, 1-74
IC 694, 2-402
IC 749, 2-407-408
IC 750, 2-407-408
IC 764, 2-219-220
IC 787, 2-95
IC 792, 2-95
IC 796, 2-95
IC 797, 2-95
IC 798, 2-95
IC 800, 2-95
IC 803, 2-95
IC 879, 2-221
IC 982, 2-31
IC 983, 2-31
IC 1014, 2-34
IC 1029, 2-25, 2-35-36
IC 1070, 2-471
IC 1158, 2-370, 2-372
IC 1178, 2-193
IC 1181, 2-193
IC 1182, 2-193
IC 1183, 2-193
IC 1185, 2-193
IC 1201, 2-421
IC 1283, 2-324
IC 1287, 2-356
IC 1295, 2-362-363
IC 1297, 2-107, 2-109
IC 1311, 2-142-143
IC 1311, 2-142-143
IC 1318b, 2-144
IC 1367, 2-187
IC 1369, 2-133, 2-157
IC 1375, 2-187-188
IC 1377, 2-188
IC 1379, 2-188
IC 1396, 1-131, 1-138
IC 1434, 1-215, 1-217-218
IC 1442, 1-219
IC 1454, 1-144
IC 1470, 1-146
IC 1559, 1-15
IC 1590, 1-117
IC 1613, 1-159, 2-487
IC 1686, 1-327
IC 1727, 1-393-394
IC 1731, 1-395
IC 1738, 1-162
IC 1747, 1-124
IC 1784, 1-397
IC 1795, 1-125
IC 1805, 1-125-126
IC 1848, 1-127-128
IC 1871, 1-129
IC 1953, 1-183
IC 2003, 1-310, 1-313
IC 2006, 1-185
IC 2062, 1-70
IC 2118, 1-190
IC 2149, 1-57, 1-61
IC 2156, 1-207
IC 2157, 1-207
IC 2163, 1-89
IC 2165, 1-89
IC 2167, 1-243

IC 2169, 1-242-243
IC 2172, 1-250
IC 2174, 1-71
IC 2177, 1-93, 1-255-256
IC 2179, 1-72
IC 2209, 1-74
IC 2247, 1-242
IC 2389, 1-74
IC 2522, 2-3, 2-5
IC 2523, 2-3, 2-5
IC 2537, 2-5-6
IC 2627, 2-128
IC 2955, 2-248
IC 2965, 2-130
IC 2995, 2-219
IC 3042, 2-429
IC 3064, 2-429
IC 3098, 2-431
IC 3136, 2-431
IC 3228, 2-95
IC 3256, 2-436
IC 3259, 2-437
IC 3260, 2-436-437
IC 3267, 2-437
IC 3290, 2-70-71
IC 3292, 2-91, 2-95
IC 3327, 2-95
IC 3340, 2-95
IC 3365, 2-95
IC 3370, 2-71
IC 3381, 2-443
IC 3392, 2-95
IC 3432, 2-95
IC 3442, 2-95
IC 3453, 2-95
IC 3473, 2-95
IC 3476, 2-95
IC 3478, 2-95
IC 3484, 2-95
IC 3500, 2-95
IC 3505, 2-95
IC 3520, 2-95
IC 3522, 2-95
IC 3528, 2-95
IC 3530, 2-95
IC 3557, 2-95
IC 3567, 2-95
IC 3568, 1-75
IC 3583, 2-449
IC 3609, 2-95
IC 3611, 2-95
IC 3612, 2-95
IC 3615, 2-95
IC 3629, 2-95
IC 3998, 2-101
IC 4237, 2-465
IC 4296, 2-77
IC 4329, 2-78
IC 4329a, 2-78
IC 4351, 2-223
IC 4402, 2-268-269
IC 4404, 2-419
IC 4406, 2-265, 2-269
IC 4444, 2-268-269
IC 4593, 2-194
IC 4604, 2-287
IC 4628, 2-342
IC 4634, 2-294
IC 4665, 2-287, 2-303
IC 4703, 2-373
IC 4725, 2-311, 2-327
IC 4756, 2-373, 2-378
IC 4812, 2-108
IC 4846, 2-18
IC 4996, 2-133, 2-147
IC 4997, 2-310
IC 5011, 2-283
IC 5013, 2-283
IC 5039, 2-284
IC 5041, 2-284
IC 5067, 2-133, 2-152
IC 5068, 2-152
IC 5070, 2-152
IC 5078, 2-65

Index

IC 5105, 2-281, 2-284-285
IC 5122, 2-66
IC 5131, 1-333
IC 5135, 1-335
IC 5146, 2-133, 2-161
IC 5152, 2-487
IC 5156, 1-336
IC 5217, 1-220
IC 5269, 1-337-338
IC 5271, 1-333, 1-338
IC 5283, 1-286
IC 5332, 1-367-368
Ihle, Abraham, 2-328
Inanna, 2-423
Ink Spot, 2-317
Innes 445, 1-161
Intergalactic Wanderer, 1-232
IRAS, 2-271
Ishara, 2-333
Ishtar, 1-315, 1-375, 2-81, 2-225, 2-423
Ixion, 2-67

J

J107, 2-355
J139, 2-307
J2143, 2-375
Jacob 16, 2-125
Jakiel, Richard W, 1-33, 1-45, 1-83, 1-90,
  1-162, 1-163, 1-178-179, 1-186, 1-188,
  1-194, 1-197-199, 1-234, 1-313, 1-329,
  1-364, 1-393, 1-395-396, 1-404, 2-4, 2-6,
  2-39, 2-44-45, 2-54, 2-179, 2-212,
  2-252-255, 2-403, 2-408, 2-443, 2-455,
  2-461, 2-471
Jason, 1-39, 1-339, 2-67, 2-169, 2-287, 2-311
Jonckheere 6, 2-314
Jonckheere 18, 2-283
Jonckheere 320, 1-263-264
Jonckheere 900, 1-209
Jones 1, 1-290
Juno, 1-77, 2-67
Jupiter, 1-85, 1-140, 1-203, 1-263, 1-375, 2-25,
  2-67, 2-117, 2-169, 2-215, 2-305, 2-399

K

Kaerius, Petrus, 1-63, 1-237
Kantheros, 2-125
Kapteyn's Star, 2-388
Keenan, P.C., 1-xv
Kemble, L.J., 1-69
Kemble's Cascade, 1-63, 1-69
Kepler, Johannes, 1-63
King 4, 1-126
King 5, 1-306
King 6, 1-66
King 8, 1-60
King 10, 1-145
King 12, 1-110, 1-112, 1-119
King 14, 1-114
King 16, 1-116, 1-119
King 18, 1-145
King 19, 1-146
King 22, 1-55-56
King Charles I, 2-41
King Lyceon, 2-385
King Minos, 2-111
Kleinmann-Low Sources, 1-268
Kochab, 2-417
Koehler, J.G., 2-308, 2-452, 2-454
Kohoutek 1-16, 2-182, 2-184
Kreimer, Evered, 1-31, 1-282, 1-405, 2-220,
  2-319, 2-326, 2-328-329, 2-331, 2-338,
  2-369, 2-410
Krueger 60, 1-133-134
Kui 51, 2-205

L

La Superba, 2-41, 2-43
La 1, 1-224
Lacerta, 1-215-216, 1-218
Lagoon Nebula, 2-313-314, 2-317-320
Lalande 153, 1-87
Lalande 1123, 2-259
Lalande 21185, 2-388
Laplace, 2-53
Laplace, nebular hypothesis of, 2-53-55
Large Magellanic Cloud, 1-391
Lavoisier, Antoine Laurant, 1-191
LDN 11, 2-295
LDN 17, 2-295
LDN 42, 2-298
LDN 66, 2-299
LDN 93, 2-317
LDN 99, 2-297
LDN 100, 2-297
LDN 108, 2-322
LDN 173, 2-297
LDN 219, 2-302
LDN 323, 2-323
LDN 327, 2-323
LDN 379, 2-358
LDN 443, 2-358
LDN 497, 2-360
LDN 531, 2-15
LDN 532, 2-361
LDN 543, 2-16
LDN 567, 2-14
LDN 582, 2-13
LDN 619, 2-18
LDN 684, 2-19
LDN 1151, 1-141
LDN 1164, 1-141
LDN 1470, 1-309
LDN 1607, 1-243
LDN 1622, 1-275
LDN 1682, 2-344
LDN 1771, 2-319
LDN 1773, 2-287, 2-298
Le Gentil, 1-58-59
Leda, 2-133
Leo, 1-77, 2-225-232, 2-234-237, 2-239, 2-241-242,
  2-244, 2-246, 2-249-251, 2-379, 2-385, 2-388,
  2-423, 2-487
Leo I Galaxy Cloud, 2-235, 2-237, 2-241-242
Leo Minor, 1-215, 2-234, 2-249-250
Leonid meteors, 2-225
Lepus, 1-85, 1-171, 1-221-223
Li 259, 2-283
Libra, 2-257-259, 2-333, 2-423
Ling, Alister, 1-98, 1-119, 1-136, 1-144, 1-240,
  1-257, 1-302-303, 1-313, 1-405
Little Dipper, 2-169, 2-224, 2-415
Little Dumbbell Nebula, 1-297
Little Gem, 2-330
Little Ghost Nebula, 2-299, 2-301
Local Galaxy Group, 1-73, 1-159, 1-191, 1-326,
  1-365, 1-389, 1-391, 2-41, 2-331, 2-385,
  2-388, 2-423, 2-445, 2-487
Lord Rosse, 1-xxxvi, 1-30, 1-391, 2-53, 2-89
Local Supercluster, 2-41, 2-81
Lupus, 2-265-266, 2-333, 2-337, 2-388
Lux, Barbara, 1-405
Lynx, 1-215, 1-229-232, 2-249
Lyra, 2-9, 2-271-272, 2-274, 2-301, 2-330
Lyrid systems, 1-xxvi

M

Magellanic Clouds, 1-159, 1-232
Maia, 1-381
Malus, 1-361
Maraldi, 1-31, 1-282
Markarian 6, 1-126
Markarian 38, 2-323
Markarian 50, 1-147
Markarian 205, 2-176
Markarian Chain, 2-437
Marling, Dr. Jack, 1-83, 1-110, 1-302-303, 1-313,
  1-403, 2-14, 2-18, 2-111
Maroldi, 1-31
Mars, 1-85, 1-149, 1-154
MCG-01-03-85, 1-160
MCG-3-26-06, 2-214
MCG-7-26-57, 2-73
McHenry, Larry E., 1-405
Mechain, Pierre, 1-29, 1-121, 1-225, 1-297,
  1-331, 2-47, 2-51-52, 2-85, 2-87, 2-89, 2-91,
  2-178, 2-220, 2-235, 2-237, 2-241-242,
  2-291, 2-331, 2-337, 2-399, 2-406, 2-412,
  2-451
Medea, 2-125, 2-169, 2-385
Medusa, 1-151, 1-279, 1-293, 1-299, 1-389
Medusa Nebula, 1-213
Mekbuda, 1-203
Melotte 15, 1-105, 1-125-126
Melotte 20, 1-293, 1-307
Melotte 22, 1-378-379, 1-381
Melotte 25, 1-375, 1-382
Melotte 28, 1-383
Melotte 71, 1-347
Melotte 72, 1-257
Melotte 111, 2-90
Menkar, 1-154
Merope, 1-380-381 1-xxxviii
Merope Nebula, 1-xxxviii
Merrill 1-1, 2-21
Merrill 2-1, 2-264
Mesarthim, 1-39
Messier 1, 1-385-386
Messier 2, 1-31-32
Messier 3, 2-55-56
Messier 4, 2-333, 2-336, 2-338
Messier 5, 2-365, 2-368-369
Messier 6, 2-333, 2-348, 2-349, 2-351
Messier 7, 2-333, 2-350-351
Messier 8, 2-318, 2-320
Messier 9, 2-287, 2-290, 2-293, 2-296-299,
  2-301, 2-303
Messier 10, 2-287, 2-291-294, 2-298, 2-302-303
Messier 11, 2-343, 2-353, 2-361-362
Messier 12, 2-291-292
Messier 13, 2-56, 2-189, 2-196-200, 2-368-369
Messier 14, 2-302
Messier 15, 1-279, 1-282
Messier 16, 2-373, 2-377-378
Messier 17, 2-324-325
Messier 18, 2-314, 2-324-325
Messier 19, 2-287, 2-293-294
Messier 20, 2-311, 2-313-314, 2-317
Messier 21, 2-311, 2-314, 2-319
Messier 22, 2-311, 2-314, 2-328
Messier 23, 2-311, 2-314-316
Messier 24, 2-311, 2-314, 2-323-324
Messier 25, 2-311, 2-314, 2-327
Messier 26, 2-353, 2-360-361
Messier 27, 1-297, 1-382, 2-315-316, 2-475,
  2-482
Messier 28, 2-311, 2-314, 2-326
Messier 29, 2-133, 2-148
Messier 30, 2-61, 2-66
Messier 31, 1-11, 1-16-18, 1-326, 1-391, 2-53,
  2-193, 2-426, 2-487
Messier 32, 1-11, 1-16-18, 1-391, 2-487
Messier 33, 1-331, 1-389, 1-391-393, 2-487
Messier 34, 1-293, 1-295, 1-301-302
Messier 35, 1-203, 1-205, 1-207-208
Messier 36, 1-47, 1-59-61
Messier 37, 1-47, 1-58 60-61
Messier 38, 1-47, 1-57-61
Messier 39, 2-133
Messier 40, 2-385, 2-410
Messier 41, 1-85, 1-87, 1-91-92
Messier 42, 1-1, 1-244, 1-259, 1-262, 1-266-269,
  1-378, 2-325
Messier 43, 1-259, 1-267, 1-269
Messier 44, 1-1, 1-77, 1-79, 1-81-82
Messier 45, 1-1, 1-375, 1-378-379
Messier 46, 1-339, 1-344, 1-348-351, 1-353
Messier 47, 1-339, 1-341, 1-344, 1-346-349
Messier 48, 2-203, 2-206-207
Messier 49, 2-423, 2-444
Messier 50, 1-237, 1-253-254
Messier 51, 2-41, 2-53, 2-55
Messier 52, 1-105, 1-108-109, 2-55
Messier 53, 2-81, 2-103
Messier 54, 2-311, 2-314, 2-329
Messier 55, 2-311, 2-314, 2-330
Messier 56, 2-271, 2-278
Messier 57, 2-271, 2-277, 2-301, 2-330
Messier 58, 2-96, 2-423, 2-450
Messier 59, 2-423, 2-452, 2-454
Messier 60, 2-423, 2-452-454

Messier 61, 2-423, 2-433, 2-436
Messier 62, 2-287, 2-292-294
Messier 63, 2-41, 2-52
Messier 64, 2-81, 2-99-100
Messier 65, 2-225, 2-235, 2-241-243
Messier 66, 2-225, 2-235, 2-241-243
Messier 66 Galaxy Group, 2-235, 2-242
Messier 67, 1-77, 1-82-83
Messier 68, 2-203, 2-220
Messier 69, 2-311, 2-314, 2-327, 2-329
Messier 70, 2-311, 2-314, 2-327, 2-329
Messier 71, 2-305, 2-308-309
Messier 72, 1-25, 1-29-30
Messier 73, 1-25, 1-30
Messier 74, 1-315, 1-330-331
Messier 75, 2-311, 2-314, 2-331
Messier 76, 1-212, 1-293, 1-297-298
Messier 77, 1-151, 1-167-168
Messier 78, 1-147, 1-259, 1-274
Messier 79, 1-221, 1-225
Messier 80, 2-333, 2-337-338
Messier 81, 2-385, 2-388, 2-392-395
Messier 81 Galaxy Group, 2-392, 2-395
Messier 82, 2-385, 2-388, 2-392-394
Messier 83, 2-73, 2-203, 2-222-223, 2-465
Messier 84, 2-95, 2-423, 2-437, 2-440-441, 2-449
Messier 85, 2-81, 2-90-91, 2-449
Messier 86, 2-95, 2-423, 2-437, 2-440-443, 2-449
Messier 87, 2-423, 2-444-445, 2-449, 2-451
Messier 88, 2-81, 2-95-96, 2-437, 2-449
Messier 89, 2-423, 2-448-449, 2-455
Messier 90, 2-423, 2-449
Messier 91, 2-96, 2-98
Messier 92, 2-189, 2-199-200
Messier 93, 1-339, 1-350-351
Messier 94, 2-41, 2-51, 2-99
Messier 95, 2-225, 2-234-236
Messier 96, 2-225, 2-235-237
Messier 96 Galaxy Group, 2-236-237
Messier 97, 2-399
Messier 98, 2-81, 2-85, 2-89
Messier 99, 2-81, 2-86-87, 2-89
Messier 100, 2-81, 2-88-89
Messier 101, 2-41, 2-178, 2-385, 2-388, 2-412, 2-414
Messier 101 Galaxy Group, 2-412, 2-414
Messier 102, 2-169, 2-177-178
Messier 103, 1-121
Messier 104, 2-423, 2-451
Messier 105, 2-225, 2-235-237
Messier 106, 2-41, 2-47
Messier 107, 2-291
Messier 108, 2-385, 2-399
Messier 109, 2-406-407
Messier 110, 1-11, 1-18, 2-487
Messier, Charles, 1-1, 1-111, 1-254, 1-301, 1-391, 2-55, 2-66, 2-230, 2-291, 2-293, 2-297, 2-325-326, 2-337, 2-437, 2-444, 2-448, 2-450
Microscopium, 2-281-283
Minerva, 2-169
Minkowski 1-4, 1-309
Minkowski 1-73, 2-22
Minkowski 1-74, 2-22
Minkowski 1-79, 2-160
Minkowski 1-80, 1-146
Minkowski 2-2, 1-69
Minkowski 2-51, 1-142
Minkowski 2-55, 1-147
Minkowski 3-1, 1-93
Minkowski 3-3, 1-257
Minkowski 3-6, 1-364
Minkowski 92, 2-137, 2-139
Minos, 1-375
Mintaka, 1-259, 1-262, 1-265
Mir 349, 2-417
Mira, 1-11, 1-151-154, 1-315, 2-225
Miram, 1-296
Mirtle, J. C., 1-81, 1-120, 1-405
Mizar, 2-169, 2-385, 2-387, 2-389
Mlbo 4, 2-335
Mlbo 6, 2-283
Monoceros, 1-93, 1-237-238, 1-240, 1-242, 1-252, 1-254, 1-257, 2-108
Morgan, W., 1-xv

Morphological Galaxy Classification System, 1-xli-xlvi
Mt. Palomar, 1-247
Musca Borealis, 1-389

N

Negy, Jr., Bennie, 1-405
Neptune, 1-259
Network Nebula, 2-150-153
New 1, 1-160
NGC 1, 1-292
NGC 2, 1-292
NGC 16, 1-292
NGC 23, 1-292
NGC 24, 1-365, 1-369
NGC 26, 1-292
NGC 40, 1-131, 1-149
NGC 45, 1-155
NGC 50, 1-328
NGC 55, 1-365, 1-369
NGC 80, 1-14-15
NGC 83, 1-14-15
NGC 91, 1-14-15
NGC 103, 1-112-113
NGC 125, 1-322
NGC 127, 1-321-322
NGC 128, 1-315, 1-322
NGC 129, 1-105, 1-112-113
NGC 130, 1-322
NGC 131, 1-370-371
NGC 133, 1-113
NGC 134, 1-365, 1-370-371
NGC 136, 1-113
NGC 139, 1-191
NGC 145, 1-155
NGC 146, 1-113-114
NGC 147, 1-113, 1-115, 2-487
NGC 148, 1-371
NGC 150, 1-371
NGC 153, 1-175
NGC 157, 1-156
NGC 160, 1-15
NGC 169, 1-15
NGC 175, 1-156
NGC 178, 1-156
NGC 181, 1-15
NGC 182, 1-321-323
NGC 183, 1-15
NGC 184, 1-15
NGC 185, 1-113-115, 2-487
NGC 188, 1-131, 1-149
NGC 189, 1-115
NGC 193, 1-321, 1-323
NGC 194, 1-322-323
NGC 198, 1-323
NGC 199, 1-323
NGC 200, 1-323
NGC 205, 1-11, 1-17-18, 2-487
NGC 206, 1-19
NGC 210, 1-156-157
NGC 214, 1-15, 1-17
NGC 221, 1-11, 2-487
NGC 224, 1-11, 1-17-19, 2-487
NGC 225, 1-105, 1-115-116
NGC 227, 1-157
NGC 233, 1-19
NGC 237, 1-157
NGC 245, 1-157
NGC 246, 1-151, 1-157-158
NGC 247, 1-151, 1-158
NGC 252, 1-19
NGC 253, 1-365, 1-370-372
NGC 254, 1-372
NGC 255, 1-156, 1-158
NGC 257, 1-323
NGC 266, 1-324
NGC 268, 1-158
NGC 274, 1-159
NGC 275, 1-159
NGC 278, 1-105, 1-117
NGC 281, 1-117, 1-361
NGC 288, 1-365, 1-372
NGC 289, 1-372
NGC 292, 2-487

NGC 300, 1-365, 1-372-373
NGC 309, 1-159
NGC 311, 1-322, 1-324
NGC 315, 1-324
NGC 318, 1-324
NGC 326, 1-324
NGC 337, 1-159
NGC 355, 1-159
NGC 357, 1-159
NGC 358, 1-118
NGC 366, 1-119
NGC 374, 1-324
NGC 379, 1-325
NGC 380, 1-325
NGC 381, 1-119
NGC 382, 1-325
NGC 383, 1-324-325
NGC 384, 1-325
NGC 385, 1-325
NGC 392, 1-325
NGC 394, 1-325
NGC 397, 1-325
NGC 398, 1-325
NGC 399, 1-325
NGC 403, 1-325
NGC 404, 1-19
NGC 407, 1-325
NGC 408, 1-326
NGC 410, 1-325-326
NGC 414, 1-326
NGC 428, 1-160
NGC 433, 1-119-120
NGC 436, 1-120
NGC 439, 1-372
NGC 441, 1-372
NGC 450, 1-160
NGC 457, 1-105, 1-120-121
NGC 467, 1-326-327
NGC 470, 1-326-327
NGC 473, 1-326-327
NGC 474, 1-326-327
NGC 483, 1-327
NGC 488, 1-315, 1-327
NGC 490, 1-327
NGC 491, 1-373
NGC 494, 1-327
NGC 495, 1-327
NGC 496, 1-328
NGC 499, 1-327-328
NGC 501, 1-328
NGC 503, 1-328
NGC 504, 1-328
NGC 507, 1-328
NGC 507 Galaxy Group, 1-328
NGC 508, 1-328-329
NGC 514, 1-329
NGC 515, 1-329
NGC 517, 1-329, 1-341
NGC 518, 1-329
NGC 520, 1-329
NGC 521, 1-160
NGC 524, 1-330
NGC 528, 1-330
NGC 532, 1-330
NGC 533, 1-160
NGC 541, 1-160
NGC 545, 1-160
NGC 547, 1-160
NGC 559, 1-120-121
NGC 568, 2-25
NGC 577, 2-471
NGC 578, 1-160-161
NGC 579, 1-389, 1-393
NGC 581, 1-105, 1-121
NGC 582, 1-389, 1-393
NGC 584, 1-161
NGC 586, 1-161
NGC 596, 1-161
NGC 598, 1-389, 1-392, 2-265, 2-487
NGC 600, 1-161
NGC 604, 1-389, 1-393
NGC 609, 1-122
NGC 613, 1-365, 1-371, 1-373
NGC 615, 1-162
NGC 628, 1-315, 1-330-332

NGC 636, 1-162
NGC 637, 1-122
NGC 650, 1-297
NGC 651, 1-297
NGC 654, 1-105, 1-123, 2-169
NGC 658, 1-332
NGC 659, 1-123-124
NGC 660, 1-331-332, 2-373
NGC 661, 1-393
NGC 663, 1-105, 1-123-124
NGC 664, 1-332
NGC 665, 2-311
NGC 669, 1-389, 1-393
NGC 670, 1-389, 1-394
NGC 672, 1-389, 1-393-394
NGC 673, 1-41
NGC 676, 1-332
NGC 678, 1-41-42
NGC 679, 1-18-19,
NGC 680, 1-41-42
NGC 681, 1-162
NGC 684, 1-389, 1-395
NGC 687, 1-19
NGC 691, 1-42
NGC 693, 1-332, 2-163
NGC 694, 1-42
NGC 695, 1-42
NGC 697, 1-41-42
NGC 701, 1-162
NGC 703, 1-20
NGC 704, 1-20
NGC 705, 1-20
NGC 706, 1-332
NGC 708, 1-18, 1-20
NGC 710, 1-20
NGC 718, 1-332
NGC 720, 1-162
NGC 735, 1-395
NGC 736, 1-395
NGC 737, 1-395
NGC 740, 1-395
NGC 741, 1-329, 1-332
NGC 742, 1-329
NGC 743, 1-124
NGC 744, 1-293, 1-298
NGC 750, 1-395
NGC 751, 1-395
NGC 752, 1-11, 1-20
NGC 753, 1-18, 1-20
NGC 759, 1-21
NGC 761, 1-395-396
NGC 770, 1-43
NGC 772, 1-39, 1-43
NGC 777, 1-396
NGC 778, 1-396
NGC 779, 1-162-163
NGC 781, 1-279
NGC 783, 1-396
NGC 784, 1-389, 1-396
NGC 785, 1-396
NGC 788, 1-163
NGC 797, 1-21
NGC 801, 1-21
NGC 803, 1-43-44
NGC 818, 1-21
NGC 821, 1-43-44
NGC 828, 1-21
NGC 833, 1-162-163
NGC 834, 1-22
NGC 835, 1-163
NGC 838, 1-164
NGC 839, 1-164
NGC 845, 1-22
NGC 864, 1-164
NGC 869, 1-293, 1-299-300
NGC 870, 1-44
NGC 871, 1-43-44
NGC 876, 1-44
NGC 877, 1-39, 1-43-44
NGC 884, 1-293, 1-299-300
NGC 886, 1-125
NGC 890, 1-397-398
NGC 891, 1-11, 1-22
NGC 895, 1-164
NGC 896, 1-125

NGC 906, 1-23
NGC 908, 1-163-164
NGC 909, 1-23
NGC 910, 1-21, 1-23
NGC 911, 1-21, 1-23
NGC 912, 1-23
NGC 913, 1-23
NGC 914, 1-23
NGC 918, 1-44-45
NGC 922, 1-191
NGC 925, 1-389, 1-397
NGC 927, 1-44
NGC 936, 1-163-164
NGC 941, 1-163-164
NGC 945, 1-165
NGC 947, 1-164
NGC 949, 1-389, 1-397-398
NGC 955, 1-165
NGC 957, 1-293, 1-300-301
NGC 958, 1-165
NGC 959, 1-398
NGC 969, 1-398
NGC 972, 1-39, 1-45
NGC 974, 1-398
NGC 978, 1-398
NGC 978a-b, 1-398
NGC 986, 1-191, 1-194
NGC 991, 1-165
NGC 1003, 1-301
NGC 1012, 1-45
NGC 1015, 1-165
NGC 1016, 1-165
NGC 1022, 1-165
NGC 1023, 1-293, 1-300-301
NGC 1024, 1-45
NGC 1027, 1-105, 1-127
NGC 1028, 1-45
NGC 1029, 1-45
NGC 1032, 1-166
NGC 1035, 1-166-167
NGC 1036, 1-45
NGC 1039, 1-293, 1-301
NGC 1042, 1-163, 1-166-167
NGC 1048a, 1-166-167
NGC 1048b, 1-166-167
NGC 1049, 1-191, 1-193-194
NGC 1052, 1-166-167
NGC 1055, 1-167-168
NGC 1058, 1-302
NGC 1068, 1-151, 1-167
NGC 1069, 1-168
NGC 1073, 1-168-169
NGC 1079, 1-193
NGC 1084, 1-178
NGC 1087, 1-168-169
NGC 1090, 1-168-169
NGC 1094, 1-169
NGC 1097, 1-191, 1-193-194
NGC 1097a, 1-193-194
NGC 1134, 1-46
NGC 1140, 1-178
NGC 1156, 1-46
NGC 1160, 1-302
NGC 1161, 1-302
NGC 1167, 1-302
NGC 1169, 1-302
NGC 1171, 1-303
NGC 1172, 1-178
NGC 1175, 1-303
NGC 1177, 1-303
NGC 1179, 1-178
NGC 1184, 1-150
NGC 1186, 1-303
NGC 1187, 1-178-179
NGC 1193, 1-303
NGC 1199, 1-179
NGC 1201, 1-194
NGC 1209, 1-179
NGC 1220, 1-306
NGC 1232, 1-175, 1-178-180
NGC 1241, 1-180
NGC 1242, 1-180
NGC 1245, 1-293, 1-306
NGC 1250, 1-305
NGC 1255, 1-194, 1-197

NGC 1257, 1-305
NGC 1259, 1-305
NGC 1260, 1-305
NGC 1264, 1-305
NGC 1265, 1-305
NGC 1267, 1-305
NGC 1268, 1-305
NGC 1270, 1-305
NGC 1271, 1-305
NGC 1272, 1-305
NGC 1273, 1-305
NGC 1274, 1-305
NGC 1275, 1-305, 1-307
NGC 1276, 1-305
NGC 1277, 1-305
NGC 1278, 1-305
NGC 1279, 1-305
NGC 1281, 1-305
NGC 1282, 1-305
NGC 1283, 1-305
NGC 1288, 1-194
NGC 1291, 1-175, 1-179-180
NGC 1292, 1-194
NGC 1293, 1-305
NGC 1294, 1-305
NGC 1297, 1-179-181
NGC 1300, 1-175, 1-179-181
NGC 1301, 1-181
NGC 1309, 1-181-182
NGC 1310, 1-197
NGC 1316, 1-191, 1-195-196
NGC 1317, 1-195
NGC 1318, 1-195
NGC 1319, 1-182
NGC 1325, 1-182
NGC 1326, 1-196
NGC 1331, 1-182
NGC 1332, 1-175, 1-182
NGC 1333, 1-308
NGC 1337, 1-182
NGC 1339, 1-196
NGC 1341, 1-196
NGC 1342, 1-293, 1-306, 1-308
NGC 1348, 1-308, 1-313
NGC 1350, 1-197
NGC 1351, 1-197
NGC 1353, 1-182
NGC 1357, 1-182-183
NGC 1358, 1-183
NGC 1359, 1-183
NGC 1360, 1-191, 1-197
NGC 1365, 1-191, 1-198
NGC 1366, 1-198
NGC 1371, 1-198
NGC 1373, 1-198
NGC 1374, 1-199
NGC 1375, 1-199
NGC 1379, 1-199
NGC 1380, 1-191, 1-199
NGC 1381, 1-200
NGC 1382, 1-200
NGC 1385, 1-200
NGC 1386, 1-200
NGC 1387, 1-200
NGC 1389, 1-200
NGC 1395, 1-183, 1-185
NGC 1398, 1-202
NGC 1399, 1-200
NGC 1400, 1-183-185
NGC 1404, 1-202
NGC 1406, 1-191, 1-199, 1-202
NGC 1407, 1-184-185
NGC 1415, 1-202
NGC 1417, 1-184
NGC 1421, 1-175, 1-184-185
NGC 1425, 1-191, 1-199, 1-202
NGC 1426, 1-184
NGC 1427, 1-202
NGC 1435, 1-378-381
NGC 1437, 1-202
NGC 1439, 1-184
NGC 1440, 1-184-185
NGC 1441, 1-185
NGC 1444, 1-293, 1-295, 1-309
NGC 1449, 1-185

NGC 1451, 1-185
NGC 1452, 1-185
NGC 1453, 1-185
NGC 1461, 1-185
NGC 1465, 1-310
NGC 1487, 1-185
NGC 1491, 1-311
NGC 1496, 1-311
NGC 1499, 1-310
NGC 1501, 1-63, 1-68
NGC 1502, 1-63, 1-69
NGC 1507, 1-185-186
NGC 1513, 1-311
NGC 1514, 1-375, 1-381-382
NGC 1518, 1-186
NGC 1521, 1-186
NGC 1528, 1-293, 1-312
NGC 1530, 1-70
NGC 1531, 1-186
NGC 1532, 1-175, 1-186
NGC 1535, 1-175, 1-187
NGC 1537, 1-187
NGC 1544, 1-150
NGC 1545, 1-293, 1-312-313
NGC 1548, 1-313
NGC 1554, 1-382
NGC 1555, 1-382
NGC 1560, 1-70
NGC 1569, 1-70
NGC 1579, 1-313-314
NGC 1582, 1-293, 1-313-314
NGC 1600, 1-187
NGC 1601, 1-187-188
NGC 1603, 1-187
NGC 1605, 1-314
NGC 1606, 1-187
NGC 1618, 1-187-188
NGC 1622, 1-187-188
NGC 1624, 1-314
NGC 1625, 1-188
NGC 1637, 1-188-189
NGC 1638, 1-188-189
NGC 1647, 1-375, 1-382
NGC 1659, 1-189
NGC 1662, 1-259, 1-263
NGC 1663, 1-263
NGC 1664, 1-51, 1-53
NGC 1666, 1-189
NGC 1667, 1-189
NGC 1699, 1-189
NGC 1700, 1-189
NGC 1720, 1-190
NGC 1723, 1-190
NGC 1725, 1-190
NGC 1726, 1-190
NGC 1728, 1-190
NGC 1744, 1-224
NGC 1746, 1-375, 1-383
NGC 1750, 1-383
NGC 1758, 1-383
NGC 1778, 1-53
NGC 1784, 1-224-225
NGC 1788, 1-264
NGC 1792, 1-171, 1-173
NGC 1798, 1-53
NGC 1800, 1-171, 1-173
NGC 1802, 1-384
NGC 1807, 1-375, 1-384
NGC 1808, 1-171, 1-173
NGC 1817, 1-375, 1-384-385
NGC 1832, 1-224-225
NGC 1851, 1-171, 1-173-174
NGC 1857, 1-54
NGC 1883, 1-56-57
NGC 1888, 1-225
NGC 1889, 1-225
NGC 1893, 1-47, 1-55-56
NGC 1904, 1-221, 1-225
NGC 1907, 1-57
NGC 1912, 1-47, 1-58
NGC 1931, 1-59-60
NGC 1952, 1-375
NGC 1960, 1-47, 1-59
NGC 1961, 1-71
NGC 1964, 1-221, 1-226

NGC 1973, 1-266
NGC 1975, 1-266
NGC 1976, 1-259, 1-267-269
NGC 1977, 1-266
NGC 1980, 1-259, 1-267
NGC 1981, 1-259, 1-266-267
NGC 1982, 1-259
NGC 1985, 1-60
NGC 1990, 1-269
NGC 1996, 1-397
NGC 1999, 1-269-270
NGC 2000, 2-419
NGC 2017, 1-221, 1-223, 1-226
NGC 2022, 1-272-273
NGC 2023, 1-271-272
NGC 2024, 1-259, 1-270, 2-103
NGC 2039, 1-273
NGC 2063, 1-273
NGC 2064, 1-274
NGC 2067, 1-274
NGC 2068, 1-259, 1-274
NGC 2071, 1-274
NGC 2090, 1-174
NGC 2099, 1-47, 1-61
NGC 2112, 1-259, 1-275
NGC 2126, 1-61-62
NGC 2129, 1-203, 1-206-207
NGC 2139, 1-227
NGC 2141, 1-273, 1-276
NGC 2143, 1-276
NGC 2146, 1-71
NGC 2149, 1-237, 1-240
NGC 2158, 1-203, 1-207-208
NGC 2168, 1-203, 1-207-208
NGC 2169, 1-259, 1-273, 1-276
NGC 2170, 1-239
NGC 2174, 1-277
NGC 2175, 1-259, 1-277
NGC 2179, 1-227
NGC 2180, 1-276-277
NGC 2182, 1-239
NGC 2183, 1-240
NGC 2185, 1-240
NGC 2186, 1-276, 1-278
NGC 2188, 1-174
NGC 2192, 1-62
NGC 2194, 1-259, 1-278
NGC 2196, 1-227
NGC 2202, 1-276, 1-278
NGC 2204, 1-88
NGC 2207, 1-89
NGC 2215, 1-237, 1-240
NGC 2217, 1-89
NGC 2219, 1-241
NGC 2223, 1-89
NGC 2225, 1-241
NGC 2226, 1-241
NGC 2232, 1-237, 1-241-242
NGC 2234, 1-210
NGC 2236, 1-237, 1-241-242
NGC 2237, 1-237
NGC 2239, 1-243
NGC 2243, 1-90
NGC 2244, 1-237, 1-243-245
NGC 2245, 1-242, 1-244
NGC 2247, 1-244
NGC 2250, 1-244
NGC 2251, 1-237, 1-241, 1-245
NGC 2252, 1-237, 1-245
NGC 2254, 1-245
NGC 2259, 1-247
NGC 2260, 1-245-246
NGC 2261, 1-237, 1-247-248
NGC 2262, 1-246-247, 1-250
NGC 2264, 1-248-249
NGC 2265, 1-210
NGC 2266, 1-203, 1-210
NGC 2268, 1-72
NGC 2269, 1-250
NGC 2270, 1-250
NGC 2276, 1-150
NGC 2280, 1-90-91
NGC 2281, 1-62
NGC 2282, 1-250
NGC 2283, 1-91

NGC 2286, 1-237, 1-250-251
NGC 2287, 1-85, 1-91
NGC 2292, 1-92
NGC 2293, 1-92
NGC 2295, 1-92
NGC 2298, 1-342-343
NGC 2299, 1-251
NGC 2300, 1-150
NGC 2301, 1-237, 1-251
NGC 2302, 1-237, 1-251-252
NGC 2304, 1-211
NGC 2306, 1-252
NGC 2309, 1-251-252
NGC 2310, 1-343
NGC 2311, 1-251, 1-253
NGC 2312, 1-253
NGC 2314, 1-71
NGC 2316, 1-253
NGC 2319, 1-252-253
NGC 2323, 1-237, 1-254
NGC 2324, 1-237, 1-252, 1-254
NGC 2325, 1-90, 1-93
NGC 2327, 1-93
NGC 2331, 1-203, 1-211
NGC 2335, 193, 1-252, 1-255
NGC 2336, 1-71-72
NGC 2339, 1-211
NGC 2341, 1-211
NGC 2342, 1-211
NGC 2343, 1-237, 1-255-256
NGC 2345, 1-85, 1-94
NGC 2347, 1-72
NGC 2350, 1-101, 1-104
NGC 2351, 1-253, 1-256
NGC 2353, 1-237, 1-253, 1-256
NGC 2354, 1-94-95
NGC 2355, 1-203, 1-212
NGC 2359, 1-95-96
NGC 2360, 1-85, 1-95-96
NGC 2362, 1-85-86, 1-88, 1-95, 1-97
NGC 2363, 1-72
NGC 2364, 1-253, 1-257
NGC 2366, 1-72
NGC 2367, 1-85, 1-97
NGC 2368, 1-257
NGC 2371, 1-211
NGC 2374, 1-85, 1-98
NGC 2379, 1-212
NGC 2383, 1-98-99
NGC 2384, 1-85, 1-98-99
NGC 2385, 1-213
NGC 2388, 1-213
NGC 2389, 1-211, 1-213
NGC 2392, 1-203, 1-214
NGC 2394, 1-101, 1-103-104
NGC 2395, 1-212-214
NGC 2396, 1-343
NGC 2401, 1-342, 1-344
NGC 2403, 1-63, 1-73
NGC 2414, 1-345-346
NGC 2415, 1-231
NGC 2419, 1-232
NGC 2420, 1-214
NGC 2421, 1-345-346
NGC 2422, 1-339, 1-347
NGC 2423, 1-339, 1-345-347
NGC 2425, 1-346, 1-348
NGC 2432, 1-348
NGC 2437, 1-339, 1-349
NGC 2438, 1-339, 1-349-350
NGC 2439, 1-339, 1-348, 1-350
NGC 2440, 1-339, 1-350
NGC 2441, 1-73
NGC 2444, 1-232
NGC 2445, 1-232
NGC 2447, 1-339, 1-350
NGC 2451, 1-339, 1-351
NGC 2452, 1-339, 1-352
NGC 2453, 1-352
NGC 2455, 1-352
NGC 2460, 1-74
NGC 2467, 1-339, 1-354
NGC 2470, 1-101, 1-104
NGC 2474, 1-233
NGC 2475, 1-233

Index

NGC 2477, 1-339, 1-353
NGC 2479, 1-354-355
NGC 2482, 1-339, 1-354-355
NGC 2483, 1-339, 1-355
NGC 2486, 1-214
NGC 2487, 1-214
NGC 2489, 1-355
NGC 2491, 1-104
NGC 2496, 1-104
NGC 2500, 1-233
NGC 2506, 1-237, 1-257-258
NGC 2508, 1-104
NGC 2509, 1-355-356
NGC 2523, 1-75, 1-357
NGC 2523b, 1-74-75
NGC 2525, 1-357-358
NGC 2527, 1-339, 1-357
NGC 2533, 1-357
NGC 2535, 1-79-80
NGC 2536, 1-79-80
NGC 2537, 1-233-234
NGC 2538, 1-104
NGC 2539, 1-339, 1-357-358
NGC 2541, 1-233-234
NGC 2544, 1-74
NGC 2545, 1-80
NGC 2546, 1-339, 1-358
NGC 2548, 2-203, 2-206-207
NGC 2549, 1-233
NGC 2551, 1-74
NGC 2552, 1-234
NGC 2554, 1-80
NGC 2562, 1-80
NGC 2563, 1-80
NGC 2567, 1-339, 1-359
NGC 2568, 1-358
NGC 2571, 1-359
NGC 2577, 1-81
NGC 2579, 1-339, 1-359-360
NGC 2580, 1-359
NGC 2587, 1-360
NGC 2588, 1-359-360
NGC 2598, 1-81
NGC 2599, 1-81
NGC 2608, 1-81
NGC 2610, 2-206-207
NGC 2613, 1-361, 1-364
NGC 2619, 1-81
NGC 2623, 1-81
NGC 2627, 1-361, 1-363-364
NGC 2632, 1-77, 1-81
NGC 2633, 1-74
NGC 2634, 1-74
NGC 2635, 1-363
NGC 2639, 2-389
NGC 2642, 2-206-207
NGC 2646, 1-74
NGC 2654, 2-388-389
NGC 2655, 1-74-75
NGC 2658, 1-364
NGC 2672, 1-82-83
NGC 2673, 1-83
NGC 2681, 2-388, 2-390
NGC 2682, 1-82-83
NGC 2683, 1-229, 1-234-235
NGC 2685, 2-388, 2-390
NGC 2693, 2-390
NGC 2701, 2-389-390
NGC 2712, 1-234
NGC 2713, 2-206, 2-208
NGC 2715, 1-75
NGC 2716, 2-208
NGC 2732, 1-75
NGC 2742, 2-389, 2-391
NGC 2744, 1-84
NGC 2747, 1-83-84
NGC 2748, 1-75
NGC 2749, 1-84
NGC 2751, 1-84
NGC 2752, 1-84
NGC 2763, 2-208
NGC 2764, 1-84
NGC 2768, 2-389, 2-391
NGC 2770, 1-234
NGC 2770b, 1-234

NGC 2775, 1-83-84
NGC 2776, 1-235
NGC 2777, 1-84
NGC 2781, 2-208
NGC 2782, 1-235
NGC 2784, 2-208
NGC 2787, 2-390-391
NGC 2793, 1-235
NGC 2798, 1-235
NGC 2799, 1-235
NGC 2811, 2-209
NGC 2815, 2-209
NGC 2818, 1-361, 1-364
NGC 2818a, 1-361, 1-364
NGC 2825, 1-236
NGC 2826, 2-133
NGC 2830, 1-236
NGC 2831, 1-236
NGC 2832, 1-236
NGC 2834, 1-236
NGC 2835, 2-209
NGC 2841, 2-385, 2-390-391
NGC 2844, 1-236
NGC 2848, 2-210
NGC 2851, 2-210
NGC 2855, 2-210
NGC 2859, 2-249
NGC 2865, 2-210
NGC 2880, 2-392
NGC 2884, 2-210
NGC 2888, 1-364
NGC 2889, 2-210
NGC 2902, 2-210
NGC 2903, 2-225, 2-229-230
NGC 2907, 2-210
NGC 2911, 2-228, 2-230
NGC 2914, 2-230
NGC 2916, 2-228, 2-230
NGC 2919, 2-230
NGC 2924, 2-211
NGC 2935, 2-211
NGC 2942, 2-249
NGC 2950, 2-390, 2-392
NGC 2955, 2-251
NGC 2962, 2-212
NGC 2964, 2-230
NGC 2967, 2-379, 2-382
NGC 2968, 2-230-231
NGC 2970, 2-231
NGC 2974, 2-379, 2-382
NGC 2976, 2-385, 2-391-392
NGC 2977, 2-172
NGC 2978, 2-381
NGC 2980, 2-381
NGC 2983, 2-211-212
NGC 2985, 2-391-392
NGC 2986, 2-212
NGC 2989, 2-212
NGC 2990, 2-381
NGC 2992, 2-212-213
NGC 2993, 2-212-213
NGC 2997, 2-1, 2-4
NGC 2998, 2-391-392
NGC 3001, 2-1, 2-3-4
NGC 3002, 2-249
NGC 3003, 2-249, 2-251-252
NGC 3005, 2-392
NGC 3006, 2-392
NGC 3008, 2-392
NGC 3016, 2-231
NGC 3018, 2-381
NGC 3019, 2-231
NGC 3020, 2-231
NGC 3021, 2-251
NGC 3023, 2-381
NGC 3024, 2-251
NGC 3027, 2-392
NGC 3031, 2-385, 2-392-393
NGC 3032, 2-231
NGC 3034, 2-385, 2-393-394
NGC 3038, 2-3
NGC 3041, 2-228, 2-231
NGC 3043, 2-394
NGC 3044, 2-382
NGC 3052, 2-212-213

NGC 3054, 2-213
NGC 3055, 2-382-383
NGC 3056, 2-3, 2-5
NGC 3065, 2-394
NGC 3066, 2-394-395
NGC 3067, 2-231
NGC 3073, 2-395
NGC 3077, 2-395, 2-397
NGC 3078, 2-213
NGC 3079, 2-395
NGC 3081, 2-213
NGC 3087, 2-3
NGC 3089, 2-3-4
NGC 3091, 2-213-214
NGC 3095, 2-4-5
NGC 3096, 2-214
NGC 3098, 2-231
NGC 3100, 2-4-5
NGC 3108, 2-4
NGC 3109, 2-214-215
NGC 3115, 2-379, 2-382-383
NGC 3124, 2-214
NGC 3125, 2-5
NGC 3145, 2-214
NGC 3147, 2-172-173
NGC 3150, 2-252
NGC 3151, 2-252
NGC 3152, 2-252
NGC 3156, 2-383
NGC 3158, 2-252
NGC 3159, 2-252
NGC 3160, 2-252
NGC 3161, 2-252
NGC 3162, 2-231-232
NGC 3163, 2-252
NGC 3165, 2-383
NGC 3166, 2-383-384
NGC 3169, 2-383-384
NGC 3175, 2-1, 2-5-6
NGC 3177, 2-232
NGC 3183, 2-172-173
NGC 3184, 2-385, 2-395
NGC 3185, 2-232-233
NGC 3187, 2-232-233
NGC 3190, 2-232-233
NGC 3193, 2-232-233
NGC 3198, 2-396
NGC 3200, 2-214-215
NGC 3203, 2-214-215
NGC 3221, 2-233
NGC 3222, 2-233
NGC 3223, 2-5-6
NGC 3226, 2-233-234
NGC 3227, 2-233-234
NGC 3241, 2-6
NGC 3242, 2-203, 2-215
NGC 3244, 2-6-7
NGC 3245, 2-252
NGC 3246, 2-384
NGC 3250, 2-6-7
NGC 3254, 2-252-253
NGC 3257, 2-6
NGC 3258, 2-6
NGC 3258b, 2-7
NGC 3259, 2-396
NGC 3260, 2-7
NGC 3267, 2-7
NGC 3268, 2-7
NGC 3269, 2-7
7NGC 3271, 2-7
NGC 3273, 2-7
NGC 3274, 2-234
NGC 3275, 2-8
NGC 3277, 2-253
NGC 3281, 2-8
NGC 3285, 2-215
NGC 3287, 2-234-235
NGC 3294, 2-253
NGC 3300, 2-234
NGC 3301, 2-234
NGC 3305, 2-216
NGC 3307, 2-216
NGC 3308, 2-216
NGC 3309, 2-216
NGC 3310, 2-396-397

NGC 3311, 2-216
NGC 3312, 2-217
NGC 3314, 2-217
NGC 3316, 2-217
NGC 3319, 2-396-397
NGC 3320, 2-397
NGC 3329, 2-173
NGC 3338, 2-232, 2-235
NGC 3344, 2-249, 2-253
NGC 3346, 2-232, 2-235
NGC 3347, 2-8
NGC 3348, 2-397
NGC 3351, 2-225, 2-234-235
NGC 3353, 2-397
NGC 3354, 2-8
NGC 3358, 2-8
NGC 3359, 2-397
NGC 3367, 2-235
NGC 3368, 2-225, 2-235
NGC 3370, 2-236
NGC 3377, 2-236, 2-238
NGC 3379, 2-225, 2-237
NGC 3384, 2-237
NGC 3389, 2-237
NGC 3390, 2-217
NGC 3395, 2-253-254
NGC 3396, 2-253-254
NGC 3403, 2-172-173
NGC 3412, 2-237-238
NGC 3414, 2-254
NGC 3415, 2-398
NGC 3418, 2-254
NGC 3423, 2-383-384
NGC 3430, 2-254
NGC 3432, 2-249, 2-254
NGC 3433, 2-238
NGC 3437, 2-238
NGC 3440, 2-398
NGC 3445, 2-398
NGC 3448, 2-398
NGC 3449, 2-8
NGC 3454, 2-238
NGC 3455, 2-238
NGC 3456, 2-127
NGC 3457, 2-238
NGC 3458, 2-398
NGC 3464, 2-217
NGC 3478, 2-398
NGC 3485, 2-238
NGC 3486, 2-249, 2-255
NGC 3489, 2-238
NGC 3495, 2-238
NGC 3501, 2-239
NGC 3504, 2-249, 2-255
NGC 3506, 2-239
NGC 3507, 2-239-240
NGC 3510, 2-255
NGC 3511, 2-125, 2-127
NGC 3512, 2-255
NGC 3513, 2-127
NGC 3516, 2-398
NGC 3521, 2-225, 2-239
NGC 3547, 2-239
NGC 3549, 2-398
NGC 3556, 2-385, 2-399
NGC 3571, 2-127-128
NGC 3577, 2-398-399
NGC 3583, 2-398-399
NGC 3585, 2-216-217
NGC 3587, 2-399
NGC 3593, 2-240
NGC 3596, 2-240
NGC 3599, 2-240
NGC 3605, 2-240
NGC 3607, 2-240-241
NGC 3608, 2-240-241
NGC 3610, 2-400
NGC 3611, 2-241
NGC 3613, 2-400
NGC 3614, 2-400
NGC 3619, 2-398, 2-400
NGC 3621, 2-217-218
NGC 3623, 2-225, 2-241-242
NGC 3625, 2-398, 2-400
NGC 3626, 2-242, 2-245

NGC 3627, 2-225, 2-242
NGC 3628, 2-225, 2-241-243
NGC 3629, 2-243
NGC 3630, 2-244
NGC 3631, 2-400-401
NGC 3636, 2-127-128
NGC 3637, 2-127-128
NGC 3640, 2-244-245
NGC 3641, 2-244-245
NGC 3642, 2-400-401
NGC 3646, 2-244
NGC 3649, 2-244
NGC 3655, 2-244
NGC 3659, 2-244
NGC 3660, 2-128-129
NGC 3664, 2-244
NGC 3665, 2-401
NGC 3666, 2-244
NGC 3672, 2-125, 2-128-129
NGC 3673, 2-218
NGC 3675, 2-401
NGC 3681, 2-244-245
NGC 3683, 2-401
NGC 3684, 2-245
NGC 3686, 2-245
NGC 3687, 2-401
NGC 3689, 2-245
NGC 3690, 2-402
NGC 3691, 2-245
NGC 3705, 2-245
NGC 3717, 2-218
NGC 3718, 2-402-403
NGC 3719, 246
NGC 3720, 2-245-246
NGC 3726, 2-402-403
NGC 3729, 2-402-403
NGC 3732, 2-129
NGC 3735, 2-173-174
NGC 3738, 2-402
NGC 3745, 2-246
NGC 3746, 2-246
NGC 3748, 2-246
NGC 3750, 2-246
NGC 3751, 2-246
NGC 3753, 2-246
NGC 3754, 2-246
NGC 3756, 2-402
NGC 3769, 2-402-403
NGC 3773, 2-246
NGC 3780, 2-403
NGC 3782, 2-403
NGC 3810, 2-246
NGC 3813, 2-403
NGC 3818, 2-427
NGC 3837, 2-247
NGC 3840, 2-247
NGC 3841, 2-247
NGC 3842, 2-247
NGC 3844, 2-247
NGC 3860, 2-247
NGC 3861, 2-248
NGC 3862, 2-248
NGC 3865, 2-128-129
NGC 3866, 2-128-129
NGC 3872, 2-248
NGC 3877, 2-403
NGC 3885, 2-218
NGC 3887, 2-125, 2-128-129
NGC 3888, 2-404
NGC 3892, 2-129-130
NGC 3893, 2-403-404
NGC 3896, 2-403-404
NGC 3898, 2-404
NGC 3900, 2-248
NGC 3904, 2-218-219
NGC 3912, 2-248
NGC 3917, 2-404-405
NGC 3923, 2-218-219
NGC 3936, 2-218-219
NGC 3938, 2-404-405
NGC 3941, 2-405
NGC 3945, 2-405
NGC 3949, 2-405
NGC 3952, 2-427
NGC 3953, 2-405-406

NGC 3955, 2-125, 2-129-130
NGC 3956, 2-129-130
NGC 3957, 2-130
NGC 3958, 2-385, 2-406
NGC 3962, 2-125, 2-130-131
NGC 3963, 2-406
NGC 3976, 2-427
NGC 3981, 2-130-131
NGC 3982, 2-406
NGC 3985, 2-406
NGC 3990, 2-408
NGC 3991, 2-406
NGC 3992, 2-406-407
NGC 3994, 2-406
NGC 3995, 2-406, 2-408
NGC 3998, 2-406, 2-408
NGC 4004, 2-248
NGC 4008, 2-248
NGC 4013, 2-407-408
NGC 4016, 2-248
NGC 4017, 2-248
NGC 4024, 2-119
NGC 4026, 2-407-408
NGC 4027, 2-117, 2-119-120
NGC 4030, 2-427-428
NGC 4032, 2-83-84
NGC 4033, 2-119-120
NGC 4036, 2-407
NGC 4038, 2-121
NGC 4039, 2-121
NGC 4041, 2-407-408
NGC 4045, 2-427
NGC 4047, 2-409
NGC 4050, 2-121
NGC 4051, 2-408-409
NGC 4062, 2-408-409
NGC 4064, 2-83-84
NGC 4073, 2-428
NGC 4085, 2-408-409
NGC 4088, 2-409
NGC 4094, 2-121-122
NGC 4096, 2-409
NGC 4100, 2-408, 2-410
NGC 4102, 2-410
NGC 4105, 2-219
NGC 4106, 2-219
NGC 4109, 2-44
NGC 4111, 2-44
NGC 4116, 2-428
NGC 4117, 2-44
NGC 4121, 2-174
NGC 4123, 2-428
NGC 4124, 2-428
NGC 4125, 2-169, 2-173-174
NGC 4128, 2-174
NGC 4129, 2-428
NGC 4133, 2-174-175
NGC 4136, 2-84
NGC 4138, 2-44
NGC 4143, 2-44
NGC 4144, 2-408, 2-410
NGC 4145, 2-44-45
NGC 4147, 2-84
NGC 4150, 2-84
NGC 4151, 2-44-45
NGC 4152, 2-84
NGC 4156, 2-44-45
NGC 4157, 2-408, 2-410
NGC 4158, 2-85
NGC 4162, 2-85
NGC 4164, 2-429
NGC 4165, 2-429
NGC 4168, 2-85, 2-429
NGC 4178, 2-429
NGC 4179, 2-429-430
NGC 4183, 2-45
NGC 4189, 2-85
NGC 4190, 2-45
NGC 4192, 2-81, 2-85
NGC 4193, 2-429
NGC 4203, 2-86
NGC 4206, 2-429-430
NGC 4212, 2-86
NGC 4214, 2-45-46
NGC 4215, 2-430

Index

NGC 4216, 2-423, 2-429-430
NGC 4217, 2-45-46
NGC 4218, 2-45
NGC 4220, 2-45-46
NGC 4222, 2-429-430
NGC 4224, 2-430
NGC 4226, 2-45
NGC 4233, 2-430
NGC 4234, 2-430
NGC 4235, 2-430-431
NGC 4236, 2-174-175
NGC 4237, 2-86
NGC 4241, 2-431
NGC 4242, 2-46, 2-54
NGC 4244, 2-41, 2-46
NGC 4245, 2-86
NGC 4246, 2-431
NGC 4247, 2-431
NGC 4248, 2-47
NGC 4250, 2-175
NGC 4251, 2-87
NGC 4254, 2-81, 2-86-87
NGC 4256, 2-169, 2-173, 2-175
NGC 4258, 2-41, 2-47
NGC 4260, 2-431
NGC 4261, 2-431
NGC 4262, 2-87
NGC 4264, 2-431
NGC 4266, 2-431-432
NGC 4267, 2-430, 2-432
NGC 4268, 2-432
NGC 4270, 2-431-432
NGC 4273, 2-432
NGC 4274, 2-81, 2-87
NGC 4277, 2-432
NGC 4278, 2-87-88
NGC 4281, 2-432
NGC 4283, 2-87-88
NGC 4284, 2-410
NGC 4286, 2-87
NGC 4290, 2-410
NGC 4291, 2-175-176
NGC 4292, 2-433, 2-436
NGC 4293, 2-88
NGC 4294, 2-432-433
NGC 4298, 2-88
NGC 4299, 2-432-433
NGC 4302, 2-88
NGC 4303, 2-423, 2-433
NGC 4303a, 2-433
NGC 4304, 2-219-220
NGC 4307, 2-436
NGC 4312, 2-88
NGC 4313, 2-432, 2-436
NGC 4314, 2-89, 2-91
NGC 4319, 2-175-176
NGC 4321, 2-81, 2-89
NGC 4322, 2-89, 2-95
NGC 4324, 2-436
NGC 4328, 2-89, 2-95
NGC 4339, 2-436
NGC 4340, 2-90-91
NGC 4342, 2-436
NGC 4343, 2-436
NGC 4346, 2-47, 2-54
NGC 4348, 2-437
NGC 4350, 2-90
NGC 4351, 2-437
NGC 4357, 2-47-48
NGC 4361, 2-117, 2-122
NGC 4365, 2-436-437
NGC 4366, 2-437
NGC 4369, 2-48
NGC 4370, 2-437
NGC 4371, 2-437
NGC 4373, 2-70-71
NGC 4373a, 2-70
NGC 4374, 2-423, 2-441
NGC 4377, 2-91
NGC 4378, 2-437
NGC 4379, 2-91
NGC 4380, 2-437
NGC 4382, 2-81, 2-90-91
NGC 4383, 2-91
NGC 4386, 2-175-177

NGC 4387, 2-440-441
NGC 4388, 2-437, 2-440-441
NGC 4389, 2-48, 2-54
NGC 4394, 2-90-91, 2-95
NGC 4395, 2-48
NGC 4396, 2-95
NGC 4402, 2-440
NGC 4405, 2-95
NGC 4406, 2-423, 2-440-441
NGC 4412, 2-441-442
NGC 4413, 2-441
NGC 4414, 2-91-92
NGC 4417, 2-442-443
NGC 4419, 2-92
NGC 4420, 2-442
NGC 4421, 2-92
NGC 4424, 2-442
NGC 4425, 2-441-442
NGC 4428, 2-442
NGC 4429, 2-442
NGC 4431, 2-442-443
NGC 4433, 2-442
NGC 4435, 2-441, 2-443
NGC 4436, 2-442-443
NGC 4438, 2-442, 2-443
NGC 4440, 2-442-443
NGC 4442, 2-443
NGC 4446, 2-95
NGC 4447, 2-95
NGC 4448, 2-92
NGC 4449, 2-41, 2-48
NGC 4450, 2-92
NGC 4452, 2-443
NGC 4454, 2-443
NGC 4455, 2-92
NGC 4457, 2-443-444
NGC 4458, 2-444
NGC 4459, 2-93
NGC 4460, 2-49, 2-54
NGC 4461, 2-444
NGC 4462, 2-122
NGC 4468, 2-93
NGC 4469, 2-444
NGC 4472, 2-423, 2-444
NGC 4473, 2-93
NGC 4474, 2-93
NGC 4476, 2-445
NGC 4477, 2-93
NGC 4478, 2-445
NGC 4479, 2-93
NGC 4483, 2-445
NGC 4485, 2-49
NGC 4486, 2-423, 2-445
NGC 4487, 2-445-446
NGC 4489, 2-93, 2-95
NGC 4490, 2-49
NGC 4491, 2-445
NGC 4494, 2-93
NGC 4496a, 2-446
NGC 4496b, 2-446
NGC 4497, 2-445-446
NGC 4498, 2-93
NGC 4501, 2-81, 2-95-96
NGC 4502, 2-95
NGC 4503, 2-446
NGC 4504, 2-446
NGC 4506, 2-95
NGC 4507, 2-70
NGC 4515, 2-95
NGC 4516, 2-95
NGC 4517, 2-446
NGC 4517a, 2-446-447
NGC 4519, 2-447
NGC 4522, 2-447
NGC 4523, 2-95
NGC 4526, 2-423, 2-436, 2-447
NGC 4527, 2-447
NGC 4528, 2-447
NGC 4532, 2-447
NGC 4534, 2-49
NGC 4535, 2-423, 2-447
NGC 4536, 2-423, 2-444, 2-448
NGC 4539, 2-95
NGC 4540, 2-96
NGC 4545, 2-177

NGC 4546, 2-448, 2-455
NGC 4548, 2-96
NGC 4550, 2-448, 2-455
NGC 4551, 2-448, 2-455
NGC 4552, 2-423, 2-448, 2-455
NGC 4559, 2-81, 2-96-97
NGC 4564, 2-448-449
NGC 4565, 2-81, 2-97-98
NGC 4567, 2-449
NGC 4568, 2-449
NGC 4569, 2-423, 2-449
NGC 4570, 2-450
NGC 4571, 2-98
NGC 4578, 2-450
NGC 4579, 2-423, 2-450
NGC 4580, 2-450
NGC 4586, 2-450
NGC 4589, 2-175, 2-177
NGC 4590, 2-203, 2-220
NGC 4592, 2-450, 2-455
NGC 4593, 2-450-451, 2-455
NGC 4594, 2-423, 2-451
NGC 4595, 2-98
NGC 4596, 2-451
NGC 4597, 2-451
NGC 4601, 2-71
NGC 4602, 2-452, 2-455
NGC 4603, 2-71
NGC 4603d, 2-71
NGC 4605, 2-411
NGC 4608, 2-451-452
NGC 4612, 2-452
NGC 4616, 2-71
NGC 4618, 2-50, 2-54
NGC 4621, 2-423, 2-452-453
NGC 4622, 2-71
NGC 4623, 2-452
NGC 4625, 2-50
NGC 4627, 2-50
NGC 4630, 2-452
NGC 4631, 2-41, 2-50
NGC 4632, 2-452, 2-455
NGC 4636, 2-452-453, 2-455
NGC 4637, 2-453
NGC 4638, 2-453
NGC 4639, 2-453
NGC 4642, 2-455
NGC 4643, 2-453
NGC 4645, 2-71
NGC 4647, 2-453-454
NGC 4648, 2-176-177
NGC 4649, 2-423, 2-453-454
NGC 4651, 2-98
NGC 4653, 2-454
NGC 4654, 2-454
NGC 4656, 2-51
NGC 4657, 2-51
NGC 4658, 2-454
NGC 4660, 2-454
NGC 4665, 2-454
NGC 4666, 2-423, 2-454-456
NGC 4668, 2-454-456
NGC 4670, 2-98
NGC 4672, 2-72
NGC 4673, 2-98
NGC 4677, 2-72
NGC 4679, 2-72
NGC 4684, 2-456
NGC 4688, 2-456
NGC 4689, 2-98
NGC 4691, 2-456
NGC 4694, 2-456
NGC 4696, 2-71-73
NGC 4696a, 2-72
NGC 4696b, 2-72
NGC 4697, 2-423, 2-455-456
NGC 4698, 2-456
NGC 4699, 2-423, 2-455-456
NGC 4700, 2-457
NGC 4701, 2-457
NGC 4706, 2-72
NGC 4709, 2-72-73
NGC 4710, 2-98
NGC 4712, 2-99
NGC 4713, 2-457

NGC 4724, 2-122
NGC 4725, 2-81, 2-99
NGC 4727, 2-122
NGC 4729, 2-73
NGC 4730, 2-73
NGC 4731, 2-455, 2-457
NGC 4733, 2-457
NGC 4736, 2-41, 2-51
NGC 4742, 2-457-458
NGC 4744, 2-72-73
NGC 4746, 2-457
NGC 4747, 2-99
NGC 4750, 2-176-177
NGC 4753, 2-457-458
NGC 4754, 2-458
NGC 4756, 2-123
NGC 4757, 2-458
NGC 4760, 2-458-459
NGC 4762, 2-423, 2-458
NGC 4763, 2-123
NGC 4765, 2-458
NGC 4766, 2-459
NGC 4767, 2-72-73
NGC 4771, 2-458
NGC 4772, 2-459
NGC 4775, 2-459
NGC 4781, 2-458-460
NGC 4782, 2-123
NGC 4783, 2-123
NGC 4784, 2-459
NGC 4786, 2-459
NGC 4790, 2-459
NGC 4791, 2-459
NGC 4792, 2-123
NGC 4793, 2-99
NGC 4795, 2-459
NGC 4796, 2-459
NGC 4800, 2-51
NGC 4802, 2-123
NGC 4808, 2-459
NGC 4814, 2-411
NGC 4818, 2-459-460
NGC 4826, 2-81, 2-99
NGC 4839, 2-100
NGC 4845, 2-460
NGC 4846a, 2-473
NGC 4856, 2-460
NGC 4860, 2-100
NGC 4861, 2-51
NGC 4864, 2-98, 2-100
NGC 4865, 2-101
NGC 4866, 2-460-461
NGC 4867, 2-100
NGC 4868, 2-51-52
NGC 4871, 2-101
NGC 4872, 2-101
NGC 4873, 2-101
NGC 4874, 2-100-101
NGC 4877, 2-460
NGC 4880, 2-460
NGC 4882, 2-101
NGC 4886, 2-101
NGC 4889, 2-100-101
NGC 4897, 2-461
NGC 4898ab, 2-101
NGC 4899, 2-461
NGC 4900, 2-461
NGC 4902, 2-461
NGC 4904, 2-461-462
NGC 4911, 2-102
NGC 4914, 2-52
NGC 4915, 2-462
NGC 4918, 2-462
NGC 4919, 2-102
NGC 4921, 2-102
NGC 4923, 2-102
NGC 4928, 2-462
NGC 4933a, 2-462
NGC 4933b, 2-462
NGC 4939, 2-461-462
NGC 4941, 2-462
NGC 4945, 2-72-74
NGC 4947, 2-73-74
NGC 4951, 2-462
NGC 4958, 2-462

NGC 4961, 2-102
NGC 4976, 2-74
NGC 4981, 2-462
NGC 4984, 2-463
NGC 4995, 2-463
NGC 4999, 2-463
NGC 5005, 2-41, 2-51-52
NGC 5011, 2-73-74
NGC 5012, 2-102
NGC 5016, 2-103
NGC 5017, 2-103
NGC 5018, 2-463
NGC 5022, 2-463
NGC 5024, 2-81, 2-103
NGC 5033, 2-51-52
NGC 5035, 2-463
NGC 5037, 2-463
NGC 5044, 2-463-464
NGC 5046, 2-463-464
NGC 5047, 2-463-464
NGC 5049, 2-464
NGC 5053, 2-103
NGC 5054, 2-464
NGC 5055, 2-41, 2-52
NGC 5061, 2-220
NGC 5064, 2-74
NGC 5068, 2-464
NGC 5072, 2-465
NGC 5074, 2-53
NGC 5076, 2-465
NGC 5077, 2-465
NGC 5078, 2-220-221
NGC 5079, 2-465
NGC 5082, 2-73-74
NGC 5084, 2-465
NGC 5085, 2-221
NGC 5087, 2-465
NGC 5088, 2-465
NGC 5090, 2-74
NGC 5091, 2-74
NGC 5101, 2-221
NGC 5102, 2-74, 2-76
NGC 5107, 2-53
NGC 5112, 2-53-54
NGC 5116, 2-104
NGC 5128, 2-67, 2-74-75
NGC 5134, 2-465
NGC 5135, 2-221
NGC 5139, 2-67, 2-75-76
NGC 5147, 2-466
NGC 5150, 2-221, 2-223
NGC 5152, 2-221
NGC 5152, 2-221
NGC 5153, 2-221
NGC 5156, 2-76
NGC 5161, 2-76
NGC 5170, 2-464, 2-466
NGC 5172, 2-102, 2-104
NGC 5180, 2-104
NGC 5188, 2-76
NGC 5193, 2-76
NGC 5193a, 2-76
NGC 5194, 2-41, 2-55
NGC 5195, 2-53, 2-55
NGC 5198, 2-55
NGC 5204, 2-411
NGC 5222, 2-466
NGC 5230, 2-466
NGC 5236, 2-203, 2-222
NGC 5247, 2-466
NGC 5248, 2-25, 2-28
NGC 5253, 2-76-77
NGC 5264, 2-223
NGC 5266, 2-77
NGC 5272, 2-41, 2-55
NGC 5273, 2-54, 2-56
NGC 5276, 2-56
NGC 5286, 2-77
NGC 5290, 2-56
NGC 5291, 2-77
NGC 5296, 2-54, 2-56
NGC 5297, 2-56
NGC 5298, 2-77
NGC 5300, 2-466
NGC 5301, 2-54, 2-56

NGC 5302, 2-78
NGC 5307, 2-78
NGC 5308, 2-411
NGC 5311, 2-56
NGC 5313, 2-56
NGC 5320, 2-54, 2-56
NGC 5322, 2-411
NGC 5323, 2-418
NGC 5324, 2-466
NGC 5326, 2-57
NGC 5328, 2-223
NGC 5330, 2-223
NGC 5334, 2-466
NGC 5338, 2-467
NGC 5347, 2-57
NGC 5348, 2-467
NGC 5349, 2-57
NGC 5350, 2-57-58
NGC 5351, 2-54, 2-57
NGC 5353, 2-58
NGC 5354, 2-58
NGC 5355, 2-57-58
NGC 5356, 2-467
NGC 5357, 2-77-78
NGC 5358, 2-57-58
NGC 5360, 2-467
NGC 5362, 2-58
NGC 5363, 2-465
NGC 5364, 2-467
NGC 5365, 2-78
NGC 5371, 2-57, 2-59
NGC 5375, 2-58-59
NGC 5376, 2-412
NGC 5377, 2-58-59
NGC 5379, 2-412
NGC 5380, 2-59
NGC 5383, 2-58-59
NGC 5385, 2-418
NGC 5389, 2-412
NGC 5394, 2-59-60
NGC 5395, 2-59-60
NGC 5398, 2-78
NGC 5406, 2-60
NGC 5409, 2-28
NGC 5411, 2-28
NGC 5412, 2-418-419
NGC 5416, 2-28-30
NGC 5419, 2-78
NGC 5422, 2-411-412
NGC 5423, 2-29
NGC 5424, 2-29
NGC 5426, 2-467, 2-469
NGC 5427, 2-467, 2-469
NGC 5430, 2-412
NGC 5431, 2-29
NGC 5434, 2-29
NGC 5434b, 2-29
NGC 5436, 2-29-30
NGC 5437, 2-29-30
NGC 5438, 2-29-30
NGC 5443, 2-412
NGC 5444, 2-60
NGC 5445, 2-60
NGC 5448, 2-412
NGC 5452, 2-418-419
NGC 5454, 2-30
NGC 5456, 2-30
NGC 5457, 2-385, 2-412-413
NGC 5460, 2-79
NGC 5464, 2-222
NGC 5466, 2-25, 2-30
NGC 5468, 2-467
NGC 5472, 2-467
NGC 5473, 2-413
NGC 5474, 2-413
NGC 5480, 2-31, 2-413-414
NGC 5481, 2-31, 2-413-414
NGC 5482, 2-31
NGC 5483, 2-78-79
NGC 5484, 2-414
NGC 5485, 2-414
NGC 5486, 2-414
NGC 5490, 2-31
NGC 5490b, 2-31
NGC 5490c, 2-31

Index

NGC 5492, 2-30-31
NGC 5493, 2-467
NGC 5494, 2-78-79
NGC 5496, 2-467
NGC 5523, 2-30-31
NGC 5525, 2-32
NGC 5527, 2-32
NGC 5529, 2-25, 2-32
NGC 5530, 2-268-269
NGC 5531, 2-32
NGC 5532, 2-32
NGC 5532b, 2-32
NGC 5533, 2-32
NGC 5534, 2-468
NGC 5538, 2-33
NGC 5542, 2-33
NGC 5543, 2-33
NGC 5544, 2-32-33
NGC 5545, 2-33
NGC 5546, 2-33
NGC 5547, 2-419
NGC 5548, 2-32-33
NGC 5549, 2-33
NGC 5556, 2-223
NGC 5557, 2-33
NGC 5560, 2-468
NGC 5566, 2-468
NGC 5569, 2-468
NGC 5570, 2-33
NGC 5574, 2-468
NGC 5576, 2-468-469
NGC 5577, 2-468
NGC 5582, 2-33
NGC 5584, 2-468
NGC 5585, 2-414
NGC 5587, 2-33
NGC 5592, 2-223-224
NGC 5595, 2-260
NGC 5597, 2-260
NGC 5600, 2-33-34
NGC 5605, 2-260
NGC 5608, 2-33
NGC 5613, 2-33
NGC 5614, 2-33-34
NGC 5615, 2-34
NGC 5627, 2-34
NGC 5630, 2-34
NGC 5631, 2-414
NGC 5633, 2-34
NGC 5634, 2-468-469
NGC 5636, 2-469
NGC 5638, 2-469
NGC 5641, 2-35
NGC 5643, 2-269-270
NGC 5644, 2-35
NGC 5645, 2-469
NGC 5653, 2-35
NGC 5656, 2-35
NGC 5660, 2-35-36
NGC 5665, 2-35-36
NGC 5666, 2-36
NGC 5668, 2-469
NGC 5669, 2-36
NGC 5673, 2-35-36
NGC 5676, 2-25, 2-35-37
NGC 5678, 2-176-177
NGC 5687, 2-36-37
NGC 5689, 2-25, 2-36-37
NGC 5690, 2-470
NGC 5691, 2-470
NGC 5693, 2-36
NGC 5694, 2-224
NGC 5695, 2-36
NGC 5701, 2-470
NGC 5707, 2-37
NGC 5713, 2-470
NGC 5714, 2-470
NGC 5716, 2-260
NGC 5719, 2-470
NGC 5728, 2-257, 2-260
NGC 5739, 2-37-38
NGC 5740, 2-470
NGC 5746, 2-470
NGC 5750, 2-471
NGC 5756, 2-261

NGC 5757, 2-260-261
NGC 5762, 2-37
NGC 5763, 2-37
NGC 5768, 2-261
NGC 5770, 2-37
NGC 5772, 2-37
NGC 5774, 2-471
NGC 5775, 2-471
NGC 5783, 2-37-38
NGC 5784, 2-38
NGC 5787, 2-38
NGC 5788, 2-38
NGC 5791, 2-262
NGC 5792, 2-257, 2-261-262
NGC 5793, 2-262
NGC 5794, 2-38
NGC 5795, 2-38
NGC 5796, 2-262
NGC 5797, 2-38-39
NGC 5804, 2-38
NGC 5805, 2-38
NGC 5806, 2-471
NGC 5808, 2-419
NGC 5812, 2-257, 2-262
NGC 5813, 2-471-472
NGC 5814, 2-471-472
NGC 5819, 2-419
NGC 5820, 2-38-39
NGC 5824, 2-270
NGC 5831, 2-472
NGC 5838, 2-471-472
NGC 5839, 2-472-473
NGC 5845, 2-472-473
NGC 5846, 2-472-473
NGC 5846a, 2-473
NGC 5850, 2-472-473
NGC 5854, 2-473
NGC 5856, 2-471
NGC 5858, 2-262-263
NGC 5861, 2-262-263
NGC 5864, 2-473
NGC 5866, 2-169, 2-177-178
NGC 5873, 2-270
NGC 5874, 2-39
NGC 5875, 2-39
NGC 5876, 2-39
NGC 5878, 2-257, 2-262
NGC 5879, 2-178
NGC 5885, 2-263
NGC 5893, 2-40
NGC 5895, 2-40
NGC 5896, 2-40
NGC 5897, 2-257, 2-264
NGC 5898, 2-263-264
NGC 5899, 2-39-40
NGC 5900, 2-40
NGC 5903, 2-263-264
NGC 5904, 2-365, 2-368-369
NGC 5905, 2-178-179
NGC 5907, 2-169, 2-178-179
NGC 5908, 2-179
NGC 5915, 2-263-264
NGC 5916, 2-263-264
NGC 5921, 2-368, 2-370
NGC 5927, 2-269-270
NGC 5929, 2-39-40
NGC 5930, 2-39-40
NGC 5936, 2-369
NGC 5946, 2-260
NGC 5949, 2-179
NGC 5951, 2-369
NGC 5953, 2-369-370
NGC 5954, 2-369-370
NGC 5957, 2-369-370
NGC 5958, 2-114
NGC 5961, 2-114
NGC 5962, 2-369-370
NGC 5963, 2-179-180
NGC 5965, 2-169, 2-179-180
NGC 5970, 2-370-371
NGC 5980, 2-371
NGC 5981, 2-180
NGC 5982, 2-180
NGC 5984, 2-370-371
NGC 5985, 2-180-181

NGC 5986, 2-265, 2-269-270
NGC 5994, 2-370
NGC 5996, 2-370-371
NGC 6001, 2-114, 2-420
NGC 6011, 2-420
NGC 6012, 2-371
NGC 6015, 2-169, 2-181
NGC 6026, 2-270
NGC 6027, 2-371-372
NGC 6027a, 2-371
NGC 6027b, 2-371-372
NGC 6027c, 2-371
NGC 6027d, 2-371-372
NGC 6027e, 2-371-372
NGC 6040, 2-193
NGC 6041, 2-194
NGC 6041a-b, 2-193
NGC 6042, 2-193
NGC 6048, 2-420
NGC 6051, 2-372
NGC 6052, 2-194
NGC 6058, 2-194
NGC 6068, 2-420
NGC 6068a, 2-420
NGC 6070, 2-370, 2-372
NGC 6071, 2-420
NGC 6072, 2-337, 2-339
NGC 6079, 2-420
NGC 6091, 2-421
NGC 6093, 2-333, 2-337-338
NGC 6094, 2-420-421
NGC 6104, 2-114
NGC 6105, 2-114
NGC 6106, 2-194-195
NGC 6107, 2-114-115
NGC 6109, 2-115
NGC 6117, 2-115
NGC 6117b, 2-115
NGC 6118, 2-370, 2-372
NGC 6119, 2-115
NGC 6120, 2-115
NGC 6121, 2-333, 2-338
NGC 6122, 2-115
NGC 6124, 2-333, 2-338-339
NGC 6129, 2-115
NGC 6131, 2-115
NGC 6131b, 2-116
NGC 6137, 2-116
NGC 6137a-b, 2-115
NGC 6139, 2-339-340
NGC 6140, 2-169, 2-180-181
NGC 6144, 2-339-340
NGC 6145, 2-186
NGC 6146, 2-195
NGC 6147, 2-195
NGC 6153, 2-339-340
NGC 6158, 2-195
NGC 6160, 2-195
NGC 6166, 2-196
NGC 6171, 2-291
NGC 6173, 2-196
NGC 6174, 2-196
NGC 6178, 2-333, 2-340
NGC 6181, 2-195-196
NGC 6192, 2-339-340
NGC 6194, 2-196
NGC 6196, 2-196
NGC 6197, 2-196
NGC 6205, 2-189, 2-196-197
NGC 6207, 2-198
NGC 6210, 2-198
NGC 6216, 2-340
NGC 6217, 2-421
NGC 6218, 2-287, 2-291
NGC 6229, 2-198-199
NGC 6231, 2-333, 2-340-341, 2-343
NGC 6235, 2-292, 2-300
NGC 6236, 2-181
NGC 6239, 2-198-199
NGC 6242, 2-333, 2-342
NGC 6249, 2-333, 2-343
NGC 6251, 2-421
NGC 6252, 2-421
NGC 6254, 2-287, 2-292
NGC 6255, 2-199

NGC 6259, 2-343
NGC 6266, 2-287, 2-292-293
NGC 6268, 2-343
NGC 6273, 2-287, 2-294
NGC 6281, 2-333, 2-339, 2-344
NGC 6284, 2-295, 2-300
NGC 6287, 2-293, 2-295, 2-300
NGC 6293, 2-295, 2-300
NGC 6302, 2-344
NGC 6304, 2-297, 2-300
NGC 6309, 2-300
NGC 6316, 2-297, 2-300
NGC 6318, 2-339, 2-345
NGC 6322, 2-333, 2-345
NGC 6324, 2-421
NGC 6325, 2-298, 2-300
NGC 6333, 2-287, 2-296-297
NGC 6334, 2-345
NGC 6337, 2-346
NGC 6340, 2-182
NGC 6341, 2-189, 2-199
NGC 6342, 2-299-300
NGC 6355, 2-287, 2-299-300
NGC 6356, 2-287, 2-299-300
NGC 6357, 2-346
NGC 6366, 2-299-300
NGC 6369, 2-287, 2-299, 2-301
NGC 6380, 2-347-348
NGC 6383, 2-333, 2-335, 2-347-348
NGC 6384, 2-301-302
NGC 6388, 2-333, 2-339, 2-348
NGC 6395, 2-181-182
NGC 6396, 2-348
NGC 6400, 2-333, 2-349
NGC 6401, 2-302-303
NGC 6402, 2-287, 2-302
NGC 6404, 2-348
NGC 6405, 2-333, 2-348-349
NGC 6412, 2-182-183
NGC 6416, 2-333, 2-349
NGC 6425, 2-339, 2-349
NGC 6426, 2-303
NGC 6440, 2-314-316
NGC 6441, 2-333, 2-339, 2-350
NGC 6444, 2-350
NGC 6445, 2-314-316
NGC 6451, 2-339, 2-350
NGC 6453, 2-333, 2-350
NGC 6469, 2-315-316
NGC 6475, 2-333, 2-351
NGC 6482, 2-200-201
NGC 6494, 2-311, 2-315-316
NGC 6496, 2-105, 2-107
NGC 6503, 2-169, 2-182-183
NGC 6506, 2-316
NGC 6514, 2-311
NGC 6517, 2-303
NGC 6520, 2-317
NGC 6522, 2-314, 2-319
NGC 6523, 2-311, 2-318
NGC 6528, 2-318-321
NGC 6530, 2-318
NGC 6531, 2-311, 2-319
NGC 6535, 2-376
NGC 6539, 2-373, 2-376
NGC 6540, 2-320-321
NGC 6541, 2-105, 2-107
NGC 6543, 2-169, 2-182-183
NGC 6544, 2-320-321
NGC 6546, 2-320-321
NGC 6553, 2-320-321
NGC 6558, 2-322
NGC 6563, 2-322
NGC 6565, 2-322
NGC 6567, 2-322
NGC 6568, 2-321-322
NGC 6569, 2-322
NGC 6572, 2-304
NGC 6574, 2-200-201
NGC 6583, 2-321, 2-323
NGC 6589, 2-324
NGC 6590, 2-324
NGC 6595, 2-324
NGC 6596, 2-321
NGC 6603, 2-311, 2-323-324

NGC 6604, 2-376-377
NGC 6605, 2-376
NGC 6611, 2-373, 2-377-378
NGC 6613, 2-324-325
NGC 6618, 2-311, 2-325
NGC 6624, 2-311, 2-326
NGC 6626, 2-311, 2-326
NGC 6629, 2-326
NGC 6631, 2-355-356
NGC 6633, 2-287, 2-303-304
NGC 6637, 2-311, 2-327
NGC 6638, 2-321, 2-326
NGC 6639, 2-355-356
NGC 6642, 2-311, 2-328
NGC 6643, 2-169, 2-182-183
NGC 6644, 2-359
NGC 6645, 2-327-328
NGC 6646, 2-274, 2-276
NGC 6649, 2-356, 2-359
NGC 6652, 2-328
NGC 6654, 2-183-184
NGC 6656, 2-311, 2-328
NGC 6664, 2-333, 2-359
NGC 6667, 2-183-184
NGC 6671, 2-274, 2-276
NGC 6675, 2-275-276
NGC 6681, 2-311
NGC 6683, 2-353, 2-360
NGC 6685, 2-275
NGC 6688, 2-275-276
NGC 6690, 2-183-184
NGC 6694, 2-353, 2-360
NGC 6695, 2-275
NGC 6702, 2-275-276
NGC 6703, 2-275-276
NGC 6704, 2-361
NGC 6705, 2-353, 2-362
NGC 6709, 2-9, 2-12
NGC 6710, 2-276-277
NGC 6712, 2-361-362
NGC 6715, 2-311, 2-329
NGC 6716, 2-330
NGC 6717, 2-330
NGC 6720, 2-271, 2-277
NGC 6723, 2-108, 2-321, 2-330
NGC 6726, 2-107-108
NGC 6727, 2-107-108
NGC 6729, 2-107-108
NGC 6738, 2-9, 2-13
NGC 6741, 2-14
NGC 6742, 2-184
NGC 6745, 2-276, 2-278
NGC 6749, 2-14
NGC 6751, 2-14-16
NGC 6755, 2-9, 2-16
NGC 6756, 2-16-17
NGC 6760, 2-17-18
NGC 6765, 2-276, 2-278
NGC 6772, 2-17
NGC 6773, 2-18
NGC 6775, 2-18
NGC 6778, 2-18-19
NGC 6779, 2-271, 2-278
NGC 6781, 2-9, 2-17, 2-19
NGC 6790, 2-19
NGC 6791, 2-271, 2-276, 2-279
NGC 6792, 2-276, 2-279
NGC 6793, 2-477
NGC 6795, 2-20
NGC 6800, 2-478
NGC 6802, 2-478-479
NGC 6803, 2-20
NGC 6804, 2-20
NGC 6807, 2-20
NGC 6809, 2-311, 2-330
NGC 6811, 2-133, 2-136-137
NGC 6813, 2-479
NGC 6814, 2-22
NGC 6815, 2-479, 2-484
NGC 6818, 2-330
NGC 6819, 2-133, 2-137-138
NGC 6820, 2-479, 2-481
NGC 6822, 2-331, 2-487
NGC 6823, 2-475, 2-479
NGC 6826, 2-133, 2-138, 2-155

NGC 6828, 2-22
NGC 6830, 2-475, 2-480-481
NGC 6834, 2-133, 2-137-138, 2-480-481
NGC 6835, 2-321, 2-331
NGC 6836, 2-321, 2-331
NGC 6837, 2-22-23
NGC 6838, 2-305, 2-308
NGC 6840, 2-23
NGC 6842, 2-139, 2-481-482
NGC 6843, 2-23
NGC 6852, 2-23
NGC 6853, 2-475, 2-482
NGC 6856, 2-139-140
NGC 6857, 2-139-140
NGC 6858, 2-23-24
NGC 6864, 2-311, 2-331
NGC 6866, 2-133, 2-141
NGC 6871, 2-133, 2-141
NGC 6873, 2-307, 2-309
NGC 6874, 2-139, 2-142
NGC 6879, 2-310
NGC 6882, 2-483
NGC 6883, 2-142-143
NGC 6885, 2-475, 2-483
NGC 6886, 2-310
NGC 6888, 2-133, 2-143-144
NGC 6891, 2-165, 2-167
NGC 6894, 2-145
NGC 6895, 2-144
NGC 6896, 2-139, 2-147
NGC 6903, 2-64
NGC 6904, 2-483-484
NGC 6905, 2-163, 2-166
NGC 6907, 2-64
NGC 6910, 2-133, 2-148
NGC 6912, 2-65
NGC 6913, 2-133
NGC 6914a, 2-149-150
NGC 6914b, 2-149-150
NGC 6921, 2-484
NGC 6923, 2-283
NGC 6924, 2-65
NGC 6925, 2-281, 2-284
NGC 6927, 2-167
NGC 6928, 2-166
NGC 6930, 2-166
NGC 6934, 2-163, 2-166-167
NGC 6936, 2-65
NGC 6938, 2-485
NGC 6939, 1-131, 1-135
NGC 6940, 2-475, 2-484
NGC 6944, 2-167
NGC 6944a, 2-167
NGC 6946, 1-131, 1-135
NGC 6949, 1-136
NGC 6950, 2-167
NGC 6951, 1-136
NGC 6954, 2-168
NGC 6956, 2-167-168
NGC 6958, 2-281, 2-284-285
NGC 6959, 1-28
NGC 6960, 2-133, 2-150-151
NGC 6961, 1-29
NGC 6962, 1-28-29
NGC 6964, 1-29
NGC 6965, 1-29
NGC 6967, 1-29
NGC 6972, 2-168
NGC 6975, 1-29
NGC 6976, 1-29
NGC 6977, 1-29
NGC 6978, 1-29
NGC 6979, 2-151
NGC 6981, 1-25, 1-29-30
NGC 6989, 2-152
NGC 6991, 2-155
NGC 6992, 2-133, 2-152-153
NGC 6994, 1-25, 1-30
NGC 6995, 2-152-153
NGC 6996, 2-153-155
NGC 7000, 2-133, 2-155
NGC 7003, 2-168
NGC 7006, 2-168
NGC 7008, 2-154
NGC 7009, 1-25, 1-30-31

NGC 7015, 2-185, 2-187
NGC 7023, 1-136-137
NGC 7024, 2-146, 2-156
NGC 7026, 2-154
NGC 7027, 2-154
NGC 7037, 2-156
NGC 7039, 2-133, 2-157
NGC 7040, 2-187
NGC 7044, 2-154, 2-157
NGC 7046, 2-187
NGC 7048, 2-154, 2-157
NGC 7050, 2-150
NGC 7052, 2-485
NGC 7055, 1-136, 1-138
NGC 7058, 2-146, 2-158
NGC 7062, 2-133, 2-158
NGC 7063, 2-133, 2-158-159
NGC 7065, 1-28, 1-31
NGC 7065a, 1-28, 1-31
NGC 7071, 2-159
NGC 7076, 1-138
NGC 7077, 1-31
NGC 7078, 1-279, 1-282
NGC 7080, 2-485-486
NGC 7081, 1-31
NGC 7082, 2-133
NGC 7086, 2-133, 2-159
NGC 7089, 1-25, 1-31
NGC 7092, 2-133, 2-159
NGC 7093, 2-146, 2-160
NGC 7094, 1-282
NGC 7099, 2-61, 2-66
NGC 7103, 2-65-66
NGC 7104, 2-66
NGC 7127, 2-146, 2-160
NGC 7128, 2-154, 2-160
NGC 7129, 1-139
NGC 7130, 1-333, 1-335-336
NGC 7133, 1-139
NGC 7135, 1-335-336
NGC 7137, 1-283
NGC 7139, 1-140-141
NGC 7142, 1-140
NGC 7150, 2-146, 2-161
NGC 7154, 1-335
NGC 7160, 1-131, 1-133, 1-141
NGC 7163, 1-335
NGC 7171, 1-28, 1-32
NGC 7172, 1-335-336
NGC 7173, 1-336
NGC 7174, 1-336
NGC 7175, 2-146, 2-161
NGC 7176, 1-335-336
NGC 7177, 1-283
NGC 7180, 1-32-33
NGC 7183, 1-32
NGC 7184, 1-32-33
NGC 7185, 1-32-33
NGC 7188, 1-33
NGC 7201, 1-336-337
NGC 7203, 1-336-337
NGC 7204, 1-337
NGC 7209, 1-215, 1-217-218
NGC 7214, 1-337
NGC 7217, 1-279, 1-283
NGC 7218, 1-32-33
NGC 7221, 1-333, 1-337
NGC 7223, 1-217
NGC 7226, 1-142
NGC 7229, 1-338
NGC 7234, 1-142, 1-144
NGC 7235, 1-131, 1-142
NGC 7240, 1-218
NGC 7242, 1-218
NGC 7243, 1-215, 1-217-219
NGC 7245, 1-215, 1-218-219
NGC 7248, 1-219
NGC 7250, 1-219
NGC 7251, 1-33
NGC 7252, 1-34
NGC 7261, 1-131, 1-142-143
NGC 7263, 1-220
NGC 7265, 1-220
NGC 7273, 1-220
NGC 7274, 1-220

NGC 7276, 1-220
NGC 7281, 1-131, 1-143
NGC 7293, 1-25, 1-34
NGC 7295, 1-220
NGC 7296, 1-220
NGC 7298, 1-220
NGC 7300, 1-33, 1-35
NGC 7302, 1-33, 1-35
NGC 7309, 1-35
NGC 7313, 1-337
NGC 7314, 1-333, 1-337-338
NGC 7317, 1-283
NGC 7318a, 1-283
NGC 7318b, 1-283
NGC 7319, 1-283
NGC 7320, 1-283-284
NGC 7331, 1-279, 1-284
NGC 7332, 1-279, 1-285
NGC 7335, 1-285
NGC 7337, 1-285
NGC 7339, 1-285
NGC 7340, 1-285
NGC 7354, 1-144
NGC 7361, 1-338
NGC 7371, 1-33, 1-35
NGC 7377, 1-33, 1-35
NGC 7378, 1-35
NGC 7380, 1-144-145
NGC 7392, 1-35-36
NGC 7394, 1-220
NGC 7416, 1-35
NGC 7419, 1-145-146
NGC 7423, 1-145-146
NGC 7428, 1-319
NGC 7429, 1-146
NGC 7443, 1-36
NGC 7444, 1-36
NGC 7448, 1-285
NGC 7450, 1-36
NGC 7454, 1-286
NGC 7457, 1-286
NGC 7463, 1-286
NGC 7464, 1-286
NGC 7465, 1-286
NGC 7469, 1-286-287
NGC 7479, 1-279, 1-287
NGC 7492, 1-339
NGC 7507, 1-367
NGC 7510, 1-131, 1-146-147
NGC 7513, 1-367
NGC 7537, 1-318-319
NGC 7538, 1-147
NGC 7541, 1-315, 1-318-319
NGC 7557, 1-318
NGC 7562, 1-318-319
NGC 7576, 1-36-38
NGC 7585, 1-36-38
NGC 7592, 1-37
NGC 7600, 1-37
NGC 7606, 1-37
NGC 7611, 1-319
NGC 7617, 1-318-319
NGC 7619, 1-288, 1-319
NGC 7623, 1-288
NGC 7625, 1-288
NGC 7626, 1-289, 1-319
NGC 7631, 1-289
NGC 7635, 1-108-110
NGC 7640, 1-23
NGC 7654, 1-105, 1-109
NGC 7662, 1-11, 1-24
NGC 7673, 1-289-290
NGC 7677, 1-289-290
NGC 7678, 1-289-290
NGC 7679, 1-320
NGC 7682, 1-320
NGC 7686, 1-11, 1-24
NGC 7704, 1-337
NGC 7713, 1-368
NGC 7714, 1-320
NGC 7715, 1-320
NGC 7716, 1-320
NGC 7721, 1-37
NGC 7723, 1-38
NGC 7724, 1-38

NGC 7727, 1-38
NGC 7741, 1-290
NGC 7742, 1-290
NGC 7743, 1-290
NGC 7748, 1-147
NGC 7750, 1-320-321
NGC 7755, 1-368
NGC 7757, 1-321
NGC 7762, 1-147-148
NGC 7769, 1-291
NGC 7770, 1-291
NGC 7771, 1-291
NGC 7772, 1-291
NGC 7778, 1-321
NGC 7779, 1-321
NGC 7780, 1-322
NGC 7781, 1-322
NGC 7782, 1-320, 1-322
NGC 7785, 1-322
NGC 7788, 1-110-112
NGC 7789, 1-105, 1-111, 2-343
NGC 7790, 1-105, 1-111-112
NGC 7793, 1-365, 1-368
NGC 7800, 1-291
NGC 7814, 1-279, 1-291-292
NGC 7822, 1-148
NGC 7839, 1-292
NGC 7876, 2-39
NGC 7881, 1-321
Nile Star, 1-85
Nimrod, 1-259
Ninurta, 2-9, 2-189, 2-203, 2-271, 2-287, 2-311, 2-333
Norma, 2-270
Norma Spiral Arm, 2-314, 2-323
North America Nebula, 2-133, 2-153, 2-155-156
Northern Cross, 2-133, 2-136
Northern Crown, 2-105, 2-111-112

O

Omega Nebula, 2-325-236
Ophiuchus, 1-259, 1-339, 2-228, 2-287, 2-289-295, 2-298, 2-302, 2-336, 2-338, 2-365, 2-373, 2-388
Oriani, 2-433
Orion, 1-47, 1-50, 1-85, 1-101, 1-131, 1-147, 1-175, 1-221, 1-237, 1-244, 1-249, 1-259-260, 1-262, 1-265-269, 1-378
Orion Association, 1-47
Orion-Cygnus Spiral Arm, 2-311, 2-314, 2-333
Orion Molecular Cloud, 1-268
Orion Nebula, 1-259, 1-262, 1-266-269
Orpheus, 2-271
Ostara, 1-221
Osypowski, Tom, 1-405
Otto Struve 147, 1-51
Otto Struve 179, 1-206
Otto Struve 300, 2-367
Otto Struve 440, 1-131
Otto Struve 475, 1-215
Otto Struve 533, 2-163
Owl Nebula, 2-399

P

Palomar 5, 2-367-368
Palomar 9, 2-330
Palomar 10, 2-307, 2-309
Palomar 12, 2-65-66
Palomar Observatory Sky Survey, 2-22
Parrot's Head Nebula, 2-319
Pattie, Steve, 1-198
Pegasus, 1-215, 1-279-280, 1-282, 1-288-289, 1-292, 1-293, 1-315, 1-319, 2-185
Pegasus I Galaxy Cluster, 1-288-289, 1-319
Peiresc, Nicholas, 1-267
Pelican Nebula, 2-152, 2-155
Per 2, 2-259
Persephone, 2-423
Perseus, 1-11, 1-124, 1-127, 1-151, 1-212, 1-279, 1-293-294, 1-296, 1-299-300, 1-305, 1-307, 1-385, 2-308
Perseus Double Cluster, 1-124, 1-293, 1-385,

2-340
Perseus OB1 association, 1-299-300
Phryxus, 1-39
Phylira, 2-67
Pickering, 2-389
Pickering's Triangular Wisp, 2-151
Pinwheel Galaxy, 1-391, 1-393, 2-412
Pipe Nebula, 2-287, 2-293, 2-295-296, 2-298-299, 2-301
Pisces, 1-25, 1-39, 1-315-316, 1-318-319
Piscis Austrinus, 1-25, 1-333-335
Pismis 1, 1-358
Pismis 24, 2-346
PK+14.1, 2-297, 2-286
PK3−4.5, 2-322
PK8+3.1, 2-315
PK2+5.1, 2-301-302
PK9−5.1, 2-326
PK11−0.2, 2-322
PK25+4.2, 2-363
PK25+40.1, 2-194
PK25−17.1, 2-330
PK27−9.1, 2-18
PK29−5.1, 2-15
PK33−2.1, 2-14
PK33−5.1, 2-17-18
PK34+11.1, 2-304
PK34−6.1, 2-19
PK36−1.1, 2-14
PK36−57.1, 1-34
PK38−25.1, 2-24
PK40−0.1, 2-15-16
PK42−14.1, 2-23
PK42−6.1, 2-20
PK43+37.1, 2-198
PK47−4.1, 2-20
PK51−3.1, 2-22
PK52−2.2, 2-21
PK52−4.1, 2-22
PK54−12.1, 2-165
PK57−8.1, 2-310
PK58−10.1, 2-310
PK59−18.1, 2-168
PK60−7.2, 2-310
PK61−9.1, 2-166
PK62+9.1, 2-278
PK64+5.1, 2-137
PK64+15.1, 2-276-277
PK64+48.1, 2-194
PK65+0.1, 2-140
PK66−28.1, 1-282
PK69−2.1, 2-145
PK72−17.1, 2-485
PK77+14.1, 2-136, 2-154
PK79+5.1, 2-143, 2-154
PK80−6.1, 2-154, 2-156
PK83+12.1, 2-138
PK84+3.1, 2-154, 2-156
PK85+4.1, 2-150, 2-154
PK89+0.1, 2-154, 2-156
PK93−2.1, 2-154, 2-160
PK93+5.2, 2-156
PK94+27.1, 2-184
PK96+29.1, 2-182-183
PK101+8.1, 1-138
PK102−2.1, 1-220
PK103+0.1, 1-142-143
PK104−29.1, 1-289
PK106−17.1, 1-24
PK107+2.1, 1-144
PK107−2.1, 1-146
PK112−10.1, 1-110
PK114−4.1, 1-110
PK116+8.1, 1-147
PK119−6.1, 1-112
PK120+9.1, 1-149
PK123+34.1, 1-75
PK130+1.1, 1-124
PK131−5.1, 1-298, 1-302
PK144+6.1, 1-68
PK144-15.1, 1-302-303
PK147+4.1, 1-69
PK147−2.1, 1-309, 1-313
PK159−15.1, 1-309, 1-313
PK161−14.1, 1-310, 1-313

PK164+31.1, 1-232-233
PK165−15.1, 1-381
PK166+10.1, 1-57, 1-61
PK170+15.1, 1-62
PK173−5.1, 1-53
PK190−17.1, 1-264
PK194+2.1, 1-209
PK196−10.1, 1-272-273
PK197+17.1, 1-214
PK205+14.1, 1-213-214
PK206−40.1, 1-187
PK208+33.1, 1-82
PK219−31.1, 1-83-84
PK220−53.1, 1-197
PK221+5.1, 1-257
PK221−12.1, 1-89
PK231+4.2, 1-349-350
PK234+2.1, 1-350
PK238+34.1, 2-211
PK239+13.1, 2-206-207
PK242−11.1, 1-93
PK243−1.1, 1-352
PK254+5.1, 1-364
PK261+8.1, 1-364
PK261+32.1, 2-215
PK294+43.1, 2-122
PK312+10.1, 2-78
PK331+16.1, 2-270
PK341+5.1, 2-339-340
PK341+13.1, 2-270
PK342+10.1, 2-337, 2-339
PK342+27.1, 2-264
PK349+1.1, 2-344-345
PK349−1.1, 2-346
PK358−7.1, 2-322
PK358−21.1, 2-109
Pleiades, 1-6, 1-8, 1-59, 1-82, 1-109, 1-375-376, 1-378-382, 2-90, 2-340, 2-342
Pleione, 1-380-381
Pluto, 2-287, 2-389
Polakis, Tom, 2-5-6, 2-64, 2-86-87, 2-167, 2-206, 2-339, 2-391, 2-401, 2-428, 2-469
Polaris, 1-150, 2-415-417
Pollux, 1-203, 1-206, 1-339, 2-133
Porrima, 2-426
Poseidon, 1-259, 1-293, 2-163
Praesepe, 1-77, 1-81
Procyon, 1-85, 1-101, 1-151, 1-259
Prometheus, 2-67
Proxima Centauri, 2-70
Ptolemy, 1-1, 2-75, 2-81, 2-328, 2-333, 2-349, 2-351
Ptolemy II, 2-81
Puppis, 1-99, 1-171, 1-339, 1-341-344, 1-347-349, 1-351-359, 1-361, 2-353
Pyxidis, 1-361, 1-363-364
Pyxis, 1-339, 1-361-362
Pz 2, 1-175

Q

Quadrans Muralis, 2-25
Quadrantids, 2-25

R

Ras Algethi, 2-191-192
RCW 127, 2-345
Regulus, 2-225, 2-228, 2-379
Rhea, 2-67
Riccioli, 2-389
Rigel, 1-xiii, 1-xiv, 1-xv, 1-xvii, 1-xxxviii, 1-190, 1-259, 1-261-262, 2-70
Ring Nebula, 2-271, 2-277-278
Ring-Tail Galaxy, 2-117, 2-120-121
Robinson, W. L., 1-405
Rosette Nebula, 1-237, 1-243-245, 1-249
Roslund 4, 2-483
Roslund 5, 2-133, 2-143
Rosse, Lord, 1-30, 1-391, 2-53, 2-89
Rst 4596, 2-355
RUGC 132, 2-350
Rumker 14, 2-67
Rumker 18, 2-69
Rumker 21, 2-268

Ruprecht 1, 1-90
Ruprecht 2, 1-91
Ruprecht 3, 1-91
Ruprecht 7, 1-92
Ruprecht 8, 1-92
Ruprecht 11, 1-94
Ruprecht 16, 1-98
Ruprecht 18, 1-99
Ruprecht 20, 1-99
Ruprecht 24, 1-345
Ruprecht 26, 1-346
Ruprecht 27, 1-348
Ruprecht 30, 1-350
Ruprecht 31, 1-350
Ruprecht 32, 1-351
Ruprecht 34, 1-351-352
Ruprecht 36, 1-352
Ruprecht 43, 1-356
Ruprecht 44, 1-356
Ruprecht 45, 1-356
Ruprecht 46, 1-356-357
Ruprecht 47, 1-356
Ruprecht 49, 1-356-357
Ruprecht 53, 1-357
Ruprecht 55, 1-358
Ruprecht 56, 1-358
Ruprecht 59, 1-359
Ruprecht 132, 2-350
Ruprecht 135, 2-375
Ruprecht 136, 2-316
Ruprecht 141, 2-357
Ruprecht 142, 2-357
Ruprecht 143, 2-358
Ruprecht 144, 2-359
Ruprecht 148, 1-314
Ruprecht 151, 1-349
Ruprecht 173, 2-133-134
Ruprecht 175, 2-146, 2-151
Russell, H.N., 244, 2-265

S

Sagitta, 1-229, 1-241, 2-21, 2-305-306, 2-475
Sagittarius, 1-123, 1-378, 2-9, 2-61, 2-81, 2-105, 2-108, 2-203, 2-287, 2-302, 2-311-312, 2-314, 2-317, 2-322-324, 2-328-329, 2-333, 2-336, 2-351, 2-353, 2-357, 2-378
Sagittarius-Carina Spiral Arm, 2-311, 2-318, 2-323, 2-333
Saiph, 1-259
Sandage, Allen, 1-73
Sandqvist-Lindroos 28, 2-348
Sargaz, 2-311, 2-333
Saturn, 1-85, 1-134, 2-169
Saturn Nebula, 1-25, 1-29-31
Savary, 2-388
Scarr, Dr. Leonard, 1-267, 2-88, 2-92
Scheat, 1-281
Schedar, 1-105
Schur, Chris, 1-148, 1-232, 1-248, 2-21, 2-341, 2-344, 2-361-362, 2-478
Scorpio-Centaurus Stellar Association, 2-265, 2-333
Scorpius, 1-95, 1-259, 2-105, 2-257, 2-287, 2-311, 2-333-336, 2-341-342
Scorpius OB1, 2-341-342
Sculptor, 1-365-374
Sculptor Galaxy Group, 1-365, 1-368-369, 1-371-372, 2-392
Scutum, 1-61, 1-215, 2-343, 2-353-354, 2-356, 2-358-361
Scutum Star Cloud, 2-15, 2-353, 2-358, 2-360-361
Secchi, Father Angelo, 2-41
See 209, 2-267
See 221, 2-267
See 428, 2-283
See 430, 2-283
See 437, 2-283
Sellinger, David T., 1-405
Semele, 2-105
Serpens, 1-268
Serpens Caput, 2-365-366, 2-373
Serpens Cauda, 2-353, 2-365, 2-373-374

Sextans, 1-215, 2-379-380
Syfert, C.K., 1-180
Seyfert's Sextet, 2-371
Shamash, 2-257
Sharpless 185, 1-118
Sharpless 2-71, 2-14
Sharpless 2-82, 2-307-308
Sharpless 2-84, 2-308
Sharpless 2-88, 2-480-481
Sharpless 2-90, 2-480-481
Sharpless 2-101, 2-138, 2-140
Sharpless 2-224, 1-57
Sharpless 2-264, 1-266
Sharpless 2-274, 1-213
Sharpless 2-282, 1-246-247
Sharpless 2-290, 1-84
Sharpless 2-301, 1-94-95
Sharpless 2-302, 1-344-345
Sharpless 2-307, 1-346
Sharpless 2-311, 1-354
Sheliak, 2-271
Sheratan, 1-39
Shupa, 2-25
Siamese Twins, 2-449
Simon Marius, 1-18
Sipazianna, 1-259
Sirius, 1-85, 1-88, 1-91, 1-101, 1-151, 1-259, 1-262, 2-197, 2-385
Skiff, Brian, 1-269
Small Sagittarius Star Cloud, 2-314, 2-322-324
Smith, Norman G., 1-405
Smyth, 1-28
Snake Nebula, 2-287, 2-298-299
So 384, 2-78
Sobieski, John III, 2-353
Sombrero Galaxy, 2-451
South 473, 1-221
South 476, 1-221
Southern Cross, 2-133
Southern Crown, 2-105-106
Spica, 2-117, 2-388, 2-423
Spindle Galaxy, 2-379, 2-382
Star Queen Nebula, 2-373, 2-377-378
Stauffer, Mark T., 1-50, 1-134, 1-296
Stellar Spectral Classes, 1-xiv-xvii
Stephan's Quintet, 1-283
Stephenson 1, 2-277
Stock 1, 2-475, 2-478
Stock 2, 1-63, 1-105, 1-124, 1-127
Stock 3, 1-119
Stock 4, 1-298
Stock 5, 1-105, 1-124
Stock 6, 1-125
Stock 8, 1-57-58
Stock 10, 1-60
Stock 11, 1-109
Stock 12, 1-105, 1-110
Stock 22, 1-119
Stock 23, 1-66
Stock 24, 1-115-116
Strabo, 2-415
Struve 11, 1-175
Struve 84, 1-151
Struve 130, 2-343
Struve 132, 2-343
Struve 231, 1-151
Struve 239, 1-389
Struve 268, 1-293
Struve 270, 1-293
Struve 295, 1-154
Struve 296, 1-293
Struve 299, 1-154
Struve 304, 1-293
Struve 307, 1-296
Struve 326, 1-39
Struve 331, 1-296
Struve 369, 1-296
Struve 390, 1-63
Struve 392, 1-297
Struve 422, 1-375
Struve 446, 1-311
Struve 452, 1-375
Struve 470, 1-175
Struve 485, 1-63
Struve 495, 1-378

Struve 516, 1-87, 1-175
Struve 518, 1-87, 1-175
Struve 519, 1-313
Struve 537, 1-88
Struve 538, 1-88
Struve 541, 1-87
Struve 543, 1-87
Struve 627, 2-426
Struve 644, 1-47
Struve 661, 1-221
Struve 668, 1-262
Struve 672, 2-259
Struve 698, 1-50
Struve 710, 2-313
Struve 715, 2-313
Struve 722, 2-313
Struve 737, 2-307
Struve 802, 1-25
Struve 808, 1-27
Struve 872, 1-50
Struve 924, 1-203
Struve 928, 1-51
Struve 924, 1-203
Struve 929, 1-51
Struve 982, 1-203
Struve 997, 1-88
Struve 1061, 1-206
Struve 1066, 1-206
Struve 1095, 1-101
Struve 1103, 1-101
Struve 1104, 1-342
Struve 1108, 1-206
Struve 1110, 1-206
Struve 1122, 1-63
Struve 1149, 1-101
Struve 1245, 1-77
Struve 1282, 1-229
Struve 1333, 1-229
Struve 1338, 1-229
Struve 1369, 1-231
Struve 1441, 2-379
Struve 1487, 2-228
Struve 1547, 2-229
Struve 1552, 2-229
Struve 1596, 2-81
Struve 1615, 2-81
Struve 1622, 2-41
Struve 1625, 1-66
Struve 1657, 2-81
Struve 1669, 2-119
Struve 1692, 2-43
Struve 1694, 1-63
Struve 1768, 2-43
Struve 1788, 2-427
Struve 1798, 2-418
Struve 1821, 2-25
Struve 1864, 2-25
Struve 1877, 2-25
Struve 1877, 2-25
Struve 1888, 2-28
Struve 1930, 2-365
Struve 1932, 2-111
Struve 1954, 2-365
Struve 1962, 2-259
Struve 1970, 2-367
Struve 1999, 2-336
Struve 2010, 2-189
Struve 2110, 2-192
Struve 2130, 2-172
Struve 2140, 2-192
Struve 2155, 2-172
Struve 2161, 2-193
Struve 2173, 2-273
Struve 2220, 2-193
Struve 2241, 2-172
Struve 2280, 2-193
Struve 2303, 2-373
Struve 2306, 2-353
Struve 2308, 2-172
Struve 2316, 2-375
Struve 2325, 2-356
Struve 2373, 2-353
Struve 2404, 2-9
Struve 2445, 2-475
Struve 2455, 2-475

Struve 2457, 2-475
Struve 2470, 2-274
Struve 2474, 2-274
Struve 2540, 2-475
Struve 2578, 2-136
Struve 2579, 2-136
Struve 2637, 2-307
Struve 2653, 2-475
Struve 2725, 2-165
Struve 2727, 2-165
Struve 2737, 2-185
Struve 2742, 2-185
Struve 2786, 2-185
Struve 2793, 2-185
Struve 2806, 1-131
Struve 2838, 1-25
Struve 2841, 1-279
Struve 2848, 1-279
Struve 2863, 1-134
Struve 2876, 1-215
Struve 2877, 1-281
Struve 2894, 1-215
Struve 2902, 1-215
Struve 2922, 1-215
Struve 3007, 1-282
Struve 3050, 1-14
Struve 3409, 1-178
Sulafat, 2-271
Summer Triangle, 2-9
Sunflower Galaxy, 2-52
Supernova 1981B, 2-448
Swan Nebula, 2-314, 2-325, 2-378

T

Tarazed, 2-9
Taurus, 1-82, 1-134, 1-259, 1-375-376, 2-225, 2-342
Taurus Dark Cloud Complex, 1-378, 1-380
Taygeta, 1-380
Thales, 2-415
Theseus, 2-111
Thessaly, 1-39
Thuban, 2-169, 2-417
Tombaugh 1, 1-92
Tombaugh 2, 1-93
Tombaugh 5, 1-68, 1-71
Trapezium, 1-261-262, 1-267-269
Triangulum, 1-375, 1-389-391
Trifid Nebula, 2-313-314, 2-317, 2-319
Tropic of Cancer, 1-77
Trumpler 1, 1-122
Trumpler 2, 1-301, 1-303
Trumpler 3, 1-129-130
Trumpler 5, 1-246, 1-248
Trumpler 6, 1-98-99
Trumpler 7, 1-99, 1-342-343
Trumpler 9, 1-355
Trumpler 24, 2-333, 2-335, 2-342-343
Trumpler 25, 2-347
Trumpler 26, 2-301-302
Trumpler 27, 2-333, 2-348
Trumpler 28, 2-333, 2-348
Trumpler 29, 2-333, 2-349
Trumpler 30, 2-333, 2-351
Trumpler 32, 2-376
Trumpler 34, 2-360
Trumpler 35, 2-360
Tyndareus, 2-133
Typhon, 2-61

U

UA 444, 1-155
UGC 203, 1-14
UGC 206, 1-14
UGC 208, 1-15
UGC 326, 1-113
UGC 326, 2-487
UGC 356, 1-15
UGC 365, 1-15
UGC 387a/b, 1-15
UGC 396, 1-114, 2-487
UGC 420, 1-323
UGC 438, 1-17

UGC 452, 1-17
UGC 454, 1-17
UGC 464, 1-19
UGC 491, 1-19
UGC 601, 1-324
UGC 686, 1-325
UGC 687, 1-325
UGC 688, 1-325
UGC 717, 1-325
UGC 718, 1-19
UGC 859, 1-326
UGC 906, 1-327
UGC 919, 1-327
UGC 920, 1-327
UGC 935, 1-328
UGC 938, 1-328
UGC 939, 1-329
UGC 956, 1-329
UGC 960, 1-330
UGC 988, 1-330
UGC 1192, 1-332
UGC 1195, 1-331-332
UGC 1210, 1-332
UGC 1248, 1-393
UGC 1249, 1-394
UGC 1283, 1-19
UGC 1348, 1-20
UGC 1349, 1-20
UGC 1414, 1-395
UGC 1421, 1-395
UGC 1431, 1-395
UGC 1437, 1-20
UGC 1440, 1-21
UGC 1497, 1-396
UGC 1501, 1-396
UGC 1541, 1-21
UGC 1633, 1-21
UGC 1655, 1-21
UGC 1672, 1-22
UGC 1676, 1-22
UGC 1744, 1-397
UGC 1831, 1-22
UGC 1866, 1-23
UGC 1868, 1-23
UGC 1875, 1-23
UGC 1878, 1-23
UGC 1887, 1-23
UGC 2002, 1-398
UGC 2105, 1-398
UGC 2124, 1-165
UGC 2510, 1-303
UGC 2608, 1-305
UGC 2626, 1-305
UGC 2639, 1-305
UGC 2654, 1-305
UGC 2665, 1-305
UGC 2673, 1-305
UGC 2696, 1-305
UGC 2809, 1-72
UGC 2847, 1-68
UGC 2947, 1-185
UGC 3013, 1-70
UGC 3429, 1-71
UGC 3653, 1-72
UGC 3708, 1-211
UGC 3709, 1-211
UGC 3798, 1-150
UGC 3879, 1-213
UGC 4016, 1-232
UGC 4017, 1-232
UGC 4036, 1-73
UGC 4097, 1-74
UGC 4114, 1-233
UGC 4264, 1-79
UGC 4271, 1-74
UGC 4312, 1-74
UGC 4313, 1-233
UGC 4359, 1-74
UGC 4362, 1-74
UGC 4443, 1-81
UGC 4509, 1-81
UGC 4574, 1-74
UGC 4604, 1-74
UGC 4619, 1-82
UGC 4620, 1-83

UGC 4708, 1-234
UGC 4759, 1-75
UGC 4763, 1-84
UGC 4772, 1-84
UGC 4806, 1-234
UGC 4818, 1-75
UGC 4823, 1-84
UGC 4825, 1-75
UGC 4832, 1-75
UGC 4838, 1-235
UGC 4894, 1-235
UGC 4942, 1-236
UGC 5364, 2-487
UGC 5470, 2-487
UGC 6253, 2-487
UGC 6697, 2-247
UGC 7118, 2-174
UGC 8091, 2-487
UGC 9275, 2-34
UGC 9286, 2-34
UGC 9610, 2-38
UGC 9749, 2-487
UGC 9851, 2-40
UGC 9920, 2-114
UGC 9948, 2-180
UGC 9977, 2-370-371
UGC 10054, 2-419
UGC 10288, 2-370, 2-372
UGC 10311, 2-114
UGC 10316, 2-115
UGC 10338, 2-115
UGC 10356, 2-115
UGC 10822, 2-487
UGC 10876, 2-182
UGC 11012, 2-183
UGC 11218, 2-181, 2-183
UGC 11238, 2-184
UGC 11269, 2-184
UGC 11300, 2-184
UGC 11325, 2-275-276
UGC 11557, 1-135
UGC 11755, 1-31
UGC 11760, 1-31
UGC 12262, 1-319
UGC 12509, 1-319
UGC 12554, 1-23
UGC 12613, 2-487
UGC 12618, 1-320
UGC 12622, 1-320
UGC 12699, 1-320
UGC 12702, 1-320
UGC 12737, 1-321
UGC 12788, 1-321
UGC-A86, 2-487
Underhay, Ernest, 1-405
Upgren 1, 2-49-50
Uranus, 1-28
Ursa Major, 1-85, 1-229, 1-235, 1-259, 1-382, 2-31, 2-41, 2-90, 2-111, 2-249, 2-385, 2-387-389, 2-392, 2-413, 2-423
Ursa Major I Cloud, 2-41
Ursa Major Moving Group, 1-85, 1-382, 2-90, 2-111, 2-385, 2-389
Ursa Major Stream, 2-385
Ursa Minor, 2-415-417

V

van den Bergh 14, 1-67
van den Bergh 15, 1-67
van den Bergh 24, 1-310
van den Bergh 31, 1-49, 1-51-52
van den Bergh 62, 1-275
van den Bergh 69, 1-239
van den Bergh 93, 1-254-255
van den Bergh 97, 1-344-345
van den Bergh 98, 1-347
van den Begrh 131, 2-149
van den Bergh 132, 2-149
van Maanen's Star, 1-317-318
Vega, 2-9, 2-271, 2-275, 2-417
Veil Nebula, 2-133, 2-150-153
Venaticorum, 2-41, 2-43, 2-47, 2-49

Vela, 1-339, 1-361
Venus, 1-85, 1-315, 1-375, 2-197
Vernal Equinox, 1-25, 1-39, 1-315
Vindemiatrix, 2-423
Virgo, 2-41, 2-67, 2-73, 2-81, 2-85, 2-95-96, 2-222, 2-257, 2-385, 2-423, 2-425-426, 2-445, 2-469
Virgo Galaxy Cluster, 2-423-426, 2-440-441, 2-444
Virgo A, 2-445
Vulpecula, 1-215, 1-297, 2-305, 2-315-316, 2-475-476, 2-478

W

Walton, Don, 1-69, 1-350, 1-405, 2-215, 2-382
Wasat, 1-206
Webb, 1-134, 1-256
Webb Society, 1-256
Whirlpool Galaxy, 2-53
Wild Duck Cluster, 2-362
Wilson, Barbara, 1-405
Winnecke 4, 2-410
Winter Triangle, 1-85, 1-101, 1-259
Witch Head Nebula, 1-190
Wolf 28, 1-317
Wolf 359, 2-228
Wolf, Max, 2-228

Z

Zeta Persei Association, 1-297, 1-309
Zeus, 1-279
Zubenelgenubi, 2-257

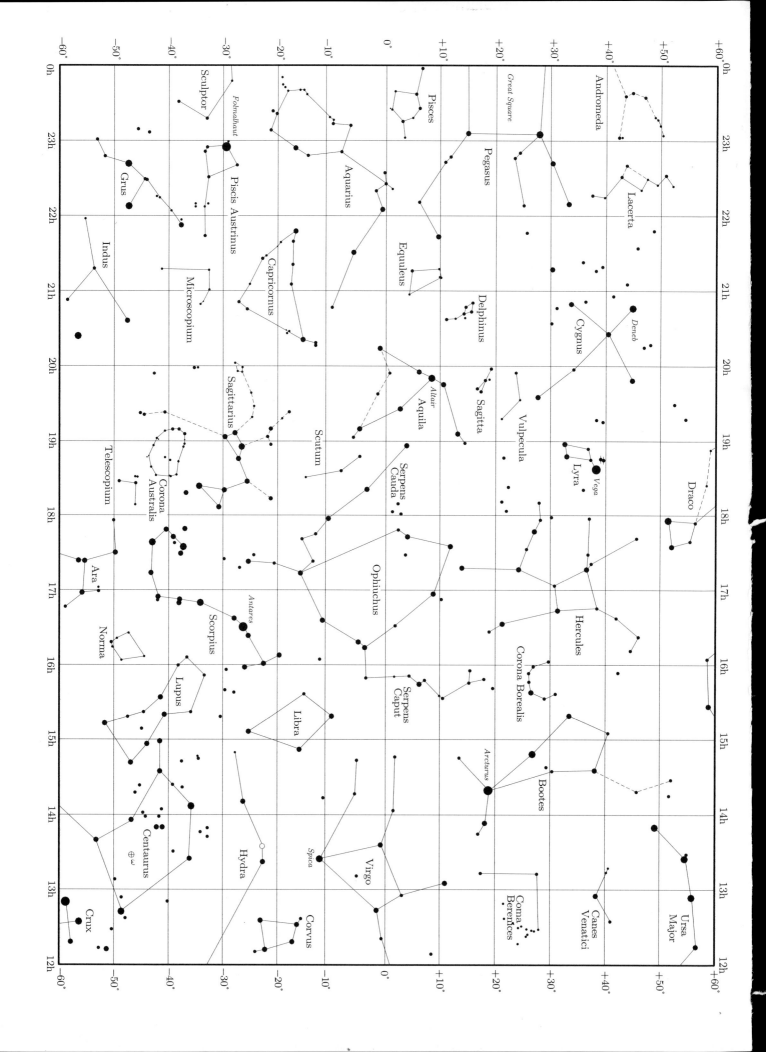